Student Study Guide
Solutions Manual

to accompany

Introduction to Genetic Principles

David R. Hyde
University of Notre Dame

Student Study Guide Prepared by
Erin N. Olson

Brian L. Olson
Saint Cloud State University

Solutions Manual Prepared by
Laurie K. Russell
St. Louis University

 **Higher Education**

Boston   Burr Ridge, IL   Dubuque, IA   New York   San Francisco   St. Louis
Bangkok   Bogotá   Caracas   Kuala Lumpur   Lisbon   London   Madrid   Mexico City
Milan   Montreal   New Delhi   Santiago   Seoul   Singapore   Sydney   Taipei   Toronto

The McGraw-Hill Companies

 Higher Education

Student Study Guide/Solutions Manual to accompany
INTRODUCTION TO GENETIC PRINCIPLES
DAVID R. HYDE

Published by McGraw-Hill Higher Education, an imprint of The McGraw-Hill Companies, Inc., 1221 Avenue of the Americas, New York, NY 10020. Copyright © 2009 by The McGraw-Hill Companies, Inc. All rights reserved.

This book is printed on acid-free paper.

1 2 3 4 5 6 7 8 9 0 QPD/QPD 0 9 8

ISBN: 978-0-07-320631-8
MHID: 0-07-320631-8

www.mhhe.com

TABLE OF CONTENTS

PREFACE

<u>**How to Use this Study Guide**</u>

Key Chapter Concepts

This section contains a summary of the important concepts presented in the chapter. It is designed to help you identify the "big picture." For example, what are the concepts presented and why are they important? Use this section to help you identify the really important ideas that you should learn and to help tie together all of the details presented in the chapter. For example, how do all the details relate to each other? You should read this section after reading the chapter in the textbook in order to identify these concepts before working the problems. Then read this section again before the exam to refresh your memory.

Understanding Key Concepts

Use this section to test your understanding of the key terms listed before this section. If you find that you cannot fill-in a blank, first look for a clue in the *key chapter concepts* section or in your textbook before looking at the answer in the *assessing your knowledge* section.

Figure Analysis

This section is designed to help you understand some of the most important or difficult figures in the chapter. These figures are included in the textbook for a reason: to help students understand an important concept. This section will help you identify the important concepts associated with these figures. A series of very simple questions are asked of several figures from the chapter that collectively guide you to the important concept being exemplified.

Use this section by answering the questions associated with each figure. Be sure to have your textbook open to the figure in order to answer the questions. After answering the questions, look at the answers in the *assessing your knowledge* section and ask yourself, "what is the major concept(s) associated with this figure?" Questions for only a few figures per chapter are included in this study guide. However, you as a student are encouraged to use a similar technique to understand the other figures in the chapter. You should be able to identify one to three major concepts that are exemplified in each figure. Identify and understand those concepts, and you will be very well prepared for an examination.

General Questions

Use this section to test your understanding of many of the important concepts found in the chapter. If you find that you cannot answer a question, first look for the answer in the *key chapter concepts* section or in your textbook before looking at the answer in the *assessing your knowledge* section.

Multiple Choice Questions

Use this section as practice for multiple choice exams. Many of these questions are challenging and require more than just memorization of facts. They are cleverly designed questions that often require analysis or interpretation in order to be answered. Thus, they are excellent practice for difficult multiple choice examinations so that you are not blindsided by unexpected questions during the exam.

Practice Problems

Much of genetics involves either quantitative or qualitative analysis of problems or situations. The problems in this section provide you the opportunity to test your understanding of these calculations and analyses. As your instructor performs a calculation or analysis in class, it may look easy or intuitively obvious. However, don't let an exam be the first time that you perform the calculation as you will likely get stuck. Even if you are confident that you understand the concept behind a problem, you *must* work through the problem before you can be sure that you can perform such a calculation on an exam. This section provides you with this practice for all of the major quantitative calculations found in the chapter and most of the qualitative analysis found in the chapter.

Assessing Your Knowledge

Use this section to check your answers to questions and problems that you worked in the chapter. Detailed answers are given that describe *how* the correct answer was reached. When using this section be sure that you learn *how* the answer was reached and not just what the right answer is.

How to Succeed in Genetics

Your genetics course will no doubt differ greatly from the other biology courses you have taken. Genetics is a problem-solving discipline that requires knowledge of many facts and a conceptualization of many ideas. This study guide emphasizes genetics as an ongoing process of inquiry and investigation and promotes the development of critical analysis skills. You cannot passively learn the subject matter presented in your textbook and study guide. You must actively participate in the learning process by working problems and analyzing genetic data. The old saying *practice makes perfect* will definitely apply to this genetics course. You should begin to study by reading the relevant chapters of your textbook. Ideally, you should read the chapter *before* the material is covered in lecture. You should always review the day's lecture as soon as possible after class; try to write down two to three important points about each figure or slide from the day's lecture. Alternatively, you may find it helpful to make an outline of important points from the lecture. After the lecture, read the *key chapter concepts* section of this study guide as a review and to help identify the most important concepts in the chapter. Do not attempt to use this section in place of reading the textbook or attending lectures. Once you feel confident that you understand the concepts covered, begin working the problems and answering the questions in the textbook and in this study guide. The *general questions* and *practice problems* found in this study guide are challenging: most will require you to apply general principles or to analyze genetic data. While detailed solutions are provided in the *assessing your knowledge* section of each chapter, don't be tempted to look at the answers until you have attempted to work the problem in its entirety. After looking at the answer, be sure to determine *why* the correct answer is correct and *how* to arrive at that answer. For multiple choice questions, you should also attempt to identify why the incorrect answers are wrong. Answering the questions and working the problems will help you to identify what concepts you are having difficulty with. You should then go back and reread the related sections in the textbook. Then read the problems or questions that you had difficulty with again to be sure that you now understand them. You should discuss any problems that you are having with your instructor. Prior to the examination, review the material in the study guide to refresh your memory. If you have followed the preceding suggestions, you should be ready for the examination.

Tips for Mastering Genetics

- **Read the assigned pages in the text.** The textbook you are using is clearly written and easy to understand. Review the textbook material before you go to class. The material will be easier to follow during the lecture if you are familiar with the subject matter to be covered. Read the study guide summary *after* the lecture and again before an examination to review the main concepts.
- **Attend all of the lectures.** Each instructor will emphasize different areas of genetics. You will want to be familiar with the course objectives and requirements so you can focus your study efforts. Your instructor may also explain concepts differently than the textbook or this study guide. Obtaining different perspectives on the same material will help you to assimilate the material and achieve a more complete understanding of the subject.
- **Ask questions.** Throughout your education you will be encouraged to ask questions. Begin asking them now! The best way to achieve a full understanding is to ask questions. You must remember that your instructor wants you to successfully complete the course. However, your instructor cannot help you if he or she is unaware that you are having problems. Before going to talk with your instructor, identify the concepts or procedures that you are having difficulty understanding or performing. In class, ask the instructor to clarify points if necessary. Don't be afraid to ask for help.
- **Learn the vocabulary.** Geneticists use a number of terms that will be new to you. You must learn this vocabulary in order to understand the concepts being discussed and to solve problems. Relevant terms appear in bold in the text.
- **Work all of the problems.** As mentioned previously, genetics involves problem solving. The study guide provides hints that you will find useful when working the problems in the textbook and study guide. Rework all problems that you had difficulty with prior to the exam.

THE PIONEERS OF GENETICS

The contributions of many scientists have created and advanced the field of genetics over the past 100 years. As you encounter the listed individuals during your studies, identify their major contribution to the current understanding of genetic principles.

G. Mendel	G. W. Beadle	E. L. Tatum	C. Bridges
M. F. Lyon	T. H. Morgan	A. Sturtevant	B. McClintock
H. B. Creighton	C. Stern	J. Lederberg	F. Griffith
E. Wollman	M. Delbruck	J. D. Watson	F. H. C. Crick
M. Wilkins	R. Franklin	O. Avery	A. Hershey
M. Chase	E. Chargaff	M. Meselson	F. W. Stahl
M. W. Nirenberg	S. Ochoa	A. Jeffreys	W. Gilbert
F. Sanger	W. Arber	H. Smith	D. Nathans
P. Berg	F. Jacob	J. Monod	S. Benzer
S. Luria	L. Margulis	J. F. Crow	R. A. Fisher
S. Wright	C. Darwin	S. J. Gould	R. C. Lewontin

CHAPTER 1 INTRODUCTION TO GENETICS

Chapter Goals
1. Know the key events in the modern history of genetics.
2. Gain an overview of the topics included in this book—the syllabus of genetics.
3. Understand why certain organisms and techniques have been used preferentially in genetic research.
4. See how genetics may be applied to human life.

Key Chapter Concepts

1.1 A Brief Overview of the Modern History of Genetics
- **Early Human History:** Scientists and philosophers have proposed many theories as to how traits are inherited from parents. For example, one early idea suggested that miniature humans were carried in the head of a sperm. Most of these early theories were refuted by later scientific experimentation. William Harvey proposed that a substance in the gametes would produce the adult structure, and this view became widely accepted. The development of light microscopy led to the discovery of protozoa and bacteria in rainwater and the discovery of the cell and the nucleus.
- **Mid-Nineteenth Century:** Gregor Mendel's crosses involving pea plants led to the discovery that heredity resulted from the inheritance of physical entities (now known as genes) passed from parents to offspring through gametes.
- **Early Twentieth Century:** Walter Sutton discovered that genes are located on chromosomes. Thomas Hunt Morgan and his colleagues demonstrated that some genes that control common traits are sex-linked.
- **Late Twentieth Century:** The era of molecular genetics began when Oswald Avery, Alfred Hershey, and Martha Chase demonstrated that DNA was the genetic material. Less than 10 years later James Watson and Francis Crick determined the structure of this amazing molecule. These discoveries provided the necessary groundwork to allow our modern understanding and manipulation of genes and organisms.

1.2 The Three General Areas of Genetics
- **Classical genetic studies** focus on the transmission of traits from parents to offspring. In this area, scientists study the behavior, inheritance, and effects of different mutations. This subdiscipline encompasses the study of Mendel's principles (chapter 2); the cell cycle, mitosis and meiosis (chapter 3); sex determination and linkage (chapter 4); genetic mapping (chapter 6); and cytogenetics (chapter 8).
- **Molecular genetic studies** focus on the genetic material. In this area, scientists study *how* specific changes in the genetic material of an organism affect its traits. This allows them to learn how cellular processes occur and how we can manipulate them in favorable ways. This subdiscipline includes studies of DNA structure, chemistry, and replication (chapters 7 and 9); gene expression (chapters 10 and 11) and its control (chapters 16 and 17); DNA cloning (chapter 12); and extrachromosomal inheritance (chapter 19).
- **Studies of population and evolutionary genetics** focus on how genes change over time in populations of organisms. This subdiscipline includes studies of quantitative traits and heritability (chapter 24); Hardy–Weinberg equilibrium, and the forces leading to evolutionary change (chapter 23); and the process of speciation (chapter 25).

1.3 Why Fruit Flies and Colon Bacteria?
- **Model organisms** help scientists perform simple experiments that can unequivocally tell us how genes, cells, and organisms function. To be useful, these organisms must possess several important characteristics. For example, a good model organism has a short generation time, produces a large number of offspring, can be easily and inexpensively reared or grown in a small space in the laboratory, exhibits interesting features that correspond to a variety of organisms, and has genomic DNA that has been completely, or almost completely, sequenced. Once an organism has been studied by one scientist, it becomes an attractive organism to be used by others since something is already

known about it. Thus, several seemingly strange organisms have become common model organisms for study in genetics.

1.4 Application of Genetics to the Human Population
- **Human diseases** often result from genetic changes. Our understanding of genes, how they function, and how they are inherited is directly responsible for our ability to treat and cure many diseases.
- **Agriculture** has also benefited from our understanding of genes and how to manipulate them. The production and composition of many of the fruits and vegetables eaten every day have been improved through genetic means.
- **Genetic counseling** is an area of genetics that allows us to predict the probability of a person developing a disease or passing a disease on to a child through a simple blood test. Whether this knowledge is desirable or detrimental is a matter of opinion to many individuals.

Key Terms

- genetics
- theory of epigenesis
- vector
- genome
- transgenic organisms
- cloned animals
- genomics
- proteomics
- gene therapy
- classical genetics
- chromosomes
- alleles
- genes
- genotypes
- phenotypes
- diploid
- haploid
- dominant
- recessive

- genetic cross
- law of segregation
- law of independent assortment
- phenotypic ratios
- chromosomal theory of inheritance
- enzymes
- one-gene–one-polypeptide
- karyotype
- cytogenetics
- molecular genetics
- mutations
- polarity
- antiparallel
- codons
- transcription
- translation
- DNA

- mRNA
- tRNA
- rRNA
- protein
- central dogma
- recombinant DNA techniques
- restriction endonuclease
- cloning
- plasmid
- complementary DNA (cDNA)
- polymerase chain reaction (PCR)
- evolutionary genetics
- Hardy–Weinberg equilibrium
- allele frequency
- punctuated equilibrium

Understanding the Key Concepts
Use the key terms or parts of their definitions to complete the following sentences.

Classical genetics is the study of (1)_____, (2)_____, and (3)_____. Gregor Mendel used pea plants to determine the basic rules of (4)_____. By observing the (5)_____ of the pea plants that resulted from a particular (6)_____, he was able to deduce which trait is (7)_____ and which trait is (8)_____ and therefore only observed when expressed in a homozygote. By looking at the segregation of two different traits determined by different genes, Mendel found that (9)_____ of different genes segregate (10)_____ of each other.

The genetic material in a cell was found to be (11)_____ and the study of its structure, replication, and expression is called (12)_____. DNA consists of repeating nucleotide subunits composed of a (13)_____ sugar, a phosphate group, and one of four nitrogenous bases, (14)_____, (15)_____, (16)_____, or (17)_____. It exists in the cell as a

(18)_____, where the two strands are (19)_____ in their polarity. The bases in one strand of the DNA are (20)_____ to the bases on the opposite strand allowing them to base-pair with each other by hydrogen bond interactions.

The (21)_____ of molecular biology describes the flow of genetic information from (22)_____ to (23)_____ which occurs by the process of (24)_____. The mRNA is then (25)_____ into (26)_____. Each protein is made up of (27)_____ that are specified in the DNA by (28)_____ consisting of groups of (29)_____ nucleotide bases.

(30)_____ can be used to manipulate the DNA of different organisms. A specific sequence of DNA can be amplified from an organism's (31)_____ by (32)_____, cut at specific DNA sequences by a (33)_____ and then (34)_____ into a (35)_____ that allows it to be propagated in bacteria. This technology has led to the discovery of a vast amount of molecular information about the genomes of different organisms.

(36)_____ is the study of changes in (37)_____ in populations over time. The (38)_____ is used to determine if evolutionary processes are acting on populations. (39)_____ occurs in a population if it is large, has random mating, and has negligible effects of (40)_____, (41)_____, and (42)_____.

Figure Analysis

Use the indicated figure or table from the textbook to answer the questions that follow.

1. Table 1.1

(a) How many different codons are there?   How many codons specify amino acids?

(b) Can codons specify more than one amino acid?

(c) How many amino acids are specified by the codons in the table?

(d) How many nucleotide combinations specify stop codons?   What are they?

(e) How many nucleotide combinations specify start codons?   What are they?

(f) How many nucleotide combinations specify the amino acid serine?   What are they?

(g) What amino acids do the following codons specify?
    ACA   CAA   AGG   GCG   CUU

(h) Why is the genetic code said to be degenerate?

2. Figure 1.13

(a) In what direction is the mRNA transcribed?

(b) What enzyme synthesizes mRNA from DNA?

(c) Indicate which nucleotides in the RNA base-pair with the following nucleotides in the DNA?
    Adenine   Cytosine   Guanine   Thymine

(d) How does this differ from base-pairing between the two strands of a DNA molecule?

3. Figure 1.16

(a) What enzymes are required for the foreign DNA to be cloned into the plasmid DNA?

(b) What molecules are cut with a restriction endonuclease and why?

(c) When the plasmid and foreign DNA are mixed together, why do they join together?

(d) If the restriction digest was not complete, when DNA ligase is added to the mixture of plasmid and foreign DNA, what molecule other than the hybrid plasmid could be formed?

(e) Could this molecule be "grown" inside bacteria? Why or why not?

General Questions

Q1-1. Classify each of the following pioneering scientific investigations according to the area of study (classical, evolutionary, or molecular genetics).

_____ a. Genetic Control of Biochemical Reactions in *Neurospora* (G. W. Beadle and E. L. Tatum)

_____ b. The Genetics of Sex in *Drosophila* (C. Bridges)

_____ c. The Chromosomes in Heredity (W. S. Sutton)

_____ d. Sex Limited Inheritance in Man (T. H. Morgan)

_____ e. Sex Chromatin and Gene Action in the Mammalian X Chromosome (M. F. Lyon)

_____ f. A Correlation of Cytological and Genetical Crossing over in *Zea mays* (H. B. Creighton and B. McClintock)

_____ g. The Effects of Unequal Crossing over at the *Bar* Locus in *Drosophila* (A. Sturtevant)

_____ h. Studies on the Chemical Nature of the Substance Inducing Transformation in *Pneumococcal* Types (O. Avery, C. MacLeod, and M. McCarty)

_____ i. Molecular Structure of Nucleic Acids: A Structure for Deoxyribonucleic Acid (J. D. Watson and F. H. C. Crick)

_____ j. Independent Functions of Viral Protein and Nucleic Acid in Growth of Bacteriophage (A. Hershey and M. Chase)

_____ k. Genetic Regulatory Mechanisms in the Synthesis of Proteins (F. Jacob and J. Monod)

_____ l. Artificial Transmutation of the Gene (H. J. Muller)

_____ m. The Role of Inheritance in Behavior (R. Plomin)

_____ n. The Genetic Changes in Populations of *Drosphila pseudoobscura* in the American Southwest (T. Dobzhansky)

_____ o. The Genetical Evolution of Social Behavior (W. D. Hamilton)

_____ p. Versuche über Pflanzen Hybriden "Experiments in Plant Hybridization" (G. Mendel)

Q1-2. Weismann performed experiments in the late 1800s with inbred strains of mice possessing normal tails. He removed the tails of a group of mice and intermated individuals from this group. All of the offspring produced by parents with cut-off tails had normal-length tails. This is a significant finding. Why?

Q1-3. Describe how genetic modifications are being used to improve crops.

Q1-4. What are the advantages of genetic testing?

Q1-5. What are the conditions required for a population to achieve genetic equilibrium?

Multiple Choice

For each of the following, circle the letter of the choice that most appropriately answers the question.

1.) Which of the following statements about the DNA double helix is *not* true?

  a. DNA polymerase replicates both strands of the double helix.

  b. The two strands of the DNA double helix are always antiparallel.

  c. DNA contains the sugar deoxyribose.

  d. Both strands of the double helix are transcribed into RNA.

  e. DNA contains four bases: adenine, cytosine, guanine, and thymine.

2.) The first genetic map demonstrating that genes exist in a linear order on chromosomes was created by

  a. Gregor Mendel

  b. Walter Sutton

  c. Alfred Sturtevant

  d. Anton von Leeuwenhoek

  e. Thomas Hunt Morgan

3.) Which of the following is *not* a characteristic of *C. elegans* that makes it a model organism for genetic research?

  a. It can be mated to generate a desired genotypic progeny.

  b. Targeted mutations in specific genes can be created.

  c. Cell movements and organ development can be visualized because it is transparent.

  d. It is a small organism with a rapid generation time.

  e. It has a small genome.

4.) William Harvey first proposed that

  a. all cells come from pre-existing cells.

  b. substances in gametes produce adult structures.

  c. microscopic particles move by Brownian motion.

  d. meiosis involves a reductional stage of cell division.

  e. genes are located on chromosomes.

5.) The first organism to have its complete genome sequenced was

  a. *Bacillus anthracis*

  b. *Escherichia coli*

  c. *Drosophila melanogaster*

  d. *Saccharomyces cerevisiae*

  e. *Haemophilus influenzae*

6.) The chromosomal theory states that chromosomes are

  a. linear arrays of genes containing the genetic information required by organisms.

  b. structures that never change from one generation to the next.

  c. found within the nucleus of every cell.

  d. factors in germ cells that produce new individuals.

  e. continually changing as organisms evolve.

7.) Which of the following is least likely to be successfully treated by gene therapy?

  a. hemophilia

  b. cystic fibrosis

  c. diseases caused by a single defective gene

  d. behavioral disorders

  e. Duchenne's muscular dystrophy

8.) Which of the following *cannot* be determined by pedigree and linkage analysis?

  a. if a gene is located on a sex chromosome

  b. if a particular chromosome has undergone a rearrangement

  c. the probability that a child will exhibit a specific phenotype

  d. whether phenotypes are dominant or recessive

  e. the probability that an individual will have a specific allele

9.) Which of the following statements regarding genetic testing is true?

  a. Genetic tests can only be performed on an unborn fetus.

  b. Genetic testing is only done for diseases where an available treatment exists.

  c. An individual's predisposition to Huntington disease, early onset familial Alzheimer disease, and several forms of cancer can be determined by genetic testing.

  d. Genetic test results should always be included as part of an individual's medical records.

  e. A positive test result indicates that the disease tested for will develop at some point during an individual's lifetime.

10.) Which of the following statements identifies a true difference between RNA and DNA?

  a. RNA contains five bases: adenine, cytosine, guanine, thymine, and uracil.

  b. RNA contains a six-carbon rather than a five-carbon sugar.

  c. RNA lacks polarity.

  d. RNA is typically single-stranded.

  e. All of the above

Assessing Your Knowledge

Understanding the Key Concepts—Answers

1–3.) genes, mutations, phenotypes; 4.) inheritance; 5.) phenotypes; 6.) genetic cross; 7.) dominant; 8.) recessive; 9.) alleles; 10.) independently; 11.) DNA; 12.) molecular genetics; 13.) deoxyribose; 14–17.) adenine, cytosine, guanine, thymine; 18.) double helix; 19.) antiparallel; 20.) complementary; 21.) central dogma; 22.) DNA; 23.) mRNA; 24.) transcription; 25.) translated; 26.) protein; 27.) amino acids; 28.) codons; 29.) three; 30.) Recombinant DNA techniques; 31.) genome; 32.) PCR; 33.) restriction endonuclease; 34.) cloned; 35.) plasmid; 36.) Evolutionary genetics; 37.) alleles; 38.) Hardy–Weinberg equilibrium; 39.) Genetic equilibrium; 40–42.) mutation, migration, natural selection.

Figure Analysis—Answers

1a.) 64; 61; b.) No; c.) 20; d.) 3; UAA, UGA, UAG; e.) 1; AUG; f.) 6; UCU, UCA, UCG, UCC, AGU, AGC; g.) ACA: threonine; CAA: glutamine; AGG: arginine; GCG: alanine; CUU: leucine; h.) The genetic code is said to be degenerate because some amino acids are specified by more than one codon. The degenerate base usually occurs in the third position of a codon, while the first two positions are the same.

2a.) 5' → 3'; b.) RNA polymerase; c.) Adenine base-pairs with uracil, cytosine base-pairs with guanine, guanine base-pairs with cytosine, thymine base-pairs with adenine; d.) In a molecule of DNA, adenine base-pairs with thymine because uracil is not found in DNA.

3a.) Restriction endonuclease, DNA ligase; b.) Both the plasmid and the foreign DNA are cut with the restriction endonuclease to make compatible ends on both molecules.; c.) The plasmid and foreign DNA join together when they are mixed because they have compatible ends formed by restriction endonuclease digestion that can base-pair with each other.; d.) If the restriction digest was not complete, the plasmid DNA may recircularize when DNA ligase is added to the mixture of plasmid and foreign DNA to form the original starting plasmid that does not contain the foreign DNA.; e.) This molecule could be "grown" inside of bacteria because it contains an origin of replication that will allow it to be propagated in the bacteria and it contains the marker that will be selected for when the bacteria are plated.

General Questions—Answers

Q1-1. (a–g and p) classical genetics; (h–l) molecular genetics; and (m–o) evolutionary genetics.

Q1-2. This result is important because it showed that acquired characteristics are *not* inherited.

Q1-3. Crops can now be genetically modified to be herbicide-resistant, allowing competing weeds to be killed by herbicides and crops to produce larger yields. Crops can also be genetically modified to delay spoiling, to grow in extreme climate conditions, and to increase their nutrient content.

Q1-4. Genetic testing can determine an individual's predisposition to developing a number of diseases and disorders. While these tests do not conclusively determine whether an individual will develop a particular disease or disorder, they can make an individual more aware of his or her susceptibility. This awareness should increase an individual's vigilance in being tested for the disease or disorder and will allow early detection if a problem does occur which allows for the greatest chance of survival. Defects in an unborn fetus can also be determined by genetic testing, making parents aware that their child may be afflicted with a disorder when it is born. This allows them to prepare for this situation or choose to end the pregnancy.

Q1-5. To achieve genetic equilibrium in a population, the population must be large, mating must be random, and there must be negligible effects from mutation, migration, and natural selection.

Multiple Choice—Answers

1.) d; 2.) c; 3.) b; 4.) b; 5.) e; 6.) a; 7.) d; 8.) b; 9.) c; 10.) d.

1

INTRODUCTION TO GENETICS

CHAPTER SUMMARY QUESTIONS

1. Both preformation and epigenesis attempt to explain the development of animals. Preformation was the belief that the gametes of each organism contained homunculi, or preformed miniature adults, that unfold into their adult shape during development. The theory of epigenesis on the other hand contends that the organism was not yet formed in the gametes. Rather, the adult structure was produced during development from substances found in the gametes.

2. 1. j, 2. d, 3. e, 4. g, 5. b, 6. i, 7. c, 8. h, 9. f, 10. a.

3. Restriction endonucleases are enzymes that cut DNA at specific sequences. Their discovery allowed for the creation of recombinant DNA molecules containing DNA from two or more different species. Recombinant DNA has proven to be very useful for research and the pharmaceutical industry.

4. a. Genomics is the study of the DNA content and gene organization in and between species.
   b. Transcriptomics is the study of all transcribed genes and how their expression changes in response to various stimuli.
   c. Proteomics is the study of all proteins expressed by a cell or individual.

5. The three general areas of genetics are classical genetics, molecular genetics, and evolutionary genetics. Classical genetics focuses on genes, mutations, and phenotypes. It deals with the transmission of the genetic material from one generation to the next. Molecular genetics is the study of the structure, replication, and expression of the genetic material. Evolutionary genetics is the study of the mechanisms of evolutionary change, or changes in the gene frequencies in populations over time.

6. Mendel's first law is the law of segregation: alleles remain distinct and are not blended together during inheritance. His second law is the law of independent assortment: alleles of different genes sort independently of each other into gametes.

7. A pedigree is a pictorial representation of related individuals and the phenotype that each individual exhibits. It can be used (a) to deduce the pattern of inheritance of a particular disease and (b) to calculate the linkage relationship or map distance between two genes. This information can be used to calculate the probability that an individual possesses a specific allele or that a child will exhibit a specific phenotype.

8. A karyotype is the representation of all the chromosomes in an individual. It reveals the sex of the individual. It can also tell whether there are changes in chromosome number and whether there are significant abnormalities in chromosome structure.

9. a. An allele is a form of a gene that has a different sequence. There can be many different alleles for a gene.
   b. Haploid is a cell or organism that possesses a single set of chromosomes.
   c. Genotype is the combination of alleles that an individual possesses.
   d. Phenotype is the characteristics that an individual exhibits.

10. 5'-P ——————————————————— OH-3'
           A   C   C   C   T   G   A   T   G

           T   G   G   G   A   C   T   A   C
       3'-OH ————————————————————— P-5'

11. Messenger RNA (mRNA) is the RNA that contains the information that will be translated into a protein's amino acid sequence. Ribosomal RNA (rRNA) forms a complex with several proteins to form the ribosome, which is the site of mRNA translation. Transfer RNA (tRNA) carries the amino acids to the ribosome for incorporation into the protein. Accuracy is ensured by the complementarity between the codons of mRNA and the anticodon sequences in tRNA.

12. The central dogma states that DNA is transcribed to mRNA, which is then translated into protein.

13. Genes consist of DNA nucleotide sequences and are located at fixed positions on the chromosomes.

14. 1. *Polymorphisms*. DNA sequencing has revealed that much more allelic variation exists within natural populations than older mathematical models could account for. Much of the polymorphism is neutral. Thus, it appears that natural selection, the guiding force of evolution, does not act differentially on many, if not most, of the genetic differences found so commonly in nature.
    2. *Punctuated equilibrium*. The fossil record had suggested that most evolutionary change was gradual. However, recent evidence revealed that evolutionary change may actually occur in short, rapid bursts, followed by long periods of very little change.
    3. *Sociobiology*. Proponents suggest that social behavior is under genetic control and is acted upon by natural selection, as is any morphological or physiological trait. This idea is controversial because it calls altruism into question and suggests, to some extent, we are genetically programmed to act in certain ways.

15. a. A cDNA is a complementary DNA that is a reverse transcribed copy of messenger RNA.
    b. PCR is polymerase chain reaction, which is a technique used to amplify desired DNA sequences.

16. (1) Has short generation time, (2) produces large number of offspring, (3) easy and inexpensive to grow, (4) exhibits interesting features found in other organisms, especially humans, (5) its genomic DNA is largely or entirely sequenced.

17.

Genus and Species	Organism
Drosophila melanogaster	**Fruit fly**
*Mus **musculus***	Mouse
***Arabidopsis** thaliana*	**Flowering plant**
*Escherichia **coli***	**Colon bacterium**
Saccharomyces cerevisiae	Baker's yeast
Danio rerio	**Zebrafish**
*Caenorhabditis **elegans***	Developing worm

18. Genetics plays an important role in the diagnosis and potential treatment of many hereditary diseases. Research in genetics has also been instrumental in developing modified plants and animals for better productivity by improving growth and disease resistance. In addition, genetic engineering has resulted in the production of pharmaceuticals from both bacteria and animals.

19. Gene therapy is the introduction of normal genes into an individual in an attempt to cure a genetic disorder or to minimize its adverse effects. Most of the problems associated with gene therapy have to do with the virus that is used to introduce the normal gene into the cells. The virus has caused a massive inflammatory response in one patient, and a leukemia-like disorder in another. Moreover, the cell with the "corrected" gene is a somatic cell and is not passed to the individual's offspring. Therefore, gene therapy must be reinitiated with every individual that suffers from the disease.

20. The results of genetic testing could be used by an insurance company to determine whether an individual would be eligible for health or life insurance. The results could also be used by prospective employers to avoid hiring an individual with a particular genotype.

CHAPTER 2 MENDELIAN GENETICS

Chapter Goals

1. Understand Mendel's first law of equal segregation and manipulate data for the inheritance of a single trait.
2. Understand Mendel's second law of independent assortment and manipulate data for the inheritance of multiple traits.
3. Be able to apply the concepts of dominance and recessiveness to interpret inheritance patterns.
4. Be able to apply the rules of probability to solve genetic questions.
5. Use the chi-square test to statistically analyze genetic data.

Key Chapter Concepts

2.1 Mendel's Experiments

- **Mendelian Inheritance:** Characteristics that are inherited in a simple Mendelian fashion are controlled by single genes with two alleles. Mendel focused on discrete traits with two alternative phenotypes in the garden pea. He first established **pure-breeding** lines by self-fertilizing individual plants. These lines would "breed true" or produce only plants with the same phenotype. Mendel was able to cross these pure-breeding lines and observe the phenotypes of the offspring produced. He then counted and classified the resulting offspring according to phenotype. This **quantitative** approach differed from that traditionally used to investigate biological questions.

- **Genetic Crosses:** The pure-breeding individuals initially crossed represented the **parental** or P_1 generation. The offspring resulting from this cross comprised the **first filial** or F_1 generation. Individuals in the F_1 generation are also referred to as **hybrids** because they are the result of a cross between two unlike parents. An F_2 generation is obtained by crossing individuals from the F_1 generation. Mendel's crosses led to the conclusions outlined subsequently. (*Note:* Modern terminology such as *gene* and *allele* will be used in the following discussion for clarity. Mendel did not use these terms.)

- **Phenotype and Genotype:** As you begin to explore the field of genetics, you will find that organisms vary in their genetic makeup as well as their physical appearance. We use the term **phenotype** to describe the individual's physical attributes. The phenotype is "what can be seen." The term **genotype** refers to the genes that an organism possesses. Remember that a **gene** is the inherited determinant of the phenotype. Each individual usually receives two copies of each gene, one from the female parent and one from the male parent. The alternative forms of the same gene are called **alleles**. For each gene, an individual may have two different alleles (**heterozygous** condition, a **heterozygote**) or two identical alleles (**homozygous** condition, a **homozygote**). The phenotype may be directly determined by the genotype, or it may result from the combined effects of genes and the environment.

- **Particulate Inheritance:** Mendel's data clearly showed that physical particles were passed from parents to offspring. We now know that these particles are genes and that they are carried on the chromosomes. Each individual possesses two alleles, while the **gametes** (sperm and egg) contain only one.

- **Complete Dominance:** While an individual has two alleles for each gene, both alleles are not necessarily visible. Mendel consistently found that the F_1 generation was composed of individuals of one phenotype while the P_1 and F_2 generations were composed of individuals with one of the two possible phenotypes. For example, crossing a plant with purple flowers to a plant with white flowers (P_1) yields all purple-flowered plants in the F_1. A 3:1 ratio (purple:white) was found in the F_2 generation. Mendel referred to the trait that was visible in the F_1 as **dominant**. The **recessive** trait was masked. Further crosses revealed that some of the F_2 purple plants produced only purple offspring when self-fertilized while others produced a 3:1 **phenotypic ratio** of purple:white plants when they were self-fertilized. We now know that purple-flowered plants in the F_2 generation are either **homozygous dominant** (*AA*) or **heterozygous** (*Aa*). White plants produced only like offspring when selfed (*aa*, **homozygous recessive**). A 1:2:1 (*AA:Aa:aa*) **genotypic ratio** is observed in the F_2 generation.

2.2 Segregation

- **Rule of Segregation:** During gamete formation, one allele from each gene pair is randomly passed into each gamete. For example, a plant heterozygous for flower color (a **monohybrid**) is as likely to form gametes containing the allele for purple color as it is to form gametes containing the allele for white color.

2.3 Independent Assortment

- **Rule of Independent Assortment:** While the rule of segregation focuses on the separation of alleles for a single gene, the rule of independent assortment considers the inheritance pattern of two or more genes simultaneously. When an individual heterozygous for two traits (*AaBb*; a **dihybrid**) forms gametes, the alleles for the first trait separate independently from those of the second trait. In this example, four different types of gametes will be formed in equal proportions: *AB*, *Ab*, *aB*, and *ab*.

2.4 Genetic Symbol Conventions

- **Allelic Designations:** Botanists and mammalian geneticists use uppercase and lowercase letters to denote dominant and recessive alleles, respectively. Microbiologists and *Drosophila* geneticists, use a plus sign to designate a **wild-type** allele and no plus sign to designate a **mutant** allele. Genes in *Drosophila* are named after the mutant phenotype, so if a mutation is dominant, it begins with an uppercase letter. If a mutation is recessive, it begins with a lowercase letter. In some cases, a gene is designated by more than one letter to differentiate it from a second gene beginning with the same letter. For genes where multiple mutant alleles exist, the mutant alleles are distinguished by either a superscript number or a superscript letter.

2.5 Probability

- **Probability Theory:** Probability is the expectation of the occurrence of a particular event. For example, the probability of a heterozygote (*Aa*) producing a gamete containing the *A* allele is 1/2 because you have one *A* allele and two possible alleles (1/2). If you were considering the probability of getting a number 3 on a six-sided die, then you would reason that there is going to be only one number 3 represented on the die with six possible different numbers total. So, you would have one of six or 1/6 chance.
- **Combining Probabilities:** Individual probabilities can be combined to obtain the probability that two or more events will both occur. We use two different rules, the sum rule and the product rule, to determine how the probabilities are combined. You can simplify these rules by phrasing the probability question in the form of an "and" or "or" statement or question. If you use "or," you will use the sum rule. If you use "and," you will use the product rule. There will be some situations where you will use both to calculate a probability.
- **Sum Rule:** The probability that one of several mutually exclusive events will occur is equal to the *sum* of the individual event probabilities.
    - What is the probability of drawing a *single card* from a standard deck of 52 cards and getting a two of clubs *or* a card with hearts on it? Notice the "or." This means we use the sum rule and *add* the individual probabilities. The chance of drawing a two of clubs is 1 chance in 52 cards or 1/52 because there is only one card that is the two of clubs in the whole deck. The chance of drawing a heart is 13 (number of cards with hearts in the whole deck) chances in 52 cards. By adding these chances together, we find that the probability of drawing a two of clubs *or* a heart is 1/52 + 13/52 = 14/52 or 0.269.
    - Now we will use a genetics example. What is the probability of producing an offspring with a dominant phenotype (homozygous dominant *or* heterozygous) from two heterozygous parents? We can use a Punnett square to set up the cross and determine what four genotypes are possible for their offspring. If we cross two heterozygous parents (*Aa*), the chance of producing a homozygous dominant offspring (*AA*) is 1/4. The chance of producing a heterozygous offspring (*Aa*) is 2/4. So, the probability of producing a homozygous dominant offspring *or* a heterozygous offspring is 1/4 + 2/4 = 3/4 or 0.75.
- **Product rule:** The probability that two independent events will both occur is equal to the *product* of their individual probabilities.

o   What is the probability of drawing *two cards* from a standard deck of 52 cards and getting a two of clubs *and* a heart, in that particular order? Notice the "and." This means we use the product rule and *multiply* the individual probabilities. The chance of drawing a two of clubs is 1/52 (see earlier). The chance of drawing a heart is 13/52 (see earlier). By multiplying these chances, we find that the probability of drawing a two of clubs *and* a heart, in that particular order, is $1/52 \times 13/52 = 13/2704$ or 0.005.

o   Now we will use a genetics example. What is the probability of producing three offspring with the genotypes homozygous dominant **and** heterozygous **and** heterozygous (*Aa*), in that particular order, from two heterozygous parents? Again, we can use a Punnett square to set up the cross and determine what four genotypes are possible for their offspring. If we cross two heterozygous parents (*Aa*), the chance of producing a homozygous dominant offspring (*AA*) is 1/4. The chance of producing a heterozygous offspring (*Aa*) is 2/4. So, the probability of producing offspring that are homozygous dominant **and** heterozygous **and** heterozygous, in that particular order, is $1/4 \times 2/4 \times 2/4 = 4/64$ or 0.063.

- **Combining the Sum and Product Rules:** There will be circumstances that require you to use both the sum and product rules together. You can recognize these because the problem will not give you a specified order that the offspring must be born in or chosen. To answer these types of questions, you can phrase the question in "and" and "or" statements or use the binomial theorem. Let's see how the problems are phrased and solved by each method using a question similar to the one we just used, but notice that *no order was specified*. What is the probability of producing two offspring with the genotypes homozygous dominant (*AA*) **and** heterozygous (*Aa*) from two heterozygous parents (*Aa*)?

   o   Use the rules in combination. You really must consider that the events have no specified order. So the order could be two different ways. So we must phrase it as, "What is the probability of producing *AA* and *Aa or* producing *Aa* and then *AA*?" The probability of producing *AA* is 1/4, and the probability of producing *Aa* is 1/2 (see earlier). So, the probability of producing two offspring with the genotypes homozygous dominant (*AA*) **and** heterozygous (*Aa*) from two heterozygous parents (*Aa*) is $1/4 \times 1/2 + 1/2 \times 1/4$ or 0.25. (You must perform the multiplication first and then the addition.)

- **Branch-Line Approach to Calculate Probabilities:** The branch-line approach is based on the product rule and Mendel's law of independent assortment. In this approach, complicated genetic questions are answered by examining each gene or trait independently. A complicated genotype is broken down into the individual genes, and the probability of the genotype or phenotype for each gene is determined and then multiplied by the probabilities of the genotypes or phenotypes for the other genes to produce the probability of obtaining the entire genotype or phenotype.

2.6 Statistics

- Statistics are used by geneticists to design experiments, summarize data, and determine whether or not the results of an experiment support the hypothesis being tested. The **chi-square test** is a statistical test frequently employed by geneticists to determine whether or not a set of data conform to an established genetic ratio. For example, a geneticist may find that a testcross yields 50 individuals with the dominant phenotype and 40 with the recessive phenotype. The question is, "Do these data conform to the 1:1 ratio hypothesized for a testcross (the **null hypothesis**)?" Geneticists perform a chi-square test and decide on the basis of the outcome of the test. The chi-square value is calculated using the formula

$$\chi^2 = \Sigma \; \frac{(O - E)^2}{E}$$

where χ is the Greek letter chi, O is the observed number for a category, E is the expected number for that category, and Σ means to sum the calculations for all categories. The calculated χ^2 values are correlated to a probability value (*p*) using a chi-square table containing theoretical values with the same **degrees of freedom**. The degrees of freedom are equal to $n - 1$, where n is the number of categories. A convention known as the **level of significance** uses a probability of 0.05 as a cutoff for rejecting a hypothesis.

Key Terms

- self-fertilization
- cross-fertilization
- parental generation, P
- first filial generation, F_1
- hybrids
- monohybrids
- second filial generation, F_2
- dominant
- recessive
- genes
- alleles
- phenotype
- genotype

- homozygous
- heterozygous
- law of segregation
- testcross
- backcross
- dihybrid
- Punnett square
- law of independent assortment
- wild-type
- mutants
- stochastic
- probability theory

- mutually exclusive outcomes
- independent outcomes
- sum rule
- product rule
- branch-line approach
- confidence limits
- testing of hypotheses
- null hypothesis
- chi-square distribution
- probability value (*p*)
- degrees of freedom
- critical chi-square
- level of significance

Understanding the Key Concepts

Use the key terms or parts of their definitions to complete the following sentences.

Gregor Mendel genetically crossed pure-breeding tall and dwarf pea plants by (1)_____ and found that all of the progeny in the (2)_____ generation were tall. Therefore, tallness was considered to be the (3)_____ phenotype. When these plants were (4)_____, the progeny in the (5)_____ generation were 3/4 tall and 1/4 dwarf. The reappearance of the (6)_____ trait indicated that the (7)_____ plants had to have the (8)_____ (*Dd*), or be (9)_____. While only the (10)_____ of the F_2 plants were visible, Mendel determined the (11)_____ of these plants by (12)_____ the F_2 individuals. Doing this, he found that all of the F_2 dwarf plants were (13)_____ for the *d* (14)_____ because they produced only dwarf plants in the F_3 generation. The tall F_2 plants produced both tall and short plants in the F_3 generation. Plants that were homozygous (15)_____ bred true and produced only tall plants, but plants that were heterozygous produced both tall and dwarf plants in a (16)_____ ratio. This result indicated that the (17)_____ ratio of the F_2 offspring is (18)_____, and thus supported Mendel's (19)_____.

Using a (20)_____ cross, Mendel looked at two traits (21)_____ and found that the two traits are inherited (22)_____ of each other because the (23)_____ of one (24)_____ assort in the (25)_____ of an individual (26)_____ of the (27)_____ of a second (28)_____. The presence of each allele in the (29)_____ occurs with (30)_____ frequency.

The (31)_____ of a given outcome can be determined by dividing the number of times an event is (32)_____ by the (33)_____ of possible cases. When events are (34)_____, the probability that one event or the other event will occur is determined by the (35)_____ of the (36)_____ of each event. When two events have (37)_____ that do not affect one another, the probability of one event and the second event occurring is the (38)_____ of their independent (39)_____. Geneticists use the (40)_____ to determine whether experimental values are (41)_____ similar or different than expected values. If the (42)_____ associated with a calculated (43)_____ value is greater than (44)_____ with the same number of (45)_____, the difference between the observed and the expected probabilities is likely due to chance.

Figure Analysis
Use the indicated figure or table from the textbook to answer the questions that follow.

1. **Figure 2.5**
   (a) What was the surprising result that Mendel observed in the F_1 progeny?

   (b) How could some of the plants in the F_2 generation be dwarf when they are the offspring of plants that are all tall?

   (c) Why is the phenotypic ratio of tall:dwarf plants in the F_2 generation 3:1?

2. **Figure 2.8**
   (a) What was the purpose of self-fertilizing the F_2 plants to produce the F_3 generation?

   (b) What did Mendel observe in the F_3 generation when he self-fertilized the dwarf F_2 plants?

   (c) What did Mendel observe in the F_3 generation when he self-fertilized the tall F_2 plants?

   (d) Demonstrate how a test cross could be used to identify whether the genotype of a tall plant is *DD* or *Dd*?

3. **Table 2.2**
   (a) For the cross *AA* × *aa*, how many F_1 gametic genotypes are there? What are they?

   (b) Draw the Punnett square for self-fertilization of the F_1 generation.

   (c) How many different genotypes are there in the F_2 progeny? How many different F_2 phenotypes are there if dominance is complete?

   (d) In the dihybrid cross, *AaBB* × *AaBb*, how many F_1 gametic genotypes are there?

   (e) What are the F_1 gametic genotypes in the dihybrid cross in part (d)?

   (f) Draw the Punnett square for self-fertilization of the F_1 generation.

   (g) In the Punnett square, indicate the different F_2 genotypes. How many are there?

   (h) For the cross, *AaBbCcDd* × *AABBccdd*, what is the proportion of recessive homozygotes among the F_2 progeny?

   (i) For the cross, *AaBbCcDd* × *AABBccdd*, break down the genotype into four independently assorting genes to determine the probability of producing the *AaBbCcDd* heterozygote from the cross. Show the Punnett square for the cross between each gene.

General Questions
Q2-1. What were the important factors in Mendel's experiments that allowed him to make reliable conclusions about how organisms inherit traits?

Q2-2. Mendel's law of segregation states that during gamete formation the two alleles separate randomly with each gamete having an equal probability of receiving either allele. Describe the experimental observations that this law explains.

Q2-3. Would it be possible to demonstrate independent assortment using a monohybrid cross? Explain.

Q2-4. Describe the difference between mutually exclusive outcomes and independent outcomes. How does one determine the probability that one of several mutually exclusive events will occur? How does one calculate the probability that two independent events will both occur?

Q2-5. From the vantage point of a scientist, what advantage does the use of the chi-square test offer as a tool in decision making?

Multiple Choice

For each of the following, circle the letter of the choice that most appropriately answers the question.

1.) When Mendel crossed a pure-breeding dominant individual with a pure-breeding recessive individual, he found that the F$_1$ progeny were which of the following?

    a. all homozygous recessive

    b. all homozygous dominant

    c. all heterozygous recessive

    d. all heterozygous dominant

    e. 1/2 homozygous dominant,
         1/2 homozygous recessive

2.) If the progeny of the dihybrid cross *WWGG* × *wwgg* are self-fertilized, how many of the F$_2$ offspring would be expected to have the genotype *Wwgg?*

    a. 0

    b. 1/16

    c. 1/8

    d. 3/16

    e. 1/4

3.) In the monohybrid cross, *Ww* × *ww*, where *W* is round and dominant and *w* is wrinkled and recessive, what is the expected phenotypic ratio of the F$_1$ offspring?

    a. 3 round : 1 wrinkled

    b. 1 round : 1 wrinkled

    c. 4 round : 0 wrinkled

    d. 1 round : 3 wrinkled

    e. 0 round : 4 wrinkled

4.) In the cross described in question 3, what is the expected genotypic ratio of the F$_1$ progeny?

    a. 1 *WW* : 2 *Ww* : 1 *ww*

    b. 2 *WW* : 2 *ww*

    c. 4 *Ww* : 0 *ww*

    d. 2 *Ww* : 2 *ww*

    e. 3 *Ww* : 1 *ww*

5.) If the F_1 generation of the dihybrid cross *WWGG* × *wwgg* (where *W* is round, *G* is yellow, and both are dominant, while *w* is wrinkled and *g* is green) is self-fertilized, what is the expected probability of getting a wrinkled, green plant in the F_2 generation?

    a. 1/4

    b. 3/16

    c. 9/16

    d. 1/16

    e. 1/8

6.) In the cross described in question 5, what is the expected probability of getting a round, green plant in the F_2 generation if the F_1 generation is self-fertilized?

    a. 1/4

    b. 3/16

    c. 1/8

    d. 9/16

    e. 1/16

7.) If you throw three dice, what is the probability that each die will be either a 3, 5, or a 6?

    a. 1/4

    b. 1/6

    c. 1/8

    d. 1/2

    e. 1/12

8.) Which of the following conventions do botanists and mammalian geneticists use to designate dominant and recessive alleles?

    a. + and − symbols

    b. different combinations of letters

    c. superscript numbers or letters

    d. uppercase and lowercase letters

    e. subscript numbers or letters

9.) In the cross, *AaBBCcDd* × *AABbccdd*, what is the probability of producing offspring that are heterozygous for all four genes?

    a. 1/8

    b. 1/16

    c. 1/32

    d. 3/32

    e. 9/32

10.) In the cross described in question 9, what is the probability of producing offspring that are phenotypically identical to one of the parents?

    a. 3/32

    b. 9/32

    c. 1/8

    d. 1/4

    e. 1/2

Practice Problems

P2-1. Consider the condition *epicanthus* in humans. This disorder, a folding of the upper eyelid, is inherited as an autosomal dominant trait. An affected male marries a normal woman and they start a family. What proportion of their children is expected to be affected by this disorder? What are the genotypes of the man and his wife?

(*Hint 1*: When determining the probability of having an offspring with a certain genotype or phenotype, look at the number of possible genotypes you observe in your Punnett square. For example, if you use a monohybrid cross and one of four is affected, that chance is 1/4 or 25%. If you consider a dihybrid cross and 3 of 16 are affected, that chance is 3/16 or 18.75%.)

P2-2. The offspring of a dihybrid cross are individually testcrossed. Predict the frequency of progeny expected in each phenotypic class.

(*Hint 2*: Diagram the cross under consideration. Use the symbols you establish consistently throughout the problem. Remember that each individual is diploid and therefore must have two alleles for each trait. Eggs and sperm normally contain only one allele for each trait. In cases involving the inheritance of two or more traits, always group alleles for the same trait together.)

P2-3. In mice, the allele for short tail (*A*) is dominant over the allele for long tail (*a*). The genotype *AA* is lethal. Predict the results of a cross between two mice with short tails.

P2-4. Consider the crosses in garden peas provided in the following table. For each of the two traits (flower color and pod position), determine which is dominant and which is recessive. For each cross, determine the genotype of both parents.

Cross	Offspring Phenotypes			
	Purple, Axial	Purple, Terminal	White, Axial	White, Terminal
Purple, axial × purple, axial	99	32	29	12
White, axial × purple, axial	112	38	114	40
Purple, terminal × white, axial	150	0	155	0
Purple, terminal × white, axial	100	99	101	104
White, terminal × white, terminal	0	0	0	192

(*Hint 3*: Examine the phenotypic ratio of the progeny to determine the genotype of the parents.
   a. A 3:1 ratio among the offspring *suggests* that both parents are heterozygous.
   b. A 1:1 ratio among the offspring *suggests* that one parent is heterozygous and the other is homozygous recessive.
   c. A 1:2:1 ratio among the offspring *suggests* that both parents are heterozygous and the mode of inheritance is incomplete dominance or codominance.
   d. A 9:7, 9:3:4, 12:3:1, 13:3, or 15:1 ratio *suggests* epistasis.
   e. A 2:1 ratio *suggests* a homozygous lethal and that both parents are heterozygous.
   f. A 9:3:3:1 ratio *suggests* that the parents are heterozygous for two traits.
   g. A 1:1:1:1 ratio *suggests* that, for each of two traits, one of the parents is heterozygous while the other parent is homozygous recessive. There are two possible crosses: *AaBb* × *aabb* and *Aabb* × *aaBb*.
   h. A 3:3:1:1 ratio *suggests* that one parent is heterozygous for two traits while the other parent is heterozygous for one gene and homozygous recessive for the second gene.)

P2-5. Maple sugar urine disease is a recessive autosomal disorder in humans. Assume that two carriers of maple sugar urine disease marry and start a family. If they have four children, what is the chance that the children will have the following phenotypes, in the following order: normal, normal, affected, normal? What is the probability that any one of the four will be affected?

[*Hint 4*: Each reproductive event is independent. The ratios given in *Hint 3* are predicted given large numbers of offspring. For example, do not mistakenly believe that a couple (each carrying a recessive disease allele) will necessarily have three normal children if their first child was affected. Rather, remember that each time the couple conceives a child, there is a 25% chance that the child will be affected regardless of the outcome of earlier pregnancies.

Hint 5: When determining the probability of having an offspring with a certain genotype or phenotype, look at the number of possible genotypes you observe in your Punnett square. For example, if you use a monohybrid cross and one of four is affected, that chance is 1/4 or 25%. If you consider a dihybrid cross and 3 of 16 are affected, that chance is 3/16 or 18.75%.

Hint 6: The product rule is used when you want to know the probability of obtaining both A **and** B.]

P2-6. A geneticist observed 46 green seedlings and 11 albino (without chlorophyll) seedlings among the progeny of a cross between two green tomato plants. What hypothesis should the geneticist test? Apply the chi-square test to the data and interpret the results.

(*Hint 7*: The critical chi-square value, the value to which a calculated chi-square value is compared, is obtained from a table. You must know the degrees of freedom and choose a level of significance to determine the critical value. The degrees of freedom represent a count of independent categories. For tests of Mendelian ratios, the degrees of freedom are equal to the number of categories minus one. The level of significance traditionally used is 0.05.

Hint 8: The null hypothesis is rejected if the calculated chi-square value exceeds the critical value obtained from the table. If the calculated value falls below the critical value, the researcher fails to reject the null hypothesis.)

P2-7. Mendel crossed pea plants with round and yellow seeds to a plant with wrinkled and green seeds. The F_1 generation was composed of plants producing round and yellow seeds. A cross of the F_1 plants resulted in 315 plants producing round and yellow seeds, 108 plants producing round and green seeds, 101 plants producing wrinkled and yellow seeds, and 32 plants producing wrinkled and green seeds. Mendel proposed that this ratio approximated a 9:3:3:1 ratio. Use the chi-square test to statistically test this hypothesis. Test the hypothesis that the data approximate a 1:1:1:1 ratio. Clearly state your conclusions.

(*Hint*: See *Hints 7* and *8*.)

Assessing Your Knowledge

Understanding the Key Concepts—Answers

1.) cross-fertilization; 2.) F_1; 3.) dominant; 4.) self-fertilized; 5.) F_2; 6.) recessive; 7.) F_1; 8.) genotype; 9.) heterozygous; 10.) phenotypes; 11.) genotypes; 12.) self-fertilizing; 13.) homozygous; 14.) allele; 15.) dominant; 16.) 3:1; 17.) genotypic; 18.) 1:2:1; 19.) law of segregation; 20.) dihybrid; 21.) simultaneously; 22.) independently; 23.) alleles; 24.) gene; 25.) gametes; 26.) independent; 27.) alleles; 28.) gene; 29.) gametes; 30.) equal; 31.) probability; 32.) observed; 33.) total number; 34.) mutually exclusive; 35.) sum; 36.) probabilities; 37.) independent outcomes; 38.) product; 39.) probabilities; 40.) chi-square distribution; 41.) statistically; 42.) probability; 43.) chi-square; 44.) *p* = 0.05; 45.) degrees of freedom.

Figure Analysis—Answers

1a.) All of the F_1 plants were tall, and the dwarf trait was absent in the F_1 generation; b.) The dwarf trait is recessive and reappears in the F_2 generation. Some of the F_2 progeny are homozygous for the *d* allele and are therefore dwarf because their parents are genotypically *Dd*, rather than being *DD*; c.) The phenotypic ratio of tall:dwarf plants in the F_2 generation is 3:1 because one-fourth of the progeny are *DD* and phenotypically tall, one-half are *Dd* and are phenotypically tall, and one-fourth are *dd* and phenotypically dwarf. The dominant allele is *D*. All plants with this allele are tall. The recessive allele is *d*. Only plants that are homozygous for this allele are dwarf. Therefore, the phenotypic ratio of tall:dwarf is 3:1.

2a.) Mendel self-fertilized the F_2 generation to produce the F_3 generation to demonstrate that the genotypic ratio of the F_2 offspring was 1 *DD*:2 *Dd*:1 *dd*; b.) When Mendel self-fertilized the dwarf F_2 plants, he found that they produced only dwarf offspring in the F_3 generation. This is because the dwarf F_2 plants produced gametes with only the *d* allele; c.) When Mendel self-fertilized the tall F_2 plants, he found that one-third bred true producing only tall offspring, but two-thirds produced both tall and dwarf offspring in a 3:1 ratio. This occurred because the tall F_2 plants had the genotype *DD* or *Dd*. The *DD* F_2 plants bred true producing only tall plants, but the *Dd* F_2 plants have the recessive dwarf allele and produce gametes bearing either the *D* allele or the *d* allele. Three of the offspring produced by the *Dd* F_2 plants are phenotypically tall (*DD* or *Dd*), while one of the offspring produced is phenotypically dwarf (*dd*); d.) In a testcross, an organism is crossed to a recessive homozygote. If the organism of unknown genotype is *DD*, the testcross will yield *DD* × *dd*, where all offspring have the genotype *Dd* and are tall. If the organism of unknown genotype is *Dd*, the testcross will yield *Dd* × *dd*, where one-half of the offspring have the genotype *Dd* and are tall and one-half of the offspring have the genotype *dd* and are dwarf.

3a.) For the cross, *AA* × *aa*, there are two F_1 gametic genotypes. They are *A* and *a*.

b.)

	A	*a*
A	*AA*	*aA*
a	*aA*	*aa*

c.) There are three different genotypes in the F_2 progeny. If dominance is complete, there are two different F_2 phenotypes; d.) In the dihybrid cross *AaBB* × *AaBb*, there are four F_1 gametic genotypes; e.) The F_1 gametic genotypes are *AB*, *ab*, *aB*, *Ab*.

f.)

	AB	*ab*	*aB*	*Ab*
AB	**AABB**	**AaBb**	**AaBB**	**AABb**
ab	*AaBb*	**aabb**	**aaBb**	**Aabb**
aB	*AaBB*	*aaBb*	**aaBB**	*AaBb*
Ab	*AABb*	*Aabb*	*AaBb*	**AAbb**

g.) There are nine different genotypes in the F_2 progeny indicated in bold in the Punnett square; h.) 1/256—use the general rule for n = 4.

i.)

	A	*a*			*B*	*b*			*C*	*c*			*D*	*d*
A	*AA*	*Aa*		*B*	*BB*	*Bb*		*c*	*Cc*	*cc*		*d*	*Dd*	*dd*
A	*AA*	*Aa*		*B*	*BB*	*Bb*		*c*	*Cc*	*cc*		*d*	*Dd*	*dd*

Aa: (1/2) × *Bb*: (1/2) × *Cc*: (1/2) × *Dd*: (1/2) = 1/16 is the probability of producing *AaBbCcDd*.

General Questions—Answers

Q2-1. Mendel crossed plants that he ensured were pure-breeding for the traits that he was attempting to look at. Therefore, his data was reliable and easy to interpret. Mendel also analyzed discrete, nonoverlapping characteristics and observed the distribution of these characteristics over the next several generations so that he could unambiguously describe each plant. Mendel used a quantitative, rather than qualitative, approach. He also conducted simple experiments and looked for statistical regularities (e.g., a 3:1 ratio).

Q2-2. This law explains why the F_1 progeny all have the dominant tall characteristic and why the dwarf phenotype reappears in the F_2 generation. All the F_1 progeny are heterozygous possessing two different alleles (Dd). The dwarf phenotype reappears in the F_2 generation because the F_1 progeny all possess the recessive allele. Each F_1 individual produces two kinds of gametes in equal frequencies D and d. When these individuals are self-fertilized, these gametes randomly fuse to produce the F_2 generation. This is diagrammed in the Punnett square:

	D	d
D	DD	dD
d	Dd	dd

From this, you can see that the 3:1 ratio of tall:dwarf plants can be explained by the hybrid nature of the F_1 individuals.

Q2-3. A monohybrid cross cannot be used to demonstrate independent assortment. Independent assortment states that segregation of a pair of alleles is independent of the separation of *other* pairs. In a monohybrid cross, only one pair of alleles is under consideration.

Q2-4. Mutually exclusive outcomes are events in which the occurrence of one possibility excludes all other possibilities, such as rolling a particular number on a die. Independent outcomes are events that do not influence one another, such as rolling two dice. The roll of the first does not affect the roll of the second die. The probability that one of several mutually exclusive events will occur is determined by the sum of the probabilities of the individual events. The probability that two independent events will occur is determined by the product of their individual probabilities.

Q2-5. Statistics provide geneticists with a consistent decision-making tool. Without statistics, a geneticist would need to rely on his or her "gut feeling" when interpreting data. While this means of examination may be effective in some cases, it does not provide a clear decision in all cases. For example, most people would agree that 49:51 closely approximates a 1:1 ratio, even without the benefit of statistics. However, the interpretation of a 60:40 ratio is substantially less clear. You are as likely to find geneticists who are willing to accept 60:40 as a 1:1 ratio as you are to find geneticists who do not accept this ratio. When statistics are used, the decision is more clear and a larger number of geneticists will agree on the interpretation.

Multiple Choice—Answers

1.) d; 2.) c; 3.) b; 4.) d; 5.) d; 6.) b; 7.) c; 8.) d; 9.) b; 10.) e.

Practice Problems—Answers

P2-1. The trait under consideration is inherited as an autosomal dominant; therefore both homozygotes and heterozygotes will be affected. The male parent in this problem is affected. His genotype is either *EE* or *Ee*. The female parent is phenotypically normal and therefore has the genotype *ee*. To answer this question, two different crosses must be performed because the specific genotype of the male is unknown.

Gametes	*E*	*E*
e	*Ee* (affected)	*Ee* (affected)
e	*Ee* (affected)	*Ee* (affected)

If the male is homozygous dominant (*EE*) and the female is homozygous recessive (*ee*), then all of the offspring produced will have the genotype *Ee* and would be affected. Male (*EE*) × female (*ee*) yields a 100% chance (4 of 4) of affected (*Ee*) children.

Gametes	*E*	*e*
e	*Ee* (affected)	*ee* (unaffected)
e	*Ee* (affected)	*ee* (unaffected)

If the male is a heterozygote (*Ee*), one-half of his sperm will contain the dominant allele (*E*) and the other half of the sperm will carry the normal allele (*e*). All of the female's eggs contain the normal allele (*e*). Male (*Ee*) × female (*ee*) yields a 50% chance (2 of 4) of affected (*Ee*) and a 50% chance (2 of 4) of unaffected (*ee*) children.

P2-2. In the case of a testcross, the individual with an unknown genotype is crossed with a homozygous recessive individual. The genotype of the unknown individual will ultimately determine the phenotype observed: a dominant phenotype can only result if the unknown individual contributes a dominant allele. First, take a look at the possible gametes produced by the parent of unknown genotype.

Unknown Parent Genotype	Gametes Produced by Unknown Parent
AABB	All *AB*
AaBB	1/2 *AB* and 1/2 *aB*
AABb	1/2 *AB* and 1/2 *Ab*
AaBb	1/4 *AB*, 1/4 *Ab*, 1/4 *aB*, and 1/4 *ab*
AAbb	All *Ab*
Aabb	1/2 *Ab* and 1/2 *ab*
aaBB	All *aB*
aaBb	1/2 *aB* and 1/2 *ab*
aabb	All *ab*

Next, add the gamete (100% *ab*) from the homozygous recessive parent (*aabb*) to create the possible offspring and examine the **phenotype** of those individuals.

Testcrosses	Frequency of Each Phenotype			
	Dominant *A*, Dominant *B*	Dominant *A*, Recessive *b*	Recessive *a*, Dominant *B*	Recessive *a*, Recessive *b*
AABB × *aabb*	100%	0	0	0
AaBB × *aabb*	50%	0	50%	0
AABb × *aabb*	50%	50%	0	0
AaBb × *aabb*	25%	25%	25%	25%
AAbb × *aabb*	0	100%	0	0
Aabb × *aabb*	0	50%	0	50%
aaBB × *aabb*	0	0	100%	0
aaBb × *aabb*	0	0	50%	50%
aabb × *aabb*	0	0	0	100%

P2-3. All mice with short tails are heterozygous because the AA genotype is lethal.

Gametes	A	a
A	AA (dies)	Aa (short tail)
a	Aa (short tail)	aa (long tail)

If you cross two heterozygotes, you expect to find a 1/4 AA:1/2 Aa:1/4 aa genotypic ratio. Elimination of the AA individuals yields 2/3 short-tailed mice (Aa) and 1/3 long-tailed mice (aa).

P2-4. To begin this problem, you need to first examine all crosses and determine which traits are dominant. Notice that the purple, axial plants in cross (a) produced a generation of plants containing all possible combinations of traits in approximately a 9:3:3:1 ratio. This ratio is expected from a cross of two dihybrids. The traits possessed by the dihybrids (purple and axial) are dominant. Once the dominant and recessive traits have been identified, fill in as many of the genotypes as possible (for example, you know that white plants are genotype pp while purple plants have at least one dominant allele P). To determine if plants with the dominant phenotype are homozygous or heterozygous, examine the phenotypes produced as well as the phenotypic ratios. For example, we can deduce that the purple plant in (b) is heterozygous because the cross produced a 1:1 ratio of plants with purple flowers to plants with white flowers. Genotypes for (b) could also be assigned by recognizing the 3:3:1:1 ratio.

    a. $PpAa \times PpAa$ 9:3:3:1
    b. $ppAa \times PpAa$ 1 purple:1 white; 3 Axial:1 terminal
    c. $Ppaa \times ppAA$ 1 purple:1 white; all axial
    d. $Ppaa \times ppAa$ 1:1:1:1
    e. $ppaa \times ppaa$ All white terminal

P2-5. In this example, the probability of producing a normal child is 3/4 and the probability of producing an affected child is 1/4. The probability of having four children (sequence: normal, normal, affected, normal) is calculated as $(3/4)^2 \times (1/4)^1 \times (3/4)^1 = 0.1055$. The probability that three will be normal and one affected, without specifying a birth order, will be $4(3/4)^3 \times (1/4)^1$ or 0.4219.

P2-6. This ratio appears to be a 3:1 (green:albino) ratio. The hypothesis to be tested is, *The ratio obtained by the geneticist approximated a 3 green to 1 albino ratio*. The chi-square test should be used. The chi-square value is calculated: $\sum[(observed - expected)^2 / expected]$.

Phenotype	Observed	Expected number
Green seeds	46	$3/4 \times 57 = 42.75$
Albino seeds	11	$1/4 \times 57 = 14.25$
Total	57	

$\chi^2 = (46 - 42.75)^2/42.75 + (11 - 14.25)^2/14.25$
   $= 0.247 + 0.741$
   $= 0.988$ (1 df)

Critical value (2 − 1 = 1 df; *Hint 4*) = 3.841
Fail to reject hypothesis because 0.988 < 3.841.

P2-7. Testing the hypothesis that the data approximate a 9:3:3:1 ratio:

Phenotype	Observed	Expected number
Round, yellow	315	$9/16 \times 556 = 312.75$
Round, green	108	$3/16 \times 556 = 104.25$
Wrinkled, yellow	101	$3/16 \times 556 = 104.25$
Wrinkled, round	32	$1/16 \times 556 = 34.75$
Total	556	

$\chi^2 = (315 - 312.75)^2/312.75 + (108 - 104.25)^2/104.25 + (101 - 104.25)^2/104.25 +$
    $(32 - 34.75)^2/34.75$

    $= 0.016 + 0.135 + 0.101 + 0.218$
    $= 0.47$ (3 df)

Fail to reject null hypothesis because $0.47 < 7.815$ (critical value for 3 df). The data approximate a 9:3:3:1 ratio.

Testing the hypothesis that the data approximate a 1:1:1:1 ratio:

Phenotype	Observed	Expected Number
Round, yellow	315	$1/4 \times 556 = 139$
Round, green	108	$1/4 \times 556 = 139$
Wrinkled, yellow	101	$1/4 \times 556 = 139$
Wrinkled, round	32	$1/4 \times 556 = 139$
Total	556	

$\chi^2 = (315 - 139)^2/1389 + (108 - 139)^2/139 + (101 - 139)^2/139 + (32 - 139)^2/139$
    $= 222.849 + 6.914 + 10.389 + 82.367$
    $= 322.519$ (3 df)

Reject the null hypothesis because $322.519 > 7.815$ (critical value for 3 df).

The data *do not* approximate a 1:1:1:1 ratio.

2

MENDEL'S PRINCIPLES

CHAPTER SUMMARY QUESTIONS

1. Cross-fertilization occurs when the pollen of one plant is used to fertilize the stigma of another plant. In self-fertilization, or selfing, the pollen is used to fertilize the stigma of the same plant. Mendel used both methods.

2. Perform a testcross by crossing the tall pea plant of unknown genotype with a homozygous recessive dwarf plant (*dd*). If the offspring consist only of tall pea plants, then the plant in question would be homozygous (cross: *DD* × *dd* → *Dd*). If the offspring consists of both tall and dwarf pea plants, then the plant in question would be heterozygous (cross: *Dd* × *dd* → 1/2 *Dd*:1/2 *dd*).

3. Mendel's second law, independent assortment, will not hold if the two genes are close to each other on the same chromosome.

4.

Cross (P generation)	$AABB \times$ $aabb$	$AABBCC \times$ $aabbcc$	$AABBCCDD \times$ $aabbccdd$
Number of different F_1 gametes	4	8	16
Number of different F_2 genotypes	9	27	81
Number of different F_2 phenotypes	4	8	16
Degrees of freedom in chi-square test	3	7	15

5. The offspring are in an approximate 3:1 phenotypic ratio of black to white. This indicates that black is dominant, white is recessive, and both parents are heterozygous. Because both parents had the same phenotype, the cross would be $Ww \times Ww$. (Remember that alleles are typically designated by the recessive trait.)

6. **a.** The offspring are in an approximate 3:1 phenotypic ratio of red eyes to brown eyes. The 3:1 ratio suggests that this is a monohybrid cross and that red eye is dominant and brown eye is recessive. The cross would be $Bb \times Bb$.

   b. There are two phenotypes in approximately a 1:1 ratio in the offspring. We don't know which allele is dominant from this cross. The two phenotypes indicate one gene, and the 1:1 ratio indicates a mating of a heterozygote and homozygous recessive individuals (the testcrossing of a heterozygote). Therefore, the cross can be depicted as $Aa \times aa$.

7. Washed eye mutant, We; wild type, We^+. (Dominant mutations are denoted with an uppercase letter, and W is already the allelic designation for the wrinkled phenotype.)

8. In the table, D = dominant; R = recessive; M = mutant; + = wild type.

Allele	Dominant or Recessive	Mutant or Wild Type	Alternative Allele	Dominant or Recessive
y^+	D	+	y	R
Hw	D	M	Hw^+	R
Ax^+	R	+	Ax	D
Co	D	M	Co^+	R
rv^+	D	+	rv	R
dow	R	M	dow^+	D
$M(2)e^+$	R	+	$M(2)e$	D
J	D	M	J^+	R
tuf^+	D	+	tuf	R
bur	R	M	bur^+	D

9. **a.** The chi square test is a statistical test used to determine the *probability* that the observed numbers deviate from those expected under a particular hypothesis.

    b. The null hypothesis assumes that the difference between the observed and expected numbers is due to chance.

    c. The level of significance is the probability level that is used as a cutoff for rejecting a hypothesis. It is typically set at 0.05.

10. 1. b, 2. a, 3. b, 4. c (one can pick the ace of diamonds).

EXERCISES AND PROBLEMS

11. **a.** Only one gamete: *AbCdE*.
     b. Two gametes: *aBCDe* and *aBcDe*.
     c. Eight gametes: *ABCDE, ABCDe, ABCdE, ABCde, AbCDE, AbCDe, AbCdE,* and *AbCde*.

12. **a.** *AA* × *Aa, aa* × *Aa, AA* × *aa,* and *Aa* × *Aa*.
     b. *AA* × *AA* and *aa* × *aa*.

13. **a.** Both the kitten and its father are long haired, so they must be homozygous for the recessive allele (*l*). The kitten could only inherit a recessive *l* allele from its father and must have obtained the other *l* allele from its mother. Therefore, she must be heterozygous (*Ll*).

     b. The kitten must be heterozygous *Ll* and could only inherit the recessive *l* allele from its homozygous recessive father. The kitten must have inherited the *L* allele from its mother. The genotype of the mother cannot be determined for sure. The mother can be homozygous for the short allele (*LL*) or heterozygous (*Ll*). In both cases, she can contribute an *L* allele to her kitten.

14. The 3:1 phenotypic ratio of the F_2 generation is composed of a 1:2:1 genotypic ratio that corresponds to 1/4 *DD*, 1/2 *Dd*, and 1/4 *dd*. The 1/4 dwarf F_2 (*dd*), when selfed, produce all dwarf progeny (*dd*). The tall F_2 (3/4 of total F_2), when selfed, fall into two categories: 1/4 (*DD*, 1/3 of the tall F_2) produces all tall, and 1/2 (*Dd*, 2/3 of the tall F_2) produces tall and dwarf progeny in a 3:1 ratio.

     We can compute the relative proportions of genotypes and phenotypes in the F_3 generation using the following table.

F$_2$ Genotype	Fraction of Genotype in F$_2$ Generation	F$_3$ Genotype		
		DD	Dd	dd
Dd (dwarf)	1/4			$1 \times 1/4 = 1/4$
DD (tall)	1/4	$1 \times 1/4 = 1/4$		
Dd (tall)	1/2	$1/4 \times 1/2 = 1/8$	$1/2 \times 1/2 = 1/4$	$1/4 \times 1/2 = 1/8$
F$_3$ genotypic ratio		3/8 DD	1/4 Dd	3/8 dd
F$_3$ phenotypic ratio		5/8 tall		3/8 dwarf

Overall, the F$_3$ are 3/8 *DD* (tall), 2/8 *Dd* (tall), and 3/8 *dd* dwarf.

15. The F$_1$ are tetrahybrids (*AaBbCcDd*). If selfed, an F$_1$ would form $2^4 = 16$ different types of gametes; $2^4 = 16$ different phenotypes would appear in the F$_2$, which would be made up of $3^4 = 81$ different genotypes; $1/(4^2)^2 = 1/256$ of the F$_2$ would have a genotype of *aabbccdd*.

16. When a decahybrid is selfed, it would produce $2^{10} = 1024$ different gametes; $1/(2^{10})^2$ or approximately 0.000001 of the F$_2$ would be homozygous recessive; $3^{10} = 59,049$ different genotypes yielding $2^{10} = 1024$ different phenotypes would appear. If the decahybrid were testcrossed, it would produce 2^{10} different gametes; $1/2^{10} = 0.00098$ of the F$_2$ would be homozygous recessive; 2^{10} different genotypes and phenotypes would appear.

17. Choice (b) is preferred because although each will reveal the correct genotype, generally, testcrossing has the greatest probability of exposing the recessive allele in a heterozygote. For example, testcrossing the *Aa* heterozygote will produce the *aa* offspring one-half of the time, while either selfing or backcrossing the *Aa* heterozygote produces the *aa* offspring one-fourth of the time. Thus, with a limited number of offspring examined per cross, testcrossing most reliably exposes the recessive allele.

18. The round, yellow F$_2$ plants are made up of four genotypes; the round, green of two genotypes; the wrinkled, yellow of two genotypes; and the wrinkled, green of one genotype. Testcrossing all these genotypes produces the following results:

W–G–:   1/16 *WWGG* × *wwgg* → all *WwGg* (round, yellow)
        2/16 *WwGG* × *wwgg* → 1/2 *WwGg* (round, yellow), 1/2 *wwGg* (wrinkled, yellow)

2/16 *WWGg* × *wwgg* → 1/2 *WwGg* (round, yellow), 1/2 *Wwgg* (round, green)

4/16 *WwGg* × *wwgg* → 1/4 *WwGg* (round, yellow), 1/4 *Wwgg* (round, green), 1/4 *wwGg* (wrinkled, yellow), 1/4 *wwgg* (wrinkled, green)

W–gg: 1/16 *WWgg* × *wwgg* → all *Wwgg* (round, green)

2/16 *Wwgg* × *wwgg* → 1/2 *Wwgg* (round, green), 1/2 *wwgg* (wrinkled, green)

wwG–:   1/16 *wwGG* × *wwgg* → all *wwGg* (wrinkled, yellow)

2/16 *wwGg* × *wwgg* → 1/2 *wwGg* (wrinkled, yellow), 1/2 *wwgg* (wrinkled, green)

wwgg: 1/16 *wwgg* × *wwgg* → all *wwgg* (wrinkled, green)

19. a. In the first cross, the yellow-to-green ratio, 120:43—almost exactly 3:1. Therefore, yellow is dominant and both parents must be heterozygous. Now look at the tall-to-short ratio, 122:41. Again, we see a 3:1 ratio, which indicates that tall is dominant and each parent is heterozygous. Thus, the cross is most likely the following dihybrid cross: *GgDd* × *GgDd*.

b. In the second cross, there are no tall progeny. Therefore, either the dwarf phenotypes are homozygous, or dwarf is dominant and at least one parent is homozygous. In the absence of the first cross, we can't determine the mode of inheritance of height. We can, however, conclude that yellow is dominant (we got a 3:1 ratio) and that each parent is heterozygous. Based on the first cross, we can conclude that this cross is *Ggdd* × *Ggdd*.

c. In the third cross, we see 41 yellow:46 green, and 45 tall:42 dwarf. Both of these ratios are approximately 1:1, and all we can conclude is that these ratios result from matings between a heterozygote and a recessive homozygote. With only this cross, we can't determine dominance. However, we can if we use all three crosses. The cross is *Ggdd* × *ggDd*.

20. Examine each trait separately. There are 104 long:34 short; and 69 brown:69 red. Length appears in a 3:1 ratio; therefore, long is dominant and the 3:1 ratio in the progeny tells us that each parent is heterozygous. Eye color appears in a 1:1 ratio. We can't conclude which allele is dominant; all we can conclude is that one parent is a recessive homozygote and that one parent is a heterozygote. (We will assume that red is the wild type.) If bw^+ = red, bw = brown, s^+ = long, and s = short, one possible way to indicate the cross is $bw^+bw\ s^+s \times bwbw\ s^+s$.

21. If you examine each trait separately, you will see in the F_1 progeny that long is dominant to short and tan must be dominant to dark. The F_2 result confirms this assumption. We see a 3:1 ratio for both tan:dark and long:short. The total number of flies is 80. An ideal 9:3:3:1 ratio would be 45:15:15:5. Our results are very close to this. Therefore, we conclude that long and tan are dominant, and that the F_1 flies were heterozygous.

22. You could set up a Punnett square and count the boxes; but unfortunately, this is an 8 × 8 matrix that yields 64 squares to count. A very tedious proposition! Set up the cross: *DdGgWw* × *DdGgWw* and look at one gene at a time.

    a. The chance of getting a dominant trait from a monohybrid cross is 3/4. Therefore, the chance of all three dominant traits together is 3/4 × 3/4 × 3/4 = 27/64.

    b. The chance of getting a recessive trait from a monohybrid cross is 1/4. Therefore, the chance for all three recessives is 1/4 × 1/4 × 1/4 = 1/64.

    c. The chance of dwarf is 1/4, the chance of green is 1/4, and the chance of round is 3/4. Therefore the total chance producing the dwarf, green, round plant is 1/4 × 1/4 × 3/4 = 3/64.

23. In cross (a), both parents have crossveins, and we see an approximate 3:1 ratio in the progeny. Crossveins must be dominant. In cross (b), both parents have red eyes, and we see an approximate 3:1 ratio in the progeny. Red eyes must be dominant.
Let or^+ = red, or = orange, cv^+ = crossveins, cv = crossveinless.

    Now that we've established dominance, let's reexamine each of the four crosses. Cross (a) involves orange-eyed parents who must be homozygous (orange is recessive). The progeny have an approximate 3 crossveins:1 crossveinless ratio, which suggests the parents are heterozygous for the *crossveinless* gene. Therefore the cross is $oror\ cv^+cv \times oror\ cv^+cv$.

    In cross (b) we see an approximate 3:1 ratio of red to orange indicating the parents are heterozygous for the *orange* gene. The progeny have an approximate 1 crossveins:1 crossveinless ratio, which suggests the parent exhibiting the crossveins phenotype is heterozygous. Therefore, the cross is $or^+or\ cv^+cv \times or^+or\ cvcv$.

    Cross (c) yields no orange phenotype indicating that at least one of the parents is homozygous dominant for the wild-type red allele (or^+). The progeny also have an approximate 1 crossveins:1 crossveinless ratio, which suggests the crossveins parent is heterozygous. Therefore, there are two choices possible for this cross: (1) $or^+or^+cv^+cv \times or^+or\ cvcv$ **or** (2) $or^+or\ cv^+cv \times or^+or^+\ cvcv$.

    In cross (d) we see four phenotypes in an approximate 9:3:3:1. This result indicates a mating between two double heterozygotes $or^+or\ cv^+cv \times or^+or\ cv^+cv$

24.   **a.** Let e = ebony, e^+ = wild type, b = black, and b^+ = wild type. The first cross is $ee\ b^+b^+ \times e^+e^+\ bb$. This will yield only $e^+e\ b^+b$. Thus, all F_1 flies will exhibit a wild type phenotype.

    b. Selfing of $e^+e\ b^+b$ results in the following:

        9/16 $e^+- b^+-$ wild type
        3/16 $e^+ - bb$ black
        3/16 $ee\ b^+ -$ ebony
        1/16 $ee\ bb$ ebony, black

    Since it is difficult to distinguish black and ebony, 7/16 will be dark-bodied and 9/16 will be normal.

    c. The first backcross is $e^+e\ b^+b \times ee\ b^+b^+$. Progeny = 1/2 ebony:1/2 normal.

The second backcross is $e^+e\ b^+b \times e^+e^+\ bb$. Progeny = 1/2 black:1/2 normal. In each case we have a testcross situation for only one gene.

25. a. 1/10,000 = 0.0001
   b. 1/10,000 = 0.0001 (Remember, *chance* has no memory!)
   c. $(1/10,000)^2$ = 0.00000001 (The product rule is used here.**)**

26. The F_1 progeny are *AaBbCcDdEe*. The chance of getting any individual with a particular homozygous genotype is $(1/4)^5$. Since we are looking for two different possibilities, we have $2(1/4)^5$ = 2/1024 = 1/512.

27. a. $1/2 \times 1/2 \times 1/2 \times 1/2$ = 1/16
   b. $1/2 \times 1/2 \times 1/2 \times 1/2$ = 1/16
   c. $(1/2 \times 1/2 \times 1/2 \times 1/2) + (1/2 \times 1/2 \times 1/2 \times 1/2)$ = 1/8 (probability of four boys or four girls)
   d. The probability of having four boys = 1/16. Therefore, the probability of having any combination other than four boys is 1 – 1/16 = 15/16.

28. This cross can be set up as a dihybrid cross: *AaPp* × *AaPp*, where *a* is the recessive albinism allele and *p* is the recessive PKU allele.

   a. The chance of a recessive genotype (*aa*) from two heterozygotes is 1/4.
   b. The probability of having albinism (*aa*) but not PKU (*P–*) is $1/4 \times 3/4$ = 3/16.
   The probability of having PKU (*pp*) but not albinism (*A–*) is $1/4 \times 3/4$ = 3/16.
   For an either/or situation, we add the probabilities, 3/16 + 3/16 = 6/16 = 3/8.
   c. For two independent events to occur simultaneously, we multiply their individual probabilities, $1/4 \times 1/4$ = 1/16.

29. a. The probability of two heterozygotes producing a child with the dominant phenotype (having no molars, *A–*) is 3/4. The probability of having four consecutive children without molars is $(3/4)^4$ = 81/256.
   b. When a particular order is specified, we multiply the probability of each event, $1/4 \times 1/4 \times 3/4 \times 3/4$ = 9/256.

30. We calculate the frequency of deaths from cancer as 300/900 = 1/3, and the deaths due to heart disease as 200/900 = 2/9.
   a. 1/3
   b. For death by cancer or heart disease, we add the probabilities, 1/3 + 2/9 = 5/9.

31. a. P(*A– B– C– D– E–*) = $3/4 \times 3/4 \times 1 \times 3/4 \times 1/2$ = 27/128.
   b. P(*A– B– C– D– ee*) or (*A– B– C– D– E–*) = 27/128 + 27/128 = 27/64.

c. The probability of having an individual with a parental phenotype is 27/64. Therefore, the remaining 1 − 27/64 = 37/64 will be phenotypically unlike either parent.

d. 1/2 × 1/2 × 1/2 × 1/2 × 1/2 = 1/32.

e. 1/2 × 1/2 × 1/2 × 1/2 × 1/2 = 1/32.

f. The probability of producing an individual with a parental genotype is 2/32. Therefore, the remaining 1 − 2/32 = 30/32 or 15/16 will be genotypically unlike either parent.

32. The cross is $Aa \times aa$. Therefore, there is a one-half chance of either a taster or nontaster offspring. We can imagine five different birth order scenarios that give one taster child (symbolized as T) and four nontasters (symbolized as N) children. T, N, N, N, N (where the taster child is the first-born); N, T, N, N, N (where the taster child is the second-born); N, N, T, N, N; N, N, N, T, N; and N, N, N, N, T. The probability of each scenario is 1/2 × 1/2 × 1/2 × 1/2 × 1/2 = 1/32. The probability that one of five children will be a taster is given by the sum rule, 1/32 + 1/32 + 1/32 + 1/32 + 1/32 = 5/32.

33. a. The selfing of a round-seeded hybrid produces nearly a 3:1 phenotypic ratio of round seeded to wrinkled seeded offspring. This suggests that round is dominant to wrinkled.

Hypothesis: $Ww \times Ww$ produces $W–:ww$ in a 3:1 ratio. Chi-square, one degree of freedom (df), = 0.263. Critical chi-square at 0.05, 1 df, = 3.841.

	Seed Type		
	Round	**Wrinkled**	**Total**
Observed numbers (O)	5474	1850	7324
Expected ratio	3/4	1/4	
Expected numbers (E)	5493	1831	7324
$O - E$	−19	19	
$(O - E)^2$	361	361	
$(O - E)^2/E$	0.066	0.197	$0.263 = \chi^2$

Fail to reject (that is, accept) the null hypothesis at the 0.05 level and therefore also at the 0.01 level.

b. Hypothesis: $Ww \times Ww$ produces $W–:ww$ in a 3:1 ratio. Chi-square, 1 df, = 0.474. Critical chi-square at 0.05, 1 df, = 3.841.

| | **Seed Type** | | |
	Round	**Wrinkled**	**Total**
Observed numbers (O)	47	13	60
Expected ratio	3/4	1/4	
Expected numbers (E)	45	15	60
$O - E$	2	−2	
$(O - E)^2$	4	4	
$(O - E)^2/E$	0.089	0.267	$0.356 = \chi^2$

Fail to reject (that is, accept) the null hypothesis at the 0.05 level and therefore also at the 0.01 level.

c. Hypothesis: The 567 F_2 plants were composed of two different genotypes in a 2:1 ratio that corresponded to Ww and WW, respectively. Chi-square, 1 df, = 0.127. Critical chi-square at 0.05, 1 df, = 3.841.

| | **Genotype** | | |
	Ww	**WW**	**Total**
Observed numbers (O)	374	193	567
Expected ratio	2/3	1/3	
Expected numbers (E)	378	189	567
$O - E$	−4	4	
$(O - E)^2$	16	16	
$(O - E)^2/E$	0.042	0.085	$0.127 = \chi^2$

Fail to reject (that is, accept) the null hypothesis at the 0.05 level and therefore also at the 0.01 level.

d. This cross produces equal numbers of violet and white flowers, which suggests that this is a heterozygote in a testcross. Similarly, there are an equal number of long and short stems that suggests a heterozygote in a testcross.

Hypothesis: $WwSs \times wwss$ produces a 1:1:1:1 of offspring. Chi-square, 3 df, = 1.084. Critical chi-square at 0.05, 3 df, = 7.815.

| | **Offspring** | | | | |
	$WwSs$	**$wwSs$**	**$Wwss$**	**$wwss$**	**Total**
Observed numbers (O)	46	40	37	41	166
Expected ratio	1/4	1/4	1/4	1/4	
Expected numbers (E)	41	41	41	41	166
$O - E$	5	−1	−4	0	
$(O - E)^2$	25	1	16	0	
$(O - E)^2/E$	0.610	0.024	0.390	0.000	$1.024 = \chi^2$

Fail to reject (that is, accept) the null hypothesis at the 0.05 level and therefore also at the 0.01 level.

34. Hypothesis: $WwGg \times WwGg$ produces $W{-}G{-}{:}W{-}gg{:}wwG{-}{:}wwgg$ in a 9:3:3:1 ratio. Critical chi-square at 0.05, 3 df = 7.815.

	Offspring				
	W–G–	**W–gg**	**wwG–**	**wwgg**	**Total**
Observed numbers (O)	315	108	101	32	556
Expected ratio	9/16	3/16	3/16	1/16	
Expected numbers (E)	312.75	104.25	104.25	34.75	556
$O - E$	2.25	3.75	–3.25	–2.75	
$(O - E)^2$	5.06	14.06	10.56	7.56	
$(O - E)^2/E$	0.016	0.135	0.101	0.218	$0.470 = \chi^2$

Since this chi-square, 0.470, is less than the critical chi-square, we fail to reject (that is, accept) our hypothesis of two-locus genetic control with dominant alleles at each locus.

35.

	Phenotype		
	Curly-Winged	**Straight-Winged**	**Total**
Observed numbers (O)	61	35	96
Expected ratio	3/4	1/4	
Expected numbers (E)	72	24	96
$O - E$	–11	11	
$(O - E)^2$	121	121	
$(O - E)^2/E$	1.681	5.042	$6.723 = \chi^2$

Critical chi-square at 0.05, 1 df, = 3.841. The chi-square for our null hypothesis was 6.723. Therefore, we reject the 3:1 ratio as an appropriate null hypothesis.

36. a. This is a dihybrid cross. From the F_1 progeny, we can determine that long-winged and tan-bodied must be dominant to short-winged and dark-bodied, respectively. Because this is a dihybrid cross, we expect the F_2 progeny should be in a 9:3:3:1 ratio.

   b. Our null hypothesis: This is a dihybrid cross with dominant alleles at each locus.

	Offspring				
	Long, Tan	**Long, Dark**	**Short, Tan**	**Short, Dark**	**Total**
Observed numbers (O)	84	27	35	14	160
Expected ratio	9/16	3/16	3/16	1/16	
Expected numbers (E)	90	30	30	10	160
$O - E$	−6	−3	5	4	
$(O - E)^2$	36	9	25	16	
$(O - E)^2/E$	0.400	0.300	0.833	1.600	$3.133 = \chi^2$

Critical chi-square at 0.05, 3 df, = 7.815. The chi-square for our null hypothesis, 3.133, is less than the critical chi-square. Therefore, we fail to reject (that is, accept) our hypothesis.

37. a. $1/6 \times 1/6 = 1/36$.

b. There are two scenarios in which a 3 and a 5 can be obtained. I. First die shows a 3, and second die shows a 5 (P = $1/6 \times 1/6 = 1/36$). II. First die shows a 5, and second die shows a 3 (P = $1/6 \times 1/6 = 1/36$). Using the sum rule (either-or), the overall probability = $1/36 + 1/36 = 1/18$.

c. P = 5/6. Simply put, there are five chances out of six that the second die will show a different number from the first.

d. P = 11/36. Simply put, there are 11 combinations out of a possible 36 that will give at least one 6. They are: 6, 1; 6, 2; 6, 3; 6, 4; 6, 5; 6, 6; 1, 6; 2, 6; 3, 6; 4, 6; and 5, 6. (We already accounted for the 6, 6 possibility and do not need to account for it a second time.)

e. The probability of rolling at least one 6 with two die is 11/36 [see part (d)]. Therefore, the probability of not rolling a 6 is $1 - 11/36 = 25/36$.

38. a. $1/2 \times 1/2 \times 1/2 \times 1/2 \times 1/2 = 1/32$.

b. $(1/2 \times 1/2 \times 1/2 \times 1/2 \times 1/2) + (1/2 \times 1/2 \times 1/2 \times 1/2 \times 1/2) = 2/32$ or $1/16$.

c. P(A– bb C– D– E–) = $3/4 \times 1/2 \times 1/2 \times 3/4 \times 1/1 = 9/64$.

d. The probability of an individual with a parental phenotype is $9/64 + 9/64 = 9/32$. Therefore, the remaining $1 - 9/32 = 23/32$ will be phenotypically unlike either parent.

e. P = 0, because all offspring of this cross will have the dominant E trait.

39. First list all possible genotypes for colored plants:
AACCRR
AACCRr
AACcRR
AACcRr
AaCCRR
AaCCRr

AaCcRR

AaCcRr

The first genotype can be eliminated because all progeny should be colored, regardless of the tester strain. *AACCRr* can be eliminated because the progeny of the first cross would have all been colored. *AACcRR* can be eliminated because the progeny of the second cross would have all been colored. *AaCCRR* can also be eliminated because the progeny of the third cross would have all been colored.

We are now left with *AACcRr, AaCCRr, AaCcRR,* and *AaCcRr*. Try *AACcRr × aaccRR* (cross 1), which will give 1/2 colored (*AaCcR–*):1/2 colorless (*AaccR–*) progeny. This could be the genotype, so try it in the second cross: *AACcRr × aaCCrr*. This too will give 1/2 colored and 1/2 colorless offspring (*AaC–rr*). Since this does not fit the observed result, the unknown genotype is not *AACcRr*. Now try *AaCCRr × aaccRR* (cross 1). This fits the results. Let's now consider another genotype: *AaCcRR*. The cross *AaCcRR × aaccRR* gives 3/4 colorless (*AaccRR, aaccRR,* or *aaCcRR*), which does not fit the results; therefore, the genotype is not *AaCcRR*. Now try *AaCcRr × aaccRR*. This, too, will give 3/4 colorless (*AaccR–, aaccR–,* or *aaCcR–*, which is not seen. Therefore, the genotype must be *AaCCRr*.

Confirm this with the other two crosses:

AaCCRr × aaCCrr → 1/4 colored:3/4 colorless, which fits.

AaCCRr × AAccrr → 1/2 colored:1/2 colorless, which fits.

40. You should change your choice because the box you chose originally has a 1/3 chance of containing the prize, whereas the remaining box has a 2/3 chance of containing the prize. The 1/3 chance of your choice is set by the fact that there were three equally likely choices at the beginning. When your friend eliminated an empty box, she left two choices: your original box and the third box. Since the probability of your original choice has not changed, the probability that the remaining box contains the prize must be 2/3 to give a combined probability of 1.0.

CHAPTER 3 MITOSIS AND MEIOSIS

Chapter Goals
1. Understand the key features of a chromosome and how chromosomes are classified.
2. Describe the key steps in the cell cycle and how the cycle is regulated.
3. Diagram and describe the relationship between mitosis and meiosis in both haploid and diploid cells.
4. Understand the relationship between Mendel's two laws and meiosis.

Key Chapter Concepts
3.1 Cell Structure and Function
- **Prokaryotes**: Prokaryotes, including bacteria, cyanobacteria, and archaebacteria have a single circular molecule of DNA, but lack a true nucleus. All other organisms, including fungi, protists, plants, and animals are eukaryotic organisms.
- **Eukaryotes**: Eukaryotes have linear DNA complexed with histones and contained within a nucleus surrounded by a nuclear membrane. In addition, eukaryotes have other membrane-bound organelles, including the **endoplasmic reticulum**, **Golgi apparatus**, **mitochondria**, and **chloroplasts** that are not common to most prokaryotes. Eukaryotes also possess structures, such as **centrosomes** that organize **microtubules**, and **spindle fibers** composed of microtubules that attach to chromosomes and ensure their proper movement during mitosis and meiosis. These structures are not present in prokaryotic cells which do not undergo mitosis or meiosis.

3.2 Chromosomes
- **Eukaryotic Chromosome Structure:** Chromosomes are structures carrying genes. After replication, chromosomes are composed of two genetically identical strands (**sister chromatids**) held together by a single **centromere**. The DNA is complexed with protein, and the chromosomes are linear arrangements of genes. Eukaryotic chromosomes are classified according to the position of the centromere: in the middle (**metacentric**), very near the end (**acrocentric**), between the middle and the end (**subtelocentric/submetacentric**), or at the end (**telocentric**) of the chromosome structure.
- **Chromosome Number (Ploidy):** Most eukaryotic cells are **diploid (2N)**. In a diploid cell, each type of chromosome is represented twice in the nucleus. One set of chromosomes is maternally derived, while the second set is paternally derived. Each pair of morphologically similar chromosomes is referred to as a **homologous pair**. **Gametes**, sperm and egg, usually contain only one set of chromosomes and are referred to as **haploid (1N)**.
- **Karyotype:** The karyotype of an organism is a photograph of the chromosomes of an organism taken during mitosis. Chromosomes can be distinguished by their size, centromere position, and banding patterns. Karyotyping allows the visualization of any gross chromosomal abnormalities as well as the determination of the sex of an organism.

3.3 The Cell Cycle and Its Control Mechanisms
- **The Cell Cycle:** When conditions are appropriate, cells undergo a continuous cycle of growth, DNA synthesis, and division. This cell cycle is divided into two main stages: interphase and mitosis. During interphase, the cell grows (**Gap 1** or G_1 **phase**), synthesizes DNA (**Synthesis** or **S phase**), and continues to prepare to divide (**Gap 2** or G_2 **phase**). Following this progression of phases (G_1 to S to G_2), the cell divides during the processes of mitosis and cytokinesis.
- **Checkpoints:** Cell cycle checkpoints, such as the G_1/S **checkpoint**, the G_S/M **checkpoint**, the **M checkpoint**, and the **spindle attachment checkpoint**, act as **surveillance mechanisms** to ensure that various events have been properly completed before the next phase of the cell cycle is initiated. The cell cycle can be delayed at any one of these checkpoints if a problem in the cell is detected, such as DNA damage, or improper attachment of the kinetochores to the spindle apparatus. Progression through these checkpoints depends upon the function of **cyclin-dependent kinases**, and the regulated **degradation** of proteins that either specifically inhibit or promote cell cycle progression.

3.4 Mitosis

- **Mitosis** is a nuclear division in which a diploid cell gives rise to two identical diploid cells. Mitosis, while not a lengthy process compared to the entire cell cycle, is further subdivided into phases: **prophase**, **metaphase**, **anaphase**, and **telophase**. During mitosis, the sister chromatids of each chromosome are separated into two different cells. In this way, each resulting cell receives an identical set of chromosomes. The **spindle apparatus** is responsible for the movement of chromosomes during mitosis. The nuclear division (**karyokinesis**) is followed by a division of the cytoplasm (**cytokinesis**). The progression of the cell cycle is genetically controlled.

3.5 Meiosis

- **Meiosis** is a nuclear division in which a diploid cell gives rise to four haploid cells. Meiosis is a reduction division in which the number of chromosomes is halved. There are two successive cell divisions: **meiosis I** and **meiosis II**. In meiosis I, homologous pairs of chromosomes are separated resulting in two *haploid* cells. This division is referred to as the **reduction division**. In meiosis II, sister chromatids are separated giving rise to four haploid cells. The second meiotic division equalizes the DNA content of the cells produced. This division is an **equational division**, the amount of genetic material per cell is reduced by half, but the chromosome number per cell is not further reduced. Both meiosis I and II are subdivided into phases: prophase, metaphase, anaphase, and telophase. Meiosis produces genetic diversity in organisms by the **random assortment** of homologous chromosomes and **recombination** in meiosis I which leads to different genotypic combinations in the gametes produced.

3.6 Comparison Between Mitosis and Meiosis

- There are similarities and differences between mitosis and meiosis. DNA replication occurs prior to both mitosis and meiosis. Mitosis involves only a single cell division, while meiosis involves two divisions. During meiosis I, homologous chromosomes pair which allows crossing over between nonsister chromatids to occur, but this pairing does not occur in mitosis. In mitosis, spindle fibers from opposite poles attach to sister chromatid kinetochores causing the sister chromatids to migrate to opposite poles in anaphase. In meiosis I, spindle fibers attach to homologous chromosome kinetochores causing homologous chromosomes to migrate to opposite poles in anaphase I, while sister chromatids remain attached and migrate to the same poles. In meiosis II, the sister chromatids separate and migrate to opposite poles in anaphase II. During mitosis, there is no change in the amount of DNA, the number of chromosomes, or allelic combinations between the parental cell and the two daughter cells. Meiosis, unlike mitosis, generates genetic diversity by independent assortment and recombination which produce new allelic arrangements.

3.7 Meiosis in Mammals

- In mammals, meiosis occurs in the gonads and results in the production of sperm or eggs. **Spermatogenesis**, occurring in male mammals, produces four equally sized **sperm cells**. In females, **oogenesis** results in the formation of four haploid cells of unequal size: a larger **ovum** (egg) and three smaller polar bodies. Meiosis is an integral component of sexual reproduction in mammals for two reasons. First, meiosis provides a mechanism for the maintenance of chromosome number. Meiosis produces haploid gametes that, at fertilization, restore the diploid number. Second, meiosis generates variation.

3.8 Life Cycles

- In most eukaryotic organisms, an alternation between a diploid and a haploid state is observed. In plants, there is an alternation of generations with important roles for both mitosis and meiosis. The flowering plant is a diploid **sporophyte**. The sporophyte gives rise to a haploid **gametophyte** through the process of meiosis. The gametophyte produces gametes by mitosis.

3.9 Chromosomal Theory of Inheritance

- This theory was proposed by Walter Sutton in 1902. The theory states that chromosomes are linear sequences of genes. The behavior of chromosomes during meiosis correlates with, and provides an explanation for, Mendel's principles.

Key Terms

- nondisjunction
- prokaryotes
- eukaryotes
- nucleoprotein
- endoplasmic reticulum
- Golgi apparatus
- mitochondria
- chloroplast
- centrosome
- centrioles
- microtubules
- spindle fibers
- spindle pole body
- chromosome
- euchromatin
- heterochromatin
- interphase
- centromere
- kinetochore
- metacentric
- telocentric
- acrocentric
- p arm
- q arm
- diploid
- homologous chromosomes
- haploid
- gametes
- karyotype
- homomorphic chromosome pairs
- homogametic
- heterogametic
- mitosis
- meiosis
- cytokinesis
- interphase
- G_1 phase

- G_2 phase
- S phase
- M phase
- proliferation
- maturation-promoting factor (MPF)
- cyclin
- kinase
- phosphorylation
- cyclin-dependent kinase
- anaphase-promoting complex (APC)
- ubiquitin
- G_1/S checkpoint
- G_S/M checkpoint
- M checkpoint
- spindle attachment checkpoint
- mitotic arrest-deficient protein 2 (MAD2)
- p53 protein
- sister chromatids
- nucleolus
- ribosomal RNA (rRNA)
- cohesin
- aster
- kinetochore
- microtubules
- polar microtubules
- prophase
- metaphase
- metaphase plate
- securin
- separase
- anaphase
- telophase
- nucleolar organizers
- meiosis I
- meiosis II
- leptonema

- zygonema
- pachynema
- diplonema
- diakinesis
- bouquet stage
- lateral element
- bivalents
- synapsis
- synaptonemal complex
- crossing over
- recombination nodules
- nonsister chromatids
- chiasmata
- tetrads
- reductional division
- monovalent
- interkinesis
- equational division
- spermatogenesis
- spermatogonium
- spermatids
- spermatozoa
- oogonia
- primary oocytes
- dictyotene
- secondary oocyte
- polar body
- ovum
- oogenesis
- parthenogenesis
- sporophyte
- gametophyte
- pollen grain
- sexual spores
- chromosomal theory of inheritance
- sex-linked

Understanding the Key Concepts
Use the key terms or parts of their definitions to complete the following sentences.

mitosis

The process of (1)__________ occurs during the (2)____S____ phase of the cell cycle to produce two identical (3)_sister chromatids_ that are separated into two different daughter cells by (4)_cytokinesis_. The first stage of this process is (5)_prophase_ when the (6)_spindle_ is formed, the chromosomes (7)_condense_ and the (8)_nuclear membrane_ breaks down. During (9)_metaphase_, the chromosomes (10)_align_ at the (11)_metaphase plate_. The two (12)_sister chromatids_ then separate and move to opposite poles during anaphase by the degradation of (13)_cohesins_ that join them along their length. The chromosomes (14)_uncoil_, and the (15)_nuclear envelope_ re-forms during telophase which is followed by the division of the other cellular components during (16)__________. Both daughter cells produced as a result of this process have genetic material that is (17)__________ to that of the (18)__________ cell.

Meiosis is separated into two divisions: one (19)__________ and one (20)__________ division. During the first meiotic division, the (21)__________ is reduced from (22)__________ to (n) in each daughter cell, when the (23)__________ are separated. Unlike (24)__________, the (25)__________ are not separated during meiosis I because (26)__________ break down everywhere, but at the (27)__________. This allows the (28)__________ to migrate to the same poles during meiosis I. During the second meiotic division, the (29)__________ are separated, but the (30)__________ per cell is not further (31)__________. Therefore, for every diploid cell that enters meiosis, (32)__________ daughter cells are produced that each have the (33)__________ number of chromosomes. Unlike mitosis which is conservative, meiosis promotes (34)__________, by allowing (35)__________ and (36)__________ between nonsister chromatids. The movement of chromosomes in meiosis is consistent with (37)__________ observations of the separation of alleles in pea plants which led to the laws of (38)__________ and (39)__________, two fundamental principles that explain genetic inheritance.

Figure Analysis
Use the indicated figure or table from the textbook to answer the questions that follow.

1. **Figure 3.8**
   (a) Which component of the MPF complex oscillates in the quantity during the cell cycle?

   (b) What is the function of the constant component of the MPF complex?

   (c) What are the two reasons why CDC2 normally does not initiate mitosis?

   (d) What protein complex breaks down cyclin B and how?

   (e) What does the degradation of cyclin B do to CDC2?

2. **Figure 3.14**
   (a) What is the function of CDC20?

   (b) What protein does the APC target for degradation?

   (c) How is separase activated?

   (d) What is the function of separase during mitosis?

3. **Figure 3.19**
   (a) What events occur during interphase?

   (b) Describe how the chromosomes move or separate during prophase, metaphase, anaphase, and telophase.

(c) What is the ultimate result of mitosis?

4. **Figure 3.28**
   (a) What two processes that occur during meiosis promote genetic diversity?

   (b) How many gametes with different genotypes are possible from independent assortment in the example shown? What are the possible different genotypes?

   (c) With one crossover, how many gametes with different genotypes are possible from independent assortment in the example shown? What are the possible different genotypes?

General Questions

Q3-1. Describe the main features of prophase, metaphase, and anaphase of meiosis I and meiosis II.

Q3-2. Describe how the processes of mitosis and meiosis explain Mendel's principles of segregation and independent assortment.

Q3-3. Would the sequence of genes found on one of the sister chromatids of a dyad necessarily be identical to the sequence of genes found on a bivalent occurring earlier in meiosis? Explain.

Q3-4. Meiosis is often termed a "reduction division." What is reduced and when?

Q3-5. Given that most mutations occur during DNA synthesis, will sperm cells or egg cells be expected to carry more new mutations? Explain.

Q3-6. Compare the processes of mitosis and meiosis with regard to each of the following features:
   a. Number of cell divisions
   b. Ploidy and number of resulting products
   c. Number of times DNA replicates
   d. Synapsis of homologous chromosomes
   e. Disjunction of homologous chromosomes
   f. Ability of products to divide by meiosis
   g. Function and significance

Multiple Choice

For each of the following, circle the letter of the choice that most appropriately answers the questions.

1.) Maturation-promoting factor (MPF) is required to progress through which of the following checkpoints?

   a. M checkpoint

   b. G_1/S checkpoint

   c. spindle attachment checkpoint

   d. G_S/M checkpoint

   e. None of the above

2.) Which of the following is *not* a true statement about the differences between prokaryotes and eukaryotes?

   a. Eukaryotic cells undergo mitosis and meiosis, while prokaryotic cells do not.

   b. Eukaryotes have membrane-bound organelles and a true nucleus containing their genetic information, while prokaryotes do not.

   c. Bacteria, cyanobacteria, and archaebacteria are prokaryotes, while all organisms classified as fungi or protists are eukaryotes.

   d. Prokaryotic cells are typically smaller in size than eukaryotic cells.

   e. Eukaryotes have numerous circular molecules of DNA, while prokaryotes have only one linear molecule.

3.) Which of the following is *not* a function of the microtubules within a eukaryotic cell?

   a. Provide the cell with shape and structure.

   b. Hold together two sister chromatids.

   c. Move sister chromatids to opposite poles during anaphase of mitosis.

   d. Move internal components of the cell to their proper destinations.

   e. Separate homologous chromosomes during anaphase I of meiosis.

4.) Crossing over between homologous chromosomes in prophase I of meiosis occurs during which of the following stages?

   a. leptotene

   b. zygotene

   c. pachytene

   d. diplotene

   e. diakinesis

5.) During metaphase I of meiosis, cohesion breaks down everywhere except for which of the following locations?

   a. in the middle of paired homologs

   b. at the centromeres

   c. at the ends of paired homologs

   d. at the chiasmata

   e. at the centrosomes

6.) Which of the following statements about meiosis is true?

   a. Following the completion of meiosis, the DNA content of a cell with 4C DNA will be reduced to one-fourth of its original content.

   b. For each cell that enters meiosis I, eight cells emerge with half the number of chromosomes of the original cell.

   c. In meiosis I, sister chromatids are pulled to different poles.

   d. In meiosis II, maternal centromeres are separated from paternal centromeres.

   e. Independent assortment of chromosomes and recombination during meiosis II produce an infinite number of different chromosomal combinations in the gametes.

7.) How many functional eggs are produced for each primary oocyte that completes meiosis?

   a. 0

   b. 1

   c. 2

   d. 3

   e. 4

8.) How many functional sperm are produced for each spermatogonium that completes meiosis?

a. 0

b. 1

c. 2

d. 3

e. 4

9.) The longest principal stage in cultured human cells is typically

a. G_1

b. S

c. G_2

d. M

e. None of the above

10.) Release of the MAD2 protein from the kinetochores of each chromosome is required for which stage of mitosis to proceed?

a. prophase

b. metaphase

c. anaphase

d. telophase

e. interphase

Practice Problems

P3-1. Consider a diploid cell with $2n = 24$. Assume that the amount of DNA in this cell at G_1 of interphase is represented by the quantity X. Complete the following table.

Stage of Division	Number of Chromosomes	Number of Chromatids	Quantity of DNA
At prophase of mitosis			
After completion of mitosis			
At prophase of meiosis I			
At prophase of meiosis II			
In sperm and egg cells			

[*Hint 1*: To determine the number of chromosomes in a cell, count the number of centromeres. One centromere equals one chromosome.

Hint 2: The amount of DNA in a cell is doubled prior to the beginning of both mitosis *and* meiosis. In mitosis, the DNA is divided equally between the two resulting cells. In meiosis, the DNA is divided equally among the four cells produced. Although there are two divisions during meiosis, DNA synthesis does not occur between divisions.

Hint 3: The terms diploid and haploid refer to the number of *chromosome sets*. A diploid number of four ($2n = 4$) indicates that there are four chromosomes (two pairs) in the cell. A haploid number of four ($n = 4$) indicates that there are four unpaired chromosomes in the cell.]

P3-2. *Drosophila melanogaster* has four pairs of chromosomes in the diploid state. Diagram the chromosome complement of a cell from an adult fruit fly and from a fruit fly gamete.

(*Hint 4*: For the chromosome complement of a cell remember that in mitosis, homologous pairs of chromosomes are present. During mitosis, these chromosomes align singly along the metaphase plate. During meiosis I, homologous pairs of chromosomes are present, but they align paired on either side of the metaphase plate. In meiosis II, the cell contains half the original number of chromosomes because only one member of each homologous pair is present.)

P3-3. Consider a cell containing a pair of metacentric chromosomes and a pair of telocentric chromosomes. Diagram the chromosome composition expected to result from (a) a mitotic division, (b) meiosis I, and (c) meiosis II.

P3-4. How many combinations of maternal and paternal chromosomes are possible in the gametes if the diploid number of chromosomes in the parent cell is (a) 2; (b) 4; or (c) 6?

P3-5. If purple endosperm (*P*) is dominant to white (*p*) in corn, give the genotype and phenotype of the endosperm cells that would result by transferring pollen from a homozygous *PP* plant to the pistils of a homozygous *pp* plant. If the pollinating plant were *pp* and the pistillate plant *PP*, would the phenotypic results be different?

Assessing Your Knowledge

Understanding the Key Concepts—Answers

1.) DNA replication; 2.) S; 3.) sister chromatids 4.) mitosis; 5.) prophase; 6.) spindle; 7.) condense; 8.) nuclear membrane; 9.) metaphase; 10.) align; 11.) metaphase plate; 12.) sister chromatids; 13.) cohesins; 14.) uncoil; 15.) nuclear envelope; 16.) cytokinesis; 17.) identical 18.) parental; 19.) reductional; 20.) equational; 21.) chromosome number; 22.) ($2n$); 23.) homologous chromosomes; 24.) mitosis; 25.) sister chromatids; 26.) cohesins; 27.) centromeres; 28.) sister chromatids; 29.) sister chromatids; 30.) chromosome number; 31.) reduced; 32.) four; 33.) haploid; 34.) genetic diversity; 35.) independent assortment; 36.) crossing over; 37.) Mendel's; 38–39.) segregation; independent assortment.

Figure Analysis—Answers

1a.) cyclin B; b.) The constant component of the MPF complex is CDC2p. It functions as a kinase to transfer a phosphate group from ATP to a specific amino acid of the protein it is acting on; c.) CDC2p normally does not initiate mitosis first, because phosphate groups block its active site, so it has to be dephosphorylated before it can be active. Second, it does not initiate mitosis because it can only function when it is combined with a molecule of cyclin B whose levels fluctuate during the cell cycle; d.) The APC breaks down cyclin B by attaching a ubiquitin molecule to it which targets the protein for proteasomal degradation; e.) The degradation of cyclin B regenerates an inactive CDC2p as the cell enters G_1.

2a.) CDC20 binds to an inactive APC activating it; b.) The APC targets securin for degradation; c.) Securin is an inhibitory protein bound to separase. Degradation of securin leads to activation of separase; d.) Separase is a protease that degrades cohesin when it is activated to liberate the sister chromatids from each other.

3a.) During interphase, the cell grows (G_1), replicates its DNA (S), and continues to prepare to divide (G_2); b.) During prophase, the chromosomes shorten and thicken, condense, and become visible. During metaphase, the chromosomes attach to microtubules at the kinetochores and become aligned at the equator of the cell in the metaphase plate. During anaphase, the chromosomes are pulled to opposite poles of the cell. During telophase, the chromosomes at opposite poles decondense as the nuclear envelope re-forms dividing the two sets of chromosomes; c.) Mitosis ultimately results in nuclear division generating two daughter cells that are genetically identical to the parental cell.

4a.) Independent assortment and crossing over between nonsister chromatids during meiosis I promote genetic diversity by producing a large number of different chromosomal combinations in the gametes of an individual; b.) Independent assortment produces gametes with four possible different genotypes. They are *ABCD, abcd, ABcd,* and *abCD*; c.) With one crossover, independent assortment produces gametes with eight possible different genotypes. They are *ABCD, abcd, abCD, AbCD, aBcd, ABcd, Abcd,* and *aBCD*.

General Questions—Answers

Q3-1.

Meiosis I	Prophase	Nuclear membrane disintegrates, spindle apparatus forms, chromosomes condense, homologous chromosomes synapse.
	Metaphase	Homologous chromosomes align on each side of the equator.
	Anaphase	Homologous chromosomes separate and move to opposite poles.
Meiosis II	Prophase	Chromosomes contract.
	Metaphase	Chromosomes align with centromeres on equator.
	Anaphase	Chromatids separate and move to opposite poles.

Q3-2. Mendel's principles of segregation and independent assortment are explained by the behavior of chromosomes during meiosis. The principle of segregation states that pairs of alleles separate from each other and randomly recombine at fertilization. The principle of independent assortment states that the separation of one pair of alleles occurs independently from the separation of other pairs of alleles. Each of these processes is demonstrated by the separation and varied alignment of homologous chromosomes during meiosis I. Assume that chromosome number 1 carries gene *A* and chromosome 2 carries gene *B*. Consider a dihybrid female in the process of meiosis. One member of the homologous pair of chromosomes 1 (1m) bears the *A1* allele. The other member of the pair (1p) contains the *A2* allele. At anaphase I, the homologous chromosomes disjoin. As a result, *A1* is partitioned into one cell while *A2* appears in the other cell. To demonstrate independent assortment, we must simultaneously consider genes *A* and *B*. Because chromosomes 1 and 2 separate from each other independently, so do the alleles of genes *A* and *B*. As a result, a gamete may contain any allele combination from the *A* and *B* genes (refer to the following figure). If the cell received chromosome 1m and 2m, the genotype would be *A1B1*. If the cell received chromosome 1m and 2p, the genotype would be *A1B2*. A chromosome complement of 1p and 2p results in the *A2B2* genotype. Finally, a gamete would have the genotype *A2B1* if it contained chromosomes 1p and 2m.

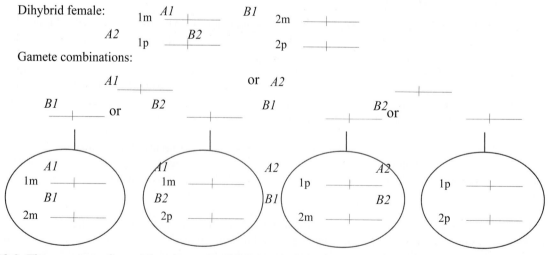

Q3-3. The sequence of genes found on one of the sister chromatids of a dyad is not necessarily identical to the sequence of genes found on a bivalent occurring in meiosis. Why? The answer: crossing over, which occurs early during prophase, results in the exchange of pieces of chromatids from homologous chromosomes. Crossing over takes place before the chromosomes have completely condensed and tetrads are visible.

Q3-4. Meiosis is a reduction division in which the number of chromosomes is reduced. The number of chromosomes is halved during the first meiotic division. The cells entering meiosis II contain one set of chromosomes and are therefore haploid.

Q3-5. Sperm cells would be expected to have more new mutations because spermatogonia are continuously dividing mitotically. Primary oocytes have already entered meiosis I before the female's birth. These cells are not continually dividing mitotically throughout the female's life.

Q3-6.

Feature	Mitosis	Meiosis
a	One	Two (meiosis I and II)
b	Two identical diploid ($2n$) cells	Four usually dissimilar haploid (n) cells
c	Once, during S phase	Once, during S phase
d	No	Yes
e	No	Yes
f	Yes	No
g	Growth, repair	Maintenance of chromosome number during sexual reproduction; generation of variation

Multiple Choice—Answers

1.) d; 2.) e; 3.) b; 4.) c; 5.) b; 6.) a; 7.) b; 8.) e; 9.) b; 10.) c.

Practice Problems—Answers

P3-1. The cell in question has 24 chromosomes (12 pairs) in the nucleus (*Hint 3*). During the division process, each chromosome is composed of two chromatids joined by a centromere. If the amount of DNA present in the cell at G_1 of interphase is X, the amount of DNA present in the cell prior to mitosis or meiosis will be $2X$ (*Hint 2*).

Stage of Division	Number of Chromosomes	Number of Chromatids	Quantity of DNA
At prophase of mitosis	24	48	$2X$
After completion of mitosis	24	0	X
At prophase of meiosis I	24	48	$2X$
At prophase of meiosis II	12	24	X
In sperm and egg cells	12	0	$1/2\,X$ or $X/2$

P3-2.   Cell from adult   Gamete

P3-3. Assume that the two circles in a–c represent the two cells resulting from a single division.

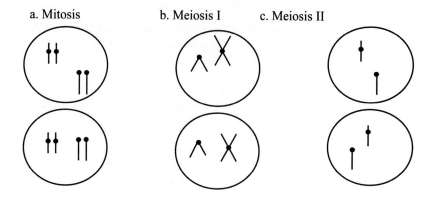

a. Mitosis   b. Meiosis I   c. Meiosis II

P3-4. To determine the number of gametes possible, use the equation 2^n, where n is the number of chromosome pairs.

    a. $2^2 = 2 \times 2 = 4$

    b. $2^4 = 2 \times 2 \times 2 \times 2 = 16$

    c. $2^6 = 2 \times 2 \times 2 \times 2 \times 2 \times 2 = 64$

P3-5. Examination of the life cycle of corn reveals that endosperm is triploid ($3n$). This ploidy level results from the fusion of two polar nuclei and one sperm nucleus (see figure 3.31 in the textbook). Therefore the endosperm genotype will be *Ppp*. If the female parent were genotype *PP*, the resulting endosperm genotype will be *PPp*. Both genotypes produce purple kernels.

3

MITOSIS AND MEIOSIS

CHAPTER SUMMARY QUESTIONS

1. 1. f, 2. h, 3. c, 4. g, 5. i, 6. j, 7. a, 8. d, 9. b, 10. e.

2. See table 3.1 for a summary answer.

3. The kinetochore is a proteinaceous structure on the centromere that attaches to the spindle fibers. The centromere is the DNA sequence at the constriction point in the eukaryotic chromosome on which the kinetochore lies.

4. When a eukaryotic chromosome replicates during the S phase of the cell cycle, one chromosome becomes two chromatids, attached near the centromere. These are sister chromatids. Chromatids of different chromosomes are nonsister chromatids.

5. Homologous chromosomes are members of a pair of *essentially* identical chromosomes. In diploid organisms, one member from each pair comes from each parent. Homologous chromosomes have the same length and centromere position and carry the same genes. However, they may possess different alleles.

6. The terms *reductional* and *equational* refer to the segregation of chromosomes during meiosis. The first division is termed reductional because the homologous chromosomes separate from each other, reducing the number of chromosomes in half. The second meiotic division is called equational because the number of chromosomes remains the same although the number of chromatids is halved.

7. The life cycle of human beings is shown in the following figure. The life cycle of the pea plant is the same as the life cycle of the corn plant that is diagrammed in figure 3.32.

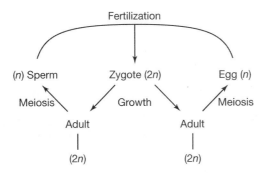

8. After the S phase, the sister chromatids are kept together by a multiprotein complex termed *cohesin*. The anaphase-promoting complex (APC) triggers the destruction of cohesin. Activated APC catalyzes the ubiquination and degradation of the protein securin, an inhibitory protein that binds and inactivates a protease called separase. Free from securin, separase can now degrade cohesin and permit the separation of the sister chromatids from each other.

9. The three major checkpoints are G_1/S, G_2/M, and M. The G_1/S checkpoint screens that the cell possesses the correct size and the DNA is not damaged. The G_2/M checkpoint evaluates if DNA replication and repair has been completed. The M checkpoint monitors proper spindle fiber assembly and attachment to kinetochores. Checkpoint failure may result in uncontrolled cell division and cancer.

10. The two meiotic mechanisms that generate genetic diversity are (1) an independent assortment of chromosomes, which starts in metaphase I and ends in anaphase I; and (2) crossing over between nonsister chromatids of homologous chromosomes, which takes place in prophase I.

11. **a.** Mitosis and meiosis I
    b. Mitosis and meiosis II
    c. Meiosis I

    d. Mitosis

    e. Mitosis, meiosis I, and meiosis II

    f. Mitosis and meiosis II

12. In diploid organisms, chromosomes occur in pairs. The haploid number, or number of *pairs* of chromosomes, may be even (4 in the fruit fly, for example) or odd (23 in humans). However, the total diploid number is given by 2 × the number of chromosome pairs, and therefore it will be an even number.

13. The difference lies mainly in cytokinesis or cytoplasmic division. In animal cells, a ring of actin forms around the cell causing a constriction of the cell membrane in the metaphase plate. In plant cells, which have cell walls, a cell plate grows in the approximate location of the metaphase plate.

14. 1. In spermatogenesis, both meiotic divisions are followed by equal cytoplasmic divisions, yielding four functional gametes. In oogenesis, unequal cytoplasmic divisions usually produce only one functional gamete and up to three polar bodies.

    2. Once spermatogenesis begins, it occurs continuously. Oogenesis, on the other hand, starts during embryonic development and then pauses until puberty, where it resumes for only one oocyte, usually, per month.

    3. Spermatogenesis occurs throughout adult life, while oogenesis ends at menopause.

    4. Spermatogenesis results in the production of hundreds of millions of sperm per day, whereas on average a human female produces about 400 ova in her lifetime.

EXERCISES AND PROBLEMS

15. a. 46 chromosomes ($2n$)

    b. 0 chromosomes (There is no nucleus in a mature red blood cell.)

    c. 23 chromosomes (n)

    d. 23 chromosomes (n)

    e. 23 chromosomes (n)

16.

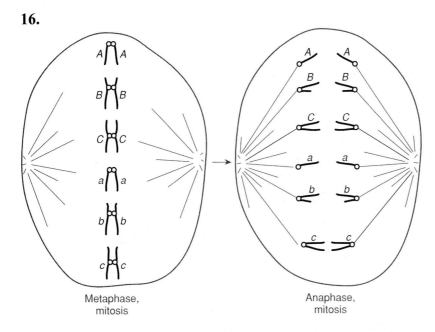

Metaphase,
mitosis

Anaphase,
mitosis

17. A total of $2^3 = 8$ different gametes can arise, each with a different combination of alleles for the three genes. A crossover between the *A* locus and its centromere does not alter gametic combinations.

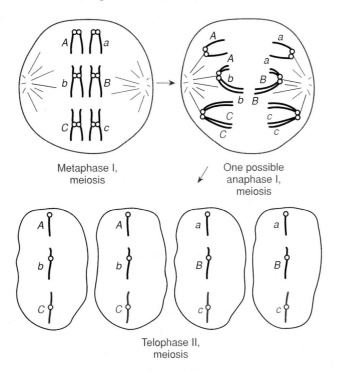

Metaphase I,
meiosis

One possible
anaphase I,
meiosis

Telophase II,
meiosis

18. a. Mitosis, prophase, $2n = 6$ (the two parallel threads represent sister chromatids); or meiosis II, early prophase, $2n = 12$ (each thread represents a chromosome, and sister chromatids are not evident yet)

b. Mitosis, metaphase, $2n = 6$, or meiosis II, metaphase, $2n = 12$

    c.   Mitosis, anaphase, $2n = 6$, or meiosis II, anaphase, $2n = 12$

    d.   Meiosis I, early prophase, $2n = 6$ (each thread represents a chromosome, and sister chromatids are not evident yet)

    e.   Meiosis I, anaphase, $2n = 6$

    f.   Meiosis II, anaphase, $2n = 6$

19. **a.**   There will be $2c$ amount of DNA in G_1 because DNA replication has not yet taken place in this diploid cell.

    b.   There will be $4c$ amount of DNA in G_2 because DNA replication has already taken place.

    c.   At the end of meiosis I, a cell will have $2c$ amount of DNA, disregarding differences brought about by differences in sex chromosomes. Since a cell entering meiosis I has $4c$ DNA, it will have $2c$ at the end of the first meiotic division and c at the end of the second meiotic division (one tetrad reduced to one chromosome).

20.

	$2n$ (= Dyads)	Bivalent (= Tetrads)
Human being	46	23
Garden pea	14	7
Fruit fly	8	4
House mouse	40	20
Roundworm	2	1
Pigeon	80	40
Boa constrictor	36	18
Cricket	22	11
Lily	24	12
Indian fern	1260	630

21. The intent of this problem is to make you think about the essential steps of meiosis, primarily the necessity to identify and properly separate members of homologous pairs of chromosomes. Presumably, any method you devise will force you through that process.

22. The number of combinations is 2^n where n = the number of different chromosome pairs. In this case, $n = 6$, so we expect $2^6 = 64$ different combinations.

23. A gamete from wheat will have 21 chromosomes, and a gamete from rye will have 7 chromosomes. Even if the 7 rye chromosomes could pair with 7 wheat chromosomes, a highly unlikely possibility, the remaining 14 wheat chromosomes could not pair (because they are different chromosomes from the same organism) and would segregate randomly during meiosis. Almost every gamete would get an incomplete set; if fertilization did occur, the zygotes would have extra chromosomes (trisomic) or would be missing some chromosomes (monosomic or nullisomic).

24.

	DNA (Number of Chromatids)	Ploidy
Spermatogonium or oogonium	2	$2n$
Primary spermatocyte or primary oocyte	4	$2n$
Secondary spermatocyte or secondary oocyte	2	n
Spermatid or ovum	1	n
Sperm	1	n

25. a. Each primary spermatocyte is diploid and will undergo meiosis to yield four cells: $4 \times 50 = 200$ sperm cells.

   b. Each secondary spermatocyte has completed one meiotic division and has one more division to go: $2 \times 50 = 100$ sperm cells.

   c. Each spermatid is haploid and will mature into a sperm cell: $1 \times 50 = 50$ sperm cells.

26. a. The primary oocyte is diploid and will undergo meiosis, but only one functional ovum results from each primary oocyte: $1 \times 50 = 50$ eggs.

   b. Each secondary oocyte will undergo meiosis II to produce an ovum and a polar body: $1 \times 50 = 50$ eggs.

27. a. 20 chromosomes ($2n$)
   b. 20 chromosomes ($2n$)
   c. 10 chromosomes (n) per nucleus
   d. 10 chromosomes (n)

28. Homologous chromosomes will pair during meiosis. Each gamete gets one of each chromosome, *A*, *B*, *C*, *D*, and *E*. Fertilization fuses two cells with the chromosome complement given. Since root cells are somatic tissue, these cells will be diploid. Therefore, the answer is **b.** *AA BB CC DD EE.*

29. The "letter" shapes have to do with the arrangement of the chromosome arms during the various mitotic stages. In metaphase, the arms of the chromosomes may protrude to the sides of each centromere, whereas during anaphase the arms are dragged behind the centromeres as they are pulled to opposite poles through the viscous cytoplasm.

30. a. 21 chromosomes (n)
   b. 42 chromosomes ($2n$)
   c. 63 chromosomes ($3n$)

31. Regardless of the number of generations, *no* offspring will have only telocentric chromosomes and *all* the progeny will have the original parental composition. Assuming meiosis occurs normally, every gamete will contain one chromosome of each of the three pairs, two different telocentric chromosomes, and one metacentric chromosome. Fertilization restores the original parental karyotype.

32. The sperm cells produced by the pollen parent have the *A* genotype. The female parent contains the egg cell and two polar nuclei, all of which have genotype *a*. One sperm cell will fertilize the egg cell resulting in a diploid zygote with genotype *Aa*. The two polar nuclei are fertilized by a second sperm cell, producing a triploid endosperm with genotype *Aaa*.

33.

Part	Chromosomes	Chromatids
a.	60	n/a
b.	60	120
c.	60	120
d.	30	60
e.	30	n/a

Chromatid refers to two DNA molecules that are joined at their centromeres. This occurs after DNA replication, prior to the start of mitosis and meiosis I. It persists at the start of meiosis II, because homologs separate during meiosis I and there is no separation of the sister chromatids. During anaphase of mitosis and anaphase II of meiosis, the DNA molecules separate from each other and the two chromatids become two chromosomes.

34. The number of different gametes produced by each parent is 2^{20}. Using the "and rule" (refer back to chapter 2), we can calculate the number of different types of offspring that can be produced from the mating: $2^{20} \times 2^{20} = 2^{40}$ possibilities!

35. a. Centromeres split during mitotic anaphase, and the separated sister chromatids are referred to as daughter chromosomes. A total of 16 chromosomes per cell means that $16/2 = 8$ chromosomes are moving toward each pole. The diploid number is therefore 8.
   b. Centromeres do not split in anaphase of meiosis I, and each centromere corresponds to a single chromosome. The diploid number is therefore 38.
   c. There are two chromatids for every chromosome, so there are 20 chromosomes at this stage. However, in meiosis II, the cells are haploid. Therefore, the diploid number is 40.

36. *AaBb* queen × *ab* drone yields for sons (or gametes): *AB, Ab, aB, ab* and for daughters: *AaBb, Aabb, aaBb, aabb*.

37. Any possible genotype, from *AAABBB* through *aaabbb* can occur in the endosperm. If at a given locus the endosperm is homozygous, so is the embryo. If the locus is heterozygous (for example, *AAa* or *Aaa*), so is the embryo. Thus, an *AAabbb* endosperm is associated with an *Aabb* embryo.

38. a. 2^{50} or about 1.1×10^{15}

  b. $2^2 = 4$

The number of gametes produced is 2^n, where n = number of independently behaving entities. If the genes are completely independent, we expect 2^{50}, and if they are completely linked, then the number of different gametes will be determined by the number of different chromosomes, so we expect 2^2. In reality, the number falls between these two extremes.

39. Both meiosis and mitosis are processes that initiate under certain circumstances of cell cycle and place. Neither is actually dependent on the chromosomal content of the cell. Thus, meiosis could begin in a haploid cell, but it would not be a successful process because there is no homolog for any chromosome to pair with. Mitosis would, however, be successful because there is no pairing (synapsis) required for successful completion of the process.

40.

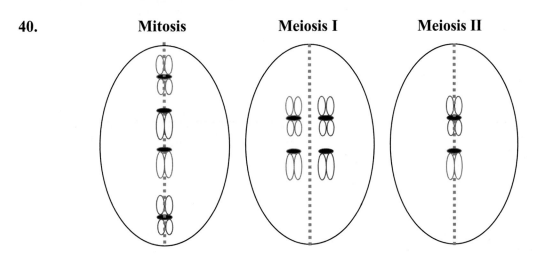

	Mitosis	**Meiosis I**	**Meiosis II**
Chromosome no.	4	4	2
Chromatid no.	8	8	4
Picograms DNA	0.6	0.6	0.3

CHAPTER INTEGRATION PROBLEM

a. The process of mitosis does not relate directly to Mendel's rules. The behavior of chromosomes during meiosis, however, explains both equal segregation (first law) and independent assortment (second law). Mendel's law of segregation can be explained by the homologous pairing and segregation of chromosomes during meiosis. Only one chromosome from each homologous pair goes into a gamete. So, for example, the *A* and *a* alleles will each segregate to a different gamete. This is also true for the *B/b*, *C/c*, and *D/d* alleles. Mendel's law of independent assortment can be explained by the relative behavior of different (nonhomologous) chromosomes during meiosis I. The pairing arrangement, separation, and migration of one homologous pair do not influence the orientation of an adjacent homologous pair. In our example, the assortment of gene *A* is independent of that of genes *B*, *C*, or *D*. In other words, a gamete is equally likely to inherit the *A* and *B* alleles or the *A* and *b* alleles. Similarly, it can obtain the *c* and *D* alleles or the *c* and *d* alleles, with equal probabilities.

b. Gene *C* is found on the *p* arm of an acrocentric chromosome, while genes *B* and *D* are found on the *q* arm of different metacentric chromosomes.

c. The cell is in prophase of mitosis. The chromosomes are condensed but not yet lined up in the center of the cell. In prophase I of meiosis there would be a synapsis between maternal and paternal homologs. In prophase II of meiosis there would only be four chromosomes (the haploid number) in the cell.

d. *AaBbCcDd.*

e. Four tetrads will form during prophase I of meiosis. These can align in eight different ways in metaphase I. To draw them all, keep one tetrad fixed and change the other three tetrads in systematic fashion.

    The following eight possible arrangements correspond to Mendel's second law of independent assortment. Each of these arrangements will occur with equal probability and each will produce two different types of gametes:

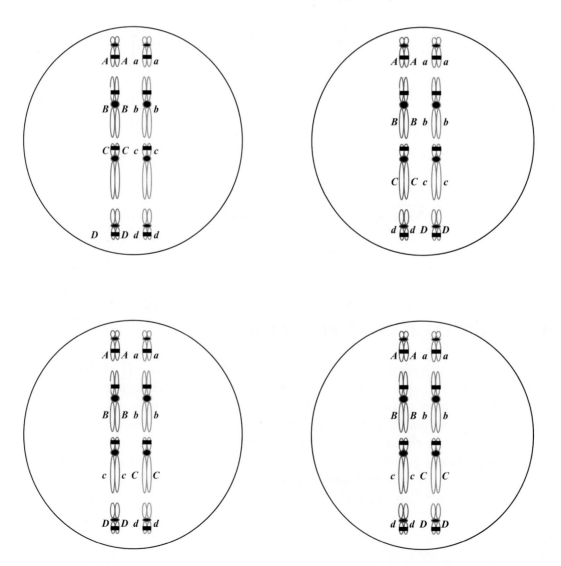

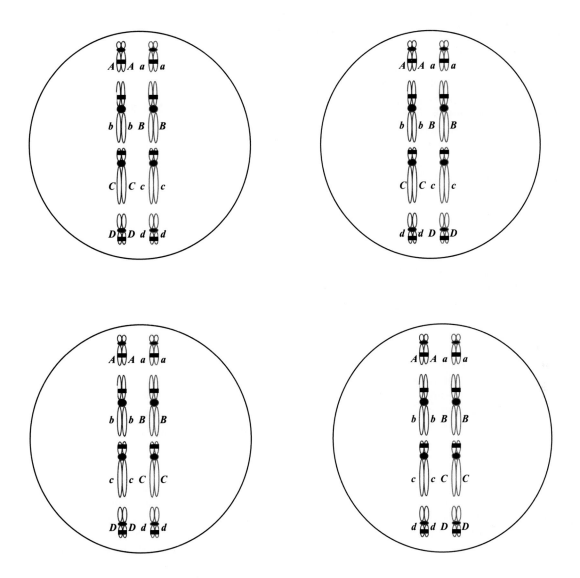

f. The number of chromosomal combinations is 2^n, where n = the number of different chromosome pairs. Thus, there are $2^4 = 16$ possible combinations. The chance of getting one member of a set is 1/2; so the chance of getting *ABCD*, or all four from the maternal side, is $(1/2)^4 = 1/16$.

g. To solve this part use an empirical approach that lists all possible combinations with two chromosomes from one parent. These are *ABcd*, *abCD*, *AbCd*, *aBcD*, *AbcD*, and *aBCd*. So the answer is 6/16 = 3/8.

h. The set of gametes with a mixture of maternal and paternal chromosomes consists of all gametes *except* those with only maternal or only paternal chromosomes. The chance of a gamete getting only maternal chromosomes is 1/16 (see part f). Similarly, the chance of a gamete getting only paternal chromosomes will be 1/16. Therefore, the chance of getting a mixture of maternal and paternal chromosomes is $1 - (1/16 + 1/16) = 14/16 = 7/8$.

CHAPTER 4 SEX LINKAGE AND PEDIGREE ANALYSIS

Chapter Goals

1. Distinguish between traits that result from genes located on the sex chromosomes and those that are present on the autosomes and whose expression are influenced by gender.
2. Explain the mechanisms of sex determination in various organisms.
3. Describe the different mechanisms organisms use to achieve balanced expression of sex-linked genes.
4. Use pedigree analysis to deduce the mode of inheritance for species that produce small numbers of offspring.
5. Calculate the probability that a particular individual in a pedigree has a specific genotype by determining the pattern of inheritance and deducing the genotypes of some of the individuals in the pedigree.

Key Chapter Concepts

4.1 Sex Linkage

- **Mechanisms of Sex Determination**: Sex determination is usually a complex process involving a series of developmental changes under genetic and hormonal control. In humans, social factors also play a significant role in the formation of a complete sexual identity. Examples of sex-determining mechanisms are discussed in the following.

- **Sex Chromosomes**: There are three basic types of sex-determining mechanisms involving genes on specific chromosomes: **XY, ZW,** and **XO**. In the **XY system**, females have a pair of morphologically identical chromosomes (a **homomorphic** pair, XX). Females produce only one type of gamete with respect to the sex chromosomes and are referred to as **homogametic**. Males have a pair of morphologically dissimilar chromosomes (a **heteromorphic** pair, XY) and therefore produce two kinds of gametes (**heterogametic**). Human beings and fruit flies have XY sex chromosomes. In the **ZW system**, female birds are heteromorphic and heterogametic (ZW) while males are homomorphic and homogametic (ZZ). In the **XO system**, which includes some species of insects, there is only one type of sex chromosome. Females are usually XX while males are X (written XO to indicate that the normal complement is a single X chromosome).

- **Genic Balance in *Drosophila***: In *Drosophila*, genes favoring maleness are found on the autosomes while genes favoring femaleness are located on the X chromosome. The ratio of X chromosomes to sets of autosomes determines if the individual is male or female. Males result if the ratio is 0.5 (1 X chromosome/2 sets of autosomes), and females result if the ratio is 1.0 (2 X chromosomes/2 sets of autosomes). This mechanism of sex determination was described by Calvin Bridges as the **genic balance hypothesis**.

- In human beings, the presence of a **sex-determining Y (SRY) gene** initiates the development of males. This gene is located on the Y chromosome. Several XX males have been identified. Genetic studies of these males have revealed that one of their X chromosomes contains a portion of the Y chromosome where the SRY gene is located.

- **Ploidy**: The number of complete sets of chromosomes an individual possesses determines sex in a group of insects which includes the bees, ants, and wasps (the Hymenoptera). In this case, females are diploid and result from fertilized eggs while males are haploid and arise from unfertilized eggs.

- **Compound Chromosomal System**: In the compound chromosomal system, several X and Y chromosomes combine to determine sex. These systems tend to be complex, and the precise mechanism of sex determination is unclear. Organisms such as the nematode *Ascaris incurva* and the duck-billed platypus have compound chromosome systems.

- **Environmental Cues**: In some species, sex is determined by the environmental cues (e.g. temperature) experienced during development. The effect of temperature on the sex of an organism can vary from species to species. Two potential target enzymes that temperature may affect to regulate sex determination are aromatase and reductase.

4.2 Proof of the Chromosomal Theory of Inheritance

- **Chromosomal Theory of Inheritance:** The chromosomal theory of inheritance states that chromosomes are linear arrays of genes that determine traits and are required to pass genetic information from one generation to the next.

- **The Case of the White-Eyed Male:** Experiments performed by Thomas Hunt Morgan in *Drosophila* suggested that the inheritance of the eye color phenotype is associated with the X chromosome. Therefore, eye color in *Drosophila* displays sex-linked inheritance. One of Morgan's students, Calvin Bridges, proved the chromosomal theory of inheritance by examining the chromosomal makeup of rare exceptional progeny resulting from his crosses of *Drosophila* that did not display the **crisscross pattern of inheritance** for eye color. Bridges was able to correctly predict what sex chromosomes would be present in these *Drosophila* mutants that experienced **nondisjunction**. These predictions were confirmed by cytology. The ability to predict the correct combination of sex chromosomes that corresponded to a specific eye color provided compelling proof that chromosomes are tightly associated with the genetic units that confer a given phenotype.

4.3 Sex Determination

- **Traits on the Sex Chromosomes:** Genes located on the sex chromosomes (X, Y, Z, and W chromosomes) follow different patterns of inheritance than genes located on the autosomes. In the XY system, this difference arises from the fact that females possess two X chromosomes while males have only a single X. Likewise, only males possess a Y chromosome. The patterns of inheritance for each type of linkage related to sex in the XY system are described in the following.
- **Sex-linked traits** also show a unique pattern of inheritance. Males inherit only a single copy of genes on the X chromosome from their mothers and are described as **hemizygous**. Since males inherit only one allele, that allele is expressed. Therefore, a single copy of a recessive allele determines the phenotype of the individual, a phenomenon known as **pseudodominance**. Males pass their single X chromosome to all of their daughters but none of their sons. Conversely, females inherit two copies of sex-linked genes: one allele from the mother and one allele from the father. Crosses involving sex-linkage can be easily identified: the phenotypes of males and females will differ in one or more generations (crisscross pattern of inheritance). In addition, switching the genotype of the male and female parent (constructing a **reciprocal cross**) will yield different results. Note that this pattern will be exactly the opposite for the ZW system in which females are ZW and males are ZZ.
- **Holandric traits** are those found only in males in the XY system. The genes controlling these traits are found on the Y chromosome. Males pass their Y chromosome to all of their sons and none of their daughters which accounts for the observed male-to-male pattern of inheritance.
- **Traits Related to Sex:** Several traits are related to sex although the genes are not necessarily located on the sex chromosomes. The genes for some traits are present in both sexes, while the trait is expressed in only one. These traits, such as breasts in females, are said to be **sex-limited**. **Sex-influenced** traits, such as pattern baldness, appear in both sexes but are expressed more frequently in one sex.

4.4 Gene Balance

- **Dosage Compensation:** Human males and females have the same number of autosomal chromosome pairs (22) although they differ in the number of X chromosomes they possess. Remember that the X chromosome is relatively large while the Y chromosome has very few functioning genes. This creates a problem: females receive two copies of the genes on the X chromosome, while males receive only one copy. How do females compensate for this dosage difference?
- **Increasing X-linked Gene Expression in Male *Drosophila*:** In fruit flies, dosage compensation is achieved by the hyperactivity of the male's single X chromosome.
- **X Chromosome Inactivation in Female Mammals:** In mammals, one of the female's X chromosomes is inactivated. Examination of the nucleus of female-derived cells reveals the presence of a darkly staining body that is absent in cells obtained from normal males. This structure, a **Barr body**, is believed to be an inactivated X chromosome. The inactivation occurs early during embryonic development and is a random process in most cases. This random inactivation leads to mosaic females that possess different patches that have an active paternal X chromosome and other patches that have an active maternal X chromosome. Recent studies have found that a few genes at the tip of the "inactivated" X chromosome remain active. These genes are required in double dose for normal female development. This idea makes sense when you consider that human females who do not possess two X chromosomes (XO) also do not undergo typical female development.

4.5 Pedigree Analysis
- **Pedigree Analysis:** Pedigrees, or family trees, represent the ancestry of a group of related individuals. Family trees are useful for examining a pattern of inheritance in a genetic study.
- **Autosomal Dominant Inheritance:** Autosomal dominant inheritance is suggested if traits do not skip generations and the trait appears in near-equal numbers among the two sexes. It is expected that if an affected person mates with an unaffected person, approximately 50% of their offspring should be affected.
- **Autosomal Recessive Inheritance:** Autosomal recessive inheritance is suggested if a trait skips a generation and the trait appears in near-equal numbers among the two sexes. If both parents are affected, all children should be affected. It is also suggested if affected individuals have unaffected parents. When affected people mate with unaffected people, all of the children in most cases are not affected. If one child is affected, this indicates that the unaffected parent is heterozygous. In this case, approximately half of the children should be affected
- **Sex-Linked Dominant Inheritance:** Sex-linked dominant inheritance is suggested if all of the daughters but none of the sons of an affected male are affected. Dominant inheritance is also suggested if an affected female produces at least 50% affected children.
- **Sex-Linked Recessive Inheritance:** Sex-linked recessive inheritance is suggested if most affected individuals are males. Affected females will produce only affected sons. Carrier females will produce a 1:1 ratio of affected: normal sons. Affected females result from carrier (or affected mothers) and affected fathers.

Key Terms

- sex chromosomes
- autosomes
- homogametic sex
- heterogametic sex
- sex-determining region Y (*SRY*)
- pseudoautosomal
- polyploids
- aneuploids
- nondisjunction
- Turner syndrome
- Klinefelter syndrome
- autosomal set
- genic balance theory
- intersex
- metamales
- metafemales
- hermaphrodite
- sex-linked inheritance
- reciprocal cross
- crisscross pattern of inheritance
- primary exceptional progeny
- secondary exceptional progeny
- gene balance
- sex-linked genes
- X-linked
- dosage compensation
- X chromosome inactivation
- Barr bodies
- mosaic
- sex linkage
- hemizygous
- pseudodominance
- carrier
- Y-linked
- holandric traits
- sex-limited traits
- sex-influenced traits
- pedigree
- affected
- proband
- autosomal dominant inheritance
- autosomal recessive inheritance
- sex-linked dominant inheritance
- sex-linked recessive inheritance

Understanding the Key Concepts
Use the key terms or parts of their definitions to complete the following sentences.

Human females have (1)_____ pairs of homologous chromosomes. (2)_____ of these pairs are (3)_____, and one of these pairs are (4)_____ chromosomes consisting of two (5)_____. This is in contrast to human males that have (6)_____ pairs of homologous chromosomes and one set of heterologous chromosomes, the (7)_____ pair. In males, the X chromosome can pair with the Y chromosome due to regions of (8)_____ that exist at the ends of the X and Y chromosomes. Pairing at these (9)_____ regions allows the X and Y chromosomes to segregate properly during (10)_____.

Maleness in human mammals is determined by the presence of the (11)_____ on the (12)_____. Mammals lacking a Y chromosome are (13)_____. In *Drosophila*, the Y chromosome does not encode a (14)_____ . Instead, sex is determined by (15)_____. This depends on a balance between alleles on the (16)_____ that favor the development of a female and (17)_____ alleles that favor the development of a male.

Birds, some fishes, and moths employ the (18)_____ of sex determination. In this system, females are the (19)_____ sex, and males are the (20)_____ sex. The (21)_____ chromosome contains the (22)_____ gene whose level of expression seems to control male development. Similarly, the (23)_____ chromosome contains the (24)_____ gene which is important in determining female development. Other organisms, such as the duck-billed platypus use several combinations of (25)_____ and (26)_____ chromosomes to determine sex, while the sex of some reptiles is largely dependent upon the environmental (27)_____.

Human females have two (28)_____ chromosomes, while males have one (29)_____ and one Y chromosome. Therefore, females will have (30)_____ copies of each gene present on the X chromosome and males will have (31)_____. For this reason, there has to be a mechanism to balance gene expression between males and females. The (32)_____ of all (33)_____ chromosomes, except for one in the human (34)_____ allows for (35)_____ in mammals. (36)_____ become (37)_____ bodies. In contrast, (38)_____ genes in the (39)_____ fruit fly are hypertranscribed relative to the (40)_____ to achieve gene balance.

The presence of only one X chromosome in human males makes them (41)_____ for the (42)_____ genes. Males inherit their X chromosome from their (43)_____. Therefore, if a female parent is homozygous for a recessive allele, (44)_____ of her sons will inherit this allele. This allele will determine their phenotype, making it (45)_____. Holandric traits are transmitted from (46)_____ to sons. Therefore these traits are (47)_____. (48)_____ traits such as breast development or facial hair distribution are expressed in one gender, while (49)_____ traits are (50)_____ in both genders, but are dominant in one and recessive in the other. The pattern of inheritance of a particular trait in a family can be diagrammed in a (51)_____. By determining whether the trait is expressed in every generation will help to determine if it is (52)_____ or (53)_____. By looking at the gender of affected individuals in each generation, it can be suggested whether a trait is likely (54)_____ or autosomal. (55)_____ traits follow the (56)_____ pattern of inheritance. If a male parent passes a trait to his (57)_____, the trait cannot be (58)_____, but must be (59)_____.

Figure Analysis
Use the indicated figure or table from the textbook to answer the questions that follow.

1. **Figure 4.7**
   (a) What does loss of the white-eyed phenotype in the F_1 progeny suggest?

   (b) Why wasn't the white-eyed phenotype equally distributed between males and females in the F_2 progeny?

2. **Figure 4.8**
   (a) What is a reciprocal cross? Describe the reciprocal cross Morgan used in terms of the sex chromosomes.

   (b) Why was the white-eyed phenotype expressed in the F_1 progeny if the trait is recessive?

   (c) Why are all of the white-eyed F_1 progeny male?

(d) Why do some of the F$_2$ females in the reciprocal cross have white eyes, while all of the F$_2$ females in the original cross had wild-type eye color?

3. **Figure 4.10**
(a) When Calvin Bridges repeated Morgan's cross of white-eyed females and red-eyed males, what rare F$_1$ progeny did he observe that Morgan had not previously described?

(b) What are the genotypes of the sex chromosomes in the primary exceptional progeny? How did Bridges explain these progeny in terms of sex chromosome inheritance?

(c) Why did Bridges hypothesize that the primary exceptional females were most likely fertilized by sperm containing a Y chromosome rather than sperm containing no sex chromosome?

(d) If the primary exceptional females had a Y chromosome, why weren't these flies male?

(e) Explain how the lethal F$_1$ progeny in this cross could've been produced in terms of sex chromosome inheritance.

(f) When the female primary exceptional progeny were crossed to wild-type males, what secondary exceptional progeny were produced?

(g) Why were the male secondary exceptional progeny fertile, while the male primary exceptional progeny were sterile?

(h) Explain how Bridges's observations proved the chromosomal theory of inheritance.

4. **Figure 4.20**
(a) What evidence in the pedigree suggests that the inheritance of polydactyly is not sex-linked?

(b) What evidence in the pedigree suggests that the inheritance of polydactyly is dominant?

5. **Figure 4.22**
(a) What evidence in the pedigree suggests that the inheritance of hypotrichosis is recessive?

(b) How are the parents of the three affected triplets related? Why do consanguineous matings often produce offspring with rare recessive traits?

6. **Figure 4.23**
(a) What evidence in the pedigree suggests that the inheritance of vitamin-D-resistant rickets is dominant?

(b) What evidence in the pedigree suggests that the inheritance of vitamin-D-resistant rickets is sex-linked?

7. **Figure 4.24**
(a) What evidence in the pedigree suggests that the inheritance of hemophilia is recessive?

(b) What evidence in the pedigree suggests that the inheritance of hemophilia is sex-linked?

General Questions

Q4-1. How was it determined whether maleness in humans is determined by the presence of a Y chromosome or by the absence of a second X chromosome?

Q4-2. Both *Drosophila* and humans utilize the XY system for sex determination. How does sex determination in humans and *Drosophila* differ?

Q4-3. If only part of one of the X chromosomes in female mammals is inactivated, how is dosage compensation between males and females achieved for the genes that are not inactivated?

Q4-4. What are the two significant inheritance patterns that are observed for X-linked recessive alleles?

Q4-5. If a woman is heterozygous for a recessive allele that causes color blindness (an X-linked phenotype), how many of her daughters and how many of her sons would be expected to be color blind if she marries and mates with an unaffected male? Explain your answer.

Multiple Choice

For each of the following, circle the letter of the choice that most appropriately answers the question.

1.) In the XO system of sex determination, which of the following determines the gender of an organism?
   a. expression of the *SRY* gene
   b. number of X chromosomes
   c. X and Y chromosome combinations
   d. absence of the Y chromosome
   e. presence of the O chromosome

2.) In the ZW system of sex determination, normal males have which of the following?
   a. two copies of the Z chromosome
   b. one copy of the Z chromosome and one copy of the W chromosome
   c. two copies of the Z chromosome and one copy of the W chromosome
   d. two copies of the W chromosome
   e. two copies of the W chromosome and one copy of the Z chromosome

3.) Calvin Bridges found that a normal male fruit fly has an X:A ratio that is which of the following?
   a. less than or equal to 0.33
   b. greater than 1.00
   c. equal to 0.67
   d. less than or equal to 0.50
   e. equal to 1.00

4.) In the XY system of sex determination, organisms that do not have a X chromosome are which of the following?
   a. normal males
   b. inviable
   c. sterile males
   d. sterile females
   e. females with Turner syndrome

5.) Temperature affects sex determination in all but which of the following organisms?
   a. Leopard geckos
   b. alligators
   c. turtles
   d. birds
   e. crocodiles

6.) Environmental sex determination (ESD), like other mechanisms of sex determination, is mediated by which of the following?
   a. inhibition of sex chromosome function
   b. loss of one of the sex chromosomes
   c. regulation of gene expression
   d. overexpression of genes on the X chromosome
   e. altering the X chromosome: autosome ratio

7.) In his experiments on sex-linked inheritance in *Drosophila*, Calvin Bridges found that primary exceptional males
   a. inherited a paternal X chromosome.
   b. occurred at the same frequency as primary exceptional females.
   c. were sterile.
   d. Both a and c
   e. All of the above

8.) Dosage compensation in mammals is achieved by
   a. decreasing the expression of genes on the sex chromosome in the heterogametic sex.
   b. increasing the expression of genes on the sex chromosome in the heterogametic sex.
   c. increasing the expression of genes on the sex chromosome in the homogametic sex.

d. decreasing the expression of genes on the
    sex chromosome in the homogametic sex.
e. duplication of critical X chromosome genes
    on the Y chromosome.

9.) The dosage compensation complex in
Drosophila is found associated with
    a. the autosomes in metamales.
    b. the X chromosome in males.
    c. one of the X chromosomes in females.
    d. the autosomes in metafemales.
    e. the Y chromosome in males.

10.) Based on the following pedigree where
shaded shapes indicate affected individuals and
nonshaded shapes indicate normal individuals,
the rare disease under consideration is most
consistent with which pattern of inheritance?
    a. autosomal dominant
    b. sex-linked dominant
    c. sex-linked dominant
    d. sex-linked recessive
    e. Unable to determine.

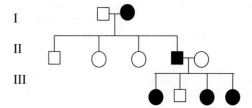

Practice Problems

P4-1. In *Drosophila melanogaster*, the allele for a miniature body is recessive and located on the
X chromosome. Design a cross of true-breeding flies that will allow the observer to distinguish males from
females based on the size of the individual.

P4-2. In the laboratory, you cross a homozygous female fruit fly with bobbed bristles and warty eyes to a
homozygous wild-type male. Assuming these two traits are recessive and sex-linked, predict the genotypes
and phenotypes of the progeny produced in the F_1 and F_2 generations. How would the results have differed
if the traits under study were sex-linked dominant? Predict the results of the cross assuming that the traits
are autosomal recessive.

[*Hint 1*: When solving sex-linkage problems, symbols are placed on the X (or Z) chromosome but not on
the Y (or W) chromosome. Place symbols for only sex-linked traits on the X chromosome. Symbols for
autosomal genes should be used as previously described in chapter 2.]

P4-3. Assume that you are a geneticist studying a new, rare genetic disorder. You are presented with a man
(A), who is affected, and his normal wife. The couple has five children: two affected daughters and three
normal sons. Each of the sons is married to a normal female and all have normal children. One of the
affected daughters married a normal man and had two sons: one with the disorder and one without the
disorder. From this information, determine the genotype of each individual represented. What can you
conclude about the parents of male A?

(*Hint 2*: Sex linkage is *suggested* if the phenotypes of males and females differ. Sex linkage is also
suggested if the results of reciprocal crosses yield different progeny phenotypes.

Hint 3: Sex linked traits often show a crisscross pattern of inheritance. Females pass an X chromosome to all of their offspring. Males pass their X chromosome to their daughters and their Y chromosome to their sons. Unique phenotypic ratios result because males are hemizygous and females are not.

Hint 4: Sex linked dominant inheritance is *suggested* if all of the daughters but none of the sons of an affected male are affected. Dominant inheritance is also suggested if an affected female produces at least 50% affected children.)

P4-4. Lisa and Tina are identical (monozygotic) twins. However, Lisa is color blind while Tina has normal color vision. Explain this difference.

(*Hint 5*: In mammals, dosage compensation alters the expression of the genotype in heterozygous females. The appearance of mosaic females may *suggest* sex linkage.)

P4-5. For each of the following sex chromosome complements, determine the sex of each individual and the number of Barr bodies present in humans. For fruit flies, provide the sex of each individual if two sets of autosomes are present and if three sets of autosomes are present.

   Chromosome complements: X, XX, XY, XXX, XXY, XYY

P4-6. Consider the plant *Lychnis alba* which has an XY sex-determining mechanism. If broad-leaved females are crossed to narrow-leaved males, the F_1 generation is composed of all broad-leaved individuals. In the F_2, all females are broad-leaved and males are either broad-leaved or narrow-leaved. Which sex is homogametic, and which is heterogametic?

P4-7. Horns in sheep are governed by a sex-influenced gene. This trait is dominant in males but recessive in females. Provide all genotypes and phenotypes possible in male sheep and in female sheep.

P4-8. Complete the following pedigree including the genotypes such that (a) Y-linked inheritance is suggested; (b) recessive autosomal inheritance is suggested and (c) recessive sex-linked inheritance is suggested.

[*Hint 6*: When working with pedigrees, write the possible genotypes under the symbols to determine whether or not the offspring could be produced from a mating between the two parents.

Hint 7: Sex-linked recessive inheritance is *suggested* if most affected individuals are males. Affected females will produce only affected sons. Carrier females will produce a 1:1 ratio of affected:normal sons. Affected females result from carrier (or affected mothers) *and* affected fathers.]

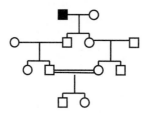

P4-9. (a.) Which modes of inheritance are consistent with the following pedigree? (b.) Which mode of inheritance is most likely?

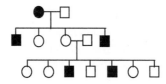

Assessing Your Knowledge

Understanding the Key Concepts—Answers

1.) 23; 2.) 22; 3.) autosomes; 4.) sex; 5.) X chromosomes; 6.) 22; 7.) XY; 8.) homology; 9.) pseudo-autosomal; 10.) meiosis; 11.) *SRY* gene; 12.) Y chromosome; 13.) female; 14.) *SRY* gene; 15.) genic balance; 16.) X chromosome; 17.) autosomal; 18.) ZW system; 19.) heterogametic; 20.) homogametic; 21.) Z; 22.) *DMRT1*; 23.) W; 24.) *ASW* or *Wpkci*; 25–26.) X; Y; 27.) temperature; 28.) X; 29.) X chromosome; 30.) two; 31.) one; 32.) inactivation; 33.) X; 34.) female; 35.) dosage compensation; 36.) Inactivated X chromosomes; 37.) Barr; 38.) X-linked; 39.) male; 40.) female; 41.) hemizygous; 42.) X-linked; 43.) mothers; 44.) all; 45.) pseudodominant; 46.) fathers; 47.) Y-linked; 48.) Sex-limited; 49.) sex-influenced; 50.) expressed; 51.) pedigree; 52–53.) dominant; recessive; 54.) sex-linked; 55.) sex-linked; 56.) crisscross; 57.) son; 58.) X-linked; 59.) autosomal.

Figure Analysis—Answers

1a.) Loss of the white-eyed phenotype in the F_1 progeny suggested that the white-eyed phenotype was recessive to the wild-type phenotype; b.) Eye color phenotype is associated with the X chromosome and white eye color is recessive. In the F_2 generation, all females inherit a wild-type allele (red eye color) on the X chromosome from their father, and either a wild-type allele (X^+), or white-eyed allele (X^W) from their mother. Since red eye color is dominant, none of the females have white eyes. The males, however, inherit the Y chromosome from their father and the X chromosome from their mother. Therefore, the males have an equal chance of inheriting either the X^+ chromosome (red eye color) or the X^W chromosome (white eye color) from their mother. For this reason, one-half of the males have red eyes (wild-type) and one-half have white eyes, but all of the females have red eyes.

2a.) A reciprocal cross is a genetic cross in which the phenotypes of the parents of a previous cross are switched.

Morgan's original cross: X^+X^+ females $\times$ X^WY males
(wild-type females $\times$ white-eyed males)

Morgan's reciprocal cross: X^WX^W females $\times$ X^+Y males
(white-eyed females $\times$ wild-type males)

b.) The recessive white allele is located on the X chromosome. Male F_1 progeny can only inherit their X chromosome from their mother because they must inherit their Y chromosome from their father. Since the mother has two X chromosomes each with the white-eyed allele, all of her sons will inherit a chromosome with a white-eyed allele; c.) The F_1 progeny exhibit a crisscross pattern of inheritance where the male passes his phenotype to his daughters and the female passes her phenotype to her sons. Since the mother is homozygous for white eye color, her sons can only inherit an X chromosome with a white eye allele. Therefore, all the F_1 males have white eyes. The females inherit one X chromosome from their father and one from their mother. Since white eye color is recessive to red eye color, all the F_1 females have red eyes due to the wild-type allele that they inherit from their father; d.) Some of the F_2 females in the reciprocal cross have white eyes because the female parent is heterozygous (X^+X^W) for the white-eyed allele and the male parent has the white-eyed allele (X^WY). The F_2 females inherit the X^W chromosome from their male parent, and one-half inherit the X^W chromosome from their female parent, so these flies will be X^WX^W or white-eyed females. In the original cross, the female F_2 progeny all inherited the wild-type allele (X^+) from the male parent (X^+Y). Since this allele is dominant, the allele inherited from the female parent did not matter because all of the females would have wild-type eye color.

3a.) In the cross of white-eyed females and wild-type males, Calvin Bridges found that 1 in approximately every 2000 males had red eyes and 1 in every 2000 females had white eyes; b.) The sex chromosome of the female primary exceptional progeny had the genotype X^WX^WY, while the males were X^+O. Since the primary exceptional females had white eyes, they had to be X^WX^W which both had to be inherited from their female parent (X^WX^W). This resulted from nondisjunction of the X chromosomes during meiosis. This X^WX^W egg was fertilized by a sperm containing a Y chromosome to produce the X^WX^WY progeny. The primary exceptional males had wild-type eye color, so their X chromosome had to have been inherited from their male parent (X^+Y). To be male, they either had to be X^+Y or X^+O; since the flies could not have inherited a Y chromosome from their female parent (X^WX^W), they received no sex chromosome from their female parent, making them X^+O. These male flies were sterile because they lacked a Y chromosome; c.) Bridges hypothesized that the primary exceptional females were most likely fertilized by a sperm containing a Y chromosome, rather than a sperm lacking a sex chromosome because for these X^WX^W eggs that resulted from nondisjunction of the female parent's X chromosomes to be fertilized by sperm lacking a sex chromosome, a nondisjunction event would've had to have occurred in the male parent as well. The likelihood of producing an individual from two gametes that have experienced nondisjunction is significantly less likely than that of producing an individual from fertilization of one gamete that has experienced nondisjunction; d.) Although the primary exceptional females had a Y chromosome, these flies were not male because the sex of a fly is determined by the X:A ratio rather than the presence of a Y chromosome which determines maleness in humans. Male flies have only one X chromosome, while the primary exceptional females have two X chromosomes; e.) The lethal $X^WX^WX^+$ progeny occurred by nondisjunction of the X chromosomes during meiosis. This X^WX^W egg was fertilized by a sperm containing an X chromosome. The lethal YO progeny occurred by fertilization of an egg lacking any sex chromosomes by a sperm containing a Y chromosome. The egg lacking any sex chromosomes was produced by nondisjunction; f.) When the female primary exceptional progeny (X^WX^WY) were crossed to wild-type males (X^+Y), the secondary exceptional progeny were red-eyed males (X^+Y) and white-eyed females (X^WX^WY); g.) The male secondary exceptional progeny were fertile, while the primary exceptional progeny were not because the male secondary progeny contained a Y chromosome; h.) Bridges was able to predict the correct combination of sex chromosomes that corresponded to a specific eye color suggesting that the chromosomes were tightly associated with genetic units that confer a specific phenotype. From the cross of the X^WX^WY exceptional white-eyed female with a wild-type male (X^+Y), he found that white-eyed female and wild-type male exceptional progeny were produced. He found that he could explain these progeny according to how the chromosomes of the female parent would segregate during meiosis to produce four possible female gametes. Once the gametes were fertilized, Bridges was able to explain what sex chromosomes were present in the progeny based on their eye color. This suggested that eye color was a phenotypic trait determined by a factor associated with the chromosomes.

4a.) The fact that the trait occurs equally in both sexes suggests that polydactyly is not sex-linked. In addition, it can be seen in the pedigree that a male passed on the trait to two of his three sons. This type of inheritance from father to son rules out X-linked inheritance; b.) Polydactyly occurs in every generation. Every affected child has an affected parent, and no generations are skipped. This pattern suggests dominant inheritance.

5a.) The inheritance pattern of hypotrichosis suggests that it is recessive because affected individuals are not found in each generation and the affected daughters come from unaffected parents. It appears for the first time in the pedigree from a consanguineous mating; b.) The parents of the affected triplets are first cousins. Consanguineous matings often produce offspring that have rare recessive traits because they share common ancestors. Therefore, a rare allele that is heterozygous in a common ancestor can be passed on to both sides of the pedigree and lead to a homozygous child.

6a.) The inheritance pattern of vitamin-D-resistant rickets suggests that it is dominant because it does not skip generations; b.) The inheritance pattern in the pedigree suggests that vitamin-D-resistant rickets is likely sex-linked because affected males pass the trait on to all of their daughters, but not to any of their sons.

7a.) The inheritance pattern in the pedigree suggests that hemophilia is recessive because the trait skips generations in affected families. There are several instances in the pedigree where an affected individual has unaffected parents and unaffected grandparents; b.) The fact that all of the affected individuals are males suggests that hemophilia is sex-linked. Because of pseudodominance, more males than females would be expected to have a sex-linked recessive phenotype. Also in the pedigree we can see that the trait is never passed from father to son; if it were, it would eliminate the possibility of the trait being X-linked. Since it is not, it suggests that hemophilia is a sex-linked trait.

General Questions—Answers

Q4-1. The existence of polyploidy and aneuploid individuals made it possible to determine if the Y chromosome is male determining. In humans, an XO individual is female, indicating that the presence of a Y chromosome is necessary for maleness. Likewise, an XXY individual is male, indicating that the presence of a Y chromosome and not a single X chromosome determines maleness.

Q4-2. In humans, the presence of a Y chromosome determines maleness, but in Drosophila, sex is determined by the ratio of autosomal alleles that favor maleness and alleles on the X chromosomes that favor femaleness. When the X:A ratio is ≥ 1.00, the organism is female. When the X:A ratio is ≤ 0.50, the organism is male.

Q4-3. All of the genes that are not inactivated on the X chromosome are also present on the Y chromosome, so both males and females have two copies of each of these genes.

Q4-4. The first pattern observed for X-linked recessive alleles is that males preferentially exhibit the recessive X-linked phenotype because they are hemizygous for the X-linked genes. Therefore, a single copy of a recessive allele determines the phenotype in males. The second pattern observed is a crisscross pattern of inheritance where the male parent passes his trait to his female offspring and the female parent passes her phenotype to her sons.

Q4-5. None of her daughters would be expected to be affected because they get one of their X chromosomes from their father who is wild-type, so the recessive allele inherited from the X chromosome of their mother would be masked. One-half of the woman's sons would be expected to be affected because they inherit their X chromosome from their mother. Since the mother is heterozygous, one-half of her sons will get the X chromosome with the recessive allele for color blindness and one-half of her sons will get the X chromosome with the wild-type allele. If the X chromosome has the recessive allele, it will be expressed in a pseudodominant fashion and the son will be color blind.

Multiple Choice—Answers

1.) b; 2.) a; 3.) d; 4.) b; 5.) d; 6.) c; 7.) e; 8.) d; 9.) b; 10.) c.

Practice Problems—Answers

P4-1. There are two possible crosses that can be designed. In the first, a miniature male is crossed with a wild-type female. In the second, reciprocal, cross, a wild-type male is crossed with a miniature female. As is obvious from the diagrammed crosses shown, cross 2 yields F_1 males and females with different phenotypes.

X^+ = chromosome bearing wild-type allele X^m = chromosome with allele for miniature body

Cross 1: P_1 **Cross 2:**

$X^mY \times X^+X^+$ $X^+Y \times X^mX^m$

Female gametes   Female gametes

		X^+	X^+	F_1			X^m	X^m
Male gametes	X^m	X^+X^m	X^+X^m			X^+	X^+X^m	X^+X^m
	Y	X^+Y	X^+Y			Y	X^mY	X^mY

Normal females and males   Normal females and miniature males

P4-2. B = allele for normal bristles, b = allele for bobbed bristles, W = allele for normal eyes, w = allele for warty eyes.

Sex-linked recessive inheritance:

	Genotypes		**Phenotypes**
P_1:	$X^{bw}X^{bw} \times X^{BW}Y$		
F_1:	$X^{BW}X^{bw} \times X^{bw}Y$		wild-type females × bobbed, warty males
F_2:	X^{BW}	X^{bw}	

	X^{BW}	X^{bw}
X^{bw}	$X^{BW}X^{bw}$	$X^{bw}X^{bw}$
Y	$X^{BW}Y$	$X^{bw}Y$

½ wild-type, ½ bobbed, warty

Sex-linked dominant inheritance:

	$X^{BW}X^{BW} \times X^{bw}Y$		
P_1:	$X^{BW}X^{BW} \times X^{bw}Y$		
F_1:	$X^{BW}X^{bw} \times X^{BW}Y$		all bobbed, warty
F_2:	$X^{BW}X^{BW}$	$X^{BW}X^{bw}$	bobbed, warty females
	$X^{BW}Y$	$X^{bw}Y$	½ bobbed, warty males and ½ wild-type males

Autosomal recessive inheritance: (refer to chapter 2)

P$_1$: *BBWW* × *bbww*

F$_1$: *BbWw* × *BbWw*   all individuals wild-type

F$_2$: 9:3:3:1 phenotypic ratio   9/16 wild-type

                                            3/16 wild-type bristles, warty eyes

                                            3/16 bobbed bristles, wild-type eyes

                                            1/16 bobbed, warty

Note: There are no differences between males and females in either generation.

P4-3. The first clue to the solution of this problem is the distribution of affected individuals. Note that a crisscross pattern of inheritance is observed: affected male to affected daughters (*Hint 3*). This indicates that the trait is sex-linked. The fact that females are affected indicates that the trait is dominant (*Hint 4*). We can conclude that the man's mother was affected. We cannot determine the genotype of his father.   *A* = allele for disorder, *a* = normal allele.

Mother	**Father**	**Man**	**His Wife**	**His Sons**	**Daughter**
$X^A X^?$	$X^? Y$	$X^A Y$	$X^a X^a$	$X^a Y$	$X^A X^a$

P4-4. Color blindness is an X-linked trait in human beings. While identical twins share all of their genes in common, differences may arise in sex-linked traits because of X inactivation. Assume that both women are heterozygotes. Lisa could be color blind if, by random chance, the X chromosome bearing the color-blind allele is active in the cells of her retina. The X chromosome possessing the normal allele is inactive in these cells. The opposite pattern would be observed in Tina: the X chromosome bearing the normal color vision allele is active.

P4-5. Recall that sex in *Drosophila* is determined by the ratio of X chromosomes to sets of autosomes. In human beings, sex is determined by the presence of the *SRY* gene on the Y chromosome. Note also that Barr bodies are found in human beings with two or more X chromosomes regardless of their sex. Barr bodies are inactivated X chromosomes; no dosage compensation occurs with the Y chromosome. Barr bodies are not found in fruit flies.

Sex determination and dosage compensation in human beings:

Sex Chromosomes	**Sex of Individual**	**Number of Barr Bodies Present**
X	Female	0
XX	Female	1
XY	Male	0
XXX	Female	2
XXY	Male	1
XYY	Male	0

Sex determination in the fruit fly, *Drosophila melanogaster*.

Sex Chromosomes	**Sex if Individual Has *Two* Sets of Autosomes**	**Sex if Individual Has *Three* Sets of Autosomes**
X	Ratio 0.5 = male	Ratio 0.33 = metamale
XX	Ratio 1.0 = female	Ratio 0.67 = intersex
XY	Ratio 0.5 = male	Ratio 0.33 = metamale
XXX	Ratio 1.5 = metafemale	Ratio 1.0 = female
XXY	Ratio 1.0 = female	Ratio 0.67 = intersex
XYY	Ratio 0.5 = male	Ratio 0.33 = metamale

P4-6. Examination of the F_1 reveals that all individuals are broad-leaved which indicates that this trait is dominant. Notice that the females in the P_1 generation possess the dominant trait. If females are the *heterogametic* sex, the female will pass an X chromosome bearing a dominant allele to all of her sons, but none of her daughters. This would result in an F_1 generation composed of broad-leaved males and narrow-leaved females. This is not the case. If the female is the *homogametic* sex, she will pass an X chromosome bearing a dominant allele to all of her offspring. In the F_1, both sexes will express the dominant phenotype. The observed F_1 phenotypes are consistent with the later hypothesis (females are homogametic).

P4-7. This problem provides you with the solution to the problem: horn presence is sex-influenced. Recall that both males and females possess the alleles for a sex-influenced trait although the trait occurs more frequently in one sex or the other. The presence of horns is dominant in males but recessive in females. The alleles are not located on the X chromosome. Genotypes and phenotypes are outlined here:

Genotype	Male Phenotype	Female Phenotype
HH	Horned	Hornless
Hh	Horned	Hornless
hh	Hornless	Horned

P4-8. Notice that a Y-linked trait is passed from male to male only. No females are affected or carry the allele. The recessive autosomal trait depicted in (b) is expressed only when two carriers produce children. Sex-linked recessive inheritance shows a crisscross pattern of inheritance. Likely carriers are noted.

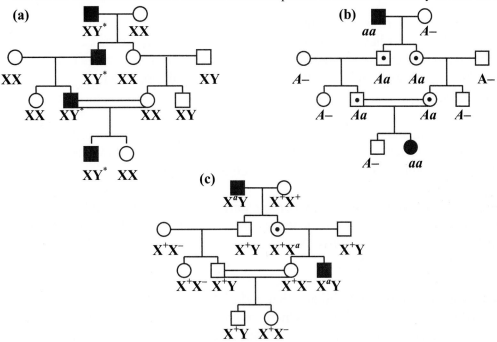

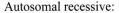

P4-9. The possible modes of inheritance are X-linked recessive and autosomal recessive with X-linked being more likely because all sons and no daughters of an affected female are also affected.

X-linked recessive:

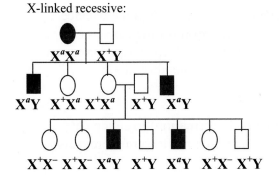

Autosomal recessive:

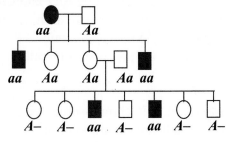

4

SEX LINKAGE
AND PEDIGREE ANALYSIS

CHAPTER SUMMARY QUESTIONS

1. 1. d, 2. i, 3. g, 4. c, 5. e, 6. f, 7. b, 8. a, 9. j, 10. h.

2. The differences are in terminology only, not in shape or size of the chromosomes. In species in which females have a homomorphic pair of sex chromosomes, the members of the pair are called X chromosomes. In species in which males have a homomorphic sex chromosome pair, the members of the pair are called Z chromosomes.

3. Sex switches are genes that determine the developmental pathway of an organism, either toward male or female, depending on the presence or state of the gene. *SRY* (sex-determining region Y) is the male sex switch in mammals.

4. Because a human male has only one X chromosome, he would be neither homozygous nor heterozygous for Duchenne muscular dystrophy. These terms only make sense in the female, where two X chromosomes are present. An affected male is said to be hemizygous.

5. Pseudoautosomal regions are found at either end of the Y chromosome in both humans and fruit flies. These regions, which are homologous to the corresponding regions of the X chromosome, allow the two sex chromosomes to pair up during meiosis. Pseudoautosomal genes are found in two copies in both males and females and, therefore, follow an autosomal mode of inheritance.

6. Both *Drosophila melanogaster* and *Caenorhabditis elegans* use a genic balance mechanism, where sex is determined by the ratio of X chromosomes to sets of autosomes (X:A). An X:A ratio of 0.5 produces males in both species, whereas an X:A ratio of 1.0 produces a female in *Drosophila* and a hermaphrodite in *Caenorhabditis*.

7. Early onset of X chromosome inactivation (Barr body formation) in calico cats; tortoiseshell cats have a later onset as seen by the fact that there were more cells present at that time and thus smaller sectors.

8. ESD refers to environmental sex determination, where the sex of the organism is determined by the environment and not by sex chromosomes of the individual. For example, sex determination in many reptilian species is determined by the environmental temperature during egg development. The mechanisms of ESD likely involve pathways for steroid biosynthesis. Two enzymes appear to play a major role: reductase, which converts testosterone into dihydro-testosterone, is found in higher levels at male-producing temperatures; aromatase, which converts testosterone into oestradiol, is found in higher levels at female-producing temperatures.

9. Exemptions should be made if the following relatives were shown to be hemophiliac: brother, sister's son, maternal uncle, maternal aunt's son, maternal grandfather, and others more distantly related on mother's side. *Note:* Hemophilia on the father's side of the family is not relevant in this case.

10. Y-linked traits share two main characteristics: (1) they are only found in males and (2) they are transmitted from an affected father to all his sons and none of his daughters.

11. Crisscross inheritance occurs when the father passes his phenotype to his daughters and the mother to her sons. X-linked genes exhibit this pattern of inheritance.

EXERCISES AND PROBLEMS

12. Because male honeybees are haploid, they produce sperm via mitosis. Meiosis would not be a successful process because there is only one copy of each chromosome and therefore no homologs to pair with. Mitosis, on the other hand, would be successful because it does not require pairing (synapsis) of homologous chromosomes.

13. **a.** 0; female
    b. 1; female
    c. 0; male
    d. 1; male
    e. 2; female
    f. 4; female
    g. 1/0 mosaic; male–female mosaic

14. In *Drosophila*, sex is determined by the ratio of X chromosomes to sets of autosomes (X:A). Males have an X:A ratio of 0.5 (if the ratio is less than 0.5, metamales are produced); females have an X:A ratio of 1.0 (if the ratio is more than 1.0, metafemales are produced). A ratio that is between 0.5 and 1.0 yields intersex individuals.

	Chromosome Composition		
	Sex Chromosomes	**Autosomes**	**Sex of Fly**
a.	X	Diploid	*Male (X:A = 0.5)*
b.	XX	Diploid	*Female (X:A = 1.0)*
c.	XY	Diploid	*Male (X:A = 0.5)*
d.	XXX	Diploid	*Metafemale (X:A = 1.5)*
e.	XXY	Diploid	*Female (X:A = 1.0)*
f.	X	Triploid	*Metamale (X:A = 0.33)*
g.	XX	Triploid	*Intersex (X:A = 0.67)*
h.	XY	Triploid	*Metamale (X:A = 0.33)*
i.	XXX	Triploid	*Female (X:A = 1.0)*
j.	XXY	Triploid	*Intersex (X:A = 0.67)*
k.	XX	Tetraploid	*Male (X:A = 0.5)*
l.	XXX	Tetraploid	*Intersex (X:A = 0.75)*
m.	XXYY	Tetraploid	*Male (X:A = 0.5)*

15. **a.** The fly is a male because the X:A ratio is 1/2 = 0.5.

   b. This abnormal sex chromosome composition was most likely the result of a nondisjunction event (failure of the chromosomes to segregate normally) during gamete formation in meiosis. Because the fly has two Y chromosomes, the meiotic nondisjunction must have occurred in the father, and specifically in meiosis II. The fly was the result of a YY sperm fertilizing an egg carrying the X chromosome.

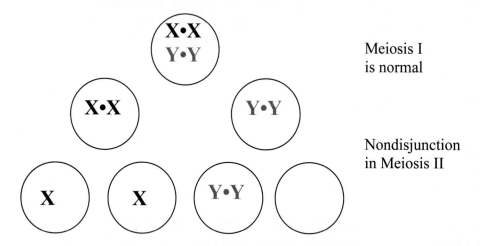

16. A grandson does not inherit a sex chromosome from his father's mother. A granddaughter, on the other hand, does not inherit any sex chromosomes from her father's father.

17. The cross, designated as XX *dsx⁺dsx* × XY *dsx⁺dsx*, produces the following Punnett square:

Female	Male			
	X dsx^+	**X dsx**	**Y dsx^+**	**Y dsx**
X dsx^+	XX dsx^+dsx^+	XX dsx^+dsx	XY dsx^+dsx^+	XY dsx^+dsx
X dsx	XX dsx^+dsx	XX $dsx\,dsx$	XY dsx^+dsx	XY $dsx\,dsx$

Therefore, the offspring consists of 3/8 females, 3/8 males, 2/8 intersexes (*dsx dsx* homozygotes).

18. a. Apparent ratio of offspring is 3:1, male to female. Fertile males are 50% *tra tra* and 50% *tra⁺tra*. Fertile daughters are all *tra⁺tra*; daughters counted as sterile males are *tra tra* = 1/4 of total.

b. If the progeny in part (a) are mated among themselves, all females will be *tra⁺tra* (*tra tra* females are sterile). Males will be *tra tra* and *tra⁺tra*. The mating of *tra⁺tra* females with *tra tra* males produces 3/4 males and 1/4 females. The mating of *tra⁺tra* females with *tra⁺tra* males produces 5/8 males (1/2 normal males plus 1/4 × 1/2 = 1/8 transformed females) and 3/8 females (1/2 females minus the 1/8 transformed). Since these matings are in equal frequencies, there are 11/16 males [(3/4 × 1/2) + (5/8 × 1/2)] and 5/16 females [(1/4 × 1/2) + (3/8 × 1/2)].

19.

		Cross	Reciprocal
P	Female	X^+X^+	$X^{lz}X^{lz}$
	Male	$X^{lz}Y$	X^+Y
F_1	Female	X^+X^{lz}	X^+X^{lz}
	Male	X^+Y	$X^{lz}Y$
F_2	Females	X^+X^+, X^+X^{lz}	$X^+X^{lz}, X^{lz}X^{lz}$
	Males	$X^+Y, X^{lz}Y$	$X^+Y, X^{lz}Y$

20. a. In chicken, the male is homogametic (ZZ) and the female heterogametic (ZW). The cross can be depicted as such:

P Z^BW × Z^bZ^b
Barred hen   Nonbarred rooster

F_1 Z^bW × Z^BZ^b
Nonbarred hen   Barred rooster

These would occur in equal proportions, so 50% of the F_1 offspring would be nonbarred hens, and 50% of the F_1 offspring would be barred roosters.

F_2 Z^BW ; Z^bW ; Z^BZ^b ; Z^bZ^b
Barred hen   Nonbarred hen   Barred rooster   Nonbarred rooster

These would occur in equal proportions, so the F_2 would consist of 1/4 barred hens, 1/4 nonbarred hens, 1/4 barred roosters, and 1/4 nonbarred roosters.

b. The cross must be $Z^BW \times Z^BZ^b$. The only nonbarred offspring would be Z^bW, and thus female.

21. P　　　$fyfy$ X^+X^+ (female)　　×　fy^+fy^+ $X^{ct}Y$ (male)

F$_1$　　　fy^+fy X^+X^{ct} (female)　　×　fy^+fy X^+Y (male)

| Female | **Male** | | | |
	fy^+ X^+	fy X^+	fy^+ Y	fy Y
fy^+ X^+	fy^+fy^+ X^+X^+	fy^+fy X^+X^+	fy^+fy^+ X^+Y	fy^+fy X^+Y
fy^+ X^{ct}	fy^+fy^+ X^+X^{ct}	fy^+fy X^+X^{ct}	fy^+fy^+ $X^{ct}Y$	fy^+fy $X^{ct}Y$
fy X^+	fy^+fy X^+X^+	$fyfy$ X^+X^+	fy^+fy X^+Y	$fyfy$ X^+Y
fy X^{ct}	fy^+fy X^+X^{ct}	$fyfy$ X^+X^{ct}	fy^+fy $X^{ct}Y$	$fyfy$ $X^{ct}Y$

F$_2$:　　　Females, 6/8 wild type, 2/8 fuzzy (or 3:1); males, 3/8 wild type, 3/8 cut, 1/8 fuzzy, and 1/8 cut and fuzzy.

22. Nothing can be inferred from the first cross. The second cross indicates that black is dominant to pink. The third cross alerts us to sex linkage, as we see a difference between males and females. Therefore, eye color in canaries is sex-linked, with black eye dominant to pink eye. Recall that in birds, the male is the homogametic (ZZ) sex. If we let B = black and b = pink, the crosses can be diagrammed as follows:

a. $Z^bW \times Z^bZ^b$

↓

Z^bW　　Z^bZ^b

All pink-eyed

b. $Z^bW \times Z^BZ^B$

↓

Z^BW　　Z^BZ^b

All black-eyed

c. $Z^BW \times Z^bZ^b$

↓

Z^bW　　Z^BZ^b

Pink-eyed female　Black-eyed male

23. **a.** X-linked
b. Gray
c. 1/2 gray:1/2 yellow in both sexes
In both crosses, we see a difference in the phenotypes of the sexes, suggesting sex linkage. The F$_1$ offspring from the first cross indicate that gray is dominant to yellow. The F$_1$ females from this cross must be heterozygous, and the two phenotypes in the F$_2$ males result from each of the X chromosomes in the F$_1$ female being hemizygous in the F$_2$ males. The first cross is therefore (calling gray the wild type):

$$X^+X^+ \quad \times \quad X^yY$$

$$\downarrow$$

$$\begin{array}{cc} X^+X^y & \times & X^+Y \\ \text{Gray} & & \text{Gray} \end{array}$$

$$\downarrow$$

X^+X^+	X^+X^y	X^+Y	X^yY
Gray	Gray	Gray	Yellow

Now diagram the second cross:

$$X^yX^y \quad \times \quad X^+Y$$

$$\downarrow$$

$$\begin{array}{cc} X^+X^y & \times & X^yY \\ \text{Gray} & & \text{Yellow} \end{array}$$

$$\downarrow$$

X^+X^y	X^yX^y	X^+Y	X^yY
Gray	Yellow	Gray	Yellow

24.

		Cross	Reciprocal
P	Female	Z^pW	Z^+W
	Male	Z^+Z^+	Z^pZ^p
F_1	Female	Z^+W	Z^pW
	Male	Z^+Z^p	Z^+Z^p
F_2	Females	Z^+W, Z^pW	Z^+W, Z^pW
	Males	Z^+Z^+, Z^+Z^p	Z^+Z^p, Z^pZ^p

25. a. We see a difference in the phenotype of male and female children suggesting some mode of sex linkage. Brown teeth cannot be Y-linked, or all of the sons would have brown teeth. Since all the daughters and one of the sons have brown teeth, brown must be an X-linked dominant trait, and white is recessive.

b. If B = brown teeth and b = white teeth, the original mating is $X^bX^b \times X^BY$. All daughters from this mating must be heterozygous. The A–B mating is therefore $X^BX^b \times X^bY$. One-half of all children will be expected to receive the X^B chromosome. Therefore, the probability that couple A–B's next child will have brown teeth is 1/2.

26. Yes. Begin by determining genotypes of the two individuals. The woman must be heterozygous X^CX^c. A man with normal vision must be X^CY, and all his daughters must receive his X chromosome and should be normal, either X^CX^C or X^CX^c. Since color blindness is recessive, the daughter must have two X^c chromosomes. *Note:* A *very* rare possibility is that the man is the father and nondisjunction occurred in *both* parents: at meiosis II in the female and either meiosis I or II in the male (see chapter 8).

27. **a.** F_1: 1 wild-type female:1 white-eyed male.
    b. F_2: 3/8 wild type; 3/8 white-eyed; 1/8 ebony; 1/8 ebony, white-eyed in both sexes.
    c. Reciprocal F_2: 6/16 wild-type females; 2/16 ebony females; 3/16 wild-type males; 3/16 white-eyed males; 1/16 ebony males; 1/16 ebony, white-eyed males.

Let X^+ = red eye, X^w = white eye, e^+ = gray body, e = ebony .

P	$X^wX^w\,e^+e^+$	×	$X^+Y\,ee$
		↓	
F_1	$X^+X^w\,e^+e$	×	$X^wY\,e^+e$
	Wild type		White-eyed

Because the two traits are independently assorting, we can use the product rule of probabilities to determine the F_2 generation. (Alternatively, we can use the Punnett square technique, as we did for problem 21.)

$(1/2)X^+$	×	$(1/2)X^w$	×	$(3/4)e^+-$	= 3/16 wild-type females
			×	$(1/4)ee$	= 1/16 ebony females
	×	$(1/2)Y$	×	$(3/4)e^+-$	= 3/16 wild-type males
			×	$(1/4)ee$	= 1/16 ebony males
$(1/2)X^w$	×	$(1/2)X^w$	×	$(3/4)e^+-$	= 3/16 white-eyed females
			×	$(1/4)ee$	= 1/16 white-eyed, ebony females
	×	$(1/2)Y$	×	$(3/4)e^+-$	= 3/16 white-eyed males
			×	$(1/4)ee$	= 1/16 white-eyed, ebony males

For the reciprocal cross,

P	$X^+X^+\,ee$	×	$X^wY\,e^+e^+$
		↓	
F_1	$X^+X^w\,e^+e$	×	$X^+Y\,e^+e$
	Wild type		Wild type

$(1/2)X^+$	×	$(1/2)X^+$	×	$(3/4)e^+-$	= 3/16 wild-type females
			×	$(1/4)ee$	= 1/16 ebony females
	×	$(1/2)Y$	×	$(3/4)e^+-$	= 3/16 wild-type males
			×	$(1/4)ee$	= 1/16 ebony males
$(1/2)X^w$	×	$(1/2)X^+$	×	$(3/4)e^+-$	= 3/16 wild-type females
			×	$(1/4)ee$	= 1/16 ebony females

$$\times \quad (1/2)Y \qquad \times \ (3/4)e^+- \qquad = 3/16 \text{ white-eyed males}$$
$$\times \ (1/4)ee \qquad = 1/16 \text{ white-eyed, ebony males}$$

28. All the offspring of the abnormal-1 cross are normal, indicating that the abnormal-1 trait must be recessive. If the trait were X-linked, the male offspring of this mating would be expected to have abnormal eyes. Since all offspring have normal eyes, the abnormal-1 trait must be autosomal and the abnormal-1 flies are homozygous. In the abnormal-2 cross, the 1:1 ratio within each sex indicates a mating between a heterozygote and a homozygote. The cross could be $X^aY \times X^AX^a$ or $aa \times Aa$. Therefore, abnormal-2 flies are heterozygous for a dominant gene that could be either autosomal or X-linked. Finally, the abnormal-3 trait must be dominant, but we cannot determine if it is X-linked or autosomal. We can tell, however, that the abnormal-3 flies are homozygous for that dominant gene.

29. We see four phenotypes, so we must have at least two genes involved. We see no difference in body color between sexes, so we can conclude that body color is autosomally controlled. Note that if body color were X-linked, F_1 males should have had dark bodies. Since all F_1 offspring have tan bodies, tan body must be dominant to dark body.

We see a difference in eye color between the sexes in the F_1 generation, suggesting that eye color is X-linked. Since the F_1 females have red eyes, red must be dominant, and hence the original white-eyed female must be homozygous for the recessive allele.

In the F_2 generation, the ratio of tan to dark is 51:16—very close to the 3:1 ratio expected for one autosomal gene. Red to white is 36:31, very close to the expected 1:1 ratio for an X-linked trait.

Let dk^+ = tan body, dk = dark body, X^+ = red eyes, X^w = white eyes. Parents are dk dk X^wX^w and dk^+dk^+ X^+Y. We can then diagram the cross between the resulting F_1 individuals: dk^+dk $X^+X^w \times dk^+dk$ X^wY. This cross yields:

$$(1/2)X^+ \times (1/2)X^w \text{ (or Y)} \times (3/4)dk^+- \ = 3/16, \text{ male or female, red-eyed, tan-bodied}$$
$$\times (1/4)dk\ dk \ = 1/16, \text{ male or female, red-eyed, dark-bodied}$$
$$(1/2)X^w \times (1/2)X^w \text{ (or Y)} \times (3/4)dk^+- \ = 3/16, \text{ male or female, white-eyed, tan-bodied}$$
$$\times (1/4)dk\ dk \ = 1/16, \text{ male or female, white-eyed, dark-bodied}$$

30. a. Let X^+ = wild-type eye, X^v = vermillion eye.

$$P \qquad X^vX^v \quad \times \quad X^+Y$$
$$\downarrow$$
$$F_1 \qquad X^+X^v \qquad X^vY$$

Therefore, a ratio of 1 wild-type female:1 vermillion male.

b. F_1 X^+X^v × X^vY

                              ↓

   F_2 X^+X^v X^vX^v X^+Y X^vY

The F_2 progeny are expected in a ratio of 1 wild-type female:1 vermillion female:
1 wild-type male:1 vermillion male.

	Offspring				
	X^+X^v	*X^vX^v*	*X^+Y*	*X^vY*	**Total**
Observed numbers (O)	58	62	60	65	245
Expected ratio	1/4	1/4	1/4	1/4	
Expected numbers (E)	61.25	61.25	61.25	61.25	245
$O-E$	−3.25	0.75	−1.25	3.75	
$(O-E)^2$	10.56	0.56	1.56	14.06	
$(O-E)^2/E$	0.172	0.009	0.025	0.229	$0.435 = \chi^2$

Critical chi-square at 0.05, 3 df, = 7.815. The chi-square, 0.435, is less than the
critical chi-square. Therefore, we fail to reject (that is, accept) the null hypothesis
and conclude that the observed numbers are consistent with the expectations of X-
linked gene inheritance.

31. **a.**   While it could be an autosomal recessive mode of inheritance, it is more likely a
   sex-linked recessive (only males are affected). If complete penetrance is
   assumed, it could not be a dominant or Y-linked mode of inheritance.

   b.   Could be an autosomal recessive, but most likely an autosomal dominant mode
   of inheritance. X- and Y-linked modes of inheritance are ruled out.

   c.   An autosomal recessive mode of inheritance is possible. It could also be an X-
   linked recessive mode of inheritance (if the allele causing the trait was brought
   into the pedigree by the maternal grandmother of the two affected individuals).
   Other modes of inheritance are ruled out.

32. Assuming 100% penetrance:

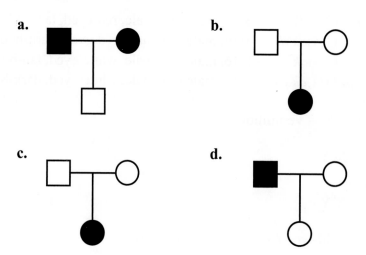

Note: Other pedigrees are possible.

33. *HH*, *Hh*, and *hh* females are all hen-feathered, whereas in males, *HH* and *Hh* genotypes produce hen-feathering, and *hh*, cock-feathering.

P	*hh*	×	*HH*
	Cock-feathered male ↓		Hen-feathered female
F₁	*Hh*	×	*Hh*
	Hen-feathered male		Hen-feathered female

This cross can be treated as two separate crosses:

	1/2 Males	**1/2 Females**
1/4 *HH*	Hen-feathered	Hen-feathered
2/4 *Hh*	Hen-feathered	Hen-feathered
1/4 *hh*	Cock-feathered	Hen-feathered

Therefore, the F₂ offspring consist of 3/8 hen-feathered males, 1/8 cock-feathered males, and 4/8 hen-feathered females.

34. a. Autosomal recessive inheritance is a possibility. However, X-linked recessive is more likely because all affected individuals are males.

    b. Autosomal dominant and autosomal recessive modes of inheritance are possible. However, the most likely mode of inheritance is X-linked dominant (affected males transmit the trait to all of their daughters but none of their sons).

    c. Autosomal dominant inheritance is possible. However, Y-linked inheritance is the most likely mode of inheritance. Only males are affected, and affected males transmit the trait to all of their sons but none of their daughters.

35. The cross can be depicted as follows

P	$X^AX^ABBCCDD$	×	$X^aYbbccdd$
		↓	
F₁	$X^AX^aBbCcDd$	×	$X^AYBbCcDd$

    a. The F₁ female produces 2^n gametes, where n = number of heterozygous gene pairs. Therefore, $2^4 = 16$ different gametes.

    b. Because the autosomal genes *B*, *C*, and *D* assort independently from each other as well as from the sex-linked gene *A*, we can treat the F₁ cross as four separate crosses.

For gene A: $X^A X^a \times X^A Y$ → 3/4 X^A– (1/4 $X^A X^A$ + 1/4 $X^A X^a$ + 1/4 $X^A Y$)
For gene B: $Bb \times Bb$ → 3/4 B–
For gene C: $Cc \times Cc$ → 3/4 C–
For gene D: $Dd \times Dd$ → 3/4 D–

We can now use the product rule to determine the combined probability of genotype X^A–B–C–D– = 3/4 × 3/4 × 3/4 × 3/4 = 81/256.

   c.   Only females can be heterozygous at all four loci:

For gene A: $X^A X^a \times X^A Y$ → 1/4 $X^A X^a$
For gene B: $Bb \times Bb$ → 1/2 Bb
For gene C: $Cc \times Cc$ → 1/2 Cc
For gene D: $Dd \times Dd$ → 1/2 Dd

The combined probability of genotype $X^A X^a BbCcDd$ = 1/4 × 1/2 × 1/2 × 1/2 = 1/32.

36. a.   The phenotype is the propensity to have twin offspring. It could be caused by a recessive or dominant, sex-linked or autosomal allele.

    b.   X-linked and Y-linked modes of inheritance are impossible. Both autosomal dominant and autosomal recessive inheritance are possible. However, autosomal dominance is more probable because autosomal recessive inheritance requires four different individuals (I-1, II-3, III-1 and IV-5) to be carriers of this rare trait.

    c.   Y-linked, X-linked dominant, and autosomal dominant modes of inheritance are impossible. Autosomal or X-linked recessive inheritance are possible. However, X-linked recessive is more probable because it requires less people to be carriers than autosomal recessive (namely, individuals II-5 and II-7).

    d.   Autosomal recessive inheritance is the only possibility.

37. a.   An autosomal dominant mode of inheritance is possible, but unlikely. Homer would have to be heterozygous for the disease (Aa), and he would have to transmit the A allele to all four of his daughters but to none of his six sons.

    b.   Autosomal recessive inheritance is possible, but unlikely. Homer would have to be homozygous for the disease (aa), and Marge would be a carrier (Aa). She would have to transmit the a allele to her four daughters and the A allele to her six sons.

    c.   Y-linked inheritance is impossible because the daughters have the disease.

    d.   X-linked dominant inheritance is very probable. Homer ($X^A Y$) gives his X chromosome to his daughters, who will all inherit the disease.

    e.   X-linked recessive inheritance is unlikely. Marge would have to be a carrier ($X^A X^a$) and transmit the X^A chromosome to her six sons and her X^a chromosome to her four daughters.

38. The genotype/phenotype designations are as follows:

Genotype	Phenotype in Males	Phenotype in Females
BB	Bald	Bald
Bb	Bald	Normal
bb	Normal	Normal

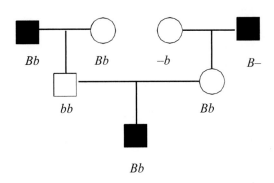

39. Assuming complete penetrance, this pedigree cannot be explained by Y-linked, autosomal or X-linked modes of inheritance. The trait is also obviously not sex-limited. So what's left? Well, sex-influenced inheritance. In this case, the trait is dominant in females and recessive in males. The genotype/phenotype designations are as follows:

Genotype	Phenotype in Females	Phenotype in Males
AA	Affected	Affected
Aa	Affected	Normal
aa	Normal	Normal

In the first two generations, the three normal males are all *Aa*, and the three affected females are all *Aa*. The genotypes of the affected females and normal male of the third generation cannot be determined for sure.

40. Because one of their daughters is albino *(aa)*, both parents must be carriers for the disease *(Aa)*. The second daughter is hemophilic (X^hX^h). Therefore, the father must have the trait (X^hY), and the mother must have at least one X^h allele. The fact that her father had normal blood clotting means that she must be normal for hemophilia (X^HX^h). Therefore, the cross can be designated as *Aa* $X^HX^h \times$ *Aa* X^hY.
Because the two traits assort independently, we can treat the cross as two separate crosses and then use the product rule to determine the combined probabilities.
a. For albinism: *Aa* $\times$ *Aa* $\rightarrow$ 3/4 *A–*
For hemophilia: $X^HX^h \times X^hY \rightarrow$ 1/4 X^HY
Therefore, the probability of genotype *A–* X^HY = 3/4 $\times$ 1/4 = 3/16.
b. For albinism: *Aa* $\times$ *Aa* $\rightarrow$ 1/4 *aa*
For hemophilia: $X^HX^h \times X^hY \rightarrow$ 1/4 X^hX^h
Therefore, the probability of genotype *aa* X^hX^h = 1/4 $\times$ 1/4 = 1/16.

   c. Only a daughter can be a carrier for both traits
       For albinism: $Aa \times Aa \rightarrow 1/2\ Aa$
       For hemophilia: $X^H X^h \times X^h Y \rightarrow 1/4\ X^H X^h$
       Therefore, the probability of genotype $Aa\ X^H X^h = 1/2 \times 1/4 = 1/8$.

41. Autosomal dominant and X-linked dominant modes of inheritance are impossible (if
complete penetrance is assumed). Autosomal recessive and X-linked recessive
modes of inheritance are possible, but unlikely (because both require that all
unaffected parents of affected children be carriers). The trait is only found in males.
However, it is not Y-linked, because the affected male in the fifth generation did not
have an affected father. The trait is most likely sex-limited.

CHAPTER INTEGRATION PROBLEM

a.

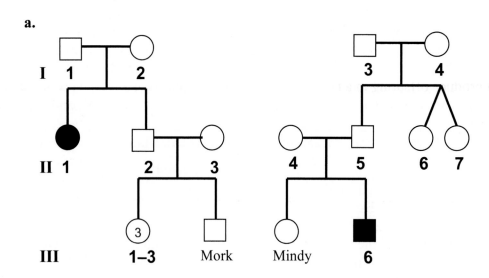

b. The only possibility for the Nanunanu syndrome is autosomal recessive. The other
four modes of inheritance are excluded.

c. Y-linked inheritance is excluded because a female (II-1) is affected. The trait cannot
be autosomal dominant or X-linked dominant because it appears in the offspring (II-
1 and III-6) of unaffected parents (I-1, I-2 and II-4, II-5, respectively). X-linked
recessive inheritance is excluded because the father (I-1) is not affected, yet his
daughter (II-1) is.

d.

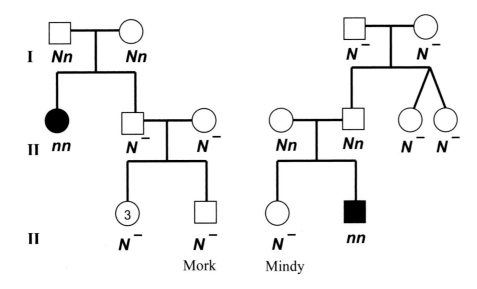

Mork   Mindy

e. For Mork and Mindy's first child to have Nanunanu syndrome, both would have to be carriers. This means Mork's father would have to be a carrier as well. We have to calculate each of these probabilities and then use the product rule. Mork's grandparents (*Nn* × *Nn*) produce offspring in a ratio of 1/4 *NN*:2/4 *Nn*:1/4 *nn*. However, we know that Mork's father is not affected, and so we are left with three possibilities: 2 *Nn* and 1 *nn*. Therefore, the probability of Mork's father being heterozygous is 2/3. Mork's father would then have a 50:50 chance of transmitting the *n* allele to Mork. Therefore, the probability of Mork being a carrier is 2/3 × 1/2 = 1/3. Mindy has a 2/3 probability of being a carrier (same reasoning as Mork's father). If Mork and Mindy are carriers, their offspring have a 1/4 chance of inheriting the Nanunanu syndrome. Thus, the overall probability is 1/4 × 1/3 × 2/3 = 1/18.

f. Two scenarios are possible: (1) Mork and Mindy are heterozygous and both children are unaffected, or (2) at least one parent is homozygous. Calculate each of these probabilities, and then use the sum ("or") rule to get to the overall probability.

Probability of scenario 1 = P(Mork heterozygous) × P(Mindy heterozygous) × P(two unaffected children) = 1/3 × 2/3 × 3/4 × 3/4 = 1/8.

The probability of both parents being heterozygous is 1/3 × 2/3 = 2/9. Therefore, the probability of at least one of them being homozygous (scenario 2) is 1 – 2/9 = 7/9.

The overall probability is thus = 1/8 + 7/9 = 65/72 or 90.28%.

g. Because their first child had Nanunanu syndrome, both Mork and Mindy must be carriers. Therefore, their next child has a 25% chance of developing the syndrome.

h. The only scenario that does *not* satisfy the "*at least* one normal child out of three" proposition is if all three children are affected. P(three affected children) = 1/4 × 1/4 × 1/4 = 1/64. Therefore, P(*at least* one normal child out of three) = 1 − 1/64 = 63/64.

CHAPTER 5 MODIFICATIONS TO MENDELIAN PATTERNS OF INHERITANCE

Chapter Goals
1. Compare the various types of dominance and how they affect the phenotype.
2. Understand how penetrance and expressivity can affect the expression of an allele.
3. Understand how epistasis and suppression of nonallelic genes affect the phenotype.
4. Using phenotypic ratios, deduce the underlying genetic interactions.
5. How does epigenetic phenomena affect the observed pattern of inheritance?

Key Chapter Concepts

5.1 Variations on Dominance

- **Incomplete Dominance:** Dominance is not universal. Incomplete dominance occurs when one allele does not completely mask the expression of the second allele. Rather, both alleles influence the phenotype of a heterozygote resulting in a unique, intermediate phenotype. If two heterozygotes are crossed, the phenotypic and genotypic ratios are the same (1:2:1). For example, assume that coat color in cattle is determined by a pair of alleles that are incompletely dominant. If you mate a bull with black hair and a cow with white hair, the resulting offspring would be gray!

- **Codominance:** Codominance is another type of inheritance pattern that results in a unique phenotype for the heterozygote. In this case, both alleles are expressed and the resulting phenotype is not an intermediate, but is a combination of the two phenotypes in the same individual. Consider the preceding cattle example. In the case of codominance, the coat of the heterozygote would be composed of both black and white hairs. If two heterozygotes are crossed, the phenotypic and genotypic ratios would again be the same (1:2:1).

- **Levels of Dominance Assignment:** It is important to note that the classification of alleles as incompletely dominant or codominant may be subject to the level of observation. For example, the codominant cow expressing both black and white hairs may appear gray from a distance and be classified by an incomplete dominance.

5.2 Multiple Alleles

- **Multiple Alleles:** Most genes have more than two alleles, although each diploid individual can only possess two of the possible alleles (one from the egg and one from the sperm). One of the most familiar examples of a gene with multiple alleles is the human ABO blood system. In this system, the A and B alleles (I^A and I^B) are dominant to the O allele (i) but codominant with respect to each other. These relationships result in the four phenotypic classes observed (blood types A, B, AB, and O). The alleles may have simple dominant recessive relationships to each other as in this blood type example, or there may be complex relationships between two or more of the alleles as in the fur color in mice example provided by your textbook.

5.3 Testing Allelism

- *Cis–trans* **Complementation Test:** Many different alleles exist for each gene, and often multiple genes are required to produce a particular trait. Therefore, it is important to be able to determine whether two alleles discovered in a population are alleles of the same gene or whether they are alleles of two different genes. The *cis–trans* complementation test is used to determine this. It is based on the fact that a recessive mutation can be complemented with a wild-type allele from another individual if that individual's mutation is present in a different gene. Therefore, if you mate two mutant individuals, wild-type offspring (complementation) indicate that the alleles are of different genes whereas mutant offspring (noncomplementation) indicate that the alleles are of the same gene.

5.4 Lethal Alleles

- **Recessive and Dominant Lethality:** Sometimes the phenotype produced by a genotype is lethal and the offspring containing the genotype are never born. These dead individuals will never be observed. Thus, when counting the number of each class of offspring in a monohybrid cross, the phenotypic ratios will deviate from the 3:1 ratio observed by Mendel. It is important to note, however, that

Mendel's principles and laws are still at work to produce the new ratios. All of the predicted genotypes are still formed; the observer just cannot see the dead progeny. A modified ratio of 2:1 instead of 3:1 will result if the mutation is recessive for the lethal phenotype (homozygotes are dead) but dominant for an observable phenotype. Sometimes, an allele is dominant for lethality. It is important to distinguish a dominant lethal allele from a recessive lethal allele that produces a dominant observable phenotype in heterozygotes.

- **Pleiotropy**: In pleiotropy, a single gene has multiple effects on the phenotype. For example, albinism is a single gene defect that produces a wide variety of phenotypic symptoms, including altered pigmentation and unsteady vision.

5.5 Penetrance and Expressivity

- **Different Phenotypes from the Same Genotype:** Individuals with identical *genotypes* do not always display the same *phenotypes*. One individual may display the mutant phenotype, and the other will retain the normal phenotype. This is an example of **reduced penetrance** because the allele fails to *penetrate* (cause the phenotype) in *all* individuals that have the allele(s). This results in an "all-or-nothing" expression pattern. The degree to which the phenotype is expressed is known as the allele's penetrance and is expressed as the percentage of individuals who display the phenotype relative to the number of individuals who have the genotype. In other instances, an array of different phenotypes can be produced from the same genotype. This is known as **variable expressivity** when the phenotype resulting from a specific genotype is expressed in varying degrees, from mild to extremely severe. Finally, there are examples of human traits, **phenocopies**, that appear to be genetically controlled, but they are actually the result of environmental influences.

5.6 Genotypic Interactions

- **Overview:** Sometimes more than one gene is responsible for a particular phenotype. This is most evident when heterozygous individuals are crossed and a 9:3:3:1 ratio or some modification of this ratio results. In these situations, the classes formed by sixteenths are different versions of a single trait. This is in contrast to the 9:3:3:1 ratio in Mendel's dihybrid cross which was different combinations of two traits. These ratios are made up of sixteenths and are a clear signal that alleles of two different genes are involved in creating the observed phenotypes.

- **Complementary and Duplicate Gene Action:** In **complementary gene action**, the action of both of two genes is necessary to produce a phenotype and a 9:7 ratio will be observed. **In duplicate gene action**, the action of either one of two different genes can produce the phenotype and a ratio of 15:1 will be observed. Thus, the genes are duplicates, or redundant. In this case, the only way to not produce the phenotype is if both genes have two mutant alleles so that neither gene can make a function protein.

- **Epistasis:** Epistasis is an example of a gene interaction in which two or more genes contribute to the same phenotype. In epistasis, one pair of alleles masks or modifies the expression of a second pair of alleles. Epistasis is similar to dominance, but it is unique in that in epistasis, one gene masks another gene while in dominance, one allele masks a different allele of the *same* gene. The **epistatic gene** is the one that masks the phenotype of the **hypostatic gene**. Epistatic interactions will produce modified 9:3:3:1 ratios in dihybrid crosses. If a recessive phenotype masks another gene (**recessive epistasis**), a 9:4:3 ratio will be observed, whereas if a dominant phenotype masks another gene (**dominant epistasis**), a 12:3:1 ratio will be observed. Notice that epistatic interactions yield three classes of progeny in dihybrid crosses; the fourth class is grouped with one of the other three. Importantly, epistatic genes usually encode enzymes that function early in biochemical pathways; loss of the earlier-acting enzyme will prevent the later-acting enzyme from producing a phenotype.

5.7 Epigenetics

- **Overview: Epigenetics** refers to changes in gene function without changes to gene sequences. An example of this is the formation of Barr bodies in female mammals. When one of the female × chromosomes is inactivated to compensate for gene dosage effects, a heterozygous individual may inactivate a different allele in different places in the body. Thus, a **mosaic** phenotype can be created where the maternal allele is expressed in some parts of the body while the maternal allele is expressed in others.

Key Terms

- epigenetic inheritance
- incomplete dominance
- codominance
- plastids
- hexosaminidase-A
- antigens
- erythrocytes
- allelic series
- complementation test
- *cis–trans* test
- *cis* configuration
- *trans* configuration

- control
- saturation
- lethal alleles
- pleiotropy
- dominant lethality
- incomplete penetrance
- deleterious alleles
- reduced penetrance
- phenocopy
- variable expressivity
- complementary gene action

- duplicate gene action
- gene families
- epistasis
- epistatic gene
- hypostatic gene
- recessive epistasis
- dominant epistasis
- suppression
- recessive suppressor
- epigenetics
- mosaicism

Understanding the Key Concepts

Use the key terms or parts of their definitions to complete the following sentences.

   In contrast to Mendel's observations, researchers studying flower color in four-o'clock plants discovered that in a monohybrid cross, the F_1 progeny were different from either parent. In fact, they displayed a phenotype (1) _____ to both of the parents. In this type of inheritance, termed (2)_____, the phenotypic ratio of the F_2 progeny of a monohybrid cross is equal to the (3)_____. The intermediate phenotype of (4) _____-zygotes usually results from the production of only (5) _____ the amount of product from a gene. A different type of inheritance, termed (6) _____, also results in an intermediate phenotype in which (7) _____ phenotypes are expressed. The ability to detect heterozygosity is important for genetic counseling and predicting the probability of having afflicted offspring. Therefore, finding a level of study at which a mutation displays (8) _____ or (9) _____ is very important.
   Most genes in a population exist in (10) _____ such that there is a great deal of variation between individuals. Furthermore, traits are often encoded by multiple genes. For example, you will learn in Chapter 6 about more than eight genes that control eye color in *Drosophila*. When new mutations are found in *Drosophila* that produce a new eye color, it is not apparent which of these eight genes the mutation resides in. A (11) _____ can be used to determine whether the mutation is (12) _____ to any of the already known eye color mutations or if it represents a new gene. To perform this test, researchers mate the new mutant fly with other flies that have other eye color mutations. The mutant fly whose cross produces (13) _____ offspring is determined to be allelic to the new mutation. In other words, both mutations reside in the same (14) _____. A (15) _____ experiment is necessary to demonstrate that (16) _____ is possible.
   When two heterozygous individuals are crossed, the phenotypic ratio of one trait is expected to be (17) _____. If a ratio of 2:1 is observed instead, it is likely that the mutation is (18) _____. Since two-thirds of the surviving progeny are heterozygous and show a different phenotype than the one-third that are homozygous, we conclude that the mutation also causes a dominant, observable phenotype. Thus, the mutation displays (19) _____ by causing two different phenotypes. Some lethal alleles are dominant. In order for them to be inherited, they must either display (20) _____ or not cause lethality until after sexual maturity.
   Anytime a ratio composed of sixteenths (9:3:3:1 or 9:4:3) is observed after crossing heterozygous individuals, it indicates that (21)_____ genes are involved. If you are observing one trait, then (22) _____ genes must be involved in producing that trait. If the action of all of these genes are required to produce a particular phenotype, it is expected that one will be (23) _____ to the other called the (24) _____. If the ratio you observe is 9:4:3, the type of gene interaction observed is

(25) _____, while a ratio of 12:3:1 indicates (26) _____. If the action of only one of the genes is required, we call this (27) _____ and the genes are said to be (28) _____.

Figure Analysis

Use the indicated figure or table from the textbook to answer the questions that follow.

1. **Figure 5.5**
(a) What are researchers trying to determine by performing the test demonstrated in figure 5.5?

(b) How can the flies in the F_1 generation of part (a) have wild-type wings?

(c) Why don't the flies in the F_1 generation of part (b) or part (c) have wild-type wings?

(d) What must be true of the mutations being tested in order for this test to work?

2. **Figure 5.8**
(a) How is the array of phenotypes different between incomplete penetrance and variable expressivity?

(b) How many of the 10 flies depicted would be predicted to have the red-eye phenotype if the white-eye mutation was 80% penetrant instead of 60%?

(c) An array of eye color phenotypes ranging from white to yellow to cream to red occurs through multiple alleles in a population. How is the array of phenotypes described in figure 5.8 by variable expressivity different from the multiple allele array?

3. **Figure 5.11**
(a) What is different between the 9:3:3:1 ratio in this figure as compared to the same ratio obtained by Mendel in dihybrid crosses?

(b) What basic piece of information is demonstrated by the fact that four phenotypes arise in sets of sixteenths from crossing two identical heterozygous individuals?

(c) In terms of the requirement for the products of genes *R* and *P*, explain how the following combs might be formed.
       Single:

       Walnut:

       Pea:

       Rose:

4. **Figure 5.16**
(a) What basic piece of information is demonstrated by the fact that *three* phenotypes arise in sets of sixteenths from crossing two identical heterozygous individuals?

(b) In terms of the requirement for the products of genes *A* and *C*, explain how the following coat colors are formed.
       Albino:

       Black:

       Agouti:

(c) Which gene is epistatic, and which is hypostatic?

General Questions

Q5-1. Compare and contrast complete dominance, incomplete dominance, and codominance.

Q5-2. Describe how genes control the fate of metabolic pathways.

Q5-3. Explain how alleles displaying reduced penetrance and/or variable expressivity can produce different phenotypes in different individuals?

Q5-4. How does the identification of epistatic interactions help researchers understand biochemical pathways?

Q5-5. How does the identification of suppressor mutations help researchers understand biochemical pathways?

Multiple Choice

For each of the following, circle the letter of the choice that most appropriately answers the question.

1.) Why are heterozygous four-o'clock flowers pink?
   a. Both alleles produce an enzyme that produces a pink pigment.
   b. The reduced amount of enzyme produced from only one red allele cannot produce enough red pigment to make a fully red flower.
   c. The R_1 allele and R_2 alleles make two different proteins that produce a pink pigment only when the proteins interact with each other.
   d. Pink flower color is dominant to both red and white flower color.
   e. None of the above

2.) At what level of observation is Tay–Sachs disease codominant?
   a. survival
   b. brain function
   c. hexosaminidase-A enzyme activity
   d. hexosaminidase-A enzyme size
   e. at all levels of observation

3.) Which of the following individuals can receive type A blood in a transfusion?
   a. type A only
   b. type AB only

c. type O only
d. type B only
e. Both type A and type AB

4.) Red eyes are dominant to brown eyes and orange eyes in *Drosophila*. If a brown-eyed male fly is mated with an orange-eyed female fly and all of the progeny have red eyes, what conclusion can be drawn from this data?
   a. The two mutations are on different chromosomes.
   b. The two mutations are in different genes.
   c. The two mutations are in the same gene.
   d. The two mutations are codominant to each other
   e. The two mutations are incompletely dominant to each other.

5.) In *Drosophila*, lobe eye shape is dominant to wild-type eye shape. If a homozygous lobe-eyed fly is crossed to a wild-type fly, all the F_1 progeny will be heterozygous. If the ratio of F_1 flies was scored and found to be 80 lobe-eyed flies and 20 wild-type flies, what should be concluded?
   a. The lobe eye allele is 80% penetrant.
   b. The lobe eye allele is 20% penetrant.
   c. The wild-type eye allele is 20% penetrant.

    d. The lobe eye allele displays variable
       expressivity.
    e. The lobe eye allele is codominant.

6.) If crossing two identical, heterozygous individuals results in a phenotypic ratio of 2:1, what should you predict about the allele in question?
    a. Two genes are required to produce the
       phenotype.
    b. It is epistatic to the wild-type allele.
    c. It is lethal in the homozygous state.
    d. It is lethal in the heterozygous state.
    e. It is displaying variable expressivity.

7.) Which of the following ratios demonstrates complementary gene action?
    a. 9:7
    b. 3:1
    c. 15:1
    d. 13:3
    e. 12:3:1

8.) Which of the following ratios demonstrates dominant epistasis?
    a. 9:7
    b. 3:1
    c. 15:1
    d. 13:3
    e. 12:3:1

9.) Which of the following is *not* characteristic of suppressor mutations?
    a. They often have no observable phenotype
       in the absence of the mutation they
       suppress.
    b. Dihybrid crosses involving a mutation and
       its suppressor produce only two different
       phenotypes.
    c. They change the phenotype of a mutant
       allele back toward the wild-type
       phenotype.
    d. They cannot suppress recessive mutations.
    e. They can be either dominant or recessive.

10.) Which of the following would be considered epigenetic inheritance?
    a. changing the sequence of a gene
    b. changing the expression of all the genes on
       a chromosome by condensing that
       chromosome
    c. creating a suppressor mutation that
       suppresses the phenotype of a mutant
       allele
    d. loss of a chromosome due to
       nondisjunction
    e. None of the above

Practice Problems

P5-1. Use the following diagrams to illustrate reduced penetrance and expressivity. Assume that each figure represents an individual female or male with the same genotype. For individuals that are unaffected, leave the circle blank. Use varying degrees of shading to illustrate (a) reduced penetrance and (b) complete penetrance with variable expressivity.

                      (a)   (b)

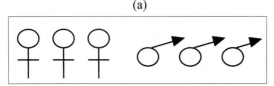 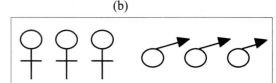

P5-2. A cross is made between a bird with blue feathers and a bird with yellow feathers. All of the offspring have green feathers. Two of the birds with green feathers are crossed, and the resulting generation consists of 10 blue birds, 20 green birds, and 9 yellow birds. How is feather color inherited? If one of the green birds is backcrossed to the yellow parent, what genotypes and phenotypes are expected in their progeny?

(*Hint 1*: To determine the pattern of inheritance, examine the phenotypes and genotypes of each generation.
    a.   The presence of an intermediate phenotype in the F_1 generation *suggests* incomplete
        dominance. The expression of both alleles in a heterozygote *suggests* codominance.

    b. Consider the *possibility* of multiple alleles if identical crosses produce different phenotypes. For example, examine the offspring produced from the mating of two parents with type A blood. In the first cross, all of the offspring have type A blood. In the second cross, three individuals have type A blood while one individual has type O blood.

    c. Gene interactions are indicated when the phenotypic ratio among offspring is expressed as a modified 9:3:3:1 ratio.

    d. A 1:2:1 ratio among the offspring *suggests* that both parents are heterozygous and the mode of inheritance is incomplete dominance or codominance.

    e. A 2:1 ratio *suggests* a homozygous lethal mutation and that both parents are heterozygous.

    f. A 9:3:3:1 ratio *suggests* that the parents are heterozygous for two genes.)

P5-3. For each of the following couples, determine the possible genotypes and phenotypes of their children:

    a. type A male and type A female   d. type AB male and type O female

    b. type B male and type A female   e. type O male and type O female

    c. type AB male and type AB female

P5-4. The Bombay phenotype in humans is a rare recessive mutation. Normally, the gene products of the *A* and *B* alleles of the ABO blood system modify the terminal sugars of a mucopolysaccharide (H structure) of the red blood cells. Individuals with the genotype *HH* or *Hh* have a complete H structure. However, individuals with the *hh* genotype produce an H substance lacking a fructose. The enzymes specified by the *A* and *B* alleles of the ABO blood system are unable to recognize and modify this incomplete H substance. As a result, the ABO system genotype is not expressed in an individual that is homozygous for the H substance abnormality (*hh*). These individuals are phenotypically type O. Predict the phenotypic ratio found in the progeny of the following couple: $I^A I^B\, Hh \times I^A I^B\, Hh$.

P5-5. In a particular species of fruit trees, two different varieties produce yellow fruit while a third variety produces red fruit. If two pure-breeding, yellow-fruited plants are crossed, all of the progeny produce red fruit. If one of these red-fruited plants is selfed, 9/16 of the progeny produce red fruit and 7/16 produce yellow fruit. (a) Suggest a mechanism of how a gene or genes control fruit color in this species of fruit trees. (b) What were the genotypes of the yellow fruit trees that were crossed? (c) How can you write the genotype of the red-fruited trees to include all possible genotypes? (d) Are any of the genes involved in producing fruit color epistatic?
(See *Hint 1*.)

Assessing Your Knowledge

Understanding the Key Concepts—Answers

1.) intermediate; 2.) incomplete dominance; 3.) genotypic ratio; 4.) hetero; 5.) half; 6.) codominance; 7.) both; 8.) incomplete dominance; 9.) codominance; 10.) multiple alleles; 11.) complementation test or *cis–trans* test; 12.) allelic; 13.) mutant; 14.) gene; 15.) control; 16.) complementation; 17.) 3:1; 18.) recessive lethal; 19.) pleiotropy; 20.) reduced penetrance; 21.) two; 22.) two; 23.) epistatic; 24.) hypostatic gene; 25.) recessive epistasis; 26.) dominant epistasis; 27.) duplicate gene action; 28.) redundant.

Figure Analysis—Answers

1a.) They are trying to determine whether the mutations that cause the two different wing phenotypes are alleles of the same gene or different genes; b.) These flies have one wild-type copy of gene 1 that they received from their father and one wild-type copy of gene 2 that they received from their mother. Since the mutations are recessive, one copy of each of these genes is sufficient to produce wild-type wings. These two mutations (alleles) are said to complement each other because they reside in different genes; c.) The alleles for the wing mutations are different alleles of the same gene. Therefore, neither parent has a wild-type copy of gene 1 to pass to the F_1 progeny who therefore can only receive two mutant copies of gene 1. Therefore, these two mutations cannot complement each other since they reside in the same gene; d.) The mutations being tested must be recessive, and thus the mutant flies that are crossed must be homozygous.

2a.) There are only two different phenotypes found in incomplete penetrance (mutant or wild-type), while a broad range of many different phenotypes is found in variable expressivity; b.) Two red-eyed flies would be found; c.) The variable expressivity array of phenotypes all result from a *single* genotype, while each phenotype in the multiple allele array is produced from a *different* genotype (or allele).

3a.) The 9:3:3:1 ratio found in figure 5.11 depicts four different *variations of the same trait*. Mendel observed four different *combinations of the phenotypes of two traits*; b.) The alleles of two different genes must be involved in producing comb morphology in chickens.
c.) Single: This must be the default morphology and does not require either of the R or P gene products. In fact, having either one or both of the R and/or P gene products will inhibit formation of this comb. Therefore, to form this comb, chickens must have *neither* the R nor the P gene products.
   Walnut: To form this comb, chickens must have the activity of *both* of the R and P gene products.
   Pea: To form this comb, chickens must have the activity of the P gene product but they cannot have the activity of the R gene product.
   Rose: To form this comb, chickens must have the activity of the R gene product but they cannot have the activity of the P gene product.

4a.) Two genes must be involved in producing coat color in mice, and one of those genes must be epistatic to the other.
b.) Albino: This coat color is produced whenever the activity of the product of gene C is missing. The presence or absence of the product of gene A is irrelevant to albino coat color.
   Black: This coat color is produced whenever the activity of the product of gene C is present and the activity of the product of gene A is absent.
   Agouti: This coat color is produced whenever the activity of the products of *both* genes A and C is present.
c.) Gene C is epistatic to gene A in a recessive manner. Therefore, we say that gene A is hypostatic.

General Questions—Answers

Q5-1. For complete dominance, one allele is completely expressed while the other is masked in a heterozygous individual. Both homozygous dominant and heterozygous genotypes produce a dominant phenotype. In contrast, for incomplete dominance and codominance, the heterozygote has its own phenotype that is different from either of the homozygotes. The heterozygous phenotype is intermediate to the parental phenotypes if the trait is incompletely dominant. For codominance, the heterozygous phenotype results from the complete expression of both alleles.

Q5-2. Genes produce proteins. Many of these proteins are enzymes that participate in metabolic pathways. A mutant gene may fail to produce a necessary enzyme, thus shutting down a specific metabolic pathway or part of the pathway.

Q5-3. Two individuals may have the same alleles at a particular locus, but they will certainly have many other genetic differences at different loci within their genome. The presence of other unique alleles from different loci can affect the penetrance or the expressivity of the allele in question. Similarly, the environment (including dietary intake) will be different for each individual. These unique environments can also influence whether a phenotype penetrates and, if it does, to what extreme the phenotype manifests.

Q5-4. Epistasis helps order gene products in a biochemical pathway. The product of an epistatic gene usually functions before the other genes in a biochemical pathway. Thus, by comparing sets of mutations in biochemical pathways, researchers can ascertain the order in which the gene products function in that pathway. This is ultimately necessary for the discovery of the function of each gene. This in turn is necessary to understand a disease's manifestation and to develop a cure.

Q5-5. Suppressor mutations usually identify two genes whose products participate in the same biochemical pathway. Often the two gene products physically interact with each other. Thus, by searching for suppressor mutations, researchers can find all the genes whose products function in a pathway.

Multiple Choice—Answers

1.) b; 2.) d; 3.) e; 4.) b; 5.) a; 6.) c; 7.) a; 8.) e; 9.) d; 10.) b.

Practice Problems—Answers

P5-1.

(a)   (b)

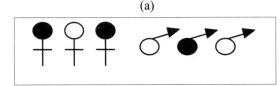   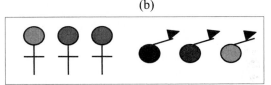

P5-2. Begin by examining the phenotypic classes and ratios produced in the F_1 and F_2 generations. The F_1 offspring were all green, a color intermediate to the parental colors of blue and yellow. In the F_2 generation, three distinct phenotypes are observed in a 1:2:1 ratio. One-half of the offspring show the intermediate phenotype (green) of the F_1 generation. This pattern indicates that feather color shows incomplete dominance. The green birds are heterozygotes.

Gametes	A	a
a	Aa (green)	aa (yellow)
a	Aa (green)	aa (yellow)

Crossing a green bird (Aa) with the yellow parent (aa) will yield a 1:1 ratio of green (Aa):yellow (aa) birds.

Note that this problem could be solved by letting AA = blue and aa = yellow *or* by letting AA = yellow and aa = blue.

P5-3. The ABO blood system displays complete dominance as well as codominance. The A and B alleles are dominant to the O allele (i) but are codominant with respect to each other. Before you begin to cross the individuals using a Punnett square, determine the possible genotypes corresponding to each blood type. Blood type A can result from genotype $I^A I^A$ or $I^A i$; blood type B can result from genotype $I^B I^B$ or $I^B i$; blood type AB corresponds to genotype $I^A I^B$; and blood type O corresponds to genotype ii.

Cross	Offspring Genotypes	Offspring Phenotypes
Cross a:		
$I^A I^A \times I^A I^A$	All $I^A I^A$	All type A
$I^A I^A \times I^A i$	1/2 $I^A I^A$, 1/2 $I^A i$	All type A
$I^A i \times I^A i$	1/4 $I^A I^A$, 1/2 $I^A i$, 1/4 ii	3/4 type A, 1/4 type O
Cross b:		
$I^B I^B \times I^A I^A$	All $I^A I^B$	All type AB
$I^B i \times I^A I^A$	1/2 $I^A I^B$, 1/2 $I^A i$	1/2 type AB, 1/2 type A
$I^B I^B \times I^A i$	1/2 $I^A I^B$, 1/2 $I^B i$	1/2 type AB, 1/2 type B
$I^B i \times I^A i$	1/4 $I^A I^B$, 1/4 $I^B i$, 1/4 $I^A i$, 1/4 ii	1/4 type AB, 1/4 type B, 1/4 type A, 1/4 type O
Cross c:		
$I^A I^B \times I^A I^B$	1/4 $I^A I^A$, 1/2 $I^A I^B$, 1/4 $I^B I^B$	1/4 type A, 1/2 type AB, 1/4 type B
Cross d:		

$I^A I^B \times ii$	$1/2\ I^A i,\ 1/2\ I^B i$	1/2 type A, 1/2 type B
Cross e:		
$ii \times ii$	All ii	All type O

P5-4. The Bombay phenotype is an example of epistasis. There are two gene pairs involved. One pair of alleles specifies the blood type. The second pair of alleles masks the expression of the first gene by interfering with the production of a complete H substance. As a result, the blood system genotype is only expressed in the presence of a dominant H allele at the second locus. The cross you are asked to consider is a dihybrid cross: both individuals are heterozygous for both genes. The phenotypes of each genotype will depend upon the interaction of the two genes.

Gametes	$I^A H$	$I^A h$	$I^B H$	$I^B h$
$I^A H$	$I^A I^A HH$ (type A)	$I^A I^A Hh$ (type A)	$I^A I^B HH$ (type AB)	$I^A I^B Hh$ (type AB)
$I^A h$	$I^A I^A Hh$ (type A)	$I^A I^A hh$ (type O)	$I^A I^B Hh$ (type AB)	$I^A I^B hh$ (type O)
$I^B H$	$I^A I^B HH$ (type AB)	$I^A I^B HH$ (type AB)	$I^B I^B HH$ (type B)	$I^B I^B Hh$ (type B)
$I^B h$	$I^A I^B Hh$ (type AB)	$I^A I^B hh$ (type O)	$I^B I^B Hh$ (type B)	$I^B I^B hh$ (type O)

Any individuals with the hh genotype are type O because they lack the ability to place the A- or B-type sugars on their red blood cells. Only when an individual has at least one dominant, wild-type allele (H) does the blood type correspond to the I locus alleles. The final phenotypic ratio observed is 3/16 type A, 3/16 type B, 6/16 type AB, and 4/16 type O.

P5-5. When looking at this data, you see a modified 9:3:3:1 ratio, immediately suggesting that two genes are interacting to control fruit color. The 9/16 are likely to be the same 9/16 as in the 9:3:3:1 ratio; all have at least one dominant allele of both genes. The 7/16 likely results from the two 3/16 classes and the one 1/16 class, none of which have a functional copy of *both* genes. Therefore you can surmise that red fruit color requires the action of *both* of the genes.

(a) It is likely that the enzyme produced by one of the two genes changes a yellow compound into another yellow compound and that the enzyme produced by the second gene changes the second yellow compound into a red compound.

(b) The selfed red fruit tree must have been heterozygous for two genes in order to produce a modified 9:3:3:1 ratio such as 9:7. Thus, together the yellow-fruited trees must have had two alleles of two different genes for a total of four different alleles. Furthermore, the two yellow-fruited trees were homozygous since they were pure-breeding. Thus the genotype options are *AABB, AAbb, aaBB,* or *aabb.* Since *AABB* would be a red plant (functional copies of both genes), we can rule it out. We can also rule out *aabb* since *four* alleles need to be present in the yellow parents. Therefore, one yellow parent must have been *AAbb* and the other *aaBB*.

(c) *A* _____ *B* _____, where _____ denotes either allele of the gene.

(d) Both genes are recessive epistatic to each other. This is a unique case of epistasis where even the gene product that acts *later* in the pathway can be epistatic. Notice that if the *aa* genotype is present, it does not matter what alleles are present in the *B* gene. Likewise, if the *bb* genotype is present, it does not matter what alleles are present in the *A* gene.

5

MODIFICATIONS TO MENDELIAN PATTERNS OF INHERITANCE

CHAPTER SUMMARY QUESTIONS

1. 1. g, 2. e, 3. a, 4. j, 5. c, 6. d, 7. h, 8. b, 9. i, 10. f.

2. In codominance, the heterozygote expresses the phenotypes of both homozygotes. In incomplete dominance, the heterozygote has a phenotype that is intermediate between those of the two homozygotes.

3. Dominance refers to the expression of an allele (and the masking of the other) when in the heterozygous condition. Epistasis is similar to dominance but is intergenic; that is, an allelic combination at one gene can mask the expression of alleles of another gene.

4. Penetrance is the proportion of individuals of a particular genotype that shows the appropriate phenotype; expressivity is the degree to which a trait is expressed.

5. Pleiotropy is the condition where a single mutant simultaneously affects several apparently unrelated phenotypic effects. It is supported when crosses cannot separate these effects and verified when a molecular or physiological mechanism of gene action is determined.

6. The disease is recessive at the level of the individual but incompletely dominant at the enzymatic level.

7. Universal donor is type O; it has no red-cell antigens and will therefore be accepted by all blood types. Universal recipient is type AB; it has no serum antibodies and can therefore accept all types of blood.

8. The I^A and I^B alleles produce glycosyl transferase enzymes that each add a different substance to the H structure, whereas the i allele produces a functionless protein that adds nothing. Thus, for example, the $I^A i$ heterozygote converts all the H structures to the A antigen making I^A dominant to i. $I^A I^B$ heterozygotes convert the H structures to approximately equal numbers of A and B antigens.

9. Gene interaction refers to the phenomenon where two or more genes have an effect on the same trait.

10. The complementation test, or *cis–trans* test, is used to determine if different recessive mutations are allelic. Mutations that are allelic (that is, in the same gene) will produce a mutant phenotype when they are in the *trans* configuration. Mutations that are non-allelic (that is, in different genes) will produce a wild-type phenotype when they are in the *trans* configuration.

EXERCISES AND PROBLEMS

11. Pink seems to be a blend of red and white. Assume that the red and white plants are homozygotes. If one of the colors were dominant, the progeny from the first cross should all be either red or white. This is not seen; therefore, neither color is dominant. Pink probably represents the heterozygote.

    Let R = gene color; R_1 = red and R_2 = white. The first cross can be indicated:

$$R_1 R_1 \quad \times \quad R_2 R_2$$
$$\text{(red)} \qquad\qquad \text{(white)}$$
$$\downarrow$$
$$\text{all } R_1 R_2$$
$$\text{(pink)}$$

    Therefore, the second cross is:
$$R_1 R_2 \quad \times \quad R_1 R_2$$
$$\text{(pink)} \qquad \text{(pink)}$$

    Any two heterozygotes will always yield a 1:2:1 genotypic ratio, or, in this case, 1/4 $R_1 R_1$:1/2 $R_1 R_2$:1/4 $R_2 R_2$. Approximately half of the progeny should be pink. We examine the cross and see that 23 out of 46 are pink.

12. We see three phenotypes in an approximate 1:2:1 ratio (3:7:4). One of the phenotypes (short) is intermediate between long ears and no ears. Therefore, we have incomplete dominance. The cross is between two heterozygotes: $L_1 L_2 \times L_1 L_2$, where L_1 = long ears and L_2 = no ears.

13. All. Since the child was type A, it must have gotten the I^A allele from its mother. The other allele in the child is either I^A or i. A man with type A ($I^A I^A$ or $I^A i$), type B ($I^B I^B$ or $I^B i$), type O (ii), or type AB ($I^A I^B$) blood could have supplied either an I^A or i allele.

14. Crosses can be $I^AI^A \times I^BI^B$, $I^AI^A \times I^Bi$, $I^Ai \times I^BI^B$, or $I^Ai \times I^Bi$.

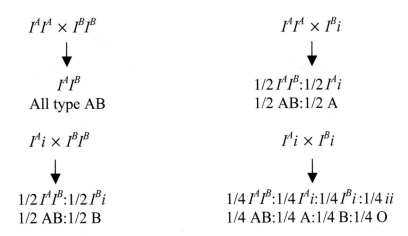

$I^AI^A \times I^BI^B$

↓

I^AI^B

All type AB

$I^Ai \times I^BI^B$

↓

$1/2\ I^AI^B : 1/2\ I^Bi$

1/2 AB : 1/2 B

$I^AI^A \times I^Bi$

↓

$1/2\ I^AI^B : 1/2\ I^Ai$

1/2 AB : 1/2 A

$I^Ai \times I^Bi$

↓

$1/4\ I^AI^B : 1/4\ I^Ai : 1/4\ I^Bi : 1/4\ ii$

1/4 AB : 1/4 A : 1/4 B : 1/4 O

15. The following table shows female and child phenotypes in the ABO system and the phenotypes of men who cannot be the father of the child.

Mother	Child	Nonfathers
A	A	—
	B	A, O
	AB	A, O
	O	AB
B	A	B, O
	B	—
	AB	B, O
	O	AB
AB	A	—
	B	—
	AB	O
O	A	B, O
	B	A, O
	O	AB

16. Constraints are set by the length of the gene (number of mutable sites) and the different possible phenotypic effects that can result from alteration of the amino acid sequence of the protein product of the gene.

17. Steve and his fiancé could be related. Both the dean and Steve's father must be I^Ai to produce O children, and each could have contributed M to produce M offspring. If the dean and Steve's father each contributed an S allele, the daughter would be SS. Note that if the daughter had B blood, she and Steve could not be related.

18. The cross is $Cc\ T_1T_2$ × $Cc\ T_1T_2$
                (purple, medium)   (purple, medium)

This problem can be worked out using the branched-line approach. To calculate the probability of each class produced, we can independently calculate the probability of each trait and apply the *product rule*.

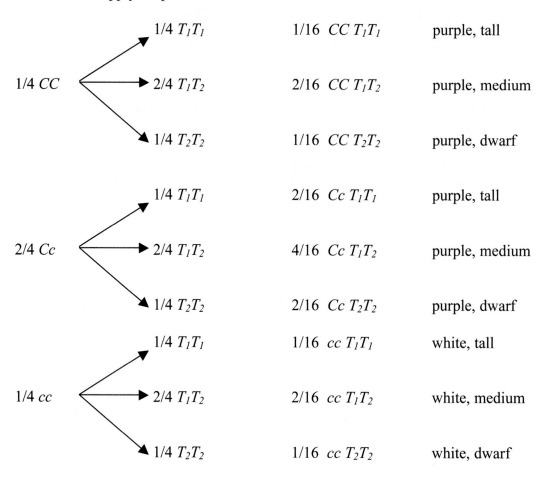

	1/4 T_1T_1	1/16 $CC\ T_1T_1$	purple, tall
1/4 CC	2/4 T_1T_2	2/16 $CC\ T_1T_2$	purple, medium
	1/4 T_2T_2	1/16 $CC\ T_2T_2$	purple, dwarf
	1/4 T_1T_1	2/16 $Cc\ T_1T_1$	purple, tall
2/4 Cc	2/4 T_1T_2	4/16 $Cc\ T_1T_2$	purple, medium
	1/4 T_2T_2	2/16 $Cc\ T_2T_2$	purple, dwarf
	1/4 T_1T_1	1/16 $cc\ T_1T_1$	white, tall
1/4 cc	2/4 T_1T_2	2/16 $cc\ T_1T_2$	white, medium
	1/4 T_2T_2	1/16 $cc\ T_2T_2$	white, dwarf

Therefore, the phenotypic ratio is
3/16 C– T_1T_1 purple, tall
6/16 C– T_1T_2 purple, medium
3/16 C– T_2T_2 purple, dwarf
1/16 $cc\ T_1T_1$ white, tall
2/16 $cc\ T_1T_2$ white, medium
1/16 $cc\ T_2T_2$ white, dwarf

19. The 15:1 ratio indicates that all genotypes except the recessive homozygote (*aabb*) produce a triangular capsule. The rounded capsule results from the recessive homozygote. One way to look at this is that the rounded form is a "default" form when neither locus has functional—dominant—alleles. However, a dominant allele at either of two loci is adequate to form the triangular seed capsule. The loci can be considered redundant in the pathway of seed capsule shape since a functional allele at one, the other, or both will provide a dominant (triangular) phenotype. At this point in time, it is impossible to know precisely what the enzymatic function of each dominant allele is.

20. a. If gene A is on an autosome, there will be 10 different genotypes and five different phenotypes.

Genotypes	Phenotypes
A_1A_1; A_1A_2; A_1A_3; A_1A_4	A_1
A_2A_2; A_2A_3; A_2A_4	A_2
A_3A_3	A_3
A_3A_4	A_3A_4
A_4A_4	A_4

b. If gene A is on the X chromosome, the alleles can now be represented as X^{A1}, X^{A2}, X^{A3}, and X^{A4}. There will be more genotypes and phenotypes, because there is a different sex chromosome composition between males and females.

Genotypes	Phenotypes
$X^{A1}X^{A1}$; $X^{A1}X^{A2}$; $X^{A1}X^{A3}$; $X^{A1}X^{A4}$	A_1 females
$X^{A2}X^{A2}$; $X^{A2}X^{A3}$; $X^{A2}X^{A4}$	A_2 females
$X^{A3}X^{A3}$	A_3 females
$X^{A3}X^{A4}$	A_3A_4 females
$X^{A4}X^{A4}$	A_4 females
$X^{A1}Y$	A_1 males
$X^{A2}Y$	A_2 males
$X^{A3}Y$	A_3 males
$X^{A4}Y$	A_4 males

21. Two loci with epistasis would explain this result. The cross $AaBb \times$ self yields 9/16 $A–B–$: 6/16 $A–bb + aaB–$: 1/16 $aabb$. The hypothesis could be verified by testcrossing the various classes.

22. The 119:32:9 ratio is very close to a 12:3:1 ratio indicating two loci with dominant epistasis. For example:

P_1 red ($AABB$) $\times$ white ($aabb$)
F_1 red ($AaBb$) $\times$ self
F_2 119 ($A–B– + A–bb$)
            32 ($aaB–$)
            9 ($aabb$)

23. Hypothesis: two-locus genetic control with recessive epistasis $AaCc \times AaCc$ produces agouti ($A–C–$), black ($aaC–$), and albino ($A–cc + aacc$) offspring in a 9:3:4 ratio. Critical chi-square at 0.05, two degrees of freedom = 5.991.

	Offspring			
	Agouti	**Black**	**Albino**	**Total**
Observed numbers (O)	28	7	13	48
Expected ratio	9/16	3/16	4/16	
Expected numbers (E)	27	9	12	48
$O - E$	1	–2	1	
$(O - E)^2$	1	4	1	
$(O - E)^2/E$	0.037	0.444	0.083	$0.564 = \chi^2$

Since this chi-square value is less than the critical chi-square, we fail to reject (that is, we accept) our hypothesis.

24. The cross is $\quad C^R C^W S^L S^S \quad \times \quad C^R C^W S^L S^S$

$\qquad\qquad\qquad$ (purple, ovoid) $\qquad$ (purple, ovoid)

This problem can be worked out using the branched-line approach. To calculate the probability of each class produced, we can independently calculate the probability of each trait and apply the *product rule*.

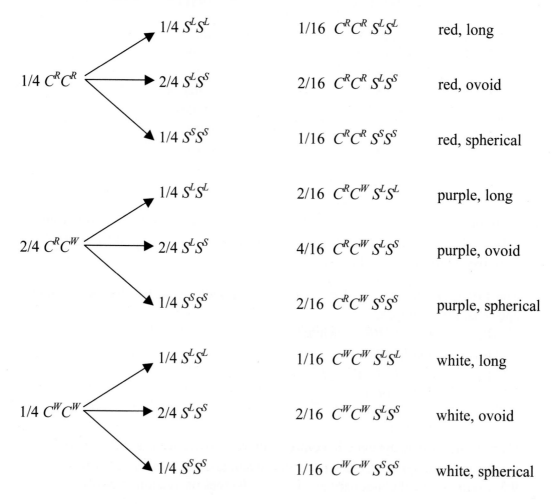

25. The animals would be bred for several generations of controlled matings. You would look for offspring types and ratios consistent with a hypothesis of the number of genes controlling the trait and the dominance relationships of alleles. One approach might be to try to establish pure-breeding lines (homozygotes), which can then be bred in a more controlled manner. If both rooster and hen were pure-breeding rose-combed types (all offspring were also rose-combed), then it would not be possible to determine the nature of the genetic control of this trait.

26. The F_1 indicates lazy is dominant; therefore, a worker must be a recessive homozygote. If one gene is involved, the cross of the F_1 female × worker male ($Ww \times ww$) should produce 1 worker :1 lazy, for this is a testcross. This result is not seen, so we must have more than one gene involved. Perhaps a worker can result from more than one gene. Let A–B– = lazy, and A–bb, aaB–, or $aabb$ = workers. The original worker is $aabb$, and the cross is

$$aabb \quad \times \quad AABB$$

$$AaBb \quad \times \quad aabb$$
(lazy)   (worker)

1/4 A–B–:1/4 A–bb:1/4 aaB–:1/4 $aabb$
 lazy   worker   worker   worker

The observed results (3 workers:1 lazy) are consistent with this explanation.

If $AaBb \quad \times \quad AaBb$

$\downarrow$

9/16 A–B–   lazy

$\left. \begin{array}{l} \text{3/16 } A\text{–}bb \\ \text{3/16 } aaB\text{–} \\ \text{1/16 } aabb \end{array} \right\}$ workers

Thus, the cross of two F_1 lazy individuals will produce 9 lazy:7 hard workers.

27. The paternal grandfather is normal (aa); therefore, the affected father must be heterozygous (Aa). The mating now becomes $Aa \times aa$ (normal mother). The probability of the father passing the polydactyly allele to his child is 50%. The probability of an affected child expressing the trait is 80%. The likelihood of these two independent events occurring together is given by the product rule: $0.5 \times 0.8 = 0.4$. Therefore there is a 40% chance that the child will exhibit polydactyly.

28. In each cross, we see that all ratios could result from a single gene (3:1, 1:2:1, and 1:1). We can easily explain the red and silver if we assume that red is dominant over silver. Thus, the first cross could be $RR \times rr$ or $Rr \times rr$. Similarly, the second cross could be $RR \times R-$ or $Rr \times Rr$. But we cannot explain any of the white progeny with only two alleles. Since all ratios are single gene ratios, we must propose a third allele that, when homozygous, produces white. Red must be dominant to white. Note that a given diploid individual can have only two alleles. To list all the possible genotypes and their phenotypes, let R = red and r = nonred. RR, Rr^S, Rr^w = red (R is dominant to all other alleles), $r^S r^S$ = silver, and $r^w r^w$ = white.

What is the phenotype of an $r^S r^w$ individual? Assume silver is dominant; that is, assume r^S is dominant to r^w. Thus, $r^S r^w$ = silver.

Now go back to the results. By our model, $r^w r^w$ is white. Therefore, in order to get white progeny, each parent must have at least one r^w allele to contribute. The cross of red by silver to produce white offspring must be

$$Rr^w \quad \times \quad r^S r^w$$
$$1/4\ Rr^S : 1/4\ Rr^w : 1/4\ r^S r^w : 1/4\ r^w r^w$$
$$\text{(red)} \quad \text{(red)} \quad \text{(silver)} \quad \text{(white)}$$

29. The cross can be diagrammed as

P   (black) $BBEE \times bbee$ (yellow)

F_1 (black) $BbEe \times$ self

F_2 9/16 $B-E-$ black
          3/16 $bbE-$ brown
          3/16 $B-ee$ ⎱
          1/16 $bbee$ ⎰   yellow

The overall ratio for this cross is 9/16 black:3/16 brown:4/16 yellow. This result is consistent with recessive epistatis. In this case, the e allele is epistatic to both the B allele and the b allele.

30. The cross is $C^B C^Y Ii \quad \times \quad C^B C^Y Ii$
                    (green)   (green)

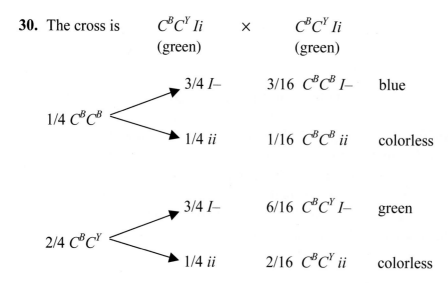

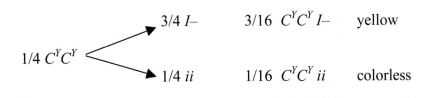

3/4 *I–*   3/16 $C^Y C^Y$ *I–*   yellow

1/4 $C^Y C^Y$

1/4 *ii*   1/16 $C^Y C^Y$ *ii*   colorless

Thus, the phenotypic ratio is 6 green:3 blue:3 yellow:4 colorless (white).

31. **a.** *AaBb* × *AaBb* yields a 9 (*A–B–*):3 (*aaB–*):4 (*A–bb, aabb*) ratio. Therefore,
AaBb × *aabb* will yield a 1 (*AaBb*):1 (*aaBb*):2 (*Aabb, aabb*) ratio.

   b. *AaBb* × *AaBb* yields a 9 (*A–B–*):7 (*A–bb, aaB–*, or *aabb*) ratio. Therefore,
AaBb × *aabb* will yield a 1 (*AaBb*):3 (*Aabb, aaBb, aabb*) ratio.

   c. *AaBb* × *AaBb* yields a 12 (*A–B–, A–bb*):3 (*aaB–*):1 (*aabb*) ratio. Therefore,
AaBb × *aabb* will yield a 2 (*AaBb, Aabb*):1 (*aaBb*):1 (*aabb*) ratio.

   d. *AaBb* × *AaBb* yields a 13 (*A–B–, A–bb, or aabb*):3 (*aaB–*) ratio. Therefore,
AaBb × *aabb* will yield a 3 (*AaBb, Aabb, aabb*):1 (*aaBb*) ratio.

   e. *AaBb* × *AaBb* yields a 1:2:1:2:4:2:1:2:1 ratio. Therefore,
AaBb × *aabb* will yield a 1 (*AaBb*):1 (*Aabb*):1 (*aaBb*):1 (*aabb*) ratio.

32. The first two crosses indicate that wild type is dominant to both oranges, and the
fourth indicates that orange-2 is dominant to pink. The fifth cross produces four
phenotypes, indicating we are dealing with at least two genes. The presence of two
genes is also suggested by the cross of orange-1 × orange-2. If these two traits were
allelic, all the progeny should have been orange. The F_1 × pink produces progeny
that resemble those from a testcross of a dihybrid. If *A–B* = wild type, *A–bb* =
orange-1, *aaB–* = orange-2, *aabb* is probably pink. The crosses in question are then

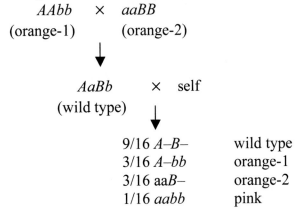

        AAbb   ×   *aaBB*
      (orange-1)   (orange-2)

            AaBb   ×   self
          (wild type)

                9/16 *A–B–*   wild type
                3/16 *A–bb*   orange-1
                3/16 aa*B–*   orange-2
                1/16 *aabb*   pink

Thus the F_2 ratio is 9 wild type:3 orange-1:3 orange-2:1 pink.

33. a.

$$Cc \times Cc$$

1/4 CC (lethal):2/4 Cc (Creeper):1/4 cc (normal)
Therefore, the ratio of surviving chicken is 2/3 creeper (Cc):1/3 normal (cc).

b. The creeper allele is dominant for leg shape because heterozygous chicken are creepers.

c. The creeper allele is recessive for lethality because only homozygous chicken die during embryological development.

34. Crosses **(a)** and **(b)** do not reveal anything about the dominance hierarchy. Each suggests a mating of a heterozygote with a homozygote. Cross **(c)** yields a higher proportion of mallard than dusky offspring, indicating that mallard is dominant to dusky. Cross **(d)** yields a higher proportion of restricted than mallard offspring, indicating that restricted is dominant to mallard. Therefore, the dominance hierarchy is $M^R > M > M^D$. The crosses are **(a)** $M^P M^P \times MM^P$; **(b)** $M^P M^P \times M^R M^P$; **(c)** MM^D $\times MM^P$; and **(d)** $MM^P \times M^R M^P$. *Note:* The dusky allele is actually designated m^d.

35. Cross **(a)** suggests a mating of true breeding black sheep, while cross **(b)** suggests a mating between a heterozygote and a homozygote sheep. Cross **(c)** yields an unusual 2:1 ratio. This is consistent with a mode of inheritance with one gene and two alleles, and where one homozygous class is inviable (lethal during embryonic development). Cross **(c)** suggests that gray coat color is dominant over black but is recessive for lethality. Let B = gray and b = black. The crosses are **(a)** $bb \times bb$; **(b)** $bb \times Bb$; and **(c)** $Bb \times Bb$. BB homozygotes die before birth, leaving a ratio of 2/3 gray (Bb):1/3 black (bb).

36. This problem can be worked out using the branched-line approach.

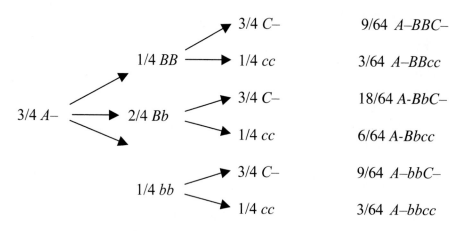

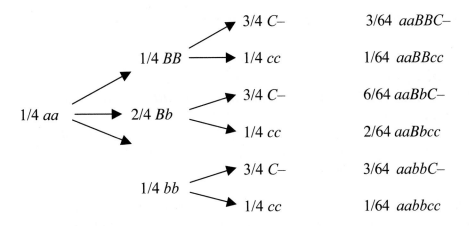

Because allele *B* is recessive for lethality, all individuals with genotype *BB* will not survive. So 16/64 (9/64 + 3/64 + 3/64 + 1/64) individuals die before birth. This leaves 48 surviving offspring. Therefore, the expected phenotypic ratio in the surviving offspring is 18 *A–BbC–*:6 *A–Bbcc*:9 *A–bbC–*:3 *A–bbcc*:6 *aaBbC–*: 2 *aaBbcc*:3 *aabbC–*:1 *aabbcc*.

37. a.   P: *AAbbCCDDII*   ×   *AABBCCDDii*

F₁: *AABbCCDDIi*   ×   *AABbCCDDIi*
        all orange

F₂:

Gametes	*ABCDI*	*ABCDi*	*AbCDI*	*AbCDi*
ABCDI	*AABBCCDDII* (orange)	*AABBCCDDIi* (orange)	*AABbCCDDII* (orange)	*AABbCCDDIi* (orange)
ABCDi	*AABBCCDDIi* (orange)	*AABBCCDDii* (purple)	*AABbCCDDIi* (orange)	*AABbCCDDii* (purple)
AbCDI	*AABbCCDDII* (orange)	*AABbCCDDIi* (orange)	*AAbbCCDDII* (yellow)	*AAbbCCDDIi* (yellow)
AbCDi	*AABbCCDDIi* (orange)	*AABbCCDDii* (purple)	*AAbbCCDDIi* (yellow)	*AAbbCCDDii* (yellow)

Therefore the F₂ ratio is 9 orange:4 yellow:3 purple.

b.   There are only two main types of flowers: colored and colorless. To have colored flowers, a plant must have a dominant allele at the *A* locus. Plants with a genotype of *aa* will have colorless flowers, regardless of the alleles at the other loci. Therefore, the probability of colored flowers is 3/4 (*A–*).

38. a. The genotypic frequencies can be obtained from a Punnett square.

Gametes	*AB*	*Ab*	*aB*	*ab*
AB	*AABB*	*AABb*	*AaBB*	*AaBb*
aB	*AaBB*	*AaBb*	*aaBB*	*aaBb*

Therefore, 1/8 *AABB*:1/8 *AABb*:2/8 *AaBB*:2/8 *AaBb*:1/8 *aaBB*:1/8 *aaBb*.
The phenotypic frequencies can be derived from the genotypic frequencies.
Therefore, 6/8 horned:2/8 hornless or a 3:1 ratio.

b. Two hornless animals can produce offspring with horns. The animals will have to be recessive for different genes. For example, if the cross is *AAbb* × *aaBB*, the offspring will all be *AaBb* and will all have horns.

39. In this example of epistasis, the *I* locus (ABO blood type) is only expressed in the presence of a dominant *H* allele at the second locus. The cross is $I^A I^B Hh \times I^A I^B Hh$.

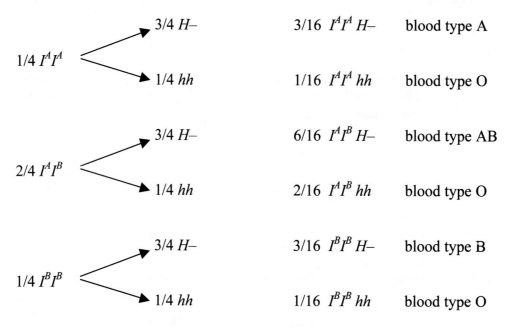

Thus, the phenotypic ratio is 3/16 A:6/16 AB:3/16 B:4/16 O.

40. Because the two genes are independently assorting, we can consider each trait separately. First, joy: the offspring are in an approximate ratio of 1 laughing:1 smiling (41:38). Therefore, one of the parents must be laughing (J_1J_1) and the other smiling (J_1J_2). Now, movement: the offspring are in an approximate ratio of 1 runner:2 joggers:1 walker (19:39:21). Thus, both parents must be joggers (M_1M_2). Combining the two traits, we can determine that the one parent is a laughing jogger and the other is a smiling one!

41. Because the four genes are independently assorting, we can consider each of them separately. The cross $Aa \times Aa$ yields two different phenotypes: $A-$ and aa. The cross $Bb \times Bb$ also yields two different phenotypes: $B-$ and bb. On the other hand, the cross $C_1C_2 \times C_1C_2$ produces three different phenotypes: C_1C_1, C_1C_2, and C_2C_2. The cross $D_1D_2 \times D_1D_2$ also produces three different phenotypes: D_1D_1, D_1D_2, and D_2D_2. The number of different phenotypes expected from the self-cross of $AaBbC_1C_2D_1D_2$ is given by the product rule: $2 \times 2 \times 3 \times 3 = 36$.

42. **a.** To be colorless, individuals must be $aabb$; the alleles at locus C do not matter. They could be either $aabbC-$ or $aabbcc$. Therefore the probability is given by $(1/4\ aa)(1/4\ bb)(1/1\ C-\ \text{or}\ cc) = 1/16$.

   b. For red probabilities, determine genotypes that will be red, calculate probability for each genotype, and add.

$A-bbC-$	$3/4 \times 1/4 \times 3/4 = 9/64$
$A-bbcc$	$3/4 \times 1/4 \times 1/4 = 3/64$
$A-B-cc$	$3/4 \times 3/4 \times 1/4 = 9/64$
$aaB-C-$	$1/4 \times 3/4 \times 3/4 = 9/64$
$aaB-cc$	$1/4 \times 3/4 \times 1/4 = \underline{3/64}$
	$33/64$

   Alternatively, calculate the frequency of blacks as $3/4\ A- \times 3/4\ B- \times 3/4\ C- = 27/64$. We calculated $1/16$ or $4/64$ colorless, so $1 - (27/64 + 4/64) = 33/64 =$ frequency of red.

43. The mating is $I^AI^B \times I^BI^O$. It will produce $1/4\ I^AI^B$:$1/4\ I^AI^O$:$1/4\ I^BI^B$:$1/4\ I^BI^O$.

   a. The probability of having a child with blood type B ($I^BI^B + I^BI^O$) is $1/2$. The probability of having two children with blood type B is $1/2 \times 1/2 = 1/4$.

   b. The probability of having a boy with blood type AB is $1/2 \times 1/4 = 1/8$. The probability of a girl with blood type A is the same. Therefore, the overall probability is given by $1/8 \times 1/8 = 1/64$.

   c. There are three different scenarios in this case. Scenario 1: First child is blood type A and the second child is not. Scenario 2: First child is not blood type A and the second child is. Scenario 3: Both children are blood type A. Calculate the probability of each scenario and then add.

Scenario 1	$1/4 \times 3/4 = 3/16$
Scenario 2	$3/4 \times 1/4 = 3/16$
Scenario 3	$1/4 \times 1/4 = \underline{1/16}$
	$7/16$

44. Using the dominance hierarchy, we can determine the potential genotypes for every phenotype. Thus, a gray rabbit can have one of four genotypes (CC, Cc^{ch}, Cc^h, and Cc); a chinchilla rabbit, three ($c^{ch}c^{ch}$, $c^{ch}c^h$, and $c^{ch}c$); a Himalayan rabbit, two (c^hc^h and c^hc); and albino rabbits only one (cc). This order of dominance allows recessive alleles to be "hidden" in various heterozygotes.

Crosses **(a)** and **(b)** suggest that all three gray rabbits are heterozygous. Cross **(c)**, gray-1 × chinchilla-1, resulted in three different genotypes, including Himalayan rabbits. Both parents must thus be heterozygous, and *at least one* of them must carry the Himalayan allele, c^h. Therefore, gray-1 = Cc^h or Cc and chinchilla-1 = $c^{ch}c^h$ or $c^{ch}c$. (*Note:* Gray-1 and chinchilla-1 cannot both carry the albino allele at the same time.) We can now go back and analyze cross **(a)**: gray-1 × gray-2. It can be represented as Cc^h (or Cc) × $C-$. The offspring include chinchillas, and so the gray-2 rabbit must be heterozygous for the chinchilla allele, c^{ch}. Therefore, gray-2 = Cc^{ch}. Let's move on to cross **(b)**: gray-2 × gray-3. It can be represented as Cc^{ch} (or Cc) × $C-$. The offspring include chinchilla, and so the gray-3 rabbit cannot be homozygous (CC). However, we cannot tell for sure what the other allele is. Indeed, it can be any of the other three alleles: c^{ch}, c^h, or c. And finally, cross **(d)**: chinchilla-2 × Himalayan, or $c^{ch}- × c^h-$. The offspring include albinos, and so both rabbits must be heterozygous for albino allele, c. Therefore, chinchilla-2 = $c^{ch}c$, and Himalayan = $c^h c$. The following table summarizes the results.

Rabbit	Genotype
Gray-1	Cc^h or Cc (only if chinchilla-1 is $c^{ch}c^h$)
Gray-2	Cc^{ch}
Gray-3	Cc^{ch}, Cc^h, or Cc
Chinchilla-1	$c^{ch}c^h$ or $c^{ch}c$ (only if gray-1 is Cc^h)
Chinchilla-2	$c^{ch}c$
Himalayan	$c^h c$

45. Two genes are involved, one autosomal recessive that produces pink eyes, and one X-linked recessive that produces white eyes. Any fly homozygous or hemizygous for white eyes will have white eyes. The fact that the F_1 flies are wild type indicates two different genes, and the different F_2 results for males and females indicate at least one X-linked gene. One-half of the F_2 males have white eyes, a ratio expected for an X-linked recessive gene; we see a 3:1 ratio for red:pink, a ratio expected for an autosomal gene. The F_2 females show a 3:1 ratio for red:pink, suggesting that all females have at least one normal X chromosome. If $X^+-A- =$ red, $X^+-aa =$ pink, $X^w X^w A- =$ white, and $X^w Y =$ white, the cross can be diagrammed as follows:

$$X^+ X^+ \, aa \quad \times \quad X^w Y \, AA$$

$$\downarrow$$

$$X^+ X^w \, Aa \qquad X^+ Y \, Aa$$
$$\text{all red}$$

$$\downarrow$$

3 X^+-A-	red	3 $X^+ Y \, A-$	red
1 X^+-aa	pink	1 $X^+ Y \, aa$	pink
		3 $X^w Y \, A-$	white
		1 $X^w Y \, aa$	white

CHAPTER INTEGRATION PROBLEM

a. The true breeding gray parent is *AABBCC*. Each of the albino parents must be homozygous recessive (*cc*) at the *C* locus. Because the *A*, *B*, and *C* genes are independently assorting, each mating can be treated as a series of monohybrid or dihybrid crosses.

Cross 1: The F_2 offspring are in a ratio of 3 gray:1 albino. The presence of no other color indicates that the F_1 rats were homozygous for the *A* and *B* genes and have genotype *AABBCc*. Otherwise, other colored types would have appeared in the offspring. Therefore, the albino-1 parent must have been *AABBcc*.

Cross 2: Without considering the *C* locus, the F_2 offspring are in a ratio of 9 gray: 3 yellow, which can be simplified to a 3:1 ratio. Since yellow offspring have the genotype *A–bb*, the preceding ratio indicates that the F_1 rats were heterozygous for the *B* gene and homozygous for the *A* gene (genotype would be *AABbCc*). Therefore, the albino-2 parent must have been *AAbbcc*.

Cross 3: Without considering the *C* locus, the F_2 offspring are in a ratio of 9 gray: 3 black, which can be simplified to a 3:1 ratio. Since black offspring have the genotype *aaB–*, the preceding ratio indicates that the F_1 rats were heterozygous for the *A* gene and homozygous for the *B* gene (genotype would be *AaBBCc*). Therefore, the albino-2 parent must have been *aaBBcc*.

b. *Cross 1:* The ratio of colored (gray, black, yellow, and cream) to albino rats is 242:83 or approximately 3:1. This indicates a mating between two heterozygotes for the *C* locus. The colors are in an approximately 9:3:3:1 ratio, indicating that the parents must also be heterozygous for the *A* and *B* genes.

Cross 2: The ratio of albino to colored (black and cream) rats is 103:99 or approximately 1:1. This indicates a mating between a heterozygote, *Cc*, and a homozygote, *cc*. The absence of gray and yellow offspring means that the parents cannot have any *A* alleles. The approximately 3 black:1 cream ratio means that the parents were heterozygous at the *B* locus.

Cross 3: There are no albino rats, meaning that the parents cannot produce the *cc* genotype. The colors are in an approximately 9:3:3:1 ratio, indicating that the parents must be heterozygous for the *A* and *B* genes.
Summary:

	Parental Phenotypes	Parental Genotypes
Cross 1	Gray × gray	*AaBbCc × AaBbCc*
Cross 2	Black × albino	*aaBbCc × aaBbcc*
Cross 3	Gray × gray	*AaBbCC × AaBbCC*, or *AaBbCC × AaBbCc*

c.   The parental genotypes can be denoted as *A–B–C–* (gray) × *A–bbC–* (yellow). The offspring include an albino (*cc*), and a cream-colored cat (*aabb*); therefore, both parents must be heterozygous for the *A* and *C* genes, and the gray parent should also be heterozygous for the *B* locus. The cross can be diagrammed as such:

AaBbCc   ×   *AabbCc*

↓

	AbC	*Abc*	*abC*	*abc*
ABC	*AABbCC* (gray)	*AABbCc* (gray)	*AaBbCC* (gray)	*AaBbCc* (gray)
ABc	*AABbCc* (gray)	*AABbcc* (albino)	*AaBbCc* (gray)	*AaBbcc* (albino)
AbC	*AAbbCC* (yellow)	*AAbbCc* (yellow)	*AabbCC* (yellow)	*AabbCc* (yellow)
Abc	*AAbbCc* (yellow)	*AAbbcc* (albino)	*AabbCc* (yellow)	*Aabbcc* (albino)
aBC	*AaBbCC* (gray)	*AaBbCc* (gray)	*aaBbCC* (black)	*aaBbCc* (black)
aBc	*AaBbCc* (gray)	*AaBbcc* (albino)	*aaBbCc* (black)	*aaBbcc* (albino)
abC	*AabbCC* (yellow)	*AabbCc* (yellow)	*aabbCC* (cream)	*aabbCc* (cream)
abc	*AabbCc* (yellow)	*Aabbcc* (albino)	*aabbCc* (cream)	*Aabbcc* (albino)

The phenotypic ratio is 9/32 gray:9/32 yellow:8/32 albino:3/32 black:3/32 cream.

d.   The polka-dot trait is dominant in males and recessive in females. This is characteristic of a sex-influenced trait. These traits are controlled by autosomal genes, so the *dotted* gene locus would be expected to be found on an autosome. Therefore, it will assort independently (Mendel's second law) from X-linked genes, which have loci on the X chromosome.

e.   Secondary spermatocytes are "future" sperm. Because secondary spermatocytes are haploid, they would have only one allele per gene. The male has a genotype of *AabbCcDd*. However, one has to consider the X and Y chromosomes as well. Therefore, the genotype can be designated as *AabbCcDdXY*. The total number of secondary spermatocytes is $2^4 = 16$ (there are four "heterozygous" positions). The genotypes can be listed systematically as such: *AbCDX, AbCDY, AbCdX, AbCdY, AbcDX, AbcDY, AbcdX, AbcdY, abCDX, abCDY, abCdX, abCdY, abcDX, abcDY, abcdX,* and *abcdY*.

f. Here again, the cross can be treated as a series of monohybrid crosses.

(1) An albino rat has to be *cc*. The probability is simply 1/4.

(2) The genotype of a yellow-colored rat is *A–bbC–*. The genotype of a polka-dotted male rat is *D–*XY. Therefore, the overall probability is $1/2 \times 1/2 \times 3/4 \times 3/4 \times 1/2 = 9/128$.

(3) The genotype of a gray-colored rat is *A–B–C–*. The dot phenotype depends on the genotype at the *D* locus and the sex of the offspring. The genotype of a male rat with no dots is *dd*XY, whereas that of a female can be *Dd*XX or *dd*XX. We can calculate each of the three possibilities, and then add them all (sum rule) to get to the overall probability. P(gray male with no dots) = $1/2 \times 1/2 \times 3/4 \times 1/4 \times 1/2 = 3/128$. P(gray female with no dots) = $1/2 \times 1/2 \times 3/4 \times 3/4 \times 1/2 = 9/128$. Therefore, the P(gray rat with no dots) = $3/128 + 9/128 = 12/128 = 3/32$.

(4) The dot phenotype depends on the genotype at the *D* locus and the sex of the offspring. The genotypes at the *A*, *B,* and *C* loci are irrelevant. A male rat is polka-dotted if he has the genotype *D–*XY, whereas a polka-dotted female has to be *DD*XX. We can calculate each of these two probabilities, and then add them up (sum rule) to get to the overall probability. P(polka-dotted rat) = $(3/4 \times 1/2) + (1/4 \times 1/2) = 4/8 = 1/2$.

CHAPTER 6 LINKAGE AND MAPPING IN EUKARYOTES

Chapter Goals
1. Understand how genes that are linked on a single chromosome affect the resulting genotypic and phenotypic ratios in the progeny.
2. Analyze and manipulate data from both two- and three-point crosses to generate linkage maps.
3. Calculate genotypic frequencies from a linkage map.
4. Understand the underlying mechanism of recombination and how haploid meiosis can be used to explain this mechanism.
5. Understand the difference between meiotic and mitotic recombination and the outcomes of both events for an organism.
6. Utilize additional methods to map eukaryotic genes.

Key Chapter Concepts

6.1 Linkage, Crossing Over, and Recombination: An Overview
- **Linkage:** Each eukaryotic chromosome is a linear sequence of genes. Genes carried on the same chromosome are physically linked together. Therefore, genes within a linkage group are inherited together unless they are separated by crossing over. Consider two linked genes, *A* and *B*, each with two alleles. Assume that an individual with genotype *AABB* is crossed to an individual of genotype *aabb* to produce a dihybrid (*AaBb*). This dihybrid will produce only gametes with the allele combination found in the original parents (*AB* and *ab*, in our example).
- **Crossing Over:** A physical exchange of DNA between nonsister chromatids of homologous chromosomes during meiosis I will generate recombinant chromosomes in which the allele distribution between chromosomes is switched. Thus, the genotype *Ab* or *aB* can appear in gametes of the preceding example if a crossover occurs between the *A* and *B* genes. Crossovers are only observed in the progeny of heterozygous individuals since the chromatids exchanging DNA must be different. The distance between two genes will determine how often new combinations of alleles (**recombination**) are formed. Crossing over will occur more often if the genes are far apart and less frequently if they are close together. Recombination frequency can therefore be used as a measure of the distance between genes.

6.2 Diploid Mapping
- **Mapping Studies:** A testcross of a heterozygous individual is used to determine the arrangement of genes on a chromosome and the distance between them. Remember that a testcross of a dihybrid yields four phenotypic classes in a 1:1:1:1 ratio and a testcross of a trihybrid yields eight phenotypic classes in a 1:1:1:1:1:1:1:1 ratio if the genes are located on separate chromosomes. In contrast, a testcross involving *linked* genes results in an excess of parental phenotypes (> 50%) and a low number of recombinant types (< 50%). Offspring that have the same allele arrangement as the original parents are referred to as **parentals** or **nonrecombinants**. Offspring with new allele arrangements resulting from crossing over are referred to as **recombinants** or **nonparentals**. Because the homozygous individual contributes only recessive alleles to the offspring, the alleles received from the heterozygous parent are the ones expressed in the offspring. This allows parental and recombinant classes to be easily identified.
- **Mapping methods:** The frequency of appearance of recombinant classes of offspring determines the distance between genes as mentioned earlier. Testcrosses involving two genes are referred to as **two-point crosses**. A cross involving three genes is referred to as a **three-point cross**. When three linked genes are considered, progeny resulting from single and **double crossovers** (those occurring between genes 1 and 2 and between genes 2 and 3) can be distinguished. The offspring resulting from double crossovers appear least frequently and indicate the order of the genes. **Positive** or **negative interference** may result if there are fewer or more double crossovers than expected, respectively. Specific details on how to solve two-point and three-point linkage map problems are given in this section of your textbook.
- **Arrangement of Alleles:** Heterozygous individuals receive different alleles from the maternal and paternal parents. When genes are linked, there are two possible arrangements of alleles in the heterozygote. First, the heterozygote can receive a chromosome bearing two dominant alleles (*AB*)

from one parent and a chromosome containing two recessive alleles (*ab*) from the other. In this case, the allele configuration in the heterozygote is described as *cis* or **coupling** because the two mutant alleles are located on the same chromosome. If one of the chromosomes of the heterozygote contains the alleles *Ab* and the other chromosome bears the alleles *aB*, the configuration is described as *trans* or **repulsion** because the two mutant alleles are found on opposite chromosomes.

- **Karyotype:** The karyotype of an organism is a photograph of the chromosomes of an organism taken during mitosis. Chromosomes can be distinguished by their size, centromere position, and banding patterns. Karyotyping allows the visualization of any gross chromosomal abnormalities as well as the determination of the sex of an organism.

6.3 Mapping DNA Sequences
- **RFLPs:** Morphological characteristics associated with specific genes are often spread far apart on chromosomes so that long stretches of DNA exist with no genes that cause observable phenotypes. This makes creating accurate maps more difficult. Fortunately, restriction enzyme recognition sequences (discussed in chapter 13) are found quite frequently in genomes. Maternal and paternal differences often exist in these sequences causing restriction enzymes to no longer be able to cut the DNA at that location. This affords researchers many more loci by which to map genes in relation to.

6.4 Haploid Mapping (Tetrad Analysis)
- **Use of Fungi:** This technique focuses on the total products of meiosis (**ascospores** retained in a sac called an **ascus**) in *Ascomycete* fungi. Mapping and recombination are easier to observe in these organisms since the haploid gametes can be grown and the phenotypes observed. Therefore, a testcross is not needed as it is in diploid organisms.
- **Unordered Spores:** Meiosis of a diploid yeast cell results in the production of three ascus types: those with spores identical to the parental spores (**parental ditype; PD**), those that contain only recombinants (**nonparental ditype; NPD**), and those that contain both classes of spores (**tetratype; TT**). The loci under consideration are linked if the parental ditypes occur more frequently than nonparental ditypes and the relative distance between genes can be determined by the frequency of appearance of NPD and TT asci.
- **Ordered Spores:** In a bread mold such as *Neurospora crassa*, meiosis produces four haploid spores that each divide mitotically, resulting in an ascus containing eight spores (**octad**). The spores are ordered according to their centromeres which are denoted by the parental mating type from which they originated (for example *A* or *a*). In the absence of crossing over between the locus and its centromere, a 4:4 pattern (*AAAAaaaa* or *aaaaAAAA*) is observed. Asci with this pattern are referred to as a **first-division segregation** (FDS) because alleles segregated during the first meiotic division. These FDS asci form when no recombination has occurred. Crossing over results in the production of asci with **second-division segregation** (SDS) patterns (those resulting from allele segregations occurring during the second meiotic division) including 2:4:2 (*AAaaaaAA* or *aaAAAAaa*) and 2:2:2:2 (*AAaaAAaa* or *aaAAaaAA*). The numbers of first- and second-division segregation asci are used to determine the distance between a gene and the centromere of its chromosome and to determine the order of genes that are linked to the same centromere.

6.5 The Mechanism of Recombination
- **Unusual Asci:** The presence of unusual orders of asci in *Neurospora* helped researchers realize that only one strand of DNA is exchanged during meiotic recombination. These asci could only be produced if **heteroduplex DNA** existed. This DNA can only occur after single-strand exchange.
- **Holliday Structures:** After heteroduplex DNA is formed, a cross, termed a **Holliday structure,** is formed between homologous chromatids. This cross structure is then resolved by cutting and reattaching two DNA strands. This **resolution** is accomplished in one of two ways. If the structure is resolved one way, flanking genes *do not* switch chromosomes while they *do* switch if it is resolved the other way.

6.6 Additional Recombination Events
- **Somatic Crossing Over:** So far, we have discussed crossovers that occur during *meiosis I*. Crossovers can also occur during *mitosis*. If this happens, the daughter cells resulting from mitosis *can* have a

different genotype than the parent cell. If this happens in one somatic cell during development of a multicellular organism, then all of the cells that come from that cell (usually a patch on the body) will have a different set of alleles than their neighboring cells. This can produce a different phenotype in that region.

- **Sister Chromatid Exchange:** Crossovers are only phenotypically observed between homologous, heterozygous, nonsister chromatids. However, crossovers do also occur between identical sister chromatids but require elaborate experiments to detect them.

6.7 Human Chromosomal Maps

- **Overview:** Maps of human chromosomes help in the elucidation of mutant genes that cause human diseases. The location of a mutant gene is mapped based on its recombination frequency relative to other known loci in the human genome. This is done by creating and studying human pedigrees of families who have inherited the disease. Once the map location is known, researchers can sequence the genes in that region to look for mutations that are found only in diseased individuals. Once the identity of the culprit gene is known, its function can be studied and a cure for the disease sought. X-linked traits are easier to map than autosomally linked traits.

Key Terms

- recombination
- recombination maps
- linkage
- chiasma
- nonrecombinants
- parentals
- recombinants
- nonparentals
- complete linkage
- linkage maps
- map unit
- centimorgan (cM)
- repulsion
- coupling
- two-point cross
- two-point testcross
- three-point cross
- reciprocal progeny
- double crossover
- measured map distance
- actual map distance
- mapping function
- positive interference

- negative interference
- coefficient of coincidence
- degree of interference
- morphological characteristic
- restriction fragment length polymorphism (RFLP)
- random strand analysis
- ascus
- tetrad
- ascospores
- minimal medium
- yeast mating types (a/α)
- parental ditype ascus (PD)
- nonparental ditype ascus (NPD)
- tetratype ascus (TT)
- octad
- first-division segregation (FDS)
- second-division segregation (SDS)
- coincidence of SDS patterns

- heteroduplex DNA
- gene conversion
- Holliday structure
- resolution
- sister chromatid exchange (SCE)
- Bloom syndrome
- G-bands
- somatic-cell hybridization
- heterokaryon
- synteny test
- assignment test
- polar body
- ovum
- oogenesis
- parthenogenesis
- sporophyte
- gametophyte
- pollen grain
- sexual spores
- chromosomal theory of inheritance
- sex-linked

Understanding the Key Concepts

Use the key terms or parts of their definitions to complete the following sentences.

Two or more genes that display (1)_____ during sexual reproduction must be on the same chromosome. These genes display (2)_____ if they never assort independently, but most genes display some degree of independent assortment. This is because (3)_____ during meiosis I can create new chromosomes that contain different combinations of alleles. These newly created chromosomes are called (4)_____, and the frequency of their appearance is related to (5)_____. The closer together two genes are on a chromosome, the (6)_____ nonparental progeny that will be observed. Thus, the frequency of nonparental appearance can be used to determine the relative distance between genes in both haploid and diploid organisms and create (7)_____. These help researchers study genes in model organisms and are essential for the discovery of the genes that cause human diseases. In diploid model organisms, these maps are developed using (8)_____ in order to study genes and inheritance, while (9)_____ analysis is used in haploid organisms to determine the mechanism of (10)_____. In humans, these maps are developed using (11)_____ analysis to identify defective genes causing inherited diseases. Creating genetic maps requires the observation of differences in an organism's DNA. Usually researchers observe (12)_____ of an organism, but sometimes DNA sequence differences can be directly determined by observing (13)_____.

In an experiment known as a (14)_____, researchers map three genes relative to each other in a diploid organism. The progeny will be found in eight classes of four sets of (15)_____ progeny. The two classes of progeny found in the greatest number demonstrate the arrangement (or distribution) of (16)_____ between the two chromosomes in the heterozygous parent. The two classes found in the fewest numbers are progeny that resulted from a (17)_____ in the heterozygous parent. This class demonstrates the (18)_____ of the genes on the chromosome; the (19)_____ gene is the one that switched relative to the other two genes. Next, the (20)_____ can be determined in (21)_____ by adding the total number of crossovers that occurred between two genes, dividing by the total number of (22)_____ and multiplying by 100. The (23)_____ can then be determined by comparing this value to a (24)_____.

The mechanism of crossing over involves the exchange of a (25)_____ strand of DNA between nonsister chromatids to form a short stretch of (26)_____ DNA and a structure named a (27)_____. (28)_____ of this structure involves cutting two strands of DNA. Depending on which two of the four strands involved are cut, you will observe a (29)_____. If the allelic difference between the two homologous, nonsister chromatids lies very near the crossover site, mitosis or DNA repair will be necessary to form a productive gamete and (30)_____ may occur. If the crossover site lies between two allelic differences, then one of the alleles may appear to switch chromatids. Often, after a crossover has occurred, a phenomenon known as positive interference (31)_____ the probability that a second crossover will occur nearby.

Figure Analysis

Use the indicated figure or table from the textbook to answer the questions that follow.

1. **Figure 6.3**
(a) Based on Mendel's predictions about independent assortment, what phenotypic ratio did researchers expect to observe in this testcross?

(b) What ratio did they actually observe?

(c) Which progeny are parentals, and which are nonparentals?

(d) Why are there very few banded, detached progeny and wild-type progeny?

(e) How were the nonparentals formed?

2. **Figure 6.7**
(a) What was the important biological principle demonstrated by Creighton and McClintock in this experiment?

(b) Why did they use maize chromosomes to demonstrate this principle?

(c) Where did the extra piece of DNA on chromosome 9 come from?

(d) What two specific characteristics of progeny did Creighton and McClintock compare in order to demonstrate their conclusion?

3. **Table 6.1**
(a) What type of cross is demonstrated in this table?

(b) Which gene is in between the other two, and how does this data demonstrate this?

(c) How many crossovers occurred between the *b* and *pr* genes in the creation of these 15,000 progeny? How many occurred between the *pr* and *c* genes? How many occurred between the *b* and *c* genes?

(d) How many map units are between the *b* and *pr* genes?

4. **Figure 6.15**
(a) List three important principles demonstrated by this graph.
    i.

    ii.

    iii.

5. **Figure 6.23**
(a) If genes *a* and *b* are linked, how many crossovers occurred between genes *a* and *b* if the following asci are observed?
    i. Parental ditype   _____
    ii. Nonparental ditype   _____
    iii. Tetratype   _____

(b) How are the numbers of each type of ascus used to calculate distances between genes?

6. **Figure 6.25**
(a) Spore formation in the organism *Neurospora crassa* is shown. What is the common name of this organism?

(b) What makes studying meiosis in this organism important or helpful to researchers?

(c) Why are there eight spores produced in the asci of this organism?

(d) If a crossover had occurred in meiosis I between the two innermost nonsister chromatids depicted in this figure, what would be the order of spores in the octad?

7. **Figure 6.30**
(a) Meiosis in what organism was studied to demonstrate that only one strand of a double-stranded DNA molecule is exchanged with a sister chromatid?

(b) Which strands are exchanged in the recombination event depicted in figure 6.30?

(c) What ordered ratio is observed in this organism that demonstrates that only one strand is exchanged?

(d) If two strands were exchanged instead of only one, what ordered ratio would be produced by a single crossover without any repair?

General Questions

Q6-1. Consider a two-point mapping problem. Can crossing over always be detected? When might it not be detected?

Q6-2. Do genes on the same chromosome always fail to assort independently during meiosis?

Q6-3. When considering a mapping function, why are larger actual map distances measured less accurately than shorter actual map distances?

Q6-4. How do you determine gene and centromere order in a *Neurospora* dihybrid mapping experiment?

Q6-5. How are 5:2 and 6:3 ratios produced in *Neurospora* octads?

Multiple Choice

For each of the following, circle the letter of the choice that most appropriately answers the question.

1.) How can genes located on the same chromosome sometimes assort independently in meiosis?
  a. because of random alignment of tetrads
  b. because of sister chromatid exchange
  c. because chromosomes have centromeres
  d. because of meiotic recombination
  e. None of the above

2.) Which of the following crossovers will *not* produce recombinant offspring in a heterozygous testcross?
  a. single crossover between two genes
  b. double crossover between two genes
  c. triple crossover between two genes
  d. double crossover between two genes with a third gene between the two crossovers
  e. None of the above are correct; all of the above can be detected.

3.) If two genes are completely linked, what percentage of offspring from a dihybrid test-cross will be recombinants?
  a. 0%
  b. between 0% and 25%
  c. between 25% and 50%
  d. 50%
  e. 100%

4.) Mapping which of the following will yield a more accurate map distance?
  a. genes on different chromosomes
  b. genes far apart on the same chromosome
  c. genes close together on the same chromosome
  d. genes on the *Drosophila* X chromosome
  e. unlinked genes

5.) In a three-point testcross, which progeny demonstrate which gene is in between the other two?
  a. parentals
  b. nonparentals that result from a single crossover
  c. nonparentals that result from a double crossover (one crossover between each pair of genes)
  d. the classes appearing least frequently
  e. Both c and d

6.) In a three-point testcross, which progeny demonstrate the configuration of alleles on the two parental chromatids?
  a. parentals
  b. nonparentals that result from a single crossover
  c. nonparentals that result from a double crossover (one crossover between each pair of genes)
  d. the classes appearing least frequently
  e. Both c and d

7.) According to figure 6.16 in your textbook, which of the following pairs of genes are *most* tightly linked?
  a. gamet eyes and miniature wings
  b. lobe eyes and curved wings
  c. orange eyes and brown eyes
  d. orange eyes and groucho bristles
  e. dwarf body and scarlet eyes

8.) Which of the following *Neurospora* octads indicate that a crossover has *not* occurred?
  a. SDS
  b. 4:4
  c. 2:2:2:2
  d. 2:4:2
  e. 6:2

9.) Somatic-cell hybridization experiments help researchers determine the location of genes on chromosomes because _____ .
  a. both homologous chromosomes will be retained.
  b. recombination will occur frequently in these hybrid cells.
  c. all but a few of the human chromosomes are lost.
  d. ordered spores will form.
  e. All of the above

10.) Why are X-linked human traits easier to map than autosomally linked traits using pedigrees?
  a. There are more RFLPs on the X chromosome than on autosomes.
  b. Men are more likely to participate in a study than women.
  c. The X chromosome is shorter than autosomes.
  d. The genotype of only one parent needs to be known to study X-linked traits.
  e. None of the above are correct; both are equally easy.

Practice Problems

P6-1. In the tomato, stems are either purple (*Aw*) or green (*aw*) and the leaves are either wooly (*Wo*) or smooth (*wo*). Given the following results of a dihybrid testcross, determine if the genes are linked. If so, how far apart are they on the chromosome? What is the linkage arrangement and the distance between genes? How would you describe the allele arrangement in the F_1 heterozygote used in the testcross?

Purple, wooly	205	Green, wooly	810
Purple, smooth	790	Green, smooth	195

(*Hint 1*: Although linkage changes the frequency of each phenotype, the number of phenotypic classes almost always remains the same. Deviations from a 1:1:1:1 ratio (for a dihybrid testcross) or a 1:1:1:1:1:1:1:1 ratio (for a trihybrid testcross) indicate linkage.

Hint 2: In terms of numbers, the parental phenotypic classes will be the most frequent and the double crossover phenotypic classes will be the least frequent.

Hint 3: In denoting linked genes, write the genotypes with a slash mark to separate maternally and paternally derived alleles (e.g., *AB/ab* rather than *AaBb*).

Hint 4: Taken together, the two parental class phenotypes represent the allele arrangement of the F_1 female that was testcrossed. For example, assume that one of the parental classes obtained from the dihybrid testcross is mutant for both traits while the other parental class is wild-type for both traits. In this case, the alleles of the F_1 dihybrid female are in the *cis* arrangement.)

P6-2. In *Drosophila*, a kidney-shaped eye (*k*), cardinal-colored eye (*c*), and ebony body (*e*) are recessive traits located on chromosome number 3. A pure-breeding wild-type male is crossed with a pure-breeding female with kidney-shaped, cardinal-colored eyes and an ebony body. All of the F_1's were wild-type. From the following F_1 testcross data, determine if the traits are linked. If so, calculate all map distances.

Kidney, cardinal, ebony	1758	Kidney	98
Wild-type (all traits)	1768	Ebony, cardinal	88
Kidney, ebony	128	Kidney, cardinal	12
Cardinal	134	Ebony	14

[See *Hints 2–4*.

Hint 5: Determination of the middle gene is essential. The parental and double-crossover phenotypic classes are used to determine the middle gene of a three-gene sequence. You start with the genotypes of the parental classes and arrange the parental genes so that each gene is in the middle. Then perform a double crossover and find a gene arrangement that fits the genotype of the double crossovers that are observed. Remember that in a double crossover, only the association of the middle gene differs when compared to the parental arrangement. For example, if the parental associations are *ABC* and *abc*, the double-crossover types will have the associations *AbC* and *aBc*. If the parental arrangement is *Abc* and *aBC*, then the double crossover types will have the associations *ABc* and *abC*. The placement of the other two genes is not critical (for example, assuming the order is *CBA* rather than *ABC* will have no effect on the calculation of map distances).

Hint 6: Single-crossover phenotypic classes cannot be reliably identified until the proper gene sequence has been established. Compare each single-crossover class to the parental classes to determine where the crossover took place.

Hint 7: When solving three-point linkage map problems, it is a good idea to use the gene symbols associated with each trait to represent the phenotypic description of testcross classes. Remember to use one symbol type (for example, *b* or *b*+) for each gene under consideration.]

P6-3. Consider three characteristics in the hypothetical animal, the zeebog from Madagascar. In the population, Madagascar zeebogs are either hairy (*H*) or hairless (*h*); crazy (*C*) or sane (*c*); and mischievous (*M*) or innocent (*m*). The results obtained from a testcross of a trihybrid female are given here.

Hairy, sane, mischievous	195	Hairy, crazy, innocent	790
Hairy, sane, innocent	205	Hairy, crazy, mischievous	816
Hairless, sane, mischievous	782	Hairless, crazy, mischievous	197
Hairless, sane, innocent	815	Hairless, crazy, innocent	200

What is the linkage arrangement and the distance between genes? How would you describe the allele arrangement in the F_1 heterozygote used in the testcross?

(See *Hints 2–7*.

Hint 8: In some cases, three traits may be studied; however, only two genes are linked, while the third is on a separate chromosome and assorts independently. This situation can be identified by examining the frequencies of each phenotypic class: four equally frequent majority classes and four equally frequent minority classes will be found. If the independently assorting gene is ignored, the problem reduces to a two-point linkage map. Statistically, genes that show map distances greater than 30 map units may actually be linked or may be independently assorting.)

P6-4. Consider three linked genes (x, y, and z) on a single chromosome. Assume that a three-point linkage map showed that genes x and z are 30 map units apart. The x–y distance is 14 map units, and the y–z distance is 16 map units. Devise a data set ($N = 1000$ individuals) with no crossover interference that supports these gene distances.

P6-5. Considering the map distances given in problem P6-4, devise a data set ($N = 1000$ individuals) that supports the given gene distances assuming 100% crossover interference.

[*Hint 9*: Crossover interference reduces the likelihood of a double crossover occurring. If crossover interference is complete ($I = 1$), then no double crossovers are observed.]

P6-6. Considering again the map distances given in problem P6-4, devise a data set ($N = 1000$ individuals) that supports the given gene distances if crossover interference is 55%.

(See *Hint 9*.)

P6-7. *Sordaria* is a mold that naturally grows on the dung of herbivores. It is an ascomycete and a close relative of *Neurospora*. In this species, spores may be black (wild-type) or tan in color. The distance between the mutant tan gene and the centromere has been reported as 26 map units. Assume that you cross a tan mutant strain to a wild-type strain and you score a total of 2000 asci. What data would you record if the given map distance is accurate?

Spore Sequence	Number of asci
4:4	_____
2:2:2:2	_____
2:4:2	_____

P6-8. Somatic cell hybridization was used to map each of four enzymes to the human chromosome on which the loci reside. The results of the analysis are given here. (a) Which chromosome contains human enzyme 1a? (b) human enzyme 2a? (c) human enzyme 3a? (d) human enzyme 4a?

	Enzyme Activity				Human Chromosome Presence				
Clone	1a	2a	3a	4a	1	2	3	4	5
A	+	--	+	--	+	--	--	+	--
B	+	--	--	+	--	+	+	+	--
C	+	+	--	+	--	--	+	+	+
D	+	+	--	--	--	--	--	+	+
E	+	--	--	--	--	--	--	+	--

P6-9. Examine the following pedigree. The traits under consideration are both X-linked and recessive. Note that the male in generation I expresses trait A. You should assume that the female in generation I carries the allele for trait B. From the information given, estimate the distance between the two genes. In what configuration (*cis* or *trans*) are the alleles of the female indicated with an arrow?

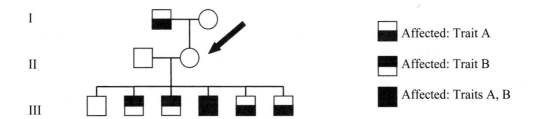

	Affected: Trait A
	Affected: Trait B
	Affected: Traits A, B

Assessing Your Knowledge

Understanding the Key Concepts—Answers

1.) linkage; 2.) complete linkage; 3.) recombination or crossing over 4.) nonparentals or recombinants; 5.) the distance between the two genes; 6.) less; 7.) recombination (or genetic) maps; 8.) testcrosses; 9.) spore, tetrad, or ascus; 10.) recombination; 11.) pedigree; 12.) morphological characteristics; 13.) RFLPs; 14.) trihybrid testcross; 15.) reciprocal; 16.) alleles; 17.) double crossover; 18.) order; 19.) middle; 20.) measured map distance; 21.) map units (or centimorgans); 22.) progeny; 23.) actual map distance; 24.) mapping function; 25.) single; 26.) heteroduplex; 27.) Holliday structure; 28.) Resolution; 29.) crossover; 30). gene conversion; 31.) reduces.

Figure Analysis—Answers

1a.) 1 banded, detached:1 banded:1 detached:1 wild type; b.) 1 banded, detached:241 banded:256 detached:1.5 wild type; c.) Parentals: banded progeny and detached progeny; nonparentals: banded, detached progeny and wild-type progeny; d.) The banded and detached mutations are present on the same chromosome, so they do not assort independently during meiosis. Furthermore, each of the parental generation chromosomes had exactly one of the two mutations. Therefore, creation of chromosomes containing both or neither of the mutations is unlikely; e.) A meiotic crossover occurred between these two mutations in the heterozygous F_1 progeny creating two new chromosomes, one with both mutations (banded, detached) and one with neither mutation (wild type).

2a.) The Creighton and McClintock experiment demonstrated that crossing over between nonsister chromatids results in a physical exchange of DNA and genes; b.) Creighton and McClintock needed an organism that had homologous chromosomes that could be visually differentiated from each other. They found a strain of maize that had an odd chromosome 9 containing unique ends that could be distinguished by microscopy; c.) The extra piece of DNA on chromosome 9 came from another maize chromosome that was accidentally attached to chromosome 9; d.) Creighton and McClintock compared the phenotype of each progeny with the morphology of its chromosome 9 to demonstrate that the creation of recombinant phenotypes correlated with the creation of recombinant morphology of chromosome 9.

3a.) Trihybrid testcross; b.) The purple eye color gene (*pr*) is in between the other two genes. The data demonstrate this, because the classes of progeny that appear in the fewest number (purple and black, curved) have a nonparental phenotype in which only the *pr* gene has been recombined (or switched) relative to the other two genes.
c.)   *b − pr*: 887 (388 + 367 + 60 + 72)
         pr − c: 2927 (1412 + 1383 + 60 + 72)
         b − c: 3550 (388 + 367 + 1412 + 1383 + 2 × 60 + 2 × 72)
d.) 5.9 map units (or cM)

4a.)   i. The observed recombination distance is smaller than the actual map distance.
        ii. The measured map distance can never exceed 50 map units.
        iii. The actual map distance can be determined from the measured map distance.

5a).   i. Parental ditype: zero crossovers
         ii. Nonparental ditype: two crossovers
         iii. Tetratype: one crossover

b.) Add the number of NPD asci and half the number of TT asci and divide this by the total number of asci. This will be your recombination frequency. Multiply this number by 100 to get distance in map units.

6a.) Pink bread mold; b.) It produces asci that have ordered spores making it possible to directly see which chromosomes segregated from each other during meiosis I and meiosis II; c.) Each of the four gametes produced by meiosis undergoes one mitosis to produce eight haploid spores; d.) *A A a a A A a a*

7a.) *Neurospora* (pink bread mold); b.) The bottom strand of the middle two chromatids; c.) 3:1:1:3; d.) 2:2:2:2:2

General Questions—Answers

Q6-1. Crossing over cannot always be detected. It is not detectable by simple methods such as observing phenotypes under the following circumstances since they will not produce recombinant offspring phenotypes. A crossover occurring outside of the region containing two genes under study will not be detected in the offspring phenotypes. A double crossover occurring between two genes will also go undetected. In fact, any even number of crossovers (four crossovers for example) occurring between two genes will go undetected. Finally, if Holliday structures are resolved in such a way as to not recombine flanking genes, this crossover will go undetected.

Q6-2. No. Genes on the same chromosome may appear to assort independently if they are located far apart on the chromosome.

Q6-3. Larger actual map distances appear shorter than they really are since many of the crossovers that occur between genes go undetected. The larger the distance, the more double or quadruple crossovers that do not get scored.

Q6-4. First determine both gene–centromere distances using ordered analysis. Then calculate gene–gene distances using unordered spore analysis. Then put the three in the order that fits both calculations the best.

Q6-5. After formation of heteroduplex DNA at a site of recombination that occurs right at an allelic difference between two nonsister chromatids, the cell repairs one or both of the mismatched nucleotides before mitosis produces the eight spores. If the mismatch in only one of the heteroduplexes is repaired, a 5:3 ratio will result, while a 6:2 ratio will result if both are repaired.

Multiple Choice—Answers

1.) d; 2.) b; 3.) a; 4.) c; 5.) e; 6.) a; 7.) c; 8.) b; 9.) c; 10.) d.

Practice Problems—Answers

P6-1. The traits are linked as evidenced by the two large and two small phenotypic classes. The two largest categories (purple, smooth and green, wooly) are parental types, while the two smallest categories (purple, wooly and green, smooth) are recombinant types. To determine the distance between genes, calculate the frequency of recombinant classes. In this example:

Recombination frequency	=	$(205 + 195) / (790 + 205 + 195 + 810)$
	=	$400/2000$
	=	$0.20 \times 100 = 20\%$ or 20 map units

Because a testcross was performed, the phenotype observed corresponds directly with the alleles received from the dihybrid (a dominant phenotype results if dominant alleles were passed to the offspring, and a recessive phenotype if recessive alleles are passed to the offspring). The two parental phenotypic classes result from a lack of crossing over between the genes of interest. Therefore, taken together, these classes represent the genotype of the dihybrid female. In this case:

Trans configuration	purple smooth	*Aw*	*wo*
	green wooly	*aw*	*Wo*

Another way of writing this is: *Aw wo/aw Wo*

P6-2. The three genes are linked on a single chromosome as suggested by the lack of a 1:1:1:1:1:1:1:1 ratio resulting from the testcross. The two largest categories represent the parentals, while the two least frequent classes result from a double crossover. Again, distances between genes are determined by calculating recombination frequencies.

- Determine the correct gene order by comparing double crossover classes to the parentals.

Parentals			**Doubles**		
k	*c*	*e*	*k*	*c*	*e+*
k+	*c+*	*e+*	*k+*	*c+*	*e*

Only the gene in the middle changes in a double crossover. Notice that all mutant traits are together on one chromosome in the parentals. In the double crossovers, ebony is located on the other chromosome. Therefore, body color is the gene in the middle.

- Rewrite each phenotypic class in the appropriate order, using symbols.

k	*e*	*c*		*k+*	*e+*	*c*		*k*	*e+*	*c*
k+	*e+*	*c+*		*k*	*e+*	*c+*		*k+*	*e*	*c+*
k	*e*	*c+*		*k+*	*e*	*c*				

- Identify each phenotypic class by comparing it to the parentals.

k	*e*	*c*	Parentals	*k+*	*e+*	*c+*
k	*e*	*c+*	Single (*e* and *c*)	*k+*	*e+*	*c*
k+	*e*	*c*	Single (*k* and *e*)	*k*	*e+*	*c+*
k	*e+*	*c*	Double	*k+*	*e*	*c+*

- Calculate map distances.

Distance between *k* and *e*
= (number of singles *k–e* + doubles / total number offspring) × 100
= [(98 + 88 + 12 + 14) / 4000] × 100
= (212/4000) × 100
= 5.3 map units

Distance between *e* and *c*
= (number of singles *e–c* + doubles / total number offspring) × 100
= [(128 + 134 + 12 + 14) / 4000] × 100
= (288/4000) × 100
= 7.2 map units

Distance between *k* and *c*
= *k–e* distance + *e–c* distance
= 7.2 + 5.3
= 12.5 map units

P6-3. When looking at this data, you find four equally frequent large categories and four equally frequent small categories. This pattern suggests that only two of the three genes being considered are linked together and the third is independently assorting. To determine which genes are linked, compare the phenotypes of the four largest classes.

h	*c*	*M*
h	*c*	*m*
H	*C*	*m*
H	*C*	*M*

Notice that hairless and sane are always together as are the alternative characteristics (hairy and crazy). These two traits are therefore linked with mischievous/innocent independently assorting. Elimination of the information about this trait yields a simple two-point linkage map problem.

h	*c*	782 + 815 = 1597
H	*C*	790 + 816 = 1606
h	*C*	197 + 200 = 397
H	*c*	195 + 205 = 400
		Total = 4000

Distance: hairless/hairy to crazy/sane $[(397 + 400)/4000] \times 100 = 19.925$ map units

Trihybrid F_1 genotype (*cis* arrangement): $\dfrac{h \quad c \quad M}{H \quad C \quad m}$

P6-4. In this problem, you are provided with map distances and asked to work back to the data. To do so, you must remember how each of the map distances provided was calculated.
- Calculate the number of double-crossover types expected. Without crossover interference, each crossover event occurs independently. Therefore, we multiply the probabilities of crossovers between each pair of genes.
  Probability of crossover between x and y = 0.14
  Probability of crossover between y and z = 0.16
  Probability of a double crossover   = $0.14 \times 0.16 = 0.0224$

  2.24% of total offspring are expected to be double-crossover types.

- Calculate the expected frequency of single-crossover types. Remember that the map distance equals the frequency of single crossovers plus the frequency of double crossovers.

  Probability of single crossover (x and y) types = 14% – 2.24% = 11.76%
  Probability of single crossover (y and z) types = 16% – 2.24% = 13.76%

- The remaining individuals are parental types.
  Parental types = 100 – 2.24 – 11.76 – 13.76 = 72.24%

- Construct the data set. Of the 1000 offspring, 72.24%, or 722, will be parentals. Fifty percent of these (361) will by xyz, while the other 50% will be x+y+z+. Assuming 1000 offspring and a *cis* allele arrangement in the F_1 trihybrid female, we obtain

x	y	z	361		x	y	z^+	69
x^+	y^+	z^+	361		x^+	y^+	z	69
x	y^+	z^+	59		x	y^+	z	11
x^+	y	z	59		x^+	y	z^+	11

P6-5. In this problem, you are asked to assume that crossover interference is 100%. In this case, the occurrence of the first crossover event completely inhibits the second crossover event. As a result, no double-crossover types are obtained.

x	y	z	350
x^+	y^+	z^+	350
x	y^+	z^+	70
x^+	y	z	70
x	y	z^+	80
x^+	y^+	z	80

P6-6. Given that the number of double crossovers is reduced by 55%, you must first calculate the number of expected double crossovers. Follow the outline from problem P6-4 to arrive at an answer.

To calculate the number of expected double crossovers:
From problem P6-4, expected number of double crossovers given no interference is 2.24%.

$$I = 1 - (\text{number of doubles observed / number of doubles expected})$$
$$0.55 = 1 - (\text{observed frequency} / 0.0224)$$
$$0.55 - 1.0 = - (\text{observed frequency} / 0.0224)$$
$$0.45 = \text{observed frequency} / 0.0224$$
$$0.45 \times 0.0224 = \text{observed frequency}$$
$$0.01008 = \text{observed frequency}$$

Observed frequency of doubles (55% interference) $\approx$ **1.0%**

x	y	z	355
x^+	y^+	z^+	355
x	y^+	z^+	65
x^+	y	z	65
x	y	z^+	75
x^+	y^+	z	75
x	y^+	z	5
x^+	y	z^+	5

P6-7. To calculate the gene–centromere distance, add one-half of all SDS asci (2:2:2:2 and 2:4:2) together and divide by the total number of asci recorded. Multiply this number by 100 to get map units.

$$0.26 \text{ map units} = 1/2(\text{number of 2:2:2:2 asci + number of 2:4:2 asci}) / 2000$$
$$0.26 = (\text{number of SDS asci}) / 4000$$
$$0.26 \times 4000 = \text{number of SDS asci}$$
$$1040 = \text{number of SDS asci (1/2 will be 2:2:2:2 and 1/2 will be 2:4:2)}$$
$$\text{number of FDS asci} = 2000 - 1040 = 960 \text{ (4:4)}$$

4:4	960
2:2:2:2	520
2:4:2	520

P6-8. For each enzyme, determine which clones produce the enzyme and which do not. Compare the enzyme presence with a chromosome presence/enzyme absence with a chromosome absence. For example, if enzyme a is produced in colony A and B but not C, the chromosome on which the gene resides should be present in colony A and B but not C.

Enzyme Activity	Chromosome Location
Human enzyme 1a	4
Human enzyme 2a	5
Human enzyme 3a	1
Human enzyme 4a	3

P6-9. This is an example of the "grandfather" method of estimating recombination from pedigree data. In this problem, the male in generation I expresses trait A while his wife carries the allele for trait B. The female in generation II is heterozygous for both traits. The alleles are in the *trans* configuration (*A b* obtained from the female, *a B* obtained from the male). Of the males in generation III, those expressing only trait A or trait B are parental types. Those expressing both or neither are the result of recombination. Male number 1 and male number 4 are recombinants. Our distance calculation will be 2/6 or ~33 map units.

6

LINKAGE AND MAPPING IN EUKARYOTES

CHAPTER SUMMARY QUESTIONS

1. 1. f, 2. g, 3. b, 4. h, 5. j, 6. i, 7. c, 8. d, 9. e, 10. a .

2. The law of independent assortment. Genes that are linked on the same chromosome are more likely to be transmitted together than would be expected according to Mendel's second law. If genes are on different chromosomes, they will assort independently, in accord with Mendel's second law. If genes are on the same chromosome but far apart from one another, they will also obey Mendel's second law.

3. **a.**   *A*   *B*
    _____

             a   *b*
    _____

    b.   *A*   *b*
    _____

             a   *B*
    _____

4. A crossover is the exchange of genetic material between the two nonsister chromatids of homologous chromosomes. It occurs at the tetrad stage in prophase I of meiosis and is evidenced by chiasmata (singular, chiasma).

5. 50%. A single crossover event involves only two of the four chromatids of a tetrad; the other two chromatids are the parental or nonrecombinant chromatids. Therefore, even if 100% of the tetrads undergo a single crossover between two linked genes, the frequency of recombinant gametes would be 50%.

6. According to the product rule, the frequency of a double crossover is equal to the frequency of the first crossover times the frequency of the second. Therefore, within

any given region, the occurrence of two events is less likely than the occurrence of a single event.

7. Double crossovers always result in switching the middle gene relative to the two flanking genes. Therefore, one can compare the two double crossover phenotypes to the two parental or nonrecombinant phenotypes. The gene that has flipped occupies the middle locus.

8. The measured map distance between two loci is the value obtained from a two-point cross. The actual map distance, on the other hand, is an idealized, more accurate value obtained from summing short distances between many intervening loci.

9. Interference is an indication of the effects of one crossover over another. In positive interference, the occurrence of the first crossover reduces the chance of the second, whereas in negative interference, the occurrence of one crossover enhances the probability that crossovers will occur in adjacent regions.

10. A RFLP, or restriction fragment length polymorphism, is a nucleotide change that results in the elimination or creation of a restriction enzyme site. RFLPs represent different alleles that can be analyzed by recombination in the same manner as morphological phenotypes. One advantage of RFLP is that it does not have to be associated with a particular gene or phenotype. Thus, there will be many more RFLPs than there are genes in the genome. This large number of RFLPs should yield at least one RFLP near every gene, which will produce a small and accurate recombination distance between the RFLP and the gene.

11. Three-point crosses capture (allow us to see) double crossovers that have taken place in the two regions defined by the three loci. However, any double crossovers that occur within one region or any crossovers involving more than two events will not be indicated correctly by random-strand analysis.

EXERCISES AND PROBLEMS

12. The cross yielded a total of 20% recombinants. Therefore, the distance between the two genes is 20 cM.

13. **a.** Four types of gametes: 25% *AB*, 25% *Ab*, 25% *aB*, and 25% *ab*.
    b. Two types of gametes: 50% *AB* and 50% *ab*.
    c. Four types of gametes: 45% *AB*, 5% *Ab*, 5% *aB*, and 45% *ab*. *Note:* 10 cM means a total of 10% recombinants.
    d. Four types of gametes: 25% *AB*, 25% *Ab*, 25% *aB*, and 25% *ab*. *Note:* Recombination frequency cannot exceed 50%. Genes that are 70 map units apart will behave as if they are unlinked.

14. This problem can be solved using a trial-and-error approach. Start with genes that are separated by the longest distance, as these must be near the ends, and then try to fit the other genes in. Begin with i–e = 27. Then consider the next longest distance that includes either i or e. In this case it is e–k = 19. Locus k could be between i and e, or e could be the middle locus. If k is in the middle, then the distance i–k would be equal to the distance i–e minus the distance e–k. This is exactly what we obtained from the crosses (27 – 19 = 8). Therefore, the order of the first three loci we have examined is i, k, e. Similar analysis yields the following map:

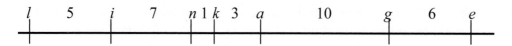

15. The recombinants in each case will add up to 20% or 0.2 frequency. Remember that there is no crossing over in *Drosophila* males.

   a. Both parents are $cu^+ e^+/cu\ e$. Female will produce 80% parental type ($cu^+ e^+$, $cu\ e$) and 20% recombinant ($cu^+ e$, $cu\ e^+$) gametes. Male will only produce two types of gametes: 50% $cu^+ e^+$ and 50% $cu\ e$. Therefore, the probability of a *cucu ee* offspring is 0.4 ($cu\ e$) × 0.5 ($cu\ e$) = 0.2 or 20%.

   b. Both parents are $cu^+ e/cu\ e^+$. Female will produce 80% parental type ($cu^+ e$, $cu\ e^+$) and 20% recombinant ($cu^+ e^+$, $cu\ e$) gametes. Male will only produce two types of gametes: 50% $cu^+ e$ and 50% $cu\ e^+$. Therefore, the probability of a *cucu ee* offspring is 0% because the male cannot provide a $cu\ e$ gamete to his offspring.

   c. The male is $cu^+ e^+/cu\ e$ and the female is $cu^+ e/cu\ e^+$. Therefore, the probability of a *cucu ee* offspring is 0.5 ($cu\ e$) × 0.1 ($cu\ e$) = 0.05, or 5%. The gamete frequencies were calculated in parts (a) and (b).

16. **a.** P   groucho   ×   rough
                       grogro ro$^+$ro$^+$ *gro$^+$gro$^+$ roro*

          F_1 female *gro$^+$gro ro$^+$ro*   ×   male *grogro roro*

          F_2 *grogro ro$^+$ro*   518
                     gro$^+$gro roro   471
                     grogro roro   6
                     gro$^+$gro ro$^+$ro   5

      (6 + 5)/1000 = 0.011 = 1.1% recombination
                         = 1.1 map units apart

   b. Given the map units, F_1 gametes are produced on the average by females as follows: *gro ro$^+$*, 49.45% = 0.4945 (98.9%/2); *gro$^+$ ro*, 49.45% = 0.4945; *gro ro*, 0.55% (1.1%/2) = 0.0055; and *gro$^+$ ro$^+$*, 0.55% = 0.0055. Males, lacking crossing over, produce only two gamete types: *gro ro$^+$* and *gro$^+$ ro*, each 50% = 0.50. Summing from the Punnett square following, the phenotypes of the offspring would be as follows: wild type, 50%; groucho, rough, 0%; groucho, 25%; and rough, 25%.

| | **Male** | |
Female	*gro ro$^+$* (0.5)	*gro$^+$ ro* (0.5)
gro ro$^+$ (0.4945)	Groucho 0.24725	Wild type 0.24725
gro$^+$ ro (0.4945)	Wild type 0.24725	Rough 0.24725
gro ro (0.0055)	Groucho 0.00275	Rough 0.00275
gro$^+$ ro$^+$ (0.0055)	Wild type 0.00275	Wild type 0.00275

17. a. The two genes are linked. The cross is a testcross. If the genes were not linked, we would expect a 1:1:1:1 ratio of offspring; we don't see that.

    b. *Trans*. The alleles that are linked will appear as the majority classes, which are Trembling, long-haired and normal, Rex. Therefore, Trembling and Rex must be in the *trans* configuration.

    c. If we let T = Trembling, and R = Rex, the cross is

$$\frac{T \quad r}{t \quad R} \quad \times \quad \frac{t \quad r}{t \quad r}$$

Recombinants are Trembling, Rex and normal, long-haired;

$$\text{Recombination frequency} = \frac{42 + 44}{300} \times 100 = 28.7\%$$

18. The linkage pattern is consistent with X linkage.

P $X^{abe,bis} X^{abe,bis}$ (female) $\times$ X^+Y (male)

F_1 $X^+X^{abe,bis} \times X^{abe,bis}Y$

F_2 Since crossovers can only occur in F_1 females, this is the same as a testcross: both sons and daughters will have the phenotype of the X chromosome inherited from the mother. Thus, in the F_2, both sons and daughters have roughly the same distribution of phenotypes. Abnormal and brown are the recombinant classes (43 + 45 + 37 + 35 = 160). The two loci are 16.0 map units apart on the X chromosome [(160/1000) $\times$ 100].

19. A dihybrid female is testcrossed (with a hemizygous male having both recessive alleles). Each recombinant class will make up about 5% of the offspring. Each parental class will make up about 45% of the offspring. Phenotypic classes will be equally distributed between the two sexes. The same results will be found for an autosomal locus if the dihybrids are females (no crossing over in males). A reciprocal cross cannot be done for X-linked genes because males cannot be dihybrid. Males dihybrid for an autosomal gene produce only two classes of offspring when testcrossed—parentals.

20. a. Female is z^+w^+/zw. She will produce gametes as follows: 35% z^+w^+ (70%/2), 35% zw, 15% z^+w (30%/2), and 15% zw^+. The male is zw Y. He will produce gametes as follows: 50 % zw, and 50% Y. Therefore, the offspring's phenotypic frequencies for both males and females is 0.35 z^+w^+, 0.35 zw, 0.15 z^+w, and 0.15 zw^+.

b. Female is z^+w^+/zw. Male is z^+w Y. He will produce gametes as follows: 50% z^+w and 50% Y. Therefore, the daughters' phenotypic frequencies are: 0.50 z^+w^+ and 0.50 z^+w. Sons' phenotypic frequencies are the same as in part (a).

c. Female is z^+w/zw^+. She will produce gametes as follows: 35% z^+w (70%/2), 35% zw^+, 15% z^+w^+ (30%/2), and 15% zw. The male is zw Y. He will produce gametes as follows: 50 % zw and 50% Y. Therefore, the phenotypic frequencies for both sons and daughters is 0.35 z^+w, 0.35 zw^+, 0.15 z^+w^+, and 0.15 zw.

d. Female is z^+w/zw^+. Male is z^+w Y. He will produce gametes as follows: 50% z^+w and 50% Y. Therefore, the daughters' phenotypic frequencies are: 0.50 z^+w^+ and 0.50 z^+w. Sons' phenotypic frequencies are the same as in part (c).

21. This is a simple three-point cross complicated by the fact that these loci are detected by electrophoretic methods. The alleles are codominant and thus, although we can't do a true testcross (no recessive homozygotes exist), we tested the trihybrid females by crossing them with males that were homozygous for the slow alleles. Thus the cross to produce the preceding data was as follows:

$$got^f got^s\ amy^f amy^s\ sdh^f sdh^s \quad \times \quad got^s got^s\ amy^s amy^s\ sdh^s sdh^s$$

In looking at the offspring, we know that each has a slow allele at each locus contributed by the homozygous males. Thus we look at the other allele in each of the offspring to know the alleles contributed by the trihybrid female. By doing that we have the same information we would have in a testcross.

The pattern of numbers among the eight offspring classes is the pattern we are used to for linkage of the three loci. We can tell from the two groups in largest numbers (nonrecombinants—classes 1 and 2) that the alleles are in the coupling (*cis*) arrangement. If we compare either of the nonrecombinant offspring with the double recombinant classes (7 and 8), we can see that the *amy* locus is in the middle. The comparisons of classes 8 with 2 or 7 with 1 should be especially clear. Again, disregard the slow alleles contributed by the male at each locus and look only at the other alleles, the ones contributed by the trihybrid. We can now infer that the trihybrid female had the following chromosomal arrangement:

$$got^f \quad amy^f \quad sdh^f$$

$$\overline{got^s \quad amy^s \quad sdh^s}$$

You should now be able to see that classes 3, 4, 7, and 8 have crossovers in the *got–amy* region and classes 5, 6, 7, and 8 have crossovers in the *amy–sdh* region. Tallying these numbers, we see that the recombinants make up $(11 + 14 + 1 + 1)/1000 = 0.027$, or 2.7%, in the *got–amy* region and $(58 + 53 + 1 + 1)/1000 = 0.113$, or 11.3%, in the *amy–sdh* region. Thus the map distances between the loci are 2.7 and 11.3 map units, respectively.

We expect $0.027 \times 0.113 = 0.00305$ double crossovers, or 3.05 per thousand. We observed only two individuals. Thus the coefficient of coincidence is $2/3.05 = 0.6557$, or about 66% of the expected double crossovers are actually observed.

22. **a.** The hotfoot locus is in the middle (compare, for example, hotfoot, a double crossover, with the wild type, a parental); there are 16.0 map units from hotfoot to either end locus: $74 + 66 + 11 + 9$ recombinants between hotfoot and waved, and $79 + 61 + 11 + 9$ recombinants between hotfoot and obese.

  b. The trihybrid parent was $o\ h\ wa/o^+\ h^+\ wa^+$.

  c. The coefficient of coincidence is $20/25.6\ [20/(0.16 \times 0.16 \times 1000)]$. Interference is $1 - (20/25.6) = 0.22$, or 22%.

23. **a.** The brittle and glossy loci are linked; ragged is assorting independently. The brittle to glossy distance is 8.0 map units ($23 + 17 + 21 + 19$ recombinants). The offspring of this cross form a pattern of four classes in high numbers, all about equal, and four classes in low numbers, all about equal. This is the pattern of simple linkage of two loci with a third locus assorting independently. There are two ways to approach this problem. By inspection you should be able to see that among the four classes in high numbers, the brittle and glossy mutants are together and their wild-type alleles are also together, with the ragged phenotype assorting independently. Thus the ragged locus is assorting independently and the four classes in high numbers are actually two classes with regard to brittle and glossy—they are the parentals. Similarly, the four classes in low numbers are the recombinants. If you cannot see this, then arbitrarily assign one reciprocal set in high numbers as parentals and one reciprocal set in low numbers as double crossovers. You will then calculate map distances such that brittle and glossy are 8 map units apart and ragged is 50 map units from the "middle" locus. In other words ragged is assorting independently.

  b. The *btl* and *gl17* alleles are in the *cis* (coupling) phase.

  c. Not relevant.

24. **a.** Work backward from the 0.61% double recombinants $(0.100 \times 0.061 \times 100)$. Thus, there would be 6 of 1000 double recombinants. In the *an-sple* region, we need the total of single + double recombinants = 100 of 1000 (10 map units). Thus, $100 - 6 = 94$; divided by 2 (two phenotypes) is 47 each. For the *sple-at* region, the total of single and double recombinants = 61 (6.1 map units). Thus,

61 − 6 = 55; divided by 2 is 27 and 28. The parentals make up the remainder for a total of 1000.

b. With a coefficient of coincidence of 0.60, only 0.366% (0.61 × 0.60) of the expected double recombinants will occur, that is, 4 instead of 6. Thus,

	Coefficient of Coincidence	
	1.0	**0.6**
Ancon, spiny, arctus oculus	422	421
Wild type	423	422
Ancon, spiny	27	28
Arctus oculus	28	29
Ancon	47	48
Spiny, arctus oculus	47	48
Ancon, arctus oculus	3	2
Spiny	3	2
Total	**1000**	**1000**

25. Notchy should be known on the X chromosome by the results of reciprocal crosses as well as linkage with other X-linked loci. If a trihybrid female is testcrossed (to an ancon, spiny-legged male fly hemizygous for the *ny* allele), the offspring will consist of, for example,

Parentals (*an sple* or *an⁺ sple⁺*)	**Males**	**Females**
Ancon, spiny legs, notchy	112	113
Ancon, spiny legs	113	112
Notchy	112	113
Wild type	113	112
Recombinants (*an sple⁺* or *an⁺ sple*)		
Ancon, notchy	12	13
Ancon	13	12
Spiny legs, notchy	12	13
Spiny legs	13	12

The genes for ancon and spiny legs are linked, separated by 10 map units, whereas the notchy gene segregates independently. Because of the genetic nature of the male parent, the sexes segregate similarly. Thus, of 1000 offspring, 900 will be in the parentals group, divided equally among sexes and genotypes. One hundred will be recombinant (10%), again, divided equally among sexes and genotypes.

26. a. $\dfrac{k \quad e^+ \quad cd}{k^+ \quad e \quad cd^+}$

b. $\underline{k \quad 6.9 \quad e \quad 5.1 \quad cd}$

The initial cross is $\dfrac{k \quad e^+ \quad cd}{k \quad e^+ \quad cd} \times \dfrac{k^+ \quad e \quad cd^+}{k^+ \quad e \quad cd^+}$

Producing a trihybrid F$_1$ female: $\dfrac{k \quad e^+ \quad cd}{k^+ \quad e \quad cd^+}$, in any order.

The last two classes (3 + 4 offspring in the F$_2$) are double crossovers and allow us to determine gene order by comparison with the parentals (880 + 887). If the order is *k cd e,* a double crossover in the F$_1$ females yields *k cd$^+$ e$^+$* and *k$^+$ cd e.* If the order is *cd k e,* a double crossover yields *cd k$^+$ e$^+$* and *cd$^+$ k e.* Therefore, the order must be *k e cd,* which gives the correct double recombinants in the F$_2$ generation. After reconstructing the trihybrid

$$\dfrac{k \quad e^+ \quad cd}{k^+ \quad e \quad cd^+}$$

and scoring each of the offspring for crossovers in the *k–e* and *e–cd* regions: map units, k–e = [(64 + 67 + 3 + 4)/2000] × 100 = 6.9; and map units, e–cd = [(49 + 46 + 3 + 4)/2000] × 100 = 5.1.

27. a. Tunicate and glossy are linked; Ligule assorts independently.
   b. *TG/tg*; *L/l.*
   c. 20.5 map units.
   We see four majority classes in equal frequencies (58, 55, 53, 59) and four minority classes in equal frequencies (15, 13, 16, 14), indicating that one gene is not linked. If two genes are linked, they should produce a 1:1:1:1 ratio in majority classes. Comparing Tunicate and nontunicate with Liguled and liguleless or Liguled and liguleless with Glossy and nonglossy produces this 1:1:1:1. Therefore, Tunicate and Glossy are linked. The nonrecombinants will appear in the majority classes: tunicate, Glossy and nontunicate, nonglossy are these classes. The alleles are in the *cis* configuration, *TG/tg.*

   Map units = $\dfrac{15 + 13 + 16 + 14}{283}$ × 100 = 20.5

28. a. The distance between *jv* (javelin) and *e* (ebony) is greater than 50 map units, so the genes should behave as if they were unlinked in females. The cross is

$$\dfrac{jv \, e}{jv^+ e^+} \qquad \times \qquad \dfrac{jv \, e}{jv^+ e^+}$$

Since there is no crossing over in males, however, the following chart can be drawn:

Female	Male jve	jv^+e^+
jve	jve/jve	jve/jv^+e^+
jv^+e^+	jve/jv^+e^+	jv^+e^+/jv^+e^+
jve^+	jve/jve^+	jve^+/jv^+e^+
jv^+e	jve/jv^+e	jv^+e/jv^+e^+

Summarizing the data in this table, we expect 5 wild type:1 javelin:1 ebony: 1 javelin, ebony (no crossing over in *Drosophila* males).

b. Calculate distances: $th - Dl = 23$; $Dl - e = 4.5$
Calculate double crossover frequency $= (0.23)(0.045) = 0.01035$
Calculate $th - Dl$ crossovers $= 0.23 - 0.01035 = 0.21965$
Calculate $Dl - e$ crossovers $= 0.45 - 0.01035 = 0.43965$
Nonrecombinants $= 1 - (0.01035 + 0.21965 + 0.43965) = 0.33035$

Each class within a category will be 1/2 of the preceding frequencies, since recombination is a reciprocal event.

c. Proceed the same way, except the frequency of double crossovers will be different.

Coefficient of coincidence = observed/expected
Therefore, (coefficient of coincidence)(expected) = observed:
$(0.4)(0.01035) = 0.00414$. This number is then subtracted from single exchange classes. (It is used above in place of 0.01035.)

a.	b.	c.
Thread, ebony	0.165	0.165
Delta	0.165	0.165
Wild type	0.005	0.002
Thread, Delta, ebony	0.005	0.002
Thread, Delta	0.110	0.113
Ebony	0.110	0.113
Thread	0.220	0.223
Delta, ebony	0.220	0.223

29. This problem requires the manipulation of equations. We know that interference = 1 – coefficient of coincidence, so coefficient of coincidence = 1 – interference = 1 – (– 1.5) = 2.5. Since

Coefficient of coincidence = $\dfrac{\text{observed double crossovers}}{\text{expected double crossovers}}$

observed double crossovers = (coefficient of coincidence) (expected double crossovers). The observed double crossover frequency is $(2.5)(0.1 \times 0.05) = 0.0125$.

30. a. First arrange the data in reciprocal classes:

a	b	+	462
+	+	c	458
a	+	c	27
+	b	+	23
a	+	+	16
+	b	c	14

Only three sets of reciprocal classes are found. Interestingly, the missing flies are those with the + + + and a b c phenotypes. These represent the double-crossover classes, which are always the least frequent. The absence of the double-crossover types could be because of the close proximity of the three loci on the chromosome, which makes double crossovers very unlikely. Another reason could be some unknown selective factor.

b. Comparing the missing double-crossover classes with the arrangement of the parental classes (most frequent), we notice that the *c* gene "switches places" between the two groups. Therefore, gene *c* must be the middle locus, and the gene order is *a–c–b*.

c. The distance between two genes can be obtained by calculating the recombination frequency between them. Therefore, the *a–c* distance = $(27 + 23)/1000 = 0.05$, or 5%; and the *c–b* distance = $(16 + 14)/1000 = 0.03$, or 3%. These values give the following map:

<u>*a*   5   *c*   3   *b*</u>

d. Interference = 1 – coefficient of coincidence. The coefficient of coincidence = observed double crossovers/expected double crossovers. Since no double crossovers were observed, the coefficient of coincidence = 0. Therefore, I = 1 (complete interference).

31. Seven recombinant classes will be expected in the offspring. These are as follows: Single recombinants for the regions *p–q*, *q–r*, and *r–s*; double recombinants for regions *p–q* and *q–r*, *p–q* and *r–s*, and *q–r* and *r–s*; and triple recombinants involving all three regions (*p–q*, *q–r*, and *r–s*).

32. Single crossover between the loci, involving only two chromatids, should produce a pattern similar to

$$ab, ab, ab^+, ab^+, a^+b, a^+b, a^+b^+, a^+b^+$$

If crossing over occurred at the two-strand stage, each crossover would involve both chromosomes and therefore all four chromatids, such as
$$ab^+, ab^+, ab^+, ab^+, a^+b, a^+b, a^+b, a^+b$$

A single crossover involves only two chromatids, but complex crossovers can involve three or four chromatids. For example, a three-strand double crossover between the a and b loci will produce a tetratype:
$$ab, ab, ab^+, ab^+, a^+b^+, a^+b^+, a^+b, a^+b$$

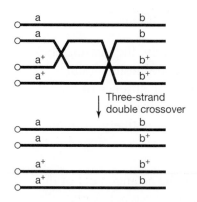

33. PD, 1, 2, 4, 6, 8–10; NPD, 3; TT, 5, 7.
Map units = {[NPD + (1/2)TT]/Total} × 100 = {[1 + (1/2)2]/10} × 100 = 20. The loci are 20 map units apart.

34. a. PD, 1, 2, 6, 8, 10, 11; NPD, 9; TT, 3, 4, 7; unscorable, 5, 12 (these could be TT or NPD).
    b. Yes, PD > NPD.
    c. Map units = {[NPD + (1/2)TT]/Total} × 100 = {[1 + (1/2)3]/10} × 100 = 25. The loci are 25 map units apart.

35. FDS, 3–5, 8, 10; SDS, 1, 2, 6, 7, 9. The distance between the *arg* locus and its centromere is (1/2)%SDS = (1/2)50% = 25 map units.

36. a. FDS, 1, 4–6, 8, 9, 11; SDS, 2, 7, 10; unscorable, 3, 12.
    b. The distance between the fuzzy locus and its centromere is (1/2)%SDS = (1/2)30% = 15 map units.

37. $ab \times a^+b^+ \rightarrow ab/a^+b^+$, which undergoes meiosis. Twelve map units means NPD + (1/2)TT = 12% of the total asci. For example, of 100 asci dissected, 6 could be NPD and 12 could be TT.

38. $a \times a^+ \rightarrow a/a^+$, which undergoes meiosis. Twelve map units means that the SDS pattern makes up 24% of the asci.

39. The best first-order estimate is that the a and b loci are on opposite sides of the same centromere. The a locus is 5.5, and the b locus is 10.5 map units from the centromere; the a–b distance, based on unordered spores is 15 map units, confirming the arrangement.

40. $ab \times a^+b^+ \rightarrow ab/a^+b^+$, which undergoes meiosis. If 1000 asci are analyzed, for example, the following results might be observed: 1, 408; 2, 42; 3, 250; 4, 250; 5, 30; 6, 5; 7, 15. To make the numbers work out, classes 4–7 must equal 300, as must classes 3 + 5–7, in order to make (1/2)%SDS = 150. Thus, if classes 3 and 4 are 250 each, 50 must be spread out among classes 5–7. NPD + (1/2)TT should equal 300 to confirm the arrangement (a–b distances). These numbers will give an a–b distance of 30.

41. Assume the order of centromere, a, b.
Class 1: No crossovers.
Class 2: A four-strand double crossover between the a and b loci.
Class 3: One crossover between the loci.
Class 4: A two-strand double crossover, one between the centromere and the a locus and the other between the loci.
Class 5: One crossover between the centromere and the a locus.
Class 6: Three crossovers. The first is between the centromere and the a locus on chromatids 2 and 3; the second is between the two loci on chromatids 1 and 3; and the third is between the two loci on chromatids 2 and 4.
Class 7: Two crossovers. The first is between the centromere and the a locus on chromatids 2 and 3; the second is between the loci on chromatids 2 and 4.

42. a. Yes; PD >> NPD (actually, no NPD).
    b. a: 2.5 map units; b: 7.5 map units. Classify each ascus—I: PD, FDS for both; II: TT, FDS for a, SDS for b; III: TT, FDS for a, SDS for b; IV: TT, FDS for a, SDS for b; V: TT, FDS for a, SDS for b; VI: PD, SDS for a and b; VII, PD, SDS for a and b; VIII: PD, SDS for a and b. We see no NPDs, so genes are linked. For gene to centromere distances, use the formula

$$\frac{1/2 \text{ (number of SDS asci)}}{100} \times 100$$

a to centromere = (1/2)(3 + 1 + 1)% = 2.5 map units
b to centromere = (1/2)(2 + 3 + 2 + 3 + 3 + 1 + 1)% = 7.5 map units

43. a. a to centromere: 6 map units; b to centromere: 0; c to centromere: 13; d to centromere: 20.5.

    b. b, c, and d are linked.

c.

$$\underleftrightarrow{\overset{6}{\hphantom{aa}}}$$

a ⟷ (6) ◦   c ⟷ (13) ◦ b ⟷ (20.5) d

$$\frac{(1/2)\,(\text{number of SDS})}{100} \times 100$$

a to centromere = $(1/2)(12) = 6$ b to centromere = 0
c to centromere = $(1/2)(26) = 13$ d to centromere = $(1/2)(41) = 20.5$

Examine each cross for PD and NPD. The second cross yields PD >> NPD, indicating that b and c are linked. The third cross indicates that c and d are linked. We know that b is located at its centromere.

Calculate the c–d distance as [number of NPD + (1/2)number TT]/100 × 100 = [3 + (1/2)50] = 28 map units. If c and d are on the same side of the centromere, most SDS for c should also be SDS for d. We see 26 SDS for c and 11 for c and d. Therefore c and d are probably on opposite sides of the centromere, consistent with the c–d distance of 28 map units calculated earlier.

44. 12.5 map units. The only genotype that grows on minimal medium is arg^+ad^+. If the two genes were unlinked, one-fourth of the progeny should have this genotype; this is not seen. The genes must be linked; wild type results from recombination between these two genes. The reciprocal class, $arg^-\,ad^-$, which has not been selected for, should be equally frequent, so map units = (2 × 25)/400 × 100 = 12.5.

45. You must analyze two genes at a time. Begin by classifying each ascus.

Ascus	pab	pk	ad	pab-pk	pab-ad	pk-ad
I	FDS	FDS	FDS	PD	PD	PD
II	SDS	SDS	SDS	PD	PD	PD
III	FDS	SDS	SDS	TT	TT	PD
IV	FDS	FDS	SDS	PD	TT	TT
V	SDS	SDS	SDS	TT	TT	PD
VI	SDS	SDS	SDS	PD	TT	TT
VII	SDS	SDS	FDS	PD	TT	TT
VIII	SDS	SDS	SDS	PD	TT	TT

We see no NPD for any of the gene pairs, so all three genes are linked. Calculate the gene–centromere distances using (1/2)(number SDS/93) × 100.

pab-centromere = $(1/2)[(35 + 2 + 2 + 1 + 3)/93] \times 100 = 23$ map units
pk-centromere = $(1/2)[(35 + 9 + 2 + 2 + 1 + 3)/93] \times 100 = 28$ map units
ad-centromere = $(1/2)[(35 + 9 + 7 + 2 + 2 + 3)/93] \times 100 = 31$ map units

Now calculate gene–gene distances using $[(1/2)TT + NPD]/93 \times 100$:

$$pab\text{–}pk = (1/2)[(9 + 2)/93] \times 100 = 6 \text{ map units}$$
$$pab\text{–}ad = (1/2)[(9 + 7 + 2 + 2 + 1)/93] \times 100 = 13 \text{ map units}$$
$$pk\text{–}ad = (1/2)[(7 + 2 + 1 + 3)/93] \times 100 = 7 \text{ map units}$$

All three genes are on the same side of the centromere.

46. Construct a pedigree of the Duffy alleles (FY^a, FY^b). Arbitrarily assign one allele to a normal chromosome 1 and the other allele to the coiled chromosome. Then, accompanying the pedigree, the alleles and their morphologically proper chromosomes would be associated.

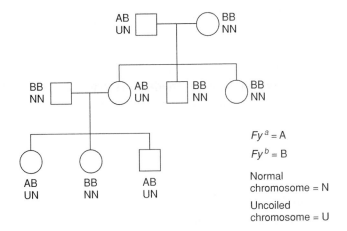

$Fy^a = A$

$Fy^b = B$

Normal chromosome = N

Uncoiled chromosome = U

47. The +/– patterns seen under the columns headed chromosome 6, 14, X, respectively.

48. These X-linked loci are 20 map units apart (2 of 10 recombinant sons according to the "grandfather method").

49. Let f = fructose-intolerance and F = Normal allele. The man's genotype must have been $I^A F/I^B f$, and the woman's is $i F/i f$. Thus, only the type A, fructose-intolerant children, are recombinant. Consequently, the genes are separated by 20 map units (2 of 10 = 20%).

50. Gluthathione reductase: 8; malate dehydrogenase: 2; adenosine deaminase: 20; galactokinase: 17; hexosaminidase: 5. For gluthathione reductase, determine what chromosomes are common to both clones A and B—this is chromosome 8. Similar logic will place the other genes on the appropriate chromosomes.

51. As we have discussed throughout this chapter, linkage has an effect on the types and proportions of gametes produced by an individual. It is, hopefully, obvious that

Minnie is the key to this problem. Crossing over during her oogenesis will determine the combination of alleles that will be transmitted to her offspring.

Let c = color blindness, C = normal vision, h = hemophilia, and H = normal clotting. The X chromosome Minnie inherited from her hemophiliac father must be $C\,h$, and the one she inherited from her color-blind mother, $c\,H$. Therefore, Minnie's sex chromosome genotype can be represented as $C\,h/c\,H$. The C and H genes are 10 cM apart; therefore, Minnie will produce gametes in the following proportions: 0.45 Ch (90%/2), 0.45 cH, and 0.05 CH, (10%/2), 0.05 ch.

Mickey's sex chromosome genotype can be designated as CH/Y. He will produce gametes in the following proportions: 0.5 CH and 0.5 Y.

a. For a child to be affected with hemophilia and not color blindness, it would have to be a son that inherited the Ch chromosome from his mother. Therefore, the probability = 0.45 (Ch X chromosome from Minnie) × 0.5 (Y chromosome from Mickey) = 0.225.

b. For a child to be affected with both hemophilia and color blindness, it would have to be a son that inherited the ch chromosome from his mother. Therefore, the probability = 0.05 (ch X chromosome from Minnie) × 0.5 (Y chromosome from Mickey) = 0.025.

c. For a child to get neither hemophilia nor color blindness, it would have to be a son that inherited the CH chromosome from his mother or be a daughter (all daughters will be normal). Calculate each probability, and use the sum rule to determine the overall probability. Probability of unaffected son = 0.05 (CH X chromosome from Minnie) × 0.5 (Y chromosome from Mickey) = 0.025. Probability of daughter = 0.5. Therefore, the probability of an unaffected child = 0.025 + 0.5 = 0.525.

52. Enzyme A on 11, B on 15, C on 18, D on 3, E on 7. Enzyme A is present in clones X and Y, and chromosome 11 is common to these two clones. Enzyme B is present only in clone X, and 15 is the only chromosome unique to clone X. Similar logic allows the assignment of the other genes.

53. Clones A and C. We need to determine the clones that have chromosome 3, because they are expected to be positive for the enzymes; we see that clones A and C have chromosome 3.

54. a. If 4% of all tetrads have a single crossover during meiosis, then 2% of all chromatids will be recombinant (half the chromatids in 4% of the cells). Thus, in fruit flies, the distance between the loci will be 2 map units.

b. In *Neurospora,* 4% of asci will exhibit second division segregation, which is also 2 map units between loci.

c. In yeast, a single crossover will generate a tetratype only, not a nonparental ditype. Again, the distance between the two loci will be recorded as 2 map units.

55. First we have to determine the linkage arrangement of Ness. Her father is a true breeding yellow (GG), fire-producer (ff) with horns (hh). Therefore, one of Ness' chromosomes has to be $G\,f\,h$, and her genotype can be represented as

$$\frac{f \quad\quad G \quad\quad h}{F \quad\quad g \quad\quad H}$$

Puff's genotype is $f\,g\,h/f\,g\,h$, and so he can only produce one type of gamete: $f\,g\,h$. Therefore, this cross is basically a testcross.

Let's now turn our attention to the "origin" of the four phenotypes in question. Offspring that have the **(a)** phenotype can arise only from a double-crossover event. Indeed, they will be half of the double-crossover class because the recombination event produces two chromosomes: $f\,g\,h$ and $F\,G\,H$. The **(b)** offspring are half the single-crossovers in the f–g region. The **(c)** offspring are half the single-crossovers in the g–h region. Offspring that are of the **(d)** phenotype will be half the parental (noncrossover) class.

First, determine the number of offspring in the double-crossover class: $0.05 \times 0.2 = 0.01$. This number will have to be subtracted from each single crossover class.

Single crossovers in the f–g region $= 0.05 - 0.01 = 0.04$
Single crossovers in the g–h region $= 0.2 - 0.01 = 0.19$
$$\text{Parentals} = 1 - (0.19 + 0.04 + 0.01) = 1 - 0.24 = 0.76$$

Therefore, the number of offspring can be calculated as follows:

a. $(0.01/2) \times 1000 = 5$ individuals that are fire-producing, green, and with horns.

b. $(0.04/2) \times 1000 = 20$ individuals that are fire-producing, green, and have no horns.

c. $(0.19/2) \times 1000 = 95$ individuals that are fire-producing, yellow, and have no horns.

d. $(0.76/2) \times 1000 = 380$ individuals that are fire-producing, yellow, and with horns.

56. a. All female offspring are wild type, yet there are eight different male phenotypes. This differential appearance of the traits indicates that the traits are sex-linked—specifically, X-linked.

The male offspring can be used to analyze the genotypes of the parents, since sons receive an X chromosome from their mother and the Y chromosome from their father. The most frequent classes (parentals) in the sons are $y\,rb^+\,m$ and $y^+\,rb\,m^+$. Because these classes correspond to the nonrecombinant X chromosomes from the mother, the mother's genotype must therefore have been $y\,rb^+\,m/y^+\,rb\,m^+$. This genotype will result in many gametes with an X chromosome carrying at least one recessive allele. However, all 1000 daughters have been observed to be wild type for all three traits. Therefore, they must have received a paternal X chromosome carrying all three wild-type alleles. Therefore, the parental cross was
$y\,rb^+\,m/y^+\,rb\,m^+$ (female) $\times$ $y^+\,rb^+\,m^+/\text{Y}$ (male).

The least frequent classes (double recombinants) are $y^+ rb^+ m^+$ and y rb m. A comparison of the parentals and the double-crossover classes reveals that gene rb has flipped and is therefore the rb gene in the middle.

b. The linkage map can be determined by an analysis of tabulated data. Only the sons will be considered because recombination cannot be scored in the female offspring.

Phenotype	Genotype	Number	Class
Ruby	$y^+ rb$ m^+/Y	326	Parental
Yellow, miniature	y rb^+ m/Y	321	Parental
Ruby, miniature	$y^+ rb$ m/Y	143	SCO ($rb - m$)
Yellow	y rb^+ m^+/Y	135	SCO ($rb - m$)
Miniature	$y^+ rb^+ m$/Y	36	SCO ($y - rb$)
Yellow, ruby	y rb m^+/Y	31	SCO ($y - rb$)
Wild type	$y^+ rb^+ m^+$/Y	5	DCO
Yellow, ruby, miniature	y rb m/Y	3	DCO
Total		1000	

RF $(y - rb)$ = $(36 + 31 + 5 + 3)/1000 = 75/1000 = 0.075$, or 7.5%
RF $(rb - m)$ = $(143 + 135 + 5 + 3)/1000 = 286/1000 = 0.286$, or 28.6%

These values give the following map:

$\underline{y\quad 7.5\quad rb\qquad\qquad\qquad 28.6\qquad\qquad\qquad\qquad m}$

c. Coefficient of coincidence = observed/expected = $(5 + 3)/[(0.075 \times 0.286) \times 1000]$ = $8/21.45 = 0.373$. Interference = $1 - \text{COC} = 1 - 0.373 = 0.627$.

57. a. In *Drosophila*, recombination does not occur in males. Because more than 50% of the offspring are wild type, the male fly's genotype must have been $+++/acb$; 50% of his gametes would receive all three wild-type alleles, and these offspring would be wild type in appearance. The other 50% of his gametes would receive all three recessive alleles. Their phenotype would therefore be determined by the genotype of the gamete received from the female parent. The offspring that received the $+++$ gametes from the male parent are wild type and are uninformative for mapping. If we subtract these 1000 offspring from the wild-type group, we are left with 321 $+++$ flies and 318 abc flies. These are the largest classes of offspring and therefore represent the parental genotypes. The genotype of the female fly must have been $+++/acb$.

b. The least frequent classes of offspring are $ab+$ and $++c$. These represent the double crossovers. The c alleles are "flipped" with respect to the a and b alleles; therefore c is the middle gene.

Recombination frequencies are calculated as follows:

RF (*a*–*c*) = (3 + 4 + 47 + 53)/1000 = 107/1000 = 0.107, or 10.7%
RF (*c*–*b*) = (3 + 4 + 131 + 123)/1000 = 261/1000 = 0.261, or 26.1%

These values give the following map:

a 10.7 *c* 26.1 *b*

c. Coefficient of coincidence = observed/expected = (3 + 4)/[(0.107 × 0.261) × 1000] = 12/27.9 = 0.430. Interference = 1 – COC = 1 – 0.430 = 0.570.

falsch

58. a. Let *g* represent the recessive allele that causes geneticitis and g^+ represent the normal allele. Both affected siblings are homozygous for *RFLP1* and for the *g* allele. Therefore, in both families, the *g* allele is in *cis* with the *RFLP1* allele and in *trans* with the *RFLP2* allele. The probability of recombination between the geneticitis gene and the RFLP is 0.07.

We do not know anything about David's genotype for the RFLP. Because he has an affected sibling, both his parents must have been heterozygous. David is not affected with geneticitis, so he could either be homozygous for the normal allele or a heterozygous carrier. The probability that he is homozygous is 1/3, and the probability that he is a carrier is 2/3. The likelihood that David is a carrier and that he contributed the *g* allele to his offspring is 2/3 × 1/2 = 1/3.

Barb is heterozygous for *RFLP1/RFLP2* and also for g/g^+. Assume that both of her parents had the genotype *RFLP1 g/RFLP2* g^+. Both parents would be expected to produce gametes in the following proportions: 0.465 *RFLP1 g*, 0.465 *RFLP2* g^+, 0.035 *RFLP1* g^+, and 0.035 *RFLP2 g*. For Barb to be a carrier of the *g* allele, she could have one of the following genotypes:

RFLP1 g/RFLP2 g^+ 0.465 × 0.465 = 0.930
RFLP1 g^+*/RFLP2 g* 0.035 × 0.035 = 0.001

The total probability that Barb is a carrier is therefore 0.930 + 0.001 = 0.931.

The probability that Barb is a carrier and passes the *g* allele on to her child is 0.931 × 0.5 = 0.4655.

The probability that the child Barb and David are expecting will be affected with geneticitis is therefore 0.4655 × 0.3333 = 0.155.

CAN SOMEONE DO MATH?!

b. If David is homozygous for *RFLP2*, if he is a carrier, his genotype must be *RFLP2* g^+*/RFLP2 g*. Based on the probabilities calculated earlier, the likelihood of this is 0.465 × 0.035 = 0.016, and the probability that a gamete will receive the *g* allele is 0.008.

Now the probability that Barb and David's child will be affected is 0.4655 × 0.008 = 0.004.

CHAPTER INTEGRATION PROBLEM

a. Crosses I and II are reciprocal crosses, and they yield different results. Therefore, the trait must be sex-linked. Moreover, females can be backflippers and they typically do not have a Y chromosome. Therefore, the *bf* locus must be on the X chromosome.

b. If the trait were dominant, the daughters of a backflipping male would be backflippers, as these would inherit his X chromosome. However, in cross I, a backflipping male produced only wild-type female offspring. Therefore, the trait must be recessive.

c. Mendel's first law is the law of equal segregation. So let's consider segregation at each locus in cross III:
   For *bf*, the ratio of backflippers to nonbackflippers is (251 + 130):(125 + 244) or approximately 1.03:1.
   For *m*, the ratio of miniature to normal wings is (251 + 125):(130 + 244) or 1:1. Therefore, there is equal segregation at each locus.

d. If epistasis was involved, there would be an effect on the expected phenotypes of Cross III's F_2 offspring. The four phenotypes (backflipper, miniature wings; backflipper, normal wings; non-backflipper, miniature wings; and non-backflipper, normal wings) that were expected to appear did show up in the offspring. Therefore, genes *bf* and *m* do not exhibit epistasis.

e. If genes *bf* and *m* are unlinked and assorting independently, a 1:1:1:1 ratio of offspring would be expected in the F_2 generation of cross III. The actual offspring are in a ratio of approximately 2:1:1:2. Therefore, it is highly unlikely that the genes are independently assorting and far more likely for them to be linked.
   A chi-square test can certainly evaluate the likelihood of independent assortment.
   Our null hypothesis: Two genes are assorting independently, and therefore produce a 1:1:1:1 ratio in the offspring.

	Offspring				
	Backflipper, miniature	**Backflipper, normal**	**Non-backflipper, miniature**	**Non-backflipper, normal**	**Total**
Observed numbers (*O*)	251	130	125	244	750
Expected ratio	1/4	1/4	1/4	1/4	
Expected numbers (*E*)	187.5	187.5	187.5	187.5	750
$O - E$	63.5	−57.5	−62.5	56.5	
$(O - E)^2$	4032.25	3906.25	3906.25	3192.25	

$(O - E)^2/E$ 21.50   17.63   20.83   17.02 $76.98 = \chi^2$

The critical chi-square at 0.05, 3 df, = 7.815. The chi-square for our null hypothesis, 76.98, is off the chart! Therefore, we reject the hypothesis of independent assortment.

f.   The genes are linked. RF ($bf - m$) = (130 + 125)/750 = 255/750 = 0.34, or 34%. Thus, genes bf and m are 34 map units apart on the X chromosome.

g.   The two recombinant classes (backflipper, normal wings, and non-backflipper, miniature wings) are the result of a crossover in the F_1 female. This crossover occurs in the tetrad stage of prophase I of meiosis.

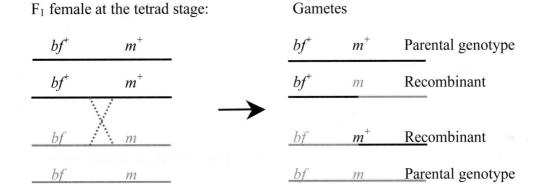

h.   The female's genotype is $bf^+ m^+/bf\, m$. The bf and m genes are 34 cM apart. Therefore, the female will produce gametes in the following proportions: 0.33 $bf^+ m^+$ (66%/2), 0.33 $bf\, m$, and 0.17 $bf^+ m$ (34%/2), 0.17 $bf\, m^+$. The male's genotype is $bf\, m^+/Y$. He will produce gametes in the following proportions: 0.5 $bf\, m^+$ and 0.5 Y.

    There are many different genotypes that will provide a phenotype that is wild type for at least one trait. For example, $bf^+ m^+/bf\, m^+$, $bf^+ m^+/Y$, and $bf\, m/bf\, m^+$. Therefore, we would have to calculate the probability of each genotype (using the product rule) and then add them all up (sum rule) to get to the overall probability.

    A much easier method is to calculate the probability of obtaining a genotype that is recessive for both traits, and then subtract that probability from 1.0 (sum of all probabilities). Daughters of this cross cannot be recessive for both traits because they inherit the $bf\, m^+$ X chromosome from their father and thus will all exhibit normal (wild-type) wings. So the only offspring that would be recessive for both traits are sons who obtained the Y chromosome from their father and the $bf\, m$ X chromosome from their mother. Probability = 0.33 ($bf\, m$ chromosome from mother) × 0.5 (Y chromosome from father) = 0.165. Therefore, the probability of an offspring that is wild type for at least one trait is = 1 – 0.165 = 0.835.

CHAPTER 7 DNA STRUCTURE AND CHROMOSOME ORGANIZATION

Chapter Goals

1. Learn the key requirements for genetic material and describe the key experiments that established the identity of the genetic material.
2. Understand the structure of DNA and its organization in chromatin.
3. Be able to distinguish the roles of the different functional units in a chromosome.

Key Chapter Concepts

7.1 Required Properties of a Genetic Material

- **The Genetic Material:** To serve as the genetic material, a molecule such as DNA must possess several characteristics. First, DNA must contain the information that controls the synthesis of enzymes responsible for cell growth, development, and functioning. Second, the genetic material must be able to self-replicate with high fidelity, yet allow a low level of mutation. Finally, the genetic material must be located in the chromosomes that carry traits from one generation to the next.

7.2 Evidence for DNA as the Genetic Material

- **Experiments of Frederick Griffith:** Frederick Griffith used two strains of *Streptococcus pneumoniae*, one virulent (S strain) and one nonvirulent (R strain). The S strain cells had an outer capsule composed of polysaccharides that protected them from the white blood cells of the host. Griffith demonstrated that the S strain could transform the R strain into virulent S strain cells by injecting the live R strain and heat-killed S strain bacteria into mice. This combination caused the mice to die, and virulent S strain cells were recovered from the dead mice.

- **Experiments of Avery, MacLeod, and McCarty:** Oswald Avery, Colin MacLeod, and Maclyn McCarty identified the transforming substance in the heat-killed S strain organisms by using different reagents in vitro to try to destroy the transforming activity of the strain. They treated the heat-killed S strain with DNase, protease, or RNase or extracted the lipids from the strain before mixing the cells with the live R strain cells and found that only treatment with DNase prevented the R strain from being transformed by the heat-killed S strain. This was the first demonstration that DNA is the genetic material of the cell.

- **Experiments of Hershey and Chase:** The experiments of Alfred Hershey and Martha Chase also supported the idea that DNA is the genetic material. They used radioactive isotopes (^{35}S) and (^{32}P) to differentially label viral proteins or nucleic acids during infection. They mixed the labeled phage particles with unlabeled bacteria and identified the phage material injected into the cell. Since the phage replicate inside the cell, the cell would contain the phage's genetic material. Hershey and Chase found that the ^{32}P label entered the cells, but the ^{35}S label did not, implying that the DNA of the phage enters the cell while the outer protein coat does not. This result suggested that the phage's DNA, not the protein, must be the genetic material.

- **RNA as the Genetic Material:** In some viruses, RNA serves as the genetic material. **Fraenkel-Conrat and Singer** showed that viruses reconstituted using RNA and proteins from different strains always produced viruses of the type associated with the RNA and not the protein after infection. This demonstrated that the RNA, not the protein, had to be the genetic material.

7.3 Chemistry of Nucleic Acids

- **Structure and Chemistry of DNA and RNA:** DNA and RNA are nucleic acids made up of **nucleotide** subunits. Each nucleotide is made of a sugar (**ribose** in RNA and **deoxyribose** in DNA), a phosphate group, and a nitrogenous base. Nucleotides are linked together by a **phosphodiester bond** between the 5′ phosphate of one nucleotide and the 3′ hydroxyl of the adjacent nucleotide. X-ray crystallography indicates that the structure of DNA is a helix. Both DNA and RNA contain the **purine** bases adenine and guanine. DNA contains the **pyrimidines** cytosine and thymine, while RNA contains the pyrimidines cytosine and uracil. In DNA, the percentage of adenine equals the percentage of thymine and the amount of cytosine equals the amount of guanine (**Chargaff's rule**) because adenine pairs with thymine and cytosine with guanine. Two hydrogen bonds form between the adenine and thymine bases, while three bonds connect the guanine and cytosine bases. **Watson** and **Crick** proposed a model structure for DNA. They described the DNA molecule as a double helix or twisted ladder with the sides of the ladder (sugar–phosphate backbones) running in opposite directions (**antiparallel**). The A=T and C≡G base pairs make up the rungs of the ladder.

- **Alternative Forms of DNA:** DNA can be either double-stranded or single-stranded and exists in three different forms: **A-DNA**, **B-DNA**, or **Z-DNA**. A-DNA and B-DNA are right-handed helices, while Z-DNA has a left-handed structure. B-DNA is the major form that is present in a cell. A-DNA is present in double-stranded RNA and in DNA–RNA hybrid molecules. Z-DNA is a unique form that may be involved in regulating gene expression in eukaryotes.

7.4 The Eukaryotic Cell

- **Complexities of the Eukaryotic Cell:** Eukaryotes are more complex than prokaryotes. Prokaryotes are primarily one-celled organisms, while eukaryotes are multicellular. Prokaryotic cells are small with little internal structure and contain a single, circular chromosome. Eukaryotes have larger cells and a number of internal organelles, and they contain a larger number of linear chromosomes with a thousand times more DNA than prokaryotes. The chromosomal DNA in a typical prokaryote, such as *E. coli*, is not complexed with proteins, and the genes are grouped in operons that are regulated together. Eukaryotic DNA is complexed with proteins, and packaged into chromatin. The genes in eukaryotes are typically separated and regulated by independent mechanisms.

7.5 The Eukaryotic Chromosome

- **Eukaryotic Chromosomes:** Each eukaryotic chromosome contains one double helix of DNA that is packed and coiled around many different proteins to compact the long DNA molecule. The first level of structure involves DNA interacting with a ball of proteins called histone proteins. The DNA wraps twice around this protein ball to form a structure called a **nucleosome**. Each DNA molecule interacts with many **histones**. This level of structure is sometime referred to as "pearls on a string" because of the way the nucleosomes appear. Between nucleosomes (pearls) are stretches of **linker DNA** (string). These structures can influence transcription. The DNA–protein structure is further coiled into a solenoid, which resembles a "slinky." This **solenoid** is then coiled into yet another solenoid or "slinky" configuration before it interacts with the scaffold. The scaffold structure of the chromosome is formed by nonhistone proteins and gives the chromosome its characteristic shape. **Polytene chromosomes** result from replication and synthesis of numerous chromatids without cell division. The dark staining regions of chromatin condensation that occur on eukaryotic chromosomes during early mitosis or meiosis are called **chromomeres**. In polytene chromosomes, several chromatids aligned laterally will appear as a dark band. The diffuse, uncoiled regions in polytene chromosomes represent regions where transcription is actively occurring and are referred to as **chromosome puffs**. **Lampbrush chromosomes** have a looped-out configuration due to the elaboration of mRNA and are found in amphibian oocytes. The DNA that is transcribed is referred to as **euchromatin**. Chromatin that does not contain active genes is called **heterochromatin**, while the repetitive DNA that surrounds the centromere is referred to as **constitutive heterochromatin**.

- **Telomeres:** The regions at the ends of a chromosome are referred to as **telomeres**. Most are repetitions of a 5–8 base sequence and are required for chromosome stability and replication. These sequences are added to chromosomes by the enzyme **telomerase**. In the cells of the majority of higher organisms, the telomeres shorten with each cell division because telomerase is not active. Telomeres are hypothesized to be the "clock" that controls the number of times a cell divides.

- **C-Value Paradox:** The C-value paradox refers to the phenomenon that DNA quantity does not correlate with organism complexity. DNA is categorized as unique or repetitive DNA. Some sources estimate that about 97% of eukaryotic DNA does not code for RNA (junk DNA) and is highly repetitive; that is, it contains many copies of the same nucleotide sequence. Some repetitive DNA contains multiple copies of nearly identical genes with similar functions, such as the β-globin gene family. Data from the human genome sequence indicate that about 1% of the human genome is in exons, 24% is in introns, and 75% is intergenic DNA.

Key Terms

- active site
- mutability
- fidelity
- transformation
- nucleotides
- nucleoside
- purines
- pyrimidines
- adenine
- cytosine
- guanine
- thymine
- uracil
- phosphodiester bonding
- tetranucleotide hypothesis
- Chargaff's rule
- X-ray crystallography
- complementarity
- antiparallel

- B-DNA
- A-DNA
- Z-DNA
- nucleosomes
- histone
- chromatosome
- chromatin
- 30-nm solenoid fiber
- radial loop–scaffold organization
- scaffold-associated regions (SARs)
- endomitosis
- polytene chromosomes
- lampbrush chromosomes
- Giemsa stain
- satellite DNA
- euchromatin
- constitutive heterochromatin

- facultative heterochromatin
- centromere
- telomere
- point centromere
- regional centromeres
- telomeres
- 3′ overhang
- G-tetraplex
- t-loop
- C-value paradox
- transposable genetic elements
- retrotransposons
- long interspersed elements (LINES)
- short interspersed elements (SINES)
- junk DNA

Understanding the Key Concepts

Use the key terms or parts of their definitions to complete the following sentences.

Three basic requirements for the genetic material of the cell are that it be associated with the (1)_____; it must contain the necessary information to direct the (2)_____ of cellular proteins; and it has to be capable of (3)_____ with high (4)_____, while allowing some (5)_____ to occur permitting the genetic material of an organism to change. Many scientists at first thought that (6)_____ were the most probable genetic material of the cell because they are more complex than (7)_____ which contains only (8)_____ different (9)_____. Experiments by Griffith, Avery, MacLeod, McCarty, and Hershey and Chase demonstrated that (10)_____ is the most common genetic material. In some cases, (11)_____ can also act as the genetic material. Without exception, the genetic material of all organisms is presumed to be (12)_____ or (13)_____, but it has been discovered that in some diseases, such as Creutzfeldt–Jakob disease, the infectious material is a (14)_____ .

According to the (15)_____ model, DNA has a (16)_____ structure consisting of two (17)_____ strands. Each strand consists of (18)_____ that are joined by (19)_____ bonds. (20)_____ between the bases on one strand and the (21)_____ bases on the opposite strand hold the two strands of the DNA molecule together. Each nucleotide of DNA consists of a (22)_____, a (23)_____, and a (24)_____ base. The backbone of the DNA double helix consists of the sugar–(25)_____ linkages, while the (26)_____ protrude to the inside of the helix. The bases can be one of two (27)_____, such as adenine or (28)_____, or one of two (29)_____, such as (30)_____ or (31)_____.

　　　　Eukaryotic DNA is packaged to reduce its length. Chromosomal DNA is wrapped around a core of (32)_____ proteins, consisting of an octamer of (33)_____, (34)_____, (35)_____, and (36)_____. The (37)_____ core particle associates with (38)_____ to form the (39)_____. It is then wrapped in a helical fashion to form a (40)_____, which is then packaged into a (41)_____ organization. This packaging of the chromatin compacts the DNA (42)_____-fold. There are (43)_____ types of chromatin. (44)_____ is associated with transcriptionally active genes. (45)_____ is located around the (46)_____ and is transcriptionally (47)_____, while (48)_____ can shift between an active and an inactive state.

　　　　The size of the genome does not always correspond to the (49)_____ of the organism. This is the basis of the (50)_____. Many organisms contain large amounts of (51)_____ that has no obvious function but is not detrimental to the organism. Some of this DNA can move around in an organism's genome. These (52)_____ introduce (53)_____ sequences into different parts of the genome. Some of these, such as (54)_____ and (55)_____ are (56)_____ that move by way of a (57)_____ intermediate. An inability to remove elements such as these may account for the large differences in the size of different organisms' genomes.

Figure Analysis
Use the indicated figure or table from the textbook to answer the questions that follow.

1. **Figure 7.4**
(a) Why does one strain (S) of *S. pneumoniae* cause a fatal bacterial infection in mice while the other strain (R) does not?

(b) Which of Griffith's injections caused the mice to die?

(c) Why did heat-treated S-type cells fail to kill the mice upon injection?

(d) Why were the heat-killed S-type cells able to kill the mice when they were injected along with live R-type cells?

2. **Figure 7.5**
(a) How did Avery, MacLeod, and McCarty confirm Griffith's result?

(b) What is a fundamental difference between how Avery, MacLeod, and McCarty demonstrated that the heat-killed S strain could transform the live R strain cells and how Griffith demonstrated it?

(c) Why were both live S strain and live R strain cells obtained in the Petri plates where heat-killed S strain cells were mixed with live R strain cells, rather than all S strain cells?

(d) What four treatments were used on the heat-killed S strain cells before they were mixed with live R strain cells to try to kill the transforming activity?

(e) What was the only reagent that killed the transforming activity? Explain why.

3. **Figure 7.6**
(a) How did Hershey and Chase differentially label viral proteins and nucleic acids?

(b) Based upon what understanding did Hershey and Chase reason that this would be possible?

(c) What was done with the labeled T2 phage?

(d) Which label was present in the progeny phage?

(e) How was this result interpreted by Hershey and Chase?

4. Table 7.5 and Figure 7.22
(a) Why are histones well-suited to bind to DNA?

(b) If chromatin is treated with large amounts of DNase, what will occur?

(c) What is the effect of denaturing DNase-digested chromatin?

(d) What would be the result of denaturing the chromatin following DNase treatment?

(e) Following DNase treatment, if chromatin is digested by mild micrococcal nuclease treatment, what will occur? What will be produced if the DNase-digested chromatin is denatured following mild micrococcal nuclease treatment?

(f) If chromatin is digested with micrococcal nuclease under harsh conditions following DNase treatment, what will occur? What would be the final products obtained following denaturation?

General Questions
Q7-1. You are provided with three samples of DNA and asked to determine which of the three has the highest G-C ratio. How can you quickly make this assessment?

Q7-2. What evidence indicates that adenine pairs with thymine and cytosine pairs with guanine in the DNA double helix?

Q7-3. Which of the following types of sequences are most common in human nuclear DNA: unique sequences functioning as genes, regulatory sequences in the DNA, or repetitive sequences?

Q7-4. Describe three telomeric structures that can be formed to protect the 3′ single-stranded overhang of a linear chromosome from degradation.

Q7-5. What is the function of the centromere?

Multiple Choice
For each of the following, circle the letter of the choice that most appropriately answers the question.

1.) In a nucleotide, a nitrogenous base is attached to the _____ of the sugar molecule?
  a. 1′
  b. 2′
  c. 3′
  d. 4′
  e. 5′

2.) Nucleotides are linked together to form DNA or RNA chains by the formation of a phosphodiester bond between a phosphate attached to the 5′ carbon of one nucleotide and a _____ of a second nucleotide.
  a. PO_4^- at the 5′ carbon
  b. OH at the 3′ carbon

c. H at the 3′ carbon

d. OH at the 2′ carbon

e. PO₄⁻ at the 3′ carbon

3.) According to Chargaff's rule, a molecule of DNA that consists of 15% adenine, must contain what percentage of cytosine?

a. 15%

b. 20%

c. 25%

d. 30%

e. 35%

4.) Which of the following did Erwin Chargaff demonstrate to be true about the base composition of the DNA of different organisms?

a. $A + T = G + C$

b. $A = C$; $G = T$

c. $A + C = T + G$

d. $A = G$; $C = T$

e. $A = C = G = T$

5.) Which of the following statements is true of Watson–Crick base-pairing?

a. Adenine forms two hydrogen bonds with cytosine; guanine forms three hydrogen bonds with thymine.

b. Adenine forms two hydrogen bonds with thymine; cytosine forms three hydrogen bonds with guanine.

c. Cytosine forms two hydrogen bonds with thymine; adenine forms three hydrogen bonds with guanine.

d. Purines base-pair with other purines to form two hydrogen bonds.

e. Any purine can base-pair with any pyrimidine to form three bonds.

6.) Which of the following is *not* true of B-DNA?

a. The bases are stacked perpendicular to the main axis.

b. The diameter of a helix is 20 angstroms.

c. It forms a right-handed helix.

d. It has 11.3 base pairs per turn.

e. It is the major form of cellular DNA.

7.) Chromosomal puffs are formed as a result of which of the following?

a. areas of Z-DNA adjacent to areas of B-DNA

b. large amounts of transcription

c. additional packaging of the chromatin

d. protection of the 3′ single-stranded overhang

e. location of the centromere

8.) Which of the following sequences is complementary to the sequence of the telomeric repeat found at the end of each human chromosome?

a. 5′-AAATCC-3′

b. 5′-TTTAAA-3′

c. 5′-TAACCC-3′

d. 5′-CCCTAA-3′

e. 5′-CCCATT-3′

9.) If the sequence of one strand of DNA is 5′-GCATCAGCTCATAG-3′, the sequence of the complementary strand is which of the following?

a. 5′-GCATCAGCTCATAG-3′

b. 5′-GATACTCGACTACG-3′

c. 5′-CTATGAGCTGATGC-3′

d. 5′-CGTAGTCGAGTATC-3′

e. None of the above

10.) The nucleosome consists of which of the following?

a. one molecule of H1, H2A, H2B, H3, H4 and 165-bp of DNA

b. two molecules of H1, H2A, H2B, H3, H4 and 165-bp of DNA

c. two molecules of H2A, H2B, H3, H4 and 145-bp of DNA

d. one molecule of H1; two molecules of H2A, H2B, H3, H4; and 145-bp of DNA

e. one molecule of H1; two molecules of H2A, H2B, H3, H4; and 165-bp of DNA

Practice Problems

P7-1. Examine the following generalized representation of a strand of nucleic acid. Determine if the strand is DNA or RNA. Identify the 5′ and 3′ ends of the molecule.

(*Hint 1*: RNA can be identified by the presence of uracil and the absence of thymine. DNA contains thymine but not uracil.)

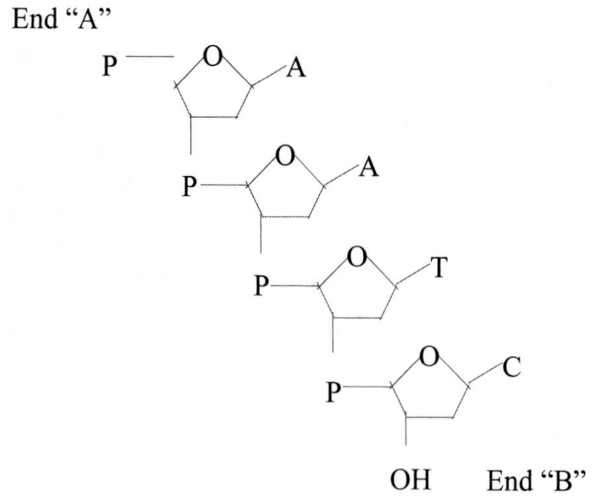

End "A"

OH   End "B"

P7-2. Chargaff's ratios were instrumental in the development of the Watson–Crick model of DNA structure. Assume that Chargaff's data were as follows: 35% A, 20% T, 20% C, and 35% G. Assume further that the data of Wilkins and Franklin were unchanged. What conclusions, if any, would have been made by Watson and Crick?

[*Hint 2*: Chargaff's rules provide a means of determining the base composition of the double-stranded DNA molecule. Knowledge of the amount of one base can be used to determine the percentage of the remaining bases in the molecule. For example, 10% adenine indicates that the molecule contains 10% thymine. The remaining 80% of the bases are guanine and cytosine, in approximately equal proportions (40% C and 40% G).]

P7-3. Consider the following DNA base compositions: $(A + G)/(T + C) = 1.0$ and $(A + T)/(C + G) = 1.5$. Is the molecule single-stranded or double-stranded?

(*Hint 3*: Chargaff's rules can be used to determine if a molecule is single-stranded or double-stranded. The amount of adenine equals the amount of thymine, and the amount of cytosine equals the amount of guanine only in a double-stranded DNA molecule.)

P7-4. Provide the strand that is complementary to the following one.
       3′ TAAGCTTAACGCGCATCAGCGCGCTCCGATCGATCGATCAGCGATCAGAGGGAAG 5′

P7-5. How would the results of Griffith, Avery et al., and Hershey–Chase have differed if the genetic material was RNA rather than DNA? How would they have differed if the genetic material was protein rather than DNA?

P7-6. Examine the base composition of each of the following DNA molecules. Order the molecules (only one strand is shown) according to the relative temperature required to denature the molecule (highest to lowest).

(*Hint 4*: The temperature at which DNA denatures, or separates into single strands, provides insights to the base composition of the molecule. The three hydrogen bonds between cytosine and guanine are more difficult to break than the two bonds between adenine and thymine. Therefore, the higher the temperature needed to denature DNA, the higher the guanine–cytosine composition.)
    a. GCTATAATAC
    b. GGCTATCGGG
    c. TTATCTAATT

P7-7. Which of the following features are associated with bacterial chromosomes? Which are found in eukaryotic chromosomes? Which of the features are characteristic of both cell types?
    a. Histone proteins   e. Nonhistone protein scaffolds
    b. Pentose sugars   f. Linear chromosomes
    c. Nitrogenous bases   g. Circular chromosomes
    d. Telomeres   h. Nucleosomes

P7-8. Consider a hypothetical organism with a haploid genome consisting of 10^8 nucleotide pairs. Eighty-five percent of the genome is unique-sequence DNA, and each gene is represented once. If the average gene sequence is 10^2 nucleotide pairs, how many different genes are potentially encoded in the region described?

P7-9. One-third of the length of a metacentric eukaryotic chromosome consists of satellite DNA. The G-band staining pattern is shown. The centromere is represented by an open circle. Show the expected R-band and C-band patterns for this chromosome.

(*Hint 5*: The chemical composition of specific chromosomal regions accounts for differences in chromosome staining.)

Assessing Your Knowledge

Understanding the Key Concepts—Answers

1.) chromosomes; 2.) synthesis; 3.) self-replicating; 4.) fidelity; 5.) mutations; 6.) proteins; 7.) DNA; 8.) four; 9.) bases; 10.) DNA; 11.) RNA; 12–13.) DNA; RNA; 14.) protein; 15.) Watson–Crick; 16.) double-helical; 17.) antiparallel; 18.) nucleotides; 19.) phosphodiester; 20.) Hydrogen bonding; 21.) complementary; 22–23.) phosphate; sugar; 24.) nitrogenous; 25.) phosphate; 26.) bases; 27.) purines; 28.) guanine; 29.) pyrimidines; 30–31.) cytosine; thymine; 32.) histone; 33–36.) H2A; H2B, H3, H4; 37.) nucleosome; 38.) histone H1; 39.) chromatosome; 40.) 30-nm solenoid fiber; 41.) radial loop–scaffold;

42.) 10,000; 43.) three; 44.) Euchromatin; 45.) Constitutive heterochromatin; 46.) centromeres; 47.) inactive; 48.) facultative heterochromatin; 49.) complexity; 50.) C-value paradox; 51.) junk DNA; 52.) transposable genetic elements; 53.) repetitive; 54–55.) LINES; SINES; 56.) retrotransposons; 57.) RNA.

Figure Analysis—Answers

1a.) The S strain of *S. pneumoniae* causes a fatal bacterial infection in mice because its cells have outer capsules composed of polysaccharide that protect them from the white blood cells. The R strain does not have this polysaccharide coat and thus can be engulfed by the mice's white blood cells; b.) The mice died when they were injected either with S-type cells alone or with R-type cells plus heat-killed S-type cells; c.) The heat-treated S-type cells failed to kill the mice because the S-type bacteria were killed by the heat treatment and thus could not propagate inside of the mice; d.) When the heat-killed S-type cells were injected along with live R-type cells, some of the DNA from the S-type cells was taken up by the R-type cells transforming them into S-type cells. These transformed R-type cells killed the mice.

2a.) Avery, MacLeod, and McCarty confirmed Griffith's result by demonstrating that live S-type and R-type cells were obtained upon plating when heat-killed S-type cells were mixed with live R-type cells; b.) A fundamental difference between Avery, MacLeod, and McCarty's experiment and Griffith's experiment was that Avery, MacLeod and McCarty's experiment was done completely in vitro. While Griffith injected the two types of *S. pneumoniae* into live mice and looked at whether the mice lived or died, Avery, MacLeod, and McCarty combined the two types of *S. pneumoniae* in vitro and plated the mixture onto Petri plates to look at the type of living bacteria; c.) Both live S strain and live R strain cells were obtained on the Petri plates because not all of the R-type cells were transformed by the S strain DNA. This is due to the fact that the transformation is a relatively rare event, so only some of the R-type cells take up the S-type DNA. For all S-type cells to be obtained on the plate, all of the live R-type cells would have had to take up the S-type DNA; d.) To try to kill the transforming activity of the heat-killed S-type cells before they were mixed with the live R-type cells, Avery, MacLeod, and McCarty treated the heat-killed S-type cells with protease (to remove proteins), Rnase or DNase, or they extracted the cells to remove lipids; e.) The only reagent that killed the transforming activity of the heat-killed S-type cells was DNase. This is due to the fact that the DNA of the S-type cells was responsible for transforming the nonvirulent R-type cells into a virulent form. Destruction of the S-type proteins, lipids, and RNA had no effect because these cellular components did not contain the genetic information that ultimately rendered the S-type cells to be virulent.

3a.) Hershey and Chase differentially labeled viral proteins and nucleic acids by infecting *E. coli* that were being grown in media containing ^{35}S or ^{32}P with T2 bacteriophage; b.) Hershey and Chase reasoned that it would be possible to differentially label viral proteins and nucleic acids on the understanding that all nucleic acids, but not proteins, contain phosphorus, while most proteins, but not nucleic acids, contain sulfur. Therefore ^{32}P would be incorporated into the nucleic acids, but not into proteins, and ^{35}S would be incorporated into proteins, but not nucleic acids; c.) The differentially labeled T2 phage were used to infect bacteria grown in nonradioactive media to determine if the labeled protein or the labeled nucleic acid entered the cell during infection; d.) The ^{32}P label was present in the progeny phage upon infection of bacteria grown in nonradioactive media with labeled T2 phage; e.) Since phage replicated inside of the bacterial cell, the phage's genetic material had to be transferred into the bacterial cell. Only the ^{32}P was found to enter the bacterial cells upon phage infection and the ^{35}S did not, implying that the outer protein coat of the phage does not enter the bacterial cell that it infects and therefore protein cannot be responsible for the production of new phages during the infection process. The ^{32}P in the DNA does enter the bacterial cell that it infects, and therefore it is the genetic material responsible for the production of new phages.

4a.) The histones are well-suited to bind to DNA because they are proteins that are lysine- and arginine-rich. Lysine and arginine are two basic amino acids that are positively charged, and therefore proteins that are rich in these residues are particularly well-suited to bind to the negatively charged DNA; b.) If chromatin is treated with large amounts of DNase, the linker DNA will be degraded leaving the chromatin fragmented into pieces of DNA wrapped around a complex of histone proteins; c.) If the DNase-digested chromatin is denatured, the DNA is released from the histone proteins; d.) If the chromatin is denatured following DNase treatment, the H1 protein, the histone core proteins, and fragments of 210-bp of DNA would result; e.) If chromatin is digested by mild micrococcal nuclease treatment following DNase treatment, additional DNA will be removed from the ends of the DNA that is wrapped around the histone core. This digestion will produce chromatosome subunits. If the DNase-digested chromatin is denatured following mild micrococcal nuclease treatment, the H1 protein, the histone core proteins, and fragments of 165-bp of DNA would result; f.) If the chromatin is digested with micrococcal nuclease under harsh conditions following DNase treatment, the H1 protein will be released and additional DNA (relative to mild micrococcal nuclease treatment) will be digested away from the ends of the DNA that is wrapped around the histone core. This will produce the nucleosome and the H1 protein. The final products obtained following denaturation would be the histone core proteins, and 145-bp of DNA.

General Questions—Answers

Q7-1. At elevated temperatures, DNA denatures. The relative temperature at which a DNA molecule denatures is related to its specific base composition (*Hint 4*). There are three hydrogen bonds joining cytosine and guanine, while only two hydrogen bonds form between adenine and thymine. Three hydrogen bonds require more energy to break than two hydrogen bonds. Therefore, G–C pairs denature at a higher temperature than A–T bonds. Determine at what temperature the three samples denature to determine which has the highest G–C content.

Q7-2. The specific pairing of bases in DNA is supported by two pieces of evidence: Chargaff's ratios and X-ray crystallography data. Chargaff found that the ratio of purines to pyrimidines was equal and that the amount of adenine equals that of thymine while the amount of guanine equals that of cytosine. This evidence suggests that the molecule is double-stranded with specific paring between adenine and thymine and between guanine and cytosine. X-ray crystallography also suggested that the structure was a double helix. The measurements made correspond with the pairing of a purine and a pyrimidine.

Q7-3. Most human nuclear DNA is made up of repetitive sequences. The function of these repeating sequences is unknown.

Q7-4. The 3′ single-stranded overhang can be protected by the formation of a planar G-tetraplex where four guanines in the guanine-rich DNA of the telomeric repeat are hydrogen-bonded to each other. The 3′ ends of telomeres can also be bound by the telomere end-binding protein (TEBP) that protects them, or they can be bound by a protein called TRF2 that can cause the 3′ overhang to loop around and interdigitate into the double helix to form a t-loop that hides the end from nucleases that may degrade it.

Q7-5. The centromere serves as the attachment point for the spindle fibers during nuclear division, and they are essential in keeping the sister chromatids together during mitosis and meiosis.

Multiple Choice—Answers

1.) a; 2.) b; 3.) e; 4.) c; 5.) b; 6.) d; 7.) b; 8.) d; 9.) c; 10.) c.

Practice Problems—Answers

P7-1. This molecule contains thymine (T) and is therefore composed of DNA. If the molecule were RNA, the base uracil (U) would have been present (*Hint 1*). End "A" contains a phosphate group indicating that this is the 5′ end of the molecule. End "B" contains an exposed –OH group indicating that it is the 3′ end of the molecule.

P7-2. The data provided indicate that the bases pair in the following way: A–G and T–C. In this case, the two pyrimidine bases pair and the two purine bases pair. This data would still indicate that the molecule is double-stranded. However, this base pairing would provide the incorrect thickness for the molecule based on the data of Wilkins and Franklin.

P7-3. In a double-stranded DNA molecule, $(A + G) = (T + C)$ because of the specific base pairings in DNA. In this case, $(A + G)/(T + C) = 1.0$. The data given are consistent with a double-stranded DNA molecule. This molecule has a high A–T content because $(A + T)/(C + G)$ exceed 1.0.

P7-4.
5′ ATTCGAATTGCGCGTAGTCGCGCGAGGCTAGCTAGCTAGTCGCTAGTCTCCCTTC 3′

P7-5. Regardless of the nature of the genetic material, the results of Griffith would not have differed. Griffith identified a "transforming" substance but did not identify which material was responsible for the transformation. In contrast, the results of Avery et al. would have differed significantly if DNA were not the genetic material. For example, if RNA had been the genetic material, the cell fraction containing RNA would have transformed bacterial cells. Likewise, the fraction containing proteins would have been biologically active if the genetic material were protein. The results of the Hershey–Chase experiments would have been identical if the genetic material were RNA because both contain phosphorus. The results would have differed if the material were protein. In this case, the progeny phage resulting from a cell provided with sodium-labeled medium would be strongly labeled.

P7-6. Molecule (b) has the highest G–C content and will denature at the highest temperature. Molecule (a) will denature at an intermediate temperature. Molecule (c) has the highest A–T content and will denature at the lowest temperature. Refer to question 7-1 for an additional explanation concerning the relationship between base composition and temperature.

P7-7. Pentose sugars (b) and nitrogenous bases (c) are characteristic of both prokaryotic and eukaryotic chromosomes. Only the prokaryotic chromosome is circular (g). Histone proteins (a), telomeres (d), scaffolds (e), linear structure (f), and nucleosomes (h) are all associated with eukaryotic chromosomes.

P7-8.　　Size of unique-sequence area　=　0.85×10^8 base pairs
　　　　Number of genes　　　　　　=　0.85×10^8 base pairs / 1×10^2 base pairs per sequence
　　　　　　　　　　　　　　　　　=　850,000 or 8.5×10^5 different sequences.

P7-9.

R-band pattern

C-band pattern

7

DNA STRUCTURE AND CHROMOSOME ORGANIZATION

CHAPTER SUMMARY QUESTIONS

1. 1. c, 2. b, 3. f, 4. e, 5. g, 6. i, 7. d, 8. j, 9. a, 10. h.

2. Avery, MacLeod, and McCarty performed experiments showing that DNA was the transforming agent, and they are thus generally given credit for formalizing the notion that DNA, not protein, is the genetic material. Chargaff, Hershey and Chase, Fraenkel-Conrat, and several others also helped shape the general view. At the time that Watson and Crick published their model, the scientific community knew that DNA was the genetic material but didn't know its structure.

3. DNA is located in chromosomes, has a structure that is easily and accurately replicated, and has the sequence complexity to code for thousands to tens of thousands of genes that a eukaryotic organism has.

4. 3'-GTAATCTGGCCATCTG-5'.

5. Sugars: DNA has deoxyribose; RNA has ribose. Bases: DNA has thymine in place of uracil; RNA has uracil in place of thymine.

6. Only proteins and nucleic acids were ever considered seriously.

7. See figure 7.18.

8. In general, prokaryotes are small, have a relatively small circular chromosome, and have little internal cellular structure compared to eukaryotes. Most prokaryotic messenger RNAs are polycistronic, under operon control; eukaryotic messenger RNAs are highly processed, monocistronic, and usually not under operon control. Prokaryotes are mostly single-celled organisms, whereas eukaryotes are mostly multicellular. Eukaryotes have repetitive DNA, absent for the most part in prokaryotes. Prokaryotic chromosomes are not complexed with protein to anywhere near the same extent that eukaryotic chromosomes are.

9. The simplest explanation is the difference in the amount of the genetic material in prokaryotes and eukaryotes. Because the average human chromosome has several centimeters of DNA, that DNA must be contracted to a size in which it can be moved during mitosis and meiosis without tangling and breaking. Nucleosomes provide the first order of coiling, and then several levels of coiling of the nucleosomal DNA bring it down to a manageable size for nuclear divisional processes.

10. The evidence for the uninemic (single DNA molecule) nature of the eukaryotic chromosome is summarized in figures 7.19 and 7.20.

11. The major protein components of eukaryotic chromosomes are the histones, which form nucleosomes, compacting the DNA, and nonhistone proteins, which form the scaffold and have other functions.

12. The length of DNA associated with nucleosomes was determined by footprinting, in which free DNA was digested, leaving only those segments protected by nucleosomes.

13. The nucleosome is composed of two each of histones H2A, H2B, H3, and H4. The H1 histone interacts with the DNA as it enters and exits the nucleosome, so it may serve to cross-link nucleosomes.

14. See figures 7.24, 7.27, and 7.28 for the relationship of the 110-Å, 30-nm, and 240-nm chromosome fibers.

15. G-bands and C-bands are illustrated in figures 7.35 and 7.36. R-bands are the reverse of G-bands. Band structures are summarized in table 7.6.

16. See figure 7.35.

17. a. A centromere is a visible constriction of the chromosome, and the point of attachment of sister chromatids.

    b. A kinetochore is a proteinaceous structure located at the centromere and that connects the chromosome to the microtubules of the spindle.

    c. A telomere is the end of a linear chromosome.

    d. A polytene chromosome is a giant chromosome consisting of bundles of chromatids formed after many rounds of endomitosis (DNA replication but no nuclear division). Polytene chromosomes are characteristic of Diptera such as *Drosophila*.

18. A chromosomal puff can be stage-specific, tissue-specific, constitutive, or environmentally induced. It is an area of active transcription in a polytene chromosome.

19. They can form complex DNA structures on the overhang such as G-tetraplexes; they can form t-loops, or they can produce end binding proteins such as TEBP.

20. The C-value paradox involves the issues of the excessive amounts of DNA in eukaryotic cells and the difference between eukaryotic species that seem to have similar complexity. It is explained by the large amount of structural DNA in chromosomes as well as the large amounts of short and long interspersed elements (SINEs and LINEs).

EXERCISES AND PROBLEMS

21. a. Double-stranded DNA, **b.** single-stranded DNA, **c.** double-stranded RNA, **d.** single-stranded DNA, **e.** single-stranded RNA, **f.** DNA–RNA hybrid. First, look at what bases are present. If T is present, the molecule is DNA, and if U is present, the molecule is RNA. Remember that for a double-stranded molecule, A must equal T (U) and G must equal C. For molecule (a), we see that A = T and G = C. In molecule (b), purines (A + G) > pyrimidines (C + T). In double-stranded molecules, purines = pyrimidines. Molecule (c) is consistent with double-stranded RNA. In molecules (d) and (e), G is not equal to C; therefore, the molecules are single-stranded. In molecule (f), T and U are present, such that T + U = A. A DNA–RNA hybrid has U opposite A in RNA and T opposite A in DNA: thus A = T + U.

22. a. 28% G, 28% C, 22% A, 22% T.

    b. Same percentages except 22% U, 0% T. Chargaff's rule states that the quantity of A = T and the quantity of G = C. If G = 28%, then C = 28% and G + C = 56%. The sum of all bases must equal 100%. Therefore, (A + T) = 100 – 56 = 44. Since A = T, (1/2)(44) = 22%. This is the amount of both A and T. For an RNA molecule, proceed the same way, except remember that U replaces T, so we have 22% U.

23. ^{32}P and ^{35}S provided a means to distinguish between bacteriophage DNA and proteins. DNA contains a lot of phosphorus (in its sugar–phosphate backbone), but no sulfur. Therefore, ^{32}P will specifically label DNA, but not proteins. Proteins, on the other hand, contain sulfur in the amino acids cysteine and methionine and usually no phosphorus (with the exception of phosphorylated proteins). Therefore, ^{35}S will specifically label proteins, but not DNA.

^{3}H and ^{14}C would not have worked in the Hershey–Chase experiment. Both DNA and proteins have large amounts of hydrogen and carbon, so it is not possible to selectively label these macromolecules using hydrogen and carbon isotopes.

24. The genetic code would somehow be read in number of tetranucleotide units, in which each unit consists of one each of the four bases (G, C, T, A). For example, one unit might be the amino acid alanine, two units might be the amino acid arginine, and so on.

25. For a double-stranded DNA molecule, A must equal T and G must equal C. There are 60 A's and 105 G's, so there must be 60 T's and 105 C's. Therefore, the molecule contains 60 + 60 + 105 + 105 = 330 bases or 165 base pairs.

26. (Lowest) 69°C, 73°C, 78°C, 82°C, 84°C (highest). Remember that a G–C pair has three hydrogen bonds and thus requires more energy to be broken than an A–T pair. Therefore, the higher the melting temperature, the higher the G–C content. Simply arrange the molecules from lowest to highest melting temperature.

27. a. Not possible because there would be a need for A's to establish complementarity.
   b. Not possible because there would be a need for T's and C's for complementarity.
   c. Possible because of the complementary relationship between C and G.

28. There must be regions of complementarity within the single-stranded regions. A melting temperature indicates some regions that are double-stranded. We can envision at least two different possible configurations:
1. Whole molecule complementary

2. Fragment complementary

In fact, most single-stranded molecules have some regions that are complementary.

29. A quick and easy way to solve this problem is by arbitrarily assigning base percentages and then substituting these values into the various equations to determine if they are true. Remember that A = T and C = G in double-stranded DNA. Let's say for example that A and T are each 20%, while C and G are 30% each. Equation **(a)** can be rewritten as (20 + 20)/(30 + 30) = 1, which is obviously incorrect. Equation **(b)** on the other hand is correct: (20 + 30)/(30 + 20) = 1. Using similar logic, equations **(c)**, **(d)**, and **(e)** will be found to be true. Equation **(f)** is incorrect.

30. In most cell types in higher eukaryotes, the number of divisions is finite. In these cells, telomerase is inactive. As the divisions proceed, the telomeres shorten until a length is reached that somehow signals the cell to stop dividing. If a mutation arose such that telomerase became active in an inappropriate cell type, the telomeres could have a functional length reestablished each generation, signaling the cell to keep dividing. The result of uncontrolled cell growth is cancer.

31. Each position can be occupied by one of four bases: (A, C, G, or T). Therefore, a five-base stretch of DNA will have $(4)^5 = 1024$ possible sequences (product rule is used).

32. If one strand contains 30% A, 20% C, 10% G, and 40% T, the complementary DNA strand will contain 30% T, 20% G, 10% C, and 40% A. Therefore, the entire molecule will contain 35% A, 15% C, 15% G, and 35% T (basically, the average of the two strands).

33. Assume that each chromosome contained two complete copies of the same DNA. Following the protocol of figure 7.19, the final results would be several chromosomes, before separation, that consisted of either two labeled chromatids or only one labeled chromatid, in a 1:1 ratio (barring sister chromatid exchanges). The labeled chromatid in chromosomes with just one chromatid labeled will have twice the label of each chromatid in the chromosomes in which both chromatids are labeled (see the following figure).

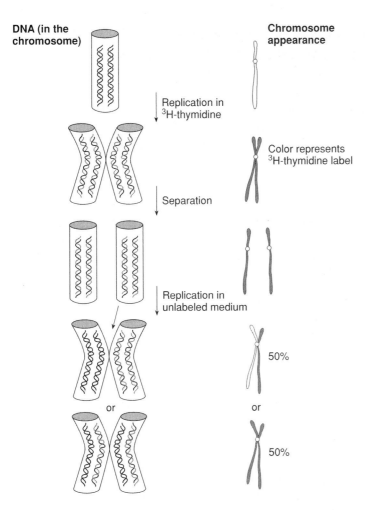

DNA (in the chromosome)

Replication in
³H-thymidine

Separation

Replication in
unlabeled medium

or

Chromosome appearance

Color represents
³H-thymidine label

50%

or

50%

34. a. The chromosome contains a total of 6×10^8 nucleotides (each base pair contains two nucleotides), and therefore 6×10^8 phosphorus atoms (one per nucleotide).

   b. B-DNA contains 10 base pairs per 3.4 nm (one helical turn). Thus, each base pair represents 0.34 nm. The length of the DNA in this chromosome would be $(3 \times 10^8 \text{ bp}) \times (0.34 \text{ nm/bp}) = 1.02 \times 10^8$ nm. To convert to micrometers, simply divide by 1000. Therefore, the genome will be 1.02×10^5 μm.

35. a. Both measurements should yield approximately the same size molecule.

   b. Method 1 will yield a DNA size larger than that determined by method 2. If the chromosome is composed of only one molecule, both methods give total DNA per chromosome. If the chromosome has more than one molecule, each molecule (method 2) will be less than the total amount of DNA per chromosome (method 1). Assume for example that the chromosome has two identically sized molecules. The molecular size determined by method 2 will be one-half the size calculated by method 1.

36. Nucleases will not digest DNA to which proteins, in this case histones, are bound. Approximately 200 base pairs must be wrapped around proteins, and there must be some unprotected regions between these 200 base-pair regions. Multiples of 200 base pairs appear because the nuclease does not cut at every unprotected sequence.

37. **a.** The total amount of DNA associated with a nucleosome = 146 bp (wrapped around the octamer) + 65 bp (linker region) = 211 bp. Therefore, the *Cryptococcus* genome will consist of 2.1×10^7 bp/211, or approximately 100,000 nucleosomes.

  b. Approximately 800,000 core histone molecules (8 per nucleosome).

38. A = 13.3%, G = 26.7%, C = 15%, T = 45%. Begin by writing the given information as equations:

  G = 2A; T = 3C; (C + T)/(A + G) = 1.5

  Now substitute for G and T in the third equation:

  (C + 3C)/(A + 2A) = 1.5

  4C/3A = 1.5

  4C = 4.5A

  C = 4.5A/4

  Now, remember that the sum of all bases must = 100%.

  A + G + C + T = 100

  Substitute equalities into this equation:

  A + 2A + (4.5A/4) + (13.5A/4) = 1

  3A + 4.5A = 1

  7.5A = 1

  A = 1/7.5 = 0.133 = 13.3%

  G = 2A = 2 (13.3) = 26.7%

  C = [(4.5)(0.133)]/4 = 15%

  T = 3C = 45%

39. Spaces between the nucleosomes must contain many promoter sequences. For the DNA to be digested, it must be unprotected. Since we see little transcription, the promoters must be missing and must have been destroyed by the nucleases.

40. **a.** B-DNA contains 10 bp per turn, and therefore this molecule will contain 100,000/10 = 10,000 turns.

  b. Z-DNA contains 12 bp per turn, and therefore this molecule will contain 100,000/12 = 8333.34 turns.

41. Yes, but only the smallest genes. The average coding part of a gene has about 1000 base pairs. A 200 base-pair region would give maximally a protein of 66 amino acids. Most proteins have much more than 66 amino acids.

42. No, because you could have a single-stranded DNA molecule with an (A + G)/(C + T) ratio of 1.0 (*example:* a DNA molecule with 30% A, 20% G, 30% C, and 20% T). To arrive at the conclusion that a particular DNA molecule is double-stranded, we must know the composition of each of the four bases.

43. Although archaeal species have attributes of both prokaryotes and eukaryotes, the only way to know would be by direct observation. It turns out that archaea do have modified nucleosomes.

44. In their 1953 paper, Watson and Crick give one reason why DNA has the deoxyribose sugar: "It is probably impossible to build this structure with a ribose sugar in place of the deoxyribose, as the extra oxygen atom would make too close a van der Waals contact." From an evolutionary standpoint, it is also useful for the cell to be able to differentiate RNA from DNA for the removal of RNA primers during DNA replication; the oxygen difference in the sugars would provide that ability. As for the difference in pyrimidines (thymine versus uracil), we believe the difference has to do with the cell's ability to repair spontaneously mutated bases. Although we will go into this in more detail in chapter 18, we can mention here that the most common spontaneous mutation of bases is of cytosine to uracil by deamination (see figure 7.8). If uracil was being created spontaneously in DNA and was a normal base in DNA, then the repair systems would not be able to differentiate a normal uracil from a mutated uracil and many mutations would go unrepaired. Hence early in evolution, thymine was probably substituted for uracil in DNA to alleviate that problem.

45. To calculate the number of base pairs, we have to know the "length" of DNA in the circular bacterial chromosome. In other words, we have to determine the "circumference" of the bacterial chromosome. This can be calculated using the formula πD, where π is 3.14 and D is 478 μ. The length of the DNA comes out to 1500 μm or 1,500,000 nm. Each base pair in B-DNA corresponds to 0.34 nm. Therefore, the genome of *Mycobacterium tuberculosis* would contain

$$1{,}500{,}000 \text{ nm}/0.34 \text{ nm} = 4.41 \times 10^6 \text{ base pairs}$$

46. Comparative DNA studies can be helpful in understanding the roles of the various types of DNA in the eukaryotic chromosomes if there are cases in which there are remarkably large differences in the amount of DNA in similar species. It can then be inferred that the basic developmental plan of an organism is contained in the one with the lower amount of DNA, and the extra DNA in the species with more DNA may be superfluous. We do have cases in which amphibians differ by as much as 100 times the amount of DNA found in similar species. The puffer fish has only one-sixth the amount of DNA as other higher eukaryotes.

47. **a.** If A = 20%, then T = 20%. Therefore, C + G = 60%. Since C = G, (1/2)(60) = 30%. The DNA is 1000 bp or 2000 bases long. Therefore, the number of cytosine bases will be 2000 × 0.3 = 600.

    b. First determine the number of base pairs. Every base pair of DNA has a linear length of 0.34 nm. A centimeter = 10^{-2} m, and a meter = 10^9 nm, and so 1 cm = 10^7 nm. Therefore, 1 cm of DNA will contain $10^7/0.34 = 2.9 \times 10^7$ bp, which equals 5.8×10^7 bases. C = 30%, and therefore the number of cytosine bases is $(5.8 \times 10^7) \times 0.3 = 1.74 \times 10^7$, or 17,400,000.

CHAPTER INTEGRATION PROBLEM

a.   First, let's review a few facts about chromosomes and cell division: (1) The number of chromosomes in a cell is equivalent to the number of functional centromeres. (2) The number of DNA molecules is equivalent to the number of chromatids. (3) The number of DNA molecules doubles during the S phase (DNA replication) of interphase; however, the number of chromosomes stays the same because the sister chromatids are still attached at the same centromere. (4) Mitosis maintains the diploid number, and so after cytokinesis, each cell will have the diploid number of chromosomes, with each chromosome consisting only of a single chromatid. (5) Meiosis I is a reductional division (2N → 1N), which yields a haploid number of chromosomes, with each chromosome consisting of two sister chromatids. (6) Meiosis II is an equational division (1N → 1N), which maintains the haploid number of chromosomes, but with each chromosome now consisting of only a single chromatid. (7) During anaphase of mitosis and anaphase II of meiosis, the number of chromosomes per cell will double, albeit temporarily, because the centromere will divide and sister chromatids will separate to opposite poles of the cell. Now, we can determine the numbers with confidence!

Stage	Number of Chromosomes per Cell	Number of DNA Molecules per Cell
i. During G1 of interphase	28	28
ii. Start of prophase of mitosis	28	56
iii. End of anaphase of mitosis	56	56
iv. After cytokinesis of mitosis	28	28
v. Start of prophase I of meiosis	28	56
vi. Start of prophase II of meiosis	14	28
vii. After cytokinesis of meiosis II	14	14

b.   At the start of mitosis and meiosis I, the cell would have already undergone DNA replication during interphase. Therefore, the nucleus will contain two times the diploid number of base pairs. At the end of mitosis, each cell will have the normal diploid number. At the end of meiosis II, each cell will have one-fourth of the amount of DNA that was present at the start of meiosis I.

The *percentage* of adenines can be determined using Chargaff's rule. If G + C = 60%, then A + T = 40%. Since A = T, then A = 20%. Therefore, the number of adenines can be calculated by multiplying the total number of base pairs by 2 (to get the overall number of *bases*) and then multiplying that by the percentage of A.

Stage	Total Number of Base Pairs per Cell	Number of Adenines per Cell
i. At start of mitosis	4×10^9	1.6×10^9
ii. At end of mitosis	2×10^9	0.8×10^9
iii. At start of meiosis I	4×10^9	1.6×10^9
iv. At end of meiosis II	1×10^9	0.4×10^9

c. **i.** The nuclei of a secondary oocyte and a first polar body are expected to have the same weight. Both cells result from a primary oocyte that has undergone meiosis I. While the division of the primary oocyte's cytoplasm is asymmetric (most of the cytoplasm goes to the secondary oocyte), the nucleus of the primary oocyte divides symmetrically yielding two equivalent daughter nuclei. The female is homogametic (XX), so the nuclei of the secondary oocyte and the polar body will each inherit an X chromosome and 22 autosomes.

ii. It depends which secondary spermatocyte is used for comparison with the secondary oocyte. Meiosis I in the male will yield two secondary spermatocytes. The male is heterogametic (XY), so the two secondary oocytes will have different chromosomal compositions in their nuclei: one will have 22 autosomes + X, and the other will have 22 autosomes + Y. The X chromosome is more than twice the size of and will be heavier than the Y chromosome. Therefore, the secondary spermatocyte with a nucleus containing the Y chromosome will be expected to be lighter than the nucleus of a secondary oocyte. On the other hand, a secondary spermatocyte with a nucleus containing the X chromosome will be expected to have the same weight as the nucleus of a secondary oocyte.

d. The proportions of bases in this genome are A = 20% (or 0.2), T = 20% (or 0.2), C = 30 % (or 0.3), and G = 30% (or 0.3). The probability of finding the sequence 5'-TAGAC-3' is given as $0.2 \times 0.2 \times 0.3 \times 0.2 \times 0.3 = 0.00072$. Therefore, the number of times the sequence is expected to occur is $0.00072 \times (2 \times 10^9) = 1,440,000$ times.

e. Consider the simplest case that will yield a strand of DNA with an (A + G)/(C + T) ratio of 0.5: this is if A + G = 0.5 and C + T = 1. The complementary strand would therefore have a content of A + G = 1 and C + T = 0.5, and a ratio of 1/0.5 = 2.

f. First, calculate the mass of a single base pair = 660 Daltons $\times$ $(1.67 \times 10^{-24}$ g/Daltons) = 1.1×10^{-21} g. The total number of base pairs in the animal's body will be $(0.33 \text{ g})/(1.1 \times 10^{-21} \text{ g/bp}) = 3 \times 10^{20}$ bp. Each base pair in the B form of DNA corresponds to 0.34 nm. Therefore, the length of the DNA will be $(3 \times 10^{20}$ bp) $\times$ (0.34 nm/bp) = 1×10^{20} nm. Because 1.0 km represents 1×10^{12} nm, this animal's DNA if stretched end-to-end would be $(1 \times 10^{20}$ nm)/$(1 \times 10^{12}$ nm/km) = 1×10^8 km.

CHAPTER 8 CHANGES IN CHROMOSOME STRUCTURE AND NUMBER

Chapter Goals

1. Understand the cytogenetic and inheritance features associated with inversions, deletions, duplications, and translocations.
2. Describe the genetic consequences of each of the four chromosomal rearrangements delineated in goal 1.
3. Define the differences between aneuploidy and euploidy.
4. Understand how aneuploidy and changes in euploidy affect mitosis and meiosis.
5. Compare how aneuploidy and euploidy affect the viability of the organism.

Key Chapter Concepts

8.1 Variation in Chromosome Structure: An Overview

- **Overview:** Various agents such as ionizing radiation, physical stress, and chemical compounds are capable of damaging chromosomes by breaking the two strands of DNA in the chromosome. Cells attempt to repair these breaks by rejoining the ends (**restitution**). Sometimes, however, cells fail to repair the mistake or rejoin inappropriate ends, resulting in the loss or rearrangement of some genetic material. These changes usually only occur to one of the two homologous chromosomes possessed by a diploid individual, yet they still can have extreme phenotypic consequences such as lethality or developmental defects such as deformity and mental retardation.

- **Single Breaks in a Chromosome:** Breaks can occur in a chromatid that can result in a fragment containing the centromere (**centric**) and a fragment lacking a centromere (**acentric**) if rejoining does not occur. During mitosis, the acentric fragment is often lost (**deletion**). Occasionally, two centric fragments join forming a **dicentric** chromosome containing two centromeres. During nuclear division, the dicentric fragment breaks as it is pulled apart by the spindle fibers in a **breakage–fusion–bridge cycle.**

- **Multiple Breaks in a Chromosome:** If this occurs, an internal piece of the chromosome can be lost (**deletion** or deficiency) or reattached in the wrong orientation (**inversion**). Breaks occurring simultaneously in two nonhomologous chromosomes can also be rejoined incorrectly. For example, the ends of two different chromosomes can be joined together resulting in a **reciprocal translocation.**

8.2 Deletions

- **Properties:** Individuals possessing deletions have a **genetic imbalance** since they have two copies of some genes and only a single copy of others. This can alter the phenotype of an individual in two major ways. First, **pseudodominance** allows recessive traits to be expressed when only one allele is inherited since no second allele exists. Second, **haploinsufficiency** allows a mutant phenotype to appear even though a wild-type allele is present since it is only present in one copy and cannot produce enough product to produce a wild-type phenotype. Cri du chat syndrome results from a deletion of part of human chromosome 5.

8.3 Inversions

- **Properties:** Inversions are a rearrangement of the DNA in a chromosome. Thus, an individual in which an inversion occurs does not gain or lose any DNA. Most often, these rearrangements do not cause any specific phenotypes. Two key points are of interest, however. First, individuals possessing inversions have reduced fertility. This is because if a meiotic crossover occurs within the inverted region, many of the gametes produced will be genetically imbalanced and will produce inviable zygotes. Second, phenotypes can arise in individuals containing inversions through one of two uncommon mechanisms. (1). If the breakpoint (end) of the inversion happens to be within a gene, the gene will be inactive and can cause a phenotype. (2). **Position effects** can change the expression of a gene depending on its location on a chromosome. For example, an inversion may bring a poorly expressed gene in proximity of a highly active transcriptional promoter.

8.4 Translocations
- **Properties:** Like individuals with inversions, individuals possessing a reciprocal translocation have a normal complement of genetic material, are often phenotypically normal, and are able to carry out mitosis normally. Fifty percent of the gametes produced by these individuals will produce genetically imbalanced gametes, unlike the lower percentage of imbalanced gametes produced by individuals possessing inversions. This is because meiotic segregation produces imbalanced gametes in translocation individuals while specific crossovers produce imbalanced gametes in inversion individuals. Reciprocal individuals thus display **semisterility**.
- **Segregation of Translocation Chromosomes:** Chromosomes containing reciprocal translocations can form a tetrad in the form of a cross during meiosis I. Different gametes are formed depending on how this tetrad segregates. In **alternate segregation**, the two chromosomes containing the translocation segregate together and the two normal chromosomes are pulled to the opposite pole. This pattern of segregation results in all balanced gametes. In **adjacent-1** or **adjacent-2 segregation**, a normal chromosome segregates with a chromosome containing a translocation, resulting in unbalanced gametes. Alternate and adjacent-1 segregation are equally likely to occur.

8.5 Centromeric Breaks
- **Robertsonian Translocations:** If a chromosome break occurs very near the centromeres of two chromosomes, the exposed centromeres can be joined to form what is called a Robertsonian translocation to form one chromosome out of two. Acrocentric chromosomes often form Robertsonian translocations by losing their short arms. After this, very little genetic material has been lost but the chromosome number has changed.

8.6 Duplications
- **Causes:** The duplication of a region of chromosomes can result from the breakage–fusion–bridge cycle and by crossovers that occur within an inversion loop as discussed previously. They can also occur by **unequal crossovers**. Crossovers are supposed to only occur between *homologous* regions of DNA. If two regions of homology exist within a single chromosome, a crossover may inappropriately occur between region 1 of one chromosome and region 2 of the homologous chromosome. This results in the duplication of the DNA between the two regions on one chromosome and deletion of this material on the homologous chromosome.
- **Three-Nucleotide Repeat Expansion:** A special type of duplication is the expansion of the number of three-nucleotide repeats found at specific places in chromosomes. Expansion of the repeats found within important genes leads to disruption of protein produced from the gene and ultimately human diseases. Fragile-X syndrome, myotonic muscular dystrophy, and Huntington disease all result from three-nucleotide repeat expansions.

8.7 Variation in Chromosome Number: An Overview
- **Overview:** Variations in chromosome number can be detrimental to the organism. In human beings, at least 50% of human miscarriages in the United States are believed to result from chromosomal abnormalities. **Euploid** cells/individuals may have additional or missing chromosome sets, while **aneuploid** cells/individuals have additional or missing chromosomes. Aneuploidy often results from **nondisjunction** during meiosis. In aneuploidy, diploid cells can be missing a single chromosome (**monosomic**), or both copies of a chromosome (**nullisomic**), or they have an extra chromosome (**trisomic**). In euploidy, cells may contain three complete sets of chromosomes (**triploid**), four complete sets of chromosomes (**tetraploid**), or more.

8.8 Aneuploidy
- **Aneuploid Phenotypes:** Human trisomics are rare because only a few chromosomes can form viable trisomics. Human monosomics are even more rare and are only found of the X chromosome or one of the two smallest autosomes (21 or 22). Aneuploidy of autosomes usually causes more severe phenotypes than aneuploidy of the sex chromosomes. Aneuploidy leads to several medically recognized syndromes in human beings including Down syndrome (trisomy 21), Edward syndrome (trisomy 18), Patau syndrome (trisomy 13), Turner syndrome (XO), the XYY and triple-X karyotype, and Klinefelter syndrome (XXY).

- **Causes of Aneuploidy**: Aneuploidy usually results from meiotic nondisjunction during either the first or the second division. By examining the allele distribution on the trisomic chromosomes, it can be determined whether the nondisjunction occurred during meiosis I or meiosis II. If two of the chromosomes are *identical*, the error must have occurred during meiosis II, while three *different* chromosomes indicates that the error occurred during meiosis I. If nondisjunction occurs during *mitosis* in a developing organism, a **mosaic** organism will result in which some of its cells will be genetically normal while others are aneuploid.

8.9 Changes in Euploidy

- **Polyploidy:** Polyploidy may result from additional sets of chromosomes from the same species (**autopolyploidy**) or from the hybridization of gametes from two different species (**allopolyploidy**). Polyploidy results in difficulty during meiotic segregation when an odd number of chromosome sets is involved since one set of chromosomes will not have a homolog to pair with. Thus, these individuals are sterile. In plants, however, cells can undergo **somatic doubling** where all the chromosomes are doubled. Then meiotic segregation is faced with an even number of chromosome sets and meiosis works.

Key Terms

- cytogenetics
- restitution
- acentric fragment
- centric fragment
- dicentric chromosome
- breakage–fusion–bridge cycle
- reciprocal translocation
- Robertsonian translocation
- fusion
- deletion
- deletion loop
- pseudodominance
- genetic imbalance
- haploinsufficiency
- inversion
- pericentric inversion

- paracentric inversion
- balancer chromosomes
- position effect
- variegation
- supergene
- alternate segregation
- adjacent-1 segregation
- adjacent-2 segregation
- semisterility
- pseudolinkage
- fundamental number (NF)
- centromeric fission
- unequal crossing over
- fragile-X syndrome
- transmitting males
- aneuploidy
- euploidy
- monosomic

- nullisomic
- trisomic
- triploids
- tetraploids
- polyploids
- sporadic Down syndrome
- familial
- isochromosome
- mosaics
- chimeras
- gynadromorph
- hermaphrodite
- autopolyploidy
- allopolyploidy
- somatic doubling
- amphidiploid

Understanding the Key Concepts

Use the key terms or parts of their definitions to complete the following sentences.

Breaks can occur in chromosomes. How these breaks are repaired, or not repaired, determines what type of chromosomal abnormality occurs. If a single break occurs and is not repaired, a (1)_____ will result. If two breaks occur within a single chromosome, an internal (2)_____ or an (3)_____ may occur. If a break occurs in two different chromosomes, a (4)_____ may occur. (5)_____ involve the loss of genetic material and, therefore, if large enough, may cause severe phenotypic defects. These defects can result from (6)_____ of some genes or by (7)_____ of some recessive alleles found on the homologous chromosome. (8)_____ and (9)_____, on the other hand, are simply a rearrangement of the DNA and do not result in the gain or loss of any genetic material. Therefore, they usually (10)_____ cause major phenotypic defects. However, these individuals usually have (11)_____ since they have a high chance of producing gametes with (12)_____. Translocation and inversion chromosomes can still (13)_____ with their homologs, but (14)_____ in regions of inversions or (15)_____ segregation patterns in translocations cause reduced fertility and (16)_____, respectively. Two unique situations can allow a translocation or inversion to create a phenotype. First, (17)_____ can cause a gene to be expressed differently when it is in a different location on a chromosome. Second, the (18)_____ may occur in the middle of a gene, disrupting its function.

Changes in the number of individual chromosomes, called (19)_____, results from nondisjunction during the first or second (20)_____ division. Resulting aneuploid gametes will be haploid for all of the chromosomes except the one that experienced the nondisjunction and will produce a (21)_____ or a (22)_____ zygote when fused with a normal gamete. In humans, most of these aneuploid zygotes are inviable and will die during embryogenesis. A few will survive until birth and even less will survive after birth. The human autosomal aneuploids that survive after birth are (23)_____, (24)_____, (25)_____, (26)_____, and (27)_____. Aneuploidy of the (28)_____ is usually more tolerable in humans, and several such disorders exist.

Changes in the whole set of chromosomes so that an organism goes from diploid to triploid or tetraploid are called (29)_____ and usually result from fusion of two nonhaploid cells. This occurs most frequently in (30)_____. These (31)_____ organisms are often sterile since the seeds produced are (32)_____. (33)_____ allows creation of genetically balanced seeds again since all chromosomes will have a homolog to synapse with.

Figure Analysis

Use the indicated figure or table from the textbook to answer the questions that follow.

1. **Figure 8.9**
(a) What is different about each of the three inversions depicted in the three columns?

(b) Which, if any, of the three inversions is likely to cause a phenotype in this individual? Which are likely to not cause a phenotype?

(c) For each of the inversions predicted to cause a phenotype, explain in general terms how that phenotype will be caused?

2. **Figure 8.14**
(a) Why does the chromosome with the inversion form a loop?

(b) Between which two genes does the crossover depicted in figure 8.14 occur?

(c) How many different gametes can be produced by the crossover depicted in figure 8.14?

(d) How many of the gametes produced by a crossover within an inversion loop will most likely produce inviable zygotes after fertilization with a normal gamete? Which ones (recombinants or nonrecombinants)?

(e) Why do individuals with a chromosome inversion have reduced fertility?

3. **Figure 8.18**
(a) Which chromosome cosegregated with the first chromosome during the first meiotic division to produce alternate segregation? Which one cosegregated with it during the first meiotic division to produce adjacent-1 segregation?

(b) Which two segregation patterns are produced most frequently? Of these two, is one of them produced more frequently?

(c) What determines which of the two most frequent segregation patterns occurs?

(d) What percentage of the gametes produced by the individual carrying this translocation will be able to produce viable offspring?

(e) Why can a completely normal gamete be produced by this individual who has a reciprocal translocation?

4. **Figures 8.26 and 8.27**
(a) What fails to segregate in nondisjunction during meiosis I? What fails to segregate in nondisjunction during meiosis II?

(b) For each of the following individuals, determine whether the nondisjunction(s) that produced them occurred in the male, the female, or both and, if determinable, during which meiotic division the nondisjunction(s) occurred?
> XX0 (two maternal X chromosomes):
> Turner syndrome with a maternal X chromosome:
> Turner syndrome with a paternal X chromosome:
> Klinefelter syndrome (only one maternal X chromosome):
> Triple-X female (only one maternal X chromosome):
> XYY karyotype:

(c) Looking at the right two panels of figure 8.26, one is unable to distinguish between XX gametes resulting from nondisjunction in meiosis I from XX gametes resulting from nondisjunction in meiosis II, but in fact they are different. What is different about the two different XX gametes, and how could you tell this?

General Questions
Q8-1. What is the difference between real suppression and apparent suppression of crossovers by a chromosomal inversion?

Q8-2. How can a large deletion in a single chromosome cause a severe phenotype or even lethality in a diploid organism since the organism has a normal copy of that chromosome?

Q8-3. If two breaks occur within one chromosome, describe several outcomes that might occur and how each is likely to affect the cell.

Q8-4. In chapter 4, you learned that calico cats are usually female and that the color pattern results from X inactivation in heterozygotes. It has been noted that calico *males* also occur, but at a very low frequency (1/3000 cats). To what do these males owe their calico coloration? If you were to karyotype one of these individuals, what would you find?

Q8-5. Compare and contrast sporadic Down syndrome versus familial Down syndrome.

Multiple Choice

For each of the following, circle the letter of the choice that most appropriately answers the question.

1.) Chromosomal breaks do *not* result from which of the following?
   a. physical stress exerted by spindle fibers
   b. adjacent-1 segregation after translocation
   c. ionizing radiation
   d. ingested chemicals
   e. dicentric chromosomes

2.) A single chromosomal break *can* result in which of the following?
   a. deletion
   b. acentric chromosome
   c. inversion
   d. translocation
   e. Both a and b

3.) Which of the following does *not* help to explain why there are very few monosomic humans?
   a. unbalanced translocations occur more frequently in monosomics
   b. haploinsufficiency
   c. pseudodominance
   d. genetic imbalance
   e. large deletions are usually lethal in humans

4.) Which of the following *cannot* result in a duplication?
   a. an unequal crossover
   b. alternate segregation of a reciprocal translocation without a crossover
   c. a crossover within a paricentric inversion
   d. a crossover within a paracentric inversion
   e. a breakage–fusion–bridge cycle

5.) Which of the following does *not* help to explain why genes distantly spaced on balancer chromosomes show very tight linkage?
   a. Inversions suppress crossovers from occurring with the balancer chromosome.
   b. Zygotes that form gametes produced by crossovers with the balancer chromosome are lethal.
   c. Crossovers occur more frequently within an inversion than outside of an inversion.

   d. Many inversions are present on the balancer chromosome.
   e. None of the above are correct; all answers help to explain why the genes show tight linkage.

6.) Pseudolinkage can help researchers to identify which of the following
   a. genes located near an inversion breakpoint
   b. genes located near a deletion
   c. genes located near a translocation breakpoint
   d. a Robertsonian translocation
   e. Both a and b

7.) If each of the following produced a phenotype, which one can revert to a wild-type phenotype through an unequal crossover?
   a. a deletion
   b. an inversion
   c. a translocation
   d. a duplication
   e. All of the above

8.) Which of the following individuals would have familial Down syndrome?
   a. 47,XY,+21
   b. 47,XX,+18
   c. 45,XX,–21
   d. 47,XXX
   e. 47,XX, t(14q;21q)

9.) Which of the following might form a mosaic organism?
   a. mitotic nondisjunction
   b. meiotic nondisjunction
   c. double fertilization of an egg
   d. triple fertilization of an egg
   e. None of the above

10.) Which of the following is likely to produce a seedless variety of fruit?
   a. tetraploid plant
   b. diploid plant
   c. amphidiploid plant
   d. triploid plant
   e. All of the above

Practice Problems

P8-1. Provide the karyotype description for the following syndromes:

(*Hint 1*: Become familiar with the nomenclature used to describe human chromosome complements. The standard "short-hand" representation of a karyotype includes information about the individual's sex, total number of chromosomes, and the abnormality observed. Specifically, the total number of chromosomes is listed first, followed by a description of the sex chromosomes, and finally a description of the autosomal abnormality, if any. For example, a female zygote with an extra copy of chromosome number 14 would be written 47,XX,+14.)

Sporadic Down syndrome:
Edward syndrome:
Patau syndrome:
Klinefelter syndrome:
Turner syndrome:
Cri du chat syndrome:

P8-2. Diagram all types of nondisjunction of the sex chromosomes that can occur during (a) meiosis I or (b) meiosis II in a normal human female and in a normal human male. Indicate the sex chromosome complement of a zygote resulting from the fusion of each gamete type produced and a normal gamete from the other parent.

(*Hint 2*: Nondisjunction can occur during mitosis, meiosis I, meiosis II, or both meiosis I and meiosis II. The likelihood of producing an abnormal gamete will depend upon when nondisjunction occurred. If nondisjunction occurs during meiosis I, all of the gametes will be unbalanced. If nondisjunction occurs during meiosis II, approximately one-half of the gametes will be normal and one-half will be unbalanced.)

P8-3. A man with brown teeth (an X-linked dominant trait) marries a woman with normal white teeth. They have a son with Klinefelter syndrome. Assume that the parents are karyotypically normal and that the boy's abnormal karyotype resulted from nondisjunction. How could you use information about the color of the boy's teeth to determine which parent experienced nondisjunction?

P8-4. Consider the following chromosomes (ovals represent centromeres). Indicate the gene order, centromere position, and banding pattern found on the resulting chromosomes if the area between the arrows is (a) deleted, (b) duplicated, or (c) inverted.

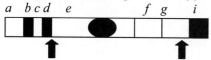

P8-5. Shown here are a pair of homologous chromosomes from a karyotype stained with the Giemsa banding procedure. The centromere is represented by an open circle. Would the banding patterns observed for this pair of chromosomes be useful in identifying potential causes of mental retardation in the individual from which they came?

Maternally derived chromosome

Paternally derived chromosome

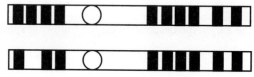

Assessing Your Knowledge

Understanding the Key Concepts—Answers

1.) deletion or deficiency; 2.) deletion or deficiency; 3.) inversion; 4.) translocation; 5.) Deletions; 6.) haploinsufficiency; 7.) pseudodominance; 8.) Inversions; 9.) translocations; 10.) do not; 11.) reduced fertility; 12.) genetic imbalances; 13.) synapse; 14.) crossovers; 15.) adjacent; 16.) semisterility; 17.) position effects; 18.) breakpoint; 19.) aneuploidy; 20.) meiotic; 21.) monosomic; 22.) trisomic; 23.) trisomy 13 (Patau syndrome); 24.) trisomy 18 (Edward syndrome); 25.) trisomy 21 (Down syndrome); 26.) monsomy 21; 27.) monosomy 22 or trisomy 8 or cat's eye syndrome; 28.) sex chromosomes or X and Y chromosomes; 29.) euploidy; 30). plants; 31.) polyploidy; 32.) aneuploid; 33.) somatic doubling.

Figure Analysis—Answers

1a.) The location of the two ends of the inversion relative to genes *a*, *b*, and *c* is different in each case; b.) The inversions in the second (middle) and third (right) columns are likely to cause a phenotype, while the inversions in the first column is unlikely to cause a phenotype; c.) Inversion 2 (middle column): One of the breakpoints for the inversion occurs in the middle of gene *b*. After inversion, gene *b* has been split so that half is in its original position but the other half has been moved far away. Thus, the activity of gene *b* will be lost causing a phenotype unless an allele on the homologous chromosome can compensate for the loss of this gene. Inversion 3 (right column): The ends of the inversion fall between two genes and their transcriptional promoters. After inversion, the genes each have a new promoter. If these genes are normally expressed differently (as they most certainly are in humans), the expression of both genes *a* and *b* will be incorrect and likely cause a phenotype.

2a.) During synapsis in prophase I of meiosis, homologous chromosomes must pair with each other. This occurs by DNA of one chromosome aligning with the homologous DNA in the other chromosome. When a region is inverted, it must loop in order for homologous DNA to align. b.) The crossover occurs between genes *E* and *F*; c.) Four; d.) Half of the gametes produced will likely produce inviable zygotes since they will either have a dicentric chromosome or an acentric chromosome, both of which will lead to a significant loss of genetic material. It is the gametes containing recombinant chromosomes that will produce inviable zygotes; e). If a crossover occurs in the inverted region of an individual's chromosome, half of the gametes produced will be unable to produce offspring. Therefore, the individual displays reduced fertility.

3a.) Alternate segregation: The first chromosome cosegregated with the fourth chromosome.
     Adjacent-1 segregation: The first chromosome cosegregated with the third chromosome.
b.) Alternate and adjacent-1 segregation patterns occur much more frequently than adjacent-2 segregation. Alternate and adjacent-1 segregation occur equally as frequent; c.) Alternate or adjacent-1 segregation is determined randomly by how the two align during metaphase I of meiosis. In other words, imagine the third and fourth centromeres rotating so that the fourth lies on top and the third on the bottom; (d) Less than 50% of the gametes will be able to produce viable offspring. This is because only those produced by alternate segregation do not contain duplications and deletions; (e) The individual has a reciprocal translocation involving only one of the two homologs of each of the two chromosomes involved. They also have a normal copy of each of these chromosomes. Since gametes only receive one copy of each chromosome, there is a chance that a gamete can receive all normal chromosomes.

4a.) Meiosis I: Homologous chromosomes failed to separate.
     Meiosis II: Sister chromatids failed to separate.

		Which sex	Which division
b.)	XX0 (two maternal X chromosomes):	Male	Cannot be determined
	Turner syndrome with a maternal X chromosome:	Male	Cannot be determined
	Turner syndrome with a paternal X chromosome:	Female	Cannot be determined
	Klinefelter syndrome (only one maternal X):	Male	Meiosis I
	Triple-X female (only one maternal X chromosome):	Male	Meiosis II
	XYY karyotype:	Male	Meiosis II

c.) XX gametes produced by nondisjunction in females differ by the alleles present on the two X chromosomes. Nondisjunction in meiosis I results in gametes with two *different* X chromosomes (each containing different alleles). Nondisjunction in meiosis II results in gametes with two *identical* X chromosomes (each containing the same alleles).

General Questions—Answers

Q8-1. Real suppression refers to the fact that crossovers do not occur in the region surrounding the breakpoints of an inversion since the DNA is bent and unaligned. Therefore, crossovers are physically inhibited. Apparent suppression refers to the fact that crossovers do not appear to happen in the region of an inversion since no recombinant progeny are observed. Importantly, these crossovers actually *do* occur; however, the recombinant progeny die before an observer can count them. Thus, they appear as if they never happened at all.

Q8-2. Many genes in animals must be present in two copies in order to produce the correct phenotype. This is because with only half the number of genes, sometimes only half the amount of protein is made from the gene. This is called haploinsufficiency. When only one copy of many genes is present (such as in the case of a large deletion), there is a good chance that one or more of these genes will display haploinsufficiency. There is also a chance that deleterious, recessive alleles will show pseudodominance when their phenotype is expressed in the hemizygous state created by the deletion.

Q8-3. Correct repair: The cell can rejoin the two breaks, and no phenotype will be observed.
Internal deletion: The internal piece of the chromosome may be lost and the outer ends rejoined. A large deletion may cause severe defects such as mental retardation, while smaller deletions may not show a phenotype.
Inversion: The internal piece of the chromosome may invert and then be rejoined to the chromosome. Inversions only rarely have phenotypic consequences. However, the inversion will increase the likelihood of producing imbalanced gametes.

Q8-4. Male calico cats must also experience the effects of X inactivation, so they must have two X chromosomes. Since they are male, they must also have a Y chromosome. Therefore, they have an abnormal number of sex chromosomes such that they are XXY. These males are heterozygous for the coat color gene which results in the same pattern effects seen in normal females. The karyotype analysis would confirm the animal's abnormal chromosome complement.

Q8-5. Both sporadic and familial Down syndrome have similar phenotypes, and both result from the possession of a third copy of the genes found on the long arm of chromosome 21. The way in which the individuals receive the third copy differentiates the two forms. In sporadic Down syndrome, an extra copy of the entire chromosome 21 is gained through nondisjunction in a parent producing the child (trisomy). The parents don't have the extra copy, and the child will not pass on this extra copy of chromosome 21. In familial Down syndrome, an extra copy of the long arm of chromosome 21 is inherited because it is attached to another chromosome (translocation). Therefore, this chromosome is inherited as an abnormal chromosome along with two normal copies of chromosome 21. The parents have the abnormal chromosome and the child can pass it on to his or her children. The key difference is that Down syndrome caused by the translocation is inherited and passed on whereas the trisomy is not.

Multiple Choice—Answers

1.) b; 2.) e; 3.) a; 4.) b; 5.) c; 6.) c; 7.) d; 8.) e; 9.) a; 10.) d.

Practice Problems—Answers

P8-1.	Sporadic Down syndrome:	47,XX,+21 or 47,XY,+21
	Edward syndrome:	47,XX,+18 (majority) or 47,XY,+18
	Patau syndrome:	47,XX,+13 or 47,XY,+13
	Klinefelter syndrome:	47,XXY
	Turner syndrome:	45,X
	Cri du chat syndrome:	46,XX,5p– or 46,XY,5p–

P8-2. (a) In females: In males:

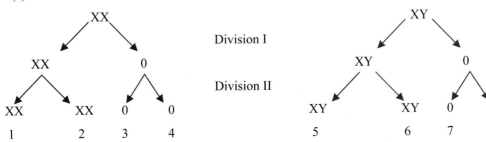

Gamete Number	Type	Normal Gamete	Zygote
1 and 2	Disomic for X	X or Y	XXX or XXY
3 and 4	Nullisomic for X	X or Y	X0 or 0Y
5 and 6	Disomic for sex	X	XXY
7 and 8	Nullisomic for sex	X	X0

(b) In females: In males:

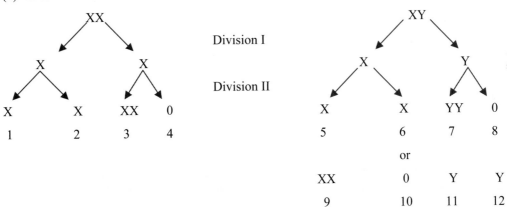

Gamete Number	Type	Normal Gamete	Zygote
1 and 2	Normal X	X or Y	XX or XY
3	Disomic for X	X or Y	XXX or XXY
4	Nullisomic for XX or Y		X0 or 0Y
5 and 6	Normal X	X	XX
7	Disomic for Y	X	XYY
8 and 10	Nullisomic for sex	X	X0
9	Disomic for X	X	XXX
11 and 12	Normal for Y	X	XY

P8-3. The sex chromosome complement of a Klinefelter male is XXY. Nondisjunction in the mother or father could produce this complement. We know that the Y chromosome was obtained from the father. However, it is unclear if the father also donated an X chromosome to his son (meiosis I nondisjunction in the father) or if the mother donated two copies of her X chromosome to her son (meiosis I or meiosis II nondisjunction in the mother). The color of the male's teeth will tell you which parent nondisjunction occurred in. Remember that brown teeth is an X-linked dominant trait. In our example, the female is normal and does not possess the allele for brown teeth but her husband does. If the male contributed an X chromosome, the Klinefelter child will have brown teeth. If, on the other hand, the female contributed two X chromosomes, the Klinefelter child should have white teeth.

P8-4.

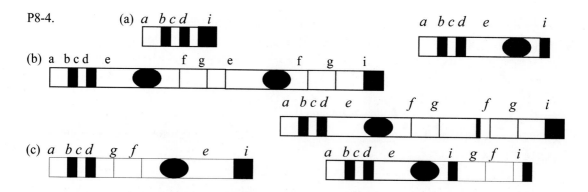

P8-5. Notice that the banding pattern of the maternally and paternally derived chromosomes are not identical: the paternal chromosome is missing the second band on the short arm. The missing band could represent a deletion on the paternal chromosome. Alternatively, the paternally derived chromosome could represent the normal situation with the extra band on the maternally derived chromosome resulting from a duplication. This information would be very useful in diagnosis of a mental deficiency or any other clinical symptom. The G-banding staining technique is the standard technique used by laboratories for cytogenetic study.

8

CHANGES IN CHROMOSOME STRUCTURE AND NUMBER

CHAPTER SUMMARY QUESTIONS

1. 1. j, 2. e, 3. b, 4. c, 5. g, 6. d, 7. f, 8. a, 9. i, 10. h.

2. A pericentric inversion is one in which the inverted DNA region contains the centromere, while in a paracentric inversion, the centromere lies outside the inverted region. Figure 8.10 provides a schematic. Both types of inversion suppress recombination in an inversion heterozygote in two ways: (1) A real mechanism due to the difficulty in complete pairing between the two homologs near the inversion breakpoints (figure 8.11), and (2) An apparent mechanism due to crossovers in the inversion loop producing gametes that will generate nonviable zygotes because of duplications and deficiencies (figures 8.13 and 8.14).

3. Both types of organisms contain three or more sets of chromosomes. Autopolyploids have chromosome sets from only one species, while allopolyploids have chromosomes from two or more different species.

4. Animals use chromosomal sex-determining mechanisms that are disrupted by polyploidy. Plants typically do not use those same mechanisms. Many plants can exist vegetatively, allowing more time for the rare somatic doubling of genome sets to occur. In addition, many plants rely on insect pollinators or wind to fertilize them, and therefore are more likely to undergo hybridization.

5. In humans, monosomy is rare, meaning that, with few exceptions, it is lethal. In fact, monosomics are also rare in spontaneous abortions, indicating that most monosomic fetuses are lost before the woman is aware of the pregnancy. The only monosomics known to be viable in human beings are Turner syndrome (45, X) and monosomics of chromosomes 21 and 22, the two smallest autosomes.

6. An acentric chromosome has no centromere, while a dicentric chromosome has two. An isochromosome is a chromosome that has a centromere and two genetically and morphologically identical arms (two long arms for example, instead of a short and a long arm). It undergoes normal segregation during cell division, whereas acentric and dicentric chromosomes do not.

7. A cross-shaped structure involving two homologous pairs of chromosomes. This structure can segregate (figure 8.17) in three different ways: (1) Alternate segregation involves the movement of homologous centromeres 1,4 and 2,3 to opposite poles. It leads to balanced gametes: one containing normal chromosomes, and the other with the reciprocal translocation. (2) Adjacent-1 segregation involves the movement of homologous centromeres 1,3 and 2,4 to opposite poles. It leads to unbalanced gametes carrying duplications and deficiencies that will usually produce nonviable offspring. (3) Adjacent-2 segregation (which is rare) involves the movement of nonhomologous centromeres 1,2 and 3,4 to opposite poles. It leads to unbalanced gametes carrying duplications and deficiencies that will usually produce nonviable offspring.

8. All chromosomes form linear bivalents. The cross-shaped figure is seen only in heterozygotes.

9. Yes, if the deletion moves a gene next to heterochromatin. Please refer to figure 8.15.

10. No, there are no inversion loops formed in homozygotes. Therefore, crossover suppression is not observed in an inversion homozygote.

11. Semisterility is the phenomenon where roughly 50% of the gametes are unbalanced and therefore cannot produce viable zygotes. It is caused by chromosomal rearrangements that result in an abnormal chromosomal complement in a gamete. For example, in individuals with heterozygous translocations, alternate segregation (which produces balanced gametes) and adjacent-1 segregation (which produces unbalanced gametes) are nearly equally likely and so cause semisterility. Heterozygous inversions may also cause semisterility.

12. A diagram will show that a crossover between a centromere and the center of the cross can change the consequences of the pattern of centromere separation. For example, the following figure diagrams a crossover between loci 4 and 5. Refer also to figures 8.17 and 8.18.

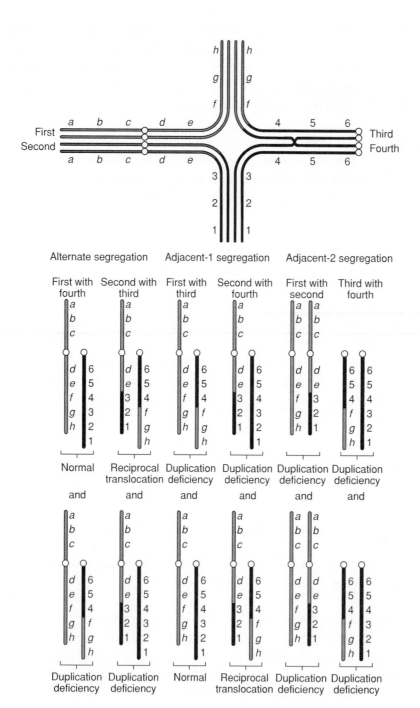

13. The following figure depicts a three-strand double crossover in a paracentric inversion loop. The first crossover occurs between E and F on the second and third chromatids. The second crossover occurs between F and G on the first and third chromatids. The products of this crossover will be (1) a dicentric chromosome with duplications and deficiencies, (2) a recombinant chromosome with normal structure, (3) a nonrecombinant inversion chromosome, and (4) an acentric chromosome with duplications and deficiencies.

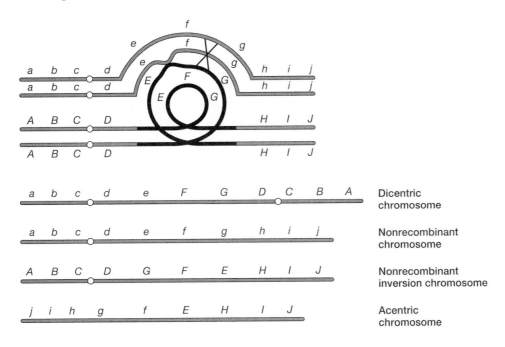

14. Problems occur during meiosis, not mitosis. Problems are worse in the odd-ploid organisms such as triploids because of the production of aneuploid gametes (see figure 8.38).

15. Odd-ploid autopolyploids have meiotic problems, producing aneuploid gametes, as do allopolyploids, because of a lack of pairing partners. Amphidiploids (allotetraploids) should have few or no meiotic problems, because each chromosome has a homolog.

16. Sporadic Down syndrome, also known as trisomy 21, is caused by a spontaneous, nondisjunction of chromosome 21. This form is random and comprises about 95% of all Down syndrome individuals. Familial Down syndrome (approximately 5% of cases) typically results from a Robertsonian translocation involving chromosome 21 and chromosome 14 (or 15). Translocation carriers do not exhibit Down syndrome; however, their children have an increased incidence of Down syndrome. (Please refer to figure 8.29.) The incidence of sporadic Down syndrome increases among children born to older mothers (because older eggs are more likely to undergo nondisjunction of chromosome 21). Familial Down syndrome does not exhibit this age-dependent expression.

17. Aneuploidy for autosomes would be more severe than that of sex chromosomes. If an individual has more than the required X chromosomes (two in females and one in males), each additional X chromosome would be inactivated in the form of a Barr body. Therefore, extra copies of the X chromosome do not cause a gene imbalance for most X-linked genes. Gene imbalances in males with additional Y chromosomes are not likely to cause severe effects either. The Y is one of the smallest chromosomes and so will contain relatively few genes. These genes are not essential for life because they are absent in females.

EXERCISES AND PROBLEMS

18. In humans, $2n = 46$. **a.** 69 ($3n$). **b.** 47 ($2n + 1$). **c.** 44 ($2n - 1 - 1$). **d.** 44 ($2n - 2$).

19. $2n = 40$. **a.** 39 ($2n - 1$). **b.** 41 ($2n + 1$). **c.** 38 ($2n - 2$). **d.** 44 ($2n + 2 + 2$). **e.** 60 ($3n$). **f.** 82. **g.** 21 ($n + 1$). **h.** 19 ($n - 1$). For parts (g) and (h), remember that a normal gamete contains a single chromosome of each pair.

20. Reciprocal translocation (some effects occur only in the heterozygous condition). Look for the cross-shaped figure at meiosis or in salivary gland chromosomes.

21. Inversion (some effects occur only in the heterozygous condition). Look for a loop at meiosis or in salivary gland chromosomes.

22. **a.**   47, XY, + 11.
     b.   44, XX, – 5 – 7.
     c.   46, XY, $20q^+$.
     d.   46, XX, t($10p^-$;$15q^+$).
     e.   45, Y, – 3 + 18.

23. **a.**   Deletion of FG.
     b.   Duplication of $BC \cdot D$.
     c.   Paracentric inversion of EFG.
     d.   Pericentric inversion of $C \cdot D$.
     e.   Deletion of C; duplication of F; then pericentric inversion of $B \cdot DEF$.

24. **a.**   The male and female have the same genotype, so the discussion applies to both. Start by giving a different designation to each of the two homologs that are carrying the same allele: ey^+–1 and ey^+–2. Therefore, the genotype of the individual is now ey^+–1 ey^+–2 ey. Three different types of segregation can occur in this trisomic during gamete formation in meiosis: ey^+–1 ey^+–2 to one pole and ey to the other pole; ey^+–1 ey to one pole and ey^+–2 to the other; and finally, ey^+–2 ey to one pole and ey^+–1 to the other. Thus, each trisomic parent can produce four different types of gametes: 2/6 ey^+ ey, 2/6 ey^+, 1/6 ey^+ ey^+, and 1/6 ey. The various possible fertilizations can be obtained by using the forked-line method for gametes.

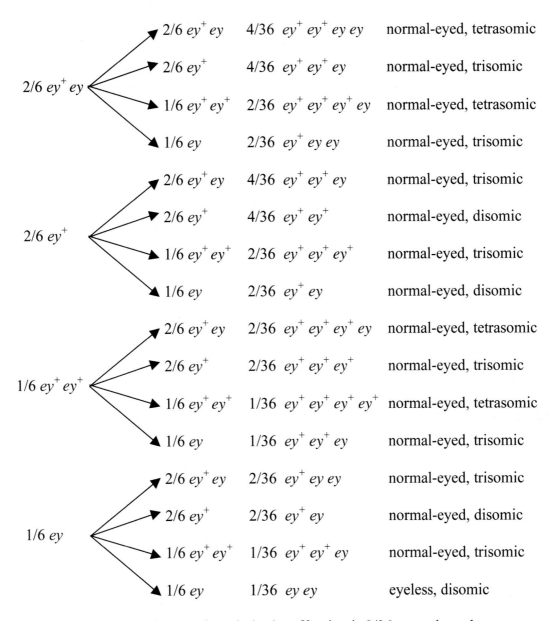

Therefore, the phenotypic ratio in the offspring is 9/36 normal-eyed, tetrasomic:18/36 normal-eyed, trisomic:8/36 normal-eyed, disomic:1/36 eyeless, disomic.

b. The male is trisomic and can produce four types of gametes, as discussed in part (a). The female is disomic and can produce two typics of gametes in equal proportions: 1/2 ey^+ and 1/2 ey.

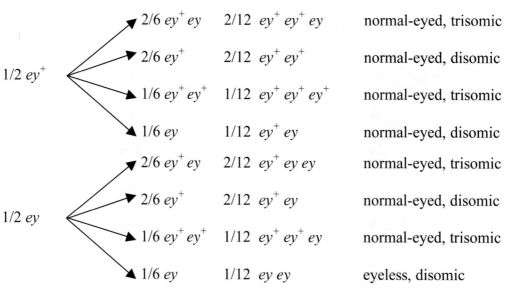

Therefore, the phenotypic ratio in the offspring is 6/12 normal-eyed, trisomic:5/12 normal-eyed, disomic:1/12 eyeless, disomic.

25. Assume crossovers as shown in the following figure:

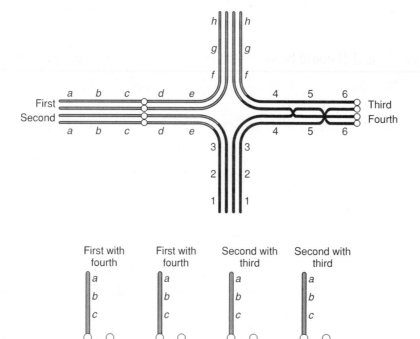

26. a. Nonreciprocal translocation of *GH* from chromosome 1 to chromosome 2.
    b. Robertsonian translocation and the loss of the acentric chromosomes.
    c. Deletion of *B* from chromosome 1; pericentric inversion of *C•DE* in chromosome 1; then a reciprocal translocation of *GH* and *TUV* between the two chromosomes.
    d. Deletion of *E* from chromosome 1; reciprocal translocation of *FGH* and *UV* between the two chromosomes; duplication of *UV* in chromosome 1, followed by an inversion of the duplicated segment.
    e. Deletion of *U* from chromosome 2; reciprocal translocation of *FGH* and *TV* between the two chromosomes; then a pericentric inversion of *C•DETV* in chromosome 1.

27. During evolution of the *Drosophila* species, each of the two large metacentric chromosomes in *D. melanogaster* underwent centric fission, thus each creating two different acrocentric chromosomes. Alternatively, two pairs of acrocentric chromosomes from *D. virilis* underwent centric fusion (Robertsonian translocation), thereby creating the two large metacentric chromosomes.

28. We have five heterozygous genes, so we expect to see $2^5 = 32$ genotypes, but we see only six genotypes. We see no exchanges between genes *B*, *C*, and *D*. These three genes could be so tightly linked that no recombination occurs between them (unlikely). Genes *B*, *C*, and *D* could be within an inversion that is heterozygous in the heterozygous parent. The recombination that does occur within this inversion results in inviable zygotes.

29. a. All females should get a wild-type X chromosome from their father. Irradiation produces chromosomal breaks, so a deletion of part of the X is possible, producing a situation of pseudodominance. (Alternatively, the offspring could have gotten a mutant X^w from the father.)

    b. Diagram the crosses (a "/" represents a deleted part of the chromosome and _____ = X chromosome).

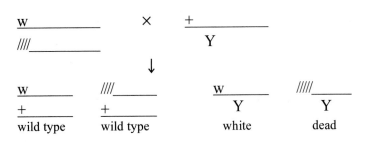

We expect all wild-type females and all white-eyed males, but in a ratio of 2 female:1 male.

30. We see about half the enzyme activity in crosses with strains B and C. Therefore, the gene must be located in the region that is common to both strains, approximately in the region located 25 to 35 map units from the left end.

31. We see that the F_2 offspring from cross A × B yields fewer progeny than the other crosses. Something unusual must be involved. One explanation is that one of the strains is homozygous for a reciprocal translocation. The translocation, when heterozygous, results in some inviable gametes or progeny, and thus reduces the number of progeny.

32.

The use of X-rays alerts us to chromosomal aberrations. The fact that a vermilion female appears when we expect all wild-type females indicates that we have a deletion. The deletion must end between v and m. If the deletion included miniature, we should have seen a vermilion, miniature female. We can draw the chromosomes of the flies in the second cross as

The question is, What is the distance between the deletion and the gene for miniature? We must look at the males from the cross. Note that we see only half as many males compared with females; those males that received the deletion X chromosome must have died. The vermilion male must have resulted from a recombination between the end of the deletion and miniature:

yielding the following chromosomes:

Therefore, map distance = (vermilion males/total males) × 100 = 1/90 × 100 = 1.1 map units.

33. We expect to see about 32% recombination between these two genes, but we see only 2%. The most likely explanation is that an inversion occurred so that these two genes came to lie close to each other; the stocks are homozygous. Since semisterility is not reported, we are probably not dealing with crossover suppression in inversion heterozygotes.

34. 3 → 4 → 1 → 2. Inversion of $u\,p\,o$ will yield (4). If this arrangement is followed by an inversion of $t\,s\,o\,p$, (1) results. Finally, an inversion of $r\,q\,p\,o$ will yield (2).

35. a.   6. There are six chromosomes per haploid set, and therefore six different trisomics are possible.

 b.   15. The chromosomes are C1, C2, C3, C4, C5, and C6. The possible double trisomics are C1C2, C1C3, C1C4, C1C5, C1C6, C2C3, C2C4, C2C5, C2C6, C3C4, C3C5, C3C6, C4C5, C4C6, and C5C6.

36.   A translocation from the tip of the normal X in the male to the Y. We expect all males to receive an X chromosome with the white-eye allele from the female. For the male to be wild type, he still must have part of the wild-type X chromosome. To test, cross this wild-type male with white-eyed females. All the female progeny should be white-eyed and all the male progeny red-eyed. Cytological examination of the chromosomes should reveal the translocation.

37.   One species contributes $n = 8$ chromosomes and the other $n = 6$ chromosomes. The hybrid will contain $8 + 6 = 14$ chromosomes. When the chromosome number is doubled, the allotetraploid (amphidiploid) will have $14 \times 2 = 28$ chromosomes.

38. a.   If the autotetraploid has 64 chromosomes, each set of chromosomes would have $64/4 = 16$ chromosomes. Therefore, there will be 16 linkage groups.

 b.   An allotetraploid is in essence a double diploid (also termed amphidiploid). This means that all 64 chromosomes exist as pairs, and so the number of chromosomes per set is $64/2 = 32$. Therefore, the number of linkage groups is 32. This is true regardless of the number of chromosome pairs contributed by the original parents. For example, if one species is $2n = 20$ and the other species is $2n = 44$, the linkage group contributions of the species will be 10 and 22, respectively. If both species are $2n = 32$, each will contribute 16 linkage groups. So the total number of linkage groups will be 32 in all such cases.

39. a.   8.

 b.   9. Two homozygotes: *AAAAAAAA*, *aaaaaaaa*, and seven types of heterozygotes, *AAAAAAAa*, *AAAAAAaa*, *AAAAAaaa*, *AAAAaaaa*, *AAAaaaaa*, *AAaaaaaa*, and *Aaaaaaaa*.

40.   Nondisjunction in one of the cells in meiosis II in the female leads to the following gametes: X, X, XX, and O (lacking a sex chromosome). The male's gametes will be X and Y.

Gametes	X	X	XX	O
X	XX normal female	XX normal female	XXX triple-X female	XO Turner female
Y	XY normal male	XY normal male	XXY Klinefelter male	YO lethal

41. Nondisjunction in meiosis I in the male (with meiosis II being normal) leads to the following gametes: XY, XY, O, and O (lacking a sex chromosome). The female can only produce X gametes.

Gametes	XY	XY	O	O
X	XXY	XXY	XO	XO
	Klinefelter male	Klinefelter male	Turner female	Turner female

42. a. These mosaics arise as a result of nondisjunction of sister chromatids in mitosis in embryonic development.

    b. The phenotypic variance can be explained by the timing of the mitotic nondisjunction. If it occurs early in embryonic development, the aneuploidy is likely to be spread throughout the body, and the phenotype may be severe; if it occurs late in embryonic development, the aneuploidy may be limited to one or a few organs or tissues, and the phenotypic effect will be milder.

43. An XO/XYY mosaic can arise from mitotic nondisjunction during early embryonic development in an XY male.

44. 32. The gamete from P will have 9 chromosomes, and the gamete from U will have 7 chromosomes. The original zygote will have 16 chromosomes, but none of these will pair. To be fertile, each chromosome must be duplicated to yield 32 chromosomes.

45. a. A male with Down syndrome will produce two types of gametes with regard to chromosome 21: 50% C21 C21 and 50% C21. The normal female will produce one type of gamete: C21. Use the forked-line method:

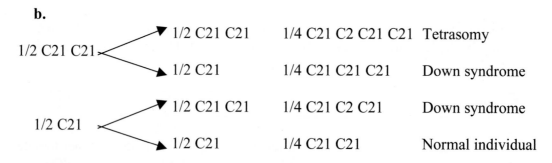

    Therefore, the phenotypic ratio is 1:1, Down syndrome to normal.

    b.

	1/2 C21 C21	1/4 C21 C2 C21 C21	Tetrasomy
1/2 C21 C21	1/2 C21	1/4 C21 C21 C21	Down syndrome
	1/2 C21 C21	1/4 C21 C2 C21	Down syndrome
1/2 C21	1/2 C21	1/4 C21 C21	Normal individual

Tetrasomy is not compatible with survival. Therefore the phenotypic ratio would be 2/3:1/3 (or 2:1), Down syndrome to normal.

46. The father. The allele for color blindness can only come from the mother. If meiosis in her is normal, an egg could get the X chromosome carrying the mutant allele. The daughter has only one X chromosome, so the sex chromosomes failed to separate in the father, and a sperm with neither X nor Y fertilized the egg.

47. Either. Let X^C = Normal vision and X^c = color blindness. The woman is $X^C X^c$, the man is $X^c Y$, and the child is $X^c X^c Y$. If meiosis is normal in the woman, the egg gets X^c, and the sperm must contain $X^c Y$. If meiosis is normal in the male, the sperm gets Y. If the chromatids of the X^c chromosome fail to separate during the second division in the female, the egg gets $X^c X^c$.

48. a. The mother. Let X^E = Faulty tooth enamel, and X^e = normal tooth enamel. The child has Klinefelter syndrome and normal teeth, so he has to be $X^e X^e Y$. The father has faulty teeth, $X^E Y$, and must have donated the Y chromosome to this child. Therefore, the child must have received both of his X chromosomes from his mother, who must be $X^E X^e$.

b. Meiosis II, because the child received two copies of the same X chromosome from his mother.

49. The first meiotic division in the father is normal, producing cells with either two X or two Y chromatids. During the second meiotic division in the cell with the two Y chromatids, both Y chromatids move to the same pole and end up in the same sperm cell.

50. Since this karyotype is a tetraploid, it is likely that after fertilization all the chromosomes duplicated. An alternative, if quite unlikely, explanation is that the fetus may have resulted from the fertilization of a diploid egg by a diploid sperm.

51. The numbers are all multiples of 14 (1, 2, 3, 4, 5, 6, 7, and 8 copies). We can thus hypothesize that the original diploid chromosome number ($2n$) is 14. The other species would be polyploids, multiples of the original 14 (tetraploid, hexaploid, and so forth). These are all the even ploids up to 112 chromosomes. As we saw, even ploids have the potential to succeed in meiosis, whereas odd ploids rarely do.

52. The key is to remember that a deletion can only cause pseudodominance of adjacent genes. For example, del 1 leads to pseudodominance of genes *e* and *j*. Therefore, genes *e* and *j* must be adjacent to each other. Using a systematic logical approach, you should come up with the following gene order: *j e l • r a d y*.

53. a. Give a different designation to each of the four alleles: A_1, A_2, a_1, and a_2. The autotetraploid can produce six possible gametes: $A_1 A_2$, $A_1 a_1$, $A_1 a_2$, $A_2 a_1$, $A_2 a_2$, and $a_1 a_2$. Therefore the ratio of gametes is 1/6 *AA*:4/6 *Aa*:1/6 *aa*.

b. While the forked-line method can be used, let's try the Punnett square approach!

Gametes	1/6 *AA*	4/6 *Aa*	1/6 *aa*
1/6 *AA*	1/36 *AAAA*	4/36 *AAAa*	1/36 *AAaa*
4/6 *Aa*	4/36 *AAAa*	16/36 *AAaa*	4/36 *Aaaa*
1/6 *aa*	1/36 *AAaa*	4/36 *Aaaa*	1/36 *aaaa*

The genotypic ratio is 1/36 *AAAA*:8/36 *AAAa*:18/36 *AAaa*:8/36 *Aaaa*:1/36 *aaaa*. The phenotypic ratio is 35/36 *A⁻*:1/36 *aaaa*.

54. If the plants were closely enough related, both in physiology and chromosome number, it is conceivable that a successful hybrid could be created. Perhaps chromosome doubling would be necessary to achieve successful meiosis. Although the odds of this are low, the odds are even lower that the hybrid would combine the desired attributes of each species. Similar experiments done in the past have rarely been successful even when viable hybrids were produced.

CHAPTER INTEGRATION PROBLEM

a. The P female has normal wings, so she must be homozygous for wild-type wing shape: N^+N^+. She is red-eyed, so she must have at least one w^+ allele. Her F_1 offspring consist of approximately equal numbers of red-eyed and white-eyed males, implying that she is heterozygous for eye color: w^+w. Therefore, her genotype is $N^+ w^+/N^+ w$.

b. The use of X-rays alerts us to chromosomal aberrations. Indeed, this is justified by the presence of 20 notched females in the F_1 offspring. These must be heterozygous for Notch: N^+N. Therefore, X-ray irradiation of the P female caused mutations (most likely deletions) in its gametes, so that some now carry the N allele. These mutant gametes, when fertilized by sperm with the N^+-carrying X chromosome, will produce zygotes that are N^+N. The Notched females produced by these zygotes have red eyes. So they could be w^+w^+ or w^+w. These females inherit the w^+ allele from their father, so they would get the w^+ or w allele from their mother. Therefore, either maternal X chromosome could have incurred the mutation.

c. Given that 4% of the F_1 offspring are Notched females, the absence of Notched males is surprising. This could be explained by the fact that Notched is lethal in the hemizygous state. Males that inherit the N allele from their mother die during embryonic development, and therefore no Notched males will survive to birth.

d. For wild-type gametes:

P $N^+ w^+/N^+ w$ × $N^+ w^+/Y$

F$_1$ 1/4 $N^+ w^+/N^+ w^+$
          1/4 $N^+ w/N^+ w^+$
          1/4 $N^+ w^+/Y$
          1/4 $N^+ w/Y$

and phenotypically

F$_1$ 1/2 normal-winged, red-eyed females
          1/4 normal-winged, red-eyed males
          1/4 normal-winged, white-eyed males

For mutant gametes:

Scenario 1 (chromosomal aberration in X chromosome carrying the w^+ allele):

P $N w^+/N^+ w$ × $N^+ w^+/Y$

F$_1$ 1/4 $N w^+/N^+ w^+$
          1/4 $N^+ w/N^+ w^+$
          1/4 $N w^+/Y$ Die in embryonic development
          1/4 $N^+ w/Y$

and phenotypically

F$_1$ 1/3 Notched, red-eyed females
          1/3 normal-winged, red-eyed females
          1/3 normal-winged, white-eyed males

Scenario 2 (chromosomal aberration in X chromosome carrying the w allele):

P $N^+ w^+/N w$ × $N^+ w^+/Y$

F$_1$ 1/4 $N^+ w^+/N^+ w^+$
          1/4 $N w/N^+ w^+$
          1/4 $N^+ w^+/Y$
          1/4 $N w/Y$ Die in embryonic development

and phenotypically

F$_1$ 1/3 normal-winged, red-eyed females
          1/3 Notched, red-eyed females
          1/3 normal-winged, red-eyed males

e. i.   White-eyed, Notched females   Red-eyed, normal-winged males
          P $N^+ w/N w$ × $N^+ w^+/Y$

F$_1$ 1/4 $N^+ w/N^+ w^+$
              1/4 $N w/N^+ w^+$
              1/4 $N^+ w/Y$
              1/4 $N w/Y$ Die in embryonic development

and phenotypically

F$_1$ 1/3 normal-winged, red-eyed females
              1/3 Notched, red-eyed females
              1/3 normal-winged, white-eyed males

ii.   White-eyed, Notched females   White-eyed, normal-winged males
         P $N^+ w/N w$ × $N^+ w/Y$

F$_1$ 1/4 $N^+ w/N^+ w$
              1/4 $N w/N^+ w$
              1/4 $N^+ w/Y$
              1/4 $N w/Y$ Die in embryonic development

and phenotypically

F$_1$ 1/3 normal-winged, white-eyed females
              1/3 Notched, white-eyed females
              1/3 normal-winged, white-eyed males

f.   For a cross to yield females that are homozygous for *Notch*, it would have to be
between a heterozygous female (N^+N) and a hemizygous male (N/Y). However, those
males are not viable. Therefore, this cross is not feasible.

g.   The cross would be between a white-eyed male and a Notched female that is
homozygous for red eyes:

         P $N^+ w^+/N w^+$ × $N^+ w/Y$

One-third of the offspring should be Notched females. If those have red eyes, their
genotype would have to be w^+w, implying that the *Notch* mutation left the w^+ allele
intact. Therefore, the mutation involved only the N gene. If on the other hand the
Notched females have white eyes, this indicates that the wild-type w^+ allele is not
being expressed in these flies. Therefore, the cause of the Notched phenotype is a
mutation that spanned the *white* locus.

CHAPTER 9 DNA REPLICATION

Chapter Goals

1. Understand and summarize the key experiments that revealed the mechanism of DNA replication in both prokaryotes and eukaryotes and what the results from those experiments would have looked like if semiconservative replication with bidirectional replication forks did not occur.
2. Diagram a DNA replication fork showing the leading and lagging strand, the location of RNA primers, Okazaki fragments, DNA ligase, helicase, single-stranded binding proteins, DNA polymerase, and the 5′ and 3′ ends of all the DNA ends.
3. Describe and diagram the major features associated with replicating a circular DNA molecule and a linear DNA molecule.
4. Understand how DNA replication resolves the issues of the leading and lagging strands.
5. Name the three different *E. coli* DNA polymerases and describe their roles during DNA replication and the activities that they possess.

Key Chapter Concepts

9.1 The Mechanism of DNA Synthesis

- **Mechanisms of DNA Replication:** Different models have been proposed for how DNA replication occurs: conservative, semiconservative, and dispersive. Each model is similar in that one double helix is used to make two identical daughter helices but differs in where the new nucleotides are added relative to the existing or "old" nucleotides.

  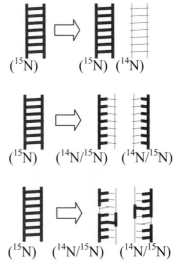

  - The **conservative model** has one daughter double helix made of completely new nucleotides and another daughter double helix made of only old nucleotides.
  - The **semiconservative model** has both helices made of one strand of new nucleotides paired with one strand of old nucleotides to make hybrid double helices.
  - The **dispersive model** also has the two daughter helices being hybrid except that any one strand of the helix may have old and new nucleotides.

- **Meselson and Stahl:** Meselson and Stahl performed the experiment which showed DNA replication occurs by a *semiconservative method* (see figure). This experiment involved growing bacteria in medium containing a dense form of nitrogen (^{15}N) for many generations, until all bacterial DNA was made with this dense nitrogen. They then switched the bacteria to medium containing normal nitrogen (^{14}N). After one, two, and three generations, they isolated DNA and centrifuged it in a density gradient to examine where the new (^{14}N) DNA was located relative to the old (^{15}N) DNA. They observed that before the DNA had replicated, the DNA localized to a band at ^{15}N (old). After generation 1, a single band between ^{14}N and ^{15}N appeared (new + old in *same* helix). After generation 2, the band at ^{14}N/^{15}N (new + old in same helix) was still there along with a new band at ^{14}N (completely new). After generation 3, the ^{14}N/^{15}N band remained, but the ^{14}N band become more prominent. *If conservative replication* was the method, then they would have observed one band for ^{14}N-containing DNA and a separate band for ^{15}N-containing DNA at generations 1, 2, and 3. *If dispersive replication* was the method, they would have observed more and more diffuse bands between the ^{14}N and ^{15}N positions with each additional generation.

9.2 The Process of Strand Synthesis

- **Prokaryotic DNA Replication:** Bacterial DNA replication proceeds from only one origin and bidirectionally. Each strand of the DNA molecule serves as a template for the synthesis of a new DNA strand. DNA replication is **semiconservative**: each strand is conserved in replication. Each nucleotide is added (polymerized) into a growing strand at the 3′ hydroxyl by a DNA polymerase. **DNA polymerase I** fills in small DNA segments during replication and repair, while **DNA polymerase II** acts as a repair enzyme. **DNA polymerase III** is the primary enzyme involved in replication: it catalyzes a bond between the 3′–OH carbon of the last nucleotide in the strand and the 5′–PO_4 carbon of the new nucleotide. Because the strands of DNA are antiparallel and the enzyme polymerizes DNA in one direction only, each strand is replicated differently.
- Replication of one strand only occurs **continuously** using the 3′ → 5′ template strand and a **primer**. A free 3′–OH in the growing strand is absolutely necessary for polymerization to occur because the enzyme adds bases in only the 5′ → 3′ direction. The newly synthesized strand is called the **leading strand**.
- DNA synthesis using the complementary strand is **discontinuous**. This **lagging strand** must be synthesized in pieces because the synthesis does not move toward the **Y-junction**. **Primase** first adds stretches of RNA **primers** that provide a free 3′–OH for the DNA polymerase. DNA polymerase adds nucleotides into the new strand until another RNA primer is encountered. These DNA fragments are called **Okazaki fragments**. DNA polymerase III extends each Okazaki fragment after priming until it reaches the primer RNA of the previous fragment, and then DNA polymerase III dissociates. The 5′ → 3′ exonuclease activity of DNA polymerase I removes the previous RNA primer and replaces it with DNA. **DNA ligase** seals the nick between the two Okazaki fragments. The 3′ → 5′ exonuclease activity of DNA polymerase removes any bases that are incorrectly paired allowing DNA polymerase to **proofread** the DNA as it is synthesized.

9.3 The Point of Origin of DNA Replication

- **Initiation of DNA Replication in Prokaryotes:** Replication is initiated in *E. coli* by binding of **initiator proteins**, the products of the *dnaA* gene, to the *oriC*, or origin of replication. This binding causes the DNA at the *oriC* to **denature**. **Helicase**, the product of the *dnaB* gene, binds to the single-stranded DNA at the *oriC* and recruits primase to form the **primosome**. The primosome synthesizes one RNA primer for leading-strand synthesis and the first primer for lagging-strand synthesis. Two copies of DNA polymerase III bind. These two copies of DNA polymerase III combine with a helicase and a primase to form a single unit called the **replisome**. Three subunits, alpha, epsilon, and theta form the **core enzyme** of DNA polymerase III, while the **holoenzyme** consists of 10 subunits. DNA polymerase III synthesizes the daughter leading strand from 5′ → 3′ in a continuous manner. The daughter lagging strand is synthesized discontinuously from 5′ → 3′. **Single-strand binding proteins** (SSB proteins) bind to the single-stranded DNA to prevent the strands from reannealing until the new strand is synthesized. As replication continues, positive supercoiling builds up in front of the replication fork. **Topoisomerases** regulate the **supercoiling** of the DNA. **Type I topoisomerases** change the **linkage number** of the DNA by one. They can relieve DNA coiling by breaking one of the DNA strands of a double helix and passing the other strand through the break. **Type II topoisomerases** break both strands of the double helix, pass another double helix through the temporary gap, and change the linkage number of the DNA in multiples of 2.

9.4 Termination of Replication

- **Termination of DNA Replication in Prokaryotes:** DNA replication in *E. coli* is terminated at the terminus region (*ter*) located 180° from the *oriC* on the circular chromosome. This terminus region contains six sites that are bound by a termination protein. The two interlinked circular daughter DNA molecules are separated by a type II topoisomerase.

9.5 Other Replication Structures

- **The Rolling-Circle Model:** This form of replication occurs in the *E. coli* F plasmid or Hfr chromosome during bacterial conjugation. Rolling-circle replication is initiated by a single-strand nick in one of the DNA strands that creates a 3′–OH that can serve to prime replication of the intact circular strand. This results in replication of a circle in the $5' \rightarrow 3'$ direction with displacement of a $5'$ tail which serves as a template for discontinuous replication using an RNA primer. When replication of the tail is complete, a nuclease cleaves the template strand from the newly synthesized leading strand producing a linear and a nicked circular DNA molecule.

- **The D-Loop Model:** The DNA in chloroplasts and in mitochondria replicates by forming a D-loop. The origin is located at different places on the two parental strands. Replication initiates at one origin and continues until it reaches the origin on the opposite strand. This allows replication to initiate on the second strand in the opposite direction resulting in unidirectional leading-strand synthesis from both origins.

9.6 Eukaryotic DNA Replication

- **Eukaryotic Polymerases and Other Enzymes:** In eukaryotes, replication occurs semiconservatively using both a leading and lagging strand as in bacteria. In addition, there are usually multiple origins of replication and additional polymerase enzymes. **DNA polymerase α-primase** generates Okazaki fragment primers. **Replication factor C** (RFC) loads a six-unit clamp called **proliferating cell nuclear antigen** (PCNA) onto the DNA. This protein serves as a processivity factor for **DNA polymerase delta** which is the major replicative polymerase in eukaryotes. RNA primers in eukaryotes are removed by RNase enzymes rather than the $5' \rightarrow 3'$ exonuclease activity of DNA polymerase I.
- **Replication Origins in Eukaryotes:** In yeast, replication originates at **autonomously replicating sequences** (ARS sites) which are bound by the **origin recognition complex** (ORC). Additional proteins are needed to initiate replication.
- **Telomeres and Termination of Replication in Eukaryotes:** To protect against loss of the $3'$ ends of linear chromosomes, the ends of the chromosomes or **telomeres** are replicated by **telomerase** that uses an associated RNA molecule complementary to the telomeric repeat as a template to extend the $3'$ ends of the telomere by **reverse transcription**.

Key Terms

- reverse transcriptase
- template
- semiconservative replication
- conservative replication
- dispersive replication
- density gradient centrifugation
- autoradiography
- theta (θ) structure
- Y-junctions
- replication forks
- replicons
- DNA polymerase I
- DNA polymerase II
- DNA polymerase III
- continuous replication
- primer

- discontinuous replication
- Okazaki fragments
- leading strand
- lagging strand
- processivity
- primase
- elongation
- nucleases
- exonucleases
- endonucleases
- proofreading
- DNA ligase
- initiator proteins
- denature
- helicase
- primosome
- holoenzyme

- core enzyme
- polymerase cycling
- single-strand binding proteins (SSB proteins)
- replisome
- supercoiled
- linkage number (L)
- topoisomers
- topoisomerase
- DNA gyrase
- rolling-circle replication
- D-loop replication
- autonomously replicating sequences (ARS)
- origin recognition complex (ORC)

Understanding the Key Concepts
Use the key terms or parts of their definitions to complete the following sentences.

Each strand of a DNA double helix serves as a (1)_____ for DNA replication. In *E. coli*, replication initiates at a single site in the (2)_____ chromosome. This site, the (3)_____ is recognized by (4)_____ that cause the DNA to (5)_____. (6)_____ binds to the (7)_____ DNA at the (8)_____, unwinds the DNA and recruits (9)_____. (10)_____ synthesizes a short strand of (11)_____ in the (12)_____ direction. This primer has a free (13)_____ group at the (14)_____ end which (15)_____ can add (16)_____, (17)_____ the replicating strand.

Only one strand of a DNA double helix can be replicated continuously because (18)_____ can only add bases to a growing DNA strand in the (19)_____ direction. The other strand is replicated (20)_____. This template strand is oriented in a 5′ → 3′ direction (21)_____ the replication fork. Replication of this strand involves the synthesis of short (22)_____ fragments. These fragments are joined together by (23)_____. As DNA replication proceeds, the 3′ → 5′ (24)_____ activity of (25)_____ ensures that bases that are incorrectly paired are removed. In *E. coli*, (26)_____ that builds up ahead of the replication fork as a result of the circular duplex winding about itself in the (27)_____ direction as the helix twists is relieved by (28)_____, which is a (29)_____. These enzymes change the (30)_____ of the DNA in multiples of (31)_____.

Eukaryotic replication of (32)_____ chromosomes occurs generally the same as replication in prokaryotes. One major difference is that there are multiple (33)_____ each with their own (34)_____ in eukaryotes. Since the eukaryotic genome is significantly (35)_____ than the prokaryotic genome, (36)_____ at multiple origins allows DNA replication to occur efficiently in eukaryotes. In addition, loss of DNA sequences at the ends of (37)_____ chromosomes has to be prevented in eukaryotes. (38)_____ uses an associated (39)_____ as a (40)_____ to add (41)_____ to the (42)_____ end of the genomic DNA.

Figure Analysis
Use the indicated figure or table from the textbook to answer the questions that follow.

1. **Figure 9.4**
(a) Describe the difference between the semiconservative, conservative, and dispersive models of replication.

(b) What experiment did Meselson and Stahl design to distinguish between these three models for DNA replication?

(c) What were the important controls in Meselson and Stahl's experiment?

(d) What would Meselson and Stahl have observed if replication occurred according to the conservative model of DNA replication?

(e) Why were Meselson and Stahl unable to distinguish between the dispersive and the semiconservative models following one round of DNA replication?

(f) How did Meselson and Stahl distinguish between the semiconservative and the dispersive models of DNA replication? What did each model predict would occur in this experiment? What did Meselson and Stahl observe?

2. **Figure 9.22**
(a) What is represented by the red DNA strands in figure 9.22?

(b) What is the polarity of the DNA strand serving as a template for leading-strand replication?

(c) In what direction are the RNA primers on the lagging strand synthesized?

(d) Why is leading-strand replication continuous, while lagging-strand replication is discontinuous?

(e) What does the replisome consist of?

(f) Describe how leading- and lagging-strand replication are coordinated.

(g) Why is the SSB protein required for DNA replication?

(h) Describe what is meant by "polymerase cycling."

(i) What is the function of DNA polymerase I?

3. **Figure 9.30**
(a) Why is it problematic to replicate the ends of linear chromosomes?

(b) What does telomerase consist of? What enzymatic activity is associated with telomerase?

(c) What is the critical function of telomerase?

(d) How does telomerase continue to add more and more repeats to the 3′ ends of linear chromosomes?

General Questions

Q9-1. Describe three characteristics of eukaryotic DNA that make its replication more complex than prokaryotic DNA replication.

Q9-2. Describe the three notable observations that John Cairns made in his experiment using autoradiography to visualize replication of a circular DNA molecule.

Q9-3. Describe how it was determined that DNA replication is bidirectional.

Q9-4. Describe the four repeating steps required for discontinuous lagging-strand synthesis and the proteins required to carry out these steps in *E. coli*.

Q9-5. Describe the functions of the α, β, ε, and θ subunits of DNA polymerase III in *E. coli*.

Multiple Choice

For each of the following, circle the letter of the choice that most appropriately answers the question.

1.) Which of the following bacterial enzymes is responsible for removing the RNA primer of an Okazaki fragment and replacing it with DNA nucleotides during lagging-strand replication?

a. primase
b. exonuclease
c. DNA polymerase I
d. RNase
e. DNA polymerase III

2.) Which *E. coli* enzyme is the main replicative polymerase in bacteria?
   a. DNA polymerase I
   b. DNA polymerase delta
   c. DNA polymerase epsilon
   d. DNA polymerase III
   e. DNA polymerase beta

3.) Which eukaryotic enzyme is considered to be the major repair polymerase?
   a. DNA polymerase beta
   b. DNA polymerase delta
   c. DNA polymerase epsilon
   d. DNA polymerase gamma
   e. DNA polymerase theta

4.) Bacterial DNA ligase seals the "nick" between two Okazaki fragments using energy from which of the following?
   a. ATP
   b. GTP
   c. NAD
   d. electron transport chain
   e. FAD

5.) Bacterial primase must remain associated with which protein to stay attached to the DNA during primer formation?
   a. helicase
   b. the clamp loader
   c. the processivity clamp
   d. SSB proteins
   e. the replicative polymerase

6.) DNA replication in eukaryotes occurs considerably slower than it does in prokaryotes because of which of the following?
   a. Eukaryotic replication can proceed in only one direction.

b. Eukaryotic DNA is packaged into chromatin.
c. The replicative polymerase is less efficient.
d. Replication of the telomeres is slow.
e. Removal of RNA primers in eukaryotes is considerably slower because it requires RNase enzymes rather than DNA polymerase alone.

7.) Eukaryotic replication occurs at a rate of approximately _____ base pairs per minute.
   a. 2000
   b. 10,000
   c. 20,000
   d. 50,000
   e. 200,000

8.) Type II topoisomerases in *E. coli* do which of the following?
   a. Change the linkage number in multiples of 2.
   b. Alleviate positive supercoiling ahead of the replication fork.
   c. Break one strand of a double helix and pass the other strand through the break.
   d. Both a and b
   e. All of the above

9.) The proofreading function of bacterial DNA polymerase ensures that an error is introduced during replication at a rate of _____.
   a. $1/10^3$ nucleotides
   b. $1/10^5$ nucleotides
   c. $1/10^7$ nucleotides
   d. $1/10^9$ nucleotides
   e. $1/10^{15}$ nucleotides

10.) Replication in *E. coli* is terminated by which of the following?
   a. telomeric repeats adjacent to the *oriC*
   b. *ter* sequences located 180° from the *oriC*
   c. *tus* sequences located 180° from the *oriC*
   d. *tus* sequences located 360° from the *oriC*
   e. an abundance of positive supercoiling ahead of the replication fork

Practice Problems

P9-1. Modern molecular genetics is based upon the assumption that DNA possesses the ability to replicate itself. Diagram and explain how DNA fulfills this assumption.

P9-2. Examine the following DNA molecule in the process of replication. Identify A–M.

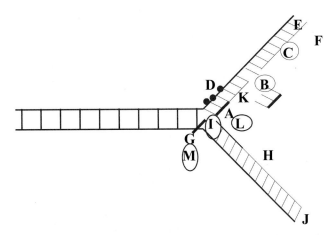

A (structure)   E (polarity)   I (enzyme)   M (enzyme creating location G)
B (enzyme)   F (polarity)   J (polarity)
C (enzyme repairing nick) G (location)   K (structure)
D (proteins)   H (strand)   L (enzyme inserting segment A)

P9-3. Would the following strand serve as the template for the leading strand or the lagging strand (assuming that the Y-junction is moving from right to left)? If it is the template for the lagging strand, what would be the first 10 bases that primase would insert?

   3'-TAAGCTTAACGCGCATCAGCGCGCTCCGATCGATCGATCAGCGATCAGAGGGAAG-5'

(*Hint 1*: DNA polymerase III requires a free 3'–OH in the growing strand to add a new nucleotide. Continuous replication can only occur in the 5' → 3' direction using the 3' → 5' template strand.)

P9-4. Consider each of the following problems (a–e) associated with prokaryotic DNA replication. For each, determine which enzyme(s) is (are) defective?
   a.   DNA synthesis is never completed because Okazaki fragments are not joined.
   b.   DNA synthesis of the lagging strand is never initiated.
   c.   Replication is never initiated although the DNA is unwound.
   d.   The newly synthesized DNA contains mismatched pairs.
   e.   New DNA strand contains some RNA bases.

P9-5. Draw a diagram of a replication fork opening from right to left. Draw the leading and lagging strands and indicate the 5' and 3' ends of each strand of DNA on the template and on the newly synthesized daughter strands.

Assessing Your Knowledge

Understanding the Key Concepts—Answers

1.) template; 2.) circular; 3.) origin of replication (*oriC*); 4.) initiator proteins; 5.) denature; 6.) Helicase; 7.) single-stranded; 8.) *oriC*; 9.) primase; 10.) Primase; 11.) RNA; 12.) 5'→ 3'; 13.) OH; 14.) 3'; 15.) DNA polymerase III; 16.) deoxyribonucleotides; 17.) elongating; 18.) DNA polymerases; 19.) 5'→ 3'; 20.) discontinuously; 21.) toward; 22.) Okazaki; 23.) DNA ligase; 24.) exonuclease; 25.) DNA polymerase;

26.) positive supercoiling; 27.) same; 28.) DNA gyrase; 29.) type II topoisomerase; 30.) linkage number; 31.) two; 32.) linear; 33.) replicons; 34.) origin; 35.) larger; 36.) initiation; 37.) linear; 38.) Telomerase; 39.) RNA; 40.) template; 41.) DNA nucleotides; 42.) 3′.

Figure Analysis—Answers

1a.) In the semiconservative model of replication, each strand of the parental DNA molecule serves as a template for the synthesis of a new strand. The daughter DNA molecules consist of one parental strand and a newly synthesized, complementary daughter strand. In the conservative model, each strand of the parental DNA molecule serves as a template for a new strand. Following replication, the parental duplex reassociates, and the newly synthesized strands associate. Alternatively, for this model, one strand of the parental DNA duplex may serve as a template for DNA replication. Following replication, the template strand reassociates with the original second strand of the parental duplex and the newly synthesized daughter strand is used as a template to synthesize the complementary strand and form the new daughter duplex. In the dispersive model of replication, only some parts of the original DNA molecule are conserved. Each daughter molecule consists of part template and part newly synthesized DNA, with neither being restricted to a single strand; b.) Meselson and Stahl used density-gradient centrifugation to look at the density of DNA obtained from *E. coli* grown in the presence of ^{15}N, or ^{14}N relative to the density of DNA obtained from *E. coli* grown in the presence of ^{15}N and then transferred and allowed to replicate in ^{14}N medium; c.) As controls, Meselson and Stahl grew *E. coli* in the presence of ^{15}N alone and in the presence of ^{14}N alone to determine where the bacterial DNA would band being allowed to incorporate only the heavy or only the light isotope following density-gradient centrifugation; d.) If replication occurred according to the conservative model of replication, Meselson and Stahl would have found that after one round of DNA replication, a new band with the density of a molecule containing ^{14}N alone would have appeared along with the original ^{15}N band; e.) Following one round of DNA replication, Meselson and Stahl found a band that was intermediate in density between light (^{14}N) and heavy (^{15}N) DNA. This was consistent with both the dispersive and semiconservative models because both predicted that the two daughter molecules would be composed of part parental DNA (^{15}N) and part newly synthesized DNA (^{14}N); f.) Meselson and Stahl distinguished between the semiconservative and the dispersive models by allowing some of the *E. coli* with heavy DNA to undergo two rounds of DNA replication in the ^{14}N medium. The semiconservative model predicted that 50% of the DNA molecules would contain two light strands, and the other 50% would contain one light and one heavy strand and be intermediate in density. The dispersive model predicted that only a band with intermediate density would appear, but this band would lie closer to the light DNA band following two rounds of DNA replication. Meselson and Stahl found that after two rounds of DNA replication, they observed two bands, one that corresponded to the light DNA (^{14}N only), and one that corresponded to the intermediate form of DNA (1/2 ^{14}N, 1/2 ^{15}N), consistent with the semiconservative model.

2a.) The red DNA strands in figure 9.22 represent the newly synthesized daughter strands; b.) 3′→ 5′; c.) 5′→ 3′; d.) Leading strand replication is continuous, while lagging-strand replication is discontinuous because DNA polymerases can only add nucleotides in the 5′→ 3′ direction and require a primer that has a free 3′–OH. The template strand that is oriented 3′→ 5′′ toward the replication fork can be used as a template for continuous replication because the primer will be synthesized 5′→ 3′′ and have a free 3′–OH group to add nucleotides onto. The template strand that is oriented 5′→ 3′ away from the replication fork must be replicated discontinuously in short fragments synthesized 5′→ 3′ and each requiring an RNA primer. In this way, replication can still proceed in the 5′→ 3′ direction on the lagging strand as well; e.) The replisome consists of two copies of DNA polymerase III, and a primosome consisting of one copy of the helicase and one copy of the primase; f.) Leading- and lagging-strand replication are coordinated by two molecules of DNA polymerase III that are attached to each other. One molecule replicates the leading strand, and the other replicates the Okazaki fragment on the lagging strand as the single-strand loop region of the lagging-strand template is pulled through the DNA polymerase III enzyme. This occurs while the replisome moves in the direction of the advancing replication fork. As one Okazaki fragment is synthesized, the primer for the next Okazaki fragment is created; g.) The SSB protein is required for DNA replication because it binds to the single-stranded DNA that is generated by the helicase and ensures that the DNA strands cannot reanneal before the new daughter strand is synthesized; h.) "Polymerase cycling" refers to the movement of DNA polymerase III off and on the DNA of the lagging-strand template as

Okazaki fragments are completed. DNA polymerase III releases one completed Okazaki fragment on the lagging-strand template and begins the synthesis of the next Okazaki fragment; i.) The function of DNA polymerase I is to complete the synthesis of Okazaki fragments after DNA polymerase III dissociates. As DNA polymerase III approaches the previous Okazaki fragment during lagging-strand replication, it dissociates. DNA polymerase I then attaches to the lagging strand and removes the RNA primer of the previous Okazaki fragment using its $5' \rightarrow 3'$ exonuclease activity. It replaces this RNA with DNA nucleotides using its $5' \rightarrow 3'$ polymerase activity.

3a.) As the replication fork moves toward the end of the chromosome, the leading strand can be replicated all the way to the 5' end of its template strand. In contrast, discontinuous replication of the lagging strand requires an RNA primer before the Okazaki fragment is synthesized. This primer can be degraded resulting in loss of the 5' end of the newly synthesized strand. This strand will serve as a template during the next round of DNA replication, and these sequences will not be able to be synthesized. Furthermore, the DNA strand will shorten more and more with each successive round of replication leading to loss of the genomic DNA sequences that are adjacent to the telomere; b.) Telomerase has both an RNA and a protein component. Reverse transcriptase activity is the enzymatic activity associated with telomerase. This prevents loss of the ends of linear chromosomes; c.) Telomerase adds copies of the telomeric repeat sequence using its RNA component as a template for the polymerization of DNA nucleotides by reverse transcription; d.) Following base-pairing of the telomerase RNA with the 3' end of the chromosome, the 3' end is extended by reverse transcription using the telomerase RNA as a template. Following synthesis of the new DNA, the telomerase translocates to base-pair with the new 3' end of the single-stranded telomere and extends the 3' end of the telomere further by reverse transcription using its RNA component as the template.

General Questions—Answers

Q9-1. Eukaryotes have linear chromosomes with telomeres. Telomerase adds copies of the telomeric repeat sequence to the ends of linear chromosomes to prevent loss of DNA sequences near the end. This is not a problem in prokaryotes because they have a single circular chromosome. Each eukaryotic chromosome has multiple replicons each with its own origin, while the *E. coli* chromosome has only one replicon. Eukaryotic DNA is packaged into chromatin by associated histone proteins, while prokaryotic DNA is not associated with histone proteins.

Q9-2. John Cairns made the observations that *E. coli* DNA is circular; the integrity of the circular DNA is maintained during DNA replication forming an intermediate θ structure; and replication initiates at a single point in the circular DNA molecule and proceeds from one or two replication forks in the circle.

Q9-3. John Cairns made the observation that a prokaryotic circular chromosome undergoing DNA replication formed a θ structure which contained two Y-junctions. To determine whether both of these Y-junctions function as replication forks, replication was allowed to begin and then tritium-labeled dTTP was added to the medium. After a short period of time, the reaction was stopped. In this experiment, the radioactive nucleotide would only be incorporated into the DNA at functional replication forks. Therefore, if replication is unidirectional, only one Y-junction will be radioactively labeled, but if it is bidirectional, both Y-junctions will be radioactively labeled. DNA was prepared and visualized by autoradiography. Both of the Y-junctions were found to be labeled indicating that replication is bidirectional.

Q9-4. The four steps required for discontinuous DNA synthesis are primer synthesis, elongation, primer removal and gap filling, and ligation. In *E. coli*, primase synthesizes an RNA primer in the $5' \rightarrow 3'$ direction de novo. DNA polymerase III adds deoxyribonucleotides to the 3' end of the RNA primer to elongate the Okazaki fragment. When it reaches the RNA primer of the previously synthesized Okazaki fragment, DNA polymerase III falls off the DNA. DNA polymerase I attaches to the lagging strand and removes the previous RNA primer using its $5' \rightarrow 3'$ exonuclease activity. It simultaneously replaces these nucleotides with DNA nucleotides using its $5' \rightarrow 3'$ polymerase activity. This leaves only a single phosphodiester bond missing between two adjacent nucleotides on the two Okazaki fragments. This nick is sealed by DNA ligase which ligates the two Okazaki fragments together.

Q9-5. The α, ε, and θ subunits make up the core DNA polymerase III enzyme. They represent the fewest number of subunits required for enzymatic activity. The α subunit has 5′ → 3′ polymerase activity associated with it, the ε subunit has 3′ → 5′ exonuclease activity associated with it, and the θ subunit is a necessary part of the polymerization core. The β subunit acts as a processivity clamp. It is present on the DNA as a dimer, and it forms a doughnut around the DNA and holds the core enzyme tightly to the DNA allowing DNA polymerase III to be highly processive during DNA replication.

Multiple Choice—Answers

1.) c; 2.) d; 3.) a; 4.) c; 5.) d; 6.) b; 7.) a; 8.) d; 9.) d; 10.) b.

Practice Problems—Answers

P9-1. DNA replication is a semiconservative process: one strand of the double helix is preserved during each replication. Each strand serves as a template for the synthesis of a new strand.

P9-2. A = RNA primer, B = DNA polymerase I, C = ligase, D = Ssb proteins, E = 5′ end, F = 3′ end, G = Y-junction, H = leading strand, I = DNA polymerase III, J = 3′ end, K = Okazaki fragment, L = primase, M = Helicase

P9-3. The DNA strand would serve as the template for the lagging strand, assuming that the 3′ end of the molecule is the Y-junction. Following the rules of complementarity, the RNA primer sequence would be 5′-AUUCGAAUUG-3′.

P9-4. Each of the problems described probably arose from a single defective enzyme.
a.   Ligase   d. DNA polymerase I or II
b.   Primase   e. DNA polymerase I
c.   DNA polymerase III

P9-5.

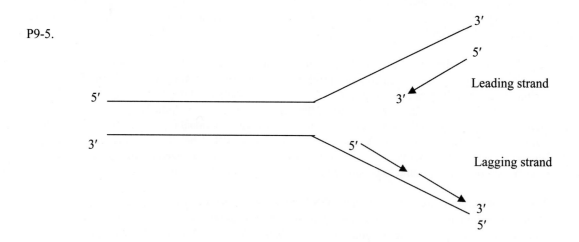

9

DNA REPLICATION

CHAPTER SUMMARY QUESTIONS

1. 1. g, 2. e, 3. a, 4. i, 5. c, 6. h, 7. d, 8. b, 9. j, 10. f.

2.

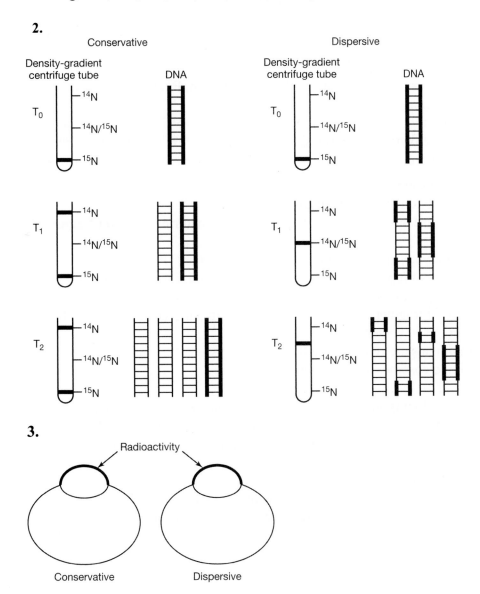

3.

4. DNA polymerases cannot initiate DNA synthesis. They require the help of the enzyme primase, a special type of RNA polymerase. Primase creates a 10–12 nucleotide RNA primer, which provides a free 3'-OH group to which DNA polymerases can add deoxyribonucleotides.

5. The 3' → 5' exonuclease activity is used to "proofread" DNA during replication. A mismatched deoxyribonucleotide at the growing end of a DNA strand can be removed before polymerization proceeds again. All three bacterial DNA polymerases possess this activity.

6. The 5' → 3' exonuclease activity is used to remove the RNA primers from the DNA after replication. Ribonucleotides are removed sequentially from the 5' to the 3' ends. Only DNA polymerase I possesses this activity.

7. DnaA proteins (c) → helicase (e) → primase (a) → DNA pol III (d) → DNA pol I (b).

8. There are many similarities and differences to choose from. First the similarities:
   (a) Replication is semiconservative.
   (b) Replication is bidirectional.
   (c) Substrates are deoxyribonucleotides.
   (d) There is a requirement for helicase, single-strand binding proteins, and primase.
   (e) DNA is synthesized in the 5' → 3' direction.
   (f) Replication is continuous on the leading strand.
   (g) Replication is discontinuous on the lagging strand, requiring Okazaki fragments.

   Replication in eukaryotes differs from that in bacteria in the following:
   (a) Multiple origins of replication per chromosome (bacteria have only one).
   (b) More DNA polymerases.
   (c) Shorter Okazaki fragments (100–200 nucleotides in eukaryotes versus 1000–2000 nucleotides in bacteria)
   (d) Okazaki fragments are removed by RNase.
   (e) Issues with nucleosomes and telomeres.

9. Both enzymes break bonds and are required for bacterial DNA replication. DNA helicase breaks hydrogen bonds, separating the two strands of DNA; DNA topoisomerase IV breaks phosphodiester bonds, separating the two completed circular daughter DNA molecules.

10. A primosome is a helicase plus a primase; it opens the DNA and creates RNA primers, and is part of the replisome. A replisome includes a primosome plus two copies of DNA polymerase III; it coordinates replication on both the leading and lagging strands at the Y-junction.

11. Continuous replication occurs 5' → 3' creating a leading strand; discontinuous replication occurs 5' → 3' backward from the replicating fork on the 5' → 3' template creating a lagging strand. Discontinuous replication is necessitated because none of the DNA polymerases acts in the 3' → 5' direction.

12. See figure 9.14.

13. Eukaryotes have linear chromosomes, whereas prokaryotes have circular chromosomes.

14. DNA polymerase III joins a 5'-triphosphate to the 3'-OH of a nucleotide already found on the DNA molecule. A pyrophosphate (P–P_i) is released in the process, leaving only one phosphate in the phosphodiester linkage. DNA ligase joins the 5'-monophosphate of one nucleotide to the 3'-OH of a neighboring nucleotide.

15. **a.** Single-strand binding proteins bind to the separated DNA strands and prevent them from reannealing until the new strands are synthesized.
    b. SSBP are not directly involved in DNA replication and will most likely be used in equal frequency during continuous and discontinuous replication.

EXERCISES AND PROBLEMS

16. Sequence II. Of the three sequences, it has the highest AT content (70%) and is therefore the easiest to "unzip."

17. Okazaki fragments would be required. Even with a unidirectional replication fork, there would still be a leading and a lagging strand.

18.

Original	5'-ATTCTTGGCATTCGC-3'
DNA complement	3'-TAAGAACCGTAAGCG-5'
RNA primer	3'-UAAGAACCGUAAGCG-5'
Replication is 5' → 3'.	

19. **a.** DNA pol III and/or DNA pol I.
    b. DNA pol I.
    c. Primase or DNA pol III.
    d. Helicase.

20. This problem can be solved using the product rule of probability. The original solution contained 80% heavy and 20% light DNA.

$$0.8 \; ^{15}N \times 0.8 \; ^{15}N = 0.64 \; ^{15}N^{15}N$$
$$0.2 \; ^{14}N \times 0.2 \; ^{14}N = 0.04 \; ^{14}N^{14}N$$
$$0.8 \; ^{15}N \times 0.2 \; ^{14}N = 0.16 \; ^{15}N^{14}N$$
$$0.2 \; ^{14}N \times 0.8 \; ^{15}N = 0.16 \; ^{14}N^{15}N$$

Consequently, the proportions are

$$0.64 \; ^{15}N^{15}N : 0.32 \; ^{15}N^{14}N : 0.04 \; ^{14}N^{14}N$$

21. **a.** The replication fork is moving right-to-left.

    b. Continuous synthesis occurs along the strand that goes in the 3' → 5' direction, and therefore, the upper template strand.

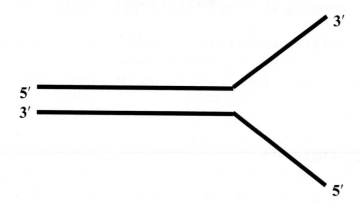

    c. Okazaki fragments form along the discontinuously replicated (5' → 3') strand, which in this case is the lower template strand.

22. Theta replication is bidirectional, whereas rolling-circle replication is unidirectional. Therefore, it will take 30 × 2 = 60 minutes to replicate the DNA via rolling-circle replication.

23. DNA gyrase creates a double-stranded break in the DNA. Another double-stranded helix is then passed through the break, which is then closed.

24. Let $^{15}N^{15}N$ DNA = HH, $^{14}N^{14}N$ DNA = LL, and $^{15}N^{14}N$ = HL.
    a. 1/4 HL:3/4 LL.
    b. 1/32 HL:31/32 LL.
    c. 1/128 HL:127/128 LL.

25. Let $^{15}N^{15}N$ DNA = HH, $^{14}N^{14}N$ DNA = LL, and $^{15}N^{14}N$ = HL.
    a. 1/8 HH:7/8 LL.
    b. 1/64 HH:63/64 LL.
    c. 1/256 HH:255/256 LL.

26. a. The two replication forks combined covered a distance of 1.8 μm in 60 minutes, and so 1.8/60 = 0.03 μm per minute. Therefore, each replication fork covers a distance of 0.03/2 = 0.015 μm per minute.

   b. Each base pair in B-DNA spans 0.34 nm. The distance 0.015 μm is equivalent to 0.015×10^3 nm. Therefore, the speed of the DNA replication fork is $(0.015 \times 10^3)/0.34 = 44$ base pairs per minute.

27. Okazaki fragments (**c**), because the assumption is that DNA polymerases can synthesize DNA in the 3' → 5' direction, as well as the 5' → 3' direction.

28. Please see figure 9.19 for the origin, figures 9.21 and 9.22 for the continuation, and figure 9.26 for the termination of DNA replication in *E. coli*.

29. Theta structure: prokaryotic DNA replication. D-loop: chloroplasts and mitochondria. Rolling circle: bacterial conjugation. Bubbles: eukaryotic DNA replication. The theta structure is an outcome of replicating a closed circle; the rolling circle is a way to transport a linear copy of a circular chromosome; bubbles are outcomes of multiple origins of replication in a linear DNA; the function of D-loops is not obvious. It may be a more efficient way of replicating a small circle without creating Okazaki fragments.

30. Finding small pieces or fragments of DNA suggests the Okazaki pieces are only slowly, if at all, joined: a function of DNA ligase. The fact that not many long DNA molecules are seen also suggests that the DNA is being broken, implicating a nuclease as well.

31. a. Because there are two replication forks, the overall rate of DNA synthesis is 2000 bp per second. Therefore, the entire *E. coli* chromosome will be replicated in 4,640,000/2000 = 2320 seconds, or 2320/60 = 38.67 minutes.

   b. *E. coli* initiates subsequent rounds of DNA replication before completely replicating the entire chromosome. Therefore, each daughter cell actually receives a chromosome that is in the process of being replicated, in addition to a fully replicated copy.

32. a. Because there are two replication forks, the overall rate of DNA synthesis is 8000 bp per minute. Therefore, the entire chromosome will be replicated in 40,000,000/8000 = 5000 minutes.

   b. Number of replicons × 8 minutes = 5000 minutes × 1 replicon. Therefore there are 5000/8 = 625 replicons.

33. a. No. There is a template but no primer.

   b. No. There is a template and a primer, but the primer does not have a free 3'-OH.

   c. Yes. There is a template and a primer with a free 3'-OH.

34. At one time molecular swivels, presumably protein in nature, located periodically along the DNA, were suggested.

35. It is unlikely that bases are added faster in developing embryos. So we must look for another mechanism. If there are more replicons, and hence more origins of DNA replication, each replicon will be shorter and be able to duplicate faster. Alternatively, and more likely, the process is regulated to slow down adult division.

36. One way to study mutations that are generally lethal is by isolating temperature-sensitive mutations. These mutations involve amino acids that disrupt the functioning of the enzymes at some critical temperature but are phenotypically normal at other temperatures. Thus, the mutant organisms can be kept alive by growing them at one temperature (the permissive temperature), but their mutant effect can be studied at the temperature in which the protein function is disabled (the restrictive temperature). (For additional discussion of these mutations, see chapter 18.)

37. You could imagine a situation in which a single molecule is replicated conservatively.

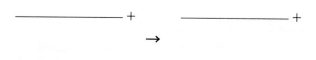

Unfortunately, all nucleic acids seem to replicate semiconservatively. Therefore, we must propose that a complementary strand (–) is made from the first:

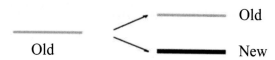

We now need to make more (+) strands from the (–). The (+) strand could separate and allow new (+) strands to be made and peeled off:

Other possibilities also exist as long as you keep in mind the complementary nature of nucleic acids.

CHAPTER INTEGRATION PROBLEM

a.

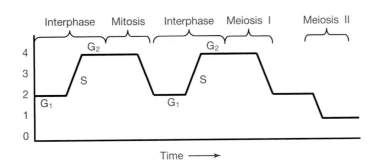

b. **i.** In prophase of mitosis, there are 10 chromosomes, with each consisting of two sister chromatids. These 20 chromatids contain $20 \times 2 = 40$ telomeres.

   ii. In prophase II of meiosis, there are 5 chromosomes, with each consisting of two sister chromatids. These 10 chromatids contain $10 \times 2 = 20$ telomeres.

c. A bivalent consists of a pair of homologous chromosomes, each of which consists of two sister chromatids. Each chromatid is a double-stranded DNA molecule. Therefore, the bivalent has $4 \times 2 = 8$ strands of DNA.

d.

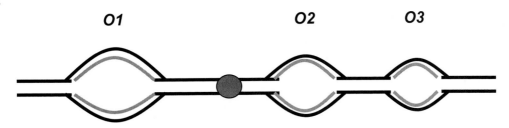

e. The speed of DNA synthesis is the same in each of the two replication forks that proceed from an origin of replication. Therefore, the smaller the amount of DNA in a replication fork area, the faster its replication.

   i. The homolog with the deletion will replicate to the short arm telomere first. This is simply because there is less DNA to replicate between *O1* and the telomere.

   ii. Because the length of DNA between *O1* and the short arm telomere is the same in the two homologs, both will replicate to the short arm telomere in roughly the same time. The same applies for **iii** and **iv**.

f. The cells that are expected to be labeled with tritium are only those that are replicating their DNA and, therefore, are in S phase. If we assume that cells of this culture are spread out in all stages of the cell cycle, then the percentage of cells in each stage should be directly proportional to the length of the stage. Therefore, $8/24 = 33\%$ of cells will be expected to have ^{3}H-labeled DNA.

g.   The order of stages in the cell cycle is: $G_1 \rightarrow S \rightarrow G_2 \rightarrow$ prophase $\rightarrow$ metaphase $\rightarrow$ anaphase $\rightarrow$ telophase. So, the acrocentric chromosome picks up the ^{3}H-label during the S phase, and would have to proceed through G_2 and prophase before it can appear as a "metaphase" chromosome. The earliest that this can happen is after a little more than 4 hours (in the case where the label was incorporated very late in S phase), and the latest is after a little more than 11 hours (in the case where the label was incorporated very early in S phase).

h.

CHAPTER 10 GENE EXPRESSION: TRANSCRIPTION

Chapter Goals
1. Be able to name and describe the different types of RNAs produced in prokaryotes and eukaryotes and what functions they have in the cell.
2. Understand and compare the sequences and mechanisms that regulate transcription in prokaryotes and eukaryotes.
3. Diagram the posttranscriptional processing events that occur in eukaryotes and describe the functions of these events.
4. Be able to discuss microRNAs and how they can regulate gene expression.

Key Chapter Concepts
- **The Central Dogma:** The **central dogma** focuses on information transfer at the molecular level. Information flows from the base sequence of DNA to an RNA molecule and then to the amino acid sequence of a protein. **Transcription** involves the transfer of information stored in DNA to an RNA molecule. The information encoded in an RNA molecule is then used to assemble a specific protein in the process of **translation**. As you will see, both processes have the same three general steps: **initiation** (start), **elongation** (proceed), and **termination** (stop).

10.1 Types of RNA
- **Protein Encoding RNAs:** DNA directs the formation of proteins, including an all-important category of proteins known as enzymes. However, DNA is located in the cell nucleus (in eukaryotes) and protein synthesis occurs in association with the ribosomes in the cytoplasm. If DNA does not leave the nucleus, how can it direct protein synthesis? The answer: **messenger RNAs (mRNAs)** carry the sequence information of DNA to the ribosomes which make proteins (**translation**).
- **Functional RNAs:** Other important RNAs are those that are not translated but function in the cell as RNAs. They are still transcribed from DNA but are never translated. **Transfer RNAs (tRNAs)** bring amino acids to the ribosome and decode the mRNA, and **ribosomal RNAs (rRNAs)** provide both structure and enzymatic activity for the ribosome. Transfer RNA is essential to the process of protein synthesis because it ensures that the correct amino acids are inserted in the proper order. This is accomplished by complementary base pairing between the sequence of three bases on the tRNA molecule (the **anticodon**) and the three base mRNA sequence (the **codon**) which specifies the amino acid to be inserted. A tRNA with a specific anticodon carries a specific amino acid. **Small nuclear RNAs** and **microRNAs** also carry out specific functions in the cell related to gene expression.

10.2 Bacterial Transcription: Preliminaries
- **RNA Polymerase and Transcription:** The core **RNA polymerase** enzyme synthesizes a single-stranded piece of RNA using one strand of the DNA as a template in the process of transcription. The core enzyme component of the holoenzyme RNA polymerase carries out the actual polymerization of nucleotides, while an associated protein, the **sigma factor**, recognizes base sequences in the DNA that direct where RNA synthesis begins.

10.3 Bacterial Transcription: The Process
- **Initiation:** RNA polymerase first interacts with the region of the DNA molecule known as the **promoter**. The promoter is a DNA sequence found **upstream** of the DNA nucleotide (start point) where transcription will be initiated, or started. Two six-base sequences aid the RNA polymerase and sigma factor in recognizing the promoter. The first sequence, known as the **–10 sequence** or **Pribnow box** (5′-TATAAT-3′ or a variation of this **consensus** sequence), is located approximately 10 bases upstream from the start point. The second sequence (5′-TTGTCA-3′) is known as the **–35 sequence** and lies even further upstream from the point where transcription will begin.
- **Elongation:** Once initiated, RNA polymerase begins adding nucleotides to the growing strand of mRNA in the 5′ → 3′ direction. Specifically, the enzyme adds cytosine when guanine is encountered on the DNA template, guanine when cytosine is encountered, adenine when thymine occurs in the template, and uracil to correspond to adenine. In this way, the mRNA has the complementary sequence

of the DNA **template strand** and the same sequence as the **coding strand** except that uracil is found in RNA in place of thymine in DNA.

- **Termination:** Synthesis of mRNA continues until one of two types of terminators is encountered. Both **rho-dependent** and **rho-independent terminators** require a sequence that will form a particular structure called a **stem loop** when synthesized in the RNA. The RNA polymerase pauses its elongation when this stem loop is formed. Rho-dependent termination requires the action of an additional protein, rho, that disrupts the DNA–RNA hybrid in the paused polymerase and causes termination. Rho-independent termination does not require an additional protein but an A–U–rich stretch that spontaneously denatures in the paused polymerase.

10.4 Ribosomes and Ribosomal RNA

- **Composition and Production:** Ribosomes are composed of both proteins and functional RNAs called rRNAs. They are assembled from two huge subunits, each containing more than 20 proteins and one or more rRNAs. The rRNAs are transcribed by RNA polymerase as a single RNA. That RNA is then cleaved into several individual rRNAs.
- **Function:** Ribosomes translate the nucleotide code of mRNAs into a chain of amino acids in the correct order to form proteins. They associate with mRNAs and recruit tRNAs with specific amino acids to the mRNA codons. The enzymatic activity of attaching amino acids is thought to result from the rRNA component of ribosomes.

10.5 Transfer RNA

- **Structure and Function:** Transfer RNAs are very small (approximately 80 nucleotides) and form a three-dimensional structure containing several important regions. All tRNAs have an overall cloverleaf structure with three stem loops that fold into an "L" shape. Each tRNA has a different **anticodon** on the anticodon loop that base-pairs with a **codon** in the mRNA being translated. Depending on the sequence of the anticodon present and a few other critical nucleotides, a specific amino acid is attached to the free 3′–OH of each tRNA by the action of **aminoacyl-tRNA synthetases**. Thus, tRNAs decode mRNAs by aligning specific amino acids with specific codons. Several of the nucleotides are modified after transcription to give tRNA's unique characteristics and functions.

10.6 Eukaryotic Transcription

- **Eukaryotic Transcription:** Transcription in eukaryotes differs from that of prokaryotes in four major ways:
    - **Localization:** Because the DNA is contained in a nucleus in eukaryotes, transcription and protein synthesis do not occur simultaneously as they do in prokaryotes. Instead, the mRNA is transcribed in the nucleus and then exported to the cytoplasm.
    - **Number of RNA Polymerases:** Unlike the single RNA polymerase in prokaryotes, there are three active RNA polymerase enzymes in eukaryotes. RNA polymerase I transcribes the rRNA gene composed of the three large rRNAs. RNA polymerase II transcribes most protein-encoding genes as well as most snRNA and miRNA genes. RNA polymerase III transcribes the tRNA genes, the smallest rRNA (5S) gene, and the U6 snRNA gene.
    - **Complexity and Number of Sequences and Associated Proteins:** The promoters for all three eukaryotic RNA polymerases have general and specific sequences that, in turn, interact with general and specific cellular proteins to promote, enhance, or even prohibit transcription. Promoters for each of the three polymerases have different types of general sequences that specify which polymerase will bind. One general sequence (5′-TATAAA-3′) referred to as a **TATA box** is found in many of these promoters and is bound by the **TATA-binding protein (TBP)** to help initiate transcription. The specific proteins available to influence transcription from a particular promoter will differ from cell type to cell type or in response to environmental conditions. These transcription factors can be classified as **activators** or **repressors** depending on whether they promote or inhibit transcription, respectively. Activators and repressors are proteins that bind DNA sequences called **enhancers** or **silencers**, respectively.
    - **Posttranscriptional Modifications:** Unlike prokaryotic RNAs, RNAs produced by RNA polymerase II in eukaryotes are extensively modified after synthesis.

- **Proofreading:** In eukaryotes, a protein complex known as TFIIS provides a proofreading function to RNA polymerase II that does not exist in prokaryotes.

10.7 Posttranscriptional Modifications in Eukaryotes

- **Overview:** Many of the RNAs produced by eukaryotic transcription are modified. As in prokaryotes, rRNAs and tRNAs are cleaved from longer RNAs and some of their ribonucleotides are modified. Unlike prokaryotes, RNAs produced by RNA polymerase II are highly processed through modification of their ends and removal of noncoding **introns** to splice together **exons** in a process called pre-mRNA splicing. In mRNAs, all of these processes occur in the nucleus and *must* be completed before the RNA can be exported to the cytoplasm for translation.

- **End Processing:** Eukaryotic RNAs produced by RNA polymerase II receive a **5′ cap** on their 5′ end and a **poly(A) tail** on their 3′ end. These modifications help to protect the RNA from degradation, promote splicing, promote export from the nucleus, and promote translation.

- **Intron Removal: Intervening sequences** in **pre-mRNAs**, called **introns**, are transcribed but are not translated and are removed by a multi-subunit enzyme called the **spliceosome** to produce mature mRNAs. The spliceosome consists of both snRNAs and proteins. The ends of introns are defined by consensus sequences in the pre-mRNA and are recognized (bound) by sequences in the snRNA components of the spliceosome. Intron removal follows two sequential transesterification reactions that are catalyzed by the spliceosome. In the first, the splice donor (5′ end of intron) attaches to the branch site adenosine to form an intron lariat. In the second, the 5′ exon attaches to the 3′ exon, releasing the intron lariat. Advantages provided by introns include alternative splicing which allows cells to produce several different (yet similar) proteins from one gene and **exon shuffling** which allows evolution of new genes more readily.

- **Self-Splicing RNAs:** Two types of introns have been discovered that can remove themselves without any help from proteins. The **group I** and **group II introns** are found in protozoans and yeast mitochondria, respectively, and are called **ribozymes** since they are RNAs that catalyze chemical reactions.

- *Trans*-**Splicing and Editing:** In a few circumstances in eukaryotes *trans*-splicing allows an exon from one pre-mRNA (called a **spliced leader**) to be spliced to the exons of different pre-mRNAs in order to produce different proteins. Also, in certain parasites, **guide RNAs (gRNAs)** are produced from genes and direct the addition or removal of nucleotides from other mRNAs in a process called **RNA editing.** RNA editing may be unique to parasites. Thus, inhibiting this process with drugs could be a potential method by which to safely treat parasite infections.

10.8 MicroRNA

- **Overview:** MicroRNAs (miRNAs) are short regulatory RNAs transcribed by RNA polymerase II from specific genes. Once produced they are cleaved into very short (usually 22 nucleotides) RNAs by the **Dicer** protein. These short RNAs then inhibit the expression of (regulate) other genes that contain a sequence complementary to the miRNA by one or more of three different mechanisms.

10.9 Update: The Flow of Genetic Information

- **Modified Central Dogma:** Since the proposal of the central dogma by Francis Crick in 1958, two new directions of information flow have been demonstrated to occur in nature. First, in opposition to transcription, viruses often produce DNA from RNA by the use of an enzyme called **reverse transcriptase.** Second, some bacterial **RNA phages** can produce RNA from RNA using a special enzyme called **RNA replicase.**

Key Terms

- central dogma
- transcription
- translation
- messenger RNA (mRNA)
- transfer RNA (tRNA)
- ribosomal RNA (rRNA)

- small nuclear RNAs (snRNAs)
- microRNAs (miRNAs)
- ribosomes
- DNA–RNA hybridization
- sense strand

- antisense strand
- coding strand
- noncoding strand
- template strand
- RNA polymerase
- sigma factor

- promoter
- footprinting
- conserved sequence
- consensus sequence
- downstream
- upstream
- Pribnow box
- −10 sequence
- −35 sequence
- upstream element
- heat shock proteins
- closed promoter complex
- open promoter complex
- rho protein
- rho-dependent terminators
- rho-independent terminators
- inverted-repeat sequence
- stem-loop structure
- open reading frame (ORF)
- leader
- 5′ untranslated region (5′ UTR)
- 3′ untranslated region (3′ UTR)
- operon
- Svedberg unit (S)
- codon
- anticodon
- aminoacyl-tRNA synthetases
- dihydrouridine (D)
- inosine (I)
- methylguanosine (MG)
- methylinosine (MI)

- pseudouridine (ψ)
- ribothymidine (T)
- transcription factors
- core promoter
- upstream promoter element
- upstream binding factor (UBF)
- TATA-binding protein (TBP)
- preinitiation complex (PIC)
- transcription factor IIIA (TFIIIA)
- transcription factor IIIC (TFIIIC)
- transcription factor IIIB (TFIIIB)
- TATA box
- TATA-less promoter
- initiator region (InR)
- downstream promoter element (DPE)
- TBP-associated factors (TAFs)
- transcription factor IID (TFIID)
- activators
- enhancers
- repressors
- silencers
- combinatorial control
- abortive RNAs
- posttranscriptional modifications
- primary transcript

- small nucleolar RNAs (snoRNAs)
- small nucleolar ribonucleoprotein particles (snoRNPs)
- pre-mRNA
- 5′ CAP
- α-amanitin
- poly(A) tail
- poly(A) polymerase
- intervening sequences
- introns
- exons
- ribozyme
- group I introns
- group II introns
- spliceosome
- small nuclear ribonucleoproteins (snRNPs)
- exon shuffling
- introns-early view
- introns-late view
- *trans*-splicing
- spliced leader (SL)
- RNA editing
- guide RNA (gRNA)
- Dicer
- microribonucleoprotein (miRNP)
- Argonaute
- gene silencing
- reverse transcriptase
- RNA phages
- RNA replicase

Understanding the Key Concepts

Use the key terms or parts of their definitions to complete the following sentences.

RNAs are produced through the process of (1)_____ in which the (2)_____ strand of a gene is read by the enzyme (3)_____ and a complementary RNA is produced. Most RNAs produced are (4)_____ containing an ORF surrounded by a (5)_____ and a (6)_____. The ORF is (7)_____ into proteins via the (8)_____. Other RNAs such as rRNAs, tRNAs, miRNAs, snoRNAs, and snRNAs do not contain an ORF but rather are functional entities within the cell. (9)_____ and (10)_____ are found in all organisms and participate in translation, while (11)_____ and (12)_____ are only found in eukaryotes where they participate in RNA processing events. Since production of these RNAs when they are not needed would be wasteful, the (13)_____ stage of transcription of individual genes is regulated. To initiate transcription, a DNA element generally called a (14)_____ is needed upstream of the transcription start site. These elements have been discovered using (15)_____ experiments and by determining (16)_____ sequences by comparing the sequences upstream of different genes. In prokaryotes, these elements are bound by the (17)_____ whose job it is to recruit the (18)_____ enzyme. This closed complex converts to an (19)_____ when the (20)_____ dissociates and the DNA in the (21)_____ denatures. Next the polymerase synthesizes a single strand of RNA in the (22)_____ direction. Transcriptional termination occurs when the polymerase pauses due to the creation of a (23)_____ in the newly synthesized RNA from an (24)_____ in the DNA.

Transcription in eukaryotes is similar to that of prokaryotes in that an RNA polymerase synthesizes a single RNA strand from a template strand of DNA. Eukaryotes, however, have (25)_____ different RNA polymerases; each transcribes a different set of genes. For instance, (26)_____ only transcribes the large rRNA genes while (27)_____ transcribes the mRNA genes and most of the snRNA and miRNA genes. The promoters for each of these polymerases have unique characteristics. In RNA polymerase I–dependent promoters, the (28)_____ binds to the (29)_____ first, followed by binding of other transcription factors such as SL1 and RNA polymerase I to the (30)_____ to form the (31)_____. RNA polymerase II-dependent promoters have TATA boxes. RNA polymerase II-dependent promoters that contain TATA boxes, however, (32)_____ is the first protein to bind, causing the DNA to (33)_____ sharply. Its binding is followed by the addition of (34)_____ to generate TFIID. The other general transcription factors and RNA polymerase II then follow to form the (35)_____ in which (36)_____ are produced. Once (37)_____ are added to the carboxyl terminus of RNA polymerase II and TFIIH (38)_____ the two strands of DNA, RNA polymerase II leaves the promoter.

While translation of newly synthesized mRNA occurs while the rest of the mRNA is still being synthesized in prokaryotes, eukaryotes do not undergo co-transcriptional translation due to the presence of a (39)_____. Instead, eukaryotic mRNAs undergo (40)_____ co-transcriptionally in order to change a (41)_____ into a mature mRNA. The first modification is the addition of a cap composed of a (42)_____ to the (43)_____ end of the mRNA. One of the functions of this cap is to stimulate (44)_____ of the first (45)_____. Similarly, addition of a poly(A) tail by (46)_____ to the (47)_____ end of the mRNA stimulates (48)_____ of the last (49)_____.

Removal of group I and group II introns found in some organisms is carried out by a (50)_____ composed of the intron itself. Group I and Group II introns differ in that group I introns utilize (51)_____ to break the phosphodiester bond at the 5′ end of the intron whereas group II introns use the 2′-hydroxyl of an internal (52)_____ to accomplish this. Nuclear pre-mRNAs follow an identical chemical reaction as group (53)_____ but use other cellular components called (54)_____ in order to form a (55)_____ that recognizes the intron–exon boundaries, aligns the two exon ends, and catalyzes the chemical reactions. The recognition of the splice donor site, which is at the (56)_____ end of the intron, is recognized by the (57)_____ through base-pairing between the mRNA donor site and the (58)_____, while the branchpoint is recognized by the (59)_____ through base-pairing. By regulating the (60)_____ of splice junctions to snRNPs, splicing enhancers and (61)_____ promote the production of (62)_____ spliced mRNAs. Thus, different mRNAs, and ultimately different proteins, can be produced in different cell types from the same gene.

The protein (63)_____ helps produce miRNAs and the protein (64)_____ helps miRNAs function to (65)_____ translation of specific genes. Thus, miRNAs are regulators of gene expression in eukaryotes. Each miRNA functions as a (66)_____-stranded RNA that is (67)_____ to a region of the specific gene that it regulates.

Figure Analysis
Use the indicated figure or table from the textbook to answer the questions that follow.

1. **Figure 10.14**
(a) What is the first region of RNA that is transcribed called?

(b) Is the sequence ATG located upstream or downstream of the promoter?

(c) Which strand of the DNA in this figure (top or bottom) is complementary to the mRNA? Which strand is the coding strand? Which strand is the noncoding strand?

(d) What would you predict to be the consequence of changing the promoter sequence (mutating it) so that TTGTCA–TATAAT became TTGTTT–TAGGAT?

(e) What would you predict to be the consequence of inverting the promoter so that TTGTCA–TATAAT was on the bottom strand of DNA when read from right to left?

2. Figure 10.17
(a) Where is the amino acid attached to this tRNA?

(b) What part of this tRNA "reads" the codons in an mRNA ORF?

(c) How is the correct amino acid attached to the correct tRNA?

(d) Why does the cloverleaf structure in part (a) look like an upside down "L" in parts (b) and (c) of figure 10.17?

(e) How can some of the nucleotides found in this tRNA not be the usual A, C, G, or U found in other RNAs?

3. Figure 10.24
(a) Which protein binds to the promoter first? What part of the DNA double helix does this protein bind? What does the binding of this protein do to the DNA?

(b) In what order do the next five protein complexes (TFIIA, TFIIB, TFIIE, TFIIH, and RNA polymerase/TFIIF) join TFIID at the promoter?

(c) What two functions does TFIIH provide to transcription? How can one complex provide more than one function?

4. Figure 10.43
(a) How can a splicing repressor protein prevent the use of a particular exon?

(b) How can a splicing enhancer promote the use of a particular splice site?

(c) If smooth muscle cells produced an mRNA containing only exons 1, 3, and 4 of the gene shown on the top panels and striated muscle cells produced an mRNA containing all four exons, what would be different between the two cells to cause this difference?

(d) How do splicing repressors and splicing enhancers know which splice sites to bind?

General Questions
Q10-1. Compare the consequences of inserting an incorrect base in the DNA template during replication and inserting an incorrect base in the mRNA during transcription.

Q10-2. If both strands of DNA of a gene are transcribed, what is the likely consequence? Why is only one or the other strand of DNA typically transcribed rather than both?

Q10-3. Other than differences in protein names, list four differences between eukaryotic and prokaryotic production of mRNAs.
i)
ii)
iii)
iv)

Q10-4. List four purposes of the 5′ cap attached to eukaryotic mRNAs.
i)
ii)
iii)
iv)

Q10-5. Describe the similarities and differences between self-splicing of group II introns and spliceosome-mediated splicing of introns found in nuclear mRNAs. Suggest a possibility of where the spliceosome may have come from during the evolution of eukaryotes.

Multiple Choice

For each of the following, circle the letter of the choice that most appropriately answers the question.

1.) Which of the following molecules is *not* produced by transcription?
   a. mRNA
   b. miRNA
   c. tRNA
   d. snRNA
   e. None of the above; all are produced by transcription.

2.) Which of the following statements is *true* concerning the template and coding strands of DNA?
   a. In a bacterial chromosome, some genes use one strand of DNA as the template and other genes use the other strand as the template.
   b. In a bacterial chromosome, all genes use the same strand of DNA as the template.
   c. The bottom strand of DNA is always the template strand for transcription.
   d. The sequence of an mRNA is complementary to the coding strand of DNA.
   e. In the small region of a single gene, two different mRNAs are made, one that is complementary to the template strand and one that is complementary to the coding strand.

3.) Which of the following most accurately describes a consensus sequence?
   a. a sequence that initiates transcription in bacteria
   b. an exact sequence that is always identical
   c. a sequence consisting of the most common nucleotides found at each position
   d. a DNA sequence that specifies the order of amino acids in a protein
   e. a sequence made of only A and T

4.) Which of the following is a function of the Pribnow box?
   a. Allow the two strands of DNA to denature.
   b. Bind the sigma factor.
   c. Terminate transcription.
   d. Bind to ribosomes.
   e. Both a and b

5.) How many subunits comprise a ribosome?
   a. one
   b. two
   c. three
   d. four
   e. five

6.) How can an enhancer located thousands of base pairs upstream of an RNA polymerase II-dependent promoter affect transcription of the distant core promoter?
   a. The DNA can loop, bringing the activator–enhancer adjacent to the core promoter.
   b. An activator bound to the enhancer can separate the two strands of DNA over the entire distance between the enhancer and the core promoter.
   c. Enhancers are mobile genetic elements that can translocate along the DNA to the promoter.
   d. TFIID, TFIIA, TFIIB, and RNA polymerase II will first bind to the enhancer and then scan along the DNA until they encounter the core promoter.
   e. Extremely large proteins bound to the enhancer can span thousands of base pairs.

7.) TBP is a component of which of the following transcription factor complexes?
   a. TFIIA
   b. TFIIB
   c. TFIID

d. TFIIE
e. TFIIH

8.) You have discovered a new type of RNA produced in eukaryotes and wish to determine which RNA polymerase synthesizes it. Which of the following would *not* help you distinguish between the three different polymerases?
   a. sensitivity of the RNAs production to different α-amanitin concentrations
   b. examination of the promoter sequences of the gene
   c. the presence of a 5′ cap on the RNA
   d. knowing the function of the RNA
   e. an in vitro transcription reaction

9.) Which of the following provided the first proof that some genes contain intervening sequences that are removed from the mRNA?
   a. Self-splicing occurred in vitro when GTP was added to RNA in a test tube.
   b. Messenger RNAs are shorter than strands of DNA.
   c. Electron microscopy of an mRNA hybridized to its corresponding gene revealed single-stranded DNA loops in the middle of the gene.
   d. A consensus sequence for 5′ splice junctions was discovered.
   e. Walter Gilbert calculated that exon shuffling could account for the evolution of all proteins from a very minimal number of original exons.

10.) How is the production of protein X inhibited in cells that express a microRNA that is perfectly complementary to the coding strand of the gene for protein X?
   a. The mRNA for protein X is cleaved.
   b. The process of transcription of the gene for protein X is inhibited.
   c. Posttranscriptional modification of the mRNA for protein X is inhibited.
   d. The process of translating the mRNA for protein X is inhibited.
   e. Both a and b

Practice Problems

P10-1. An mRNA molecule is isolated and determined to be composed of 30% A, 33% U, 18% G, and 19% C. Using this information, determine the base composition of the DNA template and the DNA coding strand.

[*Hint 1*: The coding strand of the DNA molecule has the same base sequence as the messenger RNA molecule (with U substituted for T in RNA) transcribed from the template strand.]

P10-2. Consider the following DNA molecule. What mRNA is synthesized if the top strand serves as the template? What mRNA sequence would be synthesized from the bottom strand?
   3′-GGCCATTCTACCAAAATCGC-5′
   5′-CCGGTAAGAAGGTTTTAGCG-3′

(*Hint 2*: Messenger RNA is synthesized in the 5′ → 3′ direction using a 3′ → 5′ template DNA strand. Although either strand can serve as the template, only one strand is transcribed.)

P10-3. What is the consensus sequence for the given six promoters?
   Promoter 1:   GATGTCAAACATC
   Promoter 2:   AAGGTAAAGCACT
   Promoter 3:   GAGCTCAGGCATT
   Promoter 4:   GGTGTCAAGTATA
   Promoter 5:   CAGTTCATGCGAT
   Promoter 6:   GAGGCCAAACATC
   Consensus :   _____

P10-4. Consider the following DNA molecule. Identify the Pribnow box, the direction of transcription, the start of the ORF of the gene, the mRNA sequence, the polarity of the mRNA molecule, the upstream direction, the downstream direction, the coding strand, and the template strand. The base indicated in bold and large type is the first to be transcribed.

(*Hint 3*: The base sequences given in your text for the consensus promoter region are the sequences found on the coding strand. The complementary strand is actually transcribed.)

5′-TAGGACCGCTATAATGCAGCAGAA**A**GCGATGGGCCACCCCACCAAAATCGCCTCCACGC-3′
3′-ATCCTGGCGATATTACGTCGTCTT**T**CGCTACCCGGTGGGGTGGTTTTAGCGGAGGTGCG-5′

P10-5. The sequences found in a DNA template (top) and the final mRNA product found at the ribosome (bottom) are given. How can you explain the differences in the nucleotide composition of these two molecules?

(*Hint 4*: Posttranscriptional modification of mRNA occurs before *eukaryotic* translation.)
3′-TACCGGCATCTATCCCATCCATACGGCAGGTCTTACGCGTACGCCTTTTTCAATCAT-5′
5′-AUGGCCGGUAGGUAUGCCAAUGCGCAUGCGGAAAAAUA-3′

Assessing Your Knowledge

Understanding the Key Concepts—Answers

1.) transcription; 2.) template; 3.) RNA polymerase; 4.) mRNAs; 5.) 5′ UTR; 6.) 3′ UTR; 7.) translated; 8.) ribosomes; 9.) tRNAs; 10.) rRNAs; 11.) snRNAs; 12.) snoRNAs; 13.) initiation; 14.) promoter; 15.) footprinting; 16.) consensus; 17.) sigma factor; 18.) core; 19.) open complex; 20.) sigma factor; 21.) Pribnow box; 22.) 5′ → 3′; 23.) stem-loop structure; 24.) inverted-repeat sequence; 25.) three; 26.) RNA polymerase I; 27.) RNA polymerase II; 28.) upstream binding factor or UBF; 29.) upstream promoter element or UPE; 30.) core promoter; 31.) preinitiation complex or PIC; 32.) TATA-binding protein or TBP; 33.) bend; 34.) TBP-associated factors or TAFs; 35.) preinitiation complex or PIC; 36.) abortive RNAs; 37.) phosphates; 38.) separates; 39.) nucleus or nuclear envelope; 40.) posttranscriptional modifications; 41.) pre-mRNA or primary transcript; 42.) 7-methyl-guanosine; 43.) 5′; 44.) splicing; 45.) intron; 46.) poly(A) polymerase; 47.) 3′; 48.) splicing; 49.) intron; 50.) ribozyme; 51.) GTP; 52.) adenosine; 53.) II; 54.) snRNPs; 55.) spliceosome; 56.) 5′; 57.) U1 snRNP; 58.) U1 snRNA; 59.) U2 snRNP; 60.) accessibility; 61.) splicing repressors; 62.) alternatively; 63.) Dicer; 64.) Argonaute; 65.) inhibit or repress; 66.) single 67.) complementary.

Figure Analysis—Answers

1a.) 5′ UTR; b.) Downstream; c.) In this figure, the bottom strand is complementary to the mRNA. The coding strand has the same sequence as the mRNA, and the template strand is complementary to the mRNA. Therefore, in figure 10.14, the top strand is the coding strand and the bottom strand is the noncoding or template strand; d.) Instead of a high level of transcription occurring from this promoter, many less RNA polymerases would initiate at the mutant promoter. Therefore, much less mRNA would be made from this gene. It is also likely that the mRNAs that are transcribed might initiate at a different nucleotide; e.) The promoter specifies which DNA strand is the template strand to be read by RNA polymerase and where to begin transcription. Transcription from the inverted promoter would proceed in the opposite direction (from right to left) and use the top strand as the template strand. Notice that synthesis would still occur from 5′ → 3′.

2a.) The amino acid (in this case alanine) would be covalently attached to the 3′ end of the tRNA (very top); b.) The anticodon found at the bottom of this tRNA will base pair ("read") the mRNA codon; c.) A specific aminoacyl-tRNA synthetase enzyme will bind to a specific amino acid and to a specific tRNA by recognizing the anticodon of that tRNA as well as some unique nucleotides at different places on that tRNA; d.) The T loop and the D loop fold down in the actual three-dimensional structure of a tRNA depicted in parts (b) and (c) rather than sticking straight out as depicted in part (a); e.) After the tRNA is transcribed with the usual nucleotides, specific enzymes in the cell bind to the tRNA and modify some of its nucleotides to give it specific properties and functions.

3a.) TBP is first to bind the DNA. It binds to the TATA box through the major groove of the DNA double helix causing a sharp bend in the DNA molecule; b.) TFIIA and TFIIB → RNA polymerase/TFIIF → TFIIE and TFIIH; c.) TFIIH has a kinase activity that adds phosphates to the carboxyl terminus of RNA polymerase II. It also has a DNA helicase activity that unwinds and separates the two strands of DNA ahead of RNA polymerase II. TFIIH can possess two different enzymatic activities since it is a complex composed of several different proteins. One or more of these proteins provides the kinase activity, while a different protein, or proteins, likely provides the DNA helicase activity.

4a.) A splicing repressor will bind near a 3′ splice junction and physically cover up the site. Thus, the assembling spliceosome will not see that site and will use the next available splice junction that is *after* the exon being skipped and at the beginning of the next exon; b.) A splicing enhancer will bind to a splice junction that is not normally recognized by the assembling spliceosome. Through protein–protein interactions, it will recruit the appropriate snRNPs to the appropriate splice junctions. In other words, the splicing enhancer provides an snRNP binding site when a poor consensus sequence at the splice junction is not sufficient to do so; c.) The genes present in both the smooth and striated muscle cells would be identical; both would have a gene for a splicing repressor and the gene depicted to be spliced containing all four exons. The smooth muscle cells would express the splicing repressor gene causing exon 2 of the other mRNA to be skipped. The striated muscle cells, however, would not express the splicing repressor and would include all exons in the other mRNA; d.) Different sequences lie next to specific splice junctions. These sequences are binding sites for specific splicing repressors or enhancers.

General Questions—Answers

Q10-1. Changes in the DNA sequence will persist in subsequent generations of cells. All proteins produced from this DNA will now be abnormal. Inserting an incorrect base during transcription will only result in the formation of a few abnormal proteins out of thousands of normal ones. Furthermore, this mistake will not persist and will not be transmitted since the DNA template remains unmodified.

Q10-2. The mRNA sequences produced from complementary strands of DNA are complementary to each other (5′ → 3′ mRNA pairing with the 3′ → 5′ mRNA). The two mRNAs would likely base-pair with each other, forming a double-stranded RNA molecule that could not be translated.

Q10-3. i) Prokaryotic transcription of mRNA is coupled to translation, while these two processes are independent in eukaryotes.
ii) Eukaryotic mRNA production involves posttranscriptional modifications of newly transcribed mRNAs, while prokaryotic mRNA production does not.
iii) Eukaryotic mRNA production uses a special RNA polymerase used only for mRNAs, snRNAs, and miRNAs, while prokaryotes use the same RNA polymerase for transcription of all genes.
iv) Eukaryotic transcription of mRNAs involves proofreading, while prokaryotic transcription does not.

Q10-4. i) protection from degradation; ii) promote removal of the first intron; iii) promote transport of the mRNA out of the nucleus; iv) promote translation of the mRNA by recruiting ribosomes

Q10-5. Differences: Group II introns can splice themselves, while nuclear mRNA introns require accessory proteins and RNAs to recognize and remove an intron.

Similarities: Both types of introns utilize an adenosine in the intron to break the phosphodiester bond at the 5′ splice junction followed by a second transesterification reaction in which the free end of the 5′ exon attacks the 3′ splice junction. Thus, both types of introns release the intron in the form of a lariat. Also, in both types of introns, splice junctions are recognized by RNA (snRNAs recognize nuclear mRNA introns, and the intron sequence itself recognizes group II introns).

One theory about the origin of the spliceosome is that the snRNAs of the spliceosome evolved from pieces of an ancestral group II intron that now function in *trans* to remove mRNA introns. Thus, a group II intron together with proteins, forms from several pieces to form an active splicing machine.

Multiple Choice—Answers

1.) e; 2.) a; 3.) c; 4.) e; 5.) b; 6.) a; 7.) c; 8.) d; 9.) c; 10.) e.

Practice Problems—Answers

P10-1. The mRNA has the same sequence and composition (T substituted for U) as the coding strand of DNA. Therefore, you would expect to find 30% A, 33% T, 18% G, and 19% C in the coding strand. The template strand has the complementary sequence as the mRNA. Because A and T pair, as well as C and G, you would find 30% T, 33% A, 18% C, and 19% G.

P10-2. Remember that RNA polymerase synthesizes mRNA in the 5′ → 3′ direction using a 3′ → 5′ template. The top strand in this example will be transcribed from left to right yielding the mRNA:

5′ CCGGUAAGAUGGUUUUAGCG 3′.

The bottom strand will be transcribed from right to left yielding the mRNA:

5′ CGCUAAAACCUUCUUACCGG 3′.

P10-3. Consensus: GAGGTCAAGCATT

P10-4. Remember that the Pribnow box sequence (TATAAT) occurs on the coding strand although the 3′ → 5′ strand is transcribed. The Pribnow box is in bold and demonstrates that the top strand is the coding strand while the bottom strand is the template strand. The initiation codon specifying the beginning of the ORF is indicated in bold italic.

5′ TAGGACCGC**TATAAT**GCAGCAGAA**A**GCG***ATG***GGCCACCCCACCAAAATCGCCTCCACGC 3′ - coding

3′ ATCCTGGCG**ATATTA**CGTCGTCTT**T**CGCTACCCGGTGGGGTGGTTTTAGCGGAGGTGCG 5′ - template

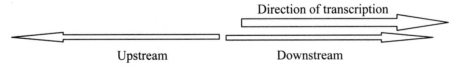

mRNA: 5′ AGCGAUGGGCCAUUCUACCAAAAUCGCCUCCACGC 3′

P10-5. First, the mRNA has a sequence that is complementary to the DNA since the DNA is the template strand. Second, the mRNA contains uracil which is not found in DNA but only in RNA in place of thymine. Third, the mRNA has been modified posttranscriptionally by splicing. The introns (bold sequences) have been removed from the originally transcribed mRNA (primary mRNA).

DNA:

3′ TACCGG**CATCTATC**CCATCCATACGG**CAGGTC**TTACGCGTACGCCTTTTT**CAATC**AT 5′

Primary mRNA:

3′ AUGGCC**GUAGAUAG**GGUAGGUAUGCC**GUCCAG**AAUGCGCAUGCGGAAAAA**GUUAG**UA 5′

Mature (spliced) mRNA:

5′ AUGGCC————GGUAGGUAUGCC ———AAUGCGCAUGCGGAAAAA ——UA 3′

10

GENE EXPRESSION: TRANSCRIPTION

CHAPTER SUMMARY QUESTIONS

1.   1. e, 2. f, 3. j, 4. d, 5. b, 6. i, 7. g, 8. a, 9. c, 10. h.

2.   The central dogma of molecular biology describes the flow of genetic information from DNA to RNA to protein. DNA controls its own replication. The transfer of information from DNA to RNA is termed *transcription*. The RNA is then converted into proteins via a process called *translation*.

3.   See figure 10.2. Complementarity is achieved between messenger RNA and ribosomal RNA and between messenger RNA and transfer RNA.

4.   Both transcription and translation occur in the cytoplasm of prokaryotes. Therefore, the mRNA can be translated before completion of its synthesis. However, the situation in eukaryotes is different: transcription occurs in the nucleus and translation in the cytoplasm. Therefore, these two processes are not coupled.

5.   **a.**

```
                                                          + 1 Transcription start site

   5'---TTGTCA-------------TATAAT-----┌─────→3'
   3'---AACAGT-------------ATATTA-----└-----5'
        −35 sequence           −10 sequence
```

b.   RNA polymerase II of humans has two types of promoters: TATA core promoters and TATA-less promoters. Both are shown in figure 10.22.

6.   The complete bacterial RNA polymerase is termed the *holoenzyme*. It consists of the core enzyme plus the sigma factor (σ). The core enzyme is composed of four subunits: two α, a β, and a β'.

7. Although the overall process is similar in prokaryotes and eukaryotes, differences exist in the number of different RNA polymerases and transcription factors in eukaryotes; exact DNA sequences recognized; polycistronic nature of prokaryotic DNA; and posttranscriptional processing in eukaryotes (splicing, capping, polyadenylation).

8. A stem-loop structure can form when a single strand of DNA or RNA has a double helical section (see figure 10.11). An inverted repeat is a sequence read outward on both strands of a double helix from a central point (see figure 10.11). A tandem repeat is a segment of nucleic acid repeated consecutively; that is, the same sequence repeats in the same direction on the same strand:

        5′-TCCGGTCCGGTCCGG-3′
        3′-AGGCCAGGCCAGGCC-5′

A DNA sequence with a seven-base inverted repeat is

        5′-ATTACCGCGGTAAT-3′
        3′-TAATGGCGCCATTA-5′

9. Footprinting is a technique in which DNA in contact with a protein is exposed to nucleases; only DNA protected by the protein is undigested. Promoters could be isolated by protection with RNA polymerase in the absence of ribonucleotides—the polymerase will not move—and then sequenced.

10. The two main types of transcription terminators in bacteria are termed rho-dependent and rho-independent. Both contain an inverted repeat sequence that generates a stem-loop structure necessary to terminate transcription. In rho-independent terminators, the stem-loop structure is followed by a run of U's which is enough to terminate transcription (U:A bonds are very weak and will denature spontaneously). Rho-dependent terminators, on the other hand, lack the U-rich sequence. They rely instead on a protein called rho (ρ), which uses ATP to denature the RNA–DNA hybrid, thereby causing termination of transcription.

11. Transcription has higher error rates. Errors of DNA polymerase tend to become permanent, whereas errors of RNA polymerase do not.

12. There are three different types of RNA polymerase III promoters. Type 1 and Type 2 promoters are rather unusual because they require sequences that are located between 55 and 80 bases downstream of the transcription initiation point and within the gene. Please refer to figure 10.21.

13. Both promoters and enhancers are DNA sequences that affect transcription. Promoters are found in both bacteria and eukaryotes, whereas enhancers are unique to eukaryotes. Promoters bind only RNA polymerase in bacteria, whereas in eukaryotes, the basal transcriptional apparatus has to bind first to the promoter before RNA polymerase can bind. Specific transcriptional activator proteins bind to enhancers to increase the transcription rate. Promoters are usually adjacent to the transcriptional start site and are highly position-dependent. Enhancers, on the other hand, function at a distance from the gene and are position-independent.

14. As a result of posttranscriptional modifications, the mature eukaryotic mRNA has features lacking in the primary transcript: (1) 5'-CAP (a 7-methyl guanosine added in the "wrong" direction—5' → 5'—by the enzyme guanylyl transferase). (2) 3'-polyA tail [a string of A residues added to the cleaved 3' end of the primary transcript by the enzyme poly(A) polymerase]. (3) Introns excised and exons spliced (by a protein–RNA complex termed the spliceosome). (4) RNA editing (by either guide RNAs inserting or deleting specific U's, or through deamination of specific C's or A's).

15. Group I introns are self-splicing introns that require a guanine-containing nucleotide for splicing. Group II introns are similar but do not require an external nucleotide for splicing. Group I and II introns are released as linear and lariat-shaped molecules, respectively.

16. **a.** The spliceosome, which is composed of several small nuclear ribonucleoproteins, functions to excise introns and splice exons in the primary mRNA transcript. It recognizes conserved sequences that flank and are within the intron.
    b. Guide RNA is an RNA molecule that guides the process of mRNA editing. It has a sequence that is complementary to the final mRNA, thereby aiding in the addition or deletion of specific U residues.
    c. Argonaute proteins bind to mature 22-nucleotide microRNAs to generate a complex termed *microribonucleoprotein*. This miRNP can perform three different functions: (1) it can bind mRNA and block its translation, (2) it can bind mRNA and result in its cleavage by endonucleases, or (3) it can enter the nucleus and repress gene transcription, a phenomenon termed *gene silencing*.

17. Alternative splicing, also termed *cis*-splicing, involves the splicing of exons that belong to the same primary mRNA transcript. *Trans*-splicing, on the other hand, joins exons from two different genes and primary transcripts into a new mRNA. Spliceosomes produce a lariat-shaped intron structure during *cis*-splicing, and a Y-shaped intron product during *trans*-splicing.

EXERCISES AND PROBLEMS

18. A consensus sequence is made up of the nucleotides that appear in a significant proportion of cases when similar sequences are aligned. The best way to get to the consensus sequence is to follow these steps:
    1. Consider each position of the sequences separately.
    2. Count the number of different bases found at that position.
    3. The base with the highest number is the one that will be included in the consensus sequence.

For example, in position 1, there are 5 G's and only 1 C and 1 A. Therefore, the consensus sequence will have a G in position 1. Similar logic provides the consensus sequence: G A T C T A G.

19. 3'—GGTAGTACTGTCTGGGAACGATTGCG—5'
    5'—CCATCATGACAGACCCTTGCTAACGC—3'

Begin by writing the strand that is complementary to the RNA. This will be the transcribed strand. Remember, U in RNA pairs with A in DNA. Since transcription proceeds in the 5' → 3' direction, the 5' end of the RNA is opposite the 3' end of the DNA.

20. With no sigma factors, there would be random starts of transcription; with no rho factor, there would be a failure to properly terminate transcription at rho-dependent termination sites.

21. The double helix must unwind in order for transcription to occur. A–T pairs, because they have only two H-bonds, are more easily disrupted than G–C pairs.

22. Hybridization of nucleic acids is carried out between species to determine the extent to which it occurs. Presumably, the greater the amount of hybridization, the more similar the genomes of two species and therefore the more closely they are related in an evolutionary sense (chapter 25).

23. Top strand: 3' —— 5' (template)
    Bottom: 5' —— 3'
    The top strand is transcribed, and the mRNA is 5'-GGGAUGCGGAAAGUCCAA-3'. Begin by writing the RNA sequence complementary to each DNA strand. Since the DNA fragment is from the beginning part of the gene, there must be a start signal, AUG, for protein synthesis. Transcription of the top strand yields such an RNA and the bottom does not. Since transcription proceeds in the 5' → 3' direction, the left end of the top strand must be the 3' end. Remember that double-stranded nucleic acids are antiparallel.

24. a. The mRNA is transcribed from the template strand based on the rules of complementarity: C (mRNA) is transcribed from G (DNA), G (mRNA) from C (DNA), A (mRNA) from T (DNA), and U (mRNA) from A (DNA). These percentages should be equivalent, and therefore the template strand of DNA would have the following percentages: 18% T, 27% G, 33% C, and 22% A.
    b. The nontemplate strand, which is the complement of the template strand, will therefore contain 18% A, 27% C, 33% G, and 22% T. Alternatively, these percentages can be derived directly by remembering that the nontemplate strand is the coding strand. Therefore, it will have the same sequence as the mRNA, substituting U for T.
    c. The base composition of the entire molecule will be the average base percentages of the two strands, or 20% A, 30% C, 30% G, and 20% T.

25. The mRNA is 5'-A U A C C G U A C C <u>A U G</u> A A G G C C-3'.
The Pribnow box is ATATTA, beginning six bases from the 3' end of the DNA. (Remember that the DNA strand shown is the template strand. So the TATAAT sequence will appear on the complementary coding DNA strand.) Transcription proceeds on the 3' → 5' strand producing a 5' → 3' RNA; therefore, it moves left to right along the given DNA molecule because double-stranded nucleic acids are antiparallel. The initiator codon is underlined in the mRNA sequence.

26. Top strand: 5' —— 3'
Bottom: 3' —— 5' (template, or transcribed strand)

Begin by writing the RNA that could be transcribed from each strand. Since the DNA represents the beginning portion of the gene, the RNA must have an AUG to start protein synthesis. Unfortunately, both strands yield RNAs with one or more AUGs. The RNA from the top strand has AUGs in both directions, but in each case the AUG is followed by a termination signal, UAA, or UAG. This RNA could not make a protein. The bottom strand produces an RNA with only one AUG. Since transcription and protein synthesis both proceed 5' → 3', the left end of the bottom strand must be 3'.

27. a.

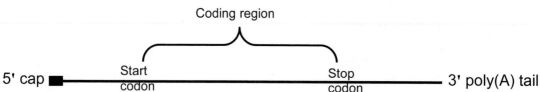

b. Each of the 129 amino acids is encoded by three bases in mRNA. There's also a stop codon which adds three more bases. Therefore, the mRNA contains at least (129 × 3) + 3 = 390 bases. The mRNA would also contain a 5' leader and a 3' trailer (see preceding figure) that add many more bases, although these are not translated.

28. If transcription of the genes is rho-dependent, the RNAs made at 40°C will be longer than those made at 30°C. Since the rho cannot function at the high temperature, RNA polymerase will read past the termination region. If transcription is rho-independent, a rho mutant will have no effect on transcription, and hence, the size of the RNAs.

29. All transcription could be rho-independent, and we would see normal transcription at both temperatures. If transcription is rho-dependent, long molecules could possibly be degraded rapidly, producing short molecules. Neither possibility is too likely since some genes seem to be rho-independent and others rho-dependent.

30.

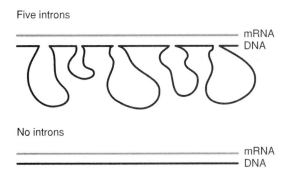

31. Introns do accumulate more mutations. Presumably, at most sites within an intron, mutations do not affect the phenotype and therefore are not selected against.

32. Enhancers bind activator proteins that also bind proteins of the RNA polymerase at the promoter. In some eukaryotic genes, there are many enhancers, allowing numerous levels of control and forcing enhancers further and further upstream. For activators bound to enhancers to bind polymerase proteins, the DNA must loop around.

33.

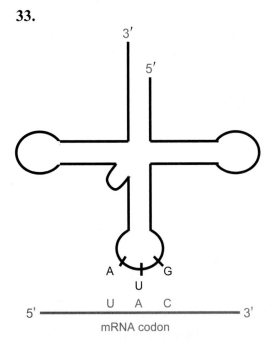

The template strand of DNA has a sequence complementary to the mRNA. Therefore, the template strand of DNA has the sequence 5'-GTA-3'.

34. The RNA is processed differently in the different tissues. In the pituitary, we see five introns and six coding regions, but in the adrenal, we see only four introns and five coding regions. The second DNA loop in the adrenal is larger than either the second or third loop in the pituitary. The third coding region must be removed in the adrenal, making the large second loop.

35.

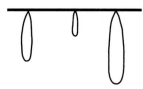

36. Prions are proteins that change the shape of proteins, not unlike many enzymes. The fact that they are mutant forms of a protein that change their unmutated equivalents into the mutated forms only makes them an oddity rather than a forbidden transfer in the central dogma. They do not represent a protein self-replication loop.

37. Given that a gene controls the production of a protein, there are both realistic and theoretical limits to the size of a gene. From what we know about the functioning of the centromere, a gene would have to occupy no more than the length of a chromosomal arm. However, given that human chromosomes must contain about 50,000 genes, it is unlikely that any gene is that large. We also know that many functioning proteins are made up of subunits, each controlled by its own gene. Thus, large proteins tend to be conglomerates of smaller ones rather than large functional units. Finally, the larger the protein, the more time it takes to transcribe and translate it, making very large size inefficient. The average protein is about 300 to 500 amino acids; with introns, and control elements, the gene for an average protein could be quite large. The largest known gene is the human dystrophin gene that codes for a cytoskeletal protein. It is 2,300,000 bases long, has 79 exons, and takes 16 hours to be transcribed.

38. a. The $(A + G)/(T + C)$ ratio of any double-stranded DNA molecule is 1.0. After all, the percentage of purines $(A + G)$ should be equal to that of pyrimidines $(T + C)$.

b. The two mRNA transcripts that result would be complementary to each other and near identical copies of the double-stranded DNA molecule that they resulted from (with U substituted for T). Therefore, the $(A + U)/(G + C)$ ratio in mRNA would be equal to the $(A + T)/(G + C)$ DNA ratio of 1.4.

c. The $(A + U)/(G + C)$ ratio would still be 1.4, regardless of which DNA strand is transcribed! The easiest way to visualize this is to draw a double-stranded DNA molecule with an $(A + T)/(G + C)$ ratio of 1.4 (that is, 14 AT pairs, and 10 CG pairs). If you transcribe either DNA strand, you will find that each RNA will have $A + U = 14$ and $C + G = 10$. Note that this will be true regardless of how you draw your DNA molecule. In other words, you could have the 14 AT pairs distributed as 14 A's on one strand and 14 T's on the other strand, or 7 A's and 7 T's on each strand, and so forth. In every single case where one strand is transcribed, the $(A + U)/(G + C)$ ratio in the mRNA will be 1.4. Try it and see!

d. As we discussed in part (b), the two mRNA transcripts would be near identical copies of the double-stranded DNA molecule (with U substituted for T), and so the (A + G)/(U + C) ratio in mRNA would be 1.0. Because it is not, the mRNA is transcribed from one strand only. This 1.2 ratio reflects the relative proportions of (A + G) and (T + C) on the transcribed DNA strand. Consider, for example, this DNA molecule which has the requisite (A + T)/(G + C) ratio of 1.4:

```
5'-TTTTTTAAAAAAAAGGGGGCCCCC-3'
3'-AAAAAATTTTTTTTCCCCCGGGGG-5'
```

Transcription of the bottom strand yields an mRNA with the sequence

```
5'-UUUUUUUAAAAAAAAGGGGGCCCCC-3'
```
and an (A + G)/(U + C) ratio of 1.18 (or approximately 1.2).

39. a.   1. The TTGTCA box is centered at –35 and so will be separated by about 25 bases from the –10 box and *not* 35 bases.
2. The second C in the AACACT sequence should be a G.
3. The enzyme is RNA polymerase and *not* DNA polymerase.
4. The coding strand and template strand should be switched.
5. The mRNA should be coming off the template strand.
6. The direction of transcription should be reversed.
7. The end of the mRNA should be 5' and *not* 3'.
8. The –10 box should *not* be transcribed!

b.

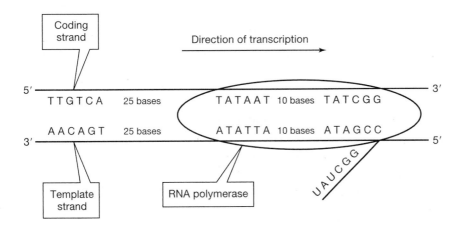

40. Transcription of all four exons produces a primary transcript that is 1000 + 500 + 1000 + 800 = 3300 nucleotides in length. If this primary transcript is cleaved 50 nucleotides before the end of the fourth exon and then a 250 nucleotide poly(A) tail is added, the size of the mature mRNA transcript will be 3300 – 50 + 250 =

3500 nucleotides long. This corresponds to the larger transcript found in unaffected individuals. The shorter transcript is 3000 nucleotides long. The shortening of this transcript by 500 nucleotides can be explained by an alternative splicing event in which exon 2 is omitted from the shorter transcript. This would result in a transcript that is $1000 + 1000 + 800 - 50 + 250 = 3000$ nucleotides in length. Another possible explanation is that exon 4 in the affected individuals might contain an alternative polyadenylation site. To produce a 3000 nucleotide transcript including exon 2, this site would have to be located 550 nucleotides before the end of exon 4. In this case, the transcript would be $1000 + 500 + 1000 + 800 - 550 + 250 = 3000$ nucleotides long.

The affected individuals in this family could have two types of mutations: (1) mutation of one of the splice junctions surrounding exon 2 or (2) a mutation in exon 4 creating a new polyadenylation site.

CHAPTER INTEGRATION PROBLEM

a. The total length of the mRNA transcript is 350 bases. However, the 50 A residues constitute the poly(A) tail which is added *after* transcription from the DNA. Therefore, the double-stranded DNA that codes for this mRNA would have to be 300 bp long. B-DNA contains 10 base pairs per 3.4 nm (one helical turn), or 1 bp per 0.34 nm. The length of the DNA would be 300 bp $\times$ 0.34 nm/bp = 102 nm.

b. The DNA coding strand of the laf^+ gene is the nontemplate strand. It is equivalent to the mRNA, with T's instead of U's. It will also lack the string of A's at the end.

c. Replication requires a primer for initiation; transcription initiation does not require a primer. Replication of the entire DNA initiates from specific sequences termed replication origins; each gene, including the laf^+ gene, is transcribed separately from its own promoter. Replication occurs in two directions using both strands of DNA as a template, while transcription is unidirectional and uses only one strand of DNA per gene.

d. The promoter for the laf^+ gene lacks a TATA-box. TATA-less promoters typically have a downstream promoter element (DPE) located at positions +28 to +34 and having the consensus sequence AGACGTG (in the coding strand). This sequence will show up in the mRNA as AGACGUG, and it does show in the laf^+ mRNA.

e. The laf^+ gene is transcribed by RNA polymerase II. RNA pol II promoters typically have an initiator region (InR) located at –3 to +5. The +1 nucleotide is usually an adenine flanked by pyrimidines. This is the case here. In addition, RNA pol II transcripts are cleaved about 15–30 nucleotides after the sequence 5'-AUAAA-3', and they pick up the poly(A) tail at the cleavage site. This is also true of the laf^+ gene.

f. The presence of two Laf proteins can be explained by alternative splicing of the *laf*⁺ mRNA. So let's first try to figure out the coding region of the *laf*⁺ gene. The initiation codon will have to be an AUG. There are two codons early on: at positions 83–85 and 118–120. The latter is almost immediately followed by a termination signal, UAA. Therefore, the initiation codon is located at positions 83–85. Now let's look for potential introns. Remember the consensus sequences of introns: GU(AAGU) at the 5' end, CAG at the 3' end, and UACUAAC in the middle. Scanning of the *laf*⁺ gene reveals two introns. A termination signal, UAA, is found at position 252–254. The start and stop signals, and introns are highlighted here:

```
1    AUCCAGUAAC UUGAUACUGA ACGAAACAGA CGUGGCCGAA CCGUACACCU

51   ACCGACUGCC UUCACGUUAC CGCGAUUAAC GAAUGAAUUA UCAUCAUUUC

101  UUUCCCGUAA GUAAACCAUG UACUAACAAU UAUCUCUCUA UCCAGGGAUU

151  CUUCGUGGAU AUGUUAAUCA CCUGGUGCGU AAGUCAGAUA UUGACUUUUA

201  UAUUUAUUCG AUUCAACGAU ACUAACUUGA CAGAAAGUG  CAAAACCUAA

251  GUAACUAUCU UCACUGCAUU AUUUGUGAAA UAAACUUCGU AAUGGGUAGC

301  AAAAAAAAAA AAAAAAAAAA AAAAAAAAAA AAAAAAAAAA AAAAAAAAAA
```

The two introns split the *laf*⁺ gene into three exons. Exon 1 extends from bases 83 to 106 and encodes 8 amino acids; exon 2 is from bases 146 to 178 and encodes 11 amino acids; and exon 3, bases 234 to 251 and 6 amino acids. The *laf*⁺ mRNA that contains all three exons will be translated to produce Laf^L which consists of 8 + 11 + 6 = 25 amino acids. The *laf*⁺ mRNA can be alternatively spliced to contain exons 1 and 3. This mRNA will produce Laf^S which consists of 8 + 6 = 14 amino acids.

g. *Mutant 1.* The failure to produce an mRNA transcript indicates that the mutation must have occurred in the promoter sequence.
Mutant 2. The failure to produce a Laf protein from a normal-length mRNA transcript indicates that the mutation likely occurred in the ribosome-binding site or in the start codon.
Mutant 3. The presence of a larger than normal protein points to translation beyond the normal termination signal. Therefore, the point-mutation converted the UAA at positions 252–254 into a triplet that codes for an amino acid. Indeed, eight codons further downstream is a UGA triplet, which is a termination codon. Therefore, the Laf^{XL} would be 25 + 8 = 33 amino acids.
Mutant 4. The Laf^{XS} is the hardest to explain. It is most likely the result of a mutation that generated a new initiation codon, which is located 18 bases upstream from a termination codon. Because initiation codons are AUG, we need to look for triplets that are AUX (where X = A, C, or U), AYG (where Y = A, C, or G), or ZUG (where Z = C, G, or U). A systematic scanning of the *laf*⁺ mRNA reveals a potential

mutation site at the CUG in position 56–58. If the coding strand of DNA had a C to A mutation at that position, the mRNA will have an A, and a translation initiation codon. This is followed by a stretch of six codons that ends with a UAA termination signal. This mRNA would be translated into a Laf^{XS} protein of seven amino acids.

h. The *laf⁺* allele encodes the normal wild-type Laf⁺ protein which is responsible for small feet. A heterozygous individual (*laf⁺laf*) possesses one *laf⁺* allele which is transcribed into normal mRNA, and then translated into the normal Laf⁺ protein. Therefore, one copy of the wild-type allele is sufficient to produce enough of the normal protein that is responsible for the wild-type phenotype of small feet. In homozygous recessive (*laf laf*) individuals, there are two mutant alleles and no wild-type allele. The absence of *laf⁺* mRNA means only nonfunctional product is produced. This leads to the mutant form of feet, or large feet.

CHAPTER 11 GENE EXPRESSION: TRANSLATION

Chapter Goals
1. Understand the one-gene/one-enzyme hypothesis and how it was elucidated.
2. Describe the general structure of amino acids and how they confer physical properties to a protein.
3. Distinguish between the levels of protein structure.
4. Describe the differences and similarities in translation initiation in prokaryotes and eukaryotes.
5. List and describe the events that occur during protein translation.
6. Give examples of posttranslational modifications.
7. Understand the experimental steps that led to solving the genetic code.

Key Chapter Concepts
11.1 The One-Gene/One-Enzyme Hypothesis
- **Overview:** Discovery of single gene mutations that each disrupted only a single enzymatic reaction in a biochemical pathway led Beadle and Tatum to propose that each gene codes for a single enzyme in the cell. This is basically true; however, a few intricacies such as alternative splicing and protein cleavage or processing can lead to multiple forms of a protein being produced from one gene.

11.2 Protein Structure
- **Amino Acid Properties:** There are 20 **amino acids** that are attached via peptide bonds to make proteins, also known as polypeptides. Each amino acid has a name, a three letter abbreviation, and a one letter abbreviation, and each has an identical common region containing a carboxyl group and an amino group. The common region is involved in attaching the amino acids together in a polypeptide. Each amino acid differs by what side group is attached to the common region. The properties of the side groups are what provide functional differences between proteins. Based on the properties of the side groups, amino acids can be divided into four groups; acidic, basic, nonpolar, and polar (uncharged).
- **Levels of Protein Structure:** Unlike DNA which almost always assumes a long double helix, proteins assume several levels of complex three-dimensional structure. This structure forms the active sites of enzymes, and the side groups of amino acids bind substrates and cofactors and carry out the reactions.
    - **Primary Structure:** The sequence of amino acids and the formation of any disulfide bridges define this level of structure.
    - **Secondary Structure:** Short stretches of amino acids form α-helices, β-sheets, extended strands, turns, and random coils through hydrogen bonding between residues.
    - **Tertiary Structure:** All of the secondary structures fold onto and around each other to form a three-dimensional structure including active sites and binding pockets.
    - **Quaternary Structure:** The binding of multiple polypeptide subunits to each other is described by this level of structure.

11.3 The Colinearity of mRNA and Protein Sequences
- **Overview:** Charles Yanofsky demonstrated that mutations at the beginning of a gene corresponded to mutations at the beginning of the protein while mutations toward the end of a gene corresponded to mutations toward the end of the protein. Thus, the important conclusion that the code in a gene was **colinear** with a protein sequence was developed.

11.4 Translation: Preliminaries
- **tRNAs:** Transfer RNA molecules carry specific amino acids to the ribosome where they are incorporated into a growing polypeptide. The amino acid is bound to the 3′ end of the transfer RNA molecule by the enzyme **aminoacyl-tRNA synthetase**. A tRNA is said to be **charged** when the amino acid is attached. The tRNA carrying an amino acid is recognized by its **anticodon** as well as other specific sequences embedded in its sequence. The anticodon ensures that the amino acids are inserted in the proper sequence: the anticodon pairs with the **codon** (sequence of three bases) within the mRNA.
- **Charging of tRNAs:** There is one aminoacyl-tRNA synthetase for each of the 20 amino acids. Each aminoacyl-tRNA synthetase binds one amino acid and charges all of the tRNAs that carry that amino

acid in a two-step reaction. First they activate the amino acid by attaching ATP to it. Second, they transfer the activated amino acid to the tRNA. Proofreading steps ensure that the correct amino acid is attached to the correct tRNA. For example, aminoacyl-tRNA synthetases discriminate between amino acids based on size in two different steps. The active site for the amino acid activation reaction only accepts amino acids as small or smaller than the correct amino acid. In the second step, all of the amino acids smaller than the correct amino acid are sent to an **editing site** and degraded. Since each amino acid has a unique size, only the correct amino acid will remain.

11.5 Translation: The Process

- **Translation:** Translation takes place in the ribosome and requires the presence of (1) an mRNA, (2) tRNAs carrying their specific amino acids, (3) the large and small ribosomal subunits which contain ribosomal RNAs (rRNAs) and form an **A site**, a **P site**, and, in prokaryotes, an **E site**, (4) protein **initiation factors**, **elongation factors**, and **release factors**, and (5) hydrolysis of GTP (guanosine triphosphate). Translation initiation involves the two ribosomal subunits and the initiator tRNA associating with the start codon of the mRNA. This initiation step is significantly different in prokaryotes versus eukaryotes. During translation elongation, the A and P sites on the ribosome position two charged tRNA molecules next to each other. The first amino acid is attached to the tRNA positioned within the peptidyl or **P site**, and the second is attached to the tRNA within the aminoacyl or **A site**. The mRNA is threaded through the ribosome to encounter the A and P sites ($5' \rightarrow 3'$). A peptide bond is then formed between the two amino acids in the adjacent sites, and then the ribosome shifts. The P site houses the tRNA that is carrying the growing peptide, while the A site houses a charged tRNA that carries the amino acid that will be added. The same three steps occur in translation as occurred in transcription: **initiation** (start), **elongation** (proceed), and **termination** (stop).

- **Initiation:** **Initiation factors** are proteins that help the two subunits of the ribosome and the initiator tRNA find and bind to the first codon. The first amino acid added to a prokaryotic protein is N-formyl methionine (methionine in eukaryotes) which is specified by the start codon (5'-AUG-3') found in the mRNA following the **Shine–Dalgarno sequence** in prokaryotes or the **Kozak sequence** in eukaryotes. The initiator tRNA enters the P site of the ribosomal complex by recognition of its complementary anticodon (3'-UAC-5').

  Prokaryotic mRNA Recognition: A consensus sequence of bases in the 5' UTR of the mRNA referred to as the **Shine–Dalgarno sequence** base-pairs with a complementary base sequence in the 16S rRNA of the 30S ribosomal subunit, thus positioning the ribosome at the correct start codon.

  Eukaryotic mRNA Recognition: Eukaryotic initiation factors bind to the 5' cap of the mRNA and assemble the initiation complex at the mRNA cap. The **scanning hypothesis** suggests that the initiation complex then scans along the 5' UTR until it encounters an appropriate start codon as determined by a surrounding consensus sequence known as the **Kozak sequence**.

- **Elongation:** Proteins known as **elongation factors** help the second (and all subsequent) charged tRNAs with an anticodon complementary to the next codon in the mRNA enter the A site of the complex. The enzyme **peptidyl transferase** catalyzes peptide bond formation between the first and second amino acids. The tRNA in the P site is now uncharged, and the tRNA in the A site carries a dipeptide (two amino acids joined by a peptide bond). The entire ribosomal complex shifts down the mRNA by one codon. In prokaryotes, the tRNA in the P site now enters the E site and is ejected. The tRNA in the A site carrying the dipeptide shifts to the P site, and the third tRNA molecule carrying an amino acid enters the A site with the help of elongation factors. The existing dipeptide is transferred to the third amino acid, the complex shifts, and the tRNA specified by the fourth codon enters. The process repeats until the entire polypeptide has been synthesized.

- **Termination:** The presence of a **stop** or **nonsense codon**, encoded by UAG, UAA, or UGA, in the A site indicates the end of the message to be translated. Nonsense codons are not recognized by a tRNA and do not specify an amino acid. Instead they interact with specific proteins known as **release factors** that release the polypeptide chain from the last tRNA and dissociate the ribosome complex from the mRNA.

- **Prokaryotic Versus Eukaryotic Translation:** In prokaryotic organisms, the processes of transcription and translation are coupled. In eukaryotic organisms, the mRNA molecule must leave the nucleus where it is synthesized and move to the ribosomes in the cytoplasm where translation occurs. A prokaryotic mRNA may also contain information for the synthesis of several proteins (**polycistronic**). Eukaryotic mRNAs are almost always **monocistronic**. Also, translation begins with a modified methionine in prokaryotes that is often removed from the mature protein, whereas eukaryotic translation begins with a normal methionine that remains as the first amino acid on the protein.

- **The Signal Hypothesis:** Some proteins being synthesized by ribosomes are destined to become a part of a cellular membrane or be exported to the outside of the cell membrane. These proteins usually contain an amino acid sequence at their very beginning called a **signal peptide**. It is the first part of the protein synthesized, and as it exits the ribosome, it is bound by the **signal recognition particle (SRP)** which translocates the entire translating complex to the endoplasmic reticulum where the newly synthesized protein can be deposited into the endoplasmic reticulum through a special pore called a **translocation channel or translocon**.

11.6 Posttranslational Changes to Peptide Chains

- **Overview:** Synthesis of a polypeptide by a ribosome is usually not sufficient to form a functional protein. The newly synthesized protein often needs to fold into a correct conformation with the help of **chaperones**. Furthermore, some proteins are cleaved after synthesis to release targeting signals such as the signal peptide or to form multiple proteins. Other proteins need to have phosphates, carbohydrates, or lipids attached to specific amino acids in order to target the protein to a particular place in the cell or to regulate the activity of the protein. These posttranslational modifications are more frequent in eukaryotes than prokaryotes.

11.7 The Genetic Code

- **Cracking the Genetic Code:** The genetic code is composed of a triplet sequence (codon) in the mRNA and is **nonoverlapping**, **unpunctuated**, **degenerate** (redundant), and nearly universal. Only rare exceptions to the code exist. Most of these exceptions are in mitochondrial translation, and most involve only a few codons. The code was determined to be read in triplets since combining two or more addition and deletion mutations (those that disrupted the reading frame) always restored the reading frame when a net loss or net gain of three nucleotides occurred. The code was finally deciphered by translating synthetically made mRNAs whose codon composition was known and by determining which charged tRNAs (20 different ones each charged with a different amino acid) bound to which known codons.

- **The Wobble Position:** The third base in the mRNA codon is the one that most often displays degeneracy in the code. For example, all codons that begin with GG code for glycine no matter what the third base of the codon is. This is important because mutations at this position are thus less likely to cause a change in the amino acid sequence of the protein. This third base is called the **wobble** base because the base in the tRNA anticodon that base-pairs with it can move slightly. This movement allows alternative base pairs to be able to form between the codon and anticodon at this position. As shown in table 11.6, G can pair with either C or U in this position. Also, if a tRNA contains the alternative base inosine (I) in the wobble position of the anticodon, the inosine can pair with A, U, or C. Thus, one tRNA with an I in the wobble position can recognize three different codons. Therefore, it is not necessary for a cell to possess 61 different tRNAs to decode the 61 amino acid encoding codons. For example, *E. coli* contains only about 50 different tRNAs.

Key Terms

- triplet-repeat diseases
- β-amyloid
- prion
- biochemical genetics
- one-gene/one-enzyme hypothesis
- amino acids
- primary structure
- secondary structure
- tertiary structure
- quaternary structure
- colinearity

- aminoacyl-tRNA synthetases
- charged tRNA
- activation site
- editing site
- initiation
- elongation
- termination
- N-formyl methionine
- initiation codon
- initiation complex
- initiation factors (IFs)

- Shine–Dalgarno sequence
- scanning hypothesis
- shunting
- 5′ untranslated region (5′ UTR)
- Kozak sequence
- open reading frames (ORFs)
- internal ribosome entry site (IRES)
- A (aminoacyl) site
- P (peptidyl) site

- E (exit) site
- elongation factors (EFs)
- peptidyl transferase
- stop codons
- nonsense condons
- release factors (RFs)
- molecular mimicry
- ribosome recycling factor (RRF)
- polyribosome
- polysome
- polycistronic
- monocistronic
- signal hypothesis
- signal peptide
- signal recognition particle (SRP)
- docking protein (DP)
- translocation channel (translocon)
- chaperones
- frameshifted
- overlapping code
- punctuation
- polynucleotide phosphorylase
- cell-free system
- degeneracy
- degenerate code
- unmixed families
- mixed families
- wobble
- site-specific variation
- reverse translation

Understanding the Key Concepts

Use the key terms or parts of their definitions to complete the following sentences.

Proteins are a chain of (1)_____ connected to each other via (2)_____. Proteins differ from one another by the composition and order of the (3)_____ different amino acids. This order and composition determines the three-dimensional structure assumed by the specific protein. The (4)_____ of each amino acid determines this structure and provides the chemical properties of the protein to aid in binding and catalysis. The (5)_____ structure of a protein can be easily determined by examining either the mRNA sequence that produced it or the actual amino acid sequence. Short motifs comprising the (6)_____ of a protein, however, are more difficult to determine but can sometimes be predicted from the amino acid sequence by knowing that alanine, glutamic acid, and methionine are more likely to be found in (7)_____ while isoleucine, tyrosine, and valine are more likely to be found in (8)_____. Other examples of motifs that comprise this level of protein structure include (9)_____, (10)_____, and random coils. How all of these motifs fold into a three-dimensional structure corresponds to the (11)_____, and the description of the number and composition of subunits constitutes the quaternary level of protein structure.

(12)_____ are enzymes that covalently attach an (13)_____ to an appropriate tRNA to form a (14)_____ tRNA. The amino acid is first selected by its small size through a sieve and activated by attachment of (15)_____ in the (16)_____ site of the enzyme. After activation, the amino acids that are too (17)_____ pass through a second sieve into the (18)_____ where they are degraded. The correct amino acid for a given aminoacyl-tRNA is thus the only activated amino acid left in the active site to be attached to the tRNA.

(19)_____ are proteins that help the ribosome and initiator tRNA assemble at the (20)_____ codon. Elongation factors are proteins that either help charged tRNAs bind to the (21)_____ or translocate the mRNA and its associated tRNAs to the next sites. (22)_____ are proteins that bind to the A site when a nonsense codon enters. In order to bind, they (23)_____ the shape of a tRNA. Once bound in the A site they release the polypeptide and (24)_____ bind to dissociate the ribosome subunits.

Synthesis of a polypeptide chain by a ribosome does not usually immediately create a functional protein. Sorting, proper folding, and in some cases proper modifications are required to make a functional protein. The (25)_____ is a peptide synthesized at the beginning of translating protein that directs the newly synthesized protein to the (26)_____ by translocating the ribosome complex via the action of the (27)_____. The newly synthesized protein is then inserted into this organelle through a pore called a (28)_____. This protein will then fold inside of the membrane in order to become part of the cells (29)_____ or to be exported to the outside of the cell. Other proteins that are not targeted to this organelle must fold the cytoplasm into the proper shape. Many (30)_____ help polypeptides fold into their proper structure. Once folded, many proteins may be cleaved or have (31)_____, (32)_____, or (33)_____ attached to certain amino acids in order to form active proteins.

The genetic code is read in triplets called (34)_____ that collectively comprise an ORF of a gene. No codons overlap each other, and no bases (punctuation) exist between codons. The gain or loss (through mutation) of one or two bases results in a (35)_____ so that every codon after the mutation will be read incorrectly. A second compensatory mutation that occurs near the first mutation and either adds or removes bases can restore the reading frame and produce a (36)_____ protein product. In an experiment designed to crack the genetic code, Matthaei and Nirenberg used a (37)_____ to translate synthetic RNAs that they made using (38)_____. They first discovered that the codon UUU codes for (39)_____. Later, Nirenberg and Leder deciphered all of the genetic code using a (40)_____. The unveiling of the genetic code revealed that the degenerate code was very organized and that half of the 16 codon families were (41)_____ families where the third position of the codon could be any of the four nucleotides. Strikingly, only (42)_____ different tRNA anticodons are required to recognize all four of these degenerate codons in this type of codon family. This is because bases in the third position, termed the (43)_____ position, can form alternative base pairs.

Figure Analysis
Use the indicated figure or table from the textbook to answer the questions that follow.

1. **Figure 11.4**
(a) Which group of amino acids carries a negative charge at physiological pH? Which group carries a positive charge?

(b) Amino acids from which group are likely to associate with the cell membrane?

(c) Which amino acid(s) can form disulfide bridges?

2. **Figure 11.17**
(a) What amino acid will be found on each of these tRNAs, and where will it be attached?

(b) What is the anticodon that "reads" the codons in an mRNA ORF in these tRNAs?

(c) Why are there two different tRNAs with the same anticodon?

(d) How can the tRNAs have the same anticodon but receive a different amino acid?

(e) Into which site on the ribosome does the tRNA on the left initially enter the ribosome complex? How about the tRNA on the right?

(f) Why can't the tRNA on the right initially enter the same site on the ribosome as the tRNA on the left?

3. **Figure 11.18**
(a) What are IF1, IF2, and IF3?

(b) In step 1, how does the 30S subunit bind to the DNA?

(c) In step 3, into which site on the ribosome does fMet-tRNA$_f^{Met}$ bind?

(d) What two purposes does IF1 serve during prokaryotic initiation?

(e) What prevents Met-tRNA$_m^{Met}$ from being used instead of fMet-tRNA$_f^{Met}$?

4. **Figure 11.21**
(a) Does figure 11.21 depict prokaryotic or eukaryotic translation initiation?

(b) Describe how translation of the ORF occurs in the scanning model shown on the top.

(c) In the shunting model shown in the middle, why is the second AUG used instead of the first?

(d) In the multiple translation start site model shown on the bottom, why is the second ORF able to be translated?

5. **Table 11.6**
(a) Why can G form a base pair with either U or C in the third position of the codon?

(b) Can G form a base pair with U in any other position in the codon?

(c) Considering these wobble rules and the genetic code shown in table 11.5, what is the *minimum* number of tRNA anticodons needed to recognize all of the codons for each of the following amino acids? What are these codons?
Proline:
Tyrosine:
Glutamine:
Isoleucine:
Arginine:
Serine:

General Questions

Q11-1. Describe how analysis of nonsense mutations demonstrated colinearity between a gene and a protein.

Q11-2. Describe the functions and actions of the prokaryotic elongation factors EF-Tu and EF-Ts and GTP during translation?

Q11-3. List five differences other than differences in numbers and names of accessory proteins between eukaryotic and prokaryotic translation.
i)
ii)
iii)
iv)
v)

Q11-4. DNA can serve directly as the template for protein synthesis only in the laboratory. Why would this mechanism of information transfer be difficult in eukaryotic cells? What difficulties would this mechanism of transfer pose?

Q11-5. You have generated a mutant strain of bacterium. When grown at room temperature, translation proceeds normally. However, when grown at 37°C, a tRNA whose anticodon is 5′ CUA 3′ and is charged with leucine is produced in the cells. What consequence will growing the cells at 37°C have on protein synthesis?

Q11-6. Describe the experimental evidence that demonstrates that the genetic code is read in triplets.

Multiple Choice

For each of the following, circle the letter of the choice that most appropriately answers the question.

1.) Which of the following amino acids is considered basic?

   a. Asp
   b. Lys
   c. Met
   d. Trp
   e. Tyr

2.) Which of the following describes the secondary level of protein structure?

   a. mRNA sequence
   b. amino acid sequence
   c. formation of pockets and binding sites on the surface of the protein
   d. assembly of two identical subunits to form an active protein
   e. folding of a short stretch of amino acids into a β-sheet

3.) Which of the following is a true statement concerning aminoacyl-tRNA synthetases?

   a. There is a different aminoacyl-tRNA synthetase for each of the 20 amino acids.
   b. All aminoacyl-tRNA synthetases bind tRNAs from the D-loop side of the tRNA.
   c. Aminoacyl-tRNA synthetases attach amino acids to tRNA in a single enzymatic step.
   d. Aminoacyl-tRNA synthetases do not have any proofreading ability.
   e. Aminoacyl-tRNA synthetases use GTP when charging a tRNA.

4.) Which of the following is *not* a regulatory element found in a prokaryotic mRNA?

   a. Shine–Dalgarno sequence
   b. initiation codon
   c. termination codon
   d. internal ribosome entry site (IRES)
   e. None of the above

5.) In what order are the three sites on a prokaryotic ribosome encountered by a tRNA?

   a. A → E → P
   b. E → A → P
   c. A → P → E
   d. P → E → A
   e. P → A → E

6.) Peptidyl transferase activity attaches the _____ group of the last amino acid incorporated into the polypeptide to the _____ of the next amino acid to be added.

   a. carboxyl, amino
   b. amino, carboxyl
   c. R, R
   d. R, carboxyl
   e. amino, R

7.) Which of the following is a true statement concerning ribosomes translating an mRNA?

   a. Ribosomes can bind to eukaryotic mRNAs before their 3′ ends have been synthesized.
   b. The elongation stage of translation does not require any energy expenditure by the cell.
   c. After one ribosome leaves the initiation codon and begins elongation, another ribosome can initiate on that same mRNA in both eukaryotes and prokaryotes.
   d. The last tRNA to bind to an elongating ribosome base-pairs with one of the nonsense codons.
   e. The polypeptide chain is never attached to a tRNA bound to the A site of the ribosome during the elongation stage.

8.) To study translation, scientists changed the amino acid cysteine to alanine *after* cysteine had been attached to its appropriate tRNA. Then, they performed translation experiments and found that alanine was incorporated into a protein where cysteine was supposed to be. What did this demonstrate about decoding of an mRNA sequence by the ribosome?

   a. The amino acid, not the anticodon, determines which amino acid is added next.
   b. The anticodon, not the amino acid, determines which amino acid is added next.
   c. Ribosomes do not allow tRNAs with the wrong amino acid attached to enter the A site.
   d. Incorporating replacing the cysteines with alanines in a protein has no effect.
   e. Alanine and cysteine are normally coded for by the same codon since the code is degenerate.

9.) Marshall Nirenberg and colleagues made a synthetic mRNA using polynucleotide phosphorylase and A and G nucleotides in a 3A:1G ratio. They then translated this mRNA into protein using a cell-free system. What percentage of the amino acids incorporated into protein would you expect to be lysine?

   a. ~66%
   b. ~56%
   c. ~26%

d. ~16%

e. ~6%

a. changing codon AAA to GAG

b. changing codon UGU to UGA

c. changing codon AGU to AGG

10.) Which of the following mutations would most likely produce the least severe phenotype?

d. changing codon CCU to ACU

e. changing codon CUU to AUU

Practice Problems

P11-1. How much energy (number of phosphate bonds) would be needed to translate a 500 amino acid protein?

P11-2. The following is the DNA sequence for the template strand of the genes for two prokaryotic proteins. The entire sequence will be transcribed as a single polycistronic mRNA. What is the sequence of the two proteins?

(*Hint 1*: The genetic code is found in the mRNA molecule. The table of codons found in chapter 11 corresponds to mRNA codons, not tRNA anticodons or DNA bases.
Hint 2: The mRNA molecule is read from 5′ → 3′ and the polypeptide is synthesized from the amino terminal to the carboxyl terminal.)

5′-TCCTCCATACCGGGCGTTTGGGGGTATCTCCTCCATACCCGACGGGATTTGGGCCGATTCCACT-3′

P11-3. How many amino acids comprise a protein whose mRNA is 369 bases in length and has 30 bases in its 5′ UTR and 30 bases in its 3′ UTR? How long would an mRNA coding for a protein with 140 amino acids be if it had 5′ and 3′ UTRs of 30 bases each?

P11-4. Complete the following table. Use a "/" to indicate either one base or the other in a position and an N to indicate any one of the four bases in that position. For tRNAs, list all tRNA anticodons that comprise the minimal number of tRNAs needed to recognize all the codons for that amino acid.

(*Hint 3*: The anticodon of the tRNA base-pairs with the mRNA codon in an antiparallel direction ensuring that amino acids are inserted in the correct order.)

Coding DNA (5′ → 3′)								
Template DNA (3′ → 5′)		CCN						
mRNA codon (5′ → 3′)	AUG			AAA/G				UAA
tRNA (3′ → 5′)								
Amino Acids	Met	Gly	Cys	Lys	Tyr	Ala	Val	Stop

P11-5. For the following mRNA sequence, provide the resulting amino acid sequence given that the genetic code is (a) a nonoverlapping triplet and (b) a triplet code overlapping by one base.
AUGCAGCAGGGCUCCGCACUGACCUAG

Assessing Your Knowledge

Understanding the Key Concepts—Answers

1.) amino acids; 2.) peptide bonds; 3.) 20; 4.) side group or R group; 5.) primary; 6.) secondary structure; 7.) α-helices; 8.) β-sheets; 9.) extended strands; 10.) turns; 11.) tertiary structure; 12.) Aminoacyl-tRNA synthetases; 13.) amino acid; 14.) charged; 15.) ATP; 16.) activation; 17.) small; 18.) editing site; 19.) Initiation factors or IFs; 20.) initiator or start; 21.) A site; 22.) Release factors or RFs; 23.) mimic; 24.) ribosome recycling factors or RRFs; 25.) signal peptide; 26.) endoplasmic reticulum; 27.) signal recognition particle or SRP; 28.) translocation channel or translocon; 29.) membrane; 30.) chaperones; 31.) phosphates; 32.) carbohydrates; 33.) lipids; 34.) codons; 35.) frameshift; 36.) functional; 37.) cell-free system; 38.) polynucleotide phosphorylase; 39.) phenylalanine; 40.) binding assay; 41.) unmixed; 42.) two; 43.) wobble.

Figure Analysis—Answers

1a.) Acidic amino acids are negatively charged, and basic amino acids are positively charged; b.) Nonpolar amino acids can insert into the hydrophobic environment of the cell membrane; c.) Only cysteine can form disulfide bridges.

2a.) N-formyl methionine will be attached to the $tRNA_f^{Met}$ in part (a) and methionine will be attached to the $tRNA_m^{Met}$ in part (b). They will be attached to the 2′–OH or the 3′–OH of the adenosine at the 3′ end of the tRNAs (very top); b.) 5′-CAU-3′; c.) The tRNA in part (a) is the initiator tRNA and recognizes the AUG initiation codon, while the tRNA in part (b) recognizes all other AUG codons in the mRNA. A different tRNA is needed for the first AUG since that tRNA must enter the ribosome through a different site than the tRNAs that recognize all the other codons; d.) An aminoacyl-tRNA will charge both of these tRNAs with methionine. A second enzyme that only recognizes the charged tRNA in part (a) modifies the methionine on that tRNA to N-formyl methionine. This modifying enzyme cannot modify the charged tRNA in part (b) because that tRNA has two regions of different sequence (highlighted in gray); e.) The $tRNA_f^{Met}$ in part (a) will initially bind to the P site, while the $tRNA_m^{Met}$ in part (b) will initially bind to the A site; f.) The $tRNA_m^{Met}$ in part (b) cannot initially enter the P site for two reasons. First, during translation initiation, an initiation factor is required to aid tRNA entry into the P site. This initiation factor cannot bind the $tRNA_m^{Met}$ in part (b) because of the sequence differences highlighted in gray. Second, during translation elongation, the P site is occupied by the peptidyl tRNA.

3a.) IF1, IF2, and IF3 are proteins called initiation factors; b.) The rRNA component (16S rRNA) of the 30S ribosome subunit forms base pairs with the Shine–Dalgarno sequence in the mRNA; c.) P site; d.) IF1 blocks the A site so that the $fMet\text{-}tRNA_f^{Met}$ cannot bind to the A site and helps IF2 and IF3 bind to the 30S ribosome subunit; e.) Sequence differences in the anticodon loop prevent the $Met\text{-}tRNA_m^{Met}$ from binding to IF2.

4a.) Eukaryotic translation initiation; b.) The preinitiation complex, including the small ribosome subunit, some initiation factors, and the initiator tRNA, forms on the 5′ cap of the mRNA. This complex then scans along the mRNA in the 5′→ 3′ direction until it encounters an appropriate Kozak sequence; c.) The second AUG is likely surrounded by a sequence that fits the consensus for the Kozak sequence better than the sequence surrounding the first AUG. Alternatively, the sequence surrounding the first AUG forms a stem loop, blocking access to the AUG. In either situation, the second AUG will be selected as the initiation codon; d.) There is an internal ribosome entry site (IRES) downstream of the first ORF. This is a sequence that allows the preinitiation complex to form at it rather than on the 5′ cap. This type of translation initiation is often found in viral mRNAs that are translated inside eukaryotic cells.

5a.) In the third position of the mRNA codon, the bases can move (or wobble) slightly to accommodate alternative base pairs such as G–U; b.) No, wobble base interactions almost always occur only in the third position of the mRNA codon.
c.) Proline: 2 (3′-GGI-5′ and 3′-GGC/U-5′) *or* (3′-GGG-5′ and 3′-GGU-5′)

Tyrosine: 1   (3′-AUG-5′)   Notice that you cannot have 3′-A U I-5′ since it would recognize the
                                     nonsense codon UAA.

Glutamine: 1   (3′-GUU-5′)

Isoleucine: 1   (3′-UAI-5′)

Arginine: 3   (3′-UCU-5′) and [(3′-GCI-5′ and 3′-GCC/U-5′) or (3′-GCG-5′ and 3′-GCU-5′)]

Serine: 3   (3′-UCG-5′ and [(3′-AGI-5′ and 3′-AGC/U-5′) or (3′-AGG-5′ and 3′-AGU-5′)]

Again, notice that you cannot have 3′-UCI-5′ instead of 3′-UCG-5′ for serine since it would also recognize
the codon for arginine.

General Questions—Answers

Q11-1. The position of a nonsense mutation in an mRNA correlated with the position of truncation of the
translated protein. Nonsense mutations that occurred near the beginning of the mRNA resulted in proteins
that were truncated near *their* beginning resulting in shorter proteins. Nonsense mutations that occurred
near the end of the mRNA resulted in truncations located near the protein's end producing nearly full-
length proteins.

Q11-2. EF-Tu binds GTP and hydrolyzes it to GDP at which point EF-Ts removes the GDP from EF-Tu
and replaces it with GTP. When EF-Tu is bound to GTP, it helps charged tRNAs bind to the A site on
ribosomes during translation elongation. Once the charged tRNA has entered the A site, EF-Tu hydrolyzes
the GTP causing EF-Tu/GDP to dissociate from the ribosome and to be recharged with GTP by EF-Ts.

Q11-3. i) Prokaryotic ribosomes contain an E site, while eukaryotic ribosomes simply eject uncharged
tRNAs from the P site after peptide bond formation.
ii) Eukaryotic preinitiation complexes assemble on the mRNA cap, while prokaryotic preinitiation
complexes form on the Shine–Dalgarno sequence in the 5′ UTR near the initiation codon.
iii) Prokaryotic translation begins by incorporating a modified methionine that is often removed from the
protein after translation, while eukaryotic begins translation with a methionine that remains as the first
amino acid in eukaryotic proteins.
iv) Eukaryotic translation occurs more slowly than prokaryotic translation.
v) Internal translation initiation is common on prokaryotic mRNAs as they are usually polycistronic.
Eukaryotic mRNAs, however, are usually monocistronic, and internal translation initiation is rare.
Other differences include the presence of a Shine–Dalgarno sequence in prokaryotic mRNAs and a Kozak
sequence or an IRES in eukaryotic mRNAs. Also, the cap and poly(A) tail play important roles in
eukaryotic translation, but they do not exist in prokaryotes.

Q11-4. In eukaryotic cells, DNA is contained within the nucleus while the site of protein synthesis is the
ribosomes which are located in the cytoplasm. If DNA directly acted as the template for protein synthesis,
the molecule would have to migrate from the nucleus to the ribosome. There are several possible problems.
First, can the molecule pass through the nuclear membrane? Second, could the molecule be damaged by
cytoplasmic enzymes? Third, what would control the DNA molecule's movement? Finally, how would
regulation be accomplished?

Q11-5. An anticodon 5′-CUA-3′ will recognize the stop or nonsense codon 5′-UAG-3′ in the mRNA.
Translation will not stop at that particular stop codon; instead, a leucine will be added to the growing
polypeptide. Amino acids will continue to be added until one of the other two stop codons (UAA or UGA)
is reached. Therefore, growing the bacterium at the high temperature will yield some proteins with
additional amino acids at the carboxyl end.

Q11-6. Francis Crick and colleagues demonstrated that the reading frame could be restored and a
functional protein produced by combining nucleotide additions and deletions that added up to a net gain or
loss of nucleotides in multiples of three.

Multiple Choice—Answers

1.) b; 2.) e; 3.) a; 4.) d; 5.) c; 6.) a; 7.) c; 8.) b; 9.) b; 10.) e.

Practice Problems—Answers

P11-1. Four high-energy phosphate bonds, such as those found in ATP and GTP, are needed for each amino acid, so 4 × 500 amino acids yields 2000 phosphate bonds required.

P11-2. Each gene is separated by a Shine–Dalgarno sequence (italic), and each has a start (bold) and stop codon (bold italic). mRNA:
*AGGAGGU***AUG**GCCCGCAAACCCCCA*UAGAGGAGGU***AUG**GGCUGCCCUAAAACCCGGCUAAG *G**UGA***

> Amino acid sequence 1:
> Ala   Arg   Lys   Pro   Pro
> Amino acid sequence 2:
> Gly   Cys   Pro   Lys   Thr   Arg   Leu   Arg

P11-3. Both UTRs (60 bases) and the stop codon (3 bases) need to be subtracted from the total number of bases to find the number of bases that code for amino acids. In this case, 369 − 60 − 3 = 306. Three base pairs equals one amino acid, so 306/3 = 102 residues. Likewise, a protein composed of 140 amino acids would be coded for by an mRNA with 140 × 3 (because the code is triplet) + 60 (5′ and 3′ UTRs) + 3 (the stop codon) or a total of 483 bases.

P11-4.

Coding DNA (5′ → 3′)	ATG	GGN	TGT/C	AAA/G	TAT/C	GCN	GTN	TAA
Template DNA (3′ → 5′)	TAC	CCN	ACA/G	TTT/C	ATA/G	CGN	CAN	ATT
mRNA codon (5′ → 3′)	AUG	GGN	UGU/C	AAA/G	UAU/C	GCN	GUN	UAA
tRNA (3′ → 5′)	UAC	CCI & CCC/U or CCG & CCU	ACG	UUU	AUG	CGI & CGC/U or CGG & CGU	CAI & CAC/U or CAG & CAU	None
Amino acids	Met	Gly	Cys	Lys	Tyr	Ala	Val	Stop

P11-5. (a) AUG   CAG   CAG   GGC   UCC   GCA   CUG   ACC   UAG
             Met   Gln   Gln   Gly   Ser   Ala   Leu   Thr   Stop

(b)

AUG   Met
  GCA   Ala
    AGC   Ser
      CAG   Gln
        GGG   Gly
          GCU   Ala
            UCC   Ser
              CGC   Arg
                CAC   His
                  CUG   Leu
                    GAC   Asp
                      CCU   Pro
                        UAG   Stop

11

GENE EXPRESSION: TRANSLATION

CHAPTER SUMMARY QUESTIONS

1. 1. c, 2. g, 3. d, 4. a, 5. f, 6. e, 7. j, 8. i, 9. b, 10. h.

2. The one-gene/one-enzyme hypothesis was proposed by George Beadle and Edward Tatum. It basically states that each gene controls the production of a single protein or enzyme. As a rule of thumb, the hypothesis still holds true today. However, there are some exceptions: (1) Some proteins are encoded by more than one gene (hemoglobin, for example), so one-gene/one-polypeptide would be appropriate. (2) Some genes encode more than one form of a protein (via alternative splicing). (3) Some genes do not encode proteins but do encode functional RNAs (rRNA and tRNA for example), so one-gene/one-RNA!

3. There are four levels of protein structure: (1) The *primary* structure corresponds to the sequence of polymerized amino acids. It is this structure that determines the higher-order levels of protein structure. Amino acids that are near each other in the primary sequence can interact to form (2) the *secondary* structure which consists of regular, repeating configurations, the most common of which are α helices and β sheets. Further folding of the polypeptide brings the secondary structures into a precise three-dimensional configuration termed (3) the *tertiary* structure. This structure creates the active site regions in many proteins. Proteins and enzymes consisting of only a single polypeptide chain possess only these three levels of structure. Proteins composed of two or more polypeptide chains have an additional level called (4) the *quaternary* structure. This is brought about by interactions of the various polypeptide subunits with each other. Many enzymes, including DNA and RNA polymerase, exhibit quaternary structure.

4. Charging of a tRNA occurs in two steps, both catalyzed by aminoacyl-tRNA synthetase:
    1. Amino acid (aa) + ATP → aa~P—adenosine + P-P
          (aminoacyl-AMP) (pyrophosphate)

    2. aa~P—adenosine + tRNA → aa~tRNA + adenosine—P
          (aminoacyl-tRNA)   (AMP)

5. In eukaryotes, translation initiation requires a 5' cap in mRNA and scanning behavior. A translation preinitiation complex forms at the 5' cap. This complex, which contains a specific cap-binding protein, eIF4F, binds a 43S protein complex composed of the small 40S ribosomal subunit, eIF3, and the eIF2 ternary complex (eIF2, GTP, and the charged initiator tRNA, Met-tRNA$_i^{Met}$). The result is a 48S complex, which is then bound by eIF1 and eIF1A. These stimulate the complex to move along the mRNA until it recognizes the initiation codon (the scanning hypothesis). When the correct initiation codon (typically the first one in the mRNA) is reached, the eIF2-GTP hydrolyzes its bound GTP to GDP, which releases eIF3 and eIF2-GDP. This allows the large ribosomal subunit to associate with the small subunit and generate the functional ribosome.

6. Methionine, with a codon of 5'-AUG-3', has two tRNAs with the same complementary anticodon, 3'-UAC-5':tRNA$_f^{Met}$, which is used only during initiation to insert N-formyl methionine, and tRNA$_m^{Met}$, which is used to insert methionine during chain elongation. The two tRNAs have nucleotide sequence differences scattered throughout them. For example at the 5' end, tRNA$_m^{Met}$ has a properly paired G, whereas tRNA$_f^{Met}$ has a mismatched C. This C is required for the methionine to be formylated. In addition, the tRNA$_f^{Met}$ has three G–C pairs in its anticodon stem that are lacking in tRNA$_m^{Met}$. These G–C pairs are required for the charged N-formyl-Met tRNA$_f^{Met}$ to enter the P site and initiate translation. The two tRNAs are shown in figure 11.17.

7. EF-Tu, bound to GTP, brings a charged transfer RNA into the A site of the ribosome. At this point, the GTP is hydrolyzed and EF-Tu/GDP is released. EF-Ts is involved in recharging EF-Tu, by first displacing the GDP to form EF-Tu/EF-Ts, and then being displaced by a new GTP. The result is the regeneration of the EF-Tu/GTP complex. EF-G, bound to GTP, catalyzes translocation, or movement of the mRNA and its associated tRNAs in relation to the ribosome. The eukaryotic equivalents are eEF1α (EF-Tu), eEF1$\beta\gamma$ (EF-Ts) and eEF2 (EF-G).

8.   **a.** P site only (initiator tRNA).
    b. P site (initiator tRNA) or A site (all other tRNAs).
    c. P site (usually) or A site (briefly before translocation).
    d. E site (usually) or P site (briefly before translocation).
    e. A site only.

9. Termination of protein synthesis in bacteria occurs when one of the three stop or nonsense codons (5'-UAA-3', 5'-UAG-3', and 5'-UGA-3') appears in the A site of the ribosome. A class I release factor (RF1 or RF2) binds to the stop codon and promotes the hydrolysis of the ester bond linking the terminal amino acid to its tRNA in the P site. This releases the polypeptide from the ribosome. RF3 (a single class II release factor), then binds to the ribosome and hydrolyzes GTP, thereby releasing RF1 or RF2 from the ribosome. Following the completion of translation, a ribosome recycling factor (RRF) binds to the ribosome and mediates the dissociation of the two ribosomal subunits.

10. The genetic code is said to be degenerate because more than one codon can specify a given amino acid. For example, serine is specified by six different codons: UCU, UCC, UCA, UCG, AGU, and AGC. The genetic code is also nonpunctuated, meaning that the codons are not separated by bases in the final mRNA transcript. Finally, the genetic code is almost universal. The meaning of the codons is not identical in all species. Exceptions have been found, most notably in mitochondrial genomes. In non-mitochondrial genomes, most of the known exceptions involve the termination codons. These encode amino acids in some organisms, such as *Paramecium* and *Mycoplasma*.

11. Proteins can undergo
    1. Cleavage of 15 to 25 amino acids from the amino terminus of membrane-bound or secreted proteins. This amino acid sequence, termed the signal peptide, is removed by the enzyme signal peptidase.
    2. Addition of phosphate groups by kinase enzymes. This leads to either the activation or inactivation of the protein.
    3. Addition of sugars or carbohydrates. These glycoproteins play a role in cell recognition.
    4. Addition of lipids. These lipoproteins allow attachment to the membrane without the protein being embedded there.

12. a. A polysome is the configuration of several ribosomes simultaneously translating the same mRNA molecule.
    b. A chaperone is a protein that aids the folding of another protein into a functional shape.
    c. Wobble refers to the reduced constraint on the third base, or 5' end, of an anticodon which allows it to base-pair with more than one base at the 3' end of codons in mRNA.

13. A signal peptide is a sequence of amino acids at the amino-terminal end of a protein that signals that the protein should enter a membrane (see figure 11.34). Although the concept is the same, the situation in eukaryotes is somewhat more complex because there are many different membrane-bound organelles, each having their own membrane-specific requirements. Signal peptides are usually cleaved off the protein after the protein enters or passes through the membrane.

14. An overlapping genetic code is one where each base is used in either two or three different codons. The two restrictions against it are (1) a mutation in one base would affect more than one codon and therefore more than one amino acid, and (2) limitations on which amino acids could occur next to each other in a polypeptide chain. For example, phenylalanine, which is encoded by UUU or UUC, could never be adjacent to glutamic acid (GAA or GAG).

EXERCISES AND PROBLEMS

15. In a biosynthetic pathway, a mutant will be able to grow if it is provided with compounds that are produced *after* the step that is defective in the mutant. Mutant 1 requires compound X and cannot grow when supplemented with any of the other compounds. Therefore, mutant 1 affects the last step of this biosynthetic pathway. Mutant 2 can grow when supplemented with compound X or with compound C, but not when supplemented with any of the other compounds. Therefore, compound C is the penultimate compound in this pathway, and mutant 2 disrupts the reaction that produces compound C. Using similar logic, the entire pathway is as follows:

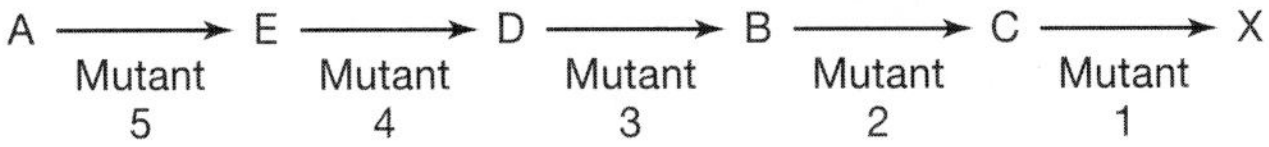

16. The messenger RNA is 5'-AUGUUACCGGGAAAAUAG-3'; the anticodons are 3'-UAC-5', 3'-AAU-5', 3'-GGC-5', 3'-CCU-5', 3'-UUU-5'; the amino acids are methionine, leucine, proline, glycine, and lysine (see the following figure).

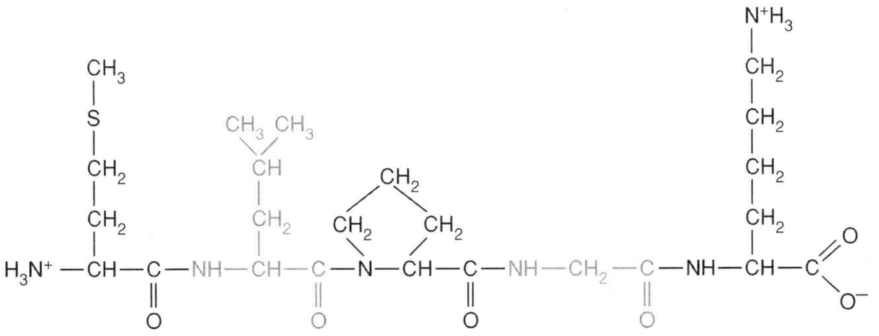

17. Please refer to figure 11.25. Use the messenger RNA of problem 16 and be sure to include EF-Tu and EF-Ts.

18. The three nonsense codons are 5'-UAA-3', 5'-UAG-3', and 5'-UGA-3'. They all begin with a pyrimidine (uracil) and have purines in the second and third positions. The theoretical anticodons are 3'-AUU/I-5', 3'-AUC/U-5', and 3'-ACU/I-5', respectively.

19. First, determine the number of codons that specify each amino acid in the pentapeptide: Ile—3, Trp—1, Gly—4, Leu—6, and Tyr—2. The number of different DNA sequences that encode these amino acids can be determined by multiplying the various possibilities (product rule of probability): $3 \times 1 \times 4 \times 6 \times 2 = 144$.

20. Please refer to figure 11.14.

21. Two. Please see figure 11.38 for an example.

22. Both spliceosomes and ribosomes are large complex structures made up of different RNA and protein molecules. The RNA in ribosomes is termed ribosomal RNA (rRNA), whereas small nuclear RNA (snRNA) is found in spliceosomes. Ribosomes are the sites of amino acid polymerization into proteins, whereas spliceosomes are involved in excision of introns and splicing of exons into mature mRNA transcripts.

23. **a.** 5'-AUG AUU GAA UGC GAG CGG AGU-3'
    b. N-met-ile-glu-cys-glu-arg-ser
    First determine the sequence of the RNA complementary to the given DNA strand. Don't forget about polarity; as the strand is written, the 5' end of the RNA will be on the left. Blocking off successive groups of three bases allows the determination of the codons. Use the code to determine the amino acid sequence.

24. The 3' end of the 16S rRNA contains the region that will base-pair with the Shine–Dalgarno sequence located right upstream of the initiation codon on mRNA molecules. This 3' region of the 16S rRNA is essential for translation initiation. Its cleavage by colicin E3 would prevent the ribosome from binding to the mRNA, and therefore translation would be inhibited.

25. 5'-UUA-3' (leucine) → 5'-UAA-3'. The consequence is that the growing polypeptide will be terminated at an improper point, probably producing a nonfunctioning enzyme or protein.

26. 5'-UAA-3' → 5'-UUA-3' (leucine). The consequence is the failure to terminate the particular protein leading to continued chain elongation to the next nonsense codon or to the end of the messenger RNA. The result is probably a nonfunctioning enzyme or protein.

27. There would be blockage of further protein synthesis because of the N-terminal formyl group that prevents a peptide bond. The growing peptide would be stopped at that point.

28. Because of wobble, only three tRNA are needed. A tRNA with anticodon 3'-UCU-5' can recognize the two codons 5'-AGA-3' and 5'-AGG-3'. A tRNA with anticodon 3'-GCI-5' can recognize the three codons 5'-CGA-3', 5'-CGU-3', and 5'-CGC-3'. The sixth and final arginine codon, 5'-CGG-3', can be recognized by a tRNA with the anticodon 3'-GCC-5'.

29. Starting with F at the NH_2 terminus, find the overlap between the various large fragments:

FGKI
  KICAB
    CABHLN
      HLNOED
        OEDJM
Therefore the sequence is NH_2-FGKICABHLNOEDJM-COOH.

30. The table could look the same (see table 11.4) except that the position would be left side = first position (5' end), top = third position (3' end), and right side = second position. For example, the codons for valine (currently 5'-GUU-3', 5'-GUC-3', 5'-GUA-3', and 5'-GUG-3') would be 5'-GUU-3', 5'-GCU-3', 5'-GAU-3', and 5'-GGU-3'.

31. GGC. The first polymer produces repeated reading frames of GCG-CGC. Therefore, GCG and CGC represent alanine and arginine, but we cannot determine which is which. The second polymer produces repeated sequences of either CGG, GGC, or GCG. We know that GCG codes for either alanine or arginine. CGC and CGG probably code for the same amino acid (wobble or unmixed family), either alanine or arginine. The final combination, GGC, is not similar to the original codons and probably codes for glycine.

32. a. The ribonucleotides A and C can combine to form eight different codons: AAA, AAC, ACA, ACC, CAA, CAC, CCA, and CCC. The frequency of each of these codons is determined by the abundance of the A and C ribonucleotides. The probability of obtaining an A at any position in a codon is 5/6, while that of C is 1/6. Using the product rule we can determine the frequencies of the codons. Then we can use table 11.5 to determine the amino acid specificities.

    AAA = (5/6)(5/6)(5/6) = 125/216 lysine
    AAC = (5/6)(5/6)(1/6) = 25/216 asparagine
    ACA = (5/6)(1/6)(5/6) = 25/216 threonine
    ACC = (5/6)(1/6)(1/6) = 5/216 threonine
    CAA = (1/6)(5/6)(5/6) = 25/216 glutamine
    CAC = (1/6)(5/6)(1/6) = 5/216 histidine
    CCA = (1/6)(1/6)(5/6) = 5/216 proline
    CCC = (1/6)(1/6)(1/6) = 1/216 proline

Therefore, there are 125/216 (or 57.87%) lysine, 30/216 (or 13.89%) threonine, 25/216 (or 11.57%) asparagine, 25/216 (or 11.57%) glutamine, 6/216 (or 2.78%) proline, and 5/216 (or 2.31%) histidine.

b. The ribonucleotides A, U, and G can combine to form $3^3 = 27$ different codons, including the three stop codons, UAA, UAG, and UGA. The frequencies of the individual bases are A = 3/5 or 0.6, U = 1/5 or 0.2, and G = 1/5 or 0.2. To list all 27 codons, we can use the following systematic approach: First, we consider two of the three bases at a time [as we did in part (a)], and then we consider the remaining codons that have all three bases together.

AAA = 0.6 × 0.6 × 0.6 = 0.216 lysine
AAU = 0.6 × 0.6 × 0.2 = 0.072 asparagine
AUA = 0.6 × 0.2 × 0.6 = 0.072 isoleucine
AUU = 0.6 × 0.2 × 0.2 = 0.024 isoleucine
UAA = 0.2 × 0.6 × 0.6 = 0.072 *Stop*
UAU = 0.2 × 0.6 × 0.2 = 0.024 tyrosine
UUA = 0.2 × 0.2 × 0.6 = 0.024 leucine
UUU = 0.2 × 0.2 × 0.2 = 0.008 phenylalanine

AAG = 0.6 × 0.6 × 0.2 = 0.072 lysine
AGA = 0.6 × 0.2 × 0.6 = 0.072 arginine
AGG = 0.6 × 0.2 × 0.2 = 0.024 arginine
GAA = 0.2 × 0.6 × 0.6 = 0.072 glutamic acid
GAG = 0.2 × 0.6 × 0.2 = 0.024 glutamic acid
GGA = 0.2 × 0.2 × 0.6 = 0.024 glycine
GGG = 0.2 × 0.2 × 0.2 = 0.008 glycine

UUG = 0.2 × 0.2 × 0.2 = 0.008 leucine
UGU = 0.2 × 0.2 × 0.2 = 0.008 cysteine
UGG = 0.2 × 0.2 × 0.2 = 0.008 tryptophan
GUU = 0.2 × 0.2 × 0.2 = 0.008 valine
GUG = 0.2 × 0.2 × 0.2 = 0.008 valine
GGU = 0.2 × 0.2 × 0.2 = 0.008 glycine

AUG = 0.6 × 0.2 × 0.2 = 0.024 methionine
AGU = 0.6 × 0.2 × 0.2 = 0.024 serine
UAG = 0.2 × 0.6 × 0.2 = 0.024 *Stop*
UGA = 0.2 × 0.2 × 0.6 = 0.024 *Stop*
GAU = 0.2 × 0.6 × 0.2 = 0.024 aspartic acid
GUA = 0.2 × 0.2 × 0.6 = 0.024 valine

The total frequency of stop or nonsense codons is 0.072 + 0.024 + 0.024 = 0.12. Therefore, the frequency of codons that specify amino acids (sense codons) will be 1 − 0.12 = 0.88. This is the number that has to be taken into account when determining the frequency or percentage of each amino acid in the protein mixture. Therefore,

Amino Acid	Frequency	Percentage
Lysine	$(0.216 + 0.072)/0.88 = 0.3272$	32.72%
Asparagine	$0.072/0.88 = 0.0818$	8.18
Isoleucine	$(0.072 + 0.024)/0.88 = 0.1091$	10.91
Tyrosine	$0.024/0.88 = 0.0272$	2.72
Leucine	$(0.024 + 0.008)/0.88 = 0.0363$	3.63
Phenylalanine	$0.008/0.88 = 0.0091$	0.91
Arginine	$(0.072 + 0.024)/0.88 = 0.1091$	10.91
Glutamic acid	$(0.072 + 0.024)/0.88 = 0.1091$	10.91
Glycine	$(0.024 + 0.008 + 0.008)/0.88 = 0.0454$	4.54
Cysteine	$0.008/0.88 = 0.0091$	0.91
Tryptophan	$0.008/0.88 = 0.0091$	0.91
Valine	$(0.008 + 0.008 + 0.024)/0.88 = 0.0454$	4.54
Methionine	$0.024/0.88 = 0.0272$	2.72
Serine	$0.024/0.88 = 0.0272$	2.72
Aspartic acid	$0.024/0.88 = 0.0272$	2.72

Note: Because 12% of all codons are nonsense codons, the polypeptide chains generated in this experiment would be relatively short.

33. We are mixing two RNA strands that are complementary; these strands will form a double-stranded RNA molecule. Since we observed the incorporation of no amino acids, the ribosome must not be able to read a double-stranded molecule.

34. Six. Begin with CUC, and then list all possible codons in which the first C is changed; then list all possible codons in which only the U is changed. Finally, list all codons that result in a change of the second C.

List 1	List 2	List 3
AUC: *ile*	CAC: *his*	CUA: *leu*
UUC: *phe*	CGC: *arg*	CUU: *leu*
GUC: *val*	CCC: *pro*	CUG: *leu*

35. Six. Three on each strand, reading in the 5' → 3' direction.

36. The likely codon cannot be determined from the information given. Either glutamate codon (GAA or GAG) can mutate in one step to valine (GUN) or lysine (AAA or AAG).

37. Because only two amino acids are incorporated, either (1) two of the three possible codons specify the same amino acid or (2) one of the codons is a stop codon. If we write out (GUA)$_n$ as GUA GUA GUA GUA . . . , we see that we could use any of three different reading frames: GUA, UAG, or AGU. If you look at the genetic code,

you will see that UAG is a stop codon, and GUA and AGU encode valine and serine, respectively.

38. Since the protein is composed of four *identical* polypeptides, only *one* gene is required for its synthesis. Each of the four polypeptides has a molecular weight of 200,000/4 = 50,000 daltons. The average molecular weight of an amino acid is 125 daltons. So, the number of amino acids in each polypeptide is 50,000/125 = 400 amino acids. These are encoded by 400 codons or 400 × 3 = 1200 bases. We also have to add the three bases that make up the stop codon. Therefore, the number of bases of exon DNA that encode this polypeptide/protein is 1203.

39. The stop codon has probably mutated to give a codon for the amino acid leucine. The longer-than-normal protein suggests that the original stop was not read. Numerous possibilities exist. If the second letter of a stop codon were changed to a U, we would have UUA or UUG leucine codons. Alternatively, the insertion of a C before the U would yield CUA, CUG, or CUA as leucine codons. Similarly, an insertion of a U next to the first G yields UUA or GUG as leucine codons. Since the next amino acid is phenylalanine, the next codon must be UUC or UUU. If a base were added, as previously, the next codon would have to begin with A or G, and phenylalanine does not begin with A or G. Therefore, the most likely explanation is a change of the second letter from an A to a U.

40. a. The 16 amino acids require at least 16 codons. However, there also has to be at least one termination signal. Therefore, the code has to consist of a minimum of 17 codons. There are only two bases, so the number of codons is given by the formula 2^n, where n = number of bases in a codon. A quadruplet code will provide 2^4 = 16 codons, which is not enough. Therefore, the minimum codon will have to be made up of *five* bases: a quintuplet genetic code.

b. This genetic code will consist of 2^5 = 32 codons. There are only two types of codons that do *not* have at least two Zs: those with no Zs, and those with only one Z. Only the RRRRR codon has no Zs. There are five codons with one Z: RRRRZ, RRRZR, RRZRR, RZRRR, and ZRRRR. Therefore, the number of codons with at least two Zs is equal to 32 − (1 + 5) = 26 codons.

41. Either CAU or CAC. Write down all possible codons for each amino acid.

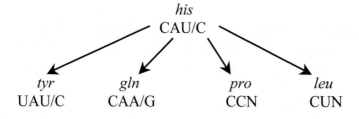

For *leu*, note that UUA/G cannot result from a single change in the *his* codon. Therefore, leucine must be CUN. All of the other amino acids could result from single changes in either the first or second base, and we are left with either codon being the one for *his*.

42. Mutant 1 is shorter than normal, indicating the cysteine codon mutated to a stop codon. Since cysteine is UGU/C, the stop must be UGA in the mutant. In mutant 2, histidine is replaced by leucine; CAU/C → CUN. Changing the A to a U will give a leucine codon. Only the third mutant allows us to determine the normal sequence. All amino acids after the first are changed, indicating a frame shift, an addition or deletion of one base. We must therefore line up the two possible sequences. Let's start with the normal sequence:

Gly	Ala	Ser	His	Cys	Leu	Phe
GGN	GCN	UCN	CAU/C	UGU/C	UUA/G	UUC/U
		AGU/C			CUN	

Then write down the possible sequences for mutant 3:

Gly	Val	Ala	Ile	Ala	Ser
GGN	GUN	GCN	AUU/C/A	GCN	UCN
					AGU/C

By aligning these sequences, the change can be seen as a deletion of the fifth base in the normal sequence (C). This causes a frame-shift mutation in mutant 3, which joins the sixth base in the normal sequence to the fourth base. So base 5 (U) in mutant 3 corresponds to base 6 (N) in the normal sequence. Therefore, the second codon (alanine) in the normal sequence must be GCU. Similar reasoning for the other positions yields the following mRNA sequence:

GGN GCU AGC CAU UGC CUC UUC/U

43.

I														II	
5' A	T	G	A	C	C	G	A	T	T	G	G	C	T	G 3'	DNA double helix
3' T	A	C	T	G	G	C	T	A	A	C	C	G	A	C 5'	
5' A	U	G	A	C	C	G	A	U	U	G	G	C	U	G 3'	Transcribed mRNA
3' U	A	C	U	G	G	C	U	A	A	C	C	G	A	C 5'	tRNA anticodons
N	Met		Thr		Asp		Trp		Leu			C			Encoded amino acids

CHAPTER INTEGRATION PROBLEM

a.   The genetic code will consist of $4^5 = 1024$ codons.

b.   The primary structure of a polypeptide refers to its amino acid sequence. The first of those 50 amino acids has to be N-formyl methionine. Each of the other 49 can be one of 21 possibilities. The number of different primary structures possible can be obtained using the product rule: 1 (N-formyl methionine) $\times$ $(21)^{49}$ = $(21)^{49}$ sequences.

c.   Scanning of the DNA sequence reveals a string of Ts (bases 262–269). Just upstream of these eight T's are two inverted repeats, GGATGAA (bases 244–250) and TTCATCC (bases 255–261), that are separated by the four-base sequence: GCTA. Transcription of this region of the DNA will form a stem-loop structure followed by a run of U's.

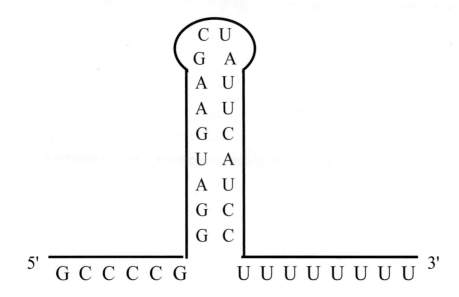

This is the hallmark of rho-independent terminators.

d.   First, let's try to look for the promoter elements in the DNA. This should be relatively easy because we are given the nontemplate or coding strand. Therefore, we can look for characteristic –35 sequences (consensus TTGACA) and –10 sequences (TATA box; consensus TATAAT). Scanning our DNA we can actually observe both types of sequences: a TTGGCA (bases 56–61) and TATAAT (bases 83–88).

   Now that we have found the promoter elements, let's figure out the mRNA sequence. The transcription initiation site is designated +1. The TATA box is centered at position –10. Therefore, the transcription intiation site will have to be about 10 bases downstream of the "middle" of the TATA box. We are also told that transcription initiation always occurs at a purine. Both of these facts point to the A at base 95 as the transcription initiation site. We have already determined the transcription terminator. To predict the entire mRNA sequence, all we have to do is take the coding DNA strand and replace its T's with U's. Therefore the *fun* mRNA is

```
+1    ACCUGGUAUG CAGGUUUAUC GCGGCUAACG GCAGUGUAGG AGGCCGGUGG

+51   AGAACAUCAA CGAUGCCAGA AAAUAAGAGA CAUCCGGGGA AACAUGAAUA

+101  GGAGGCUACU GGAUAACACA CAUAACGUAA GGGGAGAAUU AAUGCCCCGG

+151  GAUGAAGCUA UUCAUCCUUU UUUUU
```

Note: The mRNA bases have been designated with + numbers so as to differentiate them from the DNA bases.

e. We have to first find the correct translation initiation codon. This can be determined by the presence of a Shine–Dalgarno sequence a few bases upstream of it. We find such a sequence (AGGAGG) at bases +38–43. Therefore, the GUG codon (bases +47–49) is the start codon. Using the single-letter abbreviations, the amino acid sequence of the FUN protein is

```
5'-GUG GAG AAC AUC AAC GAU GCC AGA AAA UAA-3'
    M   E   N   I   N   D   A   R   K
```

f. In all six mutants, mutations converted the UAA stop codon into a sense codon. This lead to translation beyond the normal termination signal resulting in a much larger Fun protein.

g. The fact that all six mutants have the same amino acid sequence indicates that the stop codon (UAA) mutated into an amino acid that is encoded by six codons. There are only three such amino acids:

Leucine	UUA, UUG, CUU, CUC, CUA, CUG
Serine	AGU, AGC, UCU, UCC, UCA, UCG
Arginine	AGA, AGG, CGU, CGC, CGA, CGG

Mutant 1 undergoes a *single* R-to-Y base substitution. This rules out arginine because none of its six codons can be derived from UAA by a single base substitution. However, both leucine (U_U_A) and serine (U_C_A) can be the amino acid in mutant 1. In mutant 2 there is a Y insertion and an R deletion. The leucine codon CUA can be obtained from UAA by a Y insertion (C) and an R deletion (one of the two A's). However, none of the five remaining serine codons can claim this feat. (Although UCA can, it has been used for mutant 1 and so cannot be considered again. Remember that all six mutants have different DNA sequences.) Therefore, the amino acid is leucine, and the other mutant DNAs are as follows:

Mutant 3 = CUU or CUC
Mutant 4 = CUU or CUC

Mutant 5 = CUG or UUG

Mutant 6 = CUG or UUG

h. GUG GAG AAC AUC AAC GAU GCC AGA AAA UUA GAG ACA UCC GGG
    M   E   N   I   N   D   A   R   K   L   E   T   S   G

GAA ACA UGA AUA GGA GGC UAC UGG AUA ACA CAC AUA ACG UAA
  E   T   J   I   G   G   Y   W   I   T   H   I   T

Please note that this DNA sequence is that of mutant 1. The other mutants have different codon sequences for leucine (L), but they all have the same amino acid sequence for the mutant Fun protein.

CHAPTER 12 RECOMBINANT DNA TECHNOLOGY

Chapter Goals

1. Understand the basic features of restriction enzymes and how they are used in DNA cloning.
2. Compare the various types of vectors and understand how they affect DNA cloning.
3. Describe nucleic acid hybridization techniques and be able to properly interpret the data.
4. Be able to prepare a restriction map using double-digest, partial-digest, and end-labeling data.
5. Diagram how the polymerase chain reaction works.
6. Understand the methods used to sequence DNA and be able to correctly read the DNA sequence from a sequencing gel.

Key Chapter Concepts

12.1 Cloning DNA

- **Cloning:** Cloning DNA involves cleaving the DNA into specific segments, rejoining the segments, and adding a replication origin so that it can be replicated in a cell. This technique allows biologists to "cut" and "paste" DNA from various sources and to integrate "foreign" or additional genes into organisms.

- **Restriction Endonucleases:** Type II restriction endonucleases or **restriction enzymes** are valuable to the molecular geneticist because they cleave DNA at very specific places called **restriction sites**. Each enzyme recognizes and cleaves a specific sequence to create DNA ends that are either blunt (the two DNA strands are flush) or staggered (one strand is longer than the other). These enzymes occur naturally in bacteria where they protect the bacterium by cleaving foreign DNA.

- **DNA Ligation:** DNA molecules that have been cleaved can be rejoined or joined with another DNA molecule possessing the same free ends. To do this, a restriction enzyme can be used to cut the target DNA and a plasmid (one cleavage site per plasmid), creating complementary sequences on the free ends of each. The two are mixed together with ligase that will join the free ends of the target DNA and the plasmid. Some of the recircularized plasmids will have incorporated the target DNA (hybrids). When transformed into *E. coli*, these hybrids can be identified by restriction mapping following DNA isolation. These hybrids can then be used to transfer the target DNA to the host cell where it will be studied.

- **Prokaryotic Vectors: Vectors** are DNA molecules that can be maintained in a cell and can have foreign DNA cloned into them. These include plasmids, the λ phage, cosmids, bacterial artificial chromosomes, and shuttle vectors. The simplest of all vectors is a **plasmid**. Plasmids are circular DNA molecules that possess an **origin of replication** which allows the plasmid to replicate autonomously in a cell. Most plasmids also contain a **multiple cloning site region** and at least one **selectable marker** which allows bacteria carrying the plasmid to be easily identified. Usually the maximum size of a circular recombinant DNA molecule is approximately 15 kilobases. **Lambda (λ)** is a **bacteriophage** that infects bacterial cells. It allows larger DNA fragments to be cloned into it than plasmids do. Foreign DNA can be cloned between the right and left arms of λ following restriction enzyme digest of both the λ and the foreign DNA that produces compatible ends. The recombinant λ is packaged into the protein coat and used to infect *E. coli* cells. Short single-stranded sequences at the ends of the linear λ DNA called *cos* sites allow λ to form a circle in the *E. coli* cell. The DNA between the two *cos* sites in the linear concatamer is introduced into the λ particle's head. *E. coli* that have been infected with λ are killed creating a plaque on an agar plate. **Cosmids** are hybrids that contain the λ *cos* sites in a plasmid. Cosmids have an origin of replication, a selectable marker, and a single *cos* site. Large foreign DNA fragments can be ligated into a linearized cosmid. Recombinant DNA is packaged into λ protein coats and used to infect a bacterial cell. Base-pairing of the *cos* sites followed by ligation allow cosmids to behave as plasmids inside the bacterial cell. A **bacterial artificial chromosome (BAC)** allows the cloning of foreign DNA that may be several hundred to thousands of kilobase pairs in length. A BAC vector contains an origin of replication, partitioning element sequences (*par*) that ensure that a single BAC moves into each daughter cell, an antibiotic resistance gene allowing for selection of *E. coli* that contain the BAC, and a multiple cloning site. **Shuttle vectors** are circular plasmids that contain two origins of replication. One origin allows the vector to replicate in the *E. coli*, and the second origin allows it to replicate in another organism such as yeast.

12.2 DNA Libraries

- **Two Kinds of DNA Libraries:** A DNA library is composed of a pool of identical vectors that individually contain different DNA segments. These DNA segments can be pieces of genomic DNA (**genomic library**) or pieces of cDNA (**cDNA library**) that are generated from a specific tissue. Libraries, in combination with hybridization techniques, can be used to isolate genes near a known DNA sequence, or similar genes in other species. Libraries, in combination with transformation and complementation techniques, can be used to isolate the wild-type version of mutant genes.

- **Creating a Genomic Library:** **Genomic libraries** are generated by partial digestion of the genomic DNA with a restriction enzyme. Fragments generated by this digest are cloned into a vector. This ligated DNA is introduced into *E. coli*, and all of the *E. coli* cells are plated onto media that selects for the recombinant DNA. Each *E. coli* colony will have the exact same recombinant vector as the original cell. For bacteriophage-based vectors, each bacteriophage particle should contain a different recombinant genomic DNA insert. The recombinant phage amplifies itself in the infected *E. coli* cells and kills the *E. coli*. Then a new phage is released producing a plaque of lysed *E. coli* cells. The phage that is capable of infecting *E. coli* cells is maintained for the library.

- **Creating a cDNA Library:** A **cDNA library** is generated by isolating mRNA from a tissue, purifying it on an oligo(dT) column, and converting it to cDNA with **reverse transcriptase**. Once blunt-ended, double-stranded cDNA is made, it can be cloned using blunt-end linkers to generate a cDNA library. Reverse transcription uses a short single-stranded piece of DNA that is complementary to the mRNA to prime DNA synthesis. Usually in eukaryotes the sequence that is used is approximately 20 T residues, which is complementary to a portion of the mRNA's poly(A) tail. This technique can be used to examine or isolate one, or as many as all, of the expressed genes in a cell or tissue.

- **Comparing Genomic and cDNA Libraries:** The genomic library and cDNA library differ in that the genomic library contains all genes and other DNA segments such as introns and promoter regions, whereas the cDNA library contains only the expressed genes. A genomic library from an organism should be essentially the same regardless of the tissue or developmental stage at which it was isolated. In contrast, a cDNA library will be unique to the tissue and developmental stage at which the mRNA was isolated. All sequences in a genomic library will be present at nearly equal amounts, while the amount of each sequence in a cDNA library will be proportional to the amount of the corresponding mRNA in the cell.

12.3 Identifying and Isolating Specific Sequences

- **Nucleic Acid Hybridization:** Nucleic acid hybridization is one technique that can be used to identify specific DNA sequences within a population. Hybridization is the process by which one strand of a nucleic acid will bind its complementary strand, or a nearly complementary strand, via hydrogen bonding. Both strands can be DNA, but an RNA strand can hybridize to a complementary DNA strand and two complementary RNA strands can hybridize.

- **Southern Hybridization:** **Southern blotting** involves hybridization of a single-stranded nucleic acid "**probe**" to a DNA target. It is commonly used to isolate a desired gene when the entire genome is digested with a restriction enzyme. In this method, the fragments produced from the restriction enzyme digest are separated according to size by **gel electrophoresis**. While still in the gel, the separated DNA is transferred and permanently bound to a membrane. This membrane is then immersed in a solution containing a radioactively, fluorescently, or enzymatically labeled nucleic acid (the probe) possessing a base sequence complementary to one found in the target DNA. The probe will base-pair with its complementary sequence. When exposed to photographic film, the gene of interest will be "highlighted."

- **Colony or Plaque Hybridization:** Screening a library by colony or plaque hybridization involves plating the bacteria containing the library clones, transferring some of the bacterial colonies to nitrocellulose by placing the filter on top of the master plate. Cells on the filter are then lysed, and the DNA is denatured and fixed to the filter. The filter is then probed with a radioactively labeled probe that is complementary to the target sequence. After the filter is washed, it is exposed to X-ray film to identify colonies that hybridize to the probe.

- **Northern Hybridization: Northern blotting** is similar to Southern blotting except that the target is RNA, which is not treated with restriction enzymes. This analysis allows geneticists to determine whether a particular gene is being expressed (transcribed) and in what relative quantity.
- **Western blotting** is a third method that uses a type of "hybridization" which is unlike the previous two methods. For western blots, protein is the target and a special type of protein is also the "probe." In this method, an antibody that is specific to the protein of interest is used as a probe. A sample is first exposed to an antibody specific for the target protein. A second antibody carrying a fluorescent or enzymatic label is added to allow the location of the first antibody and hence the protein of interest. Western blot analysis can be used to locate specific genes by first locating the specific gene products (proteins) and to determine whether a particular gene is being expressed (transcribed and translated) and in what relative quantity.
- **Polymerase Chain Reaction:** This technique also utilizes a DNA synthetic enzyme and takes advantage of our knowledge of the replication process to amplify small samples of DNA. Primers complementary to sequences found on either side of the region of interest are created and used to initiate replication. These primers and all materials necessary for replication are mixed with a DNA sample. Multiple cycles of heating and cooling work to amplify the regions between the primers. The steps of each cycle are: (1) denaturation of the DNA helix into two separate strands, (2) hybridization of the primers to the single-stranded DNA, and (3) polymerization (DNA synthesis) of the region between the primers.

12.4 Restriction Mapping

- **Restriction Mapping:** The size of the fragments produced when a sample DNA is cleaved by restriction enzymes can be used to generate a physical map of the sample DNA. The results of a **partial digest** (one in which the process is stopped before all sites are cleaved) can be compared to those of a complete digest to determine the order of restriction sites. Two enzymes can be used simultaneously to digest a sample of DNA (**double digest**).
- **Restriction Fragment Length Polymorphisms:** Physical changes in the DNA molecule from one individual to another may change the number of restriction sites. As a result, digestion of two samples from two different individuals may yield fragments that differ in size. These differences are called restriction fragment length polymorphisms (**RFLPs**) and are detected using Southern blot analysis. Differences in the number of nucleotides between restriction sites may also result from differing numbers of repeated short segments termed **variable-number tandem repeats** (**VNTR**). VNTRs can also be mapped by polymerase chain reaction. A Southern blot using a VNTR sequence as a probe creates a **DNA fingerprint**. This DNA fingerprint may identify a single individual in a population because of the hypervariable nature of the **minisatellite** sequences. For this reason, DNA fingerprinting is a powerful tool in forensic analysis.

12.5 DNA Sequencing

- **The Dideoxynucleotide Method:** The sequence of bases in a DNA molecule can be determined by manipulating the process of DNA synthesis. In the **dideoxynucleotide method**, bases lacking a free hydroxyl group are used to terminate DNA replication when they are inserted into a growing DNA chain. A sample of DNA is divided into four sub-samples, and each is mixed with a different dideoxynucleotide base. As DNA synthesis proceeds in each sub-sample, fragments of different lengths are produced. Each fragment corresponds to the position of the complementary base in the original DNA molecule. The sequence is revealed in the pattern of fragments produced.
- **Innovations to DNA Sequencing:** Two significant modifications have simplified DNA sequencing. First, the development of methods that simplify the isolation of DNA templates, such as the generation of plasmids that permit the rapid isolation of single-stranded DNA to serve as the sequencing reaction template, and the development of PCR that allows DNA templates to be amplified. Second, DNA sequencing has been simplified by the automation of sequencing reactions that allows the results to be directly read from sequencing gels by a computer. This can be done by using dideoxynucleotides that have a different fluorescent dye attached to each of them. These dyes fluoresce at different wavelengths. After the sequencing reactions are complete, the reactions are combined and run in a single lane on a polyacrylamide gel. As the gel runs, an argon laser excites the fluorescent tags, allowing the dideoxynucleotide at the 3' end of each fragment to be identified. This automated method

yields over 1000 bases of sequence per template, compared with the several hundred bases obtained from standard ^{32}P-labeled sequencing.

Key Terms

- gene cloning
- recombinant DNA technology
- restriction endonucleases
- restriction sites
- palindromic sequences
- 5′ overhang
- 3′ overhang
- compatible ends
- linkers
- vectors
- plasmids
- multiple cloning site region
- recombinant plasmid
- bacteriophage
- *cos* sites
- cosmid

- bacterial artificial chromosome (BAC)
- F factor
- shuttle vector
- genomic library
- complete digestion
- partial digestion
- complementary DNA (cDNA)
- reverse transcriptase
- Southern hybridization
- probe
- target
- electrophoresis
- stringency
- colony hybridization
- plaque hybridization
- Northern hybridization

- polymerase chain reaction (PCR)
- Western blotting (immunoblotting)
- restriction mapping
- DNA sequencing
- double digest
- restriction fragment length polymorphisms (RFLPs)
- hypervariable loci
- microsatellite
- minisatellite
- variable number tandem repeats (VNTR)
- DNA fingerprint
- dideoxynucleotide sequencing method
- sequencing ladder gels

Understanding the Key Concepts

Use the key terms or parts of their definitions to complete the following sentences.

Type II (1)_____ recognize and cleave DNA at (2)_____, such as GAATTC or CATATG. These enzymes can cut DNA and leave either a (3)_____ end or a (4)_____ end with a single-stranded (5)_____ at the 5′ or 3′ end. If a fragment of DNA has ends that are (6)_____ with the ends of a vector, the two can be (7)_____ together producing a (8)_____ DNA molecule. Bacterial cloning vectors must have an (9)_____ so that they can reproduce (10)_____ within a bacterial cell. In addition, these vectors must possess at least one (11)_____ so that cells containing this vector can be easily identified. Plasmids are used to clone inserts that are typically (12)_____ to (13)_____ in size. Larger inserts can be cloned into lambda bacteriophage which can then be used to (14)_____ bacterial cells. The lambda vector can be selected for because it creates a (15)_____ on a plate of bacterial cells. Cosmids are used to clone even larger DNA fragments. Cosmids are like (16)_____ because they contain an origin of replication, a selectable marker, and a (17)_____, but they also contain a single (18)_____ like those found in (19)_____ that allows them to (20)_____ once they have entered a bacterial cell. In this way, cosmids can replicate and behave like a (21)_____. Bacterial artificial chromosomes can be used to clone foreign DNA that can be several (22)_____ to (23)_____ of kilobases in length.

A genomic library is created by (24)_____ digesting an organism's total (25)_____ to generate (26)_____ fragments that are approximately (27)_____ in size, and cloning these fragments into a vector. A cDNA library is generated by isolating an organism's (28)_____ which is (29)_____ to generate a (30)_____ cDNA molecule that can be cloned using blunt-end (31)_____.

A specific DNA (32)_____ sequence can be isolated in the genome, or from a library by (33)_____ using a (34)_____, radioactively labeled (35)_____ as a (36)_____. In this procedure, genomic DNA is (37)_____ with a (38)_____ and separated by (39)_____. To screen a cDNA library for a specific clone, library DNA is transformed into *E. coli* and the clones are separated by (40)_____ the bacterial colonies.

Northern hybridization is used to isolate a specific (41)_____ (42)_____ sequence. This technique differs from (43)_____ because the (44)_____ is not (45)_____ before separation by (46)_____. This technique is useful for determining (47)_____ and (48)_____ a (49)_____ is transcribed and the relative level of (50)_____.

Immunoblotting or (51)_____ blotting is used to look at (52)_____ expression in a cell. (53)_____ are separated by (54)_____ according to their (55)_____, and then transferred to a nitrocellulose filter. This filter is probed with a (56)_____ that recognizes and binds to a specific target protein. A (57)_____ coupled to an (58)_____ that catalyzes a specific chemiluminescent or colorimetric reaction is used to recognize and bind to the (59)_____, allowing a specific protein to be visualized.

The polymerase chain reaction can be used to (60)_____ a specific sequence of (61)_____ using (62)_____ that are complementary to sequences that flank the sequence of interest. In this procedure, a heat-stable (63)_____ is used, along with the four dNTPs and cycling temperatures, to amplify the DNA fragment (64)_____. A DNA molecule can be characterized by (65)_____ mapping or by DNA (66)_____.

Figure Analysis
Use the indicated figure or table from the textbook to answer the questions that follow.

1. **Figure 12.8**
   (a) What is the function of the *amp*^{r} gene on the plasmid?

   (b) What is the second method that can be used to screen for bacteria containing the plasmid with the gene of interest?

   (c) Describe how the DNA containing the gene of interest is placed into the plasmid.

   (d) How could the recombinant plasmid containing DNA lacking the gene of interest be produced?

   (e) When the bacterial cells type one, two, and three are plated on media containing X-Gal, what color will the three different types of cells be? Explain why for each type of bacterial cell.

   (f) Why are none of the bacterial type four cells found on the Petri plate when the bacterial cells are plated?

   (g) How could a scientist distinguish between the three different types of plasmids that the bacterial cells may contain?

2. **Figure 12.12**
   (a) What three fragments are generated by restriction enzyme digest of the λ DNA?

   (b) What fragments are combined and ligated to form the linear concatamers?

   (c) How is the foreign DNA packaged into the bacteriophage?

   (d) Why must the total length of the λ DNA molecule be around 49 Kb?

   (e) What is the mechanism by which the bacteriophage infects an *E. coli* cell?

   (f) Upon entering the bacterial cell, what are the two things that the λ DNA can do?

   (g) What is the advantage of using the λ vector, rather than a plasmid to clone a piece of DNA?

   (h) What is the disadvantage of using the λ vector?

3. **Figure 12.17**

(a) Why is the generation of a cDNA library particularly useful for some experiments relative to a genomic DNA library?

(b) What is the template used by reverse transcriptase? What is used to prime the reaction?

(c) What is the product of the reverse transcriptase reaction?

(d) What is the function of RNase H?

(e) What are the two functions of DNA polymerase I in generation of the cDNA?

(f) Following ligation to seal the nicks, what is the final product of the reaction? How can it be cloned?

4. **Figure 12.23**

(a) How are the proteins separated?

(b) In transferring the proteins from the gel to the nitrocellulose filter, what direction will the proteins move under an electric current?

(c) What probe is used to detect a specific protein?

(d) What is the function of the secondary antibody? Why is it necessary?

(e) How is the protein ultimately visualized?

General Questions

Q12-1. Restriction enzymes are important tools of geneticists. From what natural source are they obtained and of what value are they in nature? Why are they useful in forensic science, for example?

Q12-2. A scientist is looking for any clones in a mouse cDNA library that are related to the human *PGC1* gene. Outline the experimental protocol a scientist would use to complete this experiment.

Q12-3. Blood from two different locations is collected by police at a crime scene. Two individuals are brought in for questioning and asked to provide DNA samples. Assuming that the police have access to the victim's DNA, how could the other samples be processed to determine if one of the suspects is likely guilty of the crime?

Q12-4. A scientist has only a very small quantity of a fragment of DNA with a known sequence, but needs more to serve as a control for other experiments. Describe the simplest way that more of this fragment can be generated. How can it be verified that more of the fragment actually was generated? If more was not generated, what should the scientist check to ensure that the experiment is set up properly?

Q12-5. Describe the reasons why a restriction map for either a genomic or cDNA clone is useful.

Multiple Choice

For each of the following, circle the letter of the choice that most appropriately answers the question.

1.) The relative abundance of mRNA in a specific tissue can be most easily determined by which of the following techniques?
   a. Southern hybridization
   b. plaque hybridization
   c. Northern hybridization
   d. colony hybridization
   e. Western blotting

2.) A DNA fragment of interest can be identified by Southern hybridization using which of the following as a probe?
   a. a radioactively labeled double-stranded DNA
   b. a denatured, radioactively labeled cDNA
   c. a denatured, radioactively labeled protein
   d. a denatured, unlabeled mRNA
   e. More than one of the above is correct.

3.) Following electrophoresis, an agarose gel containing digested fragments of DNA for Southern hybridization is treated with NaOH to do which of the following?
   a. prevent diffusion of the DNA in the gel
   b. increase the efficiency of transfer
   c. visualize the DNA in the gel
   d. increase the stringency of hybridization
   e. make the DNA single-stranded

4.) Immunoblotting can be used to tell you all but which of the following about a given protein?
   a. if the protein is truncated
   b. when the protein is expressed during development
   c. in which tissues the protein is expressed
   d. the molecular weight of the protein
   e. if the amino acid sequence of the protein is altered

5.) Which of the following is *not* used in the amplification of a specific DNA sequence by PCR?
   a. a mixture of deoxyribonucleotides
   b. a thermostable DNA polymerase
   c. two single-stranded RNA primers
   d. a double-stranded DNA template
   e. All of the above are required.

6.) Which of the following is *not* required for colony hybridization?
   a. fixing the DNA to a nitrocellulose filter
   b. hybridizing the DNA with a complementary radioactively labeled probe

   c. transforming the library clones into bacteria and plating them
   d. lysing the bacterial cells containing the library clones on the master plate
   e. denaturing the DNA with NaOH

7.) Incorporation of a dideoxynucleotide into a growing DNA strand prevents it from being used for further polymerization because of which of the following?
   a. It lacks a $5'$-PO_4^- group.
   b. It lacks a $3'$-OH group.
   c. It is radioactively labeled.
   d. It has a $3'$-OH group.
   e. It has a $2'$-H group.

8.) In the standard dideoxynucleotide method of DNA sequencing developed by Sanger, which of the following is radioactively labeled?
   a. DNA primer
   b. single-stranded DNA template
   c. deoxynucleotide triphosphates
   d. dideoxynucleotide triphosphates
   e. double-stranded DNA template

9.) Mini- and microsatellites are widely varied among individuals in a population as a result of which of the following?
   a. They lead to the introduction of nonsense mutations during DNA replication.
   b. They can lead to loss of a particular restriction site.
   c. They are mobile genetic elements.
   d. They vary in the number of repeated sequences that they contain.
   e. They compromise the proofreading function of DNA polymerase.

10.) Which of the following statements regarding dideoxynucleotide sequencing with fluorescent dyes is *not* accurate?
   a. A laser is used to excite the fluorescent tags.
   b. The deoxynucleotides are each attached to a different fluorescent dye.
   c. More bases per template can be read by this method than by the standard dideoxynucleotide sequencing method.
   d. The products of the four sequencing reactions can be run in a single lane on a polyacrylamide gel.
   e. The wavelength of the fluorescence reveals the identity of the nucleotide at the $3'$ end of the DNA strand.

Practice Problems

P12-1. Examine the following autoradiogram of a gel resulting from a DNA sequencing experiment. Determine the base sequence of both strands of the DNA molecule. Provide the banding pattern that would be expected if the complementary strand of DNA were sequenced by the same method.

[*Hint 1*: The sequence of a DNA molecule (5′ → 3′) is determined by reading the gel from bottom to top in the dideoxy method.]

Target Strand:

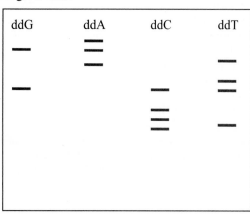

Complementary Strand:

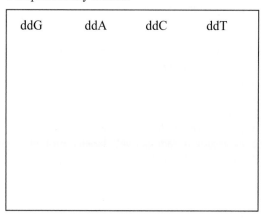

P12-2. A hypothetical gene has three different alleles (A1, A2, and A3). The RFLP sites in the region of the chromosome containing this gene are shown. The locations of sequences complementary to the probes used in the Southern blot analysis are also indicated. Predict the pattern of bands that would result from the analysis of homozygous individuals and from the analysis of individuals heterozygous for any pair of alleles.

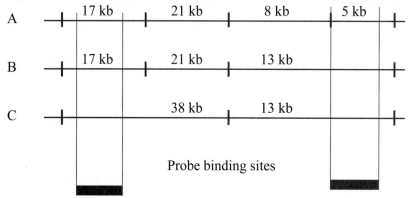

P12-3. Consider a 1250-bp-long piece of DNA that is digested with three different restriction enzymes singularly and in all combinations (double digests). The results of each single and double digest are given. Use the restriction fragment size data to construct a restriction map showing all restriction sites.

[*Hint 2*: Loss of a restriction site results in the formation of one large fragment in place of two small fragments (which together equal the size of the large fragment). Addition of a restriction site results in the formation of two small fragments in place of one larger fragment.

Hint 3: In restriction mapping, fragments appearing in a double digest should add up to a fragment from a single digest.]

Enzyme Treatment	Resulting Restriction Fragment Sizes			
Enzyme A	150	500	600	
Enzyme B	250	1000		
Enzyme C	450	800		
Enzyme A and B	100	150	400	600
Enzyme A and C	150	200	300	600
Enzyme B and C	200	250	800	

P12-4. The nucleotide sequence or a DNA sample from a hypothetical organism is given. Also given is the sequence of primers used to amplify a section of this DNA molecule. What sequence will the majority of the PCR products possess?

(*Hint 4*: PCR amplifies the DNA sequence occurring between two primers.)

DNA sequence: 5′-TCCACCCGTCGCAACAAAATTAACGTAACTTTACCCAACA-3′
                  3′-AGGTGGGCAGCGTTGTTTTAATTGCATTGAAATGGGTTGT-5′
Primers: 3′-TGGG-5′   5′-ACCC-3′

P12-5. Many of the large genome centers participating in the Human Genome Project can sequence DNA at a cost of $0.50 per base pair. If all laboratories involved in the project were as efficient, how much would it cost to sequence the entire human genome?

P12-6. For the agarose gel that follows, draw a map of the plasmid that would yield the DNA fragments shown after digesting the DNA with the indicated restriction enzymes. Show all of the *Eco*RI, *Bam*HI, and *Pst*I restriction sites in the plasmid and indicate the distance between each site.

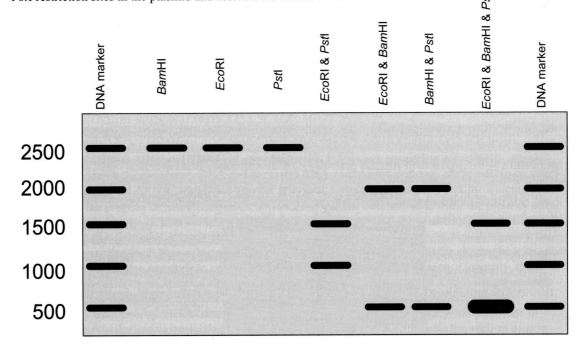

Assessing Your Knowledge

Understanding the Key Concepts—Answers

1.) restriction endonucleases; 2.) palindromic sequences; 3.) blunt; 4.) staggered; 5.) overhang; 6.) compatible; 7.) ligated; 8.) recombinant; 9.) origin of replication; 10.) autonomously; 11.) selectable marker; 12.) 100 bp; 13.) 10 Kb; 14.) infect; 15.) plaque; 16.) plasmids; 17.) multiple cloning site; 18.) *cos* site; 19.) bacteriophage lambda; 20.) circularize; 21.) plasmid; 22.) hundred; 23.) thousands; 24.) partially; 25.) genomic DNA; 26.) overlapping; 27.) equal; 28.) mRNA; 29.) reverse-transcribed; 30.) double-stranded; 31.) linkers; 32.) target; 33.) Southern hybridization; 34.) complementary; 35.) nucleic acid; 36.) probe; 37.) digested; 38.) restriction endonuclease; 39.) agarose gel electrophoresis; 40.) plating; 41–42.) RNA; target; 43.) Southern hybridization; 44.) mRNA; 45.) digested; 46.) agarose gel electrophoresis; 47–48.) when; where; 49.) mRNA; 50.) transcription or expression; 51.) Western; 52.) protein; 53.) Proteins; 54.) SDS-PAGE; 55.) molecular weights; 56.) primary antibody; 57.) secondary antibody; 58.) enzyme; 59.) primary antibody; 60.) amplify; 61.) DNA; 62.) primers; 63.) DNA polymerase; 64.) exponentially; 65.) restriction; 66.) sequencing.

Figure Analysis—Answers

1a.) The *amp*[r] gene allows bacterial cells to grow in the presence of the antibiotic ampicillin. The presence of the *amp* resistance gene on the plasmid allows only bacterial cells that contain the plasmid DNA to grow on media containing ampicillin; all other bacteria will be killed by the drug; b.) An alternate method that can be used to screen for bacterial cells containing the plasmid with the gene of interest is to plate the bacterial cells on media containing X-Gal because the plasmid contains the *lacZ* gene adjacent to the multiple cloning site. When the *lacZ* gene is expressed in *E. coli*, it produces β-galactosidase which allows cells to convert the normally colorless X-Gal into a blue precipitate. Ampicillin-resistant bacterial cells can be screened for whether or not they produce β-galactosidase. If they do, they will be blue and therefore do not contain foreign DNA cloned into the multiple cloning site; if they do not, they will be white and contain the foreign DNA; c.) DNA containing the gene of interest is digested with a specific restriction enzyme that cuts at both the 5′ and 3′ ends, or it is cut with two different enzymes that cut at each end, but not internally. The plasmid is cut with the same enzyme (or enzymes) that is used to cut the foreign DNA so that the foreign DNA and the plasmid DNA have compatible ends. Following restriction enzyme digest, the foreign DNA and the plasmid DNA are purified by agarose gel electrophoresis. The two fragments are then combined in a ligation reaction. Some of the ligation reaction is used to transform bacteria; d.) The recombinant plasmid containing the DNA other than the gene of interest could be produced because there are three restriction sites shown in the DNA fragment containing the gene of interest. A complete digest with this enzyme will yield four fragments of DNA. Some of the fragments appear to be close to the same size. When the DNA fragments are purified by agarose gel electrophoresis, they will be separated according to their size. Fragments of similar sizes will run closer to each other on the gel. If the fragment containing the gene of interest is not fully purified away from the other DNA fragments, some of the other fragments also having ends that are compatible with the digested plasmid DNA will be present in the ligation reaction. This DNA may then be ligated to the plasmid DNA, generating a recombinant plasmid containing DNA other than that containing the gene of interest; e.) The type one and type two bacterial cells will be white when plated on X-Gal media because they contain foreign DNA in the multiple cloning site which is in the middle of the *lacZ* gene. This DNA will disrupt the β-galactosidase gene, preventing production of the functional enzyme. Without functional β-galactosidase, the cells will be unable to convert X-Gal to a blue precipitate and will be white. The type three bacterial cells will be blue because they do not contain foreign DNA in the multiple cloning site, so the β-galactosidase gene is uninterrupted and fully functional. When produced, it will cleave X-Gal to form a blue precipitate; f.) None of the bacterial type four cells are found on the Petri plate when the bacterial cells are plated because these cells are not transformed with any plasmid DNA. Therefore, they do not contain an *amp*[r] gene and are killed when they are plated onto media containing ampicillin; g.) To distinguish between the three different types of plasmids that the bacterial cells may contain, X-Gal screening can initially be used. This will distinguish type one and type two bacterial cells from type three bacterial cells. To distinguish between the type one and type two cells, the scientist would have to pick the colonies off of the plate, grow them overnight in liquid media containing ampicillin, and then isolate the plasmid DNA from the cells. This DNA could then be digested with a specific restriction enzyme, or with a series of restriction enzymes, to confirm the presence or the absence of the gene of interest in the plasmid. By using enzymes that cut specifically at internal sites in the gene of interest, that either fail to cut the plasmid lacking the gene of interest, or cut at sites that will produce different size fragments in this plasmid relative to the plasmid containing the gene of interest, one should be able to determine whether the bacteria contain the recombinant plasmid with or without the gene of interest.

2a.) Restriction enzyme digest of the λ DNA generates three fragments: the left arm, the right arm, and the internal fragment; b.) To form the concatamers, the left and right arms from the digested λ DNA are combined with digested foreign genomic DNA fragments; c.) The foreign DNA is packaged into the bacteriophage by mixing the ligated DNA with the proteins that form the λ coat in vitro. By way of *cos* sites, multiple λ genomes base-pair end to end to form long linear molecules. During λ packaging, the DNA between two *cos* sites in the linear concatamer is introduced into the λ particle's head; d.) The total length of the recombinant DNA molecule between two *cos* sites must be around 49 Kb because only DNA that is close to this size will package correctly and produce an infectious λ particle. Since the λ genome is normally 48.5 Kb, you want the size to remain basically the same as this when foreign DNA is introduced. Overfilling or underfilling the head with DNA will cause changes in the outer protein covering, from expansion or collapse, and render the particle noninfectious; e.) The λ bacteriophage infects a bacterial cell by attaching to the cell wall by way of its protein coat. It then injects the λ genomic DNA inside the bacterial cell; f.) Upon entering the bacterial cell, the λ DNA can either integrate into the *E. coli* genome and reside there silently until it needs to generate more infectious particles, or it can replicate immediately to form more DNA molecules and make more coat proteins and new infectious particles. Eventually this will lead to lysis of the bacterial cell, freeing the λ particles to infect other bacterial cells; g.) Using the λ vector, rather than a plasmid to clone a piece of foreign DNA is advantageous because the efficiency of λ infection is greater than both electroporation or transformation of *E. coli*, resulting in a greater number of bacteria that will contain a recombinant λ molecule relative to a recombinant plasmid. It also may be advantageous because it permits the cloning of larger DNA fragments than plasmids do.
h.) Unlike plasmids, an λ bacteriophage cannot replicate autonomously without killing the bacterial cell it infects. This makes it harder to use it as a cloning vector.

3a.) The generation of a cDNA library, relative to a genomic DNA library is particularly useful for some experiments because much of the genomic DNA in higher eukaryotes represents sequences between genes which makes identifying the transcribed genes more difficult. Since genomic DNA contains introns and exons, the identification of open reading frames is more difficult with genomic DNA. With cDNA on the other hand, mRNA is isolated and converted to cDNA. Therefore, this is useful in determining the ORF sequence, deducing the amino acid sequence of the encoded protein, expressing eukaryotic genes in prokaryotes, and characterizing the expression of specific genes in a tissue; b.) The template used by reverse transcriptase is the poly(A) mRNA. An oligo(dT) DNA primer is used to prime reverse transcription of the template; c.) The product of reverse transcription is a RNA–DNA hybrid; d.) RNase H is used to randomly nick the mRNA strand in the RNA–DNA hybrid, providing primers for DNA polymerase I; e.) DNA polymerase I removes the RNA and replaces it with DNA reminiscent of how it acts to remove RNA primers and replace them with DNA in Okazaki fragments produced during lagging strand replication; f.) Following ligation, the final product is a blunt-ended, double-stranded cDNA. It can be cloned using blunt-end linkers that contain an internal restriction site. The cDNA with linkers attached can be digested, and the same enzyme can be used to digest a plasmid. In this way, the cDNA and the plasmid will have compatible ends and can be ligated together.

4a.) The proteins are separated on SDS-PAGE according to their molecular weights; b.) The proteins are coated with SDS and thus are uniformly negatively charged. When they are electrotransferred to the nitrocellulose filter, the proteins will move toward the anode or + charge. For this reason, the nitrocellulose filter should be placed closer to the anode and the gel should be placed closer to the cathode. In this way, the proteins will be transferred from the gel to the nitrocellulose filter; c.) A specific protein is detected on the nitrocellulose filter using a primary antibody as a probe. This antibody specifically binds to the target protein; d.) The secondary protein recognizes the first antibody and specifically binds to it. The secondary antibody is coupled to an enzyme that converts a colorless substrate to a colored precipitate, or catalyzes a chemiluminescent reaction. Therefore, when the secondary antibody binds to the primary antibody which is bound to the target protein, the target protein can be visualized by putting a substrate onto the nitrocellulose that the enzyme attached to the secondary antibody will use in a chemiluminescent reaction or that the enzyme will convert to a colored precipitate; e.) The protein is ultimately visualized on X-ray film which captures the light given off by the chemiluminescent reaction or detects the colored precipitate and indicates the location of the target protein.

General Questions—Answers

Q12-1. Restriction enzymes are isolated from bacterial cells where they normally function to cleave foreign DNA. Each restriction enzyme is named for the bacterial species from which it was isolated. Restriction enzymes are extremely useful in recombinant DNA protocols. In particular, the fact that they cleave the DNA at known sites makes them valuable for creating genetic profiles of individuals. Without a way of cleaving the DNA of every individual at exactly the same place, DNA fingerprinting would not be possible.

Q12-2. The mouse cDNA library should be transformed into *E. coli* and plated onto a Petri plate containing media to select for colonies containing the library plasmids. A nitrocellulose filter is laid over the plate once the colonies have grown up. The filter is treated to lyse the cells attached to it, releasing the DNA which is fixed to the filter and denatured with sodium hydroxide. The human *PGC1* cDNA is used as a probe. The *PGC1* cDNA is radioactively labeled, denatured, and added to the nitrocellulose filter containing the fixed library DNA under conditions of reduced stringency. The single-stranded probe will base-pair with any complementary target sequences on the membrane. The filter is then washed to remove any unbound probe and exposed to X-ray film. Any clones in the mouse cDNA library that are related to the human *PGC1* gene will hybridize with the radioactive probe. The film can then be compared to the master plate of bacterial colonies, and the related clones can be isolated from these colonies.

Q12-3. Genomic DNA is isolated from both of the suspects, the two blood samples collected at the crime scene, and the victim. Each sample of DNA is digested with the same restriction enzyme (a typical VNTR locus is flanked by *Bam*HI sites), and the DNA fragments are separated by gel electrophoresis. The DNA on the gel is denatured and transferred to nitrocellulose. A VNTR sequence is radiolabeled as a probe for Southern blotting. The probe is denatured and hybridized to the nitrocellulose. The nitrocellulose is washed to remove any unbound probe and then exposed to X-ray film. If either of the samples of blood collected from the scene has a DNA fingerprint similar to either suspect, it indicates that that suspect was present at the crime scene. If neither blood sample matches either suspect, but only matches the victim, it cannot be determined from this evidence whether either suspect is innocent or guilty. If the blood samples at the scene look completely different from the victim and both suspects, it suggests that someone else is involved in the crime.

Q12-4. The simplest way that more of the DNA fragment can be generated is by PCR. Since the sequence of the fragment is known, DNA oligos complementary to the sequences at the ends of the double-stranded DNA fragment can be designed. These primers, the template DNA, dNTPs, and the DNA polymerase are mixed. The mixture is heated to 95°C to denature the template DNA, the temperature is lowered to allow the primers to anneal to their complementary target sequences, and the temperature is then increased to approximately 72°C allowing DNA polymerase to extend each of the DNA strands. This cycle of temperatures is repeated several times during which the DNA product will be amplified exponentially. After 20 cycles, there will be approximately a million copies of the DNA fragment. To ensure that the PCR reaction worked, a small amount of the PCR product can be visualized on an agarose gel. If the reaction did not work and no product was obtained, the scientist should first check the primers to ensure that they are designed correctly and their complementary sequence is actually present in the DNA template of interest. Next, the annealing temperature of the reaction should be checked to ensure that the primers will anneal at the temperature used in the PCR reaction. The scientist should also check the length of time allowed for extension of the primers in the PCR reaction. The extension time depends on the predicted length of the amplified product. Finally, the scientist should ensure that all of the reagents required for the PCR reaction are relatively fresh and were in fact added into the reaction. Absence of the dNTPs, polymerase, either primer, or the correct concentration of the template DNA can prevent the PCR reaction from working.

Q12-5. A restriction map for either a genomic or cDNA clone can be useful because, first, it can be used to determine the amount of identity or overlap between two different sequences. In this way, a cDNA clone can be compared with a genomic clone allowing regions on the genomic clone that correspond to the exons to be identified. Second, a restriction map provides useful landmarks that help to gauge where in the genome a gene, mutation, or DNA sequence occurs. Third, a restriction map allows a researcher to correlate the genetic map with the physical map of a chromosome. This allows physical changes in the DNA, such as deletions, insertions, and loss of restriction sites to be localized on the genetic map.

Multiple Choice—Answers

1.) c; 2.) b; 3.) e; 4.) e; 5.) c; 6.) d; 7.) b; 8.) a; 9.) d; 10.) b.

Practice Problems—Answers

P12-1. Recall that reading the gel from bottom to top reveals the DNA sequence of the 5′ → 3′ strand (*Hint 1*). First, you need to determine the sequence of the target strand (5′-TCCCTTCGTAAAG-3′). Second, determine the sequence of the complementary strand (3′-AGGGAAGCATTTC-5′). Finally, determine the relative lengths of fragments that would result if a dideoxy base were inserted into the growing strand.

Target strand template: 5′-GAAATGCTTCCCT-3′

Fragments ddG sample:		Fragments ddC sample:	
5′-CTTTACG-3′	(7 bp)	5′-C-3′	(1 bp)
5′-CTTTACGAAG-3′	(10 bp)	5′-CTTTAC-3′	(6 bp)
5′-CTTTACGAAGG-3′	(11 bp)		
5′-CTTTACGAAGGG-3′	(12 bp)		

Fragments ddA sample:		Fragments ddT sample:	
5′-CTTTA-3′	(5 bp)	5′-CT -3′	(2 bp)
5′-CTTTACGA-3′	(8 bp)	5′-CTT-3′	(3 bp)
5′-CTTTACGAA-3′	(9 bp)	5′-CTTT-3′	(4 bp)
5′-CTTTACGAAGGGA-3′	(13 bp)		

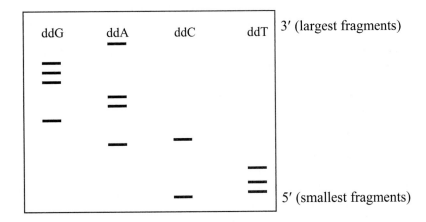

3′ (largest fragments)

5′ (smallest fragments)

P12-2.

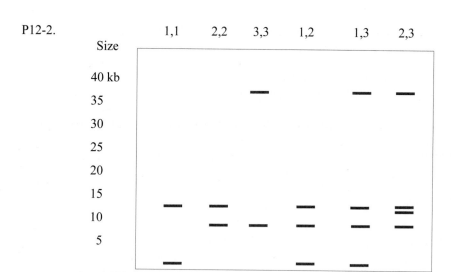

P12-3. To solve this problem, you should begin by determining the number of restriction sites that are recognized by each enzyme. The three fragments produced by enzyme A indicate that there are two restriction sites recognized by this enzyme. Enzymes B and C each produce only two fragments and thus recognize only one restriction site. Now you can compare the results from the single digests to those obtained from double digests to determine the relative distances between restriction sites and to construct the entire restriction map (*Hint 3*). Consider enzymes B and C first. The double digest produces three fragments (200, 250, and 800 bp). Notice that the 250-bp fragment appears in the single digest involving enzyme B and that the 800-bp fragment appears in the enzyme C single digest. This suggests that the 1000-bp fragment (enzyme B) contains the 800-bp fragment (from enzyme C) plus 200 bp from the 450-bp fragment (from enzyme C) (*Hint 2*). The 1000-bp fragment therefore contains a restriction site for enzyme C.

250		200		800
	B		C	

Now consider the results of digestion with enzymes A and B. The double digest reveals four fragments (100, 150, 400, and 600 bp). The 150- and 600-bp fragments are also seen in the single digests. Notice that addition of the 100- and 150-bp fragments (double digest) yields the 250-bp fragment found in digest B. This fragment contains a restriction site for enzyme A. Addition of the 400- and 600-bp fragments from the double digest yields the 1000-bp fragment from the single B digest. The second restriction site is located within this fragment. We can surmise that the 150-bp fragment is on the end, followed by the 100-bp fragment, and then the 400-bp and 600-bp fragments. We can make this determination because this arrangement is the only one possible to generate the 500-bp fragment seen in the single digest for enzyme A.

150		100		400		600
	A		B		A	

Finally, the combined map should yield the fragments found in the double digest involving enzymes A and C. The combined map is shown in the following. As you can see, this map yields fragments consistent with those reported for the double digest indicating we have correctly mapped all restriction sites.

150		100		200		200		600
	A		B		C		A	

P12-4. Remember that the sequence lying between the probes is amplified during the polymerase chain reaction (*Hint 5*). The area of probe binding is underlined. The sequence that is amplified is in bold. Very few of the final PCR products will contain the bases occurring before and after the probe binding sites (those in italic).

5'-*TCC*ACCC**GTCGCAACAAAATTAACGTAACTTT**ACCC*AACA*-3'
3'-*AGG*TGGG**CAGCGTTGTTTTAATTGCATTGAAA**TGGG*TTGT*-5'

P12-5. The entire human genome is composed of an estimated three billion base pairs. At fifty cents per base pair, sequencing the entire genome would cost approximately 1.5 billion dollars. Given that approximately 70 million dollars was invested in locating the cystic fibrosis gene alone, the international effort is very cost effective.

P12-6.

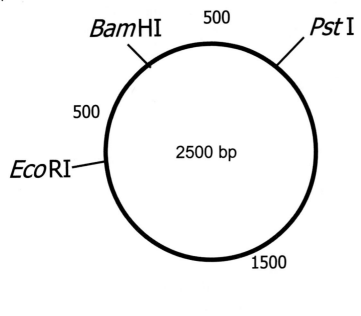

12

RECOMBINANT DNA TECHNOLOGY

CHAPTER SUMMARY QUESTIONS

1. 1. b, 2. a, 3. g, 4. h, 5. e, 6. j, 7. c, 8. d, 9. f, 10. i.

2. Restriction enzymes are naturally used by bacteria to destroy foreign DNA, presumably that of invading bacteriophages. Bacteria protect their own DNA by methylating the bases at the

recognition sequences. The methylated sequence cannot be cut by the restriction enzyme, while the unmethylated sequence (foreign DNA) is cut.

3. Type II endonucleases are valuable because they cut DNA at specific points and many leave overlapping or "sticky" ends.

4. A particular restriction endonuclease is unsuited for cloning if it destroys the DNA of interest or if it does not have an appropriate site in the vector or surrounding the foreign DNA.

5. A plasmid is a self-replicating circle of DNA found in many cells. Cosmids are plasmids that contain lambda phage *cos* sites and are useful for cloning large segments of DNA (up to 50 kb).

6. DNA can be joined by having compatible ends to begin with or by blunt-end ligation (linkers combine these ends). The appropriateness of a method depends on what DNA is to be cloned and how that DNA can be obtained. Having DNA with "sticky" ends created by the same restriction enzyme would be easiest but sometimes is not available. Adding linkers by blunt-end ligation with a particular restriction site is usually the best compromise.

7. Cloning vectors typically have
    1. An origin of DNA replication that allows them be maintained in a cell.
    2. At least one selectable marker, such as an antibiotic-resistance gene, to identify cells that contain the vector.
    3. A multiple cloning site region (containing unique restriction sites) that is used to insert foreign DNA into the vector.

8. The *lacZ* gene, which encodes the enzyme β-galactosidase, is used in plasmids as a selectable marker. Plasmid vectors are engineered so that the multiple cloning site region is located within the *lacZ* gene. So cloning foreign DNA into one of these unique restriction sites will inactivate the *lacZ* gene. The β-galactosidase enzyme can cleave the artificial compound X-Gal, which is colorless, into a molecule that forms a blue precipitate. Therefore, bacteria that turn blue when grown on X-Gal-containing agar, will have intact *lacZ* genes and therefore no foreign DNA. On the other hand, if the bacteria turn white, this means that they carry a foreign DNA insert that has disrupted the *lacZ* gene.

9. The steps in creating cDNA from messenger RNA are shown in figure 12.17. Radioactive cDNA and messenger RNA can be obtained either by using radioactive triphosphate nucleotides during the synthesis of the cDNA and messenger RNA or by end-labeling the products.

10. 1. A genomic DNA library has inserts that are much larger than those of a cDNA library.

   2. A genomic library contains essentially every DNA sequence present in the cell
       or tissue, whereas a cDNA library only contains the sequences corresponding to
       mRNA exons.
   3. An organism's genomic library will be the same regardless of the tissue or
       developmental stage, whereas a cDNA library will differ for each tissue and
       developmental stage.
   4. All sequences in a genomic library will be present at nearly equal amounts. In a
       cDNA library, on the other hand, the amount of each sequence will be
       proportional to the amount of corresponding mRNA in a cell.
   5. A genomic library can reveal information on gene organization, promoter
       sequences, and intergenic regions. In contrast, a cDNA library provides
       information about exons and open reading frames and allows expression of
       eukaryotic proteins in prokaryotic cells.

11. a. Electroporation is a technique for transfecting cells by applying a high-voltage
       electric pulse.
   b. Linkers are short artificially synthesized pieces of DNA containing a restriction
       endonuclease site. They are used in blunt-end ligation.
   c. A microsatellite is a short DNA sequence (2–5 base pairs) that is repeated tens or
       hundreds of times in tandem (side-by-side).

12. In the PCR technique, the DNA template, oligonucleotide primers, DNA
   polymerase, and dNTPs are combined. The following three steps are then repeated
   for many cycles:

   1. Denaturing the target DNA at high temperatures (for example, 95°C for
       30 seconds).
   2. Annealing the primers to complementary target sequences at lower
       temperatures (for example, 55°C for 30 seconds).
   3. Extending the target DNA at intermediate temperatures (for example,
       72°C).

   Step 2 can be altered to change the stringency between the template and
   oligonucleotide primer.

13. PCR is much more sensitive than gene cloning because it can be used to amplify
   DNA, however small in quantity or poor in quality. PCR is also far more rapid. It can
   provide millions of copies of target DNA sequences within hours, whereas many
   days are required to go through all the cloning steps. One disadvantage is that PCR
   generally requires knowledge of the sequences flanking the DNA of interest, in order
   to design the oligonucleotide primers.

14. D. Isolate DNA from organism.
    B. Cut DNA with restriction enzyme.
    F. Separate DNA by gel electrophoresis.
    A. Transfer DNA from gel to nitrocellulose filter.

E. Place filter in solution containing radiolabeled probe.

C. Expose filter to X-ray film.

15. The enzyme DNA polymerase replicates DNA by adding a new nucleotide to the 3'-OH of an existing nucleotide. Dideoxynucleoside triphosphates (ddNTPs) lack –OH groups on both their 2' and 3' carbons (please refer to figure 12.30). Therefore, if a ddNTP is incorporated into a growing DNA strand during the sequencing reaction, DNA polymerization will terminate at this point.

EXERCISES AND PROBLEMS

16. All are palindromes except for **d**. Please note that *aibohphobia* is a humorous term coined to describe the fictional fear of palindromes.

17. The DNA contain four palindromic sequences (shown in blue):

5'-TAGAATTCGACGGATCCGGGGCATGCAGATCA-3'
3'-ATCTTAAGCTGCCTAGGCCCCGTACGTCTAGT-5'

18. *Bgl*II generates 5' overhangs:

```
5'-A                G A T C T-3'
3'-T C T A G                A-5'
```
*Eco*RV generates blunt ends:

```
5'-G A T        A T C-3'
3'-C T A        T A G-5'
```

*Pvu*I generates 3' overhangs:

```
5'-C G A T            C G-3'
3'-G C            T A G C-5'
```

19. The four bases in DNA are found in equal amounts. Therefore, the frequency of each base is equal to 1/4 and the frequency of a purine (A or G) = frequency of a pyrimidine (C or T) = 1/2. To calculate the probability of a DNA sequence, we have to multiply the probabilities of each base in the sequence (product rule).
   a. P(CGCG) = $1/4 \times 1/4 \times 1/4 \times 1/4$ = 1/256.
   b. P(AGNCT) = $1/4 \times 1/4 \times 1 \times 1/4 \times 1/4$ = 1/256.
   c. P(GTPyPuAC) = $1/4 \times 1/4 \times 1/2 \times 1/2 \times 1/4 \times 1/4$ = 1/1024.
   d. P(GAATTC) = $1/4 \times 1/4 \times 1/4 \times 1/4 \times 1/4 \times 1/4$ = 1/4096.
   e. P(GCGGCCGC) = $1/4 \times 1/4 \times 1/4 \times 1/4 \times 1/4 \times 1/4 \times 1/4 \times 1/4$ = 1/65,536.

20. By convention, the first letter of a restriction enzyme is taken from the bacterial genus, and the next two letters from the species. The roman numeral refers to the order in which the enzyme was isolated. Therefore, this enzyme would be named *Big*I.

21.

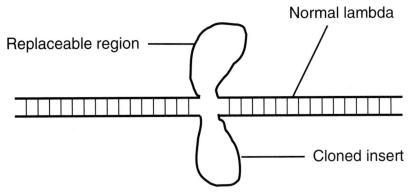

22. Use radioactive alanine transfer RNA as a probe in either a whole digest (Southern blotting) or in a genomic library (dot blotting).

23. If a messenger RNA could not be isolated, a probe could be constructed using the codon dictionary if the amino acid sequences of the protein product were known. A genomic library could be constructed with the intent of looking for the location of the gene by its expression. Expression vectors are those in which foreign DNA is expressed.

24. Partial digestion of molecule 2 leads to the following molecule:

AAAAAAAAAA
     TTTTTTTTTT

Some of these molecules will form a circle with the single-stranded T's paired with single-stranded A's. The circle eliminates the free 5' phosphate, and the enzyme can no longer work.

25. a. Because the genome is 50% GC, it will also be 50% AT, and so each base will have a frequency of 1/4 or 0.25. The probability of finding the 8-base *Not*I restriction site in this genome is $(0.25)^8 = 0.00001526$. Therefore, the number of *Not*I sites in a genome of 3.8 Mbp will be $(0.00001526)(3,800,000)$ or approximately 58 sites.

b. The average spacing between restriction sites is equal to the reciprocal of the probability of the occurrence of the restriction site. Thus, the spacing between *Not*I sites is $1/0.00001526 = 65,530$ bp.

26. In the colony 2 lane, the second band moves more slowly than the original band, so it must be larger than the original. This means that cells of colony 2 picked up two different-sized plasmids: one is a single religated plasmid (also found in colony 1) and the other is a dimer (or trimer or higher *mer*) of the plasmid. Remember that two or more molecules with the same sticky ends can combine to form one larger molecule, as shown here:

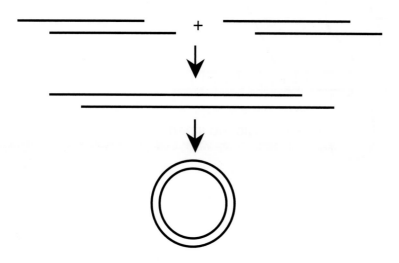

27. We must insert DNA that has no introns into bacterial plasmids. This DNA can be obtained by isolating mature, cytoplasmic messenger RNA and then using reverse transcriptase to make double-stranded cDNA. Plasmids with cDNA inserted can then be used to produce human proteins (expression vectors).

28. Electrophoretic bands of the total digest are (* indicates end label) 50, 100*, 150*, 250, and 300 bp. Bands of the partial digest are 50, 100*, 150*, 250, 300(×2), 350, 400*(×2), 450*(×2), 600, 700*, 750*, and 850* bp.

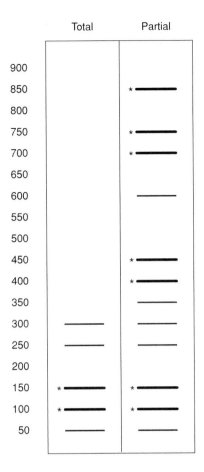

29. Mutant A: elimination of restriction site between the 300- and 50-bp segments.
Mutant B: elimination of restriction site between 100- and 300-bp segments.
Mutant C: creation of a new restriction site within the 300-bp segment, dividing it
into 75- and 225-bp segments.

30. There are four fragments in the total digest: two that are labeled (50 bp and 400 bp)
and so must be at the ends, and two that are not (100 bp and 200 bp) and so must be
in the middle. However, we cannot tell from the total digest whether the 50-bp
fragment is adjacent to the 100-bp fragment or to the 200-bp fragment. Looking at
the partial digest, we can see that there is no 150-bp fragment (labeled or not).
Therefore, the 50-bp fragment is adjacent to the 200-bp fragment, and the restriction
map of the original segment is

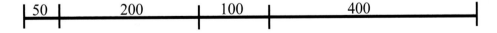

31. Alternative II is correct. Note that among the clues, alternative I requires a band of a
segment that is 200 bp long.

32. This problem begins as a trial-and-error attempt to overlay two restriction maps,
made more difficult by the fact that one enzyme, *Bam*HI, has made three cuts that are
unordered, leaving many possibilities. However, a bit of thought beforehand makes

this problem much easier. If you compare the double digest with the *Bam*HI digest, they share 200-, 250-, and 400-bp segments. The double digest has 50- and 100-bp segments replacing the 150-bp segment in the *Bam*HI digest. The inference is that there is an *Eco*RI cut in the 150-bp segment leading to the 50- and 100-bp segments, with all other segments of the *Bam*HI digestion left uncut. That leaves only two possibilities, as shown in the following; the data are insufficient to distinguish between the two choices.

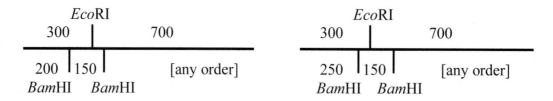

33. A linear molecule with *n* sites will produce *n* + 1 fragments, so there are two sites within the molecule. Any one of the three fragments could be in the middle.

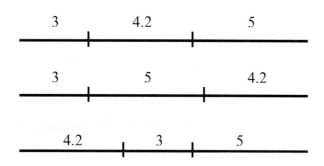

34. We know that the 6.2- and 8.0-kb *Eco*RI fragments are at opposite ends and that the 10.0- and 6.0-kb *Bam*HI fragments are at the ends. Therefore, the *Bam*HI 13.0-kb fragment must be in the middle. If the 6.2- and 6.0-kb fragments are at the same end, a double digest should produce a fragment of 0.2 kb. This is not seen, so they are at opposite ends:

| *Eco*RI | 6.2 | (2.8, 4.6, 7.4) | 8.0 |

| *Bam*HI | 10.0 | 13.0 | 6.0 |

If 7.4 is next to 6.2, we should see a 3.8 fragment in the double digest. This same-sized fragment would show up if 4.6 was next to 6.2. Because the double digest did not contain a 3.8 fragment, 2.8 is next to 6.2. The 4.6 fragment must be next.

Therefore, the restriction maps are as follows:

| *Eco*RI | 6.2 | 2.8 | 4.6 | 7.4 | 8.0 |
| *Bam*HI | 10.0 | | 13.0 | | 6.0 |

35. a. Stringency refers to the specificity of base-pairing between labeled probe and target nucleic acid. High stringency conditions, such as high hybridization temperature and/or low salt concentrations, allow for more perfect base-pairing. Low stringency conditions (low hybridization temperature and/or high salt concentrations) allow for imperfect base-pairing.

b. The appearance of three bands under low stringency conditions indicates that human brain cells contain three genes that have some homology with gene *z* of *Drosophila*. These could constitute a gene family, where only one member shows a close match with gene *z*. This is clear from the thicker band produced by the probe hybridizing to the smallest mRNA, and more importantly, by the single band formed under high stringency conditions.

36. The molecule either has no *Eco*RI sites, or it is a circle. A circle with *n* sites yields *n* molecules. If there are no *Eco*RI sites, and the molecule is linear, we expect to see the same number of *Bam*HI fragments from both the single *Bam*HI digest and from the double digest. Since we got four fragments in the double digest, but only three in the *Bam*HI digest, the molecule must be a circle. The 2.0-kb and 4.5-kb fragments appear for both *Bam*HI and *Bam*HI + *Eco*RI, so they must not contain an *Eco*RI site. The 5.5-kb fragment must contain an *Eco*RI site 2.5 kb from one end.

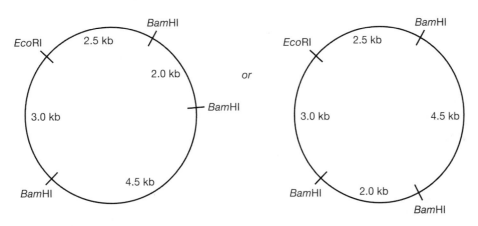

37. The first mutant will be cleaved only once, yielding bands of 1.7 kb and 2.0 kb. The second mutant will produce the 1.0-kb and the 0.7-kb fragments since this site is conserved. A new site within the 2.0-kb fragment will generate two new fragments, the sum of which is 2.0 kb. Thus the loss of a site produces two bands, and the acquisition of a new site yields four bands.

38. The second possibility predicts that a 1.5-kb fragment should be seen in the double digestion. This fragment would result from digestion of the 3.5-kb *Eco*RI fragment with *Bgl*II. Since we don't see a 1.5-kb fragment in the double digest, the second possibility does not agree with the observed results.

39. 0.5, 1.0, 2.0, 3.5 kb. The 1.0-kb band should have a higher intensity than the other bands. It helps to redraw the *Eco*RI sites:

Since *Hind*III does not cut the 3.5 band, it must cut within the 3.0 band, as drawn here. There will be two 1.0-kb bands produced as well as bands of 0.5, 2.0, and 3.5 kb.

40. 5'-CAATAGGTCGAGGTTCAATGG-3'

41. The DNA can be inserted into the M13 general sequencing vector.

42. Overlap can work for one, four, or seven bases. Lowercase letters indicate the stop sequence (tga), boldface indicates the **ATG** start sequence, and N indicates any base. One-base overlap, . . . tga**TG** . . . ; four-base overlap, . . . **At**ga . . . ; seven-base overlap, . . . **ATG**Ntga. . . .

43. Bands appear in three of the four lanes in this Northern blot, indicating that humans have mRNA that is homologous to *Drosophila* gene *z*. Thus, every human cell should have at least one gene that is homologous to gene *z*. (Remember that all cells of an organism have the same genome.) Lane 2 shows no band, so skin cells don't transcribe any gene *z* homolog. All other cells are transcribing this gene and producing an mRNA molecule that is 5 kb in length. Muscle cells produce a very large amount of this mRNA, whereas cells of the liver and lung produce a moderate amount of this mRNA. In addition, liver cells produce an mRNA that is 3 kb long. This is likely due to alternative splicing of the mRNA, which generates two different-sized mRNA molecules.

44. There is no detectable amount of protein Z in the *Drosophila* embryo. However, the amount of the protein increases during the later stages of the life cycle. It appears that gene *z* is expressed little, if any, in the embryo. In the larval stage, the expression of gene *z* becomes detectable, through its protein Z end product. Gene *z* expression increases in the pupa and reaches maximum levels in the adult fly.

45. What will appear in the gel are fragments of the newly synthesized strand. Since DNA synthesis proceeds 5' → 3', the 5' base will be T in the new strand. Proceed up the gel by indicating the bases complementary to the sequence given.

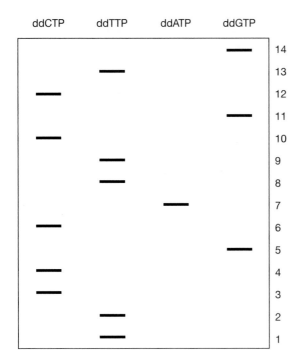

46. Recall that the dideoxy method produces bands that result from the synthesis of a new complementary band. This strand will have the following sequence:

5'-CTGATAAGGCTTTG-3'

Dideoxythymidine will produce fragments wherever a T appears in the new strand (arrows). Therefore the labeled bands will be 2-*mer,* 5-*mer,* 11-*mer,* 12-*mer,* and 13-*mer.*

47. Since *Eco*RI does not eliminate either resistance, its site must be between the *tet*ʳ and *amp*ʳ genes. The *Pst*I site must be within the *amp*ʳ gene, since insertion of DNA into this site eliminates ampicillin resistance. By similar logic, the other sites must be in the *tet*ʳ gene. In the double digests, the smaller fragment must represent the distance from the *Eco*RI site to the other site. We can draw part of the plasmid as

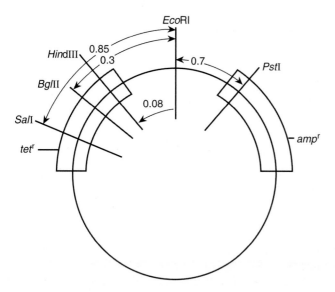

48. Mutant 1 produces mRNA and protein that are smaller than their wild-type
counterparts. This can be explained by a failure in the normal splicing mechanism of
gene *R*. A mutation occurred that likely changed the DNA consensus sequence at the
3' splice junction of an exon. The result is an alternatively spliced mRNA that lacked
at least one crucial exon. This led to a smaller nonfunctional protein.

Mutants 2 and 3 have mRNA of the same size as wild-type mRNA, but both
produce mutant proteins. The molecular defect in both mutants is most likely
attributed to base substitutions. These will not change the length of the mRNA but
may affect its translation. It appears that mutant 2 carries a mutation that converted
one sense codon into another and resulted in a nonconservative change in the amino
acid encoded. This mutation occurred in a part of the protein that is crucial for its
activity. So, although the overall size of the mutant 2 and wild-type proteins is the
same, the former has one amino acid difference in its structure, which is enough to
make it nonfunctional. The molecular defect in mutant 3 is most probably a mutation
that changed the normal termination codon into an amino acid–specifying codon.
This led to translation of a larger and therefore nonfunctional protein.

Mutant 4 also has mRNA of the same size as wild-type mRNA. However, no
R protein is detectable. This means that the mRNA is not translated, and so the
defect must have occurred during or before translation initiation. Likely explanations
include mutations in the 5'-CAP or the translation initiation codon in mRNA.

No detectable mRNA or protein is seen in the mutant 5 lanes. Thus, gene *R* is
"turned off." This is caused by a mutation in the promoter region of the gene, which
inhibits binding of RNA polymerase. The end result is no mRNA, and therefore no
protein.

49. a. We know that the 6.0- and 4.5-kb *Eco*RI fragments are at the ends of the
genomic clone, and so are the 6.0- and 3.5-kb *Bam*HI fragments. Moreover, the
two 6.0-kb fragments must be on opposite ends since the double digest did not
produce any 6.0-kb fragment. So a partial restriction map would look like this:

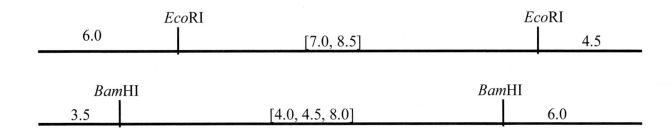

At this point, we can begin a trial-and-error procedure to place the middle fragments in their right order. Let's start with the 8.0-kb *Bam*HI fragment. If it is next to 3.5, we should see a 5.5-kb fragment in the double digest, regardless of which of the two middle *Eco*RI fragment is next to 6.0. If, on the other hand, the 8.0-kb *Bam*HI fragment is next to 6.0, we should see one of two different-sized fragments depending on which of the two middle *Eco*RI fragments is next to 6.0: if it is the 7.0 fragment, then a 5.5-kb fragment should be seen in the double digest; if it is the 8.5 fragment, then a 7.0-kb fragment would be generated. We see neither a 5.5- nor a 7.0-kb fragment. Therefore, the 8.0-kb *Bam*HI fragment should be in the middle. Let's now consider the 4.0-kb *Bam*HI fragment. If it is next to 3.5, we should see a 1.5-kb fragment in the double digest. We do. However, that 1.5 is already accounted for by the *Eco*RI cut of the 6.0-kb *Bam*HI fragment. Since we cannot have two different 1.5 fragments, the 4.0-kb *Bam*HI fragment cannot be next to 3.5 and must therefore be adjacent to the 6.0-kb fragment. Similar trial-and-error reasoning will place the two middle *Eco*RI fragments: the 7.0 next to the 6.0, and the 8.5 next to the 4.5. Therefore, an overlay of the two restriction maps is

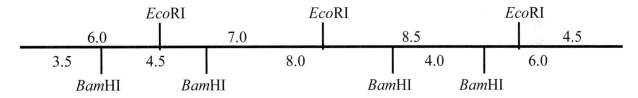

b. Digestion of an end-labeled 7.0-kb *Eco*RI fragment with *Xba*I produced 6.0- and 1.0-kb fragments, both of which are end-labeled. This means that there is an *Xba*I site 1 kb from one of two ends of the fragment. But which end? Digestion of the labeled 6.0-kb fragment with *Bam*HI yielded a 5.0-kb labeled fragment. Therefore, the *Xba*I site must be on the left end of the *Eco*RI fragment, just 1 kb from the *Bam*HI site. The *Hin*fI sites, on the other hand, are on the right end of the *Eco*RI fragment. The following map incorporates the *Xba*I (*X*) and *Hin*fI (*H*) sites into the earlier restriction map:

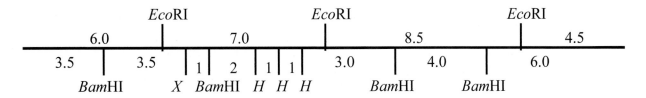

CHAPTER INTEGRATION PROBLEM

a.   A circular molecule that contains *n* sites for restriction enzymes will produce *n* fragments when digested. In contrast, a linear molecule will generate $n + 1$ fragments. Digestion of the viral genome with *Bgl*II yielded three fragments, as did the *Eco*RI digestion. If the genome is circular, there must be three restriction sites for each enzyme, whereas a linear genome would only contain two sites for each restriction enzyme. Therefore, a double digest of the viral genome would yield five fragments if the genome is linear (4 restriction sites + 1), and six fragments if the genome is circular (6 restriction sites). We see that the actual double digest produced six fragments, and so the viral genome must be circular.

b.   The 4.0-kb fragment appears in both the *Bgl*II and *Bgl*II + *Eco*RI digests, so it is a *Bgl*II fragment that lacks an *Eco*RI site. Similarly, the 20-kb *Eco*RI fragment should not contain a *Bgl*II site. The other four fragments (17, 5, 3, and 1) are only found in the double digest and are therefore *Bgl*II/*Eco*RI fragments. Let's consider different combinations of these four fragments to see if any two add up to the single *Bgl*II and *Eco*RI fragments. We find two such combinations: $17 + 5 = 22$ kb, and $17 + 1 = 18$ kb. Therefore, the 22-kb *Bgl*II fragment contains an *Eco*RI site and the 18-kb *Eco*RI fragment contains a *Bgl*II site.

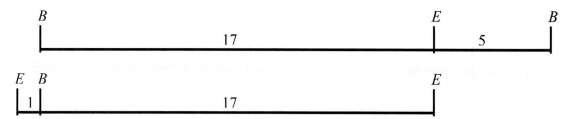

Because they share the 17-kb fragment, the two maps can be combined:

This leaves the 3 kb as the only fragment from the double digest that has not been mapped. However, this fragment cannot be adjacent to either end of the preceding map because all of its possible combinations have been accounted for. The next logical step is to place the 20-kb *Eco*RI fragment on the left of the mapped segment and the 4-kb *Bgl*II on the right. Now we can place the remaining 3-kb fragment in between the 20-kb and 4-kb fragments, thereby completing the viral genome's circular map.

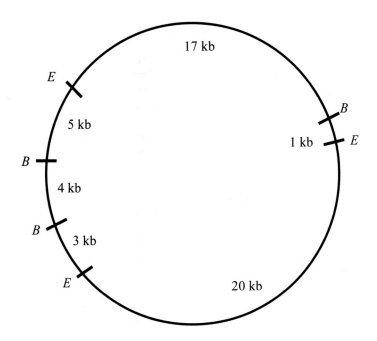

c. First let's calculate the frequency of each base. The genome has a GC content of 70%, and so the frequency of G and C is 0.7/2 = 0.35 each. Therefore, the frequency of A and T will be (1 − 0.7)/2 = 0.15 each. As mentioned earlier, the probability of the occurrence of a restriction site by chance is the product of the probabilities of each base in the sequence. Therefore, the probability of the occurrence of the *Bgl*II site AGATCT is 0.15 × 0.35 × 0.15 × 0.15 × 0.35 × 0.15 = 0.0000620, and that of the *Eco*RI site GAATTC is 0.35 × 0.15 × 0.15 × 0.15 × 0.15 × 0.35 = 0.0000620. To estimate the number of restriction sites for each enzyme, we simply multiply the probability of the site by the genome size. Therefore, each of the two enzymes is expected to have 0.0000620 × 50,000 bp = 3.1 sites. The actual number of sites for each enzyme was 3. Therefore, the viral DNA appears to have a random distribution of nucleotides.

d. P(TTAA) = 0.15 × 0.15 × 0.15 × 0.15 = 0.000506.
    Therefore, the number of expected *Mse*I restriction sites is 0.000506 × 50,000 = 25.3.
    P(CCGG) = 0.35 × 0.35 × 0.35 × 0.35 = 0.01500
    Therefore, the number of expected *Hpa*II restriction sites is 0.01500 × 50,000 = 750.
    P(CCNNGG) = 0.35 × 0.35 × 1 × 1 × 0.35 × 0.35 = 0.01500
    Therefore, the number of expected *Bsa*JI restriction sites is 0.01500 × 50,000 = 750.
    P(GTMKAC) = 0.35 × 0.15 × 0.5 × 0.5 × 0.15 × 0.35 = 0.000689
    Therefore, the number of expected *Acc*I restriction sites is 0.000689 × 50,000 = 34.4.
    P(RGCGCY) = 0.5 × 0.35 × 0.35 × 0.35 × 0.35 × 0.5 = 0.00375
    Therefore, the number of expected *Hae*II restriction sites is 0.00375 × 50,000 = 187.5.

P(TTAATTAA) = 0.15 × 0.15 × 0.15 × 0.15 × 0.15 × 0.15 × 0.15 × 0.15 = 0.00000025.

Therefore, the number of expected *Pac*I restriction sites is 0.00000025 × 50,000 = 0.0125. In other words, it is highly unlikely for this viral genome to have a *Pac*I site.

e. The viral DNA cut with *Dpn*II (^GATC) will have the following 5' overhangs:

```
5'-                    G  A  T  C-3'
3'-C  T  A  G                   -5'
```

The pEN DNA cut with *Bgl*II (A^GATCT) will have the following 5' overhangs:

```
5'-A                  G  A  T  C  T-3'
3'-T  C  T  A  G                  A-5'
```

Cloning of the viral DNA into the pEN plasmid will produce the following recombinant DNA molecule:

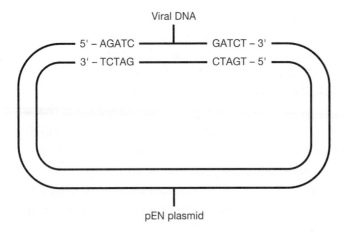

As you can see from the figure, the left end of the viral DNA has the sequence AGATC and the right end has the sequence GATCT. For the cloned viral DNA to be cut out of the plasmid with *Bgl*II, both ends will have to have the restriction site AGATCT. The probability that AGATC is followed by a T is 0.15 (which is the frequency of T in the viral DNA). The probability that GATCT is preceded by an A is also 0.15 (the frequency of A in the viral DNA). Therefore, the probability of finding *both* restriction sites is 0.15 × 0.15 = 0.0225. So *Bgl*II can be used to cut out the viral DNA from the pEN plasmid. However, it is only successful about 2.25% of the time.

f. The proteins have a total molecular weight of 90 × 25,000 = 2,250,000 daltons. The average molecular weight of an amino acid is 125 daltons. So, the number of amino acids encoded by this virus is 2,250,000/125 = 18,000 amino acids. These are specified by 18,000 codons or 18,000 × 3 = 54,000 bases. However, the viral genome is 50,000 bp in size, and this includes many DNA sequences that don't encode amino acids (promoters, 5' and 3' untranslated sequences, and so forth). The only way that the virus can contain 54,000 codons is through overlapping genes.

g. The smallest set of degenerate probes will be derived from the sequence of five amino acids that is encoded by the smallest number of codons. First, we need to refer to the genetic code (table 11.5) to determine the coding degeneracy of each amino acid in protein T. The result is

1	Met	Ala	Thr	Asp	Gly	Ser	Trp	Val	Leu	Cys	Asn	Leu	Met	Ile	Tyr
	1	4	4	2	4	6	1	4	6	2	2	6	1	3	2

16	Trp	Ser	His	Leu	Gly	Glu	Gln	Trp	Thr	Ser	Ile	Leu	Gly	Trp	Glu
	1	6	2	6	4	2	2	1	4	6	3	6	4	1	2

31	Ile	Met	Arg	Asp	Pro	Asn	Leu	Trp	Trp	Leu	Gly	Phe	Phe	Ile	Ser
	3	1	6	2	4	2	6	1	1	6	4	2	2	3	6

The number of different codons specifying a sequence of five amino acids can be obtained by multiplying the number of codons of each amino acid in the sequence (product rule again). Scanning the various combinations reveals that the sequence *Gly Trp Glu Ile Met* (positions 28–32) yields the smallest number of degenerate probes: $4 \times 1 \times 2 \times 3 \times 1 = 24$ probes.

The codons specifying the five amino acids are

Glycine	Tryptophan	Glutamic acid	Isoleucine	Methionine
GGA	TGG	GAA	ATA	ATG
GGC		GAG	ATC	
GGG			ATT	
GGT.				

The 24 combinations can be arrived at by a systematic approach are:

GGA TGG GAA ATA ATG	GGG TGG GAA ATA ATG
GGA TGG GAA ATC ATG	GGG TGG GAA ATC ATG
GGA TGG GAA ATT ATG	GGG TGG GAA ATT ATG
GGA TGG GAG ATA ATG	GGG TGG GAG ATA ATG
GGA TGG GAG ATC ATG	GGG TGG GAG ATC ATG
GGA TGG GAG ATT ATG	GGG TGG GAG ATT ATG
GGC TGG GAA ATA ATG	GGT TGG GAA ATA ATG
GGC TGG GAA ATC ATG	GGT TGG GAA ATC ATG
GGC TGG GAA ATT ATG	GGT TGG GAA ATT ATG
GGC TGG GAG ATA ATG	GGT TGG GAG ATA ATG
GGC TGG GAG ATC ATG	GGT TGG GAG ATC ATG
GGC TGG GAG ATT ATG	GGT TGG GAG ATT ATG

h.   The *Bgl*II + *Eco*RI double digest contains six fragments: 20, 17, 5, 4, 3, and 1 kb. The mixture of degenerate probes hybridized to a DNA molecule that migrates far faster than the 20-kb marker and just a little slower than the 1-kb marker. Thus, the 20-kb and 1-kb fragments are ruled out as possible locations for gene *T*. Because the band forms toward the bottom of the gel, the 17-kb fragment may be eliminated as well. However, in the absence of other DNA markers, we cannot narrow gene *T*'s location down to only a single fragment because the 3-, 4-, and 5-kb fragments will run to relatively the same vicinity as the fragment that hybridized to the probe.

i.   Mutant 1 has undergone a mutation that eliminated the *Bam*HI restriction site between the 3- and 4-kb fragments. Thus, the double digest produced a 7-kb fragment that combines the two smaller fragments.

   Mutant 2 has undergone a 1-kb deletion in the *Bam*H/*Eco*RI 3-kb fragment yielding a 2-kb fragment in the double digest.

   In the Southern blot, the probe hybridized to a DNA fragment in mutant 1 that is larger than that of the wild type. This indicates that in this mutant, gene *T* is located on a larger DNA fragment than in the wild type. Because the mutation did not affect the 5-kb fragment, it can be ruled out as the location for gene *T*. In mutant 2, the probe hybridized to the same DNA fragment that it bound to in the wild type. Thus, gene *T* must be located on the 4-kb fragment. If it were on the 3-kb fragment in the wild-type virus, it should be found on the 2-kb fragment in mutant 2, and therefore would have migrated faster.

CHAPTER 13 APPLICATION OF RECOMBINANT DNA TECHNOLOGY

Chapter Goals
1. Understand how DNA markers are used in mapping human mutations.
2. Describe how mutant genes in humans are identified and cloned.
3. Discuss the different eukaryotic vectors and their unique features and applications.
4. Understand the various methods used to generate transgenic organisms.
5. Discuss the important considerations when developing gene therapy for a human disease.

Key Chapter Concepts
13.1 Mapping Mutations in Eukaryotes

- **DNA Markers:** DNA markers, such as restriction fragment length polymorphisms (RFLPs), variable number tandem repeats (VNTRs), microsatellites, and single-nucleotide polymorphisms (SNPs) can be used to localize mutations to chromosomal regions, to find and analyze DNA segments, and for recombination mapping.

- **Restriction Fragment Length Polymorphisms (RFLPs):** RFLPs are nucleotide changes that result in the elimination or the creation of a restriction site. It can be detected by restriction enzyme digest of the genomic DNA with a specific enzyme followed by Southern hybridization using a probe that will detect a single RFLP locus at a single site in the genome. In the case of sickle cell anemia, individuals with the disease lack a specific restriction site in the β-globin gene. Therefore, the RFLP can be used to diagnose the disease. RFLPs are advantageous for gene mapping because they are scattered throughout the genome, but they can only distinguish between two alleles.

- **Variable Number Tandem Repeats (VNTRs):** VNTRs are short sequences (10–100 bp) that are repeated in a tandem array and vary in size among different individuals in a population. They are detected by restriction enzyme digest with an enzyme that does not cut within the repeat, followed by Southern hybridization using the repeat sequence as a probe.

- **Microsatellites:** Microsatellites are very short sequences (2–5 bp) that are repeated in a tandem array from 100–1000 bp in length. Some microsatellites are directly associated with the **triplet repeat diseases** that result from an increase in the number of triplet nucleotides within the ORF of a gene. Analysis of individuals with one of these triplet repeat diseases, Huntington disease, revealed that the number of CAG repeats correlated with the severity or with the age of onset of the disease. Microsatellites are typically examined by PCR amplification using primers that flank the tandem repeats.

- **Single-Nucleotide Polymorphisms (SNPs):** SNPs are single-nucleotide changes in the DNA sequence that may or may not affect a restriction enzyme site. It is less likely that an individual will be heterozygous for a specific SNP than it is that they will be heterozygous at a specific VNTR or microsatellite locus because there are only four potential alleles at any given site of an SNP. SNPs can be identified by denaturing and annealing genomic DNA to a radioactively labeled DNA oligo complementary to a specific SNP sequence and then treating with S1 nuclease which will cleave the DNA at any single-stranded sequence leading to the visualization of a shortened probe. SNPs can also be identified by hybridization of the genomic DNA to oligonucleotides that correspond to the different SNP variant sequences.

- **LOD Scores:** Calculating lod (**lo**g of **od**ds) scores is another method that is becoming more common for mapping human genes and quantitative trait loci (QTL) in other organisms. For example, a gene for schizophrenia was recently mapped in humans and a DNA region involved in aggressive stinging behavior was recently located in bees. Mapping by LOD score circumvents the problem of humans having small numbers of offspring for genetic study by combining data from many small families into one large pool. A good LOD score, usually 2 or 3, indicates, respectively, a 100 to 1 or 1000 to 1 probability favoring linkage between the gene of interest and the marker locus at the map distance in the calculation.

13.2 Cloning Eukaryotic Genes

- **Positional Gene Cloning:** Positional gene cloning involves identifying a gene base on its location in the genome. When a mutation is defined by a region in the genomic DNA, the gene that is mutated can be identified and cloned by positional cloning. If multiple genes are present in the same region, additional research will be necessary to identify the affected gene. This can be done by Northern hybridization or Western blot to determine which genes exhibit an expression pattern that matches either the gene of interest or the disease symptoms, or by using genomic DNA sequence information from several different organisms.

- **Chromosome Walking:** Chromosome walking uses hybridization techniques and DNA libraries to identify overlapping DNA segments. One DNA segment (probe) is used to screen a genomic library. The target is then used as the probe to screen the genomic library again. Eventually, this technique yields a very large segment of cloned DNA for study. In addition, this technique is also useful for locating genes that lie next to genes whose sequences are already known.

- **Haplotype Map:** The haplotype is the specific set of SNP alleles on a chromosome for a given individual. The human HapMap project involves analyzing one SNP for every 5 kb across the human genome. This has led to the identification of tag SNPs which are SNP alleles that are unique to a specific haplotype. The analysis of tag SNPs in individuals can be used to determine their haplotype based on the haplotypes of other individuals that possess the same SNPs. This analysis will simplify the localization of human disease genes. In addition, it will allow the identification of these disease genes in undiagnosed individuals by comparing the haplotypes of undiagnosed individuals with those of individuals diagnosed for a specific disease.

13.3 Eukaryotic Vectors

- **Yeast Vectors:** The **2-micron plasmid** is a naturally occurring plasmid in baker's yeast, *Saccharomyces cerevisiae*. The **yeast episomal plasmid** (**YEp**) has both an *E. coli* origin of replication and a yeast 2-micron origin. In addition, it has two selectable markers, one for *E. coli* (conferring antibiotic resistance) and one for yeast (a gene required for amino acid biosynthesis). The **yeast integrative plasmid** (**YIp**) lacks the yeast 2-micron origin and must integrate into the host yeast genome in order to be stably expressed. A wild-type yeast gene cloned into a YIp can be disrupted by cloning another gene inside of it. When this gene recombines with the wild-type genomic copy, it will inactivate the genomic copy. **Yeast centromere plasmids** (**YCp**) have a yeast centromere and a chromosomal origin of replication different from the 2-micron origin which ensures that there is one copy of the plasmid per cell. **Yeast artificial chromosomes** (**YACs**) are linear versions of YCp that contain telomeres. These vectors allow large inserts to be cloned into them.

- **Plant Vectors:** The **tumor-inducing (Ti) plasmid** is a 200-kb bacterial plasmid from the soil bacterium *Agrobacterium tumefaciens*. Upon infection, this bacterium transforms plant cells by the random integration of the Ti plasmid into the chromosome of the plant host causing a crown gall tumor to form. This is the most studied system for introducing foreign genes into plants. Foreign genes can be introduced into plants using the Ti plasmid either by (1) the integrating vector strategy which requires recombination between the T-DNA and a defective Ti plasmid within the *A. tumefaciens* or (2) the binary vector method in which the foreign DNA is cloned into a Ti plasmid lacking the virulence genes which is then introduced into *A. tumefaciens* cells that contain a defective Ti plasmid that contains the virulence genes.

- **Transposable Elements:** Transposable elements are segments of DNA that are able to move from one DNA molecule or region to a new molecule or location. **Transposase** is an enzyme encoded by the transposable element that acts on the terminal inverted repeats at both ends of the transposable element catalyzing the movement of the entire element. **P elements** are transposable elements identified in *Drosophila* that are commonly used in generating transgenic organisms.

- **Viral Vectors:** Viruses are commonly used vectors in higher organisms because they permit the replacement of part of their DNA with foreign DNA, and they can integrate their nucleic acid sequences into the host genome permitting foreign DNA to be stably inherited. SV40, a DNA tumor virus, is commonly used as a vector.

- **Use of Vectors to Express Foreign Genes:** Vectors must be capable of certain functions in order for the genes that they carry to be successfully expressed. For eukaryotic genes to be expressed in a prokaryote, for example, the gene must have the proper transcriptional, posttranscriptional, and translational signals, such as having a proper promoter and a 5′ cap, 3′poly(A) tail, being properly spliced, and having the appropriate signals for translation initiation.

13.4 Site-Specific Mutagenesis
- **Mutations:** Two types of mutations are typically generated and analyzed: (1) insertional mutations in which the central portion of a gene is replaced with another selectable gene, and (2) site-specific mutations that change one or more amino acids in the encoded protein. This site-directed mutagenesis is accomplished by using an oligonucleotide containing the desired mutation that is otherwise complementary to the cDNA that we want to mutate. The cDNA is denatured and allowed to anneal to the mutant oligo. DNA polymerase extends the sequence from the primer to create a double-stranded DNA that is transformed into *E. coli*. As the plasmid replicates in *E. coli*, one-half will contain a wild-type copy of the gene, and one-half will contain a mutant copy of the gene.

13.5 Introducing Foreign DNA into Cells
- **Transformation/Transfection: Transformation** refers to the introduction of foreign DNA into yeast and prokaryotes. Two techniques are primarily used to introduce plasmid DNA into *E. coli*; **chemical transformation** and **electroporation**. Several methods are used to introduce foreign DNA into eukaryotic host cells. Among the techniques used are **chemical transfection, liposome-mediated transfer**, and **biolistic transfer** which is used to deliver foreign DNA into mitochondria and chloroplasts. Transfection refers to the introduction of foreign DNA into a higher eukaryotic cell. Organisms that have taken up foreign DNA are referred to as **transgenic**.

13.6 Mouse Genetics
- **Transgenic Mouse:** One type of genetically altered mouse that results from the random insertion of DNA into the genome of an oocyte is a transgenic mouse. The transgene must confer a dominant phenotype because two endogenous copies of the same gene are also present in the mouse. Thus, the transgene could be the wild-type gene introduced into a homozygous mutant background, or a mutant allele with a dominant phenotype.
- **Knockout Mouse:** A knockout mouse results from the physical exchange of the transgene with the endogenous genomic copy. This procedure involves three major steps. First, a vector carrying the modified gene is created. The center of the gene to be modified is replaced with a gene conferring neomycin resistance. Outside the flanking gene regions, the gene for thymidine kinase from herpes simplex virus is cloned. Second, the targeting vector is introduced into mouse **embryonic stem (ES) cells**. Cells in which homologous recombination between the targeting vector and one copy of the endogenous gene has taken place are selected for. These cells will contain the neo^r gene and lack the tk^{HSV} gene. These cells are injected into a blastocyst-stage mouse embryo, which develops into a chimeric mouse. First a determination is made of whether any of the ES cells are incorporated into the germ line of the chimeric mouse, and mice that are homozygous for the knockout allele are then generated by mating two heterozygous mice.

13.7 Human Gene Therapy
- **Human Gene Therapy:** Human gene therapy involves the introduction of transgenes into human somatic cells to correct an inherited disease.
- **Treatment of Severe Combined Immunodeficiency:** Severe combined immunodeficiency was the first human disease treated by gene therapy. Several features of this disease made it plausible to attempt to treat it by gene therapy. SCID is caused by a recessive mutation in a single gene, the defect is restricted to the lymphocytes, and the B and T lymphocytes are relatively easy to isolate and reintroduce into the patient. Treatment involved infecting the T cells of the patient with a wild-type *ADA* gene and then reintroducing them into the patient. This therapy has been successful in some patients leading to a reduction or eradication of their symptoms, but affected individuals still have the same probability of passing the disease allele onto their offspring. Care must be taken in using retroviral vectors because the insertion site cannot be controlled and in some instances their use has led to tumor development in gene therapy patients due to the inactivation or overexpression of a protooncogene. Adenovirus, which infects lung epithelial cells, is a promising vector for gene therapy to treat cystic fibrosis.

13.8 Cloned Organisms

- **Dolly:** The sheep Dolly is the first mammal to be cloned by the replacement of a haploid egg nucleus with the diploid nucleus from the cell of an adult sheep. Since then, several other animals, including cats, pigs, rabbits, mice, and cattle, have been cloned.
- **Concerns:** The success of cloning animals remains questionable because current technology requires that a large number of nuclear transfers are required to produce a viable clone. In addition, cloned animals have typically been observed to have shortened life spans. Dolly died of cancer and arthritis at 6 years of age, considerably younger than most sheep that live 11 or 12 years. In addition, some researchers have reported that cloned animals have shown a propensity for disease and have exhibited developmental and physiological abnormalities potentially due to a differentiated cell being "reprogrammed."

13.9 Practical Benefits from Gene Cloning

- **Medicine:** Basic knowledge about how genes work, the identification of genes associated with genetic diseases, and knowledge of how these genes function has increased tremendously due to genetic engineering. Recombinant DNA technology has made available large quantities of biological substances that were previously in short supply and has made it possible to combat or immunize against certain diseases. Transgenic animals can be used as models to develop and test pharmaceutical or gene therapies for diseases.
- **Agriculture:** Genetically modified crops that are resistant to insect pests, frost, and premature ripening have been approved for planting. Many are also being made resistant to herbicides allowing farmers to kill unwanted weeds without killing their crops. This has led to an increase in crop yield which is necessary to feed the ever-growing population.
- **Industry:** Engineering bacteria to break down toxic wastes, modifying yeast to use cellulose to produce glucose and alcohol for fuel, and developing better food-processing methods and waste conversion are some of the industrial applications of biotechnology.

13.10 Ethical Considerations

- **The Green Revolution and Genetically Modified Organisms (GMOs):** There is much debate about the risk of genetically modified plants, not only whether they are safe to consume, but whether they are safe to generate and grow as well. The potential dangers of GMOs, which we do not realize now, may cause significant problems in the future when it may be too late to remove the effects.
- **Cloning Organisms and Individuals:** The ability to move genes at will and to analyze the genomes of individuals for the presence of alleles associated with specific diseases brings up questions regarding the moral and ethical nature of recombinant DNA technology and its use. Each day we are presented with more options which continue to fuel the ethical debate over how far science should go.

Key Terms

- transgenic animals
- transgenic plants
- genetically modified organisms (GMOs)
- restriction fragment length polymorphisms (RFLPs)
- variable number tandem repeat (VNTR)
- microsatellite
- triplet repeat diseases
- single-nucleotide polymorphism (SNP)
- logarithm of odds score (lod score)

- positional gene cloning
- chromosome walking
- haplotype
- 2-micron (2-μm) plasmid
- yeast episomal plasmid (YEp)
- yeast integrative plasmid (YIp)
- merodiploid
- yeast centromere plasmid (YCp)
- yeast artificial chromosome (YAC)
- rescue

- tumor-inducing (Ti) plasmid
- transposable elements
- transposase
- P element
- helper element
- site-specific mutagenesis
- oligonucleotide (oligo)
- chemical transformation
- electroporation
- transformed
- transfection
- transgenic
- liposomes
- biolistic

- transgenic mouse
- knockout mouse
- homologous recombination
- knockout-targeting vector

- embryonic stem (ES) cells
- chimeric
- severe combined immunodeficiency (SCID)

- Green Revolution
- preimplantation genetic diagnosis (PGD)

Understanding the Key Concepts
Use the key terms or parts of their definitions to complete the following sentences.

Restriction fragment length (1)_____ are DNA markers in which specific (2)_____ in the genomic DNA have been (3)_____ or (4)_____ among different individuals in a population. These sites can be used to reliably predict the (5)_____ of a particular disease allele by (6)_____. A variable number tandem repeat is another DNA marker that can vary in the (7)_____ that are present at a particular site in the genome in different individuals. The number of (8)_____ at a VNTR locus can be determined by (9)_____ the DNA with a specific (10)_____ that cleaves the DNA at sequences (11)_____ the repeat and then using (12)_____ hybridization with the (13)_____ sequence as a probe to detect the relative (14)_____ of the fragment. Since VNTRs can include only one or two repeats, but can be as large as several (15)_____ repeats, there can be (16)_____ of possible (17)_____ at a VNTR locus. A (18)_____ is a (19)_____ repeated sequence than a VNTR, consisting of only 2 to 5 base pairs. Some of these sequences are associated with (20)_____, such as Huntington disease or fragile-X syndrome. In these diseases, the (21)_____ in the disease gene correlates with the (22)_____ of the disease symptoms and the (23)_____ of onset. These sequences in the genome are typically detected by (24)_____ using (25)_____ that are (26)_____ to sequences flanking the repeat. In this way, a single (27)_____ at a specific genetic locus can be (28)_____.

A single-nucleotide (29)_____ occurs approximately once every (30)_____ base pairs in genomes of most organisms. Only (31)_____ alleles are present at each SNP making it (32)_____ that an individual will be (33)_____ for a specific SNP than it is that it will be heterozygous at a (34)_____ or (35)_____ locus. Clusters of SNPs are present in (36)_____-free regions flanked by (37)_____. The (38)_____ of an individual can be determined by analyzing the SNP alleles on a chromosome. The human HapMap project has allowed the identification of (39)_____ that will simplify the localization of human (40)_____ genes by looking for (41)_____ between these genes from different individuals with the presence of specific (42)_____.

If the chromosomal location of a specific mutation that is associated with a particular disease is known to be between two DNA markers, the mutated gene can be cloned by (43)_____ gene cloning using a technique known as (44)_____. In this technique, two (45)_____ DNA sequences flanking each of the DNA markers are used to (46)_____ a (47)_____ library. The (48)_____ clones are isolated and (49)_____ by restriction enzyme digest. The two DNA fragments that extend the farthest in each direction are used as (50)_____ to (51)_____ the library again. This is repeated until the clones from each side (52)_____ each other. The gene of interest should then be present in the isolated genomic clones. Since several open reading frames will likely be present, the mutated gene associated with the disease symptoms is typically identified by looking at the (53)_____ of the candidate genes.

Different eukaryotic (54)_____ are used to introduce DNA into different eukaryotic cells. The (55)_____ plasmid can be used to (56)_____ a specific gene in yeast to determine if a loss of function mutation has an associated (57)_____. The (58)_____ plasmid from (59)_____ can be used to introduce foreign DNA into damaged plant cells. The growth of these transformed plant cells leads to the formation of a callus that can be used to generate a (60)_____ plant. When this plant is crossed with other plants, the (61)_____ will be inherited along with the other plant genes. (62)_____ vectors allow foreign DNA to be stably inherited because they (63)_____ into the host genome.

Mice can be genetically altered by either the random insertion of foreign DNA into their genomes to create a (64)_____ mouse, or by (65)_____ between a transgene and the endogenous genomic copy to generate a (66)_____. The expression of transgenes in mice allows the phenotypes associated with specific genes or loss of function mutations in specific genes to be determined.

Figure Analysis

Use the indicated figure or table from the textbook to answer the questions that follow.

1. **Figure 13.5**
(a) How many CAG repeats do normal individuals have in the Huntington gene? How many do individuals with Huntington disease symptoms typically have?

(b) How does having more CAG repeats in the DNA affect the Huntington protein?

(c) What is the larger number of CAG repeats found to be associated with?

(d) How many repeats are individuals that exhibit Huntington symptoms later in life typically found to have?

(e) How many repeats are individuals that express Huntington symptoms earlier in life typically found to have?

(f) How can you tell from the data shown in figure 13.5 that the Huntington disease is dominant?

2. **Figure 13.7**
(a) In this method of SNP detection, how are small fragments of genomic DNA produced?

(b) Why is the genomic DNA denatured?

(c) What is being used as a probe? What is the function of the S1 nuclease?

(d) What is the expected product if the probe and the target are different at the SNP site?

(e) What is the expected product if the probe and the target are complementary at the SNP site?

(f) How is the sequence of the SNP site determined?

3. **Figure 13.8**
(a) In this method of SNP detection, how can the SNP of interest be looked at specifically?

(b) What is the purpose of the internal primer?

(c) What is the single-stranded probe hybridized to?

(d) How is the sequence of the SNP identified?

(e) How will the pattern of hybridization differ among homozygous and heterozygous individuals?

4. **Figure 13.9**
(a) What fragments of microsatellite 1 and microsatellite 2 are present in the mother?

(b) What fragments of microsatellite 1 and microsatellite 2 are present in the father?

(c) How many alleles did each child inherit from their mother? How many from their father?

(d) For microsatellite 1, what fragment do all of the children with the dominant trait have in common?

(e) For microsatellite 2, what fragment did all of the children without the dominant trait inherit from their mother?

(f) For microsatellite 2, what fragment do all of the children with the dominant trait, except one, have in common? Which one is the exception?

(g) What can be said about the seventh child who has the dominant trait, but not the fragment associated with the dominant trait? What fragment did she inherit from her father? What fragment did she inherit from her mother?

5. **Figure 13.9**
(a) From the four candidate genes located in this region of chromosome 7, how was the gene responsible for cystic fibrosis identified?

(b) How was the identity of this gene confirmed?

(c) Identify the three locations where mutations that cause cystic fibrosis could be present relative to the candidate gene.

(d) Describe how these mutations would affect the expression of the chloride transporter protein that was found to be encoded by the cystic fibrosis gene accounting for the disease symptoms.

6. **Figures 13.29–13.31**
(a) What are the critical features of the knockout targeting vector?

(b) What are the three possible outcomes of the targeting vector in mouse ES cells?

(c) What type of medium should the transfected ES cells be plated on to distinguish between these three possible outcomes? Describe what type of cells will live on this medium and what type of cells will die. Explain why.

(d) What is the next step for ES cells that have undergone homologous recombination?

(e) How can a researcher determine if the ES cells are incorporated into the developing mouse?

(f) Why is it important for some of the knockout ES cells to incorporate into the germ line of the chimeric mouse?

(g) Why are the chimeric mice mated to a homozygous recessive black mouse?

(h) How are the mice that contain the knockout allele identified?

(i) What two mice should be mated to produce the knockout mouse?

(j) What proportion of the offspring from this mating are expected to be homozygous for the knockout?

General Questions

Q13-1. How is a RFLP detected? Describe how this strategy can be used to diagnose sickle cell anemia.

Q13-2. Describe how a researcher can go about cloning a gene that has only been mapped to a specific chromosomal location between two DNA markers.

Q13-3. What is the purpose of the human HapMap project? How will this project simplify the localization of human disease genes?

Q13-4. Describe the difference between the yeast episomal, yeast integrative, and yeast centromere plasmids. How can a yeast integrative plasmid be used to disrupt a chromosomal gene in yeast?

Q13-5. Describe the two methods that can be used to introduce a foreign gene into a plant by way of infection with *Agrobacterium tumefaciens*.

Multiple Choice

For each of the following, circle the letter of the choice that most appropriately answers the question.

1.) A minisatellite is typically a _____ sequence repeated in a tandem array that is found an average of once every _____ in the human genome.
   a. 2–5 bp; 1 kb
   b. 10–100 bp; 100 kb
   c. 200–300 bp; 100 kb
   d. 2–5 bp; 30 kb
   e. 10–100 bp; 30 kb

2.) Which of the following are directly associated with the inherited triplet repeat diseases?
   a. SNPs
   b. VNTRs
   c. microsatellites
   d. RFLPs
   e. transposons

3.) A strong linkage between two loci exists when the composite lod score is _____ or more.
   a. 1
   b. 3
   c. 5
   d. 7
   e. 10

4.) Which of the following are true statements about SNPs?
   a. Most RFLPs are SNPs, but most SNPs are not RFLPs.
   b. By identifying only a few SNPs in a given region, the remainder of the SNPs in the same region can be predicted.
   c. Tag SNPs identified by the human HapMap project simplify the determination of the haplotypes of other individuals.
   d. Only a and c
   e. All of the above

5.) Which of the following is required for insertion of the T-DNA into the plant genome?
   a. L border sequences
   b. presence of foreign DNA
   c. P element
   d. virulence genes
   e. transposase

6.) Which of the following on the Ti plasmid is required for recombination to occur between two Ti plasmids?
   a. L border sequences
   b. T-DNA
   c. auxin, cytokinin, and opine synthesis genes
   d. virulence genes
   e. origin of replication

7.) Site-specific mutagenesis relies on which of the following?
   a. loss of the proofreading activity of DNA polymerase
   b. incorporation of modified deoxynucleotides
   c. design of a mutagenic primer
   d. utilization of a specific selectable marker to isolate mutagenized plasmids from nonmutagenized plasmids
   e. use of specialized *E. coli* cells that tolerate DNA mismatch mutations during DNA replication

8.) Foreign DNA can be introduced into eukaryotic cells by all but which of the following methods?
   a. viral infection
   b. electroporation
   c. bacterial transformation
   d. liposome-mediated injection
   e. calcium phosphate transfection

9.) ZnSO$_4$ was added to the diet of transgenic mice containing the *RGH* gene to do which of the following?
   a. increase the number of offspring that they produced
   b. induce the expression of the promoter controlling the *RGH* gene
   c. promote homologous recombination between the *RGH* gene and the mouse genome
   d. select for mice with high *RGH* expression
   e. retard their growth

10.) Which of the following is *not* a moral or an ethical question that has been raised as a result of advances in the field of recombinant DNA technology?
   a. Can we use a cloned copy of a specific human gene to rescue the loss of function of a homologous gene in another organism?
   b. What long-term effects may result from preimplantation genetic diagnosis?
   c. What will prevent parents from producing children with a desired tissue type?
   d. Is it appropriate to plant and market crops that encourage the use of herbicides?
   e. What conditions are considered to be genetic diseases?

Practice Problems

P13-1. Researchers wish to find the genomic location of a gene responsible for causing a particular autosomal dominant disease. The results of PCRs of three different microsatellites in a family afflicted with the disease are shown. Both parents are heterozygous for all three microsatellites as are all of their children. The father is heterozygous for the gene that causes the disease. Individuals afflicted with the disease are shown in bold. Which microsatellite most likely resides nearest the gene that causes the disease?

[*Hint 1*: Alleles of microsatellites differ in the number of repeats present. Therefore, a PCR of a microsatellite region will give a different size product (band on a gel) for each allele.

Hint 2: Children will receive one allele of each gene from each parent. Unlinked genes will assort their alleles independently of each other, while completely linked genes will always assort two alleles together.]

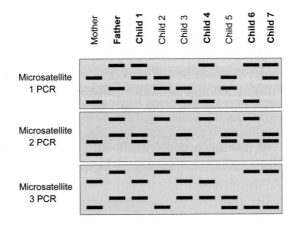

P13-2. A region of a human chromosome was discovered to contain an RFLP by Southern blot analysis. A *Bam*HI restriction digest of homozygous and heterozygous individuals was performed, and a probe was made that is complementary to that region of the chromosome. A Southern blot using that probe revealed the bands depicted. Draw a map indicating the position of and distances between each *Bam*HI site in each allele. Also indicate where the probe must anneal.

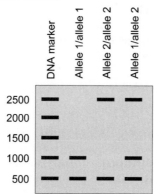

P13-3. A region of a human chromosome depicting five *Eco*RI restriction sites (E1–E5) is shown. The distances in base pairs is given between each of the sites, and the region of complementarity of a Southern blot probe is indicated. Three different RFLP alleles exist in this region. For each of the individuals described, draw the bands that you would expect to see on a Southern blot using the indicated probe after digestion with the *Eco*RI enzyme.

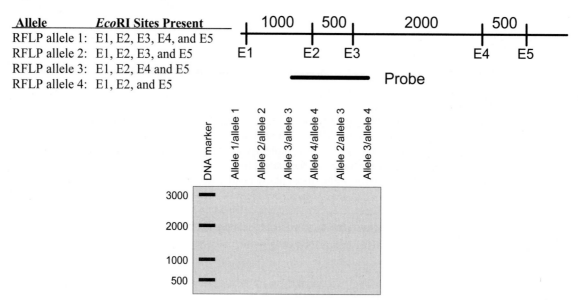

Allele	*Eco*RI Sites Present
RFLP allele 1:	E1, E2, E3, E4, and E5
RFLP allele 2:	E1, E2, E3, and E5
RFLP allele 3:	E1, E2, E4 and E5
RFLP allele 4:	E1, E2, and E5

P13-4. You are trying to map the gene for a newly identified human genetic disorder. You find and analyze the DNA from a large number of small families. The following is a table of lod scores that you have calculated. Which locus is closest to your gene of interest?

Locus	lod
DS199	2.0
DS180	2.0
DS662	2.6
DS455	1.3
DS207	3.0

P13-5. You are a researcher attempting to determine the function of two different proteins. The following DNA sequences represent the coding strand of the two genes that code for the proteins you are interested in. Notice that each sequence given begins with the start codon and a space is provided between each codon.

(a) You hypothesize that a phosphate is added to the threonine at position 8 (bold codon) of the following protein and that this posttranslational modification is important for the function of this protein. To test this, you plan to mutate this threonine to alanine (an amino acid that cannot be phosphorylated) using site-specific mutagenesis of its gene. Write out the primer sequence that you would create in order to produce the desired Thr→Ala mutation with as few nucleotide changes as possible.

```
Met Gly Val  His Ile Leu Pro Thr Pro Pro Asn Ser Gly Val Ile Phe
ATG GGG GTG  CAC ATA CTC CCA ACC CCT CCG AAT TCA GGA GAT ATT TTC
```

(b) You hypothesize that the acidic residue glutamic acid near the beginning of the protein whose gene sequence follows is important for the function of the protein. To test this hypothesis, you wish to change this negatively charged amino to an amino acid with a positive charge. Which amino acid would you change glutamic acid to so that you would only have to change a single nucleotide? Write out the primer sequence that you would create in order to produce the desired mutation.

```
ATG GCT TCT TTG CGA ACG CTC GGC CAG GAG AAT TCT CCG CTA CAG CGG
```

Assessing Your Knowledge

Understanding the Key Concepts—Answers

1.) polymorphisms; 2.) restriction sites; 3–4.) lost or eliminated; created 5.) probability; 6.) recombination mapping; 7.) number of repeats; 8.) repeats; 9.) digesting; 10.) restriction enzyme; 11.) flanking; 12.) Southern; 13.) repeat; 14.) size; 15.) hundred; 16.) hundreds; 17.) alleles; 18.) microsatellite; 19.) smaller or shorter; 20.) triplet repeat diseases; 21.) number of repeats; 22.) severity; 23.) age; 24.) PCR; 25.) primers; 26.) complementary; 27.) microsatellite; 28.) amplified; 29.) polymorphism; 30.) 1000; 31.) four; 32.) less likely; 33.) heterozygous; 34–35.) VNTR; microsatellite; 36.) recombination; 37.) recombination hotspots; 38.) haplotype; 39.) tag SNPs; 40.) disease; 41.) associations; 42.) tag SNPs; 43.) positional; 44.) chromosome walking; 45.) unique; 46.) screen; 47.) genomic DNA; 48.) positive; 49.) mapped; 50.) probes; 51.) screen; 52.) overlap; 53.) expression patterns; 54.) vectors; 55.) yeast integrative; 56.) disrupt or knockout; 57.) phenotype; 58.) Ti; 59.) *Agrobacterium tumefaciens*; 60.) transgenic; 61.) T-DNA; 62.) Viral; 63.) integrate; 64.) transgenic; 65.) homologous recombination; 66.) knockout.

Figure Analysis—Answers

1a.) Normal individuals can have up to 34 CAG repeats in the Huntington gene; individuals with symptoms of Huntington disease usually have more than 42 CAG repeats; b.) More CAG repeats in the DNA increases the length of the polyglutamine tract in the Huntington protein; c.) The larger number of CAG repeats in the Huntington gene was found to be associated with the severity of the disease, or with an early age of onset; d.) Individuals that exhibit Huntington symptoms later in life are typically found to have between 50 and 75 repeats; e.) Individuals that exhibit Huntington symptoms earlier in life are typically found to have over 100 repeats; f.) The genomic Southern blot shows that individuals that exhibit Huntington symptoms are heterozygous and have one allele with around 30 CAG repeats, and a second allele with 50 or greater CAG repeats, while the normal individual has only a single band indicating that both alleles must be around 30 repeats. Since the early-, and late-onset individuals are heterozygous, but display Huntington symptoms, the disease must be dominant because these individuals do have one normal allele.

2a.) Small fragments of genomic DNA are produced by restriction enzyme digest or by randomly shearing the DNA; b.) The genomic DNA is denatured so that it can anneal to the labeled, single-stranded probe; c.) The probe is a radiolabeled, single-stranded DNA oligonucleotide that is completely complementary to one of the SNP sequences. S1 nuclease is used to cleave any single-stranded DNA after the genomic SNP and the probe are annealed to each other; d.) If the probe and the target are not completely complementary at the SNP site, then both strands are cleaved at the site. One would expect to see a shortened probe after gel electrophoresis and exposure to X-ray film; e.) If the probe and the genomic DNA are completely complementary at the SNP site, the probe will not have any single-stranded section and will not be cleaved by S1 nuclease, so it will remain intact after electrophoresis; f.) The sequence of the SNP site is determined by finding a probe that is completely complementary to the SNP sequence. For example, if the first probe has a T in the varying position, but is not completely complementary, a second probe that differs at that site in the SNP sequence can be tried by replacing the T with a C in the probe. If this probe is not completely complementary and is still cleaved by S1 nuclease, a third probe can be tried with an A in the varying position. If this probe is not completely complementary and is still cleaved by S1 nuclease, a fourth probe can be tried with a G in the varying position. One of these probes should be completely complementary and not cleaved by S1 nuclease following annealing to the SNP genomic target, while the other three should be cleaved by S1 nuclease. In this way, the sequence of the SNP can be determined.

3a.) The SNP of interest can be looked at specifically by using primers that flank the SNP to amplify it from the genomic DNA by PCR; b.) An internal primer is used to synthesize and radioactively label only one DNA strand. This is then used as a probe; c.) This probe is hybridized to four different short oligonucleotides, each containing a different variation of the SNP sequence, which are attached to a solid support to create a spotted SNP array; d.) The hybridization pattern revealed by X-ray film identifies the sequence of the SNP. The single-stranded, radiolabeled probe will only hybridize to oligonucleotides with which it is completely complementary under stringent hybridization conditions; e.) The pattern of hybridization will differ among heterozygous and homozygous individuals because individuals who are heterozygous for the SNP will show hybridization to two different oligonucleotides, while individuals who are homozygous for the SNP will show hybridization to only a single oligonucleotide.

4a.) For microsatellite 1, the mother has fragments A and D. For microsatellite 2, the mother has fragments A and C; b.) For microsatellite 1, the father has fragments B and C. For microsatellite 2, the father has fragments B and D; c.) For both loci, each child inherited one allele from their mother and one allele from their father; d.) For microsatellite 1, all of the children with the dominant trait do not have a fragment in common. Half of the afflicted children received fragment A from their mother, and the other half received fragment D; e.) For microsatellite 2, all of the children without the dominant trait inherited fragment C from their mother; f.) For microsatellite 2, all of the children with the dominant trait have fragment A, except the seventh daughter; g.) The seventh child who has the dominant trait, but not fragment A, is likely the result of a recombination event occurring at microsatellite 2 in the mother between fragments A and C. The child then inherited a recombined fragment C from her mother. This event allowed fragment C to contain the necessary information from fragment A to confer the dominant trait in this child. This child inherited fragment C from her mother and fragment B from her father.

5a.) The cystic fibrosis gene was identified because it was the only one of the four candidate genes that was expressed in all three of the expected tissues, the lungs, pancreas, and sweat glands, which are the three tissues that are affected by cystic fibrosis; b.) The identity of this gene was confirmed by cloning it from both unaffected and affected individuals in several different pedigrees and looking for mutations in the gene from the affected individuals that were not present in the unaffected individuals; c.) Mutations in the cystic fibrosis gene that lead to the disease could be present in the ORF itself, in the promoter, or in an intron; d.) If the mutation is present in the ORF itself, it could be an inactivating mutation that causes the protein to misfold or misfunction. If the mutation is present in the promoter, it could affect the transcription of the gene and lead to either greater or reduced levels of the mRNA relative to the wild-type gene. If the mutation is present in the intron, it could affect the splicing of the final mRNA that would produce mRNAs with an altered size, amino acid sequence, or domain structure relative to wild-type.

6a.) The knockout targeting vector should carry the gene to be targeted with the center of the gene being replaced by the gene conferring neomycin resistance. Outside of the flanking gene regions, the gene for thymidine kinase from the herpes simplex virus (tk^{HSV}) should be cloned; b.) In ES cells, the targeting vector may undergo one of three events. First, that vector may not insert into the genomic DNA. Second, the vector may insert randomly in the genome. Third, the vector may undergo homologous recombination and replace a genomic copy of the gene with the knockout copy of the gene.

c.) To distinguish between these three possible outcomes, the ES cells should be plated on a medium that contains both neomycin and gancyclovir. Cells that did not have the vector insert into the genomic DNA will be killed by neomycin. For cells where the vector inserted into the genome randomly, the entire targeting vector, including the neo^r and tk^{HSV} genes will recombine into the genome. These cells will be gancyclovir-sensitive, although they are neo^r and will die. If the vector undergoes homologous recombination with the genomic copy of the targeting gene, a neo^r insertion lacking the tk^{HSV} gene will be produced. Therefore, these cells will be both neo^r and gancyclovir-resistant; d.) ES cells that have undergone homologous recombination are then injected into a blastocyst-stage mouse embryo which is homozygous for the recessive allele that produces the black fur color phenotype; e.) If the ES cells are incorporated into the developing mouse, it will have both black and brown fur and be chimeric because the ES cells that the targeting vector were introduced into were derived from a mouse line with brown fur which is dominant to black fur, while the ES cells that underwent homologous recombination were injected into a mouse with black fur; f.) Some of the knockout ES cells must be incorporated into the germ line of the chimeric mouse in order for the knockout allele to be passed on to the next generation; g.) The chimeric mice are mated to a homozygous recessive black mouse to determine if any of the knockout ES cells did incorporate into the germ line of the chimeric mice. If they did, this mating should produce heterozygous brown mice; h.) Mice that contain the knockout allele are identified by analyzing genomic DNA isolated from a clipped section of a brown-furred mouse's tail by Southern hybridization or by PCR amplification; i.) To generate a knockout mouse, two mice (one male, one female) that are both heterozygous for the knockout allele should be mated; j.) From this mating, 25% of the offspring would be expected to be homozygous for the gene knockout.

General Questions—Answers

Q13-1. A restriction fragment length polymorphism (RFLP) is detected by digestion of the genomic DNA with a specific restriction enzyme, followed by electrophoresis and Southern hybridization using a probe that will detect a single RFLP locus at a single site in the genome. This strategy can be used to diagnose sickle cell anemia because this disease is associated with a single nucleotide change in the β-globin gene that eliminates an *Mst*II restriction site. When genomic DNA from a normal individual is digested with *Mst*II, two bands are detectable by Southern hybridization using a probe that is internal to the β-globin gene and spans the internal *Mst*II site. In contrast, when genomic DNA from an individual with sickle cell anemia is digested with *Mst*II, only a single band is detected by Southern hybridization using the same probe because the internal *Mst*II site is missing. Therefore, this restriction site polymorphism is diagnostic of sickle cell anemia.

Q13-2. To clone a gene that has been localized to a specific chromosomal location between two DNA markers, unique sequences flanking each of the DNA markers can be used as probes to screen a genomic DNA library. Positive clones that hybridize to these sequences can be isolated and characterized by restriction enzyme digest. The overlapping restriction maps are arranged to determine the clones that extend the farthest in both directions or in one particular direction from the original probe. DNA fragments corresponding to the ends of the clones that extend the farthest in the correct direction are used as probes to screen the genomic DNA library again. A new group of overlapping clones is identified and arranged by restriction mapping. This process is continued until the region between the two walks from each side overlap. The gene of interest must then be located within the genomic clones isolated in the walk.

Q13-3. The purpose of the human HapMap project is to examine the haploid genotypes (haplotype), or the specific set of single-nucleotide polymorphisms on a chromosome for the individuals in a population. This project will simplify the localization of human disease genes because researchers can use the HapMap to look for associations between tag SNPs and specific diseases. For example, researchers can determine if a specific tag SNP is closely associated with a specific disease gene. Tag SNPs that are associated will be found in individuals that have the disease, but are less likely to be found in individuals that do not have the disease. Tag SNPs that are not associated with the disease will be randomly distributed. This will suggest whether a gene associated with a tag SNP may be involved in the disease associated with it. Likewise, the presence of a particular tag SNP in an individual can suggest the presence of the disease allele.

Q13-4. The yeast episomal plasmid contains the yeast 2-micron origin; an *E. coli* origin of replication; and two selectable markers, one for in *E. coli* typically conferring drug resistance and one for in yeast typically a gene required for amino acid biosynthesis. The yeast integrative plasmid lacks the 2-micron origin of replication and cannot replicate autonomously in yeast. The plasmid must recombine with the yeast genome to be stably expressed. The yeast centromere plasmid contains a yeast chromosomal origin of replication other than the 2-micron origin and a centromeric sequence that results in one copy of the plasmid per cell. To disrupt a chromosomal gene in yeast using a yeast integrative plasmid, the gene to be disrupted is cloned into YIp. Another selectable gene is then cloned into the middle of the gene to be disrupted leaving the sequences at the end of the gene that is being targeted intact. When transformed into yeast, homologous sequences at both ends of the gene being targeted in YIp will recombine with the chromosomal copy of the gene. This recombination integrates the disrupted copy of the gene into the yeast chromosome in place of the wild-type gene.

Q13-5. To introduce a foreign gene into a plant by the integrating vector strategy, the foreign gene is cloned into a plasmid containing the T-DNA. This recombinant plasmid is introduced into *A. tumefaciens* cells that contain a defective Ti plasmid. The two plasmids recombine inside the *A. tumefaciens* cells to generate a complete Ti plasmid. These *A. tumefaciens* cells are then used to infect damaged plant cells. The *A. tumefaciens* will transfer the T-DNA into the plant genome. To introduce a foreign gene into a plant by the binary vector method, the foreign gene is cloned into a Ti plasmid that lacks the virulence genes. This plasmid is introduced into *A. tumefaciens* cells that contain a defective Ti plasmid that contains the virulence genes. When this *A. tumefaciens* cell infects a damaged plant cell, the virulence genes on the second plasmid act on the right and left borders of the first plasmid to mediate the insertion of the T-DNA into the plant genome.

Multiple Choice—Answers

1.) b; 2.) c; 3.) b; 4.) e; 5.) d; 6.) a; 7.) c; 8.) c; 9.) b; 10.) a.

Practice Problems—Answers

P13-1. Microsatellite 1 most likely resides closest to the gene responsible for the disease. To determine this, you must compare the microsatellite alleles that the afflicted children received from their *father*. Notice that every afflicted child received the *large* paternal allele of this microsatellite while every nonafflicted child received the *small* paternal allele. Thus, this microsatellite is tightly linked to the gene causing the disease. In the father, the large microsatellite allele must be linked to his mutant gene while the small microsatellite allele must be linked to his normal gene. The paternal alleles of microsatellites 2 and 3 assorted independently of the mutant gene indicating that they are unlinked to the gene causing the disease. Since the mother is nonafflicted, it is irrelevant which maternal microsatellite alleles the children received.

P13-2.

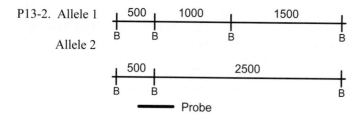

P13-3.

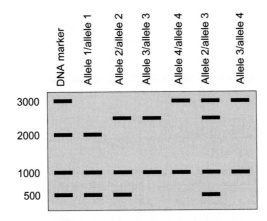

P13-4. A good lod score is 2 or 3, with 3 being better. A number of loci fit this description indicating that these loci have a high probability of being close to your gene of interest. Since *DS207* has the best LOD score (3.0), it is assumed to be the closest to your gene.

P13-5. (a) Either the coding strand or the template strand can be changed because the final product of mutagenesis will be double-stranded DNA produced by replication. The two possible primer sequences are shown here where the nucleotide change is underlined.

```
5'-CAC ATA CTC CCA GCC CCT CCG AAT TCA-3' or
3'-GTG TAT GTG GGT CGG GGT GGC TTA TGA-5'
```

(b) The only codon shown that encodes glutamic acid is the tenth codon GAG. This glutamic acid must be changed to one of the basic amino acids (Lys, His, or Arg). Changing it to Lys requires only one nucleotide (the GAG → AAG) to be changed, whereas changing it to His or Arg requires two nucleotides to be changed. Therefore, you should change glutamic acid to Lys. The two possible primer sequences are shown here where the nucleotide change is underlined.

```
5'-ACG CTC GGC CAG AAG AAT TCT CCG CTA-3' or
3'-TGC GAG CCG GTC TTC TTA AGA GGC GAT-5'
```

13

APPLICATION OF RECOMBINANT DNA TECHNOLOGY

CHAPTER SUMMARY QUESTIONS

1. 1. j, 2. i, 3. h, 4. g, 5. f, 6. e, 7. d, 8. c, 9. b, 10. a.

2. A RFLP (restriction fragment length polymorphism) is a nucleotide change that results in either the elimination or creation of a restriction enzyme site. A SNP (single nucleotide polymorphism) is a single nucleotide change in the DNA sequence that does not necessarily involve a restriction enzyme site. Most RFLPs are SNPs, but most SNPs are not RFLPs. SNPs are more randomly and densely distributed throughout the genome.

3. Minisatellites are also termed variable number tandem repeats (VNTRs). They are short sequences (10 to 100 bp) that are repeated in tandem array hundreds of times. This provides a great advantage over RFLPs in genetic mapping studies. With RFLPs only two alleles can be distinguished; presence or absence of the restriction enzyme site. With VNTRs there can be hundreds of alleles at a given locus. This large number of alleles increases the likelihood that an individual will be heterozygous. This high percentage of heterozygosity increases the probability that any mating will generate useful recombination mapping data.

4. Chi-square analysis requires a fairly large number of progeny. This is the case in *Drosophila*, but not humans. LOD (logarithm of odds) analysis was devised by Newton Morton to deal with the smaller data sets in human pedigrees.

5. Two major techniques are used to introduce plasmid DNA into bacterial cells: (1) chemical transformation, which involves treating the cells with a calcium chloride solution, and (2) electroporation, which involves briefly exposing the cells to high-voltage electricity. Both techniques make the cell permeable to DNA in its environment.

6. Foreign DNA can be introduced into eukaryotic cells by infection with animal virus vectors as well as several other techniques including electroporation, liposomes, and biolistics. The general term used for these processes is transfection.

7. The YCp, or yeast centromere plasmid, contains a yeast centromere and a chromosomal origin of replication other than the 2-μm plasmid origin. These two features result in usually only one copy of the plasmid per cell. The YAC, or yeast artificial chromosome, is a linear version of YCp that contains telomeres. It can be used to clone larger inserts.

8. Chromosome walking is a technique for cloning overlapping chromosomal regions starting from an arbitrary point (see figure 13.10). It is useful for determining relative locations of genes in uncharted regions as well as cloning regions too big to fit in a single vector.

9. *Agrobacterium tumefaciens* can infect many dicotyledonous plants, causing tumors known as crown galls. The bacterium possesses a 200-kb plasmid called the tumor-inducing (Ti) plasmid. The plasmid contains a 15-kb region termed T-DNA (for transferred DNA). Transformation of plant cells occurs when the T-DNA randomly integrates into the chromosome of the plant host. This leads to uncontrolled growth and crown gall formation.

10. The most commonly used technique for site-specific mutagenesis involves oligonucleotides (oligos). These consist of 17–25 bases and are designed to contain the desired mutation in the middle of the sequence. The target cDNA is cloned into a plasmid vector. The oligo is then hybridized to denatured target cDNA template. Addition of DNA polymerase and the four dNTPs extends the oligo sequence in the 5' → 3' direction. The end result is a double-stranded DNA plasmid that contains a single nucleotide mismatch at the site to be mutated. The plasmid is then used to transform *E. coli* cells. As the bacteria grow, the plasmid replicates, such that half the plasmids contain the mutation and half possess the original wild-type sequence.

11. **a.** The haplotype, or *haplo*id geno*type*, is the specific set of SNP alleles on a chromosome of a given individual.
   b. The YIp, or yeast integrative plasmid, is a yeast plasmid that contains a bacterial origin of replication but no yeast plasmid origin. Therefore, it cannot autonomously replicate in yeast cells, and so genes that are cloned in it must recombine with the genome to be stably expressed.
   c. The P element is a transposable element in *Drosophila* that is commonly used to create transgenic organisms.

12. After infecting a mammalian cell, SV40 recombinant virus can (1) replicate and complete its life cycle with the help of nonrecombinant viruses, (2) exist as a circular plasmid in the host's cytoplasm, or (3) integrate into the host's chromosome. In (2) and (3), the virus replicates its genome without making active virus particles.

13. Both types of mice are genetically altered for use in genetic studies. A transgenic mouse results from the random insertion of DNA into the mouse genome, whereas a knockout mouse results from the physical exchange of the transgene with the endogenous genomic copy. Transgenic mice provide model systems for human biological conditions and for gene therapy in humans. Knockout mice are especially useful for studying a number of processes, such as development, neuronal function, and immunology. Some have also been generated to model human genetic diseases such as cataracts and asthma.

14. Gene therapy is the treatment of inherited diseases using wild-type copies (transgenes) of defective genes. Unlike the transgenic and knockout mice techniques, the transgenes are only introduced into somatic cells. Offspring of these individuals, therefore, will not inherit the transgene and will have a certain probability of getting the disease.

15. The term *clone* refers to the creation of a genetically identical individual. To create Dolly, scientists transferred the nucleus of a 6-year-old ewe's udder cell into an enucleated egg cell from another sheep. Therefore, Dolly contained an exact copy of the ewe's nuclear genome. However, mitochondria also contain a genome, and

Dolly's was derived from the egg. Therefore, Dolly is not truly a genetic clone, because her total genotype was not completely identical to the 6-year-old ewe.

16. Genetic engineering has provided numerous benefits. Here's a sampling:

    1. Medicine
       • Advancement in the basic knowledge of how genes work (and don't work).
       • Cloning and sequencing of many genes involved in hereditary disorders.
       • Production of large quantities of biological substances previously in short supply (insulin, blood-clotting factors).
       • Creation of transgenic animals to serve as models for human genetic disorders.
    2. Agriculture
       • Creation of transgenic crops that resist pests, frost, and premature ripening.
       • Modification of rice to maintain its vitamin A potential even after husk removal.
    3. Industry
       • Engineering bacteria to break down toxic materials.
       • Modifying yeast to digest wood (cellulose) and convert it directly to alcohol for use as fuel.

17. Concerns include the following: (1) Are GMOs safe for human and animal consumption? (2) Are we able to control the spread of GMOs? (3) Is it ethical or moral to tamper with an organism's genetic material?

EXERCISES AND PROBLEMS

18. RFLPs can be generated by base substitutions, insertions, or deletions. They could also be caused by inversions and translocations.

19. A clone will hybridize to all fragments overlapping its location. Therefore, gene number 1 will hybridize to fragments 3, 4, and 5, while gene number 2 will hybridize to fragment 6 only.

20. Let's begin with designating the genotypes of all individuals in the family.

    | Jack: 2, 5 | Child I: 1, 5 | Child III: 2, 7 | Child V: 2, 4 |
    | Jill: 4, 7 | Child II: 3, 6 | Child IV: 7, 7 | |

    Children conceived by Jack and Jill should have an allele from each parent. A child conceived by Jack and a woman other than Jill should have an allele from Jack and no allele that matches Jill's. Similarly, a child conceived by Jill and a man other than Jack should have an allele from Jill and no allele that matches Jack's. An adopted child would most likely have two alleles that are different from those of Jack and Jill.

Now let's examine the genotypes of the children. Child I has an allele from Jack (5) and none from Jill. Child II does not share any allele with either Jack or Jill. Child III has an allele from Jack (2) and one from Jill (7). Child IV has two 7 alleles: one comes from Jill, the other from his father, who also happens to have a 7 allele. Child V has one allele from Jack (2) and the other from Jill (4).

Therefore, children III and V were conceived by Jack and Jill, child I is Jack's, child IV is Jill's, and child II was adopted.

21. RFLPs, SNPs, and VNTRs are not always associated with a visible change in the cellular phenotype. For example, they can occur in noncoding regions of the genome. RFLPs and SNPs can still be silent, even if they do occur in coding regions. This is due to the degeneracy of the genetic code: a single-base substitution does not necessarily change the specified amino acid.

22. a. The probe will hybridize to a 4-kb restriction fragment in DNA molecule 2 and a 5-kb fragment in molecule 4. In both molecules 1 and 3, the probe will bind to a 3-kb fragment, and so these two molecules cannot be distinguished from each other using this probe. Therefore, the sample would contain three alleles: 3, 4, and 5 kb.

b. There are six possible genotypes: three homozygotes ("3/3," "4/4," "5/5") and three heterozygotes ("3/4," "3/5," "4/5").

23. For bacteria to make large quantities of a human protein, the protein must be expressed from cDNA rather than from genomic DNA. Because cDNA is derived from mature mRNA, it lacks introns and can therefore produce a suitable transcript for bacteria.

24. a. Both John and Jane have a 7-kb fragment, which constitutes the entire region of the chromosome (3.3 + 1.1 + 2.6). John also has three distinct fragments, which correspond to all three restriction fragments. Jane, on the other hand, has a 2.6-kb fragment, and a 4.4-kb fragment which is equivalent to 1.1 + 3.3. Therefore, the chromosomal arrangement of the RFLP is as follows:

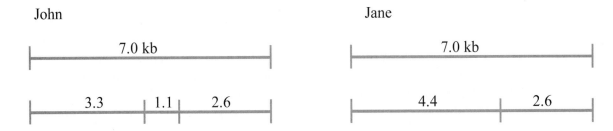

John

Jane

7.0 kb

3.3   1.1   2.6

7.0 kb

4.4   2.6

b. John and Jill's children will inherit one allele (chromosomal region) from each parent. Therefore, to determine the children's RFLP patterns we can consider all the various ways the parental chromosomes can combine. For simplification, let's designate John's alleles as A_1 (top DNA segment) and A_2 (bottom), and Jane's alleles as A_3 (top DNA segment) and A_4 (bottom). Therefore, the possible genotypes and RFLP patterns of the children are A_1A_3, which produces only the 7 kb band; A_1A_4, which produces bands of 2.6, 4.4, and 7 kb; A_2A_3, which produces bands of 1.1, 2.6, 3.3, and 7 kb; and A_2A_4, which produces bands of 1.1, 2.6, 3.3, and 4.4 kb.

25. The probe reveals three distinct bands in this pedigree: a high band, an intermediate band, and a low band, which correspond to three different alleles. Individuals I-1, I-2, II-1, and II-4 are affected, and they all have the high band. In addition, this band is absent in individuals II-2 and II-3 who are unaffected. Therefore, the high band is associated with the disorder.

Let's designate the high band, intermediate band, and low band as RFLP alleles A, B, and C, respectively. The genotypes of the individuals are as follows: I-1 = AC (heterozygote), I-2 = AB (heterozygote), II-1 = AA (homozygote), II-2 = BC (heterozygote), II-3 = BC (heterozygote), and II-4 = AC (heterozygote).

26. **a.** Because the gene for color blindness is on the X chromosome, females will have two alleles for the gene (and can be homozygous or heterozygous), while males will only have one (and are said to be hemizygous). This explains the presence of two bands in females and only one in males. Carla shows only one thick band because she is a homozygote.

   b. Ethan is the only color-blind individual in the family. Therefore, the low band that he displays is the one associated with the disorder.

   c. Ethan must have inherited the color-blind allele from his mother, Beatrice, who is a carrier. Carla must be a carrier as well because she inherited the color-blind allele (low band) from her mother, and the normal allele from her father (also a low band). Dora, however, is not a carrier. She must have inherited the high band, with a normal allele, from her mother. Her father contributed the low band and a normal allele.

27. If the normal DNA is cut with *Pvu*II, five restriction fragments would be expected. The order from left to right would be 400, 200, 800, 50, and 550 bp. All three individuals show less than five fragments and therefore exhibit polymorphisms for this region. Individual I shows a total of four fragments, three of which (400, 200, and 550 bp) are expected. The fourth fragment, which is 850 bp, corresponds to 800 + 50 in the normal DNA sequence. Therefore, in individual I, site 4 on the map has changed so that *Pvu*II does not cut at it anymore. Individual II shows two bands: a thin one of 800 bp and a thick one of 600 bp. The bottom band must correspond to two different 600-bp fragments. The 800-bp fragment is expected, but what about the two 600-bp fragments? Looking at the map, we can see how these two fragments can arise. In individual II, sites 2 and 5 on the map have changed so that they are unrecognizable by *Pvu*II. The fragment in this individual will be 400 + 200 = 600 bp, 800 bp, and 50 + 550 = 600 bp. Individual III shows three fragments, two of which (200 and 400 bp) are expected. The third fragment is 1300 bp. If sites 4 and 5 on the original map are changed, the resulting fragment would be 1400 bp (800 + 50 + 550 bp). Therefore, individual III also carries a 1-kb deletion. Indeed, it is very likely that this deletion includes restriction sites 4 and 5 on the original map.

28. The probability that an individual has these five alleles can be obtained by multiplying together the frequencies of each allele in the general population: $0.03 \times 0.11 \times 0.05 \times 0.07 \times 0.57 = 0.0000066$. The odds that the suspect is innocent are 1 in 151,515 (the reciprocal of the probability).

29. The number of possible banding patterns is determined by the various polymorphisms present. With such a large population, it is highly likely that all possible polymorphisms are found. We can list them all, using a systematic approach. Let's first designate the E sites as $E1$, $E2$, and $E3$, from left to right.

DNA Polymorphism	Fragments Generated by *Eco*RI Digestion (kb)	Fragments Detected by Probe (kb)
All three sites present	4, 6, 2, 3	6, 2, 3
Site $E1$ missing	10, 2, 3	10, 2, 3
Site $E2$ missing	4, 8, 3	8, 3
Site $E3$ missing	4, 6, 5	6, 5
Sites $E1$ and $E2$ missing	12, 3	12, 3
Sites $E1$ and $E3$ missing	10, 5	10, 5
Sites $E2$ and $E3$ missing	4, 11	11
All three sites missing	15	15

30. **a.** Because the DNA is digested with *Xho*I only, the *Acc*I restriction sites are irrelevant. Individuals I and II have all three X sites, and so *Xho*I digestion will produce three DNA fragments: one of unknown size (at the left end of the map), 7 kb, and 5 kb. The genomic DNA region that is homologous to the probe overlaps the last two fragments. Therefore, the probe will hybridize to the 7-kb and 5-kb fragments. Individual III lacks site X'', and so *Xho*I digestion will produce only two DNA fragments: one of unknown size and another 12 kb (7 + 5). The probe will hybridize only to the 12-kb fragment.

b. Individual I has all six restriction sites. Digestion by both *Xho*I and *Acc*I will produce five DNA fragments: 3, 2, 5, 1, and 4 kb (from left to right). The genomic DNA region that is homologous to the probe overlaps the last three fragments. Therefore, the probe will hybridize to the 5-, 1-, and 4-kb fragments. Individual II lacks the A''' site, and so the double digestion will produce four DNA fragments: 3, 2, 5, and another 5 kb (1 + 4). The probe will hybridize to both 5-kb fragments. Individual III lacks the X'' site, and so the double digestion will produce four DNA fragments: 3, 2, 6 (5 + 1), and 4 kb. The probe will hybridize to the 6- and 4-kb fragments.

31. There are four possible alleles for a coding SNP: A, C, G, or T. At a single locus, these four alleles can combine to form 10 different genotypes: 4 that are homozygous (AA, CC, GG, TT) and 6 that are heterozygous (AC, AG, AT, CG, CT, and GT).

Therefore, for two coding SNPs of two different genes, there would be $10 \times 10 = 100$ different genotypes.

32. a. There are four unaffected children, so the father must be heterozygous. The mother and all unaffected children will be homozygous recessive. All affected children will be heterozygous.

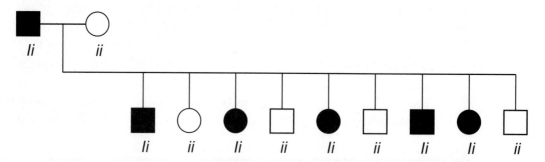

b. For microsatellite 1, the affected father has PCR products A and B. He had five affected children: two received the A fragment and three received the B fragment. Furthermore, half of the unaffected children received "A" and the other half "B." Therefore, we cannot detect any linkage between microsatellite 1 and the *illness* locus.

A completely different situation is apparent with microsatellite 2. All the unaffected children inherited PCR product A from their father, and four of the five affected children inherited PCR product C from their father. The first child (II-1) is affected but possesses PCR product A. This result suggests that eight of the nine children inherited one of the two paternal chromosomes, and child II-1 must have inherited a recombinant chromosome.

c. The lod score = log (probability of linkage at *X* map units/probability of no linkage). If the *I* locus and microsatellite 2 are unlinked, then the genotype of each child occurs at a probability of 1/4 due to independent assortment. Therefore, the probability of the specific outcomes for all nine children is $(1/4)^9 = 0.0000038$. At a distance of 12 map units between the two loci, recombinant gametes would occur at a frequency of 12% and nonrecombinant (parental) gametes at a frequency of 88%. Each of the two recombinant gametes would therefore occur at a frequency of $12/2 = 6\%$, and each of the two parental gametes at $88/2 = 44\%$. The probability of eight nonrecombinant gametes and one recombinant gamete is $(0.44)^8 (0.06)^1 = 0.0000843$.

Therefore, the lod score = log $(0.0000843/0.0000038)$ = log 22.18 = 1.346.

33. For several years the premise seemed reasonable. Scientists began to extract DNA from fossils, including the DNA from termites and other organisms in amber. The DNA was then amplified with the polymerase chain reaction. Fossil DNA in excess of 30 million years old was reported. However, recently, scientists have cast doubt on these claims. DNA begins to break down immediately on the death of the organism from various processes; the possibility of DNA millions of years old and still intact is now considered highly unlikely, although one group claims that DNA preserved in amber retains its integrity much longer than under other preservation processes. The DNA results reported for the fossil DNA have been ascribed to modern-day contaminants and to artifacts of the polymerase chain reaction.

34. a. By looking for regions of overlap, the following map can be produced. Note that because partial digests are used to generate the library, fragments at the end of a clone may overlap only partially with a fragment in the next clone.

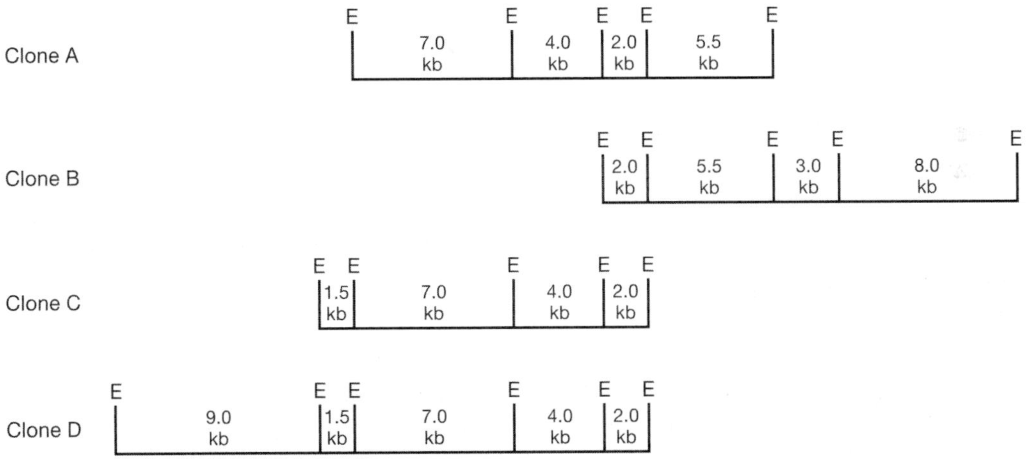

b. The minimum *Eco*RI fragment that is hybridizing to the 1.0-kb cDNA is the 2.0-kb fragment. This is the only fragment found in all four clones.

c. To extend the walk, the end fragments should be used. The 3.0- or 8.0-kb fragment from clone B would extend the walk in one direction. To extend the walk in the other direction, the 9.0-kb fragment from clone D should be used.

d. There is not enough information provided to determine which direction is toward the centromere and which is toward the telomere.

CHAPTER INTEGRATION PROBLEM

a. Yodeling cannot be a Y-linked trait because there are female yodelers. It cannot be inherited in an autosomal recessive or X-linked recessive manner because two affected individuals, II-6 and II-7, produce an unaffected child, III-4. X-linked dominant inheritance can also be excluded because an affected male, I-1, produced an unaffected daughter, II-4. Therefore, all modes of inheritance are excluded, except for autosomal dominance.

b. Let's first consider their affected daughter (individual II-4). Her chromosomal arrangement A3B1 cannot be associated with the yodeling locus because it is also found in her sister (II-5) who is a nonyodeler. Therefore, the affected daughter must have inherited A3B1 from her mother and, consequently, A2B3 (which is associated with yodeling) from her father. It follows that the unaffected daughter must have received arrangement A1B4 from her father. Only one arrangement, A4B2, is given for the affected son (II-6). Since we have established that A2B3 is associated with yodeling, this son should also have this chromosomal arrangement, and so he must have inherited A4B2 from his mother. Therefore, the linkage arrangement in the parents is as follows: A2 B3 Y/A1 B4 y (father) and A3 B1 y/A4 B2 y (mother).

c. Individuals II-2 and II-3 (unaffected sons) must have received A1B4 from their father. We cannot determine the chromosome that II-3 received from his mother, but we can for his brother. The marriage of II-2 and II-1 produced a child (III-2) that is homozygous for A4B2, and so each of the parents must possess that A4B2 chromosome. The older child (III-1) received the A1B4 chromosome from his father, and consequently the A3B1 from his mother. Now let's move to the couple that is affected. II-6 and II-7 are assumed to have identical genotypes: A4B2/A2B3. They have one unaffected daughter (III-4), so she must be homozygous for A4B2. Their affected son (III-3) must have at least one A2B3 chromosome; however, we cannot tell for sure what the other chromosome is. Their daughter (III-5) must carry the A2B3 chromosome (she is affected) and the A4B2 chromosome (she has children of her own that are unaffected). The marriage of III-5 and III-6 produced an unaffected daughter who is assumed to be homozygous (A4B2), and so the father must carry the A4B2 chromosome. He also must have the A1B4 chromosome found in his unaffected son (IV-1). Finally, we can tell that the affected son IV-3 carries the A2B3 chromosome; however, we cannot establish the identity of his other chromosome.

To summarize, here are the genotypes:

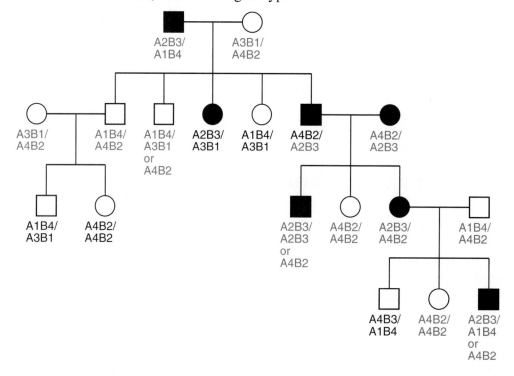

d. Individual IV-1 has an unusual genotype. He carries the chromosomal combination A4B3, which is not found elsewhere in the pedigree. In addition, the B3 locus is associated with yodeling, yet IV-1 is yodeling-challenged! Therefore, the arrangement on that particular chromosome is A4B3y.

e. The genetic mechanism underlying IV-1's unusual genotype is crossing over during meiosis I in his mother. Indeed, a double crossover event allows the coming together of alleles A4, B3, and y.

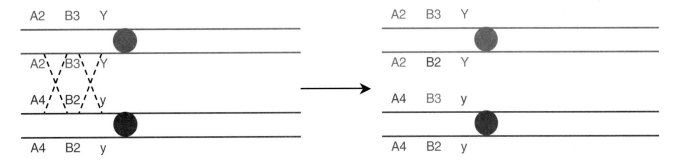

f. Begin by writing the RNA that could be transcribed from each strand. Since the DNA represents the beginning portion of the gene, the RNA must have an AUG to start protein synthesis. Transcription from the bottom strand produces an mRNA with AUGs in both directions, but in each case the AUG is followed by a short open reading frame and a termination signal: UAA (from left to right) and UAG (from right to left). Thus the bottom strand cannot produce the Yodeling mRNA. The mRNA transcribed from the top strand also contains AUGs in both directions. The one going from left to right is followed by a UGA stop signal only six codons later. However, the AUG going from right to left (positions 7–9) starts an open reading frame that extends beyond the given DNA fragment. Therefore, the top strand is the template strand and its right end must be 3'. The polarity of the DNA segment will be:

```
5'-ACACGGTATG GATCATACGG ATTGCTAATG ACTGTGATAG GCATGTACCA-3'
3'-TGTGCCATAC CTAGTATGCC TAACGATTAC TGACGGCGAT CGTACATGGT-5'
```

The coding DNA sequence and the first 15 amino acids of the Yodeling protein are:

```
5'-ATG CTA GCG GCA GTC ATT AGC AAT CCG TAT GAT CCA TAC CGT GT-3'
   Met Leu Ala Ala Val Ile Ser Asn Pro Tyr Asp Pro Tyr Arg Val
```

g. There are two codons that specify histidine: CAC and CAT. The asparagine codon is AAT, and so a change of the first A to a C would generate a histidine codon. This mismatched base should fall in the middle of the oligonucleotide sequence:

```
5'-AGTCATTAGCCATCCGTATGA-3'
```

h. There are three nonsense codons: TAA, TAG, and TGA. An inspection of the coding DNA sequence reveals that there are only two codons that are one base off the nonsense codons: TAT and TAC, both of which encode tyrosine. The TAC codon falls toward the end of the 50-bp region of the given DNA. Because the oligonucleotide is 21 bases long and the mismatched base should be in the middle of its sequence, the TAC codon cannot be the target of the mismatch: there is not enough DNA provided to figure out the sequence. So we should focus our attention on the TAT codon. If the second T is changed to an A or a G, a nonsense codon will be formed. Therefore, two different oligonucleotide sequences are possible:

         5'-GCAATCCGTA<u>A</u>GATCCATACC-3'

or   5'-GCAATCCGTA<u>G</u>GATCCATACC-3'

CHAPTER 14 GENOMICS AND BIOINFORMATICS

Chapter Goals
1. Describe the two different approaches to sequence a genome.
2. Discuss how the different genetic maps (sequence, linkage, cytogenetic) correlate with one another and why this is important.
3. Describe the methods used to annotate the genomic DNA sequence.
4. What is a microarray, how do you use it, and what information does it provide?
5. What is two-dimensional gel electrophoresis, how do you perform it, and what information does it provide?
6. Outline a yeast two-hybrid experiment, how do you perform it, and what information does it provide?
7. Compare and define genomics, transcriptome, proteomics, and interactome.

Key Chapter Concepts
14.1 Approaches to Sequencing a Genome
- **Benefits of Sequences:** Knowing the sequence of the entire genome of an organism is extremely useful for science and society. It helps us understand why an organism looks different than another organism, discover the organisms' evolutionary relationships to each other, elucidate the origins of life, discover new genes and genes that cause diseases, and clone and manipulate genes in order to understand diseases and potentially develop cures. Two common methods exist for the sequencing of a genome, ordered contig sequencing and shotgun sequencing. The latter method was used to sequence the human genome. A **draft sequence** comprises a DNA sequence that still contains many gaps between scaffolds of sequence and may still have some errors since individual nucleotides may have only been found in three to five independent **reads**. Still, this sequence is very useful as it is. In a **finished sequence,** a lot more work is required to fill in these gaps and 10 reads per nucleotide are needed in order to be confident that each nucleotide is correct.
- **Ordered Contig Sequencing:** This method requires creating a collection of clones of very large, random fragments of the genome to be sequenced. Overlapping clones are then placed in order by the use of several mapping techniques. Any redundant clones are thrown out, and the nonredundant clones form a **minimal tiling path**. Each clone is subcloned and then sequenced. This method minimizes the number of sequencing reactions and simplifies the assembly of the final sequence because redundant DNA need not be sequenced and the order is known. It, however, requires vast amounts of work to order the clones prior to sequencing.
- **Shotgun Sequencing:** This method also requires creating a collection of clones of random fragments of the genome to be sequenced; however, these clones are not ordered before sequencing. One **read** is performed from each end of every clone. Therefore, all sequencing can be performed using only two primers that anneal to vector sequence. Because which DNA fragments are sequenced is random, some regions may be sequenced several times while other regions are never sequenced. This leaves gaps between long stretches of known sequence. These gaps are filled in by several techniques including using the **paired-sequence reads** and primer walking. The assembly of all the sequences requires computers to align overlapping sequences to form a **sequence contig**.

14.2 Correlating Various Maps
- **Benefits of Correlating Maps:** Before the sequence of a genome is determined, various genomic maps will undoubtedly have already been generated. It is important to correlate the information in these maps with the newly determined genomic sequence. These maps can be helpful both for the assembly of the sequence to determine which contigs are connected and to determine which sequences correspond to which chromosomes. Importantly, these various maps enable our understanding of the genomic sequence, making it more useful. The information in these maps may help to distinguish gene sequences from noncoding or regulatory sequences as well as to identify genes that cause disease when disrupted.
- **Types of Maps:** Some of the following maps are actual physical maps that directly correlate with distances in a genomic sequence. Others are theoretical maps that represent relative distances that

do not always directly correlate with physical distances. Still theoretical maps do effectively
define order and chromosome location.

- o **Linkage Maps**: Determined by recombination frequencies, these maps demonstrate the
  order of and approximate distances between markers although they do not always *directly*
  correlate with distance. Maps of VNTRs, microsatellites, and SNPs provide much higher
  resolution than mutation maps and thus provide an excellent way to find the location of
  mutant genes that cause diseases.
- o **Cytogenetic and Chromosome Rearrangement Maps:** These maps correlate with exact
  distances between markers. By looking at what genes reside in the DNA sequence near
  the position where a cytogenetic map has identified the breakpoint of a chromosomal
  rearrangement that causes a disease, it is possible to find the culprit gene whose
  misregulation causes the disease.
- o **Expressed-Sequence Tags:** An **expressed-sequence tag (EST)** is a sequence discovered
  by sequencing random cDNAs from an organism. They are useful because they indicate
  that a region of the genome with identical or complementary sequence must be expressed.
  Therefore, the collection of ESTs helps researchers to determine what sequences in the
  genome correspond to genes.

14.3 Annotating the Genomic Sequence

- **Overview:** Determining what DNA sequences encode genes and what sequences simply lie
  between genes is sometimes quite difficult. In higher eukaryotes it is even more challenging
  because the coding sequence of most genes is interrupted by many long introns. Computer
  programs have been developed to search for *probable* gene sequences in newly sequenced
  genomes.
- **Methods:** The first step to annotating a genome is to identify all ORFs. Those that encompass an
  EST sequence are most likely to be genes. Then, surrounding gene landmarks, such as sequences
  that promote transcription, translation, or splicing, are sought since actual genes must have these
  sequences. Finally, degenerate codon usage is determined in the ORF and compared to codon
  biases typically found in that organism. For example, some organisms tend to use one or two
  codons of unmixed families of codons more frequently than the other two codons for a particular
  amino acid such as glycine. An ORF that does not follow the codon bias of that organism is not
  likely to be an actual expressed gene.

14.4 Transcriptomics

- **Overview:** Since the complex phenotypes of organisms and diseases often result from the
  collaboration of many alleles, it is important to consider all the genes expressed in an organism or
  tissue rather than just a single gene. Since testing thousands of genes individually would take too
  long, methods have been developed to analyze the expression of all the genes of an organism
  simultaneously. These methods can then determine (at the level of gene expression) why two
  tissues are different from each other. The two tissues may be different states of the same tissue or
  different tissues in an organism.
- **Methods**
  - o **SAGE Analysis**: In a **serial analysis of gene expression (SAGE)** experiment, a
    collection of cDNAs is made from different samples and the amount of each mRNA
    present is determined relative to the amount in the other sample. This is accomplished by
    cleaving off and saving an 11 nucleotide **TAG** from every cDNA produced. These
    TAGS are then linked together, cloned, and sequenced. The more frequently a TAG
    corresponding to a particular mRNA appears, the greater the mRNAs level of expression.
  - o **Microarray Analysis:** A **microarray** or **gene chip** experiment also seeks to determine
    the relative amount of each mRNA in two different samples. In these experiments,
    however, cloning and sequencing is not necessary. Instead robots spot tiny amounts of
    oligonucleotides (or sometimes cDNAs) onto a glass slide. Oligonucleotides
    complementary to each gene to be analyzed are placed in unique, known locations on the
    slide. Then fluorescently labeled cDNAs are prepared from the samples to be analyzed.
    The cDNAs of each sample are each labeled with a different color (usually red and
    green). The labeled cDNAs are then allowed to hybridize to the sequences on the slide,

and nonhybridizing cDNAs are washed away. The amount of each color present on a gene spot correlates with the amount of its mRNA present in the respective sample.

14.5 Proteomics

- **Overview: Transcriptomics** attempts to understand gene expression simply by measuring the amount of each mRNA produced in a sample. Phenotypes, however, usually require proteins and posttranslational modifications of proteins. Therefore, the most direct measure of gene expression and how it affects phenotypes requires analysis of all the proteins produced, their activity, and their interactions and modifications. This field is known as **proteomics**.

- **Methods**
    - **Two-Dimensional Gel Electrophoresis:** In chapter 12 you learned that **SDS gel electrophoresis** is used to separate proteins based on size. **Isoelectric focusing** gels separate proteins based on charge. To separate nearly every protein in a cell to a unique location, a two-dimensional gel is used. In the first step, proteins are separated based on charge, and then in a second step, negatively charged SDS is added to every protein to equalize the charge on each protein and then they are separated based on size. Thus, each protein will migrate to a unique spot on a gel. The protein at each spot can be identified by various techniques, and the amount of the protein (size and intensity of the spot) can be compared between different samples.
    - **Mass Spectrometry:** A **mass spectrometer** is used to determine the amino acid sequence of a protein and thus its identity. Posttranslational modifications can also be identified. For example, a single spot can be removed from a two-dimensional gel, subjected to mass spectrometry, and its identity and any modifications determined.
    - **Coimmunoprecipitation:** Antibodies are used to bind to a specific protein of interest and purify that protein in a coimmunoprecipitation experiment. Any proteins that bind to the protein of interest will copurify with the protein of interest. These proteins can then be identified by mass spectrometry as described earlier to determine what proteins in a cell interact with the protein of interest. Proteins that interact with each other are considered to be part of a complex.
    - **Yeast Two-Hybrid System:** Like coimmunoprecipitations, this assay is designed to determine what proteins interact with a protein of interest. It is based on the principle that transcription factors are composed of separable domains. The transcription factor can function as long as the two domains are at the promoter whether they are part of the same protein or not. Thus, the ORF for a protein of interest is fused to the ORF for one domain of a transcription factor that promotes DNA binding, and a library of cDNAs is fused to the ORF for the other domain. The only way that the second transcription factor domain can be brought to the promoter is if the two proteins fused to the two domains bind to each other. When the two domains are brought to a promoter, they activate transcription of a reporter gene whose transcription can easily be assessed.

14.6 Ethical Considerations

- **Overview:** The Ethical, Legal, and Social Issues (**ELSI**) Program is an American program that seeks to determine and clarify how the information gained through genetics and the recent Human Genome Project should be used by science, society, businesses, and government. Its role is to begin considerations of serious questions before we actually need to make decisions on these questions. These questions include whether determining genetic predispositions is a right and if so, whose right it is (individuals, parents, employers) and what are the legal responsibilities of an individual with a particular genetic makeup who commits a crime?

Key Terms

- genomics
- proteomics
- bioinformatics
- read
- ordered contig sequencing

- contig
- shotgun sequencing
- physical map
- sequence tagged sites (STSs)

- minimal tiling path
- paired-end reads
- sequence contig
- scaffolds
- consensus sequence

- draft sequence
- finished sequence
- in situ hybridization
- fluorescent in situ hybridization (FISH)
- expressed-sequence tag (EST)
- annotation
- codon bias

- transcriptomics
- serial analysis of gene expression (SAGE)
- linker
- TAG
- microarray
- gene chip
- promoter reporter
- SDS gel electrophoresis

- isoelectric focusing
- two-dimensional gel electrophoresis
- mass spectrometer
- yeast two-hybrid system
- fusion protein
- ELSI Program

Understanding the Key Concepts
Use the key terms or parts of their definitions to complete the following sentences.

The sequence of the human genome was determined by the (1)_____ method, and a (2)_____ that still contained many gaps was released to the public in the year 2000. The finished sequence took another 3 years to complete to ensure the (3)_____ of each base and to eliminate the (4)_____. Since moving to a finished sequence requires so much extra time and effort, science is moving toward generating a (5)_____ sequence of many different organisms rather than finished sequences of only a few organisms. These are sufficiently complete to perform analyses and will undoubtedly be improved by researchers as they work on individual genes.

To generate a sequence, (6)_____ of approximately 600 to 1000 bases are assembled into (7)_____ based on overlapping sequences. To ensure accuracy, the identity of each base should ideally be determined in 5 to 10 independent sequences. Once large scaffolds are found, separated by gaps, (8)_____ and primer walking can be used to fill in the gaps and connect the scaffolds. The next task is to determine which sequences reside on which chromosomes and (9)_____ the genome to determine which sequences encode genes. This is accomplished by looking for ORFs that are surrounded by gene landmarks, encompass (10)_____ and follow any known (11)_____ of the organism.

The field of transcriptomics uses (12)_____ and (13)_____ to determine the amount of all (14)_____ present in a sample and compares this between samples, while the field of (15)_____ compares the presence of (16)_____ between samples. Experiments such as (17)_____ allow researchers to compare the amount of protein in one sample versus another. To perform this experiment, researchers must perform (18)_____ followed by (19)_____. Each spot then indicates a different protein or a different modification of a protein. Experiments such as (20)_____ and (21)_____ allow researchers to determine which proteins (22)_____ with each other to form a complex. Once this has been determined between each pair of proteins in the proteome of an organism, we can describe the (23)_____.

Figure Analysis
Use the indicated figure or table from the textbook to answer the questions that follow.

1. **Figure 14.12**
   (a) How many genes are bound by this probe in the chromosomes depicted in figure 14.12?

   (b) In what phase of the cell cycle were the cells from which these chromosomes came?

   (c) Why do the chromosomes have two dots instead of just one?

   (d) When analyzing a genome that has recently been sequenced, what information can be gained by performing FISH?

2. **Figure 14.22**
   (a) What is the purpose of performing SAGE analysis? In other words, what are researchers trying to learn when they perform SAGE?

   (b) How is biotin used in SAGE analysis?

(c) What is the purpose of the streptavidin beads in SAGE analysis?

(d) What two roles do the linkers perform in SAGE analysis?

3. **Figure 14.32**
   (a) What proteins are precipitated in step 3?

   (b) How are the precipitated proteins identified?

   (c) How could you use this experiment to test if two specific proteins interact if you have antibodies against each of the proteins?

4. **Figure 14.35**
   (a) What are researchers trying to determine by performing this experiment?

   (b) Describe how a fusion protein is generated from the bait plasmid.

   (c) What in the lower panel indicates that the two proteins are interacting?

   (d) If a library of cDNAs is cloned into the prey plasmid, will every fusion protein correspond to a protein expressed in the organism from which the cDNA library came?

   (e) Why are we able to test the interaction between two *human* proteins by putting their DNA into a *yeast* cell?

General Questions

Q14-1. Discuss the advantages and disadvantages of shotgun sequencing versus ordered contig sequencing.

Q14-2. How are fluorescent cDNAs created during microarray analysis?

Q14-3. Briefly describe four approaches used in determining which parts of a genome correspond to real genes.
i)
ii)
iii)
iv)

Q14-4. What useful information can be gained by performing microarray experiments in which gene expression is compared between tumor tissue and normal tissue in several patients with a particular form of cancer? If a proteomic approach were just as feasible as the microarray approach, why might that approach be more informative?

Q14-5. When performing ordered contig sequencing, why must at least two different restriction digests be performed when subcloning the contigs into smaller clones to be sequenced?

Multiple Choice

For each of the following, circle the letter of the choice that most appropriately answers the question.

1.) Why is it important to know which two reads are paired with each other when performing shotgun sequencing?
   a. It helps to assemble sequence contigs near repetitive sequences found often in a genome.
   b. Paired reads usually overlap each other, thus allowing easy sequence contig assembly.
   c. It helps to identify the ends of eukaryotic chromosomes.
   d. It minimizes the number of reads that need to be compared when assembling the sequence.
   e. All of the above

2.) In a finished eukaryotic genome sequence, which of the following can be used to determine which sequence contig corresponds to which chromosome?
   a. Compare maps of known markers to the sequence contigs.
   b. Perform FISH with sequences found on the sequence contig.
   c. Sequence the telomeres.
   d. Look for chromosomal rearrangements.
   e. Both a and b

3.) Which of the following is likely to show the most similarity between humans and worms?
   a. nucleotide sequence of two related genes
   b. DNA sequence of the promoter of two related genes
   c. amino acid sequence of two related proteins
   d. nucleotide sequence of the mRNAs from two related genes
   e. entire genome sequence of the two organisms

4.) Which of the following techniques would likely generate data most similar to the data generated by SAGE analysis?
   a. DNA microarray analysis
   b. DNA sequencing
   c. two-dimensional gel electrophoresis
   d. coimmunoprecipitation followed by mass spectrometry
   e. yeast two-hybrid analysis

5.) In an isoelectric focusing gel, why do the proteins move through the gel when electric current is applied?

   a. The proteins are coated with negatively charged SDS.
   b. The proteins' modifications and amino acid side groups impart an overall negative charge on the proteins at low pH.
   c. Each protein is attracted to a specific pH.
   d. The backbone of proteins, composed of peptide bonds, is negatively charged.
   e. Proteins do not move through an isoelectric focusing gel.

6.) Which of the following methods can be used to determine all of the proteins that interact with a specific protein of interest?
   a. SAGE analysis
   b. two-dimensional gel electrophoresis
   c. DNA sequencing
   d. yeast two-hybrid analysis
   e. Both a and b

7.) From where do we get the knowledge of an expressed sequence tag (EST)?
   a. observing that ribosomes can bind to eukaryotic mRNAs before their 3′ ends have been synthesized
   b. sequencing a genome
   c. sequencing a cDNA
   d. sequencing a protein
   e. observing promoter reporters

8.) In a yeast two-hybrid analysis, what type of protein domain should be fused to the bait?
   a. transcriptional activation domain
   b. DNA binding domain
   c. fusion domain
   d. translation domain
   e. ATP binding domain

9.) Which of the following can be determined by using a tandem mass spectrometer (MS/MS)?
   a. presence of a specific protein in a sample
   b. position of a posttranslational modification within a protein
   c. codons used by each amino acid in the protein
   d. relative amount of a specific protein present in each of two different tissues
   e. Both a and b

10.) Which of the following is most often the first thing done to get an idea about the possible function of an unknown ORF found in a newly sequenced genome?

a. perform FISH to determine where it is located on a chromosome and if it lies at a breakpoint of a chromosomal rearrangement known to cause a disease

b. perform a promoter reporter assay to determine the expression pattern of the ORF

c. compare the predicted amino acid sequence of the ORF to proteins of known function in other organisms

d. perform two-dimensional gel electrophoresis to determine the expression level of the protein

e. run an isoelectric focusing gel to determine the isoelectric point of the protein

Practice Problems

P14-1. The following five sequencing reads were found in a shotgun sequencing project. Use the sequences to assemble a sequence contig.

(*Hint 1*: During shotgun DNA sequencing, *random* clones are sequenced and may correspond to either of the two strands of DNA. Look for regions of identity or complementarity between different sequences to align them.)

5'-TTTCGGGATCGATGATTACCGCTAGGCAAT-3'
5'-AAGGGGGCAGTGACAAATCACGTAATTTGACTGCCCATTGAAGGTAGTTCC-3'
5'-CTCTCATCGGATGCTTAG-3'
5'-AAGCATCCGATGAGAGTGATACGATTGCCTATAGC -3'
5'-TCACTGCCCCCTTAAAGGTAATTTCGGGATCGAT-3'

P14-2. The following bacterial DNA sequence of 200 base pairs contains a short gene. Can you find it using the common methods used to annotate genes? Notice that the sequence begins in the first row and continues in the subsequent rows. The number at the beginning of each row corresponds to the first nucleotide in that row. You are given that an expressed sequence tag has been found in this organism with the sequence 5'-ACGCCTGCTCCCCAATTTACAAG-3'.

```
  1  5'-GCCATTCATGCATTGCCAGTCATTAGGCCAGTAGCCAGCTATAGCCATAG-3'
     3'-CGGTAAGTACGTAACGGTCAGTAATCCGGTCATCGGTCGATATCGGTATC-5'

 51  5'-CTTGTAAATTGGGGAGCAGGCGTGATGAAGAGTCTCTGGGGCCTATATGC-3'
     3'-GAACATTTAACCCCTCGTCCGCAGTACTTGTCAGAGACCCCGGATATACG-5'

101  5'-GTAGGCTTGTTCCATCAACCTCCTTATGATCGCTGGCAACCATTATAACC-3'
     3'-CATCCGAACAAGGTAGTTGGAGGAATACTAGCGACCGTTGGTAATATTGG-5'

151  5'-GCCAATTCCGGCCTGACAAGAGGATTCCAGCATGCTTAGCCCAGTAACTT-3'
     3'-CGGTTAAGGCCGGACTGTTCTCCTAAGGTCGTACGAATCGGGTCATTGAA-5'
```

P14-3. A microarray experiment was performed to determine which genes are expressed differently in normal breast tissue versus cancerous breast tissue. Red-labeled cDNAs were made from a breast tumor, and green cDNAs were made from normal breast tissue surrounding the tumor. The following results were obtained on a human microarray chip. Which genes are expressed *more* highly in breast tumors? Which genes are expressed *less* in breast tumors versus normal breast tissue?

Results: GRN = green; YEL = yellow   Key: Location of genes A–N

P14-4. Two possible ORFs almost completely overlap each other on opposite strands of DNA found in the sequenced genome of *E. coli*. When you look at the codon for each glycine found in each ORF, you find the numbers of each codon given in the following table. Knowing the codon bias of *E. coli* shown in table 14.1 of your text, which ORF is more likely to be a real gene?

Codon	ORF1	ORF2
GGA	7	0
GGG	10	1
GGU	9	17
GGC	7	14

P14-5. Researchers wished to determine the identity of all the human proteins that bound to a human protein known as Cdc4. They performed a yeast two-hybrid analysis in which they screened a library of cDNAs cloned from human skin cells. The researchers mistakenly cloned the library into the bait plasmid and Cdc4 into the prey plasmid. After the analysis, they found hundreds of cDNA clones that activated expression of the reporter. After sequencing the positive clones, the majority of them turned out to be transcription factors. Do these transcription factors really interact with Cdc4, or did something go wrong in the experiment and the transcription factors represent false-positives? What is the most likely explanation for the identification of so many transcription factors? If the cDNA library would have been cloned into the prey plasmid and Cdc4 into the bait plasmid, what other types of proteins might express the reporter but not bind to Cdc4?

Assessing Your Knowledge

Understanding the Key Concepts—Answers

1.) shotgun sequencing; 2.) draft sequence; 3.) accuracy; 4.) gaps; 5.) draft; 6.) reads; 7.) sequence contigs; 8.) paired-end reads; 9.) annotate; 10.) expressed-sequence tags or ESTs; 11.) codon bias; 12.) serial analysis of gene expression or SAGE; 13.) microarrays or gene chips; 14.) mRNAs; 15.) proteomics; 16.) proteins; 17.) two-dimensional gel electrophoresis; 18.) isoelectric focusing; 19.) SDS gel electrophoresis; 20.) coimmunoprecipitation; 21.) yeast two-hybrid analysis; 22.) interact or bind; 23.) interactome.

Figure Analysis—Answers

1a.) Four copies of one gene are seen in this analysis; b.) Mitosis; c.) Each chromosome is composed of two identical sister chromatids since the cell was in mitosis and had already replicated its DNA; d.) It can be determined on which chromosome a particular sequence contig or scaffold resides.

2a.) Researchers use SAGE analysis to determine the relative expression level of all genes in a particular tissue. This analysis can compare two different tissues or two different states of one tissue; b.) Biotin is a natural molecule that binds very tightly to a protein known as streptavidin. The oligo-dT primer used to synthesize cDNA has biotin attached to it. Thus, all of the cDNAs synthesized can easily be attached to beads coated with the streptavidin protein; c.) Binding of the cDNAs to beads through the biotin–streptavidin interaction allows researchers to retain the ends of the cDNAs and wash away the fragment of the cDNA that is not wanted; d.) The linkers provide a restriction enzyme recognition sequence that allows a cut to be made 14 bases away (11 bases into the cDNA), thus generating the 11-base TAG. They also contain a sequence necessary to anneal primers to for performing PCR in a later step.

3a.) Any protein that either binds to the antibody or binds to a protein that binds the antibody will be precipitated; b.) The precipitated proteins are subjected to mass spectrometry which will determine the amino acid sequence of short peptides generated from these proteins. The sequence of these peptides identifies the proteins that they came from; c.) You could use one antibody to precipitate the first protein. If the two proteins interact, then the second protein should coprecipitate and be present in the precipitate. Its presence can be detected by performing a Western blot (discussed in chapter 12) of the precipitate using antibodies against the second protein. If it is present, you can conclude that the two proteins interact. A necessary control would be to demonstrate that the second protein does not precipitate if the first protein is absent. This would rule out the possibility that the second protein binds to the precipitating antibody instead of the first protein.

4a.) The yeast two-hybrid experiment determines whether two proteins bind to each other. Alternatively, it can be used to determine all of the proteins that bind to a protein of interest; b.) Two ORFs are placed in line with each other (no stop codon after the first ORF and no start codon at the beginning of the second ORF) on a plasmid such that a continuous reading frame is generated. The reading frame of the first ORF must be maintained into the second ORF. Transcription generates a single mRNA that is translated into a single protein containing amino acid sequences of both proteins fused together; c.) A blue precipitate is produced by the enzyme produced from the reporter gene. Thus, blue cells indicate that the two proteins interact and the reporter gene is thus being expressed; d.) Not every cDNA clone will produce a fusion protein that corresponds to a protein in the organism from which the cDNA came. The cDNA must be in the correct orientation relative to the transcriptional activation domain ORF, and the reading frame must be maintained. Most of the fusion proteins will in fact not satisfy these criteria and will produce nonsense proteins. These will not interact and will be ignored; e.) The genetic code is universal. Therefore, a human gene that is transcribed and translated in yeast will make the same amino acid sequence that it would if it were made in human cells.

General Questions—Answers

Q14-1. **Advantages of shotgun sequencing:** Less effort is needed to order the contigs by mapping different genes and markers prior to sequencing making it less time-consuming. Thus, since there are less steps involved, it is more easily amenable to automation. Also, a meaningful sequence is generated immediately upon beginning the process, while there is a significant delay in ordered contig sequencing. **Disadvantages of shotgun sequencing:** Assembling the sequence reads into sequence contigs is more difficult especially because of repeat sequences scattered throughout the genome. Since the clone to be sequenced is random, many more sequencing reads must be performed and some sequences will be sequenced many times in order to reach a point where all the sequence has been sequenced at least once.

Q14-2. Oligo-dT primers are annealed to the poly(A) tail of all the eukaryotic mRNAs, and reverse transcriptase creates a cDNA using three nonlabeled deoxynucleotides and one deoxynucleotide that has a fluorescent label attached to it.

Q14-3. i) You can look for ORFs that are realistically long enough to make a protein.
ii) You should be sure that the codons in the ORF fit with any codon biases known in that organism.
iii) You can look for the presence of gene landmark consensus sequences such as promoter elements, translation regulatory sequences, or posttranslational modification consensus sequences. These should be positioned in appropriate locations relative to the start and stop codons of the ORF.

iv) The presence of one or more ESTs located near the beginning or the end of the ORF supports the idea that the ORF is a real gene.

Q14-4. This experiment will demonstrate which genes are expressed inappropriately in tumors, aiding in our understanding of how cancer develops. It will also describe different expression patterns in different tumors. If the treatment and outcome of each of these individuals is monitored, it can be determined which expression pattern responds best to which treatment. Then when a new person is diagnosed with the cancer, the tumor's gene expression profile can be determined in order to know how best to treat the person. The best method of profiling a tumor would involve a proteomic approach where it is determined what proteins and what posttranslational modifications are altered. This is only inferred by microarray analysis. Just because an mRNA is more abundant, however, does not necessarily mean that more active protein is made from it. Similarly, no change in the levels of an mRNA does not mean that the activity of the encoded protein is not changed since regulation of the protein may be altered. Unfortunately, such a simple, comprehensive proteomic approach does not currently exist.

Q14-5. If only one restriction digest were performed, then sequencing reads from every subclone would end in the restriction enzyme recognition sequence and contain no overlapping sequence with other subclones. Thus, the order of the subclones could not be determined. Creating two or more sets of subclones produces sequence reads with overlapping sequences.

Multiple Choice—Answers

1.) a; 2.) e; 3.) c; 4.) a; 5.) b; 6.) d; 7.) c; 8.) b; 9.) e; 10.) c.

Practice Problems—Answers

P14-1.
5′-GGAACTACCTTCAATGGGCAGTCAAATTACGTGATTTGTCACTGCCCCCTTAAAGGT
AATTTCGGGATCGAT GATTACCGCTAGGCAATCGTATCACTCTCATCGGATGCTTAG-3′

P14-2. Begin by locating the DNA that matches the EST. This sequence will be contained within the ORF. Look for any ORFs that contain this sequence, and then look for appropriate regulatory sequences surrounding this ORF. The ORF for the gene producing this EST is highlighted in gray as are the −10 and −35 sequences that promote its transcription. The initiation and termination codons are bolded and the Shine–Dalgarno sequence and the first nucleotide of the mRNA are shown in bold italic.

```
5′-GCCATTCATGCATTGCCAGTCATTAGGCCAGTAGCCAGCTATAGCCATAG-3′
3′-CGGTAAGTACGTAACGGTCAGTAATCCGGTCATCGGTCGATATCGGTATC-5′

5′-CTTGTAAATTGGGGAGCAGGCGTGATGAAGAGTCTCTGGGGCCTATATGC-3′
3′-GAACATTTAACCCCTCGTCCGCAGTACTTGTCAGAGACCCCGGATATACG-5′

5′-GTAGGCTTGTTCCATCAACCTCCTTATGATCGCTGGCAACCATTATAACC-3′
3′-CATCCGAACAAGGTAGTTGGAGGAATACTAGCGACCGTTGGTAATATTGG-5′

5′-GCCAATTCCGGCCTGACAAGAGGATTCCAGCATGCTTAGCCCAGTAACTT-3′
3′-CGGTTAAGGCCGGACTGTTCTCCTAAGGTCGTACGAATCGGGTCATTGAA-5′
```

P14-3. **Higher in breast tumor:** Genes H, D, M
      Lower in breast tumor: Genes A, I, E, L

P14-4. ORF2 is likely to be an actual gene. This is because its glycine codon usage is similar to the average glycine codon usage in all of *E. coli*'s genes. ORF1 has a very different distribution of glycine codons.

P14-5. The cDNA library should be cloned into the prey plasmid where the proteins would be fused to the transcriptional activation domain. Many proteins exist that can activate transcription. When they are fused to the DNA binding domain, in the bait plasmid, they will express the reporter regardless of whether it binds Cdc4 or not. These transcription factors most likely expressed the reporter because they have their own transcriptional activation domain and the GAL4 DNA binding domain brought them to the promoter of the reporter. In the correct orientation (cDNA library in the prey plasmid), very few proteins will naturally have a DNA binding domain that recognizes the promoter of the reporter. Therefore, almost all proteins will require binding to Cdc4 in order to be brought to the promoter. Still, false-positives can result. Any protein that binds to the GAL4 DNA binding domain itself will express the reporter as will any proteins that can bind the DNA found in the promoter of the reporter. These false-positives are usually identified by screening the GAL4 DNA binding domain in the absence of fused Cdc4 as a control.

14

GENOMICS AND BIOINFORMATICS

CHAPTER SUMMARY QUESTIONS

1. 1. i, 2. g, 3. j, 4. b, 5. a, 6. d, 7. e, 8. c, 9. f, 10. h.

2. Genomics is the study of mapping and sequencing genomes. Proteomics is the study of all proteins in an organism. Genomics deals with the genomic DNA sequence of organisms, which is the same for all cells within an organism. Proteomics is based on which proteins are expressed; different tissues express different subsets of proteins.

3. The DNA sequence alone is not very useful. Linkage maps help to provide landmarks, or points of reference, within the sequence. This is helpful for locating the positions of genes within the sequence.

4. The assembly of repetitive DNA sequences is a significant problem for shotgun sequencing because the sequences were not localized to a physical map. The problem was overcome by generating scaffolds, sequence contigs joined by known gaps. By sequencing the clones linking the contigs up to the point of the repetitive sequence, the size of the gap could be determined. Knowing the size of the gap and the size of the repeat sequence, it is straightforward to calculate the number of copies of the repeat within a gap.

5. FISH stands for fluorescent in situ hybridization. In this technique, fluorescent labels are attached to a nucleic acid probe, which is then hybridized to metaphase chromosomes. FISH allows the identification of the chromosomal location of the DNA sequence complementary to the probe. Use of this technique allows the linkage and sequence maps to be correlated to the cytological map of the chromosome.

6. ESTs, or expressed sequence tags, are valuable in genomic mapping because they serve as landmarks that identify the location of a gene within the genomic DNA sequence. ESTs are portions of cDNA sequences but are not always useful in identifying ORFs in the genome because they may be derived from untranslated portions of the gene.

7. The annotation of the genome involves trying to identify all of the genes and their corresponding ORFs for translation. It is important to allow understanding of the function of the genome.

8. Alternative splicing is a situation in which the exons of a single gene are differentially incorporated or removed during RNA processing to produce different mRNAs. It complicates genomic analysis because there may be more than one mRNA derived from a single gene. When a gene is alternatively spliced, it may be necessary to isolate multiple cDNAs to account for all of the exons of the gene.

9. A sequence "read" is the DNA sequence derived from a single oligonucleotide primer and is typically 600–1000 bases of sequence. A contig is the compiled sequence of multiple, overlapping sequence reads.

10. In the ordered contig sequencing technique, genomic clones are identified and arranged in an overlapping series. The minimal set of overlapping clones to represent the entire genome is selected for sequencing, and the mapping information is used to aid assembly of the sequence data. In the shotgun sequence technique, random clones are selected for sequencing, and the sequence data from the random clones are assembled into a contiguous DNA sequence. Ordered contig sequencing requires additional effort to create a map of the clones; however, assembly of the sequence data is easier because the placement of each clone in the physical map is known. The shotgun sequencing technique is faster because time is not spent creating a physical map; however, assembly of the sequence DNA is more difficult without a map.

11. Tags used to order contigs include known genes, microsatellites, and sequence-tagged sites. Known genes can be used as probes for Southern hybridization very efficiently. Microsatellites can be used in PCR amplification. Sequence-tagged sites are generated from the ends of genomic clones. These types of tags are all informative and provide the most information when used in combination.

12. A genome project is "complete" when it contains no gaps and is as accurate as possible. This usually requires a minimum of 3 to 5 reads for several nucleotide positions for a "draft" sequence and 10 reads for a "finished" sequence. It is not necessary to know every nucleotide of an organism's genome to end a project. Scientists are increasingly satisfied with "draft" sequences because these sequences are generally sufficient for researchers to identify genes and begin analyses. There is a greater emphasis now on obtaining draft sequences from a large number of organisms.

EXERCISES AND PROBLEMS

13. Based on the gel pattern shown, clones 1, 3, and 4 are likely to be related. These fragments share restriction fragments that are the same size, suggesting that these clones overlap. Clone 2 does not share restriction fragments with the other clones; therefore, it is unrelated.

14. To assemble sequencing runs into a sequencing contig, look for overlapping sequences. For example, the first two sequences can be aligned as follows:

    CTGGGTGGAAACCCATGTTGGGGAAGCACCCGGGGCCCTTTTTC
                        GGAAGCACCCGGGGCCCTTTTTC . . . . . .

    The same strategy can be used to align the remaining sequences. The sequence contig is

    CGGCAGGCAGTCTGGGTGGAAACCCATGTTGGGGAAGCACCCGGGGCCCTTTTTCC
    TCTATCTCTATCGTGGACCTGCCTTTTCCGAACCCC

15. These three runs can be aligned as follows:

    GAGTGACGAGACCTATAGGATCCTGGATAGGCTGAACAACCAGAGTGCCGCTGATT
    GCCCTACATGGGCAAGGATAATGGCGGCAAGGGTGG

    Additional DNA sequence needs to be generated to resolve discrepancies in the runs. The underlined positions represent nucleotides that are a G in one sequence and a C in another sequence. Additional data are required to determine which sequence is correct.
        If this DNA sequence corresponds to part of an open reading frame, it is written in the 5' → 3' direction. Two start codons are present in the forward direction; neither of these is followed by a stop codon. In the reverse direction, one start codon is present (TAC in this direction), but it is followed immediately by a stop codon.

16. 3×10^9 bases/60,000,000 bases per day = 50 days. Theoretically, this is the fastest possible time required for sequencing the human genome at this rate. However, this number only allows single coverage of the genomic sequence. When random clones are used for sequencing, it is necessary to sequence several times more than the number of nucleotides in the genome to guarantee complete coverage.

17. The normal karyotype for *Fauna hypotheticus* is 28, XY. Therefore, the organism would have 26 autosomes (13 pairs) plus the two sex chromosomes X and Y. The nuclear genome refers to the 13 autosomes (one of each pair) plus the X and Y chromosomes. Assuming that these 15 chromosomes are of equal length, the entire genome could be sequenced in
    $$(2 \times 10^9 \text{ bases})/(700 \text{ bases per read}) = 2.86 \times 10^6 \text{ reads}$$

$(2.86 \times 10^6$ reads$)/15$ chromosomes read simultaneously $= 1.9 \times 10^5$ reads per chromosome.

Reading from both chromosome ends requires $(1.9 \times 10^5$ reads$)/2 = 9.5 \times 10^4$ reads per chromosome end. Each read cycle takes one day, so 9.5×10^4 days are required. Therefore, 9.5×10^4 days$/365$ days per year $= 260$ years are needed to sequence the *Fauna hypotheticus* genome.

18. **a.** The oligonucleotide primers used in the DNA sequencing reactions bind to the vector sequences adjacent to the cloning sites. Therefore, only two different oligos are required to generate the reads from both ends of all the clones in a single vector. Since, two different vectors are used, a total of four oligos would be needed for the project.

b. A DNA sequence of $50 \times 2 \times 10^9 = 10^{11}$ base pairs has to be generated. There are 80 sequencing machines, so a total of $80 \times 700 = 56,000$ bp can be read every 2 minutes.

   The entire sequence would require $(2$ minutes $\times 10^{11}$ bp$)/(56,000$ bp$) = 3.57 \times 10^6$ minutes. A year has 60 minutes $\times$ 24 hours $\times$ 365 days $= 525,600$ minutes. Therefore, the sequencing project would require $(3.57 \times 10^6$ minutes$)/(525,600$ minutes per year$) = 6.8$ years or approximately 6 years, 9 months, and 18 days.

19. **a.** The principle in ordering the STS markers is that those that are positive in each YAC must be adjacent to each other. YAC A contains STS markers 1, 3, and 9, so these must be adjacent. However, STS 6 must be near 1 and 9 (YAC B) but not near 3, so the order of these markers can be 6, 1, 9, 3 or 6, 9, 1, 3. YAC D reveals that markers 9, 3, and 4 should be adjacent, so the order should be 6, 1, 9, 3, 4. Similar logic can be used to place STSs 8 and 2 to the right of marker 4. This leaves STSs 5 and 7. We do know from YAC E that they must be adjacent to marker 6. However, their order with respect to 6 is ambiguous: 7–5–6 or 5–7–6. Therefore, the physical map is

Please note that the map with STS 2 on the left and markers 5, 7 on the right is equally valid.

b. The YAC clones can be aligned from left to right as follows:

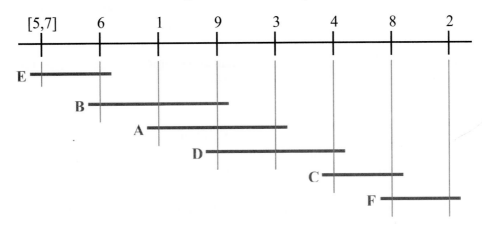

20. a. There are five N's in the middle of the sequence; each can be occupied by any of the four bases (A, C, G, T). Therefore, there are $4^5 = 1024$ different restriction sites for *Sfi*I.

b. The best enzyme would be the one that cuts the DNA most infrequently. It would produce fewer fragments, thereby making the generation of the map much easier.

   Let's start by calculating the frequency of each base. The genome has a GC content of 66%, and so the frequency of G and C is 0.66/2 = 0.33 each. Therefore, the frequency of A and T will be (1 – 0.66)/2 = 0.17 each. The probability of the occurrence of a restriction site by chance is the product of the probabilities of each base in the sequence. To estimate the number of restriction sites for each enzyme, we simply multiply the probability of the site by the genome size.

   P(GGCGCGCC) = $(0.33)^8$ = 0.000140; and so the number of expected *Asc*I restriction sites is $0.000140 \times 4.1 \times 10^6 = 576$.
   P(TTAA) = $(0.17)^4$ = 0.000835; and so the number of expected *Mse*I restriction sites is $0.000835 \times 4.1 \times 10^6 = 3423$.
   P(GGCGCC) = $(0.33)^6$ = 0.001291; and so the number of expected *Nar*I restriction sites is $0.001291 \times 4.1 \times 10^6 = 5295$.
   P(TTATAA) = $(0.17)^6$ = 0.000024; and so the number of expected *Psi*I restriction sites is $0.000024 \times 4.1 \times 10^6 = 99$.
   P(GGCCNNNNNGGCC) = $(0.33)^4(1)^5(0.33)^4 = (0.33)^8$ = 0.000140; and so the number of expected *Sfi*I restriction sites is $0.000140 \times 4.1 \times 10^6 = 576$.
   Therefore, the enzyme *Psi*I is the best choice for generating the restriction map.

c. The length of each restriction fragment is equal to the reciprocal of the probability of the occurrence of the restriction site. Thus, the length of *Asc*I restriction fragments is 1/0.000140 = 7143 bp. The number of *Mse*I restriction sites in this fragment would be $0.000835 \times 7143 = 6$. Therefore, each linear *Asc*I fragment will generate seven fragments, each of about 1020 bp.

21. Starting with larger BAC and YAC clones first makes assembly easier. All of the smaller fragments from a single BAC or YAC should be assembled into a single contig. Using several different size inserts ensures that there will be overlap among fragments. The smaller inserts can be sequenced completely, or nearly completely, in a single read; the larger 10-kb fragments are useful for identifying sequences flanking repeats.

22. The evolution of complex organisms does not require a dramatic increase in the number of genes. Rather, using alternative splicing allows organisms to become more complex and create more gene products without increasing the number of genes.

23. Many of the amino acid differences between human and chimpanzee sequences must be important for function. For example, differences in the *FOXP2* gene between humans and chimps are thought to be responsible for the development of speech.

24. If GC content is 50%, then the likelihood of obtaining any given 20-nucleotide sequence is $(1/4)^{20}$, or 9.09×10^{-13}. The human genome contains approximately 3×10^9 bases. Therefore, the expected frequency of a given 20-nucleotide sequence is $9.09 \times 10^{-13} \times 3 \times 10^9$, or 0.003 copies per genome. A 20-nucleotide sequence should therefore be sufficient to uniquely identify a gene.

25. Oligonucleotides that are 65 nucleotides long are sufficiently long to hybridize specifically to transcripts but less likely than a full-length cDNA to hybridize nonspecifically with related transcripts.

26. Some examples of comparisons include analysis of gene expression patterns in two different tissues from the same organisms to identify tissue-specific expression patterns, analysis of gene expression patterns at different developmental time points to reveal the developmental timing of gene expression, and analysis of healthy versus diseased tissue to examine gene expression patterns in disease.

27. The scientist should focus her attention on the 150 genes with either increased or decreased expression in the cancerous tissue. She should examine these genes to identify candidate genes that might be likely to play a role in the development of cancer. These candidate genes would become the focus of further studies to characterize the development of cancer and could become therapeutic targets.

28. This experiment allows identification of the DNA sequences that control expression of this gene. Experiments such as this allow scientists to understand how gene expression is controlled and to examine the effects of different treatments or conditions on gene expression. For example, a scientist might investigate the effects of drugs on gene expression, using the reporter gene to monitor expression levels.

29. **a.** A protein with a lower molecular weight in the mutant cell line is likely the result of a nonsense mutation, which would produce a premature termination codon and a truncated protein.

b. A protein with a higher molecular weight in the mutant cell line may be the result of a mutation changing the normal stop codon to an amino acid codon, causing read-through and production of a larger-than-normal protein.

c. A protein with the same molecular weight but a different isoelectric point probably contains a missense mutation resulting in substitution of an amino acid with a different charge.

d. A protein with the same molecular weight and isoelectric point but different abundance may be the result of a mutation in regulatory sequences, such as the promoter or enhancer.

e. A protein with the same molecular weight and isoelectric point but greater abundance could be the result of a mutation in a regulatory sequence that normally decreases gene expression.

30. Posttranslational modifications such as phosphorylation or glycosylation can affect protein function. Identification of these changes can help us to understand changes in protein function that result from differences in these posttranslational modifications, even in the absence of changes in sequence and expression levels.

31. The group of proteins labeled (A) differ in charge but not in molecular weight. These spots are likely due to a difference in phosphorylation, which will change the charge of the protein with insignificant effects on molecular weight. The group of proteins labeled (B) differ in molecular weight but not in charge. This group of spots represents changes in glycosylation, which will affect molecular weight but not charge. The group of proteins labeled (C) differ in both molecular weight and charge and are most likely unrelated proteins.

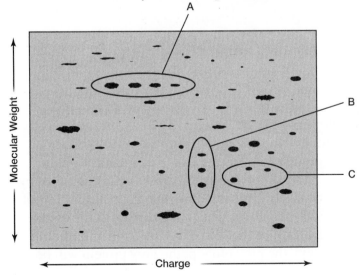

32. The changes are marked in the following figure. In treatment A, the circled protein has changed in charge but not in molecular weight, indicating that it has likely been phosphorylated in response to the treatment. In treatment B, the circled protein has increased in molecular weight, but its isoelectric point is unchanged; this indicates that the protein may have been glycosylated in response to treatment B. In treatment C, a protein spot is absent at the circled position. This indicates that the expression level of this protein has been changed in response to treatment C. This change could be indicative of a decrease in transcription of this gene or an increased rate of degradation of the protein.

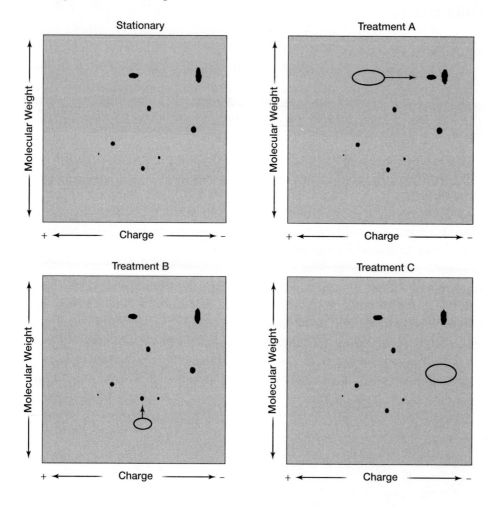

33. The yeast two-hybrid system employs two fusion proteins. The bait is a fusion protein containing the DNA-binding domain of the transcription activator Gal4p fused to a protein of interest. The prey fusion protein contains the Gal4p activation domain fused to a target protein. A reporter plasmid containing a reporter gene controlled by Gal4p is also used in this system. The Gal4p DNA-binding domain, which is part of the bait protein, interacts with the UAS$_g$ sequence, which is upstream of the reporter protein. The DNA-binding domain by itself cannot activate transcription of the reporter. If the bait protein interacts with the prey protein, the prey protein will be recruited to the reporter gene, and the activation domain of Gal4p will activate transcription of the reporter gene. Activation of the reporter gene in this system is indicative of an interaction between the bait protein and the prey protein.

The yeast two-hybrid system may fail to identify interactions if two proteins are part of a multiprotein complex. In this situation, more than two proteins may be required to form stable protein–protein interactions.

34. If one of the proteins is a transmembrane receptor, this interaction will not be detected. A protein with a transmembrane domain will be localized to the plasma membrane rather than the nucleus. The protein–protein interaction must occur in the nucleus in order to activate the reporter gene and be detected. Another situation in which the yeast two-hybrid system might fail to detect a known protein–protein interaction is if phosphorylation of one of the proteins is required for the interaction. If the required phosphorylation event does not occur in yeast, the proteins will not be able to interact.

35. The colonies that grow on the +Trp +Leu +Ura plate may or may not contain any of the vectors. This is a control to assess the viability of the colonies. On the +Trp +Ura plate, the colonies contain the prey vector (can grow without leucine). On the +Leu +Ura plate, the colonies contain the bait vector. On the +Ura plates, the colonies contain both the bait and prey vectors. On the plate with no additions, the colonies that grow contain all three vectors, and the uracil reporter gene has been activated. The two colonies growing on the plate with no additions contain candidate cDNAs that may interact with the transcription factor being studied

If TFIIA is cloned into the bait vector and TFIIB is cloned into the prey vector, colonies receiving the TFIIA bait vector should grow in either the presence or absence of tryptophan, and colonies receiving the TFIIB prey vector should grow in the presence or absence of leucine. All colonies that are able to grow on the +Ura plate should also grow on the plate with no additions if the reporter vector was successfully transformed into the yeast. Because these two proteins are known to interact, whenever both the bait and prey vector are present, the proteins should interact and activate transcription of the reporter gene, *URA3*.

36. The brother who possesses the allele associated with colon cancer has an increased risk of developing colon cancer.

37. The gene should be mapped using linkage analysis of DNA markers. Once the chromosomal location of the disease gene is known, the annotated sequence data should be examined to identify candidate genes known to map to that part of the chromosome. After candidate genes are identified, the scientist should look for mutations in the gene in patients who have the disease.

38. These three ESTs map are very close to one another in the genomic sequence and appear to hybridize to the same mRNA. It is likely that these three ESTs represent sequences derived from different exons of the same gene. To test this hypothesis, one would screen a cDNA library to isolate a full-length cDNA clone and sequence this clone. If the hypothesis is correct, the full-length cDNA clone should contain all three cDNA sequences.

39. There are likely to be many differences in the genomes of the 10 individuals with the disease and the 10 individuals who do not have the disease. Sequencing these genomes will yield many differences and is likely to be uninformative. A better approach would be to use linkage analysis to identify the chromosomal location of the gene causing this disease and then to look for candidate genes that map to that region. These candidate genes could be selected for sequencing to identify a mutation.

40. Another explanation for the results is that the two proteins interact with one another and therefore coimmunoprecipitate. One way to test this hypothesis would be to test protein–protein interactions in a yeast two-hybrid assay. Because one of the proteins is known to be a transmembrane receptor, the yeast two-hybrid construct should be designed to include the intracellular domain of the receptor but not the transmembrane domain so that the fusion protein will not be localized to the plasma membrane.

41. Begin by assigning numbers to each of the fragments. Note that the numbers assigned here do not correspond to the sizes of the fragment but are intended to uniquely identify each fragment. Beginning with the number 1 for the largest fragment, the genomic clones contain the following fragments:

      Clone A: 1, 5, 8, and 12
      Clone B: 2, 9, and 10
      Clone C: 1, 5, 8, and 13
      Clone D: 2, 9, and 11
      Clone E: 3, 7, 8, and 12
      Clone F: 4, 6, 11, and 14
      Clone G: 2, 5, 10, and 13

Next look for overlaps between fragments. For example, clone F contains fragments 4, 6, 11, and 14. Clone D contains fragments 2, 9, and 11. Clones D and F must overlap in region 11. Fragments 4, 6, and 14 are not found in any clones other than F; therefore, clone F lies at one end of the contig. There is not enough information to order fragments 4, 6, and 14 with respect to one another. Fragments 2 and 9 from clone D are also found in clone B; fragment 2 but not fragment 9 occurs in clone G.

Using this logic, the following map of the contig can be constructed:

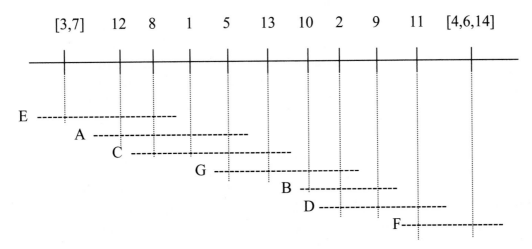

[3,7]   12   8   1   5   13   10   2   9   11   [4,6,14]

E
A
C
G
B
D
F

42. a. The clones should be arranged in the following order: ECADB.

b. The *sme* gene is most likely in clone B or C, as it lies between probes 2 and 3 on the map. The *nb* gene is most likely in clone A or clone E, as it lies between probes 4 and 5 on the map.

c. (1) Sequence the two genes in wild-type flies and in the *sme* mutant flies. Identification of a mutation in one of the genes would support identification of the gene as *sme*. (2) Perform genetic crosses in which strains carrying deletions of the genes are crossed with the *sme* mutant. Lack of complementation in the offspring indicates that *sme* lies within the deletion. (3) Analyze the expression of the two candidate genes in the *sme* mutant at the RNA and protein levels. Reduced or absent expression in the mutant compared to a wild-type strain would support identity between the candidate gene and *sme*.

CHAPTER INTEGRATION PROBLEM

a. The long-bristle phenotype is dominant to wild type. The loss of bristles on the thorax is recessive to the wild-type allele. If the two alleles are designated Lb (long bristles) and Lb^+ (wild-type bristle length), the original mutant fly had the genotype Lb/Lb^+, and crosses can be diagrammed as follows:
P: Lb/Lb^+ (long-bristle mutant) $\times$ Lb^+/Lb^+ (wild type)
F_1: 1/2 Lb/Lb^+ (long bristles) and 1/2 Lb^+/Lb^+ (wild type)
Cross two F_1 flies with long bristles (Lb/Lb^+ $\times$ Lb/Lb^+)
F2: 1/4 Lb^+/Lb^+ (wild type)
   1/2 Lb/Lb^+ (long bristles on abdomen, normal thorax bristles)
   1/4 Lb/Lb (long bristles on abdomen and loss of bristles on thorax)

b. EST 3 most likely corresponds to a mutation in this region. This is the only EST probe that hybridizes to bands of different sizes in the wild-type and mutant flies.

c. The fact that in the mutant, the larger probe hybridizes to both the centromere and a location near the telomere of chromosome 3 suggests that the mutation is due to an inversion that results in the production of a shorter transcript. Because the 5' end of the gene is near the end of the chromosome in both the wild-type flies and the mutant flies, the gene must be oriented so that the 5' end of the coding (nontemplate) strand of gene is nearest the end of the chromosome, and the 3' end of the coding strand is more proximal to the centromere. Transcription would therefore proceed toward the centromere.

d. Mutations in this gene are affecting transcription patterns of other genes. The fact that this mutation results in a dominant phenotype in the abdomen but a recessive phenotype in the thorax suggests that it may be dimerizing with different partners in the different segments. For example, if the protein forms a homodimer in the abdomen but a heterodimer in the thorax, the mutation should be dominant in the abdomen but recessive in the thorax.

e. The presence of a slightly smaller protein suggests that the smaller mRNA on the Northern blot was missing only a small part of the coding region. A Western blot would be expected to look like this:

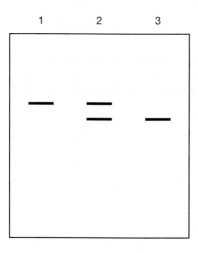

CHAPTER 15 GENETICS OF BACTERIA AND BACTERIOPHAGES

Chapter Goals
1. Describe the different types of bacterial and phage mutants.
2. Identify the different forms of the F plasmid DNA, and explain how each form transfers bacterial genomic DNA and is used for gene mapping.
3. Understand how transformation is used for gene mapping.
4. Explain the two different types of transduction and how each can be used for gene mapping.
5. Describe how genes in the phage genome are mapped.
6. Summarize how Benzer assigned complementation groups in the *rII* gene and how he mapped mutations within an *rII* cistron.

Key Chapter Concepts
15.1 Bacteria and Bacteriophages in Genetic Research
- **Characteristics that make them suitable for genetic research:** Several properties of bacteria and bacteriophages make them particularly suitable for genetic research. They have a short generation time which allows for the production of large amounts of DNA for analysis, and the large number of daughter cells allows for the isolation of rare recombination events. They also have much less genetic material than eukaryotes do, and the organization of the material is much simpler. Finally, they are easy to handle.

15.2 Morphological Characteristics
- **Bacteria:** Bacteria lack a true nucleus, have no nuclear membrane, and have only a single "naked" circular chromosome, although they may contain **plasmids**.
- **Bacteriophages:** Bacteriophages consist exclusively of genetic material surrounded by a protein coat. They are classified according to what type of genetic material they have and by the structural features of their protein surfaces. Bacteriophages are obligate parasites. Once their genetic material penetrates a host cell, the bacteriophage can take over the metabolism of the cell, allowing them to make multiple copies of themselves.

15.3 Techniques of Cultivation
- **Bacterial Culturing and Screening Techniques:** Bacteria are excellent model organisms for genetic analysis of biochemical traits. Strains with nutritional requirements can be generated and studied to determine which genes in a biosynthetic pathway are nonfunctional or defective. A bacterial strain that grows to form colonies on medium containing only the minimal nutritional necessities (**minimal medium**) is referred to as a **prototroph**. Strains of bacteria with a defect in a biosynthetic pathway have additional nutritional requirements and are referred to as **auxotrophic**. Auxotrophic strains can grow only on medium containing the nutrient they are unable to synthesize due to a genetic mutation. Mutant strains that can grow *only* on a supplemented or (**enriched**) or **complete medium** are called **conditional-lethal mutants**. Various kinds of **selective medium** (minimal medium plus one additional substance) are used to identify conditional-lethal mutants. Bacteria may also differ in their **resistance** and **sensitivity** to antibiotics. Drug sensitivity can be used to select colonies with particular mutations.
- **Bacteriophage Culture:** Phages grow only in living cells. **Bacterial lawns** cultivated on Petri plates serve as a medium for the growth of bacteriophages. Addition of the phage causes **lysis** of the bacterial cells that it infects and produces a clear **plaque** on the Petri plate. Large quantities of bacteriophages can also be grown in flasks of bacterial suspensions.

15.4 Bacterial Phenotypes
- **Colony Morphology:** This class includes the form, color, and size of the colony that grows from a single cell. This is the first of three general classes of bacterial phenotypes.
- **Nutritional Requirements:** These are the compounds that must be added to minimal medium so that bacteria will grow. These requirements reflect mutations that disrupt one or more enzymes in the bacteria's biosynthetic pathway. Wild-type bacteria (prototrophs) have no nutritional requirements.

This is the second of the three classes of bacterial phenotypes. **Replica-plating,** which is used to determine if a bacterial strain is auxotrophic, involves growing bacterial colonies on complete medium and then transferring the colonies to selective medium to determine whether they are able to continue growing. Based on the media on which the bacteria can and cannot grow, the bacteria are given special designations to indicate their genotype. For example, a *met⁻pro⁺* bacterial strain requires methionine for growth on minimal medium, but does not require proline. In terms of energy sources, the plus or minus notation has a different meaning. For example, a strain that can utilize the sugar galactose as an energy source is designated as *gal⁺*, whereas a *gal⁻* strain needs a carbon source other than galactose for growth.

- **Resistance and Sensitivity:** The third common classification of bacterial phenotypes involves resistance and sensitivity to drugs, phages, and other environmental insults. Antibiotic-resistant genes can be found on either plasmids or in the bacterial genome. Wild-type bacteria are typically sensitive to antibiotics and to specific types of bacteriophages, but mutants that are resistant to either a specific antibiotic or a specific phage can be screened for by identifying colonies capable of growing in the presence of a specific antibiotic or phage.

15.5 Bacteriophage Life Cycles and Phenotypes

- **Phage Life Cycles:** Bacteriophages can only replicate within a bacterial cell, and their genetic material is dedicated to accomplishing the task of infection and replication. All phages infect bacterial cells by attaching to the cell surface and injecting their genetic material into the cell. This genetic material can either immediately begin to produce new phages leading to lysis of the bacterial cell (**virulent phage**) or can integrate into the bacterial genome (**prophage**) or into a plasmid where it can remain dormant until a stimulus to the bacterial cell induces the phage to excise itself and enter the lytic cycle (**temperate phage**).

- **Phage Phenotypes:** Phage phenotypes include two categories: plaque morphology and growth characteristics on different bacterial strains. Plaque size results from the rate of lysis of the bacterial cell, while the host range of a bacteriophage refers to the bacterial strains that are sensitive to it. Both of these characteristics are encoded by the bacteriophage genome.

15.6 Overview of Sexual Processes in Bacteria

- **Bacterial Sexual Processes:** Bacteria can undergo sexual processes involving the recombination of genetic material between two different individuals or genotypes. This allows genetic studies to be conducted on bacteria. Bacteria can undergo recombination with another cell by one of three methods: **transformation**, **conjugation**, and **transduction**. Phages can also exchange genetic material when a bacterium is infected by more than one phage particle.

15.7 Transformation

- The first process, **transformation**, involves the uptake of DNA found in the environment and the incorporation of this genetic material into the cell's genome through recombination. Cells that undergo transformation have a surface protein called a **competence factor** which is needed to bind the extracellular piece of DNA. Two crossovers are needed to incorporate a piece of linear foreign DNA into the cell's genome. The remaining exogenous DNA is degraded.

- Transformation mapping provides a measure of the relative distance between two loci called a **co-occurrence** or **cotransfer index**. A large co-occurrence index indicates that the genes under study are close together, while a small co-occurrence index indicates a greater map distance. The number of recipient bacteria that incorporate one or the other (single transformants) or both (double transformants) of the donor alleles is counted. The cotransfer index is calculated by dividing the number of double transformants observed by the total number of transformants observed.

15.8 Conjugation

- **Conjugation:** Conjugation in bacteria such as *E. coli* involves the transfer of genetic material from one cell (the donor) to another (the recipient). Cells carrying a **fertility** or **F factor** (F⁺ strain) are capable of transferring a single strand of DNA to a recipient cell lacking an F factor (F⁻ strain). The F factor is an independently replicating piece of DNA called a **plasmid**. In another strain, **Hfr**, the F factor has become part of the bacterial chromosome. During conjugation, genetic material from the donor cell passes to the recipient cell through a protein channel called the **F-pili** or **sex pili** which connects the two cells. The foreign DNA is incorporated into the genome of the recipient through the process of crossing over.

- **Using Time of Transfer and Recombination Distances to Map Genes:** The processes of conjugation can be exploited to map the relative positions of genes on the bacterial chromosome. During conjugation, bacterial genes are transferred sequentially over time with the F factor being transferred last. An **interrupted mating** technique can be used to map genes. At specific time intervals, "mating" cells are disrupted, ending the transfer of genetic material. The genotypes of the F$^-$ cells are then determined by replica-plating. This technique involves selecting the first allele that enters the recipient and then determining the amount of time required for subsequent alleles to enter the F$^-$ cell. This technique provides estimates, in units of time, of the distance between genes and is most accurate for genes that are distantly located. The distance between prokaryotic genes in a circular chromosome can also be determined by recombination mapping of linked genes after conjugation. This method involves selecting the last gene to enter the recipient cell so that all of the desired genes are present in the recipient cell. Recombination distances between pairs of genes can then be determined by determining the percentage of all of the crossover events between the two genes.

- **F′ Conjugation: F-duction or Sexduction:** In addition to the F$^+$ and Hfr strains, the F plasmid can also spontaneously recombine between transposable element sequences to integrate into the bacterial genome within an F$^+$ culture. The F plasmid can also spontaneously excise itself from the bacterial chromosome. At a very low frequency, this excision is imprecise, resulting in an F plasmid that also contains some of the bacterial genomic sequences. This does not produce an F$^+$ cell, but produces a new class of donor cell called an **F′**. The passage of this F′ factor to an F$^-$ cell is called **F-duction** or **sexduction,** and the recipient cell usually becomes an F′ merozygote. Genes transferred from the bacterial genome of the donor strain to the recipient strain are stably maintained without recombination because they are present on the F plasmid.

15.9 Transduction

- The final sexual process in bacteria is **transduction**. In this case, a bacterial virus or **bacteriophage** transfers bacterial DNA between cells. The bacteriophage replicates within a bacterial host cell. An occasional error results in the incorporation of bacterial DNA into the virus. This bacterial genomic DNA can be physically joined to either the phage DNA (**specialized transduction**) or a separate fragment of DNA (**generalized transduction**). When this virus infects another cell, bacterial DNA carried by the virus may be recombined into the new host.

- **Specialized Transduction:** The usefulness of specialized transduction in mapping genes is limited because it involves the integration of a defective λ genome that contains flanking bacterial genomic DNA into a single location in the *E. coli* genome.

- **Generalized Transduction:** Following infection of some virulent phage in bacterial cells, the bacterial genomic DNA is digested to reduce the replication, transcription, and translation of bacterial genes. Generalized transduction involves the incorporation of random bacterial genomic DNA fragments into new phage particles during packaging. Genes that are closer together are more likely to be cotransduced.

- **Establishing Gene Order and Map Distances:** Transduction can also be used to establish gene order and map distances between bacterial genes. Two-factor transduction experiments can be used to establish gene order. The greater the frequency that two alleles cotransduce, the closer they are relative to each other. Three-factor transduction experiments are the most useful in mapping studies. A prototroph is used to make transducing phages which are then provided with auxotrophic host cells. The number of cells that are transduced for one, two, or three loci is determined. The cotransduction frequency is the percentage of total transductants that contain both of the alleles of interest. **Cotransductance** values are then used to establish relative distances between genes.

15.10 Phage Genetics

- **Mapping Genes by Recombination:** Recombination in a phage genome can be studied by coinfecting a bacterial cell with two genetically distinct phage particles that possess different alleles. In these cells, the different phage genomes may either remain intact to yield the two parental genomes or they may cross over to yield two recombinant genomes. Each of the four phage genotypes produces a phenotypically distinct plaque phenotype allowing both recombinant and parental phage genotypes to be distinguished. The recombination distance or frequency between phage genes is the percentage of recombinant phages in the total number of phages produced.

- **Fine-Structure Mapping:** Seymour Benzer used classical recombination and mutation techniques with bacteriophages to provide reasonable estimates on the details of the size and number of sites within a gene that are capable of mutation and recombination. Benzer isolated a large number of independently derived *rII⁻* mutants and crossed them among themselves to determine their **complementation groups**. From this data, Benzer defined the **cistron** as the smallest functional genetic unit that exhibits a *cis–trans* position effect. By examining an extremely large number of mutations in different regions defined by deletion mutations, Benzer found that there was a minimum recombination frequency which led Benzer to define the **recon** as the minimal recombination unit which was found to consist of a single base pair. In addition, Benzer found that mutations are not evenly distributed throughout the cistron.

Key Terms

- bacteriophages
- phenomics
- plasmids
- capsids
- capsomeres
- heterotrophs
- autotrophs
- synthetic medium
- colony
- prototrophs
- auxotrophs
- conditional-lethal mutant
- enriched medium
- selective media
- bacterial lawn
- lysis
- plaques
- colony morphology
- nutritional requirements
- replica-plating
- virulent phages
- lysate
- temperate phages
- prophage

- lysogen
- transformation
- conjugation
- transduction
- chromosome
- competence factor
- heteroduplex region
- cotransformation
- reversion
- cotransformation frequency (cotransfer index)
- fertility factor, F
- F-pili
- Hfr
- merozygote
- exogenote
- endogenote
- interrupted mating
- F-duction (sexduction)
- specialized transduction
- generalized transduction
- transducing particle
- cotransduction
- selected locus

- multiplicity of infection
- fine-structure mapping
- *cis–trans* complementation
- complementation groups
- cistron
- titer
- recon
- hotspots

Understanding the Key Concepts
Use the key terms or parts of their definitions to complete the following sentences.

Bacteria are typically grown in either liquid or solid (1)_____ medium. Prototrophs can grow on (2)_____ medium that contains a (3)_____ source, but auxotrophs are (4)_____ bacteria that have specific (5)_____. Typically, these bacteria can only grow on (6)_____ medium. (7)_____ that have a single mutation in an enzyme in the biosynthetic pathway of a particular amino acid can grow on (8)_____ medium that consists of (9)_____ medium supplemented with the specific required amino acid. Unlike bacteria, bacteriophages can only grow in (10)_____. (11)_____ cultures are infected with bacteriophages that attach to the (12)_____ of the bacterial cells and inject their (13)_____ into the cell. (14)_____ phages can integrate into the bacterial genome and enter the (15)_____ cycle in which they remain in a quiescent state until they are induced to enter the (16)_____ cycle. (17)_____ phages immediumtely begin the (18)_____ cycle upon entering the bacterial cell which typically results in (19)_____ of the bacterial cell.

Bacteria take up foreign DNA from their environment by (20)_____ which involves binding of the DNA and (21)_____ of one strand by a membrane-bound (22)_____. Linear, single-stranded foreign DNA can be incorporated into the bacterial chromosome by (23)_____ involving two (24)_____ events that regenerate a (25)_____ chromosome capable of replication. The (26)_____ frequency of two alleles from different genetic loci can be used to determine the relative (27)_____ between two genes. The higher the (28)_____, the (29)_____ the two loci are.

Conjugation is the (30)_____ transfer of genetic material from one (31)_____ to another. This process requires the direct (32)_____ between two cells. The (33)_____ of the donor cell form a connecting bridge with the recipient cell. If the donor cell is F^+, the (34)_____ is transferred to the F^- cell making it (35)_____. If the donor cell is from the Hfr strain, the recipient cell remains (36)_____ after conjugation because not all of the (37)_____ is transferred. The piece of DNA that is transferred from the Hfr cell to the recipient cell cannot be stably (38)_____ without recombining into the genome. (39)_____ is a technique that can be used to measure the (40)_____ of transfer of different alleles into the recipient cell. This technique allows the relative (41)_____ between genes to be determined according to how much (42)_____ it takes for each gene to be transferred to the recipient cell. Genes can also be mapped by conjugation based on the (43)_____ frequency between two genes. The (44)_____ the number of recombinants, the greater the physical (45)_____ between two genes.

Transduction is the (46)_____ transfer of bacterial genomic DNA. (47)_____ transduction is dependent on the faulty excision of a (48)_____ from the bacterial genome which leads to the packaging of some of the adjacent bacterial genomic DNA into the phage (49)_____. (50)_____ transduction allows a (51)_____ piece of bacterial genomic DNA to be packaged into the phage. By looking at the (52)_____ frequency of different genes, the relative map distance and (53)_____ of the genes can be determined. In terms of transduction, the closer two genes are together, the (54)_____ their (55)_____ frequency.

Figure Analysis
Use the indicated figure or table from the textbook to answer the questions that follow.

1. **Figure 15.9**
   (a) What is the purpose of replica-plating the bacterial colonies?

   (b) What is the difference between a prototroph and an auxotroph?

   (c) How could the nutrient needs of colonies that grow on the complete medium, but not on the minimal medium, or the minimal medium supplemented with methionine, be determined?

(d) How are bacteria that are incapable of utilizing lactose as an energy source designated?

(e) Which medium will an auxotroph that is not *met⁻* grow on? Which medium will it fail to grow on?

(f) What is the likely deficiency in an auxotroph that is *met⁻*?

(g) How is a strain that can utilize galactose to grow designated? How is a strain that requires leucine to grow designated?

(h) What does it mean to say that a strain is *gal⁻*? What does it mean to say that it is *met⁻*?

2. **Figure 15.14**
    (a) How does a bacterial cell take up DNA by the process of transformation?

    (b) What bacteria are competent to be transformed? What extracellular DNA is competent to cause transformation?

    (c) How many crossover events are required for the linear single-stranded foreign DNA to recombine into the bacterial genome? Why?

    (d) What does the heteroduplex region in the circular chromosome consist of?

    (e) What is the origin of the DNA that is being degraded in figure 15.14?

    (f) Following recombination of the foreign DNA and the bacterial chromosome, what is produced when the bacterial chromosome is replicated and cell division occurs?

3. **Figures 15.16 and 15.17**
    (a) For the recombinant transformants to be *tyrA⁺ cysC⁺*, where does the crossover event have to occur relative to the *tyrA* and *cysC* genes in the bacterial chromosome?

    (b) What will be the genotype of the transformant if one crossover event occurs downstream of the *tyrA* gene in the bacterial chromosome and the other occurs downstream of the *cysC* gene in the bacterial chromosome?

    (c) Which media can the *tyrA⁺ cysC⁻* recombinants grow on? Which media do they fail to grow on?

    (d) If 15 *tyrA⁺ cysC⁺* colonies, 43 *tyrA⁺ cysC⁻*, and 55 *tyrA⁻ cysC⁺* transformants are observed, what is the cotransformation frequency of the two loci?

    (e) If the cotransformation frequency for *tyrA⁺ cysC⁺* is 0.17, for *bio⁺ tyrA⁺* is 0.32, and for *bio⁺ cysC⁺* is 0.05, which genes are closest together? Which genes are farthest apart? What is the order of the genes?

4. **Figures 15.18 and 15.19**
    (a) How were the *met⁺ bio⁺ thr⁺ pro⁺* colonies generated in Lederberg and Tatum's experiment?

    (b) How did Lederberg and Tatum know that these colonies were not generated by spontaneous reversion?

    (c) After mixing, how many colonies did Lederberg and Tatum find that were *met⁺ bio⁺ thr⁺ pro⁺*?

    (d) How did Bernard Davis eliminate the possibility that the *met⁺ bio⁺ thr⁺ pro⁺* prototrophs were produced by transformation?

    (e) Why did colonies appear on the minimal medium in Bernard Davis' experiment if the filter was removed?

5. **Figures 15.40 and 15.41**

(a) Why are there no plaques produced upon infection of the *E. coli* K12 strain with the r_a^-/r_b^+, r_a^-/r_b^+ mutant phage?

(b) Why were plaques produced upon infection of the *E. coli* K12 strain with the r_a^+/r_b^-, r_a^-/r_b^+ mutant phages? By what two processes could these plaques have been produced?

(c) Why did Benzer isolate phages from individual plaques following infection of the K12 strain with the r_a^-/r_b^+, r_a^-/r_b^+ mutant phages and use them to infect *E. coli* strains B and K12 individually? What did he expect?

(d) What did Benzer find when he crossed an *rII* point mutant with an *rII* deletion mutant located in a different region of the gene?

General Questions

Q15-1. Compare transformation, transduction, and conjugation with respect to the source of foreign DNA, the mechanism of foreign DNA acquisition and incorporation, and the mapping information provided.

Q15-2. Describe the similarities and the differences between the three different F strain donors with respect to how they affect the F⁻ recipient strain, whether they can transfer genomic DNA, and how.

Q15-3. Explain how transformation can be used to determine the order of prokaryotic genes and the relative map distance between two genes.

Q15-4. Briefly explain the two methods by which genes can be mapped by conjugation.

Q15-5. Explain why generalized transduction is more useful than specialized transduction for mapping genes.

Multiple Choice

For each of the following, circle the letter of the choice that most appropriately answers the question.

1.) Which of the following characteristics can be used to treat nonnaturally competent bacteria to make them more competent?
    a. potassium chloride
    b. sodium chloride
    c. sodium hydroxide
    d. calcium chloride
    e. sodium phosphate

2.) Which of the following is true of a bacterial strain with the genotype *tyrA⁻ bio⁻ thr⁻ gal⁻ met⁺*?
    a. It can only grow on minimal medium supplemented with methionine and galactose.
    b. It can only grow on medium containing tyrosine, threonine, biotin and galactose.
    c. It can only grow on medium containing methionine and glucose.

d. It can grow on medium containing tyrosine, threonine, biotin and glucose.

e. It can only grow on complete medium.

3.) Which of the following is *not* a true statement about prokaryotic crossing over?

a. Cell division following a crossover event between the *E. coli* chromosome and a piece of linear foreign DNA results in two cells that contain the original double-stranded DNA with a region of foreign DNA.

b. Crossing over in prokaryotes is not a reciprocal process.

c. Two crossover events between the circular chromosome and a linear piece of foreign DNA are required to regenerate a circular chromosome capable of replication.

d. Part of the bacterial chromosomal DNA is degraded by exonucleases following heteroduplex formation.

e. A single crossover event between the circular chromosome and the F plasmid results in a circular molecule.

4.) If a *tyrA⁻ cysC⁻* strain of *B. subtillis* is transformed with DNA from a *tyrA⁺ cysC⁺* strain, the majority of the cells will have which of the following genotypes?

a. *tyrA⁺ cysC⁻*

b. *tyrA⁺ cysC⁺*

c. *tyrA⁻ cysC⁻*

d. *tyrA⁻ cysC⁺*

e. It cannot be determined because the incorporation of the donor DNA in the recipient bacteria will be different in every experiment.

5.) As a result of bacterial conjugation, which of the following can occur?

a. An F⁺ cell becomes F⁻ and an F⁻ cell becomes F⁺.

b. Genes in the donor's genome can be transferred to a recipient cell.

c. Free DNA in liquid culture can produce recombinant prototrophs.

d. Only b and c.

e. All of the above

6.) Conjugation in the Hfr strain differs from that of a normal F⁺ strain because of which of the following?

a. The F plasmid in the Hfr strain lacks an origin of replication.

b. The Hfr strain transfers the F factor 1000 times faster to render the recipient cell F⁺.

c. The donor Hfr strain does not transfer DNA through the F-pilus.

d. The F plasmid in the Hfr strain cannot integrate into the genome.

e. The DNA transferred from the Hfr strain is unable to replicate autonomously in the recipient

7.) In Jacob and Wollman's experiment of interrupted mating, it was ensured that the alleles being tested were transferred from the Hfr cells (*strˢ aziʳ tonʳ leu⁺ galB⁺ lac⁺*) to the F⁻ recipient cells (*strʳ aziˢ tonˢ leu⁻ galB⁻ lac⁻*) by doing which of the following?

a. Infecting the cells with T1 phages to eliminate the Hfr cells

b. Plating the cells on medium containing streptomycin

c. Replica-plating the cells from complete medium to minimal medium supplemented with galactose

d. Adding sodium azide to the cell mixture before plating

e. Plating the cells on selective medium lacking leucine

8.) If an A⁺B⁺C⁺ strain of bacteria is infected with phages and the lysate is used to infect an A⁻B⁻C⁻ strain and the following results are obtained for the number of transductants in each class when A⁺ bacteria are selected for, what is the cotransduction frequency of A and C?

A⁺B⁺C⁺	32
A⁺B⁺C⁻	48
A⁺B⁻C⁺	3
A⁺B⁻C⁻	267

a. 8%

b. 9%

c. 10%

d. 12%

e. 15%

9.) Using an A⁺B⁺C⁺ Hfr exogenote or transducing phage in the bacterial strain A⁻B⁻C⁻, it is essential to know the number of recombinants having the genotype A⁺B⁺C⁻ to determine the _____ recombination distances in conjugation mapping, while in cotransduction mapping, it is essential to know this genotype to determine _____ cotransduction frequency.

a. B to C; the A to B
b. A to B; the A to B and A to C
c. A to B and B to C; the A to B
d. B to C; no
e. A to B and B to C; the B to C

10.) According to Benzer's experiments, if the titer of a phage on *E. coli* strain B is determined

to be 4×10^8 plaques per milliliter and 18 plaques per milliliter are found on the *E. coli* K12 plate, what is the recombination frequency?

a. 4.5×10^{-6}
b. 9.0×10^{-6}
c. 4.5×10^{-8}
d. 9.0×10^{-8}
e. 7.2×10^{-9}

Practice Problems

P15-1. The gene order of several Hfr strains was determined by means of interrupted mating. Using this information, construct a map of the bacterial chromosome. Indicate the location of each F factor.

(*Hint 1*: The F factor integrates into the chromosome at different places in different bacterial strains of the same species. The order and direction of transfer is consistent within a strain, but the direction may differ between strains. For example, strain A may transfer genes in the following order: *A B C D E*, while strain B transfers gene *D* first, followed by *C B A* and finally *E*. The F factor, though, is *always* the very last gene that is possible to be transferred. So, the first gene transferred is near the F factor because the chromosome is circular.)

Strain	Order of Transfer (First → Last)							
1	*C*	*O*	*N*	*J*	*U*	*G*	*A*	*T*
2	*U*	*G*	*A*	*T*	*I*	*O**	*N**	*C*
3	*T*	*I*	*O**	*N**	*C*	*O*	*N*	*J*
4	*N*	*J*	*U*	*G*	*A*	*T*	*I*	*O**
5	*G*	*U*	*J*	*N*	*O*	*C*	*O**	*N**

P15-2. A prototrophic strain (*cys⁺ leu⁺*) was used to transform an auxotrophic strain (cys⁻ leu⁻). Examine the patterns of bacterial colony growth on the following different medium plates. Identify the genotype of each colony. Which are transformed cells, and which are untransformed cells?

(*Hint 2*: Bacterial genotypes are determined by growth on selective medium. For example, if a colony is growing on medium that lacks tyrosine, then the bacteria are tyrosine-positive or *tyr⁺*. However, if a colony does not grow on medium that lacks tyrosine, then the bacteria have a mutation that does not allow them to synthesize tyrosine and are therefore *tyr⁻*.)

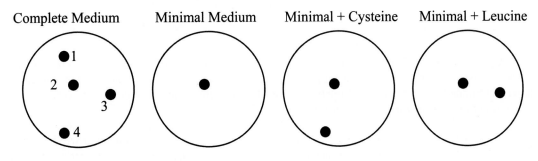

Complete Medium   Minimal Medium   Minimal + Cysteine   Minimal + Leucine

P15-3. A bacterial cell with genotype a^+b^+ served as the donor strain in a transformation experiment. The recipient strain had the genotype a^-b^-. The transformed classes obtained included: 400 a^+b^+; 250 a^+b^-, and 304 a^-b^+. What is the cotransfer index?

(*Hint 3*: Double transformant or transduction classes are those in which two of the donor markers have been transferred to the recipient cell. Single transformant or transduction classes are those in which only one of the donor markers has been transferred to the recipient cell. Cells that were not transformed or transduced are not considered in mapping experiments.)

P15-4. A transduction experiment was conducted using the donor strain $a^+b^+c^+$ and the recipient strain $a^-b^-c^-$. The a^+ allele was used as the selection variable. From the transductant classes obtained, determine the cotransductance frequencies: 30 $a^+b^+c^+$ colonies, 44 $a^+b^+c^-$ colonies, 60 $a^+b^-c^-$ colonies, 10 $a^+b^-c^+$ colonies.

P15-5. An interrupted mating experiment was conducted with the donor Hfr strain $a^+b^+c^+d^+e^+f^+g^+$ and the recipient strain $a^-b^-c^-d^-e^-f^-g^-$. The appearance of donor genetic markers in recipient cells is given in the following table. What is the gene order and distance between genes?

(*Hint 4*: Conjugation is used to determine relative map distances. Remember that distances obtained from interrupted mating studies are expressed in time. This technique is most effective for genes that are distantly located on the chromosome.)

Time (mins)	a^+	b^+	c^+	d^+	e^+	f^+	g^+
5	−	−	+	+	−	−	−
10	−	−	+	+	−	−	−
15	−	−	+	+	+	−	−
20	−	−	+	+	+	−	−
25	−	−	+	+	+	−	−
30	+	−	+	+	+	−	−
35	+	−	+	+	+	+	−
40	+	−	+	+	+	+	+
45	+	−	+	+	+	+	+
50	+	+	+	+	+	+	+

P15-6. A lys^+ his^+ val^+ strain is used to transduce a lys^- his^- val^- strain. Transductants were selected for synthesis of lysine by plating cells on minimal medium. The genotypes found among the transformants follow. Determine the gene order and relative cotransductance frequencies. Which of the cotransductance frequencies represents the greatest distance between genes?

(*Hint 5*: If two genes are close together, a high frequency of cotransfer or cotransductance will be observed. A low frequency of cotransfer or cotransductance indicates that the genes are distant. Remember that the cotransfer/cotransductance index is a measure of how often two genes are transferred together: it is the opposite of a recombination frequency.

Hint 6: Gene order can be established in transformation and transduction experiments by examining the genotype of the least frequent class. The least frequent class is that in which the gene in the middle is the *only* gene that is *not* transferred to the recipient cell. To transfer the genes at the left and the right only, two independent events must have transferred each gene individually. For example, donor genotype $a^+b^+c^+$ is used to transform recipient genotype $a^-b^-c^-$. The least frequent class has the genotype $a^-b^+c^+$. The gene order is *bac* or *cab*.)

$lys^+ his^- val^-$	800	$lys^+ his^- val^+$	10
$lys^+ his^+ val^-$	450	$lys^+ his^+ val^+$	200

Assessing Your Knowledge

Understanding the Key Concepts—Answers

1.) synthetic; 2.) minimal; 3.) carbon or energy; 4.) mutant; 5.) nutritional requirements; 6.) complete; 7.) Auxotrophs or Bacteria; 8.) selective; 9.) minimal; 10.) living cells; 11.) Bacterial; 12.) surface; 13.) genetic material; 14.) Temperate; 15.) lysogenic; 16.) lytic; 17.) Virulent; 18.) lytic; 19.) lysis; 20.) transformation; 21.) digestion; 22.) nuclease; 23.) recombination; 24.) crossover; 25.) circular; 26.) cotransformation; 27.) map distance; 28.) cotransfer; 29.) closer together; 30.) one-way; 31.) strain of bacteria; 32.) physical contact; 33.) F-pili; 34.) F plasmid; 35.) F$^+$; 36.) F$^-$; 37.) F plasmid; 38.) expressed; 39.) Interrupted mating; 40.) time; 41.) distance; 42.) time; 43.) recombination; 44.) larger; 45.) distance; 46.) phage-mediated; 47.) Specialized; 48.) prophage; 49.) head; 50.) Generalized; 51.) random; 52.) cotransduction; 53.) order; 54.) greater; 55.) cotransduction.

Figure Analysis—Answers

1a.) The bacterial colonies are replica-plated to determine the specific nutrient needs of the auxotrophic bacteria; b.) A prototroph is a colony that can grow on the minimal media (wild-type), while an auxotroph is a mutant bacteria that has specific nutritional requirements, such as methionine in figure 15.9; c.) The nutrient needs of colonies that grow on the complete medium, but not on the minimal medium or the minimal medium supplemented with methionine, could be determined by replica-plating the colonies on the complete medium to plates containing minimal medium that is supplemented with other nutrients or with different combinations of other nutrients; d.) *lac⁻*; e.) An auxotroph that is not *met⁻* will grow on complete medium, but fail to grow on both minimal medium and minimal medium supplemented with methionine; f.) The likely deficiency in an auxotroph that is *met⁻* is a mutation in a gene that encodes an enzyme that is required for one step in the methionine biosynthetic pathway; g.) *gal⁺*; *leu⁻*; h.) To say that a strain is *gal⁻* means that it requires a carbon source other than galactose for growth; that is, the strain will not grow on galactose. To say that a strain is *met⁻* means that it requires methionine for growth; that is, it will only grow on medium that contains methionine.

2a.) The cell binds the DNA, and one strand of the DNA is digested by a membrane-bound nuclease as the other strand enters the bacterial cell; b.) Bacteria that are competent to be transformed either have a surface competence factor protein that can bind to the extracellular DNA during transformation, or they are made competent by either growing them in an enriched medium, or by treating them with calcium chloride to make them more permeable. Extracellular DNA that is competent to cause transformation must be double-stranded; c.) Two crossover events are required for the linear single-stranded foreign DNA to recombine into the bacterial chromosome. Two crossover events are required to regenerate a circular chromosome capable of proper replication. An odd number of crossover events between the circular chromosome and a linear molecule produces a linear molecule that cannot properly replicate in a prokaryotic cell; d.) The heteroduplex region in the circular chromosome consists of one host DNA strand and one foreign DNA strand that may not be completely complementary; e.) The DNA that is being degraded was originally part of the bacterial chromosome; f.) Following recombination between the foreign DNA and the bacterial chromosome, each strand of the chromosome serves as a template for a new strand when the bacterial chromosome is replicated. After cell division, one cell will contain the original double-stranded chromosome, and the other cell will contain a double-stranded DNA molecule with a region of double-stranded foreign DNA.

3a.) For the recombinant transformants to be $tyrA^+ cysC^+$, the two crossover events have to occur in sequences flanking the outside segments of both genes in the bacterial chromosome; b.) $tyrA^- cysC^+$; c.) The $tyrA^+ cysC^-$ recombinants grow on complete medium and minimal medium supplemented with cysteine. They fail to grow on minimal medium and minimal medium supplemented with tyrosine; d.) The cotransformation frequency is 0.13; e.) The *bio tyrA* genes are closest together. The *bio cysC* genes are farthest apart. The gene order is *bio tyrA cysC*.

4a.) The $met^+ bio^+ thr^+ pro^+$ colonies were generated by recombination between the $met^- bio^- thr^+ pro^+$ strain with the $met^+ bio^+ thr^- pro^-$ strain; b.) Lederberg and Tatum ruled out spontaneous reversion in their experiment by plating the two strains separately on minimal medium and found no colonies that were capable of growing suggesting that there were no spontaneous revertants in their experiment. Each auxotrophic strain had two mutations in different biosynthetic pathways; the probability that both loci in one strain would simultaneously and spontaneously revert is approximately $10^{-7} \times 10^{-7} = 10^{-14}$; c.) Lederberg and Tatum found 1 in 10^7 cells plated were $met^+ bio^+ thr^+ pro^+$; d.) Bernard Davis eliminated the possibility that the $met^+ bio^+ thr^+ pro^+$ prototrophs were produced by transformation by placing the two different bacterial strains on opposite sides of a U-tube so that they were separated by a filter that had pores large enough to allow DNA to pass through it, but not large enough to allow whole cells to pass. When the cells in the two arms of the U-tube were plated, Davis found that no prototrophs were produced in either arm even though the DNA from each strain would have been allowed to pass to the other strain. This experiment indicated that the transfer of genetic material was not due to transformation, but required cell-to-cell contact; e.) When the filter was removed, the bacterial strains could mix allowing cell-to-cell contact that then permitted prototrophs that could grow on the minimal medium to be produced.

5a.) Infection of the *E. coli* K12 strain with the r_a^-/r_b^+, r_a^-/r_b^+ mutant phage produced no plaques because these two phages together were unable to produce a functional R_a protein. Thus, these two different *rII* mutant phages failed to complement each other because they had mutations in the same complementation group; b.) Infection of the *E. coli* K12 strain with the r_a^+/r_b^-, r_a^-/r_b^+ mutant phages produced plaques because these two phages together could produce both a functional R_a protein and a functional R_b protein since the mutations in each phage were in two different genes. The plaques may have been produced by either recombination or complementation; c.) Benzer isolated phages from individual plaques following infection of the K12 strain with r_a^+/r_b^-, r_a^-/r_b^+ mutant phages and used them to infect *E. coli* strains B and K12 individually to determine if the plaques were produced by either complementation or recombination. He expected that if the original plaque was due to recombination, then the subsequent infections would yield approximately an equal number of plaques on both *E. coli* strain B and K12. If the original plaque was due to complementation, then the subsequent infections would yield plaques only on strain B, but not on the K12 strain. This was based on the fact that the rII^- mutants do not grow on *E. coli* strain K12, whereas the wild-type rII^+ phages can, and both can grow on *E. coli* strain B; d.) Benzer found that when one *rII* point mutant was coinfected with an *rII* deletion mutant located in a different region of the gene at high multiplicity of infection into *E. coli* strain B then recombination between the two mutant sites could produce some wild-type (rII^+) phages. When the resulting lysate was used to infect the *E. coli* K12 strain at low multiplicity of infection, only the wild-type recombinant phages produced plaques.

General Questions—Answers

Q15-1.

Characteristic	Transformation	Transduction	Conjugation
Source of DNA	Exogenous	Phage	Another bacterial cell
Mechanism of foreign DNA acquisition	Competence factor on recipient cell surface	Infection by phage	Donor with F factor, cell-to-cell contact mediated by sex pili
Mechanism of foreign DNA incorporation	Crossing over	Crossing over	Crossing over
Mapping information	Gene order and map distance	Gene order and map distance	Gene order and *relative* distance

Q15-2. Transfer of the F factor from an F$^+$ donor cell to an F$^-$ recipient cell renders the recipient cell F$^+$. The F plasmid in the F$^+$ donor strain can recombine with the circular genome and integrate into it at a variety of sites in the genome containing transposable elements. When an integrated F plasmid starts to transfer from the donor to the recipient cell, it takes with it the genomic DNA that it is physically linked to, but the efficiency of genomic transfer is low. Transfer of the F factor from the Hfr donor strain is extremely low such that the F$^-$ recipient cell remains F$^-$ after conjugation, although the Hfr strain transfers genomic DNA at a rate about 1000 times higher than the normal F$^+$ strain. All of the cells in the Hfr strain have the F plasmid integrated at the same location. Some of the F plasmid enters the recipient cell, followed by the donor's genomic DNA, and then the remainder of the F DNA. Since all of the F DNA typically does not transfer to the recipient cell, the cell remains F$^-$. The efficiency of genomic transfer by the Hfr cell is high, but the transferred piece of plasmid is unable to replicate autonomously and must recombine with the recipient genome to be stably expressed. The F$'$ cell, which is generated when an F plasmid is imprecisely excised from the bacterial chromosome such that some of the bacterial genomic DNA is also excised, is the third class of donor cell. Conjugation between an F$'$ cell and an F$^-$ recipient cell renders the recipient cell an F$'$ merozygote because it has two copies of some genes. Unlike the Hfr cell, the genomic DNA transferred to the recipient cell is stably maintained without recombination because it is present in the autonomously replicating F plasmid.

Q15-3. The general idea of transformation mapping is to add DNA from a donor strain with a known genotype to a recipient strain with a known genotype that has different alleles at two or more loci and look for the incorporation of the donor alleles into the recipient strain of bacteria. The more frequently alleles from two different loci are incorporated into the recipient strain together, the more closely together these loci are located to each other. By counting all of the transformed cells and determining their genotypes, it is possible to calculate the cotransformation frequency. This indicates the relative distance between these two loci on the bacterial chromosome. A higher cotransformation frequency corresponds to a smaller distance between two genes. By systematically examining many loci, their relative order can be established.

Q15-4. Genes can be mapped by conjugation according to time of transfer or by recombination distances. In the time of transfer method, the first allele that enters the F$^-$ recipient cells is selected for at various time intervals following conjugation. At each interval, the cells are replica-plated to determine the other alleles that have transferred along with the first allele from the Hfr donor strain. Based on time of transfer relative to the first allele, a map of the order and time that each subsequent gene transferred is established. To map genes by recombination after conjugation, the last gene to enter the recipient cell is selected for so that all of the desired genes are present. Recombinants are then tested by replica-plating for the presence of additional donor alleles that were transferred earlier, and the genotypes of the recombinants are determined. The observed number of each recombinant class is used to determine the recombination distances between pairs of genes. The distance between two genes is equal to the percentage of all of the crossover events between two genes. This yields a genetic map that is in map units.

Q15-5. Specialized transduction involves the production of defective λ genomes that are attached to flanking genomic DNA from *E. coli* by the faulty excision of λ out of the bacterial genome. λ is known to recombine at a single site in the *E. coli* genome between the *gal* and *bio* genes. Since mapping through transduction is based on the cotransduction frequencies, or the frequency of two genes recombining into the recipient genome together, and only loci adjacent to the prophage can be transduced to the recipient cell by specialized transduction, this method has not proven to be particularly useful for mapping. Generalized transduction is much more useful for mapping gene order and relative map distance between genes. In this process, random bacterial genomic DNA is packaged into the phage head allowing virtually any bacterial locus to be transduced by this mechanism. Only genes that are very close to each other will be cotransduced. By determining the percentage of total transductants that contain both of the alleles of interest, the cotransduction frequencies and relative map distances can be determined. This method is more useful for gene mapping because it will include a greater number of loci than specialized transduction will.

Multiple Choice—Answers

1.) d; 2.) d; 3.) a; 4.) c; 5.) b; 6.) e; 7.) b; 8.) c; 9.) a; 10.) b.

Practice Problems—Answers

P15-1. Diagram each strain separately, noting similar positions of genes in each. The symbol [F] is used to denote the position of the F factor.

Strain
1   [F]*CONJUGATION* or *CONJUGATIO*N**[F]
2   [F]*UGATIO*N*CONJ* or *UGATIO*N*CONJ*[F]
3   [F]*TIO*N*CONJUGA* or *TIO*N*CONJUGA*[F]
4   [F]*NJUGATIO*N*CO* or *NJUGATIO*N*CO*[F]
5   [F]*GUJNOCN*O*ITA* or *GUJNOCN*O*ITA*[F]; notice this gene order is the same as that of strains 1 to 4, but the direction is *opposite* from that of strains 1 to 4.

Order: ***CONJUGATION***

P15-2. All colonies grow on complete medium, but only colony 2 is able to grow on the minimal medium. Colony 2 is prototrophic, while colonies 1, 3, and 4 are auxotrophic. Examination of the selective medium plates indicates that colony 4 grows in the presence of cysteine indicating this colony lacks the gene for cysteine synthesis. Colony 3 grows in the presence of leucine indicating that the gene for leucine synthesis is defective in this colony. Colony 1 is unable to grow on either of the selective medium plates indicating that it has both defective genes.

Colony	Genotype	Transformed?
1	$cys^- leu^-$	No
2	$cys^+ leu^+$	Yes
3	$cys^+ leu^-$	Yes
4	$cys^- leu^+$	Yes

P15-3. The cotransfer index is calculated by dividing the number of double transformants by the total number of transformants.

r = number of a^+b^+ transformants/total number of transformants
$r = 400/(400 + 250 + 304)$
$r = 400/954 = 0.4193$ or ~42%

P15-4. In this problem, marker a^+ is used as the selection variable. We can therefore calculate only the $a^+ b^+$ and a^+c^+ cotransductance indexes.

a^+b^+ cotransductance frequency = number of $a^+ b^+$ transductants/total transductants obtained
= (30 + 44)/144
= 0.514 or ~51%

a^+c^+ cotransductance frequency = number of $a^+ c^+$ transductants/total transductants obtained
= (30 + 10)/144
= 0.278 or ~28%

P15-5. To determine the gene order, examine the time chart provided. The first gene transferred (earliest time) is the first gene in the sequence, while the last to be transferred is the last gene. The distance between genes can be calculated by subtracting the time of the previous entry from the time of the next gene entry. Notice that genes c and d cannot be ordered because they were both transferred within the first 5 minutes.

Gene	Time First Observed
a^+	30
b^+	50
c^+	5
d^+	5
e^+	15
f^+	35
g^+	40

Gene order: c^+,d^+ _____ e^+ ____ a^+ ___ f^+ ___ g^+ ___ b^+
Distances: <5   10   15   5   5   10

P15-6. Notice that only the cells prototrophic for lysine are reported. To determine the gene order, examine the class with the lowest frequency: lys^+ his^- val^+. Notice that his^+ was not transferred indicating that it is the middle gene. The gene order is *lys his val*. Calculate cotransductance indexes as before. Note that all distances must be in reference to *lys*.

lys^+ val^+ cotransductance frequency = number of lys^+ val^+ transductants/total transductants obtained
= (10 + 200)/1460
= 210/1460 = 0.1438 or 14.38%

lys^+ his^+ cotransductance frequency = number of lys^+ his^+ transductants/total transductants obtained
= (450 + 200)/1460
= 650/1460 = 0.4451 or 44.51%

The 0.1438 cotransductance frequency represents the greatest actual distance between genes. The further apart two genes are the less likely they are to be cotransduced.

15

GENETICS OF BACTERIA AND BACTERIOPHAGES

CHAPTER SUMMARY QUESTIONS

1. 1. e, 2. d, 3. c, 4. b, 5. a, 6. j, 7. i, 8. h, 9. g, 10. f.

2. The prokaryotic chromosome is a double-stranded DNA circle that is small compared with most eukaryotic chromosomes. Viral chromosomes can be DNA or RNA, single- or double-stranded, linear or circular.

3. A minimal medium contains only those substances required by the wild-type organism to grow; a complete medium is a minimal medium that has been enriched with a complete array of organic compounds including amino acids and nucleic acid subunits. An enriched medium is the minimal medium to which nutrients have been added. A selective medium is a particular case of an enriched medium in which only one or a few items have been added.

4. There are three main properties that make bacteria and bacteriophages very suitable for genetic research: (1) They have a short generation time, which allows for the production of a large amount of DNA for analysis. (2) They have much less genetic material with much simpler organization than eukaryotes do. (3) They are easy to grow in the laboratory.

5. The three general classes of bacterial phenotypes are colony morphology, nutritional requirements, and resistance to drugs or infection by bacteriophage. (1) Colony morphology relates simply to the size, color, and form of the colony that grows from a single cell. The wild-type form of a bacterium exhibits a specific colony morphology, which is typically under genetic control. Mutants with altered colony morphologies can be readily identified and studied. (2) Nutritional requirements are the compounds that bacteria cannot synthesize, and therefore must be added to minimal media in order for the bacteria to grow. Wild-type bacteria (prototrophs) have no nutritional requirements. Nutritional mutants are termed auxotrophs, and they typically have defects in one or more enzymes in their biosynthetic pathways. (3) Resistance to drugs and infection by bacteriophages is typically the result of mutations because wild-type bacteria are usually sensitive. The resistance genes can be borne on plasmids or the bacterial genome.

6. The two general classes of phage phenotypes are plaque morphology and host range. (1) Plaque morphology is dictated by the rate of bacterial lysis. Wild-type bacteriophage lyse bacteria slowly and produce small plaques with fuzzy edges. Rapid-lysis mutants produce large smooth-edged plaques. (2) Host range refers to the bacteria that are sensitive to the bacteriophage. Phage T4, for example, will infect and lyse *E. coli* strain B, but not strain K12.

7. Replica-plating is a rapid screening technique for nutritional mutants or auxotrophs. A culture of bacterial cells is spread on a petri plate of complete (nonselective) medium. Both prototrophs (wild-type bacteria) and auxotrophs will grow, and their colonies will appear in a random arrangement. A piece of sterilized velvet cloth is gently pressed onto the plate, picking up some cells from each colony. The velvet is then pressed onto a number of replica plates, thereby reproducing the pattern of colonies from the original plate. Each of the replica plates contains complete medium that lacks a particular amino acid or vitamin. These plates are called selective plates. If a colony appears on the original master plate but fails to appear at the same location on a selective plate, that colony would be auxotrophic for the compound that is missing from the replica plate. The individual colonies of interest can then be isolated from the master plate and grown out. Replica-plating can also be used to screen for antibiotic resistance.

8. A colony is a visible mass of cells derived usually from a single progenitor. A plaque is the equivalent growth of phages on a bacterial lawn, producing a cleared area lacking intact bacteria.

9. A plasmid is an independent genetic entity within a cell, usually a small extrachromosomal double-stranded circle of DNA. It integrates by a single crossover between the plasmid and the bacterial chromosome. It leaves by a reverse process of "looping out" (see figure 15.22).

10. a. Prophage is a phage that has integrated its DNA into the bacterial chromosome.
  b. Lysate is the material released from a lysed cell.

c. Lysogeny is the bacterium–phage interaction in which the phage has become a prophage that can be induced at a later date.

11. Transduction is the transfer of DNA from one bacterium to another via a bacteriophage. Generalized transduction involves the random inclusion of a piece of the bacterial DNA in a phage coat. Any piece of DNA has a low but equal chance of being transduced. Specialized or restricted transduction comes about through incorrect looping out of a prophage. Only genes from a particular locus on the bacterial chromosome are transduced, but do so at a high frequency.

12. **a.** MOI, or multiplicity of infection, refers to the ratio of phage per bacterium. It only applies to transduction.
    b. Competence is the ability of a bacterium to pick up extraneous DNA from its environment. It applies to transformation only.
    c. All three processes increase genetic diversity.
    d. None of the three processes involve a reciprocal DNA exchange between two bacteria.
    e. Only transformation is sensitive to extracellular DNases.
    f. Both transformation and transduction can still occur in a U-tube apparatus.
    g. Sexduction is another term for $F' \times F^-$ conjugation.

13. In conjugation, distances can be calculated based on either time of transfer or recombination distance. Time of transfer mapping is performed via the interrupted mating technique. In the case of recombination distances, the last gene to enter from the donor is selected and the recombination frequency between the pairs of genes is determined.

    Mapping through either transformation or transduction is performed by measuring the cotransformation or cotransduction frequencies. These represent the frequencies of two genes recombining into the recipient genome together. Thus, they represent the frequency of recombination events that *flank* the pair of genes rather than the frequency of recombination events *between* the pair of genes. In addition, a higher recombination distance or time of transfer represents genes that are farther apart, while a larger cotransformation of cotransduction frequency represents genes that are close together.

14. For transformation, please see figures 15.13 and 15.14; for conjugation, figures 15.21, 15.23, and 15.25; for transduction, figures 15.31, 15.33 to 15.35.

EXERCISES AND PROBLEMS

15. **a.** *pen^r*, **b.** *azi^s*, **c.** *his$^-$*, **d.** *gal$^-$*, **e.** *glu$^+$*, **f.** *ton^s*.

16. The bacterium could have survived and produced a colony if it was on a λ-free area, it became lysogenic (and thus resistant to further phage attack), or it was genetically resistant to phage λ.

17. The original strain was *arg⁻ his⁻*. Colonies on minimal medium plus histidine are *arg⁺ his⁻*. Colonies on minimal medium plus arginine are *arg⁻ his⁺*. Colonies on both are wild type (*arg⁺ his⁺*).

18. Where phages cannot grow: *E. coli ton^r*, phage *h⁺*. Where phage can grow: *E. coli ton^s*, phage *h⁺* or *h,* or *E. coli ton^r*, phage *h*.

19. Colony 1 grew on the minimal medium plus histidine and arginine, so cells must be *his⁻* and/or *arg⁻*. However, it did not grow on medium that contains one of these two amino acids and not the other. Therefore, the genotype of colony 1 must be *his⁻ arg⁻*. Colony 5 also grew on the minimal medium plus histidine and arginine. However, because colony 5 grew on plates supplemented with arginine and not histidine, the genotype of these cells must have been *arg⁻*. Colony 2 grew on minimal medium plus lysine, but not on minimal media supplemented with leucine. So the genotype of colony 2 should be *leu⁻*. Similarly, the genotype of colony 3 is *lys⁻*. Colony 4 grew on a minimal medium supplemented with both histidine and methionine, but not on one with histidine. Therefore, the genotype can be either *his⁻ met⁻* or *his⁺ met⁻*. Note that it cannot be *his⁻ met⁺* because it should have grown on the histidine-containing plate.

20. Colony 1, *glu⁻ xyl⁺ arg⁻*; colony 2, *glu⁺ arg⁻ his⁻*; colony 3, prototrophic; at least *glu⁺* (may be unable to use other sugars); colony 4, *glu⁺ arg⁻*; colony 5, not just *thr⁻ val⁻* or *thr⁻* or *val⁻*. May be just *ile⁻* or *thr⁻ ile⁻ val⁻* or some combination; colony 6, *glu⁻ gal⁺ his⁻* or *glu⁻ gal⁺ his⁺*.

21. Conjugation experiments are designed so that gene recombination in the F⁻ recipient cells can be detected. After DNA is transferred and conjugation is interrupted, Hfr donor cells still contain copies of the genes in question. Therefore, these Hfr cells would have to be eliminated from the mixed culture, so that only F⁻ cells remain. This can be accomplished using Hfr donor strains with sensitivity genes, F⁻ recipient strains with resistance genes, and a medium containing the selective agent. The sensitivity gene should be far from the origin of transfer point of the Hfr chromosome. This would allow the genes under study to enter the recipient cell and recombine before the sensitivity gene does. If the sensitivity gene was located near the origin of transfer, it would pass into the F⁻ cell very early during conjugation. Consequently, there is a great reduction in the recovery of loci distal to the selective marker because both the Hfr and F⁻ members of a conjugation event can be killed by the selective agent (for example, an antibiotic such as streptomycin).

22. When the drug sensitivity locus passes into the F⁻ strain, there is a general decline in the recovery of recombinants after selection because both the Hfr and F⁻ members of a conjugation event can be killed.

23. The order is *trp purB pyrC* since the class in lowest numbers (4) represents the double recombinants and hence exchanged the outside loci but not the middle one (*purB*). The relative co-occurrence of *trp* and *purB* is $r = (86 + 67)/171 = 0.89$. The relative co-occurrence of *trp* and *pyrC* is $r = (86 + 4)/171 = 0.53$.

24.

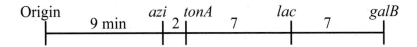

25. *lac* to *gal*, 9 minutes; *gal* to *his*, 27 minutes; *his* to *argG*, 24 minutes; *argG* to *xyl*, 11 minutes; *xyl* to *ilv*, 4 minutes; *ilv* to *thr*, 17 minutes; *thr* to *lac*, 8 minutes. For every interruption, a complete medium plate with selecting antibiotics to kill Hfr cells and one each of the seven selected plates are needed.

26. For example, use an Hfr that is wild type but str^s with the F factor integrated at minute 20. Use an F⁻ strain that is *pyrD⁻*, *trp⁻*, *man⁻*, *uvrC⁻*, *his⁻*, and str^r. Interrupt mating at 1-minute intervals; plate cells on complete medium with streptomycin to kill Hfr cells and grow recombinant and nonrecombinant F⁻ cells. The next day, after colonies have grown up, replica-plate onto selective media. The following data would be generated:

	Colony Growth on Media Selective for				
Minute	$pyrD^+$	trp^+	man^+	$uvrC^+$	his^+
0	−	−	−	−	−
1	+	−	−	−	−
8	+	+	−	−	−
16	+	+	+	−	−
24	+	+	+	+	−
25	+	+	+	+	+

27.

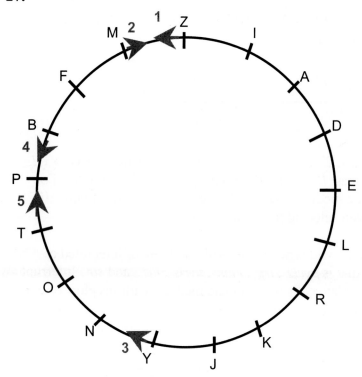

28. The class in lowest frequency (trp^+ his^- tyr^+, 107) identifies *his* as the middle locus (*trp his tyr*); relative cotransformation frequencies are (total = 2600 + 418 + 685 + 1180 + 107 + 3660 + 11,940):

$$trp \text{ to } his: \quad r = \frac{1180 + 11{,}940}{\text{total}} = \frac{13{,}120}{20{,}590} = 0.64$$

$$his \text{ to } tyr: \quad r = \frac{3660 + 11{,}940}{\text{total}} = \frac{15{,}600}{20{,}590} = 0.76$$

$$trp \text{ to } tyr: \quad r = \frac{107 + 11{,}940}{\text{total}} = \frac{12{,}047}{20{,}590} = 0.59$$

29. The order is *a c b,* and *c* is close to *a.* Genes *c* and *a* are cotransformed 76% of the time, suggesting that these two genes are very close and *b* is far away (*a-b* cotransduction is 10%, or 0.1). Two orders are possible: *a-c—————b* and *c-a—————b.* If the first order (*a-c—————b*) is correct, a^+ b^+ c^- results from a double crossover; this class should be the least frequent, and it is.

30. Try a gene order and draw the exchange(s) necessary to produce $a^+ b^+ c^+$. Remember that the least frequent class is the one that requires four crossovers to bring in the outer markers, but not the center one. Try *abc*:

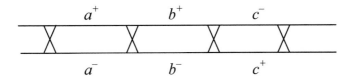

This yields $a^+ b^- c^-$, if *b* is in the middle. If the gene order is *abc*, only two crossovers would be needed to produce the class $a^+ b^+ c^+$. Therefore, *b* is not in the middle. Let's try *c* in the middle:

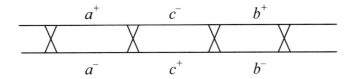

This order produces the desired class ($a^+ c^+ b^+$ is the same as $a^+ b^+ c^+$) as a consequence of four crossovers, which switch the middle locus.

31. *thr leu pro his*. We see that cells that are thr^+ are the most frequent. The chance of interruption in the conjugation increases with the length of time for the mating. Therefore, genes farther from the origin of transfer appear less frequently. We can order the genes based merely upon the frequency of genotypes seen. The order must be *thr leu pro his*. Since we see no his^+, and since we stopped the mating at 25 minutes, his^+ must be after minute 25 on the map of this Hfr strain.

32. a. Since 40% are $trp^+ arg^+$, *arg* and *trp* must be closer than *trp* and *leu*. Therefore, *leu* cannot be in the middle. The two gene orders are *arg trp leu* or *trp arg leu*.

    b. Select for leu^+ transformants and then score $leu^+ arg^+$ and $leu^+ trp^+$. If the first order is correct, $leu^+ arg^+ < leu^+ trp^+$. Since *trp* and *arg* are cotransduced 40% of the time, they must be close and adjacent. If the order is *arg trp leu*, the genotype $leu^+ arg^+ trp^-$ requires two sets of exchanges (total of four crossovers), whereas $leu^+ arg^+ trp^+$ requires only one set of crossovers (two total). If the order is *trp arg leu*, $leu^+ arg^+ trp^-$ requires only one set of exchanges and should be more frequent than $leu^+ arg^- trp^+$ which requires two sets of exchanges.

33. The numbers of the first three transformant classes indicate that each gene, by itself, is readily transformed. We notice that classes with *b* and any other gene are quite rare, a situation that indicates *b* is far from *a* and *c*. We notice that *a* and *c* are cotransformed about 13% [(93 + 6)/775] of the time, *b* and *c* about 2.5% [(14 + 6)/775] of the time, and *a* and *b* about 2% [(11 + 6)/775] of the time. Therefore, *a* and *c* are close; *b* is farther away; *c* is probably in the middle.

34. Based on the time of first appearance of the markers, the order is origin, *a, d, b, c,* and *e*. Let the F factor be located at position 0. Gene *a* appears after only 10 minutes. Therefore *a* is 10 minutes from the origin. Similar logic allows the placement of the other genes.

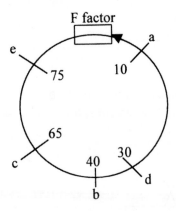

35. a. Since there is no lysine in the medium, lys^+ must be present to allow growth. Genotypes that will grow are
$lys^+ his^+ val^+$
$lys^+ his^+ val^-$
$lys^+ his^- val^+$
$lys^+ his^- val^-$

b. Both lys^+ and val^+ must be present to allow growth.
$lys^+ his^+ val^+$
$lys^+ his^- val^+$

c. Both lys^+ and his^+ must be present to allow growth.
$lys^+ his^+ val^+$
$lys^+ his^+ val^-$

d. No growth on the minimal medium means there are no $lys^+ val^+ his^+$ cells. So, on minimal medium + histidine, only $lys^+ his^- val^+$ will grow; and on minimal medium + valine, only $lys^+ his^+ val^-$ will grow.

e. lys^+ and val^+ are close together; they are cotransformed 75% of the time. The order could be *lys val his* or *val lys his*.

f. *val lys his*. If the order is *val lys his*, $val^+ lys^- his^+$ should be rare, since this genotype results from a double exchange; and indeed, this class is the least frequent.

36. Mix the phages together with bacteria with increasing quantities of the two phages. Knowing the numbers of each in a particular case, it is possible to predict the proportion of cells doubly infected (product of probabilities). The recovery of recombinants should increase with that probability. In other words, recombination should occur only in doubly infected cells.

37. a. A double crossover results in $r^+ t z^+$ and $r t^+ z$ phages. The frequency of double crossovers is $0.02 \times 0.008 = 0.00016$. There are 50,000 phages, so 0.00016

× 50,000 = 8 are expected to be double recombinants. Therefore, 8/2 = 4 are expected to be $r^+ t z^+$.

b. The coefficient of coincidence = (observed double crossovers)/(expected double crossovers) = 30/8 = 3.75. Therefore, interference = 1 – 3.75 = – 2.75. This negative interference implies that one crossover increases the likelihood of an adjacent crossover.

38. This problem is done just like a eukaryotic mapping problem (for example, *Drosophila*). The data are equivalent to offspring from a testcross (really just the gametes from the trihybrid). For example, the 3467 and 3729 classes are the parentals or nonrecombinants; the 162 and 172 classes are double crossovers. Comparison of these two classes reveals that locus *r* is in the middle, and the linkage relationship is *m r tu*. The *m-r* distance = *m-r* recombinants/total offspring = (162 + 172 + 474 + 520)/10,342 = 0.128, or 12.8%. The *r-tu* distance = (162 + 172 + 853 + 965)/10,342 = 0.208, or 20.8%. Expected double crossovers = 0.128 × 0.208 × 10,342 = 275.3. Coefficient of coincidence = (observed double crossovers)/(expected double crossovers) = 334/275.3 = 1.21.

39. Small recombination frequencies should be approximately additive. Note that recombination distances are twice the value of wild-type plaques since the double mutant recombinants were not counted. Thus, the data table should be

Cross	Percent Wild-Type Plaques	Percent Recombinants
a × *b*	0.3	0.6
a × *c*	1.0	2.0
a × *d*	0.4	0.8
b × *c*	0.7	1.4
b × *d*	0.1	0.2
c × *d*	0.6	1.2

The largest distance is between *a* and *c;* therefore *a* and *c* must be at opposite ends. Since *a-b* = 0.6, *b* must be 0.6 units to the right of *a*. This position gives *b-c* as 1.4, the observed distance. We now have the following map:

If *d* is to the left of *b*, then *d-c* should be greater than 1.4, a result not seen. Therefore, *d* is 0.2 units to the right of *b*.

40. Treat this cross like a three-point eukaryotic cross and group reciprocal classes together. The parentals are $a^+ b^+ c^+$ and $a\, b\, c$, whereas the double crossovers are $a^+ b\, c$ and $a\, b^+ c^+$. Since double crossovers rearrange the middle alleles, a is in the middle. Proceed to calculate map distances as follows:

$$b - a = [(371 + 309 + 160 + 178)/6502] \times 100 = 15.7 \text{ map units}$$

$$a - c = [(954 + 879 + 160 + 178)/6502] \times 100 = 33.4 \text{ map units}$$

The complete map is

41. To calculate map distance, you must have the number of recombinations and the total number of progeny. Since all phages grow on strain B, this number must equal the total number of progeny; this is 250×10^7. Since only wild-type phages grow on K12, and since wild types result from recombination between two genes,

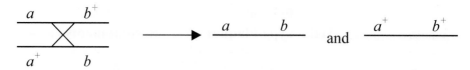

the number that grow on K12 must be recombinants. But this number represents only half of the recombinants, for the double mutant will not grow on K12. Total recombinants are:

$1 \times 2 \quad (2 \times 50)(10^4) = 10^6$
$1 \times 3 \quad (2 \times 25)(10^4) = 5 \times 10^5$
$2 \times 3 \quad (2 \times 75)(10^4) = 1.5 \times 10^6$

Map distance 1–2 $= [(10^6)/(250 \times 10^7)] \times 100$
$\qquad\qquad = (0.4 \times 10^{-3})(100) = 4 \times 10^{-2} = 0.04$
Map distance 1–3 $= [(5 \times 10^5)/(250 \times 10^7)] \times 100$
$\qquad\qquad = (0.2 \times 10^{-3})(100) = 2 \times 10^{-2} = 0.02$
Map distance 2–3 $= [(1.5 \times 10^6)/(250 \times 10^7)] \times 100$
$\qquad\qquad = (0.6 \times 10^{-3})(100) = 6 \times 10^{-2} = 0.06$

42. Gene pairs that are cotransduced must be closely linked together, whereas those that are not cotransduced must be farther apart on the phage chromosome. The data show that gene *a* cotransduces with both *d* and *f*, but *d* and *f* do not exhibit cotransduction. Therefore, *a* and *d* are more closely linked than *d* and *f*. Indeed, gene *f* does not cotransduce with any other gene, indicating that *f* is on one end of the phage genome. In contrast, gene *d* cotransduces with gene *b*. Furthermore, gene *b* does not cotransduce with gene *a*. Therefore, the order of the four genes considered so far must be *f* – *a* – *d* – *b*. Gene *b* cotransduces with both genes *c* and *e*. In addition, genes *c* and *e* exhibit cotransduction, and neither of them cotransduces with *d*, *a*, or *f*. Consequently, the orientation of genes *c* and *e* relative to *b* cannot be determined from the data given.

43. Use replica-plating on selective media with arabinose as the sole carbon source, thus selecting for *ara*⁺ cells. Although all three loci can be cotransduced, the rarity of $ara^+\ leu^-\ ilvH^+$ indicates *leu* is the middle locus *(ara leu ilvH)*. Cotransductance frequencies:

ara to $leu = (9 + 340)/(9 + 340 + 32) = 0.92$
ara to $ilvH = (0 + 340)/(340 + 32 + 9) = 0.89$

44. Examine those crosses that yield a large difference between $ara^+\ leu^+$ and $ara^+\ thr^+$: crosses 1, 4, and 6. Begin with cross 1 and arbitrarily choose a gene order for the two mutants; for example, assume the order is *thr ara-1 ara-2 leu*. Cross 1 can then be diagrammed as

$$thr^+\ ara\text{-}1^+\ ara\text{-}2^-\ leu^+ \quad \text{Donor}$$

$$\overline{\qquad\qquad\qquad\qquad\qquad\qquad}$$

$$thr^-\ ara\text{-}1^-\ ara\text{-}2^+\ leu^- \quad \text{Recipient}$$

Draw the crossovers needed to produce $ara^+\ leu^+$ and $ara^+\ thr^+$. We see that with the preceding order, $ara^+\ leu^+$ results from two sets of exchanges or two pairs of double crossovers. Since double crossovers are rare, such transductants should also be rare. The $ara^+\ thr^+$ class would come about by only one pair of crossovers, and should be found in larger numbers than $ara^+\ leu^+$. However, the results show that $ara^+\ leu^+ \gg ara^+\ thr^+$. Therefore, our initial order is wrong, and the correct order should be *thr ara-2 ara-1 leu*. If you diagram the second order, the numbers are consistent. In cross 4, by similar logic, $ara^+\ leu^+ \ll ara^+\ thr^+$, so the order is *thr ara-1 ara-3 leu*. Since we know *ara-1* is closer to *leu* than *ara-2*, and that *ara-3* is closer to *leu* than *ara-1*, the order becomes *thr ara-2 ara-1 ara-3 leu*. Cross 6 confirms that *ara-3* is closer to *leu* than *ara-2*. Therefore the gene order is *thr ara-2 ara-1 ara-3 leu*.

45. The cotransduction frequency of *leu* and *azi* indicates *leu* is closer to *azi* than to *thr*. Thus, two orders are possible: *leu azi thr* or *azi leu thr*. If the first order is correct, and *leu*⁺ is selected, $leu^+\ azi^r > leu^+\ thr^+$; this prediction fits. However, with the same gene order, and selecting for *thr*⁺, $thr^+\ azi^r > thr^+\ leu^+$; this result is not seen, and the order must be $azi^r\ leu^+\ thr^+$. This second order does predict that $thr^+\ leu^+ > thr^+\ azi^r$. Therefore, the correct gene order is *azi leu thr*.

46. Recall that if two deletion mutations overlap, they cannot give wild-type recombinants; and conversely, wild-type recombinants can be produced when two deletion mutations do not overlap.

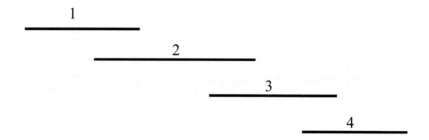

Note that the data do not provide the actual endpoints of the deletions, so other maps can be drawn as well.

47. Failure to produce wild-type recombinants indicates that the mutation must have occurred in a region of the gene that corresponds to the deletion(s) in question. The overlap patterns of the deletions define seven regions: region 1 only, overlap of 1 and 2, 2 only, overlap of 2 and 3, 3 only, overlap of 3 and 4, and 4 only. We can thus localize mutations M to S according to these seven areas:

Mutation M: region 3 only.
Mutation N: region of 2 and 3 overlap.
Mutation O: region of 4 only.
Mutation P: region of 3 and 4 overlap.
Mutation Q: region of 2 only.
Mutation R: region of 1 and 2 overlap.
Mutation S: all regions except for unique and overlap areas of 4.

The following figure shows where each of these mutations map to.

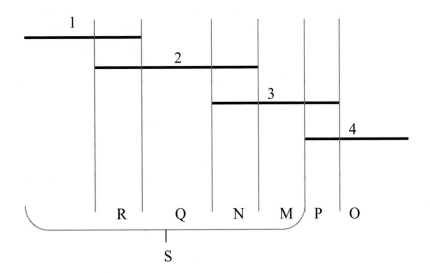

48. It may seem odd that bacteria have evolved the ability to take up DNA from the environment when that DNA came from a dead bacterium. However, we believe that many processes have evolved because of their ability to generate variability in an organism, variability that might make an organism more capable of adapting to a changing environment. This is especially true for haploid organisms that generally accrue variability by mutation. Thus taking up foreign DNA could be a quick way to generate a lot of variability. Also, this DNA may not be "bad" DNA, even though arising from dead organisms. In fact, DNA may only be available in the environment if there is a lot of it. That would indicate an environment in which there is a lot of bacterial growth. That would indicate that the DNA is from successful organisms rather than failed ones.

CHAPTER INTEGRATION PROBLEM

a. The gene order can be determined from the sequence of colony appearance. The first colonies appeared at time "x" on plate 3, which lacks nutrient C. Therefore, these colonies must have a genotype of c^+. The next colonies arose at time "$x + 5$ minutes" on plate 1, which selects for a^+. Colonies that are e^+ (plate 5) appeared at "$x + 10$ minutes," followed by b^+ colonies (plate 2) at "$x + 12.5$ minutes." Gene d^+ is the last to enter the F$^-$ cell, about 5 minutes after gene b^+. The following figure gives the order and relative location of the genes:

b. x is the approximate time at which gene c enters the F⁻ cell. However, F plasmid DNA has to enter first. The *oriT* is in the middle of the F plasmid DNA in the Hfr chromosome, so it is located approximately 250 kbp/2 = 125 kbp from gene c. This length of DNA is equivalent to x and is equal to (125,000 bp × 70 min)/2,500,000 bp = 3.5 minutes.

c. In the first 3.5 minutes of conjugation, only plasmid DNA is transferred. Therefore, the amount of chromosomal DNA transferred is equivalent to 21.5 minutes, which is (21.5 min x 2,500,000 bp)/70 min = 767,857 bp. B-DNA contains 10 base pairs per 3.4 nm (one helical turn). Thus, each base pair represents 0.34 nm. The length of this DNA would be (767,857 bp) × (0.34 nm/bp) = 261,071 nm. To convert to micrometers, simply divide by 1000. Therefore, the length of the transferred chromosomal DNA is approximately 261 μm.

d. Agar type 4 lacks nutrient D and so selects for cells with genotype d^+. One reason for the very few colonies is that gene d^+ is very far from the origin of transfer in the Hfr chromosome. This makes it unlikely that gene d^+ will be transferred to the F⁻ cell. Alternatively, d^+ could be relatively close to the origin, but tightly linked to the str^s allele. Therefore, the cells will pick up both d^+ and str^s. However, recombination between the exogenote and endogenote will render the F⁻ cell sensitive to streptomycin, and thereby unrecoverable.

e. I.   True. 1×10^{-7} is equivalent to 1 in 10 million, and e^+ cells are prototrophic.

    II. False. The frequency of a double revertant is the product of the two individual reversion frequencies. Therefore, $a^+ b^+$ revertants are $(1 \times 10^{-5})(2 \times 10^{-6}) = 2 \times 10^{-11}$ and $c^+ d^+$ revertants are $(3 \times 10^{-5})(1 \times 10^{-6}) = 3 \times 10^{-11}$.

f. The DNA segment has a palindrome (shown in red) and an inverted repeat (indicated by arrows).

```
        ⟹
5'-CAGGTTGGTG CTTCTCACCA CCAAAAGCAC CACACCGGTC-3'
3'-GTCCAACCAC GAAGAGTGGT GGTTTTCGTG GTGTGGCCAG-5'
                              ⟸
```

g. The nick at *oriT* generates the following structure:

```
5' – CTTCTCACCACCAAAAGCAC – 3'

3' – GAAGAG      3' – GTTTTCGTG – 5'
              ╲TGGTG – 5'
```

Rolling circle replication begins at the 3' end of the broken strand, using the unbroken strand as a template. The displaced 5' end of the broken strand (underlined) will enter the F⁻ cell first.

h. Heavy DNA (^{15}N-labeled) is shown in bold, and light DNA (^{14}N-labeled) is shown in regular font.

5'-**CAGGTTGGTG CTTCTCACCA CCAAAAGCAC CACACCGGTC**-3'
3'-GTCCAACCAC GAAGAGTGGT GGTTTTCGTG GTGTGGCCAG-5'

CHAPTER 16 GENE EXPRESSION: CONTROL IN BACTERIA AND PHAGES

Chapter Goals

1. Distinguish between inducible and repressible operons and describe the difference between positive and negative regulation.
2. Diagram how the *lac* operon is regulated by *cis*- and *trans*-acting functions. Describe how expression of the *lac* operon will be affected by different genotypes in merozygotes.
3. Understand the similarities and differences in regulation of the *lac*, *ara*, and *trp* operons and how these differences relate to the biochemical pathway encoded by the structural genes.
4. Understand how phage λ regulates the choice between lysogeny and lysis.
5. Describe additional transcriptional, translational, and posttranslational regulatory mechanisms observed in bacteria and bacteriophages.

Key Chapter Concepts

16.1 The Operon Model

- **Control of Transcription:** While some proteins are produced at all times, most proteins are produced only as required by specific environmental conditions. Bacteria often perform this by coordinately regulating related genes in an **operon.** Proteins needed only when a substrate is present (enzymes that *break down* a substance) are generally controlled by **inducible operons**. Transcription of these genes is "turned on" when their substrate is present. Transcription of genes encoding proteins needed only when a substance is not present (enzymes that *produce* a substance) are generally controlled by **repressible operons**. Transcription of these genes is "turned off" when the cellular levels of the substance are sufficiently high. Both inducible and repressible systems can be under positive regulation, negative regulation, or both. **Positive regulation** involves a transcriptional activator that promotes transcription, while **negative regulation** involves a transcriptional repressor that inhibits transcription. Thus, inducible versus repressible describes how the operon responds to environmental signals, while positive versus negative describes what type of transcription factor controls the operon. Importantly, a particular operon may have multiple and/or redundant levels of control.
  - **Inducible Systems:** Inducible systems are those in which transcription is "turned on" when a particular substance is encountered in the environment. The *lac* operon is the classic example of an inducible system.
  - **Repressible Systems:** Repressible systems are those in which transcription is "turned off" when a particular substance is encountered in the environment. The *trp* operon is the classic example of a repressible operon.
  - **Positive Regulation:** In positive regulation, a **positive regulator** or activator binds to the operon and *helps* RNA polymerase bind, thus increasing transcription.
  - **Negative Regulation:** In negative regulation, a **negative regulator** or repressor binds to the operon and *prevents* RNA polymerase from binding, thus decreasing transcription.

16.2 The Lactose Operon (Inducible System)

- *Lac* **Operon Mechanisms (Negative Regulation):** Three proteins needed for acquiring and metabolizing the sugar lactose are encoded by genes that share transcriptional control. Production of the three proteins occurs only when lactose is available. This operon is inducible since lactose induces its transcription, and it is under negative regulation since a repressor protein controls the induction.
  - **Uninduced**, lactose absent and operon genes *not* transcribed ("turned off"): When lactose is not present, the lactose metabolism proteins are not needed. Therefore, it is advantageous to stop transcription of these three genes by "hiding" the promoter from the RNA polymerase. A DNA sequence located near the promoter (the **operator**) and a protein that is encoded elsewhere in the bacteria's genome (the *lac* **repressor**) work together to hide the promoter. If the promoter is "hidden" from the RNA polymerase by the repressor protein binding to the nearby operator, RNA polymerase cannot initiate transcription; the genes will be "turned off."

o **Induced**, lactose present and operon genes transcribed ("turned on"): When lactose is present, some is converted to allolactose that can bind to an **allosteric site** on the *lac* repressor protein. This binding causes an **allosteric** change in the *lac* repressor structure, reducing its **affinity** for the operator sequence. Thus, the *lac* repressor "falls off" the operator. Without the *lac* repressor hiding the promoter, RNA polymerase can freely bind to the promoter. The promoter functions to position the RNA polymerase and aid in initiating transcription of the three downstream genes whose expression is needed in the presence of lactose.

- *Lac* **Operon Mutations:** Mutations can alter the function of the individual genes of the operon, the promoter of the operon, the repressor protein, or the operator DNA sequence. Operator mutations are **cis-dominant** since they affect only the operon in which they are *physically* located. Mutations in the repressor gene (*lacI* gene) are **trans-acting** since the mutant protein produced can diffuse throughout the cell and affect any *lac* operon. Repressor mutations are generally recessive to a wild-type regulator gene regardless of the *physical* location of the operon. One exception is a dominant repressor mutation called the **superrepressor mutation** or *lacI^S* that binds to the operator regardless of whether or not lactose is present. This is because its mutation is in the allosteric site and thus prevents allolactose from ever binding the *lac* repressor.

16.3 Catabolite Repression
- **Positive Regulation of the *Lac* Operon:** The *lac* operon, as well as most other operons (including the *ara* operon) involved in catabolizing sugars, is also under positive control through the transcriptional activator known as **catabolite activator protein (CAP)**. CAP binds to a CAP-binding site found near the promoters of these operons and helps RNA polymerase to bind to the promoter of the operon. Since bacteria prefer to utilize glucose over any other sugar, the CAP protein ensures that activation of operons such as the *lac* and *ara* operons are only "turned on" in the *absence* of glucose. This regulation occurs because CAP can only bind the CAP site when the allosteric regulator **cyclic-AMP (cAMP)** is bound to CAP. Importantly, cAMP levels are only high when glucose is *absent* in the environment. Thus, CAP–cAMP can only bind operons when glucose is absent. This is known as **catabolite repression**.

16.4 The Arabinose Operon
- *Ara* **Operon Mechanisms:** The *ara* **operon**, like the *lac* operon is inducible when glucose is absent and a particular sugar (arabinose in this case) is present in the environment. The CAP protein provides positive regulation, and the AraC protein (produced from a neighboring gene) is a *trans*-acting factor that acts as both a positive and negative regulator. AraC binds to two different sites in the *ara* operon (inducer and operator sites), and induction in the presence of arabinose is outlined in the following.
    o **Uninduced**, arabinose absent and operon genes *not* transcribed ("turned off"): The AraC proteins bound to the two sites bind to each other causing the DNA to loop, blocking access to the promoter by RNA polymerase. Thus, the genes are "turned off."
    o **Induced**, arabinose present and operon genes transcribed ("turned on"): When present, arabinose can bind to the AraC proteins. This binding prevents the AraC proteins bound at the two sites from interacting with each other. Thus, the DNA loop is released and the *ara* operon promoter is accessible to RNA polymerase.

16.5 The Tryptophan Operon (Repressible System)
- *Trp* **Operon Mechanism (Negative Regulation):** The genes in the *trp* **operon** encode enzymes that are necessary for the synthesis of the amino acid tryptophan. Operons such as this are known as **anabolic operons** since their enzymes synthesize molecules. The *trp* operon is repressible since its transcription is "turned off" in the presence of tryptophan and it is under negative regulation since a repressor protein controls it.
    o **Repressed**, tryptophan present and operon genes *not* transcribed ("turned off"): In repressible systems, the repressor protein does not recognize and bind the operator unless it is in combination with a **corepressor**. Here, the corepressor is tryptophan. An excess of tryptophan "turns off" transcription by forming a corepressor–repressor complex that binds to the operator and "hides" the nearby promoter from RNA polymerase.

    ○   **Derepressed**, tryptophan low and operon genes transcribed ("turned on"): When tryptophan levels are low inside the cell, very little tryptophan will be available to bind the *trp* repressor. An allosteric change then reduces the affinity of the repressor for the operator sequence. The repressor alone cannot bind the operator and "falls off" the DNA so that the nearby promoter is left freely available to RNA polymerase.

16.6 The Tryptophan Operon (Attenuator-Controlled System)

- **Attenuation Systems:** A common method of regulating amino acid biosynthetic operons involves attenuation systems in which translation of the **leader peptide gene** of an mRNA from the operon requires a specific amino acid. If the amino acid is present, translation occurs efficiently, causing termination (attenuation) of transcription. If the amino acid is not abundant, translation "stalls" allowing transcription to continue.

- *Trp* **Operon Mechanism (Attenuation):** The *trp* operon is also regulated by a mechanism called attenuation. The **attenuator region** is a nucleotide sequence between the promoter and the genes of an operon. The **leader transcript** is the beginning of the mRNA transcribed from the operon that contains the attenuator region. The mRNA sequence of the leader transcript can form two different RNA structures as it is being transcribed by RNA polymerase. One structure is tolerated (**antiterminator stem**), while the other structure (**terminator** or **attenuator**) causes the RNA polymerase to stop transcription. The rate of *translation* of the **leader peptide gene** is actually what determines which of the two RNA structures are made, and thereby, whether or not the *trp* operon genes are transcribed. When levels of tryptophan are low, translation is slowed because the ribosomes "stall" at tryptophan codons. This stalling allows the "tolerated" RNA structure to form, and transcription of the operon is completed. However, with sufficiently high levels of tryptophan, translation speed is high (no stalling of the ribosomes) allowing formation of the RNA structure that causes the RNA polymerase to stop transcription. This method of control is partially redundant with the *trp* repressor control but also allows regulation in response to shortages in other amino acids.

- **TRAP Control:** In a different bacteria, *Bacillus subtilis*, attenuator formation in the leader transcript, and thus inhibition of transcription, only occurs when a protein known as *trp* **RNA-binding attenuation protein (TRAP)**, together with tryptophan, binds to the leader transcript. If no tryptophan is present, TRAP will not bind the leader transcript and the attenuator structure will not form.

16.7 Lytic and Lysogenic Cycles in Phage λ

- **Lytic and Lysogenic Cycles in Phage λ:** Control of gene expression is also important in bacteriophages. The bacteriophage λ (lambda) can enter either a **lysogenic** or **lytic** cycle. The lytic cycle is favored when the bacteria are actively growing, while the lysogenic cycle is favored when bacterial growth is limited. Which cycle is entered depends upon which transcriptional repressor accesses two critical operator sites in the phage genome known as the left and right operators. Lysogeny is promoted by the CI repressor, while the lytic cycle is promoted by the Cro repressor. CI accumulates and promotes lysogeny when the levels of the bacterial protease FtsH that degrades CII are low. Thus CII, a transcriptional activator of CI, can accumulate. A complete list of the gene products involved in phage λ infection including promoters, terminators, and antiterminators is provided in Table 16.1 of your textbook.

16.8 Other Transcriptional Control Systems

- **Promoter Sequences:** Bacterial promoters may have slightly different sequences in their promoters. Different promoter sequences may bind to different sigma factors of RNA polymerase. For instance, the promoters of heat shock genes are only recognized by σ^{32} which is only expressed after heat shock. Furthermore, the affinity or "tightness" of binding of RNA polymerase or other DNA binding proteins, such as a repressor, can also influence transcription. If the promoter sequence is the exact sequence of nucleotides (consensus) with which the DNA binding protein interacts, then the protein will bind the DNA tightly and perform its job efficiently, be it attracting RNA polymerase or initiating transcription. Deviations from the consensus sequence reduce the efficiency of binding because the interaction is imperfect.

16.9 Translational Control

- **Translational Polarity:** The position of a gene within a polycistronic mRNA can influence the degree to which it is translated. The genes that are the last to be transcribed are translated less efficiently than those that are transcribed first and thus do not produce as much protein. It is believed that this phenomenon is a result of the following: (1) genes that are transcribed first are available to be translated relatively longer, (2) mRNAs are degraded from the 3′ end, and (3) genes whose products are not needed in high amounts means that they are generally located at the 3′ ends of operons.
- **Antisense RNA:** RNA–RNA hybridization can also function as a regulatory mechanism. If a complementary RNA (**antisense RNA**) anneals to the 5′ end of an mRNA, translation can be inhibited since the Shine–Dalgarno sequence is "hidden."
- **Riboswitches:** The 5′ UTR of a bacterial mRNA may contain a **riboswitch** whose secondary structure can change depending on whether a small molecule is bound. If present, the small molecule binds the **aptamer domain** of the mRNA, changing the secondary structure in the **expression domain**. Some expression domains can alter transcription by altering the creation of a Rho-independent terminator, while others can alter translation by altering the accessibility of the Shine–Dalgarno sequence.
- **Ribosome-Binding Efficiency:** The binding of the ribosome to the mRNA can influence translation efficiency. Recall that the Shine–Dalgarno sequence is the part of the mRNA where the ribosome binds due to an interaction of the mRNA with the 16S ribosomal RNA. If the Shine–Dalgarno sequence is the exact complement (consensus) to the ribosomal RNA, the mRNA will be translated efficiently. Deviations from the Shine–Dalgarno sequence reduce the efficiency of translation because the binding of the ribosomal RNA is imperfect.
- **Codon Preference:** The nucleotide sequence found in the mRNA molecule can also influence the rate of translation. Some codons for a given amino acid are used more frequently than others in an organism (known as **codon bias**). The tRNAs for the frequently used codons are abundant in cells, while the tRNAs for rarely used codons are rare. Therefore, a greater percentage of "preferred" codons results in increased translation speed, while the process is slowed by an abundance of codons specifying rare tRNAs.
- **Amino Acid Starvation:** A final translational control mechanism is the **stringent response** which regulates translation in response to a lack of amino acids. Since amino acids are missing, there is no need to produce more ribosomes. Thus, if uncharged tRNAs begin entering the ribosome (a sign of amino acid starvation called the **idling reaction**), a protein (**stringent factor**) produces a signaling molecule (guanosine tetraphosphate) that interacts with RNA polymerase causing it to decrease transcription of rRNA genes and to increase transcription of amino acid biosynthetic operons.

16.10 Posttranslational Control

- **Feedback Inhibition:** In this case, the product of a biosynthetic pathway turns off the pathway by binding and allosterically inactivating an enzyme in its own biosynthetic pathway. This is usually the first enzyme in the pathway. An example of this mechanism can be observed when looking at the pathway that synthesizes cytidine triphosphate (CTP). When excess CTP is available in the cell, it can bind and temporarily inactivate aspartate transcarbamylase, the first enzyme in the pathway that synthesizes CTP.
- **Protein Degradation:** Posttranslational control can also be achieved through the differential degradation of proteins. Proteins with certain amino acids at the amino terminal end have longer lives than proteins with other amino acids at the amino terminal end (**N-end rule**). For example, a protein with methionine, serine, alanine, threonine, valine, or glycine at the amino end will have a longer life in the bacterial cell (> 20 hours) than the same protein with an arginine at the amino end (2 minutes). Another mechanism of controlling protein degradation known as the **PEST hypothesis** depends on whether the protein has a particular arrangement of proline (P), glutamic acid (E), serine (S), and threonine (T) amino acids. Proteins that are rich in these PEST elements are short-lived.

Key Terms

- *lac* operon
- inducible genes

- inducer
- repressible gene

- repressor
- positive regulation

- positive regulator
- negative regulation
- negative regulator
- allosteric site
- operon
- β-galactosidase
- *lac* permease
- transacetylase
- *lac* repressor gene
- operator
- allosteric protein
- affinity
- *cis*-dominant
- *trans*-acting
- *cis*-acting
- *lacI*^{S} superrepressor mutation
- *lacI*$^-$ promoter mutation
- catabolite repression

- cyclic-AMP (cAMP)
- catabolite activator protein (CAP)
- arabinose operon (*ara* operon)
- anabolic operons
- tryptophan operon (*trp* operon)
- corepressor
- derepressed
- attenuator region
- leader transcript
- leader peptide gene
- terminator
- attenuator
- antiterminator stem
- *trp* RNA-binding attenuation protein (TRAP)
- left operon
- right operon

- late operon
- repressor region
- site-specific recombination
- antiterminator protein
- cooperativity
- anti-sigma factor
- translational polarity
- antisense RNA
- riboswitches
- aptamer domain
- expression domain
- codon bias
- stringent response
- idling reaction
- stringent factor
- feedback inhibition
- N-end rule
- PEST hypothesis

Understanding the Key Concepts

Use the key terms or parts of their definitions to complete the following sentences.

Bacterial genes are often expressed as a (1)_____ mRNA transcribed from an operon. The promoters of these operons are controlled by (2)_____-acting factors that bind to operators which are (3)_____-acting elements near the promoter. If the protein that binds the operator is a repressor of transcription, or (4)_____, the operon is under (5)_____ regulation while operons that are bound by transcriptional activators, or (6)_____, are under (7)_____ regulation. Operons expressing catabolic enzymes are usually (8)_____ by the substrate metabolite, or a derivative of it, while operons expressing anabolic enzymes are usually (9)_____ by the (10)_____ of the biosynthetic pathway. For instance, the (11)_____ is repressed by binding of the (12)_____ and tryptophan to the operator, while the (13)_____ is induced by the presence of (14)_____ through the removal of the *lac* repressor protein from the (15)_____. In the latter case, (16)_____ acts as the inducer, and tryptophan acts as the (17)_____ in the former case. Another way that anabolic pathways are regulated is through (18)_____ of the first enzyme in the biosynthetic pathway. In these situations, the small molecules elicit (19)_____ effects on their respective proteins by binding to one site, called the (20)_____, and causing an effect at a different place on the protein through an induced conformational change.

A second method exists in bacteria for the regulation of amino acid biosynthetic operons. This involves the creation of either (21)_____ or (22)_____ stem-loop structures in the beginning of the mRNAs called the (23)_____ in order to regulate (24)_____. Which structure is formed in this region depends on whether or not the (25)_____ is sitting on region 2. If it is, due to (26)_____ tryptophan levels, then the (27)_____ can form.

After an mRNA is transcribed, the expression of an active protein may still be regulated in several ways. First, (28)_____ may be inhibited due to an (29)_____ that is complementary to the (30)_____UTR of an mRNA and blocking the Shine–Dalgarno sequence. Alternatively, the formation of a secondary structure in the 5′ UTR by a (31)_____ may inhibit either transcription or translation depending on what sequences are present in the (32)_____ domain. Finally, the (33)_____ ORFs of a polycistronic mRNA generally produce more proteins as do ORFs that mostly contain codons of abundant tRNAs, thus fitting the organism's (34)_____. Finally, in the (35)_____, a stringent factor produces (36)_____ when (37)_____ levels are low in the cell. This inhibits transcription of (38)_____ and activates transcription of genes encoding enzymes that synthesize amino acids. After a protein is synthesized, its degradation may be regulated such that proteins with a (39)_____ sequence are degraded more quickly than those lacking this sequence. Similarly, the identity of the amino acid at the (40)_____ terminus of the protein also dictates the rate at which the protein is degraded. Having an (41)_____ at this position will cause the protein to be degraded the most rapidly.

The bacteriophage λ can enter either a (42)_____ or a lysogenic life cycle after infecting bacteria. The lysogenic life cycle occurs when bacterial growth is limited and the other life cycle choice would not leave any living bacteria for propagation. In this cycle, an integrase protein promotes (43)_____, causing the phage genome to integrate into the bacteria genome. At a later time, it can be excised out of the genome by (44)_____. The decision to enter a particular life cycle is mainly regulated by binding of the (45)_____ and (46)_____ repressors to operators in the phage genome that control transcription of specific genes. The (47)_____ repressor promotes lysogeny by binding the right and left operators. It only accumulates to high enough levels to do so when the (48)_____ protein inhibits the degradation of the (49)_____ protein which can then activate expression of the *cI* gene.

Figure Analysis
Use the indicated figure or table from the textbook to answer the questions that follow.

1. **Figure 16.7**
   (a) Use the terms *positive regulation* or *negative regulation* and *inducible* or *repressible* to describe the system depicted in this figure.

   (b) Which gene depicted in figure 16.7 is not part of the *lac* operon?

   (c) Which component of the figure binds to the *lac* operator sequence?

   (d) Which gene encodes the repressor protein?

   (e) Which component of the figure is considered the inducer?

   (f) Under what conditions does transcription of the operon occur? Under what conditions does the *lac* repressor protein bind to the *lac* operator?

2. **Figures 16.10 and 16.11**
   (a) What was the purpose of performing the experiments depicted in these figures?

   (b) Why was one copy of the *lacZ* and *lacY* genes mutated in this experiment?

   (c) Which constitutive mutation is *cis*-acting and which is *trans*-acting?

3. **Figures 16.23 and 16.24**
   (a) Under what two conditions does the rho-independent transcription termination signal form?

   (b) Which two regions base-pair to form the terminator? Which two base-pair to form the antiterminator stem?

(c) Describe why the sequence of U's after the terminator/attenuator is important. Why can't this string of U's still provide the same function when the antiterminator stem forms?

(d) How does the antiterminator stem prevent formation of the terminator/attenuator?

(e) How does the ribosome prevent formation of the antiterminator stem when tryptophan levels are high? Why can't the ribosome do this when tryptophan levels are low?

General Questions
Q16-1. Why is there a need to regulate gene expression in organisms?

Q16-2. Describe two types of mutations in the *lacI* gene that could cause the effects that the *lacI*^{S} superrepressor mutation displays.

Q16-3. Why are attenuation systems not found in eukaryotic organisms?

Q16-4. Why does catabolite repression exist in bacteria?

Q16-5. Arabinose and allolactose are both inducers of metabolic operons. Describe the effect that each has on its respective regulatory protein *lac* repressor and AraC protein, respectively.

Q16-6. What aspect of the bacteria determines whether the lytic or lysogenic cycle develops after infection by phage λ? If a cell is infected with phage λ possessing a defective *cI* gene, will lytic or lysogenic development occur? Which will occur if the phage possesses a defective *cro* gene?

Multiple Choice
For each of the following, circle the letter of the choice that most appropriately answers the question.

1.) Which of the following operons would be considered to be under positive regulation?
   a. An operon where a transcriptional activator binds to the operator only when it is bound to the sugar raffinose.
   b. An operon where a transcriptional activator binds to the operator only when it is not bound to cAMP.
   c. An operon where a transcriptional repressor binds to the operator only when it is bound to the sugar raffinose.
   d. An operon where a transcriptional repressor binds to the operator only when it is not bound to cAMP.
   e. Both a and b

2.) Which of the following operons would most likely be considered inducible?
   a. an operon that contains genes that encode enzymes that synthesize uracil
   b. an operon that contains genes that encode enzymes that synthesize histidine
   c. an operon that contains genes that encode enzymes that metabolize raffinose
   d. an operon that contains genes that encode enzymes that transport leucine into the cell
   e. Both a and b

3.) Allolactose is derived from (i) _____
and controls the *lac* operon by binding directly to
the (ii)_____.

    a. i) galactose   ii) operator
    b. i) lactose   ii) operator
    c. i) galactose   ii) *lac* repressor
    d. i) lactose   ii) *lac* repressor
    e. i) lactose   ii) β-galactosidase
                                      enzyme

4.) Which of the following is an inducer?
    a. tryptophan
    b. allolactose
    c. *lac* repressor
    d. CAP
    e. *lac* operator

5.) The phenotype of which of the following
mutations cannot be restored (complemented) by
a wild-type operon on an F′ plasmid?
    a. operator mutation
    b. dominant *lac* repressor mutation
    c. recessive *lac* repressor mutation
    d. recessive *lacZ* mutation
    e. Both a and b

6.) When are cAMP levels low in a bacterial
cell?
    a. when glucose concentrations are high
    b. when glucose is absent
    c. when lactose concentrations are high
    d. when lactose is absent
    e. when allolactose concentrations are higher
       than glucose concentrations

7.) Consider the hypothetical *lys* operon. This
operon encodes three genes required for the
synthesis of the amino acid lysine. What effect
would lysine most likely cause if it bound to the
allosteric site on a transcriptional activator of the
lys operon?
    a. It would cause the activator to bind to the
       lys operator.
    b. It would cause the activator to "fall off" the
       lys operator.

    c. It would cause the activator to recruit RNA
       polymerase to the *lys* promoter.
    d. It would cause the activator to form an
       antiterminator stem in the leader
       transcript.
    e. None of the above are likely to occur.

8.) Which of the following would inhibit entry
into the lysogenic cycle after infection of
bacteria by phage λ?
    a. high levels of the bacterial FtsH protease
    b. amino acid starvation of the bacteria
    c. limited bacterial growth
    d. a mutation causing constitutive activation
       of the CII transcription factor
    e. mutation of the Cro protein

9.) Which of the following does *not* help to
ensure that heat shock genes are expressed only
after heat shock in bacteria?
    a. The promoter of the σ^{32} gene is activated
       by heat.
    b. The σ^{32} protein is stabilized by heat.
    c. The Pribnow boxes of heat shock genes
       contain a sequence that is not recognized
       by σ^{70}.
    d. The σ^{32} protein only binds to the promoters
       of heat shock genes at elevated
       temperatures.
    e. The Pribnow box of heat shock genes is
       bound by σ^{32}.

10.) Imagine that the *ura* operon encodes the
three genes *uraA*, *uraB*, and *uraC* that are all
required for the biosynthetic pathway for the
production of uracil. UraA is required first,
followed by UraB, and finally UraC.
Transcription of the *ura* operon is controlled by
the repressor protein UraR binding to the *ura*
operator. Which two proteins would most likely
be subject to allosteric regulation by uracil?
    a. UraA and UraB
    b. UraA and UraR
    c. UraC and UraR
    d. UraB and UraC
    e. UraA and UraC

Practice Problems

P16-1. What are the consequences of these *lac* operon mutations when glucose is absent?

[*Hint 1*: A mutated repressor can disrupt expression of any susceptible promoter in the genome
(*trans*-acting), while a mutation in a promoter or operator will affect only genes on the same chromosome
(*cis*-dominant).

Hint 2: When determining operon function, assign phenotypes to the genomic DNA and F′ plasmid DNA
separately before considering their joint action.]

a.) an operator that is unable to bind the repressor

b.) a promoter that is unable to bind RNA polymerase

c.) a repressor that cannot bind allolactose

d.) a repressor that cannot bind the operator

e.) mutant in part (a) + an F′ plasmid-encoded operator

f.) mutant in part (a) + an F′ plasmid-encoded repressor

g.) mutant in part (c) + an F′ plasmid-encoded repressor

h.) mutant in part (d) + an F′ plasmid-encoded repressor

i.) $lacI^S$ $lacO^+$ $lacZ^+$ $lacY^-$ $lacA^+$ and F′ $lacI^+$ $lacO^+$ $lacZ^-$ $lacY^+$ $lacA^-$

j.) $lacI^-$ $lacO^+$ $lacZ^+$ $lacY^-$ $lacA^+$ and F′ $lacI^+$ $lacO^+$ $lacZ^-$ $lacY^+$ $lacA^-$

k.) $lacI^S$ $lacO^+$ $lacZ^+$ $lacY^-$ $lacA^+$ and F′ $lacI^+$ $lacO^c$ $lacZ^-$ $lacY^+$ $lacA^-$

P16-2. Consider a pool of mutations that are identified as A, B, C, and D in a hypothetical inducible operon under negative regulation. Given the consequences of each mutation (listed in the given table), identify in which of the following locations it is possible for the mutation to be:

(*Hint 3*: Mutations of structural genes result in the production of defective products.)

Promoter
Operator
Repressor allosteric site
Repressor DNA binding domain
Gene encoding enzyme 1
Gene encoding enzyme 2

Mutation	Enzyme 1	Enzyme 2	Possible Locations
A	Inducible	None	
B	Constitutive	Constitutive	
C	None	Inducible	
D	None	None	

P16-3. Consider the mutations described in the left-hand column of the following table. Identify whether the *lac* operon will be *on* or *off* in each of the four different environments listed across the top.

Mutation	No Lactose or Glucose	Lactose Only	Lactose and Glucose	Glucose Only
Wild-type operon				
CAP cannot bind CAP site				
CAP cannot bind cAMP				
Lac repressor cannot bind allolactose				
Lac repressor cannot bind operator				

P16-4. On the *lac* operon molecules provided, indicate the location of the RNA polymerase, the repressor protein, allolactose, and the mRNA molecule during induced transcription of the *lac* operon and during uninduced transcription.

a.) Induced

| Repressor gene | Promoter | Operator | β-galactosidase gene |

b.) Uninduced

| Repressor gene | Promoter | Operator | β-galactosidase gene |

P16-5. Which of the genes in each of the following pairs would most likely produce more protein?

(*Hint 4*: Codon biases for MS2 bacteriophage are given in table 16.2 of your text.)
  a.) *lacZ* or *lacA*
  b.) *trpE* or *trpA*
  c.) **Gene A** with Pribnow box 5′-TATACT-3′ or **gene B** with Pribnow box 5′-TATAAT-3′
  d.) **Gene C** with 17 base pairs between the –35 and –10 sequences or **gene D** with 13 base pairs between the –35 and –10 sequences
  e.) **Gene E** with Shine–Dalgarno sequence 5′-AGGAGGU-3′ or **gene F** with Shine–Dalgarno sequence 5′-AGCAGGA-3′
  f.) MS2 bacteriophage **gene G** with a Gly codon distribution of
       20% GGU, 23% GGC, 27% GGA, and 30% GGG or MS2 bacteriophage **gene H** with a Gly codon distribution of
        48% GGU, 37% GGC, 7% GGA, and 8% GGG

Assessing Your Knowledge

Understanding the Key Concepts—Answers

1.) polycistronic; 2.) *trans*; 3.) *cis*; 4.) negative regulator; 5.) negative; 6.) positive regulators; 7.) positive; 8.) inducible or induced; 9.) repressible or repressed; 10.) end product; 11.) *trp* operon; 12.) *trp* repressor or TrpR protein; 13.) *lac* operon; 14.) lactose; 15.) operator; 16.) allolactose; 17.) corepressor; 18.) feedback inhibition; 19.) allosteric; 20.) allosteric site; 21.) terminator or attenuator; 22.) antiterminator; 23.) attenuator region or leader transcript; 24.) transcription; 25.) ribosome; 26.) high or sufficient; 27.) terminator or attenuator; 28.) translation; 29.) antisense RNA; 30.) 5′; 31.) riboswitch; 32.) expression; 33.) first; 34.) codon bias; 35.) stringent response; 36.) guanosine tetraphosphate; 37.) amino acid; 38.) rRNA genes; 39.) PEST; 40.) amino or N-; 41.) arginine or Arg; 42.) lytic; 43.) site-specific recombination; 44.) excisionase; 45.) CI; 46.) Cro; 47.) CI; 48.) CIII; 49.) CII.

Figure Analysis—Answers

1a.) Inducible, positive regulation; b.) *lacI*; c.) The *lac* repressor protein binds to the *lac* operator. d.) *lacI*; e.) Allolactose, which is derived from lactose; f.) Transcription of the operon only occurs when lactose is present *and* glucose is not present. The *lac* repressor protein binds to the operator whenever lactose is *not* present, regardless of whether glucose is present or not.

2a.) Researchers were trying to determine whether mutations that constitutively activated the *lac* operon were in the *lacI* gene or in the *lac* operator. These experiments also corroborated the prediction by Jacob and Monod that the *lacI* gene encoded a *trans*-acting factor while the *lac* operator was a *cis*-acting element; b.) This was necessary in order to be able to determine whether the genomic operon or the plasmid-encoded operon was being expressed. One operon only expressed a functional LacZ protein, while the other would only express a functional LacY protein; c.) The constitutive *lac* operator mutation is *cis*-acting and the *lacI* mutation is *trans*-acting.

3a.) The terminator forms under high levels of tryptophan or during amino acid starvation; b.) Regions 3 and 4 base-pair to form the terminator; regions 2 and 3 base-pair to form the antiterminator; c.) The string of U's forms a very weak interaction with the template strand of DNA during transcription. When the

RNA polymerase is "stalled" over this region (because of the terminator), the interaction is too weak to sustain itself and the RNA and RNA polymerase "fall off" the DNA and transcription is terminated. When the antiterminator forms, the string of U's is still present. The terminator, however, is absent so that the polymerase does not "stall" on the string of U's. Thus, the U-rich RNA and T-rich DNA do not need to hold together for as long; d.) If region 2 is accessible when region 3 is synthesized, the two will base-pair with each other (forming the antiterminator stem) since region 4 has not yet been synthesized. Once region 4 is transcribed, region 3 is already base-paired to region 2 and cannot form a third interaction; e.) When tryptophan levels are high, the ribosome can quickly translate through the leader peptide gene because tRNA^{trp} is readily available. Thus, at the time that region 3 is being synthesized, the ribosome has moved past region 1 and is covering region 2, preventing region 2 and 3 from base-pairing. When tryptophan levels are low, however, the ribosome stalls at Trp codons in region 1, preventing it from reaching region 2. Thus region 2 *is* available to base-pair with region 3 to form an antiterminator stem.

General Questions—Answers

Q16-1. Control of expression is important for several reasons. First, without control processes, the cell would produce all of its proteins in large quantities whether they were needed or not. This unnecessary, energy-consuming process would be very costly to the cell. Second, there may be times when certain proteins are needed in greater or lesser quantity and there must be a way of suppressing or enhancing the activity of specific genes. Finally, control of gene expression in eukaryotic organisms is an important aspect of cell differentiation and tissue specialization (see chapter 17).

Q16-2. The simplest mutation that would cause the superrepressor effect would be a mutation in the allosteric site of the *lac* repressor such that allolactose could not bind. Thus, the *lac* repressor could never be removed from the operator by allolactose. The second type of mutation would be a mutation that still allowed allolactose to bind the allosteric site but prevented the conformational changes in the *lac* repressor that normally alters its DNA binding ability. Thus, even when allolactose was bound, the *lac* repressor could still bind the operator.

Q16-3. In attenuation systems, the speed of translation controls transcription. This requires the processes of transcription and translation to be coupled, as they are in prokaryotic cells. In eukaryotic cells, transcription takes place in the cell nucleus while translation occurs in association with the ribosomes in the cytoplasm.

Q16-4. In order to utilize sugars other than glucose for energy, bacteria must expend energy to first convert them to glucose. Therefore, if glucose is already present in the environment, bacteria prefer to simply use that glucose and stop making glucose from other sugars. Catabolite repression ensures that bacteria do not even waste energy making the enzymes that produce glucose from other sugars when glucose is present.

Q16-5. Allolactose binding to the *lac* repressor causes a conformational change in the DNA binding domain of the *lac* repressor, reducing its affinity for the operator. Thus, it falls off the DNA. Arabinose binding to the AraC protein causes a conformational change in a protein binding site that disrupts the interaction between AraC proteins. Thus, the interaction between proximal and distal AraC binding sites is released and the DNA loop is released. The AraC protein, however, remains bound to the DNA.

Q16-6. The healthiness, or growth state, of the bacteria helps to determine whether the lytic or lysogenic life cycle develops. Lysogeny is preferred when bacterial cell growth is limited, and the lytic cycle is preferred when the bacteria are actively growing. The *cI* gene is associated with a lysogenic cycle, while the *cro* gene is associated with a lytic cycle. If the *cI* gene is defective, lytic development will occur. Lysogenic development will occur if the *cro* gene is defective.

Multiple Choice—Answers

1.) e; 2.) c; 3.) d; 4.) b; 5.) e; 6.) a; 7.) b; 8.) a; 9.) d; 10.) b.

Practice Problems—Answers

P16-1. a.) Always on. The mutation prevents binding of the repressor protein even when lactose is absent. Thus, it cannot be turned off (uninduced) and transcription of the *lac* operon is constitutive; b.) Always off. A mutated promoter that does not allow binding of the RNA polymerase will not allow transcription of these genes to occur even if lactose is present. As a result, transcription is prevented and the operon is always "off" (cannot be induced); c.) Always off. A mutated repressor that cannot bind allolactose will result in no transcription even when lactose is present because allolactose can *never* remove the repressor and, therefore, RNA polymerase can never access the promoter. Thus, the *lac* operon cannot be induced; d.) Always on. A mutated repressor cannot bind to the operator when lactose is absent. This results in constitutive expression of the *lac* operon; e.) Always on. Constitutive expression from this operon will still occur. Since the operator is *cis*-acting, only the operator that is *physically* attached to the operon (mutant operator) can control the operon; f.) Always on. Constitutive expression from this operon will also still always occur since the genomic operon already has a wild-type repressor; g.) Always off. Transcription from this operon cannot be induced because the genomic mutant repressor is a dominant mutant and thus cannot be complemented. The mutant repressor that is forever bound (also known as a superrepressor) will never allow RNA polymerase to bind the promoter; h.) Inducible. This situation will allow the operon to be inducible and under control again. The plasmid-encoded wild-type repressor will be transcribed and translated in the bacteria, as will the mutant genome-encoded repressor. The wild-type repressor protein can then diffuse to and bind the operator if no lactose is present. This wild-type repressor can then be released if the inducer (allolactose) is present. The mutant repressor is recessive and can be complemented; i.) All genes Always off. The genomic dominant superrepressor will diffuse throughout the cell and repress both the genomic and the F′ plasmid-encoded operons even when lactose is present. Thus, the LacZ, LacY, and LacA proteins will never be made; j.) All genes inducible. Both operons will be inducible and under normal control. The wild-type repressor protein can diffuse to and bind the operators of both operons if no lactose is present. This wild-type repressor can then be released if the inducer (allolactose) is present. The genomic mutant repressor is recessive and can be complemented. Thus, the LacZ, LacY, and LacA proteins will all be made only when lactose is present; k.) LacZ and LacA Proteins never made and LacY protein always made. The genomic repressor is a dominant mutant that will always inhibit transcription of any operon in the cell with a *normal* operator. Thus, the genomic operon will never be expressed and the LacZ and LacA proteins will never be made. The F′ plasmid-encoded operon, however, has a constitutively active operator mutation that will prevent the superrepressor (and the normal repressor) from binding. Thus, the F′ plasmid-encoded operon can never be turned off, even in the absence of lactose. Since the only functional gene in the F′ plasmid-encoded operon is *lacY*, LacY will be the only protein produced and it will be constitutive.

P16-2. A: Enzyme 2
B: Operator or repressor DNA binding domain
C: Enzyme 1
D: Promoter or repressor allosteric site

P16-3.

Mutation	No Lactose or Glucose	Lactose Only	Lactose and Glucose	Glucose Only
Wild-type operon	Off	On	Off	Off
CAP cannot bind CAP site	Off	Off	Off	Off
CAP cannot bind cAMP	Off	Off	Off	Off
Lac repressor cannot bind allolactose	Off	Off	Off	Off
Lac repressor cannot bind operator	On	On	Off	Off

P16-4. a.) Induced

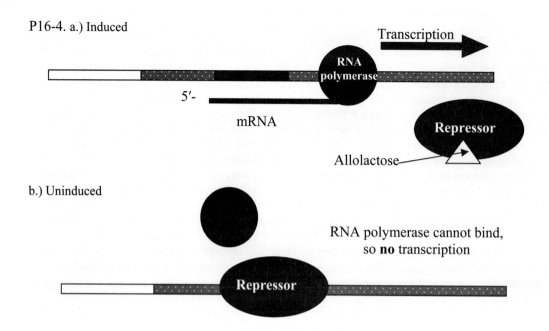

b.) Uninduced

RNA polymerase cannot bind,
so **no** transcription

P16-5. a.) *lacZ*; b.) *trpE*; c.) **gene B**; d.) **gene C**; e.) **gene E**; f.) **gene H**.

16

GENE EXPRESSION: CONTROL IN BACTERIA AND PHAGES

CHAPTER SUMMARY QUESTIONS

1. 1. d, 2. j, 3. g, 4. e, 5. h, 6. a, 7. i, 8. f, 9. b, 10. c.

2. The enzyme adenylyl cyclase catalyzes the conversion of ATP to cAMP. Adenylyl cyclase is inhibited glucose. Therefore, when glucose levels are high, adenylyl cyclase is inhibited, and cAMP levels in the cell are low. When glucose levels are low, adenylyl cyclase is active, and cAMP levels in the cell are high. Thus, cAMP levels are inversely correlated to glucose levels. The catabolite activator protein (CAP) binds cAMP; this binding is required for CAP to activate the transcription of operons encoding enzymes that metabolize other sugars. When glucose levels are high, cAMP is low, CAP is inactive, and operons that encode enzymes for the metabolism of sugars other than glucose are not activated. This ensures that the cell will metabolize glucose when it is available. When glucose levels are low, cAMP is high, CAP is active, and operons encoding enzymes that metabolize other sugars may be activated.

3. For an inducible operon to work through positive control only, an activator would have to be unable to bind to an inducer sequence in the DNA in the absence of a metabolite. In the presence of the metabolite, the activator protein would bind to the metabolite, undergo a conformational change, and then bind to the inducer sequence to activate transcription.

4. In the *trp* operator, the TrpR repressor protein cannot bind to the operator sequence in the absence of tryptophan; in this situation, the lack of repression permits transcription of the genes of the *trp* operon. When tryptophan is present, it acts as a corepressor; in other words the TrpR protein binds to tryptophan and then binds to the operator to repress transcription.

   The attenuator system is also responsive to tryptophan concentrations. When tryptophan is scarce, the levels of charged tRNA^{Trp} are low. In the leader peptide, 2 out of 14 amino acids are tryptophan. When tRNA^{Trp} levels are low, the ribosome pauses at the two tryptophan codons, allowing the mRNA to form the antiterminator stem. Formation of this mRNA secondary structure prevents formation of a terminator structure in the mRNA, and the remaining genes of the *trp* operon will be transcribed. When tryptophan is abundant, tRNA^{Trp} levels are not limiting, so the ribosome will not pause at the tryptophan codons. In this situation, the terminator stem-loop structure forms. This structure is a rho-independent transcription terminator, and transcription of the operon will be terminated before the structural genes of the *trp* operon are transcribed. (See figures 16.21, 16.22, 16.23, and 16.24.)

   In the *trp* operon the operator and attenuator systems work together. When tryptophan levels are low, the operon is derepressed because the TrpR repressor cannot bind to the operator, and transcription is not attenuated. Both mechanisms ensure that the enzymes for tryptophan synthesis will be transcribed when tryptophan levels are low. When tryptophan levels are high, the operon is repressed because tryptophan acts as a corepressor for TrpR, and transcription is attenuated. Again, these systems work together to repress transcription of the structural genes when tryptophan concentrations are high.

   When other amino acids are scarce, transcription of the *trp* operon is attenuated, as shown in figure 16.24.

5. Begin by constructing a table including the genotypes of the merodiploids and the synthesis of functional TrpD and TrpE proteins in the absence or the presence of tryptophan. For clarity, assume that when both the operator and attenuator systems are working properly, there is no synthesis of the structural genes of the *trp* operon in the presence of tryptophan (indicated by a – symbol) and high levels of synthesis (indicated by +++) in the absence of tryptophan. When only one of these systems is working properly, the system will be leaky, so in the absence of tryptophan, there is a low level of synthesis (+).

The *trpP, trpO,* and *trpL* mutants will act in *cis*, and the *trpR* mutant will act in *trans*. The following table summarizes the expected results of these mutants.

The repressor–corepressor system can work with any type of operon. However, the attenuator system will only work with operons that control amino acid synthesis because it involves charged tRNAs.

Genotype	Without Tryptophan		With Tryptophan	
	TrpE	TrpD	TrpE	TrpD
$trpR^+trpP^+trpO^+trpL^+trpE^-trpD^+$ / $trpR^+trpP^+trpO^+trpL^+trpE^+trpD^-$	+++	+++	−	−
$trpR^-trpP^+trpO^+trpL^+trpE^-trpD^+$ / $trpR^+trpP^+trpO^+trpL^+trpE^+trpD^-$	+++	+++	+	+
$trpR^+trpP^-trpO^+trpL^+trpE^-trpD^+$ / $trpR^+trpP^+trpO^+trpL^+trpE^+trpD^-$	+++	−	−	−
$trpR^+trpP^+trpO^-trpL^+trpE^-trpD^+$ / $trpR^+trpP^+trpO^+trpL^+trpE^+trpD^-$	+++	+++	−	+
$trpR^+trpP^+trpO^+trpL^-trpE^-trpD^+$ / $trpR^+trpP^+trpO^+trpL^+trpE^+trpD^-$	+++	+++	−	+

6. Antisense RNA is RNA complementary to the 5' end of an mRNA. Antisense RNA works by binding to the mRNA through complementary base pairing and preventing translation of the mRNA. The most obvious source of the regulatory RNA is transcription of the nontemplate strand of the gene. Antisense RNA could be used to treat disease by preventing transcription of disease alleles.

7. A riboswitch is an RNA sequence that regulates gene expression by binding small molecules that control the secondary structure of the RNA. Riboswitches are very effective at controlling gene expression because the RNA directly binds to a metabolite without the need for other factors such as proteins. This allows the RNA to respond directly to small changes in metabolite concentration.

8. The λ repressor favors lysogeny and represses lysis; therefore, a λ phage that enters an *E. coli* cell containing high concentrations of the λ repressor will likely undergo lysogeny. The *cro* gene product favors lysis and represses lysogeny, so if the same phage enters an *E. coli* cell containing high concentrations of Cro, it will probably enter the lytic cycle.

9. *cI*: The CI repressor represses the lytic cycle and favors lysogeny. Therefore, if *cI* is mutated, the phage will enter the lytic cycle.

   cII: The CII protein promotes transcription of the *int* gene, which is required for integration, from the P_{RE} promoter, as well as transcription of the *cI* gene from the P_{RE} promoter. If *cII* is mutated, *int* and *cI* will not be transcribed as well; both of these genes encode proteins required for lysogeny. Therefore, the *cII* mutant will be more likely to undergo lysis.

cIII: The CIII protein inhibits a bacterial protease that degrades the CII protein. If *cIII* is mutated, the CII protein will be degraded, and the phage will be more likely to enter the lytic cycle, as described above.

N: The N protein is an antiterminator that allows transcription to continue from both the left and right promoters. If *N* is mutated, the phage will be more likely to enter the lytic cycle.

cro: The Cro repressor represses lysogeny and favors the lytic cycle. If *cro* is mutated, the phage will be more likely to enter the lytic cycle.

att: The *att* sequence is involved in site-specific recombination between the λ prophage and the *E. coli* genome. If the *att* site is mutated, it will affect site-specific recombination, making lysis more likely.

Q: The *Q* gene product is required for complete transcription of the late operon. If *Q* is mutated, transcription of the late operon will terminate prematurely, and lysis will not occur. Therefore, the phage will undergo lysogeny.

10. O_{R1}: The O_{R1} site is the operator site with the highest affinity for the repressor; binding at this site enhances binding at the O_{R2} site. If O_{R1} is mutated, the repressor will not bind efficiently, and the right operon will fail to be repressed, making lysis more likely.

O_{R3}: This site has the lowest affinity for the repressor; binding of the repressor to O_{R3} results in repression of the *cI* gene itself. If O_{R3} is mutated, *cI* will fail to be repressed and levels of the repressor will rise, so lysogeny will be favored.

P_L: Mutation of the left promoter will prevent transcription of all genes transcribed from the left promoter. The genes transcribed from the left promoter are required for recombination and integration; in the absence of these gene products, lysis will occur.

P_{RE}: This promoter site is required for transcription of the repressor at high levels. If this promoter is mutated, the repressor will not be transcribed, and lysis will be favored.

P_{RM}: This promoter is required for repressor maintenance. If it is mutated, levels of the repressor will fail to be maintained, and lysogeny will occur.

t_{L1}: This sequence acts as a terminator, so transcription will continue, and *cIII* will be transcribed. The phage will be more likely to undergo lysogeny.

nutL: As in the preceding, transcription will continue through t_{L1}, which will favor lysogeny.

nutR: Transcription will continue through t_1 to the *cII* gene, which will favor lysogeny.

11.   λ phage DNA exists in two linear forms and one circular form. When packed in the bacteriophage head, λ-DNA is linear, with the *cos* sites at the two ends. Shortly after infecting a bacterial cell, the linear phage DNA circularizes. To integrate into the host cell genome as a prophage, the circular DNA is broken at a different site to create the second linear form.

12.   UV light induces the SOS repair response in bacterial cells. One of the components of this response is conversion of the RecA protein (which is normally involved in recombination) into a protease (RecA*). RecA* cleaves the λ repressor, which makes operator sites available for the Cro protein. This response eventually leads to transcription of the genes for the lytic cycle.

13.   Feedback inhibition occurs when a product of a metabolic pathway inhibits an enzyme that acts at an earlier step in the pathway. Binding of the end metabolite to the regulated enzyme induces a conformational change in the enzyme that inactivates it.
         Allosteric proteins also play a role in regulating transcription. For example, TrpR represses the *trp* operon only when it is bound to tryptophan.

14.   One mechanism that controls the rate of protein degradation is the N-end rule. Under this mechanism, identity of the N-terminal amino acid determines the half-life of the protein. Regulation of the removal of the N-terminal methionine or other N-terminal amino acids can regulate the half-life of the protein. A second model for control of the rate of protein degradation is the PEST hypothesis. Under this hypothesis, protein degradation is regulated by regions of a protein that are rich in proline, glutamic acid, serine, and threonine. Proteins rich in these regions tend to be degraded more rapidly than proteins that do not have these regions.

EXERCISES AND PROBLEMS

15.   Recall how the different parts of the *lac* operon work. The *lacI* gene encodes the repressor; in the *lacI⁻* mutant, no repressor is made, and the operon is constitutively active. The *lacOᶜ* mutant is constitutive because the repressor cannot bind the operator sequence. The *lacIˢ* mutation is always repressed because the superrepressor fails to bind the inducer and therefore constantly represses the operon. Therefore, the following phenotypes will be observed:

    a. Inducible
    b. Constitutive
    c. Constitutive
    d. Constitutive
    e. Always repressed
    f. Constitutive

16. Consider one DNA molecule at a time. If a molecule cannot make the functional *lacZ* gene product because it is *lacZ⁻*, it can be ignored. The O^c mutant acts in *cis*, so *lacZ⁺* alleles that are on the same DNA molecule as the O^c allele will be expressed constitutively. The *lacI* gene encodes a diffusible protein, so it will act in *trans* and affect both alleles. With these concepts in mind, the merozygotes will have the following phenotypes:

    a. Inducible
    b. Inducible
    c. Inducible
    d. Inducible
    e. Constitutive
    f. Constitutive

17. Begin with the structural genes (*lacZ* and *lacY*). If a structural gene is mutated, then that DNA molecule can be disregarded for expression of that gene product. Next consider mutations that act in *cis*. If the promoter is mutated (P^-), none of the gene products on that DNA molecule will be transcribed. If the operator is mutated to O^c, then that DNA molecule will be constitutively transcribed. Finally, consider the repressor, which acts in *trans*. I^s is dominant to I^+, and I^+ is dominant to I^-. The superrepressor fails to bind the inducer even at high concentrations and will constantly repress a wild-type operator sequence; however, the superrepressor cannot bind to O^c. The I^- mutant cannot bind to the operator, so it cannot repress the operon regardless of whether the inducer is present or not. In the following table, a + symbol indicates that functional gene product is produced, and a − symbol indicates that functional gene product is not produced.

| | **No IPTG present** | | **IPTG present** | |
Genotype	**β-Galactosidase**	**Permease**	**β-Galactosidase**	**Permease**
lacZ⁺Y⁻/F' lacZ⁻Y⁺ *lacIˢZ⁺Y⁻/*	−	−	+	+
F' lacI⁻Z⁻Y⁺ *lacIˢOᶜZ⁺Y⁻/*	−	−	−	−
F' lacP⁻Z⁻Y⁺ *lacZ⁺Y⁻/*	+	−	+	−
F' lacP⁻Z⁻Y⁺ *lacIˢZ⁺Y⁻/*	−	−	+	−
F' lacOᶜI⁻ZY⁺ *lacI⁻OᶜZ⁺Y⁻/*	−	+	−	+

$F'\ lacZ^-Y^+$	+	–	+	–
$lacP^-O^cZ^+Y^-/$				
$F'\ lacI^-Z^-Y^+$	–	–	–	+

18. The *lacY* gene encodes the permease that is responsible for transporting lactose into the bacterial cell. If the permease were nonfunctional, no lactose would enter the cell, and the inducer would never be synthesized. Therefore, the genes of the *lac* operon would be constitutively repressed, and the cell would fail to respond to the addition of lactose.

19. Both alleles could contain a mutation in the *lacZ* gene. Another possibility is that the original *E. coli* strain contains the I^s superrepressor mutation.

20. **a.** They could be defective in either the repressor or the operator. Both the I^- mutant and the O^c mutant are constitutively active.
   b. If the mutation is I^-, the merodiploid should be inducible because the wild-type repressor will bind to the operator on either DNA molecule. If the mutation is O^c, the merodiploid will still be constitutive because the operator mutation is *cis*-dominant.

21. The operator sequence is probably located in region S, which is absent in deletions 3 and 4. If the operator sequence is deleted, the mutant should be constitutive, as seen in deletions 3 and 4. The promoter sequence is probably located in region 1. Deletion of the promoter will eliminate all mRNA expression from this DNA molecule.

22. The mutations could affect adenylyl cyclase or the catabolite activator protein (CAP). If adenylyl cylcase is mutated, then no cAMP will be made, and all four operons will be uninducible. If the gene encoding CAP is mutated, the cells will be unresponsive to cAMP.

23. **a.** The *cis*-acting mutation is in the *araO* site and causes a lack of repression in the absence of arabinose. Even though the repression is lost, only partial activation is observed because CAP binding is required for full activation. The *trans*-acting mutation is in the *araC* gene, which encodes the inducer.
   b. If a merodiploid is generated, partial activation will still be observed in the absence of arabinose because one of the mutations is *cis*-acting.

24. This mutant will be constitutive for *lacZ* because the constitutive operator sequence cannot be bound by any form of the repressor.

25. Begin by considering the single mutants. A mutation in the gene for tryptophan synthetase should result in no tryptophan synthase activity regardless of the absence or presence of tryptophan. Examination of the three single mutants (strains 1, 2, and 3) shows that the only mutant that fits this requirement is strain 3, in which gene *b* is mutated. Therefore, *b* encodes the enzyme for tryptophan synthase.

Both the operator and repressor mutations are expected to result in the expression of tryptophan synthase activity in both the absence and presence of tryptophan activity. Therefore, single mutations cannot distinguish between these two types of mutations, and we must examine the merodiploids. An operator mutation is expected to act in *cis* and should result in tryptophan synthase activity whenever the operator mutant is in *cis* to b^+, the wild-type tryptophan synthase allele. The repressor mutation, on the other hand, will be recessive to the wild-type repressor allele, so a merodiploid that is heterozygous for the wild-type and mutant repressor allele should be repressed in the presence of tryptophan.

We can now predict the results for the merodiploids for the two possibilities remaining. If *a* is the operator and *c* is the repressor, then we would predict that tryptophan synthase activity will be positive in the absence or the presence of tryptophan. If *a* is the repressor and *c* is the operator, we predict the same results. Thus, strain 4 is uninformative. Similarly, for strain 5, tryptophan synthase activity should be positive when tryptophan is absent and negative when tryptophan is present for both possibilities; this strain too is uninformative. Strains 6 and 7 can distinguish between the two possibilities. If *a* is the operator and *c* is the repressor, then strain 6 should be positive in the absence of tryptophan and negative in the presence of tryptophan; strain 7 should be positive under both circumstances. If *a* is the repressor and *c* is the operator, then strain 6 should be positive under both circumstances, and strain 7 should be positive in the absence of tryptophan and negative in the presence of tryptophan. The results from strains 6 and 7 match the first possibility; therefore, *a* is the operator and *c* is the repressor.

26. The attenuator system requires only one binding event (charged tRNA) rather than 11; therefore, it can respond more sensitively to changes in tryptophan concentration.

27. **a.** If RNA polymerase cannot bind the promoter, the genes of the operon will not be transcribed. A mutation in the promoter will be *cis*-dominant in a partial diploid.

   b. If the repressor–corepressor cannot bind the operator, the genes of the operon will be transcribed constitutively. A repressor mutation will not be *cis*-dominant in a partial diploid.

   c. If the repressor cannot bind the corepressor, the genes of the operon will be transcribed constitutively. In a partial diploid, this mutation will not be *cis*-dominant.

28. Repression and attenuation allow the cell to monitor both amino acid levels and charged tRNA levels.

29. If the Cro protein has a high affinity for O_{R1}, then the Cro protein will likely bind O_{R1} before the CI repressor protein does. Therefore, the phage will more likely cause lysis of the infected bacteria.

30. The mutant strain could have a mutation in the CI repressor. This would result in the phage always causing lysis of the host cells.

31. Mutations in *cI, cII,* or *cIII* would cause the λ phage to follow the lytic cycle more frequently. Mutations in the *CI* repressor would cause the phage to follow the lytic cycle always.

32. When the bacteria are shifted to 42°C, the CI repressor will denature and no longer function, which will make the operator sites available for the Cro protein. The phage DNA will be excised from the bacterial genome, and lysis will occur.

33. When the lysogen is heated to 42°C, the CI repressor unfolds and cannot bind the λ O_R, which permits transcription of the *cro* gene. Expression of the Cro protein resulted in the excision of the λ prophage. Before, the prophage could enter the lytic cycle, the temperature was lowered to 30°C and the CI repressor refolded and bound O_R and O_L. This prevented both the replication of the λ-DNA and the reintegration of the λ-DNA into the bacterial genome. After several cell divisions, the λ-DNA is lost from many of the bacterial cells. Raising the temperature to 42°C again would result in lysis of the bacterial cells that contained the λ prophage, but would not affect the bacterial cells that had lost the prophage. These latter cells would then be able to form colonies.

34. Transcriptional control is more efficient than translational control because it ensures that a cell does not expend energy transcribing a gene that is not needed. Translational control can be used to directly respond to levels of a molecule such as a metabolite without the need for other protein factors.

35. Heat activates transcription of the *htpR* gene, which encodes an alternative sigma factor, σ^{32}. The σ^{32} protein recognizes the promoters of the heat shock genes. A GC-rich sequence rather than an AT-rich sequence is used for heat shock promoters. GC-rich sequences are harder to melt; AT-rich sequences in these promoters would tend to be melted at lower temperatures, while GC-rich promoter sequences would tend to melt only at higher temperatures, which is appropriate for a response to heat.

36. The data are consistent with the presence of two promoters: an early promoter that produces a polycistronic message including both *rIIA* and *rIIB* and a late promoter that transcribes only *rIIB*. A nonsense mutation in *rIIA* could cause ribosomes to dissociate from the polycistronic transcript, meaning that early expression of the *rIIB* gene would be reduced. The late promoter presumably overlaps with the 3' end of the *rIIA* gene, so that a small deletion near the end of the *rIIA* gene would delete the late promoter and prevent transcription of *rIIB* from the late promoter. Transcription of *rIIB* from the early promoter would not be affected. The following figure diagrams the structure of these genes.

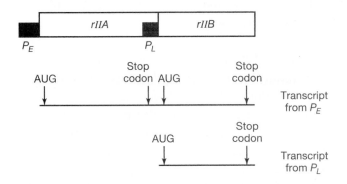

37. Because mRNA and protein levels are normal, the mutation in this strain likely affects a posttranslational control mechanism involving feedback inhibition. For example, if an enzyme in the pyrimidine synthesis pathway is normally inhibited by high levels of cytosine and thymidine, a mutation in the regulatory domain of this enzyme could render it insensitive to cytosine and thymidine. Therefore, it would not be inhibited by cytosine and thymidine and would continue synthesizing pyrimidines.

38. First, it is more efficient to utilize one sugar preferentially and to express the genes for the metabolism of other sugars only when the first sugar is absent. Second, many of the alternative sugars (such as lactose) are converted to glucose or its derivatives before metabolism. For example, lactose is cleaved by β-galactosidase to produce glucose and galactose. The galactose is then converted to glucose 6-phosphate, which is the first compound formed when metabolizing glucose. Thus, metabolizing lactose still requires all the enzymes that are used in glucose metabolism. Because these sugars pass through the glucose metabolic pathway, the cell prefers to metabolize any glucose (if it is present) before expressing additional enzymes that will convert other sugars to glucose or one of its derivatives.

39. The short half-life of mRNAs in bacteria ensures that transcription regulation will be efficient and able to respond rapidly. If mRNAs persisted in the cell for hours instead of minutes, then genes could not be turned off rapidly when necessary.

40. The mutation is probably in one of the tryptophan codons in the leader peptide. The tryptophan codon is TGG; TGT and TGC are both cysteine codons. If one of the tryptophan codons is mutated to a cysteine codon, the bacteria would be less sensitive than usual to tryptophan starvation; however, when starved for both of these amino acids, the response would be normal.

41. The mutation in this strain likely involves codon bias. One of the glycine codons may have been mutated from a common codon to a rare codon. When the common codon is used, the charged tRNA should be abundant when glycine is present. However, if a rare codon is used, the charged tRNA will be less abundant even when glycine is present. Because of the normally lower availability of the tRNA for the rare glycine codon, the ribosome would likely pause at this codon, which could

allow the antiterminator structure to form. In this situation, the full-length RNA would be transcribed despite the presence of glycine.

42. The terms *positive* and *negative regulation* refer to the effect that a regulatory protein has on transcription. The terms *inducible* and *repressible* describe whether a system is regulated by a substrate or a product. An inducible system is regulated by a substrate or a molecule related to a substrate. If the activator protein cannot bind DNA in the absence of the inducer but can bind DNA to activate transcription in the presence of the inducer molecule, the inducible system will exhibit positive regulation. If a repressor protein blocks transcription in the absence of the inducer molecule and fails to block transcription when the inducer molecule is present, the inducible system will show negative regulation (see figure 16.3).

A repressible system will exhibit negative regulation if a co-repressor molecule is required to block transcription, as is the case in the *trp* operon. However, a repressible system could also exhibit positive regulation if an activator protein promotes transcription in the absence of the co-repressor molecule but fails to activate transcription when the co-repressor molecule is present.

43. Allolactose is produced only 1% of the time, so its concentrations will tend to be low even when it is present. If lactose were the inducer, its concentrations would be much higher, and the system would tend to be leaky.

44. The LacI protein binds to two of the three operator sites as a tetramer. Maximum repression is achieved only when all three sites are bound by a repressor. When the repressor is bound to two of these sites, the intervening DNA, which includes the promoter, is looped out and is inaccessible for RNA polymerase. Use of sites that are separated from one another allows DNA looping to occur. The use of multiple sites also allows the system to be more sensitive to change in levels of the inducer.

45. **a.** The wild-type strain (number 1) shows active enzymes in the presence of Luv, and no enzymes when Luv is absent. Therefore, the operon is expressed in the presence of Luv and repressed in its absence. This is the hallmark of an inducible operon.

b. Strain 2 (v^-) is associated with an inactive enzyme 2, but no other abnormality. Therefore, *v* is the structural gene for **enzyme 2**. Because an x^- mutant (strain 4) prevents the transcription of the operon, *x* must be the **promoter** element. Strains 3 (w^-) and 6 (z^-) are both associated with constitutive synthesis of the Luv operon. Therefore, one of these two is the operator region, and the other is the gene for the repressor protein. But which is which? Recall, that the repressor protein diffuses through the cytoplasm (*trans*-acting), whereas an operator region can only affect adjacent genes (*cis*-acting). Furthermore, a mutant operator is dominant to its wild-type allele, whereas a mutant repressor is recessive. The z^-/z^+ merozygote (strain 11) still showed constitutive synthesis, indicating that z^- is dominant over z^+. Therefore, *z* is the **operator** region, and *w* is the **repressor** gene. And by elimination, *y* must correspond to the structural gene for **enzyme 1**.

c.

Strain	Genotype	Luv Not Found		Luv All Around	
		Enzyme 1	Enzyme 2	Enzyme 1	Enzyme 2
1	$v^+ w^+ x^+ y^+ z^+$	N	N	A	A
2	$v^- w^+ x^+ y^+ z^+$	N	N	A	I
3	$v^+ w^- x^+ y^+ z^+$	A	A	A	A
4	$v^+ w^+ x^- y^+ z^+$	N	N	N	N
5	$v^+ w^+ x^+ y^- z^+$	N	N	I	A
6	$v^+ w^+ x^+ y^+ z^-$	A	A	A	A
7	$v^- w^+ x^+ y^+ z^+/$				
	F' $v^+ w^+ x^+ y^+ z^+$	N	N	A	A + I
8	$v^+ w^- x^+ y^+ z^+/$				
	F' $v^+ w^+ x^+ y^+ z^+$	N	N	A	A
9	$v^+ w^+ x^- y^+ z^+/$				
	F' $v^+ w^+ x^+ y^+ z^+$	N	N	A	A
10	$v^+ w^+ x^+ y^- z^+/$				
	F' $v^+ w^+ x^+ y^+ z^+$	N	N	A + I	A
11	$v^+ w^+ x^+ y^+ z^-/$				
	F' $v^+ w^+ x^+ y^+ z^+$	A	A	A	A

46. **a.** This is the coding strand of DNA and therefore the nontemplate strand.

b. There are 20 amino acids, so the probability of finding a phenylalanine at any position would be 1/20. The probability of the tripeptide *phe-phe-phe* will be (1/20)(1/20)(1/20) = 1/8000. Therefore, a polypeptide of 8000 amino acids will be expected to have one *phe-phe-phe* sequence.

c. This discrepancy could be related to the levels of charged (aminoacylated) tRNAs in the cell. Amino acids, such as tryptophan, that normally have low cellular levels of charged tRNAs are likely to have few codons in the leader mRNA. If instead they had many codons, ribosome stalling would take place even at the normal low levels of charged tRNAs. In contrast, amino acids, such as phenylalanine, that normally have high cellular levels of charged tRNAs are likely to have many codons in the leader mRNA. If they had few codons, ribosome stalling would not take place as often as needed.

d. i. Attenuation will not occur. When charged phe-tRNAs are scarce, the ribosome would stall on the *phe* codons in region 1, which is prevented from base-pairing with region 2. Region 2 is then free to base-pair with region 3, thereby precluding the formation of the terminator stem loop (between regions 3 and 4).

ii. Attenuation will occur. No translation means that stem loops are formed between regions 1 and 2, and between regions 3 and 4. Stem loop 3–4 will cause rho-independent termination of transcription.

iii. Attenuation will not occur. Region 1 cannot base-pair with region 2, and so stem loop 2–3 (antiterminator) forms.

iv. Attenuation will occur. The ribosome would stall on the *phe* codons. However, region 2 cannot base-pair with region 3, thereby causing stem loop 3–4 to form.

v. Attenuation will not occur, simply because the stem loop 3–4 transcription terminator cannot form.

e. The mutant leader peptide sequence and encoded amino acids are given here:

```
5'-ATG AAA CAC ATA CCG TTT TCT TCG CAT TCT TTT TTA CCT TCC CCA TGA-3'

   Met Lys His Ile Pro Phe Ser Ser His Ser Phe Leu Pro Ser Pro
```

Two *phe* codons remain in the leader peptide sequence, so low levels of phenylalanine may still prevent attenuation by causing the ribosome to stall in region 1. However, the mutant sequence now contains four *ser* and three *pro* codons. Therefore, low levels of serine and proline would also prevent attenuation in this mutant.

CHAPTER INTEGRATION PROBLEM

a. There may be a λ prophage inserted between the *met*⁺ and *trp*⁺ genes. As the single-stranded DNA is preparing to enter the recipient cell, the λ CI protein would be removed (it only binds to the double-stranded O_R and O_L sequences). The transferred DNA is converted to double-stranded DNA in the recipient cell. If the entire λ genome transferred into the recipient cell (*met*⁺ transferred, but not *trp*⁺), the absence of the CI protein would allow the double-stranded prophage to excise from the bacterial genome and enter the lytic cycle in the recipient cell.

b. At 42°C, the λ CI repressor would denature and the prophage would excise from the bacterial genome. With the λ-DNA being absent from the bacterial genome, it could not be transferred by conjugation and the recipient cell could not enter the lytic cycle. The absence of the λ-DNA would also result in a normal conjugation distance between the *met*⁺ and *trp*⁺ genes.

c. The kanamycin-resistant strain already contains a λ prophage. This would result in λ CI repressor protein already being present in the recipient cell. Thus, when the donor prophage enters the recipient cell and becomes double-stranded, the endogenous

CI protein would bind to the newly entered O_R and O_L sites and prevent the excision of the entering prophage.

d. The *leu*⁺, *met*⁺, *bio*⁺, *trp*⁻ cells would grow on plates that lacked leucine, would grow on plates that lacked methionine, but could not grow on plates that lacked tryptophan.

e. The *trp*⁺ gene was deleted in strain F.

f. An inversion may have switched the order of the *trp*⁺ and *met*⁺ genes.

g. Strain F appears to contain a deletion of much of the 10-kb region. Several bands are missing entirely from the Southern blot. Strain M has two bands that are smaller than those in the original strain, again consistent with a deletion. The results are also consistent with an inversion with breakpoints located in the two bands of altered size.

CHAPTER 17 GENE EXPRESSION: CONTROL IN EUKARYOTES

Chapter Goals

1. Understand the organization of the *cis*-acting control elements and their various roles in regulating transcription.
2. Understand the different types of *trans*-acting elements, the common motifs that they share, and how they function to regulate gene expression.
3. Describe a complex regulatory network and identify the steps that could be involved in regulating gene expression.
4. Enumerate the different methods used to modify chromatin that result in changes in gene expression.
5. Describe the posttranscriptional and translational regulatory mechanisms that affect protein expression.
6. Diagram the common themes in regulating gene expression in prokaryotes and eukaryotes.

Key Chapter Concepts

17.1 The *cis* Regulatory Elements: Control of RNA Polymerase II Action

- **RNA Polymerase II Promoter:** The RNA polymerase II promoter consists of two elements: the **initiator region (*InR*)** which flanks the transcriptional start site, and the **TATA box** located at -25 which is similar to the prokaryotic -10 region. Binding of RNA polymerase II to the promoter requires the formation of the **preinitiation complex (PIC)** which consists of six general transcription factors, including **TFIID**. Formation of the preinitiation complex allows basal-level transcription to occur. TFIID is composed of 10 different proteins including the **TATA-binding protein (TBP)**, $TAF_{II}150$, and $TAF_{II}250$. TBP binds the TATA box, while $TAF_{II}150$ and $TAF_{II}250$ bind to the initiator region. Binding of TFIID causes the DNA to bend and leads to further acetylation of histones H3 and H4.
- **Proximal-Promoter Elements:** Proximal-promoter elements are sequences found in the first 200 nucleotides upstream of the promoter that bind specific proteins that either activate or repress transcription. Proximal-promoter elements can either be generic or cell- and tissue-specific. Generic elements are required for regulating the amount of transcription, while cell- and tissue-specific elements are required for controlling when and where the gene is expressed. Two generic proximal-promoter elements are the **CCAAT box** and the **GC box**. The CCAAT box is bound by the CCAAT-binding transcription factor (CTF), and the GC box is bound by the SP1 protein. These proteins aid in the binding of additional proteins in the transcriptional complex. The CCAAT box and GC box, along with the TATA box, are the only critical *cis*-acting elements required for transcription. The location of these elements relative to the TATA box is critical for the level of transcription of a gene. Cell- and tissue-specific proximal-promoter elements differ from generic elements in that the number of cell-specific elements in the upstream region of a gene can be significantly greater than the number of generic elements, and the proteins that bind to these tissue-specific elements are only expressed in a subset of cells at specific times during development allowing for complex regulation of different genes.
- **Enhancers:** Enhancers are DNA sequences that either bind proteins that act as transcriptional **activators** or proteins that act as transcriptional **repressors**. Enhancers affect the maximal level of transcription and control the temporal and spatial expression patterns of genes. The distance of an enhancer from a gene and the orientation of the enhancer do not affect its ability to regulate transcription. Enhancers either change the conformation of the chromatin making it more accessible for binding transcription factors and RNA polymerase II, or they bind **architectural proteins** that are capable of bending the DNA allowing other proteins bound at the enhancer region to physically interact with RNA polymerase II or the transcription factors.
- **Silencers:** Silencers provide a negative form of regulation in eukaryotes. They either act on a specific gene or negatively affect transcription on a global scale. Gene-specific silencers bind proteins that change the conformation of the DNA at a specific site in the chromatin. Global silencing involves binding of the same protein to a sequence that is found in the promoters of many different genes. Like enhancers, silencers can act either upstream or downstream of a gene.

- **Insulators:** Insulators are DNA sequences that are bound by specific proteins that can shield a gene from the effects of a neighboring enhancer or silencer. These elements ensure that enhancers and silencers affect only the proper gene.
- **Interactions Within and Between Enhancers:** The interaction of different proteins binding within an enhancer and between different enhancers and silencers determines the final expression pattern of a gene and the level of expression.

17.2 The *trans* Regulatory Elements: Proteins
- **DNA-Binding Motifs:** Proteins bind to specific DNA sequences through motifs found in the DNA-binding region of the proteins. The amino acid sequence of these motifs dictates what DNA sequence a protein will bind to.
    - **Helix-Turn-Helix:** The helix-turn-helix DNA-binding motif consists of two α-helices that are connected by a "turn" that consists of several amino acids. One helix binds in the major groove of the DNA, and the other interacts with the sugar–phosphate backbone. The **homeobox** proteins, first identified in *Drosophila*, contain the helix-turn-helix domain.
    - **Zinc Finger:** The zinc finger motif is composed of two antiparallel β-strands and an α-helix. Two histidine or cysteine residues in the α-helix and two cysteine residues in the β-strands covalently bind a zinc atom. The α-helix of the zinc finger binds in the major groove of the DNA.
    - **Basic Helix-Loop-Helix:** The basic helix-loop-helix motif consists of two amphipathic α-helices having all of the charged amino acids on one side of the helix. The basic region is an extension of an α-helix that contacts the DNA in the major groove of the DNA. The charged region of the α-helices allows proteins that contain this domain to dimerize.
    - **Basic Leucine Zipper:** The leucine zipper motif is an α-helix composed of four or more leucine residues that are seven amino acids apart. This motif allows proteins containing it to form homo- or heterodimers where the leucines on the one helix fit in between adjacent leucines on the opposite helix. The two α-helices wrap around each other to form a **coiled coil** structure.
- **Transcriptional Activation Versus Repression:** Regulatory proteins can function as either activators or repressors. Formation of a homodimer or a heterodimer, such as in the Myc-Max system, can determine whether a protein will activate or repress transcription. The Max protein can homodimerize with itself and repress transcription, or it can heterodimerize with the Myc protein and activate transcription. Repressors can also inhibit transcription by binding to activators to prevent their ability to bind DNA or to block their transcriptional activation domain.

17.3 The *trans* Regulatory Elements: RNAs
- **MicroRNAs (miRNAs):** miRNAs regulate the expression of a gene by disrupting translation through either the cleavage of the corresponding mRNA or by base-pairing with the mRNA to block translation. Either of these events results in a decreased amount of protein that is translated from the mRNA.
- **Antisense RNAs:** Antisense RNAs are complementary to a sequence within a target RNA. Base-pairing between the regulatory RNA and the target RNA can block translation of the mRNA or affect the stability of the mRNA, while base-pairing at an intron–exon junction can block the splicing of a pre-mRNA.

17.4 Example of a Complex Regulatory Network: The Dorsal Protein
- **Signal Transduction:** Not only does a cell respond to intracellular demands for altering how its genes are expressed, it also responds to environmental cues to change gene expression. Generally, the pathway starts by the interaction of a protein receptor at the cell surface with a hormone or protein. This interaction initiates a complex cascade of multiple protein–protein interactions that will eventually activate a specific transcription factor. These cascades can be highly complex and branching, involve many interactions, and involve numerous protein modifications, such as protein phosphorylation, to affect protein activity.

17.5 The Role of Chromatin in Gene Regulation

- **Chromatin Remodeling:** The eukaryotic chromosome is tightly compacted around proteins. The basic unit of chromatin structure is the nucleosome, which is an impediment to transcription. In order for RNA polymerase to be recruited to the promoter, the chromatin state must be modified so that the genomic DNA is accessible to the transcription machinery. Chromatin remodeling can be achieved by acetylation of the histons, the incorporation of variant histones into the nucleosome, or by the enzyme-mediated movement of the nucleosomes.

- **Detection of Nucleosome Position:** The chromatin status near a gene is examined by using the enzyme DNase I to digest genomic DNA. At low concentrations, DNase I digests DNA that is free of nucleosomes (**DNase I hypersensitive sites**). DNA that is associated with nucleosomes is protected from digestion. DNase I hypersensitive sites are often found upstream of genes that are transcriptionally active.

- **Histone Acetylation and Deacetylation:** Histones can exist in either an acetylated or a deacetylated form. Acetylation is performed by histone acetyltransferases (HATs) which add acetyl groups to the amino-terminal lysines of histones H3 and H4. This decreases the interaction between the histone core and the negatively charged backbone of the DNA and is typically associated with transcriptional activation. Histone deacetylases (HDACs) reduce the number of acetyl groups present on the H3 and H4 histones and lead to transcriptional repression. Binding of some activators may recruit HATs, while binding of some repressors may recruit HDACs.

- **Histone Methylation:** The lysines and arginines on histones H2A, H2B, H3, and H4 can all by methylated by histone methyltransferase (HMTase). Methylation of histones can either activate or repress transcription. Complex regulation exists where the acetylation status of the histones can affect their methylation status and vice versa.

- **Enzyme-Mediated Movement of Nucleosomes:** The location of the nucleosomes on the DNA can also be altered by the **ISWI** or **SWI–SNF protein complexes**. The ISWI protein complex hydrolyzes ATP to slide the DNA across the surface of the histone core, while the SWI–SNF protein complex uses ATP to dissociate the DNA from the histone core and reassemble the histone core in a new location on the DNA thereby allowing increased transcription of the nearby genes.

- **Incorporation of Variant Histone Proteins into Nucleosomes:** Variant forms of the histone proteins can be incorporated into the nucleosomes to alter the chromatin state. Some H1 histone variants and the mH2A variant are associated with transcriptional repression, while the histone H3.3 and the H2AZ variants are associated with transcriptional activation.

17.6 Methylation of DNA

- **DNA Methylation:** In eukaryotes, methyl groups can be added to cytosine bases when they are next to guanine bases (CpG). The promoter of an inactive gene is more heavily methylated than a promoter for an active gene. This modification of the cytosine bases is reversible and may change as a cell differentiates or undergoes other cellular changes. Methylation is postulated to affect transcription by prohibiting transcription directly, affecting regional chromatin structure, binding of **methyl-CpG-binding proteins** to 5-methyl-cytosine that recruit HDACs, or affecting DNA structure.

17.7 Imprinting

- **Imprinting:** Parental influences may also lead to the differential expression of alleles, depending on their parental origin. This phenomenon is referred to as **imprinting** and affects at least 20 known human genes. Some genes are **maternally imprinted** (expressed from the paternal allele), while other genes are **paternally imprinted** (expressed from the maternal allele). The imprint, which is believed to be site-specific methylation, is removed and reprogrammed during gametogenesis. For example, if a gene is *maternally imprinted*, a daughter will express only the allele she received from her father in all her *somatic* tissues. When her *reproductive* tissues were forming, though, the imprint was removed and *both* alleles were imprinted according to *her* gender. Therefore, she can only pass on an imprinted allele to her children. Her mate will give the allele that is expressed in her offspring. Conversely, for paternally imprinted genes, she would give the allele that is expressed in her offspring. The *Igf-2* gene in mice is imprinted in the female, while the *H19* gene located 70 kb downstream is imprinted in the male.

17.8 Posttranscriptional Regulation

- **Proteins Binding to the 3′ Untranslated Regions:** One mechanism of posttranscriptional regulation involves the binding of a protein to the 3′ untranslated region (3′ UTR) of an mRNA to modify the mRNA or repress its translation. This can lead to deadenylation of the 3′ poly(A) tail which decreases the stability of the mRNA and reduces the efficiency of translation.
- **Alternative Splicing:** Alternative splicing is another mechanism of posttranscriptional regulation in which multiple mRNAs are produced from a single primary transcript. In this way, several different proteins can be produced by the same gene.
- **Nonsense-Mediated mRNA Decay:** Nonsense-mediated mRNA decay eliminates mRNAs that possess either a premature translational termination codon or a translation termination codon that is too far from the poly(A) tail of the mRNA. It plays an important role in the degradation of mRNAs that contain mutations and in the maintenance of the correct levels of specific wild-type mRNA transcripts. Nonsense-mediated mRNA decay occurs by removal of the 5′ cap followed by 5′ → 3′ exonuclease degradation, 3′ deadenylation followed by 3′ → 5′ exonuclease degradation, or internal endonuclease cleavage followed by exonuclease degradation in both directions.

17.9 Translational Control

- **General Translation Regulation: Phosphorylation:** Translation can be regulated in a general manner by modifying the translation initiation factors.
- **Specific Translation Regulation:** Translation of specific mRNAs can be regulated by its interaction with specific proteins. An RNA-binding protein may interact with a specific protein involved in translation initiation and thereby prevent the ability of the protein to interact with the other proteins required for translation initiation. An RNA-binding protein can have different effects on the translation of two different mRNAs to regulate a physiological process in a coordinated manner such as is found in binding of the IRP RNA-binding protein to the IRE sequence present on the ferretin and transferrin receptor mRNAs.

17.10 Posttranslational Control

- **Protein Modifications:** Posttranslational modifications are covalent changes that occur to proteins to modify their activity within the cell. This may involve methylation, phosphorylation, or acetylation of a protein which can affect the activity, localization, or the stability of the protein. In some cases, these modifications affect the ability of a protein to function in a signal transduction cascade.
- **Protein Degradation—Ubiquitination:** Another method of posttranslational regulation is ubiquitin-mediated protein degradation. This regulated degradation involves the attachment of one or more molecules of **ubiquitin** to a protein which targets the protein for proteosomal degradation. Misfolded and mutant proteins are degraded through this pathway, as are proteins that need to be expressed for only a short time in the cell, such as transcription factors, or proteins that play a role in cell cycle regulation.

17.11 Interaction of Multiple Regulatory Mechanisms

- Multiple regulatory mechanisms in the eukaryotic cell control the amount of each protein that is present at any given time and its activity. Regulatory mechanisms control the amount of mRNA that is being transcribed, how it is being processed, its translation into a protein, and the activity of the protein.

17.12 Eukaryotic Versus Bacterial Gene Regulation

- **Similarities in the Regulation of Gene Expression:** Both bacteria and eukaryotes bind activator and repressor proteins to regulate transcription. Some of the basic DNA-binding motifs are conserved, and the activities of DNA-binding proteins can be modified by binding coregulatory molecules. The expression of genes needs to be regulated in response to the environment and to the needs of the organism.
- **Differences in the Regulation of Gene Expression:** Bacterial DNA-binding sites are clustered around the promoter, while those in eukaryotes can be hundreds of kilobases upstream or downstream of the transcription unit. Chromatin organization plays a critical role in the regulation of gene expression in eukaryotes, but is absent in bacteria. There is a larger variety of *cis*-acting elements present in

eukaryotes than in bacteria, and the DNA-binding proteins in eukaryotes can be covalently modified to regulate their activity. Posttranscriptional regulatory processes, such as mRNA splicing, and RNA editing are absent in prokaryotes. mRNA stability plays a less significant role in the regulation of gene expression in prokaryotes than in eukaryotes because transcription and translation are coupled in prokaryotes. DNA methylation has a profound effect on repressing transcription in eukaryotes, but not in prokaryotes. Eukaryotes, unlike prokaryotes, use miRNAs in addition to DNA-binding proteins to regulate gene expression.

Key Terms

- preinitiation complex (PIC)
- transcription factor IID (TFIID)
- TATA-binding protein (TBP)
- proximal-promoter elements
- enhancers
- activators
- repressors
- architectural proteins
- silencers
- insulators
- motifs
- helix-turn-helix
- homeobox

- zinc finger
- basic helix-loop-helix (bHLH)
- basic leucine zipper (bZIP)
- coiled coil
- basic helix-loop-helix/leucine zipper
- miRNAs
- antisense RNAs
- signal transduction pathway
- DNase I hypersensitive sites
- ISWI
- SWI–SNF protein complexes

- CpG islands
- methyl-CpG-binding proteins
- imprinting
- epigenesis
- alternative splicing
- nonsense-mediated mRNA decay (NMD)
- posttranslational modifications
- ubiquitination
- proteosome
- small nucleolar RNAs (snoRNAs)

Understanding the Key Concepts

Use the key terms or parts of their definitions to complete the following sentences.

The transcription of a gene by (1)_____ requires the following (2)_____ regulatory elements: the (3)_____, the proximal-promoter, and (4)_____ and silencers. Binding of RNA polymerase II to the (5)_____ requires the formation of the (6)_____, which consists of (7)_____ general (8)_____, including TFIID. (9)_____ is a component of TFIID that binds to the TATA box located at approximately (10)_____ relative to the transcription start site. Upstream of this element is the (11)_____ which contains the (12)_____ box and the (13)_____ box. These two elements are typically located between (14)_____ and (15)_____ relative to the transcription start site. Mutations in these two elements, along with those in the (16)_____ significantly decrease the (17)_____. (18)_____ or (19)_____ bind to enhancer elements which can be located several kilobases (20)_____ or (21)_____ of a gene. Activators may recruit (22)_____ that reduce the association between the (23)_____ and the DNA, allowing the chromatin to open up, and (24)_____ transcription. Repressors may recruit (25)_____ that reduce the number of (26)_____ present on histones (27)_____ and (28)_____ to (29)_____ transcription. (30)_____, like activators and repressors, typically contain one of several common motifs in which an (31)_____ in the protein binds in the (32)_____ of the DNA. Silencers are DNA sequences that are bound by proteins that change the (33)_____ of the DNA and (34)_____ transcription. Insulators prevent (35)_____ or (36)_____ from affecting (37)_____ genes. Methylation of the DNA at (38)_____ is also associated with transcriptional repression. Housekeeping genes are typically (39)_____, while genes that are not expressed are typically (40)_____. Imprinted genes are (41)_____ as a result of (42)_____. The *H19* gene in mice is imprinted in the (43)_____, such that progeny of this mouse express the (44)_____ inherited allele and fail to exhibit the expected Mendelian (45)_____ ratios.

        In addition to the transcriptional regulation of genes, gene expression can also be regulated at the (46)_____, translational, and (47)_____ levels. Alternative splicing and nonsense-mediated mRNA decay are forms of (48)_____ regulation in eukaryotes. Posttranslational regulation is manifested through the (49)_____ of proteins. Protein degradation can be mediated by (50)_____ or by (51)_____-mediated proteosomal degradation. (52)_____ is a (53)_____-amino-acid peptide that is attached to a target protein to (54)_____ it for degradation by the (55)_____. The (56)_____ delivers the target protein to the (57)_____ which transfers ubiquitin to the target protein. Once ubiquitinated, the (58)_____ of the proteosome allows the protein to enter the (59)_____ cylindrical central chamber. (60)_____ forms of proteins and proteins that are only required for a (61)_____ period of time in the cell are degraded by this method.

Figure Analysis

Use the indicated figure or table from the textbook to answer the questions that follow.

1. **Figure 17.25**
   (a) How is the Dorsal protein held in an inactive state?

   (b) How is the kinase activity of the Toll transmembrane receptor protein activated?

   (c) What is the effect on the Pelle protein of the Pelle–Tube heterodimer binding to the intracellular side of the Toll receptor?

   (d) What protein is a substrate of Pelle?

   (e) How does phosphorylation of the Cactus protein activate the Dorsal protein?

2. **Figure 17.26**
   (a) What is the function of DNase I?

   (b) Why is the DNA digested with a restriction enzyme following DNase I digestion?

   (c) On the Southern blot, why are the fragments of DNA obtained from the transcriptionally active tissue smaller than those obtained from the transcriptionally repressed tissue in this assay?

   (d) Where are DNase I hypersensitive sites typically found?

3. **Figure 17.28**
   (a) What enzyme acetylates newly synthesized histones? Where does it do this?

   (b) Which histones are acetylated?

   (c) What amino acids are acetylated on these histones? Where are these amino acids located in the protein?

   (d) What is the function of HAT-A?

   (e) What is the effect of histone acetylation?

   (f) What is HDAC? What is its function?

   (g) Which proteins recruit HATs? Which proteins recruit HDACs?

4. **Figures 17.34 and 17.37**

(a) What is the major difference between the maternal and the paternal chromosomes in terms of *Igf-2* and *H19* expression in mice?

(b) Why is *H19* transcription repressed on the paternal chromosome?

(c) Which gene is maternally imprinted? Why? What is the result of this imprinting?

(d) Which gene is paternally imprinted? Why? What is the result of this imprinting?

(e) If two heterozygous mice, both with the genotype *Igf-2/Igf-2m* are mated, what will determine the size of the progeny?

5. **Figure 17.39**

(a) Why is the Sxl protein present in the early embryo of the female flies but not in the early embryo of the male flies?

(b) What is the function of the Sxl protein in females in the late embryo?

(c) What is the effect of loss of Sxl protein activity in females? Why?

(d) What mutation in the *Sxl* gene would be lethal to males? Why?

(e) Describe how alternative splicing produces two different forms of the Sxl protein in male and female flies.

6. **Figure 17.42**

(a) What is the role of transferrin?

(b) What is the role of ferritin?

(c) When would the cell want to express more ferritin? Why?

(d) When would the cell want to express more transferrin receptor? Why?

(e) When is the ferritin mRNA translated? Explain how its translation is regulated.

(f) When is the transferrin receptor mRNA translated? Explain how its translation is regulated.

(g) Why is the transferrin receptor mRNA degraded when intracellular iron levels are high?

(h) Why is translation of the ferritin mRNA blocked when intracellular iron levels are low?

General Questions

Q17-1. Explain why transcription initiation in eukaryotes requires modification of the chromatin. Describe the three mechanisms by which chromatin remodeling can be achieved to allow transcriptional activation.

Q17-2. Describe the three general mechanisms by which repressors act to inhibit transcription.

Q17-3. Describe how the competition for dimer formation involving the Myc and Max proteins affects transcription in proliferating and nonproliferating cells.

Q17-4. What three features of enhancers distinguish these DNA elements from the other *cis*-acting regulatory elements?

Q17-5. Compare and contrast the mechanisms by which microRNAs and antisense RNAs control the translation of a specific target RNA.

Multiple Choice

For each of the following, circle the letter of the choice that most appropriately answers the question.

1.) Which of the following DNA-binding motifs found in several different DNA-binding proteins does *not* contact the DNA by way of one or more α-helices binding to the major groove of the DNA?
   a. zinc finger
   b. leucine zipper
   c. helix-turn-helix
   d. basic helix-loop-helix
   e. None of the above

2.) Which of the following proteins has been shown to have histone acetyltransferase activity?
   a. Histone H3
   b. $TAF_{II}250$
   c. RNA polymerase II
   d. TBP
   e. $TAF_{II}150$

3.) Proteins that bind to enhancers can exert their effect over tens of kilobases upstream or downstream of a gene by which of the following mechanisms?
   a. Induce bending of the DNA to bring the proteins bound at enhancers near the promoter.
   b. Activate signal transduction cascades that result in the binding or the prevention of binding of general transcription factors at the promoter.
   c. Introduce a change in the conformation of the chromatin in the region of a gene.
   d. Both a and c
   e. All of the above

4.) Mutations in which region of the promoter or proximal-promoter do *not* reduce the level of transcription?
   a. adjacent to the CCAAT box
   b. within the CCAAT box
   c. within the TATA box

   d. within the GC box
   e. None of the above; mutations in all of these will reduce the level of transcription.

5.) Which RNAs will be eliminated by nonsense-mediated mRNA decay?
   a. microRNAs
   b. mutant mRNAs with premature translation termination codons
   c. alternatively spliced mRNAs
   d. Both b and c
   e. All of the above

6.) Which of the following activities is *not* associated with the proteosome?
   a. trypsin
   b. proteinase K
   c. caspase
   d. ATPase
   e. chymotrypsin

7.) Angelman syndrome results from which of the following?
   a. a truncated SNRPN protein
   b. mutation of a snoRNA
   c. loss of the UBE3A protein
   d. loss of function mutations in the proteosome
   e. binding of a protein to an insulator

8.) DNA methylase modifies which of the following?
   a. sugar–phosphate backbone
   b. adenine
   c. cytosine
   d. guanine
   e. thymine

9.) If a gene is paternally imprinted,
   a. the progeny will all express the paternal allele.
   b. the male progeny will express the paternal allele, and the female progeny will express the maternal allele.

c. 50% of the progeny will express the
   paternal allele, independent of the sex of
   the progeny.
d. the progeny will all express the maternal
   allele.
e. None of the above

10.) Z-DNA can be stabilized by which of the
following?
   a. movement of the nucleosomes by the ISWI
      or SWI–SNF complex
   b. bending of the DNA by transcriptional
      activators
   c. variant forms of histone proteins
   d. methylation
   e. histone acetylation

Practice Problems

P17-1. On the following piece of eukaryotic genomic DNA, label the following *cis*-regulatory elements: the
promoter, proximal-promoter, potential locations of enhancers and silencers for gene A and potential
locations for an insulator for gene B. Label the *InR*, TATA box, CCAAT box, and GC box for genes A and
B, and clearly indicate whether each element is part of the promoter, proximal-promoter, or enhancer. For
all of the *cis*-regulatory elements, indicate the location of each (in base pairs) relative to the transcription
start site.

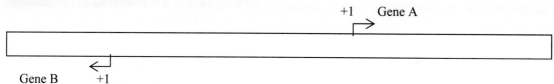

Indicate which protein binds to each of the DNA elements listed.
a.) TATA box
b.) CCAAT box
c.) GC box
d.) Enhancer

P17-2. Indicate the effect that each item listed on the left has on gene expression. Use a (+) symbol to
indicate activation, a (–) symbol to indicate repression, or a (±) symbol if the item listed can have
differential effects in different situations.
a.) Histone acetylation

b.) Hypermethylated CpG islands

c.) Exchange of the H2AZ–H2B dimer for the
   H2A–H2B dimer in the nucleosomes

d.) Histone methylation

e.) Histone deacetylation

f.) Hypomethylated CpG islands

g.) A repressor binding to an activator

h.) Binding of the SWI–SNF complex

i.) A protein binding to a silencer

j.) Myc–Max heterodimer

k.) Dorsal–Cactus heterodimer

l.) Incorporation of the histone H3 variant
   CENPA into the nucleosomes

P17-3. A scientist is studying the expression of gene *E* in different tissues. The following figure diagrams what the scientist knows about the genomic location of gene *E* and the probes available to her for analysis.

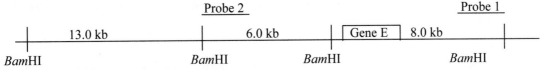

Describe an experiment the scientist could perform to determine if gene *E* is actively expressed in brain, heart, kidney, lung, or spleen cells. Outline the steps of the experimental procedure, and indicate what the experimental result will look like and the size of the fragments if gene *E* is not actively transcribed and what the result could look like if gene *E* is actively transcribed.

P17-4. For each of the following biological molecules or processes, indicate whether each occurs in prokaryotes, eukaryotes, or both to identify how gene expression is similar or different in these two groups of organisms.

a.) mRNA splicing

b.) MicroRNAs

c.) Operons with polycistronic mRNAs

d.) Covalent modification of proteins

e.) Small promoters with nearby operators

f.) Nucleosomes

g.) DNA-binding proteins acting as activators or repressors

h.) Alternative splicing

i.) Coupling of transcription and translation

j.) DNA methylation

k.) RNA editing

l.) Enhancers several kilobases away from the transcription start site

m.) Expressing a gene in the right cell at the right time

P17-5. A scientist cloned a 15-kb region upstream of a *Drosophila* gene that is actively expressed during embryonic development. Several large deletions in the promoter region were then generated, and either the full-length region or each of the deletion constructs was fused to a β-galactosidase reporter gene. Each construct was introduced individually into *Drosophila*. The 15-kb region exactly reproduced the expression pattern of the gene in the embryo which looks like:

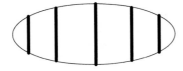

Draw what the expression pattern in the embryo would be expected to look like for each of the deletion constructs shown according to the description of the elements that each promoter region contains:

a.) The region from 12.5 to 15 kb contains an enhancer element that is required for maximal expression of stripe 2.

b.) The region from 10 to 12.5 kb contains a silencer element that is required to restrict stripe 3 to its proper width.

c.) The region from 7.5 to 10 kb contains a silencer element that is required to restrict stripe 4 to its proper width.

d.) The region from 5 to 7.5 kb does not contain an element that controls the expression of this gene based on the reporter gene expression.

e.) The region from 2.5 to 5 kb contains an enhancer element that is required for maximal expression of stripes 1 and 5.

f.) The region from 0 to 2.5 kb contains the basal promoter required for any expression in all five stripes.

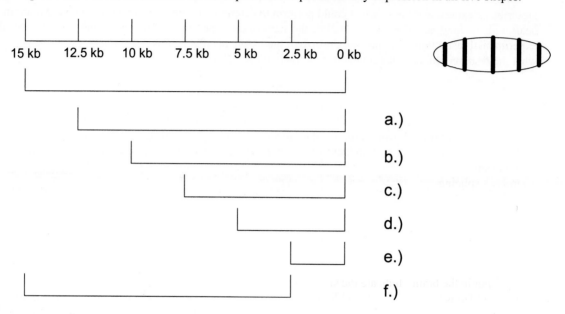

P17-6. Indicate the function of each of the following in ubiquitin-mediated proteosomal degradation. Number each protein or process to indicate the order in which it is required during proteosomal degradation.

Component	Function
Two 19S proteosomal caps	
E1 enzyme	
Ubiquitin	
20S proteosomal chamber	
E2 enzyme	
ATPases	
ATP hydrolysis	
E3 enzyme	

P17-7. Consider a paternally imprinted gene. You received allele A from your mother and allele B from your father. You pass the A allele to both of your two children, a boy and a girl. Your spouse has two C alleles. Which allele will your daughter express?

P17-8. Identify each of the following as a mechanism of posttranscriptional, translational, or posttranslational regulation that is utilized in eukaryotic organisms.
a.) Cleaving the poly(A) tail of an mRNA

b.) Phosphorylation of a protein to activate it

c.) MicroRNAs

d.) Phosphorylation of translation factors

e.) Producing multiple RNAs from a single gene

f.) Ubiquitin-mediated degradation

g.) Nonsense-mediated mRNA decay

h.) Binding of the translational machinery to the 3′ UTR of an mRNA

P17-9. A scientist wishes to analyze the methylation pattern of the promoter region of a gene. The locations of the 5′-CCGG-3′ sequences in the promoter, the distances between these sites, and the probes available for Southern blot analysis are shown in the following diagram:

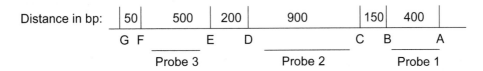

Genomic DNA is isolated from both brain and heart tissue, knowing that the gene is transcribed in the brain, but not in the heart. The genomic DNA is divided in two, and half is digested with *Hpa*II and half with *Msp*I. *Hpa*II cleaves the 5′-CCGG-3′ sequence only if it is not methylated, while *Msp*I cleaves the 5′-CCGG-3′ sequence whether the second cytosine is methylated or not. The digested DNA is analyzed by Southern blot with each of the three probes.

    a.) Draw what the Southern blot using probe 1 will look like if site B is methylated in the heart but not in the brain. Indicate the size of the fragments.

    b.) Draw what the Southern blot using probe 2 will look like if site C is not methylated in either tissue, site D is methylated in both the heart and the brain tissue, while site E is methylated in the brain tissue but not in the heart tissue. Indicate the size of the fragments.

    c.) Draw what the Southern blot using probe 3 will look like if site E is methylated in brain tissue but not in heart tissue, while site F is not methylated in either tissue. Indicate the size of the fragments. (Remember, site D is methylated in both tissues.)

a.) Probe 1

      Brain   Heart

*Msp*I   *Hpa*II   *Msp*I   *Hpa*II

b.) Probe 2

      Brain   Heart

*Msp*I   *Hpa*II   *Msp*I   *Hpa*II

c.) Probe 3

      Brain   Heart

*Msp*I   *Hpa*II   *Msp*I   *Hpa*II

d.) From this analysis, which site would you propose to be critical in regulating the tissue-specific expression of the gene?

Assessing Your Knowledge

Understanding the Key Concepts—Answers

1.) RNA polymerase II; 2.) *cis*; 3–4.) promoter; enhancers; 5.) promoter; 6.) preinitiation complex; 7.) six; 8.) transcription factors; 9.) TBP; 10.) –25; 11.) proximal-promoter; 12–13.) CCAAT; GC; 14.) –35; 15.) –200; 16.) TATA box; 17.) level of transcription; 18–19.) Activators; repressors; 20–21.) upstream; downstream; 22.) histone acetyltransferases; 23.) nucleosomes; 24.) activating; 25.) histone deacetyltransferases; 26.) acetyl groups; 27–28.) H3; H4; 29.) repress or inhibit; 30.) DNA-binding proteins; 31.) α-helix; 32.) major groove; 33.) conformation; 34.) repress; 35–36.) enhancers; silencers; 37.) neighboring; 38.) CpG islands; 39.) hypomethylated; 40.) hypermethylated; 41.) inactivated; 42.) DNA methylation; 43.) male; 44.) maternally; 45.) phenotypic; 46–47.) posttranscriptional; posttranslational; 48.) posttranscriptional; 49.) stability; 50.) proteases; 51.) ubiquitin; 52.) Ubiquitin; 53.) 76; 54.) mark; 55.) proteosome; 56.) E3 ubiquitin ligase; 57.) E2 conjugating enzyme; 58.) 19S; 59.) 20S; 60.) Mutant; 61.) short.

Figure Analysis—Answers

1a.) The Dorsal protein is held in an inactive state by binding of the protein to the Cactus protein; b.) The kinase activity of the Toll transmembrane receptor is activated by Spätzle binding to the extracellular portion of the receptor; c.) Binding of the Pelle–Tube heterodimer to the intracellular portion of the Toll receptor phosphorylates Pelle and releases it from the Toll receptor; d.) The Cactus protein is a Pelle substrate that is phosphorylated when Pelle is activated; e.) Phosphorylation of the Cactus protein activates the Dorsal protein because it disrupts the Cactus–Dorsal dimer which releases free Dorsal. Phosphorylated Cactus is degraded, and the free Dorsal subunit moves to the nucleus where it can bind to specific enhancers to activate or repress gene expression.

2a.) DNase I is an enzyme that digests DNA that is free of nucleosomes; b.) Following DNase I digestion, the DNA is digested with a restriction enzyme that will produce a predictable pattern of fragments on the Southern blot. By comparing the pattern obtained from DNA isolated from repressed tissue to that isolated from transcriptionally active tissue, it is possible to determine if the DNase I was able to digest any of the DNA and to map the DNase I hypersensitive sites; c.) The fragments of DNA obtained from the transcriptionally active tissue are smaller than those obtained from the transcriptionally repressed tissue because regions of the DNA that were free of nucleosomes in the transcriptionally active tissue were accessible to DNase I. At these sites, DNase I degraded the DNA. In the transcriptionally repressed tissue, these nucleosome-free sites were not present preventing degradation by DNase I; d.) DNase I hypersensitive sites are typically found upstream of genes that are transcriptionally active.

3a.) HAT-B acetylates newly synthesized histones in the cytosol; b.) Histones H3 and H4 are acetylated; c.) Lysine residues in the amino terminal tails of histones H3 and H4 are acetylated; d.) The function of HAT-A is to acetylate H3 and H4 histones that are already incorporated in a nucleosome in the nucleus; e.) Histone acetylation decreases the interaction between the histone core and the negatively charged sugar–phosphate backbone of the genomic DNA; f.) HDAC is a histone deacetyltransferase. HDACs reduce the number of acetyl groups present on histones H3 and H4 to repress transcription; g.) Activators recruit HATs, while repressors recruit HDACs.

4a.) The major difference between the maternal and the paternal chromosomes in terms of *Igf-2* and *H19* expression in mice is that the maternal chromosome lacks methylation in the insulator region between the two genes allowing the CTCF protein to bind to the insulator such that the enhancers downstream of the *H19* gene cannot activate *Igf-2* transcription. In the paternal chromosome, the insulator and the *H19* upstream region are methylated, preventing CTCF protein from binding to the insulator and also repressing *H19* transcription; b.) The *H19* upstream region on the paternal chromosome is methylated repressing its transcription; c.) *Igf-2* is maternally imprinted because the absence of methylation in the insulator region of the DNA allows the CTCF protein to bind to the insulator preventing the enhancer from activating transcription of *Igf-2*. The result is that progeny express the *Igf-2* phenotype that corresponds to the paternally inherited allele; d.) *H19* is paternally imprinted because methylation of the upstream region of the gene prevents its transcription in male mice. The result is that progeny express the *H19* phenotype that corresponds to the maternally inherited allele; e.) If two heterozygous mice both with the genotype *Igf-2*/*Igf-2m* are mated, the size of the progeny will be determined according to which of these two alleles is inherited from their father.

5a.) The Sxl protein is not present in the early embryo of male flies because these flies lack the *sxl* transcription factors so the gene cannot be transcribed from the P_e promoter; b.) The Sxl protein in the late embryo of female flies binds to the exon 3 splice acceptor site of the *Sxl* pre-mRNA produced from the P_L promoter. This binding prevents exon 3 from being spliced into the final mRNA; c.) Loss of Sxl protein activity in females is lethal because it results in hypertranscription of X-linked genes from both X chromosomes; d.) Mutation of the premature translational stop codon in the *Sxl* gene of male flies would result in the production of a wild-type functional Sxl protein which would be lethal because it would prevent hypertranscription of the X-linked genes; e.) Alternative splicing produces two different forms of the Sxl protein in female and male flies because female flies express the Sxl protein transcribed from the P_e promoter in the early embryo. This protein binds to the exon 3 splice acceptor site of the *sxl* pre-mRNA produced from the P_L promoter in the late embryo blocking the inclusion of exon 3 in the final mRNA in females. Male flies do not have the *Sxl* protein in the early embryo because it cannot be transcribed from the P_e promoter. Therefore, exon 3 is included in the mature mRNA. This exon contains a premature translational stop codon which produces a truncated, nonfunctional Sxl protein in the male flies.

6a.) Transferrin binds extracellular iron (Fe^{3+}), and by binding to the transferrin receptor, it moves Fe^{3+} into the cell; b.) Ferritin sequesters iron in cells to keep iron levels from becoming too high which could be toxic; c.) The cell would want to express more ferritin when intracellular iron levels are high to reduce the amount of free Fe^{3+} in the cell; d.) The cell would want to express more transferrin receptor when intracellular iron levels are low to bring more Fe^{3+} into the cell; e.) Ferritin mRNA is translated when intracellular iron levels are high. The IRE in the 5′ UTR of the ferritin mRNA is not bound by the iron regulatory protein (IRP) at this time because the IRP is bound to iron (Fe^{3+}). Lack of binding of the IRP to the 5′ UTR permits translation of the ferritin mRNA; f.) When intracellular iron levels are low, the transferrin receptor mRNA is translated because the mRNA is stabilized by binding of the iron regulatory protein (IRP) to the IRE in its 3′ UTR; g.) The transferrin receptor mRNA is degraded when intracellular iron levels are high because the IRE in the 3′ UTR of the mRNA is not bound by the iron regulatory protein (IRP) leading to the destabilization and degradation of the mRNA. By regulating receptor synthesis, the cell can reduce Fe^{3+} endocytosis; h.) The translation of the ferritin mRNA is blocked when intracellular iron levels are low because the IRE in the 5′ UTR of the ferritin mRNA is bound by the iron regulatory protein blocking its translation. When iron levels are low, ferritin is not required to bind intracellular Fe^{3+}.

General Questions—Answers

Q17-1. In eukaryotes, genomic DNA is tightly associated with histone proteins which play a major role in transcriptional repression. For RNA polymerase II, the promoter-recognizing TFIID transcription factor, and transcriptional activators to have access to the DNA to allow the initiation of transcription, the chromatin state must be modified. Chromatin remodeling can be achieved by the acetylation of histones H3 and H4 by histone acetyltransferases (HATs) that either acetylate newly synthesized histones or add additional acetyl groups to nuclear acetylated histones. Acetylation of the histones reduces the association between the nucleosomes and the genomic DNA and opens the chromatin to allow additional activators to bind and the preinitiation complex to be formed. Chromatin remodeling can also be achieved by enzyme-mediated movement of the nucleosomes to make the promoter and enhancer sequences available for protein binding. This movement of the nucleosomes is achieved by either the ISWI protein which hydrolyzes ATP to slide the histones along the DNA, or by the SWI–SNF protein complex which utilizes ATP to disassemble nucleosomes in one location on the DNA and reassemble them at another location. The third mechanism of chromatin remodeling involves the incorporation of variant histone proteins into nucleosomes. The incorporation of these histone variants can be involved in either the repression or the activation of transcription.

Q17-2. Repressors can act to inhibit transcription by competing with activators for binding to the same or an overlapping enhancer binding site or by binding directly to an activator protein to either block its DNA-binding domain, which prevents it from binding to the enhancer site, or to block its transcription activation domain, which allows it to bind to the enhancer sequence but renders it unable to activate transcription.

Q17-3. The Max protein is a basic helix-loop-helix/leucine zipper protein that lacks a transcriptional activation domain. This protein is expressed in all cells and can form either a homodimer in the absence of the Myc protein or a heterodimer with the Myc protein. The Myc protein is also a basic helix-loop-helix/leucine zipper protein, but unlike the Max protein, the Myc protein possesses a transcriptional activation domain. The Myc protein is only expressed in proliferating cells. Therefore, in all nonproliferating cells, the Max protein homodimerizes and binds to specific enhancer sequences where it blocks transcriptional activation because it lacks the transcriptional activation domain. In proliferating cells, where the Myc protein is expressed, the Max protein preferentially binds to the Myc protein, rather than homodimerizing. Therefore, in these cells, the Myc and Max proteins form a heterodimer that binds to an enhancer and can now activate transcription of target genes due to the presence of the transcriptional activation domain on the Myc protein.

Q17-4. Three features of enhancers distinguish these DNA elements from the other *cis*-acting regulatory elements. Enhancers can be moved up to tens of kilobases upstream or downstream of a gene and can still exert their effect on transcription. The orientation of an enhancer can also be reversed without affecting transcriptional activation unlike the other *cis*-acting elements. Finally, unlike the promoter and the proximal-promoter elements, enhancer elements determine the maximal level of transcription and control the spatial and temporal expression of specific genes, rather than controlling whether or not transcription occurs or affecting the basal level of transcription.

Q17-5. MicroRNAs control the translation of the corresponding mRNA by either cleaving the mRNA or base-pairing with it to block translation. The pri-miRNA forms a stem-loop structure that is cleaved into a pre-miRNA. The stem-loop region of the cleaved pre-miRNA is removed to form a duplex miRNA. One RNA strand forms a microribonucleoprotein with a protein of the RNA-induced silencing complex (RISC). If the miRNA is completely complementary to the target mRNA, endonuclease activity of the RISC cleaves the mRNA and it is degraded. If the miRNA is not completely complementary to the target mRNA, the base-paired regions block translation without cleaving the mRNA. Antisense RNAs are usually longer than miRNAs, and they do not form a stem-loop structure or a ribonucleoprotein with the RISC complex. Antisense RNAs are complementary to a target RNA sequence which allows the two RNA molecules to base-pair. Base-pairing between the antisense RNA and the target RNA can block translation of the mRNA similar to the miRNAs, or affect the stability of the mRNA. Base-pairing at an intron–exon junction can block the splicing of the pre-mRNA.

Multiple Choice—Answers

1.) e; 2.) b; 3.) d; 4.) a; 5.) d; 6.) b; 7.) c; 8.) c; 9.) d; 10.) d.

Practice Problems—Answers

P17-1.

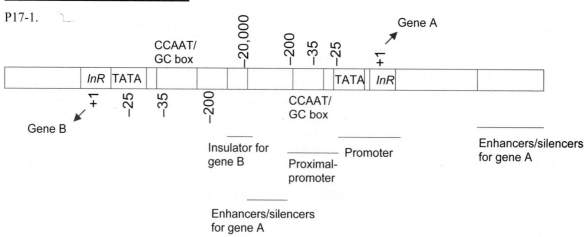

a.) TBP; b.) CTF; c.) SP1; d.) Activators or repressors

P17-2.
a.) +; b.) –; c.) +; d.) ±; e.) –; f.) +; g.) –; h.) +; i.) –; j.) +; k.) –; l.) –

17.3. To determine if gene *E* is actively being expressed in these cells, the scientist can perform a DNase I hypersensitivity assay. In this assay, genomic DNA is isolated from the brain, heart, kidney, lung, and spleen cells and is treated with a low concentration of DNase I. The histone proteins are removed, and the DNA is digested with *Bam*HI. The DNA is then run on an agarose gel and Southern blot hybridization is performed using the two available probes which are radioactively labeled. If the gene is not actively transcribed, two fragments, 8.0 kb and 6.0 kb, will be detected. If the gene is actively transcribed, the two probes will detect fragments smaller than 8.0 and 6.0 kb.

P17-4.
a.) Eukaryotes; b.) eukaryotes; c.) prokaryotes; d.) eukaryotes; e.) prokaryotes; f.) eukaryotes;
g.) prokaryotes/eukaryotes; h.) eukaryotes; i.) prokaryotes; j.) prokaryotes/eukaryotes; k.) eukaryotes;
l.) eukaryotes; m.) eukaryotes

P17-5.

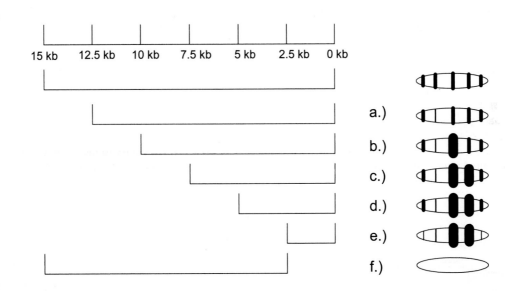

P17-6.

	Component	Function
6.	Two 19S proteosomal caps	Permit ubiquitinated proteins to enter the central chamber; denature the proteins
3.	E1 enzyme	Transfers ubiquitin to the ubiquitin-conjugating enzyme
2.	Ubiquitin	A peptide attached to a protein marking it for degradation
8.	20S proteosomal chamber	Degrades proteins into peptides
4.	E2 enzyme	Ubiquitin conjugating enzyme
7.	ATPases	Unfold ubiquitinated proteins as they enter the central core
1.	ATP hydrolysis	Attaches ubiquitin to the ubiquitin-activating enzyme (E1)
5.	E3 enzyme	Ubiquitin ligase; delivers target protein to the E2 enzyme and attaches ubiquitin to it.

P17-7. If you are female, all of your children will express the allele you gave them (allele A) because the allele your spouse gave them (allele C) is paternally imprinted. However, if you are male, all of your children will express the allele your spouse gave them (allele C) because the allele you gave them (allele A) is paternally imprinted.

P17-8. a.) Posttranscriptional; b.) posttranslational; c.) posttranscriptional; d.) translational;
e.) posttranscriptional; f.) posttranslational; g.) posttranscriptional; h.) translational

P17-9.

a.)

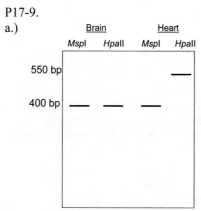

b.)

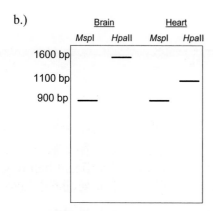

c.)

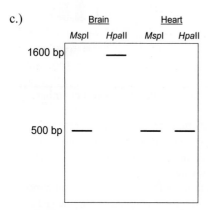

d.) Site B

17

GENE EXPRESSION: CONTROL IN EUKARYOTES

CHAPTER SUMMARY QUESTIONS

1. 1. h, 2. e, 3. g, 4. a, 5. j, 6. d, 7. i, 8. c, 9. f, 10. b.

2. The formation of the preinitiation complex (PIC) is a prerequisite for the binding of RNA polymerase II to the promoter region. The PIC is a group of six different general transcription factors. One of those is transcription factor IID (TFIID), which is composed of more than 10 different proteins, including the TATA binding protein (TBP). The TBP binds the TATA (Hogness) box in the minor groove of DNA. This causes significant bending of the DNA, thereby permitting the binding of other proteins, called TBP-associated factors (TAFs).

Another protein in the TFIID complex is the TAF_{II} 250 protein, which is a histone acetyltransferase. This enzyme binds DNA that is associated with acetylated histones and then adds more acetyl groups to the lysine residues found on the amino ends of histones H3 and H4. The DNA bending induced by TBP leads to further acetylation of histones H3 and H4 by TAF_{II} 250 protein. This may enhance the binding of additional transcription factors.

The formation of the PIC allows RNA polymerase II to bind to the promoter. This permits basal-level transcription to occur. Maximal transcription requires the binding of proteins to proximal-promoter elements and enhancers.

3. A general transcription factor facilitates transcription in all cells, whereas a specific transcription factor directs cell- or tissue-specific transcription.

4. Enhancers are DNA sequences that bind proteins and control gene transcription in a particular cell or tissue. Enhancers have two main features that distinguish them from other *cis*-acting elements: (1) they are position-independent (can be found upstream, downstream, or even within a gene), and (2) they are orientation-independent (reversing their polarity/direction does not affect their ability to regulate transcription). Enhancers affect the maximal level of transcription and control the gene's expression pattern both temporally (in time) and spatially (in tissue).

5. DNA-binding motifs include the helix-turn-helix and zinc finger. The helix-turn-helix is made up of two α helices that are connected by a "turn" consisting of several amino acids. The zinc finger is composed of a zinc atom that is covalently bound to one α helix (containing two histidine residues) and two antiparallel β sheets (containing two cysteine residues).

Protein dimerization motifs include the leucine zipper and helix-loop-helix. The leucine zipper consists of an α helix with four or more leucines that are evenly spaced at seven-amino-acid intervals. The helix-loop-helix consists of two α helices that are separated by a nonhelical loop.

6. Transcriptional repressors can bind to an enhancer site and prevent the binding of the transcriptional activator to it. Repressors can also bind to the activator directly and prevent it either from binding to DNA or from activating transcription.

7. Mechanisms that are commonly used to regulate transcription factors include: (1) binding of ligands, (2) protein–protein interactions, (3) posttranscriptional modifications, (4) chromatin remodeling, and (5) DNA methylation.

8. Both microRNA (miRNAs) and antisense RNAs are *trans*-acting factors that regulate gene expression. MicroRNAs are transcribed as a pri-mRNA (primary miRNA) that folds into a stem-loop structure which is then processed into a short RNA. This processed RNA associates with the Argonaute protein to form a microribonucleoprotein (miRNP) that constitutes part of the RNA-induced silencing complex (RISC). The miRNA has the potential to base-pair with a specific mRNA and block its translation or target its cleavage. Antisense RNAs, on the other hand, are not processed nor do they form a ribonucleoprotein. They are usually longer than miRNAs. They bind to complementary target RNA and block its splicing, repress its translation, or decrease its stability.

9. The Nanos protein represses the translation of Hunchback via a posttranscriptional mechanism involving the Pumilio protein. The 3' untranslated region of *hunchback* mRNA contains a binding site for the Pumilio protein. If the Nanos protein is absent, the *hunchback* mRNA is unmodified and is translated. However, if the Nanos protein is present, it binds to the Pumilio protein, and the Pumilio–Nanos complex then deadenylates the 3' poly(A) tail of the *hunchback* mRNA. The loss of the poly(A) tail decreases the stability of the mRNA and the efficiency of its translation. Both of these mechanisms lead to a decreased amount of the Hunchback protein.

10. a. HAT-A, or type A histone acetyltransferases, are nuclear enzymes that bind acetylated lysines and add additional acetyl groups to the amino-terminal lysines on histones H3 and H4. They are associated with transcriptional activation.
    b. HAT-B, or type B histone acetyltransferases, are cytosolic enzymes that acetylate newly synthesized H3 and H4 histones. They too are associated with transcriptional activation.

    c.   HDAC, or histone deacetyltransferases, reduce the number of acetyl groups present on the H3 and H4 histones. They repress transcription.

    d.   HMTases, or histone methyltransferases, add methyl groups to the lysine and arginines on all four nucleosome histone proteins. They can either activate or repress transcription.

11. Sex determination in *Drosophila* involves the alternative splicing of the *Sex lethal* (*Sxl*) gene. The *Sxl* gene is required for all aspects of sexual dimorphism in fruit flies. Early in embryonic development, the *Sxl* gene is transcribed from the P_e promoter in female flies, but not in males. Therefore, females will have the Sxl protein, but males will not. Later in development, the *Sxl* gene is transcribed from the P_L promoter in both males and females. In males, the primary *Sxl* transcript is spliced to include exon 3, which contains a translational stop codon. This results in a truncated and nonfunctional Sxl protein in males. In females however, the Sxl protein that was formed earlier in development binds to the splice acceptor site of exon 3. This binding masks the site during splicing, thereby preventing exon 3 from being included in the mature mRNA. The *Sxl* mRNA in females is then translated to produce a full-length and functional Sxl protein.

12. ISWI and SWI-SNF are protein complexes that can change the location of nucleosomes so that the promoter and enhancer sequences are available for protein binding. Both protein complexes can hydrolyze ATP for energy. They differ in the mechanism of action. The ISWI complex slides the histone core along the DNA, whereas the SWI-SNF complex disassembles and then reassembles the histone core.

13. Nonsense-mediated mRNA decay (NMD) is a process that eliminates mRNAs that possess a translation termination codon that is either premature or too far from the poly(A) tail of the mRNA. NMD may occur through three different pathways: (1) removal of the 5' cap followed by 5' → 3' exonuclease degradation, (2) removal of the 3' tail followed by 3' → 5' exonuclease degradation, and (3) endonucleolytic cleavage near the premature termination codon followed by 5' → 3' exonuclease degradation toward the 3' poly(A) tail and 3' → 5' exonuclease degradation toward the 5' cap. The third mechanism is used by *Drosophila*.

14. Methylation of cytosine may block transcription by preventing activators from binding to their DNA target sequences. Methylation also allows the binding of methyl-CpG-binding proteins that can recruit either transcriptional repressors or HDACs that deacetylate histones and therefore repress transcription.

15. There are several fundamental differences in the control of gene expression between bacteria and eukaryotes.

1.   In bacteria, genes are often grouped in operons and are coordinately expressed through the transcription of a single polycistronic mRNA. In contrast, eukaryotic genes are usually single transcriptional units.

2.   Bacterial DNA is not organized into nucleosomes. Therefore, bacterial genes are usually directly accessible to transcriptional activators or repressors. Eukaryotic DNA on the other hand is organized into nucleosomes, which play an important role in gene regulation. Transcription of eukaryotic genes is influenced by covalent modifications of histone proteins, by repositioning of nucleosomes, and by methylation levels.

3.   Transcription initiation is much more complex in eukaryotes. Eukaryotic structural genes have large promoters that have to bind several general transcription factors before RNA polymerase II can activate transcription. Bacterial genes have small promoters, and the RNA polymerase holoenzyme is either blocked or stimulated by the actions of repressors or activators, respectively. Moreover, in eukaryotes, transcriptional activators can bind to enhancers which can be located at large distances from the transcriptional start site. These distant enhancers occur much less frequently in bacterial cells.

4.   Eukaryotes have a more extensive array of posttranscriptional regulation mechanisms than bacteria. For example, they utilize alternative splicing of mRNAs and mRNA editing. In addition, mRNA stability is more important in eukaryotes than bacteria because in eukaryotes transcription and translation are not coupled.

16. eIF2, or eukaryotic initiation factor 2, is involved in the regulation of gene expression at the level of translation. Phosphorylation of eIF2 inhibits the binding of the initiator tRNA^{Met} to the 40S ribosomal subunit. This leads to the blocking of translation of all mRNAs.

17. The order of events in the Dorsal signal transduction pathway is: 5. Cleavage of Spätzle, 2. Binding of Spätzle to the Toll receptor, 1. Phosphorylation of Pelle, 3. Phosphorylation of Cactus, 6. Release of Dorsal from its heterodimer partner, 4. Entry of Dorsal into the nucleus.

18. 1.   False. The *myc* gene is a <u>proto</u>-oncogene that may cause cancer when inappropriately expressed.

2.   False. The *myc* gene is <u>not</u> expressed in cells that are terminally differentiated. (It is expressed in cells that are actively dividing.)

3.   True.

4.   False. The <u>Max</u> homodimer represses gene transcription. (Myc cannot form a homodimer.)

5.   True.

6.   False. The <u>Myc</u> protein contains a transcription activation domain. (The Max protein can bind to DNA as a homodimer; however, it lacks the ability to activate transcription.)

    7.   True.

    8.   True.

19. 1.   False. The addition of the <u>polypeptide</u> ubiquitin to proteins targets them for destruction by the proteasome.

    2.   True.

    3.   False. ATP + ubiquitin + UAE-E1 → UAE-E1-ubiquitin + <u>AMP + PPi</u>.

    4.   False. UAE-E1-ubiquitin + U<u>C</u>E-E2 → U<u>C</u>E-E2-ubiquitin + UAE-1. (Enzyme 2 is the <u>u</u>biquitin-<u>c</u>onjugating <u>e</u>nzyme.)

    5.   True.

    6.   False. The proteasome complex is 26S and consists of a 20S cylindrical chamber and <u>two 19S caps</u>. (Remember that sedimentation units are not additive.)

    7.   False. The proteins are degraded in the central core by three protease activities: chymotrypsin, trypsin, and <u>caspase</u>.

    8.   False. The proteasome is involved in the degradation of mutant proteins, as well as proteins that are needed for <u>short</u> periods of time (transcription factors, for example).

EXERCISES AND PROBLEMS

20. Both eukaryotic basal transcription factors and bacterial sigma factors bind promoters to stimulate transcription by RNA polymerase. They differ in the way they attach to the promoter. Sigma factors cannot bind to the promoter on their own. Rather, they are complexed with the RNA polymerase core enzyme to form the holoenzyme. Eukaryotic basal transcription factors, however, can bind to promoters independently of RNA polymerase.

21. The thyroid hormone response element (THRE) is the binding site of the thyroid hormone receptor. It functions as an enhancer (in the presence of the thyroid hormone) and as a silencer (in the absence of the thyroid hormone). If the thyroid hormone receptor cannot bind to THRE, it will not be able to influence (activate or repress) the transcription levels of the downstream genes. Therefore, the genes will exhibit basal (low) transcriptional activity, regardless of thyroid hormone levels.

22. If a gene is actively expressed, its DNA should be accessible and thus free of nucleosomes. Therefore, this genomic DNA would be susceptible to digestion by DNase I. On the other hand, genes that are transcriptionally inactive are expected to be in a compacted state due to their association with nucleosomes, and thus the genomic DNA would be resistant to DNase I digestion. In the heart and liver, only a 4-kb fragment is visible. This suggests that gene *T* is inactive in those tissues. However, two smaller fragments are visible in the kidney, indicating active transcription of gene *T* in that tissue.

23. In mice, the *Igf-2* gene exhibits genomic imprinting. In an individual, the maternal allele is silenced through DNA methylation while the paternal allele is expressed.

Cross	Genotypic Ratio	Phenotypic Ratio
A. *Igf-2 Igf-2m* female × *Igf-2 Igf-2* male	50% *Igf-2 Igf-2* 50% *Igf-2 Igf-2m*	100% normal size
B. *Igf-2 Igf-2m* male × *Igf-2 Igf-2* female	50% *Igf-2 Igf-2* 50% *Igf-2 Igf-2m*	50% normal size 50% small size
C. *Igf-2m Igf-2m* female × *Igf-2 Igf-2* male	100% *Igf-2 Igf-2m*	100% normal size
D. *Igf-2m Igf-2m* male × *Igf-2 Igf-2* female	100% *Igf-2 Igf-2m*	100% small size

24. The TfR-IRE-b sequence contains a stretch of complementary ribonucleotides that are separated by six ribonucleotides. This sequence will form a secondary stem-loop structure, as shown in the following. Note that there is a single bulge (C) on the 5' side of the stem.

```
       G   U
    A        G
  C            C
    G        C
    A        U
    A        U
    G        C
    G        C
C
    U        A
    A        U
    U        A
    U        A
    A        U
    A        U
    5'       3'
```

The fact that the five TfR sequences are highly conserved indicates that they have an important function. To test whether the TfR-IRE-b sequence can bind proteins, a gel retardation assay can be performed. The premise is that an RNA to which a protein binds would move more slowly in a gel and is retarded compared to an RNA with no bound protein. In addition, one could use synthetic TfR-IRE-b sequences containing base substitutions that prevent stem-loop formation. These RNAs would not allow protein binding. Furthermore, substitutions that restore stem-loop formation would also restore protein binding.

25. Nucleosome structure is highly conserved throughout eukaryotic organisms. Therefore, the role of nucleosomes in regulating gene expression must also be conserved in eukaryotes.

26. The *UBE3A* gene encodes the ubiquitin protein ligase 3E. Its lack of expression results in Angelman syndrome. The *UBE3A* gene is imprinted in the male, and so an individual only expresses the maternal allele of this gene. Therefore, a child (*boy* or *girl*) would exhibit Angelman syndrome if the child inherits a mutated allele from his or her mother. The pedigree of this family is shown in the following. For simplicity, U = *UBE3A* gene, + = normal, and – = mutant.

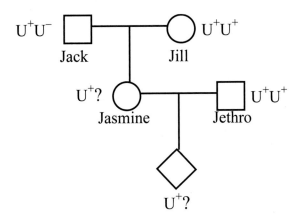

Jack has a 50% chance of transmitting his mutated *UBE3A* allele to Jasmine. Jasmine, in turn, has a 50% chance of transmitting this mutated *UBE3A* allele to her child. Therefore, the probability that the child will exhibit Angelman syndrome is $1/2 \times 1/2 = 1/4$ (product rule of probability). This 25% chance is the same for a son or daughter.

27. An active *Sxl* gene is required for female, not male, development. Therefore, XY embryos in which the *Sxl* gene is deleted are expected to become phenotypically normal males.

28. Let *A* = transcriptional activator 1, and *B* = transcriptional activator 2. The cross can be represented as *AaBb* × *AaBb*. This dihybrid cross yields offspring in the following proportions: 9/16 *A–B–*:3/16 *A–bb*:3/16 *aaB–*:1/16 *aabb*.
   a. Since black fur requires the expression of both activators, only the *A–B–* genotypic class will produce animals with black fur. Therefore, the ratios would be 9/16 black-furred:7/16 white-furred animals.
   b. Since black fur requires only one transcriptional activator, three of the four genotypic classes will produce animals with black fur. The only white-furred animal would have the genotype *aabb*. Therefore, the ratio would be 15/16 black-furred:1/16 white-furred animals.

29. The Northern blot shows an mRNA transcript for all three tissues. However, in the Western blot a protein is visible only in the muscle tissue. Therefore, gene *Z* must be regulated at the translational level.

30. Genes that are normally regulated by Myc–Max will be regulated inappropriately. In nondividing cells, which express only Max, these regulated genes will be activated instead of repressed. In dividing cells, which express both Myc and Max, the regulated genes will be repressed instead of being activated.

31. In prokaryotes, transcription and translation are coupled. In eukaryotes, transcription occurs in the nucleus and translation in the cytoplasm, so these two processes are not linked together. The coupling of transcription and translation in prokaryotes is an important part of the operon strategy.

32. This problem can be solved using a systematic approach that lists all possible combinations.

Three Exons	Four Exons	Five Exons
E1–E2–E3	E1–E2–E3–E4	E1–E2–E3–E4–E5
E1–E2–E4	E1–E2–E3–E5	
E1–E2–E5	E1–E2–E4–E5	
E1–E3–E4	E1–E3–E4–E5	
E1–E3–E5	E2–E3–E4–E5	
E1–E4–E5		
E2–E3–E4		
E2–E3–E5		
E2–E4–E5		
E3–E4–E5		

Therefore, a total of 16 different proteins are possible from the alternative splicing of this mRNA.

33. Six different gene expression control elements were identified here. The first is in the region between 15.0 and 20.0 kb and is a DNA element involved in repressing transcription in stripe 1. When this silencer is deleted, an increase in expression is observed in stripe 1. The region between 12.5 and 15.0 kb includes an element that activates expression in stripe 2; deletion of this enhancer element results in a decrease in expression in stripe 2. The third element is between 10.0 and 12.5 kb and is an enhancer of transcription in stripe 3. The fourth element is between 7.5 and 10.0 kb and is an enhancer of transcription in stripe 1. The fifth element is between 5.0 and 7.5 kb and is also an enhancer of transcription in stripe 1. The sixth element is between 0 and 2.5 kb and is an enhancer of transcription in stripe 4.

34. a. Both Max–Max homodimers and Myc–Max heterodimers will be unable to bind DNA; therefore, both repression and activation will be blocked. Genes regulated by Myc–Max will be expressed at low levels in both (**i**) nonproliferating and (**ii**) proliferating cells.

b. The Myc–Max heterodimer will fail to bind DNA. (**i**) Myc is normally not expressed in nonproliferating cells, so the mutation will have no effect in nonproliferating cells. (**ii**) In proliferating cells, genes normally activated by Myc–Max will fail to be activated.

c. If Max cannot bind to DNA, genes normally regulated by the Myc–Max system will not be repressed or activated. Therefore, genes regulated by Myc–Max will be expressed at low levels in both (**i**) nonproliferating and (**ii**) proliferating cells.

d. If Myc is unable to bind Max, Myc–Max heterodimers will not form, so Max–Max homodimers will be found in both nonproliferating and proliferating cells. Therefore, (**i**) nonproliferating cells will be normal, but (**ii**) in proliferating cells, genes normally activated by Myc–Max heterodimers will be repressed by Max–Max homodimers.

e. If the transcriptional activation domain of Myc is mutated, Myc–Max heterodimers will form, but transcription activation will not occur. Therefore, (**i**) nonproliferating cells will be normal, but (**ii**) in proliferating cells, genes normally activated by Myc–Max heterodimers will be repressed.

35. If Myc is expressed inappropriately in nonproliferating cells, its expression will activate expression of target genes involved in cell division. This will make the cells more likely to divide and will contribute to the development or progression of cancer.

36. a. Both *Msp*I and *Hpa*II recognize the sequence CCGG. *Hpa*II is sensitive to methylation and will cut this sequence only when it is not methylated. *Msp*II is not sensitive to methylation and will cut this sequence regardless of whether it is methylated or not. In liver, only the third *Msp*I site is methylated. In kidney, the second, third, and fifth sites are methylated. The restriction maps for *Msp*I and for *Hpa*II are shown in the following. Notice that the *Hpa*II maps are different for liver and kidney because the methylation patterns are different in these tissues.

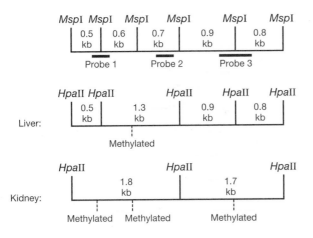

b. There may be other methylated CpG sequences in this region. If a CpG that is not part of the recognition sequence for these enzymes is methylated, it will not be detected in this assay.

c. The two sites that are methylated in the kidney but not in the liver (that is, the second and fifth *Msp*I sites) are most likely associated with repression of expression of this gene.

d. If the DNA methylase that acts in this region were inactive, then none of these sites would be methylated and *Hpa*II would cut all possible sites. Therefore, in both liver and kidney, the *Hpa*II pattern would be identical to the *Msp*I pattern. Expression of this gene would increase in the kidney but may or may not be affected in the liver.

CHAPTER INTEGRATION PROBLEM

a. In each lane, the fragments should add up to 4 kb in size. This is true of both *Hpa*II lanes, so no fragment is missing from these digests. However, this is not the case for the *Msp*I fragments, which add up to 2.75 kb in the kidney digest and 1.25 kb in the liver digest. Because the restriction enzyme *Msp*I cuts at both methylated and unmethylated 5'-CCGG-3' sequences and because kidney and liver DNA are expected to have the same sequence (barring a mutation), the *Msp*I digests in both tissues should yield identical results. Therefore, the 0.75- and 0.50-kb fragments found in the liver digest are missing from the kidney digest, while the 1.50-, 1.0-, and 0.25-kb fragments found in the kidney digest are missing from the liver digest. The full gel pattern would be

Kidney DNA		Mol. Wt. Markers	Liver DNA	
*Msp*I	*Hpa*II		*Msp*I	*Hpa*II
		— 3.0 kb		—
	—	— 2.0 kb		
—		— 1.5 kb	—	
—	—	— 1.0 kb	—	—
—	—	— 0.75 kb	—	
—		— 0.50 kb	—	
—	—	— 0.25 kb	—	

b. *Msp*I digestion of the 4-kb region yields five different fragments with sizes of 0.25, 0.50, 0.75, 1.0, and 1.5 kb. The partial *Msp*I restriction map accounts for the three midsized fragments and for a fragment that is 1.75 kb long. This fragment should be the source of the smallest (0.25 kb) and largest (1.5 kb) bands in the *Msp*I digest, and so it should contain an *Msp*I restriction site. Therefore, there are two possible restriction maps for this 4-kb region:

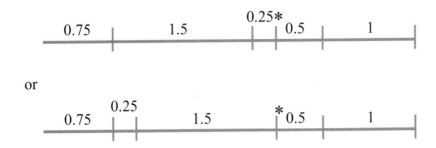

or

But which map is correct? To answer this question, we have to examine the *Hpa*II kidney digests. Remember that the site indicated by an asterisk is methylated in all tissues, and so *Hpa*II will not cut at this site. If the first map is correct, then *Hpa*II digestion should yield two different 0.75-kb fragments: the left-most fragment and the 0.5 + 0.25 fragment. Because the *Hpa*II digest only produced one 0.75-kb fragment, the first map cannot be correct. Indeed, that digest did produce a 2-kb fragment. This is clearly consistent with the second map. Therefore, the complete *Msp*I restriction map is

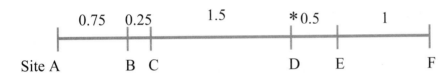

Site A   B C   D   E   F

c.   Site D is methylated in all tissues. The presence of other methylated restriction sites can be determined by analyzing the *Hpa*II digests, since this endonuclease only cuts 5'-CCGG-3' sequences that are not methylated. The *Hpa*II digests of kidney genomic DNA revealed four fragments on the Southern blot, as would be expected from the restriction map. Therefore, in kidney DNA, only site D is methylated.

Digestion of liver genomic DNA by *Hpa*II produced only two fragments: 3 kb and 1 kb. The 1-kb fragment would result if sites E and F are unmethylated. The 3-kb fragment consists of four smaller fragments: 0.75 + 0.25 + 1.5 + 0.5. This large fragment can be accounted for if *Hpa*II cuts at sites A and E, but not at sites B and C. Therefore, in liver DNA sites B, C, and D are methylated.

d.   The degree of methylation of DNA is related to silencing of a gene. The gene has more methylation sites in liver DNA than in kidney DNA. Therefore, it is likely that the gene is actively transcribed in the kidney but not in the liver.

e.   To determine the relative locations of the exons and introns, you would isolate the corresponding cDNA from a kidney cDNA library, digest the clone with a restriction enzyme, and electrophorese the restriction digest through an agarose gel. This gel could then be used in a Southern blot with smaller genomic DNA fragments from this region being used as the probe. The probes could also be hybridized to a Southern blot containing the restriction digested genomic DNA. A probe hybridizing to both the genomic DNA and cDNA would correspond to an exon; a probe hybridizing to the genomic DNA but not to the cDNA would correspond to an intron. Alternatively, the same genomic DNA probes could be hybridized to a Northern blot that contains kidney mRNA. A probe hybridizing to both the genomic

DNA and mRNA would correspond to an exon; a probe hybridizing to the genomic DNA but not to the mRNA would correspond to an intron. Sequencing genomic and cDNA would confirm these relationships.

f.   To determine the direction of transcription, you could sequence the gene or cDNA and look for open reading frames, which would be followed by a polyadenylation signal sequence. This would reveal the direction of transcription. The approximate location of the promoter could be identified by first locating the 5'-UTR (untranslated region), which is the sequence in either the cDNA or genomic DNA that precedes the translation initiation codon. The sequence that is then upstream of the 5'-UTR sequence in the genomic DNA, which is absent in the cDNA, would correspond to the relative location of the promoter. (Remember, that the RNA polymerase II promoter is located upstream of where transcription begins and will not be present in the mRNA or the corresponding cDNA.)

g.   *Msp*I generates 5' overhangs

```
5'–C                          C  G  G–3'
3'–G  G  C                          C–5'
```

h.   The *Ren*I restriction site includes an *Msp*I site. Therefore, every *Ren*I site is also an *Msp*I site.

i.   Because the *Msp*I restriction site is smaller than that of *Ren*I, not every *Msp*I site would be a *Ren*I site. For an *Msp*I site to be cut by *Ren*I, the bases at opposite ends of the site have to be 5'-AY and RT-3'. The probability of the chance occurrence of this stretch of bases is the product of the probabilities of each base in the sequence. First let's calculate the frequency of each base. The mouse genome has a GC content of 42%, and so the frequency of G and C is $0.42/2 = 0.21$ each, and the frequency of A and T will be $(1 - 0.42)/2 = 0.29$ each. Therefore, the probability of the occurrence of the sequence 5'-AY. . . RT-3' is $0.29 \times 0.5 \times 0.5 \times 0.29 = 0.021$. This means that only 2.1% (approximately 1 in 50) of *Msp*I sites in the mouse genome are also *Ren*I sites.

CHAPTER 18 DNA MUTATION, REPAIR, AND TRANSPOSITION

Chapter Goals
1. Learn the major mechanisms by which spontaneous mutations are induced.
2. Describe the various types of mutagens and how they produce mutations in DNA.
3. Understand the various mechanisms by which a cell detects mutations and repairs them.
4. Describe the various classes of transposons and their mechanisms of transposition.
5. Be able to diagram how transposable elements can produce mutations.

Key Chapter Concepts

18.1 Mutation
- **Mutation:** The term **mutation** refers to the process by which a gene changes structurally. An individual that expresses the effects of this change is called a **mutant**. Nucleotides are the smallest units of mutation. For every mutational change in the DNA, there may be a corresponding change in the protein product of the mutant gene. It is important to realize that not all mutations produce an altered phenotype and that more than one mutation can produce the same phenotypic effect. Geneticists can determine if two mutations are alleles of the same gene by creating a heterozygote and noting its phenotype (**complementation test**). A mutant phenotype indicates that the mutations occurred in the same gene. Examination of the offspring of this heterozygote suggests whether or not the mutations are structural, as well as functional, alleles. If the wild-type phenotype occurs at the background mutation rate, the mutations occur at the same place on the cistron (the smallest unit coding for a single polypeptide).
- **Phenotypic Classes:** Different phenotypes that are affected by mutations are **morphological**, **nutritional** or **biochemical**, and **behavioral**. A **lethal phenotype** results in the death of an organism, while a **deleterious phenotype** results in reduced viability.
- **The Fluctuation Test—A Turning Point for Bacterial Genetics:** Luria and Delbrück studied the generation of *ton^r* mutants from a wild-type *ton^s E. coli* strain. They demonstrated by the **fluctuation test** that phenotypic variants found in bacteria are the result of mutations rather than induced physiological changes.

18.2 Types of Mutations and Their Effects
- **Types of Mutations: Spontaneous mutations** are produced naturally in the life cycle of an organism, while **induced mutations** are generated by treating the organism with high-energy radiation or different chemical **mutagens**. Mutations that consist of single changes in the nucleotide sequence are **point mutations**. **Nonsense** mutations result in the formation of an inappropriate stop codon, while a **missense** mutation changes an amino acid specified in the code. Point mutations that add or delete bases cause a shift in the reading frame (a **frameshift** mutation) thus potentially altering the amino acid sequence of a protein encoded for by a mutated gene. Point mutations that do not produce a detectable phenotypic change are **silent mutations**. Subsequent mutations can restore gene function (**intragenic suppression**) or the original gene sequence (**back mutations**).
- **Conditional Mutations:** As discussed in chapter 15, mutations can be **conditional-lethal** or **conditional mutant** where an organism has the wild-type phenotype in the **permissive condition** but is inviable or has a mutant phenotype in the **restrictive condition**. Examples of these are **nutrient-requiring mutants** and **temperature-sensitive mutants**.
- **Mutation Rates:** The **mutation rate** is the number of mutations that arise per gene for a specific length of time in a population. These rates vary tremendously. The **reversion rate** corresponds to the change from the mutant phenotype back to the wild-type phenotype.

18.3 Mutagenesis: Origin of Mutation
- **The Origin of Mutations:** Mutations can occur spontaneously during DNA replication or can be induced by a variety of environmental factors or by exposure to certain chemicals.
- **Spontaneous Mutations:** A purine or pyrimidine can be replaced by another base of the same class (a **transition**). If a purine is replaced with a pyrimidine or vice versa, a **transversion** is said to occur.

Depurination, deamination, and mutations arising from the interaction of DNA with active oxygen species are three types of spontaneous mutations that affect the DNA of an organism.

- **Chemical Mutagenesis:** Certain chemicals can cause mutations at rates that are considerably higher than the rates at which spontaneous mutations arise. The **Ames test** is a routine screening test used to identify mutagens. **Base analogs** are chemicals that mimic the structure of a base, such as 5-bromouracil or 2-aminopurine. Incorporation of these molecules into the DNA produces transition mutations. Nitrous acid and heat both lead to transition mutations by two different mechanisms. Nitrous acid replaces amino groups on nucleotides with keto groups, while heat deaminates cytosine to form uracil. Ethyl methanesulfonate (EMS) and ethyl ethanesulfonate (EES) are two chemicals that can generate transversion mutations. EMS and EES add an ethyl group to any of the four bases. Acridine dyes can cause insertion and deletion mutations in the DNA.

- **Misalignment Mutagenesis:** Insertions and deletions in the DNA can also be produced by the misalignment of a template strand and a newly synthesized strand in a region containing a repeated sequence. This is the source of the mutations that generate the abnormal phenotypes of the **triplet diseases**, such as fragile-X syndrome, and the **polyglutamine diseases**.

- **Intergenic Suppression:** In addition to a reversion mutation, or intragenic suppression, intergenic suppression can also restore the wild-type phenotype in an organism with a mutant protein. This occurs by a mutation in a suppressor gene which typically encodes a tRNA or an interacting protein. Mutation of a tRNA can change the way in which a codon is read. Suppressor tRNAs can correct for missense mutations, nonsense mutations, and frameshift mutations. Mutation of an interacting protein can restore the interaction between it and the original mutant protein. This allows the two mutant proteins to generate a functional multiprotein complex and restores the wild-type phenotype.

- **Mutator and Antimutator Mutations:** While intergenic suppressors represent mutations that restore the wild-type phenotype, mutator and antimutator mutations cause an increase or decrease in the overall mutation rate of the cell.

18.4 DNA Repair

- **DNA Repair:** Cells normally have the ability to correct mistakes found in the DNA molecule. Loss of this ability may result in mutations, cell death (**apotosis**), or cancer. **DNA photolyase**, in the presence of light, breaks the linkages of adjacent pyrimidines (**dimers**) caused by exposure to UV light. Damaged portions of DNA are removed by the general mechanism of excision repair. Enzymes remove damaged bases in the process of **base excision repair**. Dimers may be removed in the process of **nucleotide excision** repair. **Mismatch repair** removes mismatched bases and is the most common type of DNA repair.

- **Double-Strand Break Repair:** DNA damage that breaks both strands of the DNA double helix, such as that induced by exposure to ionizing radiation, is repaired by **nonhomologous end-joining** or by **homology-directed recombination**. Nonhomologous end-joining involves the trimming of the two broken DNA ends so that they can be ligated by DNA ligase IV. This can result in the loss of some of the information contained in the broken region of the DNA. Homology-directed recombination involves the use of the Rad51 protein to invade the sister chromatid at a homologous sequence and use it as a template to repair the DNA double-strand break.

- **Recombination Repair:** Recombination repair is a **postreplicative repair** process by which gaps in one of the DNA strands created by a failure of DNA replication are fixed. The RecA protein in *E. coli* is used for filling a postreplicative gap in newly replicated DNA with a strand from the undamaged sister duplex.

- **The SOS Response:** The **SOS response** is induced by the presence of single-stranded DNA resulting from postreplicative gaps in *E. coli*. RecA interacting with single-stranded DNA activates a protease activity that cleaves specific proteins, such as the LexA protein which normally represses the transcription of several genes that have an **SOS box** in their promoters. Cleavage of LexA leads to the transcription of genes that inhibit cell division, and SOS repair polymerases that can replicate damaged DNA.

18.5 Transposable Genetic Elements

- **Transposons** are mobile genetic elements that possess the ability to disrupt the expression of a gene and produce a mutant phenotype.

- **Bacterial Transposable Elements:** Insertion sequences (IS elements), composite transposons, and complex transposons are three different classes of transposable elements in bacteria.
- **IS Elements:** IS elements are the simplest transposons. They consist of a gene that encodes **transposase** flanked by two terminal inverted repeat sequences.
- **Composite Transposons:** A composite transposon consists of a central region that usually contains bacterial genes flanked by two similar IS elements.
- **Complex Transposons:** Complex transposons contain genes other than transposase, and they do not contain flanking IS elements.
- **Mechanisms of Transposition:** There are two major mechanisms of transposition: **conservative replication** and **replicative transposition**. In conservative replication, the transposon moves to a new location without copying itself. The donor site retains the duplicated flanking sequence of the transposon. Replicative transposition involves the duplication of the transposon in which a copy is inserted in a new location and the original transposon remains at the donor site.
- **Eukaryotic Transposable Elements: Corn Ac–Ds Elements:** Barbara McClintock received a Noble Prize for her pioneering research on transposons in corn. McClintock examined a mutation in maize that causes a kernel to be purple instead of the wild-type color, white. However, she found that not all kernels with the mutation expressed it: the gene causing purple color was sometimes suppressed. McClintock hypothesized that a second gene, which she termed the dissociator or *Ds*, was responsible for the suppression. The revolutionary idea that McClintock proposed was that the *Ds* gene moved around the genome, "turning on" or "turning off" adjacent genes. The movement of the *Ds* gene was controlled by a third gene, the activator (or *Ac*) gene.
- *Drosophila* **Transposable Elements:** Two classes of transposable elements in *Drosophila* are *P* **elements** and **copia elements**.
  - *P* **Elements:** The *P* element contains terminal inverted repeats that flank four exons. When the *P* element RNA is transcribed in somatic cells, the third intron containing a translation termination codon is not spliced out producing a 66 kDa repressor of transposase. In germ cells, all three introns are removed when the *P* element is transcribed producing an mRNA that encodes the 87 kDa transposase protein. This difference in splicing is due to the presence of a protein in somatic tissues that recognizes and binds a sequence in the third intron. *P* elements are routinely used by scientists to introduce foreign DNA into flies.
  - **Copia Elements:** Copia elements are defined by long terminal repeats that contain one ORF that is homologous to the *gag*, *int*, *env*, and *pol* genes in RNA retroviruses. These elements represent a **retrotransposon** because they move by way of an RNA intermediate.
  - **Transposable Elements in Humans: Long interspersed elements (LINES)** and **short interspersed elements (SINES)** are two common classes of human transposons. LINES encode the reverse transcriptase gene, but lack LTRs unlike retrotransposons. SINES are primarily composed of the *Alu* sequence. These elements lack LTRs and do not encode a reverse transcriptase, but are flanked by 7- to 21-bp direct repeats of the insertion sequence.

Key Terms

- mutation
- morphological phenotypes
- nutritional phenotypes (biochemical phenotypes)
- behavioral phenotypes
- lethal phenotype
- deleterious phenotype
- fluctuation test
- spontaneous mutations

- induced mutations
- point mutations
- silent mutations
- missense mutations
- nonsense mutations
- frameshift mutation
- back mutation (reverse mutation)
- intragenic suppression

- conditional mutant
- permissive condition
- restrictive condition
- nutrient-requiring mutants
- temperature-sensitive mutants
- mutation rate
- reversion rate
- mutation frequency

- tautomeric shift
- transition mutation
- transversion mutation
- depurination
- deamination
- base analogs
- triplet diseases
- anticipation
- polyglutamine diseases
- intergenic suppression
- suppressor gene
- missense mutations
- mutator
- antimutator mutations
- damage reversal
- dimerization

- photoreactivation
- excision repair
- base excision repair
- DNA glycosylase
- AP endonuclease
- base flipping
- nucleotide excision repair
- mismatch repair
- nonhomologous end-joining
- homology-directed recombination
- postreplicative repair
- recombination repair
- SOS response
- transposable elements

- insertion sequences (IS elements)
- transposase
- composite transposon
- complex transposons
- conservative replication
- replicative transposition
- cointegrate
- selfish DNA
- *P* element
- hybrid dysgenesis
- retrotransposons
- long interspersed elements (LINES)
- short interspersed elements (SINES)

Understanding the Key Concepts

Use the key terms or parts of their definitions to complete the following sentences.

Mutations in the genome of an organism can arise (1)_____ or can be (2)_____ by chemical (3)_____ or by exposure to X-rays or UV light. Mutations in the DNA can lead to (4)_____, (5)_____, or (6)_____ changes in the phenotype of an organism. Some mutations are (7)_____ and lead to the death of an organism, while other mutations are (8)_____ allowing the organism to have the wild-type phenotype under (9)_____ conditions and a (10)_____ phenotype under (11)_____ conditions.

Point mutations are single (12)_____ changes in the DNA sequence. Point mutations can be (13)_____ mutations resulting in the (14)_____ of one amino acid for another in a protein. Point mutations can also be located in the (15)_____ position of a codon such that they are silent. Point mutations can be nonsense mutations which introduce a premature (16)_____ codon and lead to the production of a (17)_____ protein, or they can involve the (18)_____ or (19)_____ of a base in a DNA sequence. This type of point mutation will result in a (20)_____ which causes all of the bases (21)_____ of the mutated site to be read (22)_____. Transition mutations exchange a (23)_____ for a pyrimidine, or a (24)_____ for a purine in the normal DNA sequence. Transversion mutations exchange a (25)_____ for a pyrimidine. Spontaneous mutations can be produced by the incorporation of a base that has undergone a (26)_____ shift, or by the (27)_____ or (28)_____ of a base that has already been incorporated into the DNA. The addition or deletion of bases in the DNA can occur due to the presence of a (29)_____ sequence in the DNA which can cause the (30)_____ of the template strand and the newly synthesized daughter strand during (31)_____. Mutations like these are associated with the (32)_____ diseases.

Some DNA mutations can be (33)_____ by a second (34)_____ that (35)_____ the mutant phenotype. Intragenic (36)_____ are mutations that occur in (37)_____ as the first mutation, while (38)_____ are mutations that occur in a (39)_____ than the original mutation. Both of these second site mutations restore the wild-type phenotype. (40)_____ mutations can occur at the same site as the first mutation and also restore the wild-type phenotype. The spontaneous (41)_____ can vary greatly from gene to gene and from organism to organism. The (42)_____ rate of these mutations is typically much less than the (43)_____ rate because it is dependent on the occurrence of specific mutations at specific sites.

Certain chemicals that act as (44)_____ can be used to introduce mutations into the DNA at a rate that is much (45)_____ than the rate at which spontaneous mutations occur. Base (46)_____, such as 5-bromouracil and 2-aminopurine are mutagenic because they are incorporated into the DNA and cause (47)_____ mutations. Nitrous acid can be used to (48)_____ bases and produce (49)_____ mutations as well.

Damaged DNA can be repaired by (50)_____ of the damage or by the (51)_____ of the damaged base or bases. The (52)_____ repair enzymes are carried along with the (53)_____ machinery to ensure that errors escaping the (54)_____ activity of DNA polymerase are repaired. In eukaryotes, nucleotide excision repair is linked to (55)_____. In humans, this repair process removes a stretch of DNA approximately (56)_____ in length and then fills the gap left in the DNA. (57)_____, such as those caused by exposure to ionizing radiation, are repaired by either (58)_____ or (59)_____ which uses the (60)_____ as a template to repair the break.

(61)_____ possess the ability to cause (62)_____ in the genome of an organism because they are genetic elements that can move around the genome of an organism. These elements can (63)_____ into a gene, or into the (64)_____ region of a gene affecting its transcription.

Figure Analysis

Use the indicated figure or table from the textbook to answer the questions that follow.

1. **Figures 18.6 and 18.7**
   (a) In what two conformations can adenine and cytosine exist? What is the normal conformation?

   (b) In what two conformations can guanine and thymine exist? What is the normal conformation?

   (c) What base normally base-pairs with the amino form of adenine in a molecule of DNA? How many hydrogen bonds are formed between these two bases?

   (d) What base after undergoing a tautomeric shift can base-pair with the amino form of adenine? How many hydrogen bonds are formed between these two bases?

   (e) What base normally base-pairs with the keto form of guanine in a molecule of DNA? How many hydrogen bonds are formed between these two bases?

   (f) What base after undergoing a tautomeric shift can base-pair with the keto form of guanine? How many hydrogen bonds are formed between these two bases?

   (g) What is the result of a tautomeric shift in either the incoming base or a base in the template strand during DNA replication?

   (h) In general, what type of mutation is produced by the replacement of one base pair with a different base pair through a tautomeric shift?

2. **Figures 18.9 and 18.10**
   (a) Describe the two mechanisms by which transversion mutations can occur.

   (b) What is the intermediate in the transversion of a GC base pair to a TA base pair by base modification?
   (c) How can an AA base pair form? How many hydrogen bonds are formed in this base pair?

   (d) Why are transversion mutations typically more serious than transition mutations?

3. **Figure 18.21**

(a) What enzyme removes the damaged base from a nucleotide within the DNA?

(b) How does this enzyme gain access to the damaged base in the DNA double helix?

(c) Following removal of the base, what type of site is created in the DNA?

(d) What is the function of the AP endonuclease?

(e) What other enzymes are required to complete the repair of the DNA?

4. **Figure 18.23**

(a) What is the function of the $UvrA_2$–UvrB complex?

(b) What happens to the $UvrA_2$–UvrB complex if it encounters a site of DNA damage?

(c) Where do the UvrB and UvrC subunits nick the DNA?

(d) What is the function of UvrD?

(e) What final two enzymes are required to fill the gap and seal the nick in the DNA?

5. **Figures 18.25 and 18.26**

(a) How are mismatched base pairs following DNA replication detected?

(b) How do the mismatch repair proteins recognize which base is wrong in the mismatched pair?

(c) What is the function of the MutS, MutL, MutH, and MutU enzymes in *E. coli*?

(d) MutU is the functional homolog of which enzyme in the nucleotide excision repair system?

6. **Figure 18.27**

(a) What is the function of the Ku70/Ku80 heterodimer in nonhomologous end-joining?

(b) Why would the cell want to initiate a mechanism by which a double-strand break will be repaired, but some DNA sequences may be lost?

(c) What could happen if more than two broken ends are present?

7. **Figure 18.28**

(a) How is homology-directed recombination different than nonhomologous end-joining in terms of maintenance of the original DNA sequence?

(b) How is single-stranded DNA generated in the process of homology-directed recombination?

(c) What is the function of the Rad51 protein?

(d) What is heteroduplex DNA? How is it generated?

General Questions

Q18-1. Consider a gene A in human beings. How can mutations of this gene cause several different genetic disorders or no disorder at all?

.

Q18-2. Explain the following observation: RNA viruses accumulate mutations at a faster rate than DNA viruses.

Q18-3. High dosages of X-rays produce dominant lethals but do not affect fertility in a number of insect pest species. Individuals possessing one copy of a dominant lethal are not viable. If the female of a pest species mates only once, how can the relationship between X-rays, mutation, and fertility be used to design an eradication program?

Q18-4. Describe why a frameshift mutation in a specific ORF is potentially more devastating to a cell than a missense mutation in the same ORF? If this ORF encodes a protein kinase, a missense mutation in what part of the protein could be devastating to the cell?

Q18-5. Describe the mechanisms by which a wild-type phenotype can be restored when a mutation results in a mutant protein and phenotype.

Q18-6. Describe how the SOS response in *E. coli* is activated and the effects of stimulating this pathway.

Q18-7. Describe how transposable genetic elements can cause deleterious mutations in bacteria.

Multiple Choice

For each of the following, circle the letter of the choice that most appropriately answers the question.

1.) Luria and Delbrück's experiment to account for phenotypic variants in *E. coli* was designed to identify which of the following?
   a. type of *E. coli* mutations that cause a wild-type ton^s *E. coli* strain to become ton^r
   b. suppressor mutations of the *E. coli* ton^r phenotype
   c. which chemicals could cause ton^s *E. coli* to become ton^r
   d. whether ton^r bacteria are the result of random mutations or physiological adaptations
   e. inheritance pattern of X-ray–induced mutations in *E. coli*

2.) A second mutation within a mutant gene can have all but which of the following outcomes?
   a. It can restore the wild-type function without restoring the original sequence.
   b. It can restore the original sequence and function of the gene.

   c. It can occur at a different site than the first mutation and cause a phenotype that is significantly worse than the phenotype of the first mutation alone.
   d. It can occur at a different site than the first mutation and cause a phenotype that is no worse than the first mutation.
   e. None of the above; all of these outcomes are possible.

3.) Which of the following is an example of a silent mutation?
   a. AUU → AUG
   b. UAU → UGU
   c. AGU → AGC
   d. AAC → AAG
   e. UGG → UGA

4.) For the DNA sequence,
ATGCAT**TG**GAAAGCTGGA,
Met His Trp Lys Ala Gly
encoding the peptide shown, insertion of a G
between the two bases indicated in bold, changes
the original amino acid sequence of the peptide
to which of the following?
   a. Met-His-Trp-Gly-Lys-Ala
   b. Met-His-Trp-Asp-Leu-Trp
   c. Met-His-Cys-Gly-Ser-Trp
   d. Met-His-Trp-Glu-Ser-Trp
   e. Met-His-Cys-Gly-Leu-Ala

5.) For the DNA sequence,
ATGTACGTCCCCGACGCTTA**T**ACGCAG,
deletion of the base indicated in bold results in
which type of mutation?
   a. nonsense
   b. transition
   c. missense
   d. silent
   e. transversion

6.) Loss of the amine group from a cytosine base
produces which type of mutation?
   a. nonsense
   b. transition
   c. missense
   d. silent
   e. transversion

7.) In eukaryotes, which of the following DNA
repair processes is linked with transcription?

   a. base excision repair
   b. nucleotide excision repair
   c. mismatch repair
   d. nonhomologous end-joining
   e. recombinational repair

8.) In *E. coli*, thymine dimers are reversed by
which of the following?
   a. base excision repair
   b. mismatch repair
   c. DNA photolyase
   d. nucleotide excision repair
   e. DNA glycosylase

9.) Of the errors introduced during DNA
replication in *E. coli*, 99% are corrected by
_____.
   a. the mismatch repair system
   b. nonhomologous end-joining
   c. homology-directed recombinational repair
   d. $3' \rightarrow 5'$ exonuclease activity of DNA
      polymerase
   e. the nucleotide excision repair system

10.) Which of the following transposable
elements move by way of an RNA intermediate?
   a. Ac–Ds elements
   b. *P* elements
   c. IS elements
   d. complex transposons
   e. copia elements

Practice Problems

P18-1. Sickle-cell anemia results from an abnormal hemoglobin molecule that causes the red blood cells of
homozygotes to sickle. In sickle-cell hemoglobin, a single amino acid, glutamic acid (glu), is replaced by
valine (val). Can you identify the change that probably occurred?

(*Hint 1*: When determining the base sequence of an mRNA molecule given the amino acid sequences of the
normal and mutant protein, write down all of the possible codons corresponding to each amino acid. To
distinguish between alternate bases in the third position of the codon, compare the normal sequence with
each mutant sequence.

Hint 2: When determining the nature of a base change, consider the fewest changes in the normal mRNA
sequence necessary to produce the mutant sequence.)

P18-2. Consider the following mRNA sequences from a hypothetical organism. Identify each mutational
type illustrated, and provide the amino acid sequence corresponding to each mutant mRNA.

(*Hint 3*: The number of altered amino acids in a mutant protein suggests the type of mutation that has
occurred. A missense mutation generally produces an incorrect amino acid, while a frameshift mutation
produces a protein with several to many incorrect amino acids. A protein with an amino acid sequence that
is correct but terminates prematurely suggests the occurrence of a nonsense mutation.)

Normal: AUGUUCAGCUGGGGGGGUGUAA

Mutant 1:   AUGUUCAGCUGAGGGGUGUAA

Mutant 2:   AUGUUCAGCUGGCGGGUGUAA

Mutant 3:   AUGGUUCAGCUGGGGGGGUGUAA

P18-3. You have isolated a new UV-sensitive mutant strain of bacteria. A number of colonies from the mutant strain and a wild-type strain were cultured. Individual plates of each strain were exposed to the same dosage of UV radiation and then assigned to one of two treatments: light-incubated or dark-incubated. The percentage survival for each is given. Explain the differences observed between the mutant and wild-type strains and the light and dark treatments.

	UV Dosage Results
Wild-type, light	85%
Mutant, light	80%
Wild-type, dark	60%
Mutant, dark	9%

P18-4. In *Drosophila*, two recessive mutations affecting eye color, white and cherry, are sex-linked. Given the following results, determine if the two mutations are located in the same or in different genes.

    Cross: White-eyed female × cherry-eyed male
    Offspring: White-eyed males and light cherry-eyed females

(*Hint 4:* To verify allelism, examine the phenotype of a heterozygote for both mutations. If wild-type, the mutations are not alleles of the same gene. If mutant, the mutations are presumed to be functional alleles.)

P18-5. *Sordaria* is a fungus that can exist as a haploid or a diploid similarly to baker's yeast. You have isolated six recessive mutant strains of *Sordaria* that require leucine in the medium to live. You mate each mutant haploid strain to each of the others. The following results note whether the new diploid cell lives (+) or dies (−) on medium that doesn't contain leucine.

	M1	M2	M3	M4	M5	M6
M1	−	+	−	+	+	+
M2	+	−	+	+	+	+
M3	−	+	−	+	+	+
M4	+	+	+	−	+	+
M5	+	+	+	+	−	−
M6	+	+	+	+	−	−

a. How many complementation groups have you isolated?
b. Why does a diploid cell derived from two of the same mutant cells (e.g., M1 fused to M1) never show complementation?

Assessing Your Knowledge

Understanding the Key Concepts—Answers

1.) spontaneously; 2.) induced; 3.) mutagens; 4–6.) morphological; nutritional or biochemical; behavioral; 7.) lethal; 8.) conditional; 9.) permissive; 10.) mutant; 11.) restrictive; 12.) nucleotide; 13.) missense; 14.) replacement; 15.) wobble; 16.) translation termination; 17.) truncated; 18–19.) insertion; deletion; 20.) frameshift; 21.) downstream; 22.) out of frame; 23.) pyrimidine; 24.) purine; 25.) purine; 26.) tautomeric; 27–28.) depurination; deamination; 29.) repetitive; 30.) misalignment; 31.) DNA replication;

32.) triplet; 33.) suppressed; 34.) mutation; 35.) masks; 36.) suppressors; 37.) the same gene; 38.) intergenic suppressors; 39.) different gene; 40.) Back or reverse; 41.) mutation rate; 42.) reversion; 43.) mutation; 44.) mutagens; 45.) higher; 46.) analogs; 47.) transition; 48.) modify; 49.) transition; 50.) reversal; 51.) removal; 52.) mismatch; 53.) replication; 54.) proofreading exonuclease; 55.) transcription; 56.) 27 to 29 base pairs; 57.) DNA double-strand breaks; 58–59.) nonhomologous end-joining; homology-directed recombination; 60.) sister chromatid; 61.) Transposons; 62.) mutations; 63.) insert; 64.) promoter.

Figure Analysis—Answers

1a.) Adenine and cytosine can exist in the normal amino form or in the rare imino form; b.) Guanine and thymine can exist in the normal keto form or in the rare enol form; c.) The amino form of adenine normally base-pairs with the keto form of thymine in DNA; two hydrogen bonds are formed between these two bases; d.) The imino form of cytosine can base-pair with the normal amino form of adenine after undergoing a tautomeric shift; two hydrogen bonds are formed between these two bases; e.) The keto form of guanine normally base-pairs with the amino form of cytosine in DNA; three hydrogen bonds are formed between these two bases; f.) The enol form of thymine can base-pair with the normal keto form of guanine after undergoing a tautomeric shift; three hydrogen bonds are formed between these two bases; g.) The result of a tautomeric shift in either the incoming base, or a base in the template strand during DNA replication is mispairing; h.) Replacement of one base pair with a different base pair through a tautomeric shift produces transition mutations.

2a.) Transversion mutations can occur by a tautomeric shift of one base followed by the rotation of the other base around its glycosidic bond to form the *syn* configuration. They can also occur by base modification; b.) The intermediate in the transversion of a GC base pair to a TA base pair by base modification is the formation of a base pair between adenine and 8-oxoguanine which is formed if guanine is oxidized; c.) An AA base pair can form if one base undergoes a tautomeric shift and the other rotates about its glycosidic bond. Two hydrogen bonds are formed in the AA base pair; d.) Transversion mutations are typically more serious than transition mutations because mutating a purine to a purine or a pyrimidine to a pyrimidine in the wobble position of a codon typically does not change the encoded amino acid in a protein, whereas mutation of a purine to a pyrimidine or vice versa at the wobble position is more likely to change the encoded amino acid in a protein.

3a.) DNA glycosylase removes the damaged base from a nucleotide within the DNA; b.) It gains access to the damaged base by flipping the base out; c.) Following removal of the base, an AP site lacking either a purine or a pyrimidine is created in the DNA; d.) The AP endonuclease cleaves the phosphodiester backbone of the DNA on the 5′ side of the AP site; e.) To complete the base excision repair, DNA polymerase inserts a nucleotide at the 3′ end of the nicked strand. An exonuclease, lyase, or phosphodiesterase enzyme removes the base-free nucleotide or single-stranded DNA, and DNA ligase seals the nick.

4a.) The $UvrA_2$–UvrB complex moves along the DNA looking for damage; b.) If the $UvrA_2$–UvrB complex encounters a site of DNA damage, the $UvrA_2$ dimer dissociates. The UvrB subunit remains bound to the DNA causing it to bend and recruiting the UvrC protein; c.) UvrB nicks the DNA four to five nucleotides on the 3′ side of the lesion. UvrC nicks the DNA eight nucleotides on the 5′ side of the lesion; d.) UvrD unwinds the DNA to release the 14- to 15-base oligonucleotide and UvrC; e.) DNA polymerase I and DNA ligase are the final two enzymes required to fill the gap and seal the nick in the DNA.

5a.) Mismatched base pairs following DNA replication are detected by the mismatch repair system which follows behind the replication fork and recognizes any mismatch errors; b.) The cell assumes that the parental DNA strand is correct when two bases are mismatched. It recognizes this strand based on its methylation state. The daughter strand is typically not methylated at this stage since it was just replicated; c.) The MutS protein detects the mismatch and finds the methylation signal along with the MutL protein. MutS and MutL activate MutH which is an endonuclease that nicks the unmethylated strand. The MutU protein is a helicase that unwinds the nicked oligonucleotide containing the mismatched base pair; d.) MutU is the functional homolog of the UvrD protein in nucleotide excision repair.

6a.) The Ku70/Ku80 heterodimer binds to the broken chromosomal ends, recruits a protein kinase, and trims the ends at the break site so that they can be properly ligated together; b.) The cell would want to initiate a mechanism by which the double-strand break will be repaired, although some DNA sequences may be lost because this is more favorable than loss of a larger portion of the chromosome. Loss of a small region of the DNA provides the cell with a better chance of survival; c.) If more than two broken DNA ends are present in the cell, incorrect attachments can take place as a result of nonhomologous end-joining. This can lead to translocation events.

7a.) Homology-directed recombination is different than nonhomologous end-joining in terms of maintenance of the DNA sequence because it restores the entire DNA sequence by using the sister chromatid as a template to synthesize a DNA strand that transverses the break; b.) Single-stranded DNA is generated during homology-directed recombination because a $5' \rightarrow 3'$ exonuclease digests the DNA at the break site; c.) The Rad51 protein coats the single-stranded DNA that is generated at the break site and catalyzes the invasion of the sister chromatid at a homologous sequence; d.) Heteroduplex DNA is a region of DNA formed when one single-stranded DNA molecule invades a sister chromatid at a homologous sequence displacing the second strand of the original duplex so that DNA strands from two different DNA duplexes are paired.

General Questions—Answers

Q18-1. Not all mutations result in the production of an abnormal protein (see *Hint 4* for example). Therefore, not all mutations lead to the expression of a disorder. Several disorders may arise from mutations in the same gene provided each alters a different amino acid(s).

Q18-2. RNA viruses must undergo reverse transcription to form DNA which is then used as a template to synthesize the RNA genome. This extra step in the process of reproduction provides an additional opportunity for mistakes to be made in the fidelity of replication.

Q18-3. Males of the pest species could be exposed to high levels of radiation and then released in large numbers. Because fertility is unaffected, these males could presumably mate normally. However, the offspring they produce will show marked decreases in viability. If a large population can be released, females will be more likely to first encounter and mate with a treated male. This approach will only work for a species where the females mate only once but males mate with several females. In addition, the treated males must be as attractive as the wild-type males occurring in the field naturally.

Q18-4. A frameshift mutation in a specific ORF is potentially more devastating to a cell than a missense mutation in the same ORF because it changes the reading frame from the site of the mutation onward, whereas a missense mutation changes only a single amino acid in a protein to a different amino acid. Therefore, a frameshift mutation may affect several of the necessary domains in the protein and is more likely to abolish or change its activity. A missense mutation would affect only one domain in the protein and is far less likely to affect the activity of the protein. However, if an amino acid that is critical to the function of the protein is mutated as a result of a missense mutation in the gene, this single change can inactivate the protein. For example, if the ORF encodes a protein kinase, a missense mutation that disrupts its kinase activity, or its ability to bind to cofactors, or target substrates can have devastating effects on the cell.

Q18-5. A back, or reverse, mutation at the same site in the mutant gene as the original mutation will restore the wild-type phenotype. A second mutation in the same gene as the first but at a different site can mask the effect of the original mutation (intragenic suppression). Finally, a mutation in a second gene can mask the effect of the first mutation (intergenic suppression). When mutated, intergenic suppressors, such as tRNAs or interacting proteins, can change the way in which a codon is read.

Q18-6. The SOS response in *E. coli* is activated by the RecA protein binding to single-stranded DNA that is generated when the cell is exposed to UV light, or other DNA damage-inducing agents, or when DNA replication is inhibited. RecA binding to single-stranded DNA stimulates its protease activity which cleaves specific proteins including LexA. Cleavage of LexA activates the transcription of genes with a

SOS box in their promoters. These genes encode proteins that facilitate replication of damaged DNA and inhibit cell division to increase the time that the cell has to repair the DNA before the next round of DNA replication.

Q18-7. Transposable genetic elements can insert into a bacterial gene to produce an aberrant mRNA or in to the promoter of a gene affecting its level of transcription. Transposons can also cause genomic rearrangements such as deletions, inversions, and translocations due to recombination between transposable elements located on the same chromosome (deletions, inversions), or on distinct chromosomes (reciprocal translocations).

Multiple Choice—Answers

1.) d; 2.) e; 3.) c; 4.) d; 5.) a; 6.) b; 7.) b; 8.) c; 9.) d; 10.) e.

Practice Problems—Answers

P18-1. The codons corresponding to glu are GAA and GAG. The codons corresponding to val are GUU, GUC, GUA, and GUG. When determining the nature of a base change, consider the fewest changes in the normal mRNA sequence necessary to produce the mutant sequence. In the sickle-cell hemoglobin, a transition probably occurred: the adenine (GAG or GAA) was probably replaced by uracil (GUG or GUA) resulting in the codon for valine.

P18-2. To determine what type of mutation has occurred we must compare the mutant mRNA sequence with the normal mRNA sequence. The consequences of this change will be revealed upon examination of the amino acid sequence coded for by the mutant mRNAs.

Normal:	AUG	UUC	AGC	UGG	GGG	GUG	UAA	
	Met	Phe	Ser	Try	Gly	Val		
Mutant 1:	AUG	UUC	AGC	**UGA**	GGG	GUG	UAA	
	Met	Phe	Ser					
Nonsense mutation								
Mutant 2:	AUG	UUC	AGC	UGG	**CGG**	GUG	UAA	
	Met	Phe	Ser	Try	**Arg**	Val		
Missense mutation								
Mutant 3:	AUG	**GUU**	**CAG**	**CUG**	**GGG**	**GGU**	**GUA**	A
	Met	**Val**	**Gln**	**Leu**	**Gly**	**Gly**	**Val**	. . .
Frameshift mutation (Addition)								

P18-3. It appears that the mutant strain has a defective base and/or nucleotide excision repair system. Notice that both the wild-type and mutant strains have a high survival rate in the presence of light. This indicates that the mutant strain, like the wild-type, has an active DNA photolyase enzyme that repairs dimers induced by UV radiation when light energy is available. In the dark, the wild-type survival is reduced but is still high because an alternative repair mechanism can be used. A mutation in this system would account for the very low survival of the mutant strain incubated in the dark.

P18-4. To determine if two mutations are in the same gene we must examine the phenotype of the heterozygote. In the case of sex linkage, females are heterozygous. The females resulting from this cross are mutants indicating that the two mutations are in the same gene.

P18-5. a. All cells are recessive mutant for leucine metabolism, but there are many genes in the pathway to synthesize leucine. Any one of those enzymes can be mutant. For example:

$$A- \qquad B- \qquad C- \qquad D-$$

$$M \;\rightarrow\; N \;\rightarrow\; O \;\rightarrow\; P \;\rightarrow\; \text{leucine}$$

Let's say gene *A* encodes the enzyme that is necessary to convert M to N. If a haploid cell is blocked in the pathway by a mutation in any of the genes (*A*, *B*, *C*, or *D*), the cell will not be able to make its own leucine. That cell will die if leucine is not provided in the medium. If a recessive mutant (*a*) cell acquires a normal version (*A*) of the mutant gene, it will now be able to make leucine and will appear phenotypically normal, just like a heterozygote for a recessive genetic trait (*Aa*). The mutant cell (*a*) can acquire this normal version by being fused with a cell that contains the normal version of the gene (*A*). However, if a mutant cell (*a*) is fused with a cell that has the same mutant gene (*a*), it will not be able to make the enzyme that converts M to N and will be unable to grow on medium that lacks leucine, just like a homozygous recessive (*aa*). For example, when we fuse M1 to itself or M3, we see no growth (no complementation). This means both have the same mutant leucine gene, so we've identified the first complementation group. When we fuse M1 to M2, M4, M5, or M6, we see growth (complementation). This means each of these mutant cells (M2, M4, M5, and M6) has a *different* mutant leucine gene than cells M1 and M3. In the end, we have four complementation groups, so we are working with four *different* leucine genes. Group 1 contains M1 and M3. Group 2 contains M2. Group 3 contains M4. Group 5 contains M5 and M6.

b. If a mutant cell is fused with a cell that has the same mutant gene, it will still be phenotypically mutant just like a homozygous recessive (*aa*) cell.

18

DNA MUTATION, REPAIR, AND TRANSPOSITION

CHAPTER SUMMARY QUESTIONS

1. 1. d, 2. c, 3. b, 4. a, 5. f, 6. e, 7. j, 8. i, 9. h, 10. g.

2. Silent mutations are those that do not produce a detectable phenotypic change. They can occur in the region between genes, in introns, in the 5' and 3' untranslated regions of a gene, and in the wobble position of a codon.

3. Salvador Luria and Max Delbrück developed the *fluctuation* test. They used this test to demonstrate that bacteria underwent random mutagenesis rather than physiological induction.

4. Both types of mutation restore the wild-type function of a gene without restoring its original sequence. An intragenic suppression occurs at a different site in the *same* gene than the first mutation, while an intergenic suppression occurs in a *different* gene, called a suppressor gene (which is typically a tRNA gene).

5. There are two different conformations for each of the four DNA bases. Adenine and cytosine can exist in either the normal amino form or the rare imino form, while guanine and thymine can exist in either the normal keto form or the rare enol form. A tautomeric shift occurs when the normal form of each base undergoes a proton shift to produce its rare tautomeric form. This leads to inappropriate base pairing, which ultimately results in mutations during DNA replication.

6. The mutation rate is the number of mutations that arise per gene for a specific length of time in a population. The mutation frequency for a given gene is the number of mutant alleles per the total number of copies of the gene within a population. Therefore, the mutation rate represents newly generated mutations, whereas the mutation frequency represents the prevalence of a specific mutation, both old and newly generated.

7. Triplet diseases are inherited diseases that are caused by expansions of three-nucleotide repeats. These repeats may be present anywhere within the gene. Triplet diseases have two main features: (1) They usually affect neurons, producing symptoms such as mental retardation or neurodegeneration. (2) They exhibit anticipation, which means that the severity of the disease increases in subsequent generations due to the increasing number of trinucleotide repeats.

8. Both chemicals are base modifiers. Ethylethane sulfonate is an alkylating agent that can add an ethyl group to any of the four bases. It can promote transitions and transversions. Nitrous acid replaces amino groups ($-NH_2$) on nucleotides with keto groups ($=O$). It causes transitions by converting cytosine into uracil, which acts like thymine, and adenine into hypoxanthine, which acts like guanine.

9. Both types of mutation arise from tautomeric shifts during DNA replication. A substrate transition occurs as a result of tautomerization in the incoming base, whereas a template transition results from tautomerization in the base on the template strand.

10. Ultraviolet light induces cross-linking, or dimerization, between adjacent pyrimidines in DNA. The principal products of UV irradiation are thymine–thymine dimers, although cytosine–cytosine and cytosine–thymine dimers are occasionally produced. The dimers distort the DNA and result in the failure of the pyrimidines to base-pair with their complementary purines. This type of DNA damage can be repaired by three different mechanisms: (1) photoreactivation, which involves photolyase-mediated reversal of the thymine dimer without removing the bases, (2) nucleotide excision repair, which involves removal and replacement of a short stretch of DNA that includes the pyrimidine dimer, (3) postreplicative repair, which involves a specific group of DNA polymerases that are capable of replicating damaged DNA.

11. The mismatch repair system recognizes and repairs base-pair errors that are not corrected by the 3' → 5' exonuclease proofreading activity of DNA polymerases. The repair system assumes that the parental or template strand is correct and that the error occurred in the daughter strand. The template strand is recognized by the methylation of the adenine in 5'-GATC-3' sequences. The base on the daughter strand (nonmethylated) is removed thereby preserving the base in the parental DNA strand. In *E. coli*, the mismatch repair system consists of proteins encoded by the *mutH*, *mutL*, *mutS*, and *mutU* genes. A MutS protein homodimer detects the mismatch because of distortions to the DNA structure. A MutL protein homodimer then binds. The two homodimers find the methylation signal and activate the MutH endonuclease, which nicks the unmethylated strand at the 3'-CTAG-5' recognition site. The MutS-MutL tetramer then loads the MutU helicase, which unwinds the nicked strand. Any one of at least four different exonucleases degrades the unwound oligonucleotide. The gap is then filled by DNA polymerase III, and DNA ligase seals the final phosphodiester nick.

12. The SOS response is a complex, inducible repair system in *E. coli* that is used to repair DNA when excessive damage has occurred. The RecA protein interacts with single-stranded DNA, which stimulates a protease activity that is normally silent in the RecA protein. The RecA* protease activity cleaves the LexA protein. LexA is a transcriptional repressor of about 18 genes, many of which are involved in DNA repair and the inhibition of cell division. Therefore, the inactivation of the LexA repressor by RecA* protease allows the transcription of these genes.

13. Double-stranded DNA breaks can be repaired by one of two mechanisms: (1) nonhomologous end-joining, which brings the broken chromosomal ends back together, and (2) homology-directed recombination, which relies on the nucleotide sequences of a homologous piece of DNA, such as a sister chromatid or homologous chromosome. Nonhomologous end-joining may result in the loss of some sequence information, whereas homology-directed recombination usually restores the entire DNA sequence.

14. *IS* elements, or insertion sequences, are relatively small transposable elements in bacteria. Unlike other transposons, they carry no bacterial genes. They are sometimes referred to as "selfish DNA" because they may exist without a noticeable benefit to their host cell.

15. The direct repeats flanking transposons result from the transposition mechanism. The target site in the host chromosome is cleaved in a staggered manner, leaving single-stranded ends. The transposon is then inserted between the free ends. The single-stranded regions are then converted into double strands by DNA polymerases, thereby generating two directly repeated sequences flanking the transposon.

16. Transposons may induce deletions and inversions through pairing and recombination between copies of the transposable elements. If the copies of a transposable element are present in a directly repeated orientation, pairing followed by a single crossover

event between the two copies can result in the deletion of one copy and the region between transposons. If the copies of the transposable element are present in an inverted orientation, pairing followed by a single crossover event between the two copies can result in an inversion of the region between the transposable elements. Refer to figure 18.41.

17. The three different classes of transposable elements in bacteria are *IS* elements, composite transposons, and complex transposons. The simplest in structure are *IS* elements which encode a transposase enzyme flanked by terminal inverted repeats. Composite transposons contain *IS* elements at the end and usually encode bacterial genes, such as antibiotic resistance genes, in the middle. Complex transposons contain terminal inverted repeats and a transposase gene. They may also contain other genes. Indeed, complex transposons are similar to composite transposons, except they lack the *IS* elements at the end. Each of these three transposable elements can employ one of two major mechanisms for transposition. In conservative transposition, the transposon moves to a new location without copying itself (although the donor site retains the duplicated flanking sequence). In replicative transposition, the transposon is duplicated and the copy is inserted into the recipient site, while the original transposon remains at the donor site.

18. Hybrid dysgenesis is the production of sterile *Drosophila* offspring due to transposable element activity in germ cells. When a male of a P stock is crossed to a female of an M stock, the resulting F_1 hybrid offspring are sterile. On the other hand, the reciprocal cross between an M male and a P female yields normal F_1 offspring.

Analysis of the stocks revealed that P flies contain one or more copies of a transposon, the *P* element, while the M flies either lacked or contained shorter nonfunctional versions of the *P* element. The sterility in the F_1 hybrids resulted from transposition of the *P* element within the genomes of germ cells, causing chromosomal abnormalities, insertions into essential genes, and an overall disruption of gamete formation. Therefore, the cross of a P male with an M female results in sterile hybrid progeny. Hybrid dysgenesis does not occur if the cross is between a female fly from a P stock and a male fly from an M stock. The *P* element encodes a transposase enzyme as well as a repressor of transposase. The egg cells from a P female *Drosophila* contain enough of this repressor protein in the cytoplasm to prevent the transposition of the *P* element. Therefore, a zygote produced from an M male and a P female will have the repressor and so the offspring will be normal. On the other hand, a zygote produced from a P male and an M female (who lacks a full-length *P* element) would not have the repressor. The *P* elements passed on through the sperm would transpose in the germ cells of the F_1 hybrid, causing it to become sterile.

19. LINES (long interspersed elements) and SINES (short interspersed elements) are two common classes of human transposons. SINES, which are 150 to 500 base pairs in length, are much smaller than LINES, which have a full length of 6.5 kilobase pairs. LINES resemble retrotransposons, whereas SINES resemble short RNAs, such as the 7SL cytoplasmic RNA, small nuclear RNAs and tRNAs. Both LINES and SINES transpose through RNA intermediates. They differ in that LINES encode a reverse transcriptase enzyme, whereas SINES do not and are therefore completely dependent on the transposase being provided by another element.

20. Bacterial and eukaryotic transposable elements typically have inverted repeats at their ends and at least one gene in the middle. The transposons are also similar in that they generate direct repeats during their insertion at target sites.

EXERCISES AND PROBLEMS

21. There are six codons for arginine: CGU, CGC, CGA, CGG, AGA, and AGG. A single base substitution could occur at any of the three positions in each codon thereby resulting in a different codon.

  a. The possible transition mutations, mutant codons, and encoded amino acids are presented in the following table:

Wild-Type Codon	Mutant Codon (Encoded Amino Acid)
CGU	UGU (cys), CAU (his), CGC (arg)
CGC	UGC (cys), CAC (his), CGU (arg)
CGA	UGA (stop), CAA (gln), CGG (arg)
CGG	UGG (trp), CAG (gln), CGA (arg)
AGA	GGA (gly), AAA (lys), AGG (arg)
AGG	GGG (gly), AAG (lys), AGA (arg)

  b. The possible transversion mutations, mutant codons, and encoded amino acids are presented in the following table:

Wild-Type Codon	Mutant Codon (Encoded Amino Acid)
CGU	AGU (ser), GGU (gly), CCU (pro), CUU (leu), CGA (arg), CGG (arg)
CGC	AGC (ser), GGC (gly), CCC (pro), CUC (leu), CGA (arg), CGG (arg)
CGA	AGA (arg), GGA (gly), CCA (pro), CUA (leu), CGC (arg), CGU (arg)
CGG	AGG (arg), GGG (gly), CCG (pro), CUG (leu), CGC (arg), CGU (arg)
AGA	CGA (arg), UGA (stop), ACA (thr), AUA (ile), AGC (ser), AGU (ser)
AGG	CGG (arg), UGG (trp), ACG (thr), AUG (met), AGC (ser), AGU (ser)

22. UV light causes pyrimidine dimers, the most common of which are thymine–thymine dimers. Therefore the DNA molecule with the highest percentage of thymine will be expected to be the most sensitive to UV light damage, while that with the smallest thymine percentage should be the least sensitive. (*Note:* The thymines have to be adjacent to each other on the same DNA strand.) The known % GC content of each molecule allows us to calculate the percentage of each of the four bases.

	% Content			
Molecule	**G**	**C**	**A**	**T**
I	25	25	25	25
II	20	20	30	30
III	15	15	35	35

Therefore, DNA molecule I is the least sensitive, while molecule III is the most sensitive.

23. To determine the type of mutation that has occurred, we must compare the mutant coding sequence to the wild-type sequence.

 a.   5'-ATG CCG ACT AAC TAT AGG AAA . . .-3'
          Met Pro Thr Asn Tyr Arg Lys

 The mutation is a C to T transition. It is a silent mutation because it resulted in a codon specifying the same amino acid.

 b.   5'-ATG CCG ACT AAC TAA AGG AAA . . .-3'
          Met Pro Thr Asn Stop

 The mutation is a C to A transversion. It is a nonsense mutation because it resulted in a stop codon and a shortened polypeptide.

 c.   5'-ATG CCG ACT AAC TAC AAG AAA . . .-3'
          Met Pro Thr Asn Tyr Lys Lys

 The mutation is a G to A transition. It is a missense mutation because it resulted in a codon specifying a different amino acid.

 d.   5'-ATG CCG ACT CAA CTA CAG GAA . . .-3'
          Met Pro Thr Gln Leu Gln Glu

 The mutation is an insertion of a C, and therefore a frame-shift mutation.

 e.   5'-ATG CCG ACT AAC TAA GGA AAC . . .-3'
          Met Pro Thr Gln Stop

 The mutation is a deletion of a C. This is a frame-shift mutation that generates an immediate stop codon downstream.

24. Frameshift mutations are caused by insertions or deletions of bases (that are not multiples of 3). These will shift the reading frame for all codons downstream from the mutation. Single base-substitutions, on the other hand, only affect a single codon. Therefore, frameshift mutations will have a more deleterious effect than base-substitution mutations because they affect a larger proportion of amino acids in the polypeptide.

25. Transitions are base-substitutions from a purine to another purine or from a pyrimidine to another pyrimidine. Transversions involve a change of a purine to a pyrimidine or *vice versa*. The table lists all possibilities:

Transition	$A \to G, G \to A, C \to T, T \to C$
Transversion	$A \to C, C \to A, A \to T, T \to A,$
	$G \to C, C \to G, G \to T, T \to G$

Therefore, the ratio of transitions to transversions is 4 to 8 or 1 to 2.

26. A genetic system that correctly repaired every damaged base would not be advantageous to an organism. New mutations are necessary to increase genetic diversity within a species. While it is important that most damaged bases and mutations be repaired, a low mutation rate is desirable because of the advantages of genetic diversity.

27. A point mutation can be either a missense (new amino acid), nonsense (stop codon created), or frameshift mutation (addition or deletion of a number of bases not divisible by 3). These can all be corrected by simple reversion. Intragenic suppression can rectify the frameshift mutation. Intergenic suppression (transfer RNA change) can rectify all three mutations.

28. *IS* elements have terminal inverted repeats. The pair in **(a)** is a direct repeat. The pair in **(b)** is inverted; however, it is not complementary. The pair in **(c)** is complementary but not inverted. The pair in **(d)** is complementary and inverted, and could therefore be found at the ends of an *IS* element.

29. The mutagen 2-aminopurine can only induce transition mutations: mainly AT to GC, but also GC to AT. Of the five amino acid replacements, three can result from transitions: **(b)** Ala → Thr, **(c)** Lys → Arg, and **(e)** Gly → Glu. The other two (Pro → Gln, and Ser → Tyr) are the result of transversions and therefore cannot be induced by 2-aminopurine.

30. The histidine auxotroph probably contains a deletion. If a few bases are missing, nothing is available to cause transitions or transversions. It is highly unlikely that the correct number of missing bases could be spontaneously and correctly inserted.

31. We know that the anticodon pairs with the codon, and we expect nonsense suppressors to contain an altered anticodon. The fact that the nonsense codon can be read by a transfer RNA with a normal anticodon but altered dihydrouridine loop suggests that the way in which this loop interacts with the ribosome causes the anticodon sequence to be misread.

32. **(a)** Glu → Asp, and **(d)** Ala → Gly could result from a single transversion event.

33. As mentioned in solved problem 3, hydroxylamine causes only GC to AT transition mutations in DNA. The three nonsense codons are 5'-UAA-3', 5'-UAG-3', and

5'-UGA-3'. Hydroxylamine can be used to induce nonsense mutations in wild-type genes. For example, the codon 5'-UGG-3', which specifies tryptophan, can be changed to 5'-UAG-3' or 5'-UGA-3'. Hydroxylamine, however, cannot reverse nonsense mutations. It can only act on the G in the stop codons and, therefore, can only result in a different stop codon. It is important to remember that the mutations occur at the DNA level. The respective double-stranded DNA sequences of 5'-UAG-3' and 5'-UGA-3' are

Coding strand: 5'-TAG-3'   5'-TGA-3'
Template strand: 3'-ATC-5'   3'-ACT-5'

Any transition mutation in DNA produces:

Coding strand: 5'-TAA-3'
Template strand: 3'-ATT-5'

resulting in an mRNA with a 5'-UAA-3' nonsense codon.

34. This event would not be considered a mutation because it did not involve a "permanent" change in the DNA.

35. The term *jumping gene* is not an accurate description for all transposable elements. While some transposons excise from the DNA and insert themselves in a new location, others insert copies of themselves. This second type of transposable element remains at the original site and therefore does not "jump."

36. 1. False. The *his⁻* culture used in the Ames test is that of the <u>bacterium *Salmonella typhimurium*</u>.
   2. False. The revertant colonies were formed on agar <u>lacking</u> histidine. (The bacterial cultures are plated on minimal media.)
   3. True.
   4. False. Tube <u>III</u> approximates what would happen in the human body.
   5. False. Compound A is not mutagenic <u>nor are</u> its metabolites. (The number of colonies is roughly the same in all three tubes.)
   6. False. Compound B is <u>not</u> mutagenic but its metabolites are. (Tube III has nine times more colonies than Tube II.)
   7. True.
   8. False. Compound C <u>could be</u> carcinogenic. (Mutagenic agents are potentially, but not necessarily, carcinogenic.)

37. a. Nitrous acid is a base modifier that converts adenine into hypoxanthine and cytosine into uracil. Therefore, treatment with nitrous acid will produce the following DNA sequence:

5'-HTUG-3'
3'-THGU-5'

b. Hypoxanthine and uracil have the base-pairing properties of guanine and thymine, respectively.

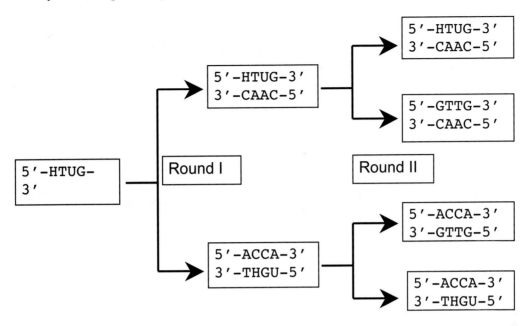

38. The two mutagens are base analogs: 2-aminopurine substitutes for adenine during DNA replication, but may base-pair with cytosine, while 5-bromouracil substitutes for thymine, but may base-pair with guanine. Therefore, after one round of DNA replication, the two DNA molecules would be

5′-ATCG-3′   and   5′-PBCG-3′
3′-BPGC-5′   3′-TAGC-5′

The second and third rounds of DNA replication would yield the following sequences:

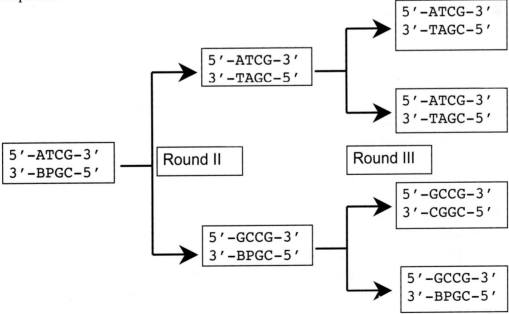

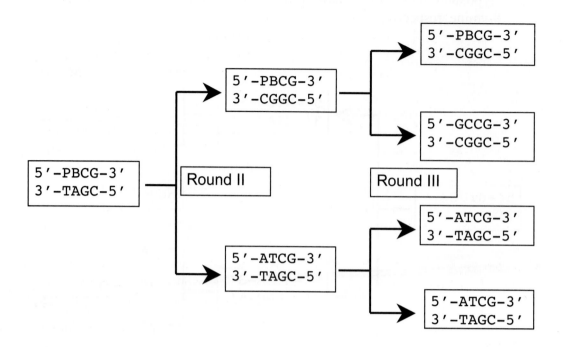

39. For example, if the 20 individual cultures of table 18.1 had values of 15, 13, 15, 20, 17, 14, 21, 19, 16, 13, 27, 14, 15, 26, 12, 21, 14, 17, 12, and 14, then the mutation theory would not have been supported because the variation between the individual and bulk cultures would not have been different.

40. **a.** The mutation rate in this population is the number of *new* mutations per gamete. The number of new mutations is $21 - 9 = 12$. Two gametes are required to produce an individual, and therefore the mutation rate is $12/(420,316 \times 2) = 12/840,632 = 1.43 \times 10^{-5}$.

     The mutation frequency is the number of mutant alleles divided by the total number of gene copies in this population. (There is no need to differentiate between new and old mutations.) Therefore, the mutation frequency is $21/840,632 = 2.5 \times 10^{-5}$.

b. When calculating the mutation rate and frequency we have assumed that the mutant gene is completely penetrant and that it does not lead to the death of the individual in utero. These assumptions may not be true, and therefore the estimate would be inaccurate.

41. Duchenne muscular dystrophy is an X-linked recessive disorder. It is typically not passed from father to son, because males receive the Y chromosome from their father and the X chromosome from their mother. Unless the son has Klinefelter syndrome (XXY), it is not possible for him to have inherited the condition from his father Franz. Rather, the son must have inherited the disease from his mother, who must be a carrier. The mutation could have occurred in her germ line or that of her mother's. A radiation-induced mutation in Franz's germ-line cells cannot explain the appearance of Duchenne muscular dystrophy in his son. Therefore, the nuclear power plant should not be held responsible.

Achondroplasia, an autosomal dominant disorder, could be transmitted from a father to his son. The achondroplastic gene mutation could have been induced by exposure to radiation. Therefore, Klaus's lawsuit has merit.

42. a. Let X* = mutant X chromosome in irradiated male fly. The parental cross is thus: X*Y × XX. This will produce an F_1 offspring of 1/2 X*X:1/2 XY, or 1 female:1 male.

b. The F_1 cross (X*X × XY) produces the following progeny: 1/4 X*X:1/4 XX:1/4 X*Y:1/4 XY. The X*Y males will die in utero because they are hemizygous for an X-linked recessive lethal mutation. Therefore, the female-to-male ratio in the F_2 will be 2:1.

43. The presence of temperature-sensitive *dnaE* mutants that exhibit conditional lethality indicates that the encoded gene product is essential for viability. Therefore, the α subunit is responsible for the 5' → 3' polymerase activity of DNA pol III. The ε subunit, on the other hand, catalyzes the 3' → 5' exonuclease activity. In its absence (*dnaQ* mutants), there is a 10,000-fold increase in the rate of mutations from the wild type. Finally, deletion mutants of *holE* are fully viable, indicating that the θ subunit is not required for the polymerase or exonuclease activity of the DNA pol III holoenzyme. (However, it does stimulate the exonuclease activity of the ε subunit.)

44. The preferential binding of the Tn10 transposase to hemi-methylated DNA is similar to the binding of the MutS and MutL proteins to hemi-methylated sequences during mismatch repair. This mechanism ensures that the Tn10 element has just been replicated, and the replication machinery is nearby and available to assist in the transposition of the Tn10 element.

45. The number of depurinations in a human diploid cell is 19,500/(18 hours × 60 minutes) = 18 per minute. Because each base pair contains one purine (and one pyrimidine), the total number of purines in a human diploid cell is 6×10^9. Therefore, the rate of spontaneous depurination per purine per minute is $18/(6 \times 10^9)$ = 3×10^{-9} depurinations per purine per minute.

46. The wild-type DNA sequence of this mRNA is (with the coding strand on top):

```
5'-CCGAC ATG TGG ACA AGT GAA CCG TCA GCA TAA GCACG-3'
3'-GGCTG TAC ACC TGT TCA CTT GGC AGT CGT ATT CGTGC-5'
```

Proflavin is an intercalating agent that causes insertion or deletion mutations in DNA. It typically causes frameshift mutations. The fact that there was a silent mutation with only a single base difference between the mutant and wild-type DNA sequences suggests that the insertion or deletion affected the stop codon TAA. An insertion of a G at the second or third position of this stop codon would produce TGA or TAG stop codons, respectively. Insertion of an A at the second or third position of the stop codon would produce TAA stop codons. Insertion of a T at the second or third position of this stop codon would produce TAA or TAG stop codons, respectively. A deletion could also explain the proflavin-induced mutation. The base immediately following the TAA codon is a G. Thus, a deletion of either A would produce a TAG stop codon.

47. The *IS* element is still capable of transposition. However, it has to rely on the transposase enzyme of another element.

48. **a.** The probability that *either* gene will undergo a mutation in a single generation is given by the "or" (sum) rule of probability: P(mutation in *T*) + P(mutation in *Z*) = $(7 \times 10^{-5}) + (3 \times 10^{-6}) = 0.000073 = 7.3 \times 10^{-5}$.

    b. The probability that gene *T* does *not* undergo a mutation in one generation is $1 - (7 \times 10^{-5}) = 0.99993$. Similarly, the probability that gene *Z* does *not* undergo a mutation in one generation is $1 - (3 \times 10^{-6}) = 0.999997$. Therefore, the probability that *neither* gene will undergo a mutation in a single generation is given by the "and" (product) rule of probability: P(no mutation in *T*) × P(no mutation in *Z*) = $0.99993 \times 0.999997 = 0.999927 = 9.99927 \times 10^{-1}$.

49. Postreplicative repair is crucial to multicellular eukaryotic organisms. Many cells in these organisms (humans, for example) are fully differentiated and no longer undergo cell division or DNA replication. However, these cells are still subjected to a variety of spontaneous and induced mutations. Although these cells do not transmit the mutations to future generations, they must still transcribe mRNAs from the mutant genes. If these mRNAs contain mutations present in the gene (such as through a base modification) or if the transcription of a gene is blocked (such as through a pyrimidine dimer), then a functional protein cannot be produced. Therefore, a mechanism to repair DNA damage after DNA replication is very important.

50. There are six mRNA codons specifying arginine: 5'-CGU-3', 5'-CGC-3', 5'-CGA-3', 5'-CGG-3', 5'-AGA-3', and 5'-AGG-3'. The wild-type arginine codon could not be 5'-AGA-3' or 5'-AGG-3' because the 5' A base would have to undergo a *transversion* in order for the arginine codon to be converted to a stop codon in two steps. Similarly, the wild-type arginine codon could not be 5'-CGU-3' or 5'-CGC-3' because the 3' pyrimidine base would have to undergo a *transversion* before the arginine codon can be converted to a stop codon. Therefore, the only possibilities for the arginine codon are 5'-CGA-3' and 5'-CGG-3'. The missing amino acids and stop codons are given in the following table:

Arginine Codon	Mutant Codon (Encoded Amino Acid)	Stop Codon
CGA	CAA (glutamine)	UAA
CGG	UGG (tryptophan)	UAG
	UGG (tryptophan)	UGA
	CAG (glutamine)	UAG
	CGA (arginine)	UGA

51. The *P* element is activated in the germ line. Therefore, the scientist must cross these flies and look at the F_1 offspring.

52. a. We can begin by writing the sequence of the wild-type mRNA, using the symbols N = any nucleotide, R = any purine (A or G), and Y = any pyrimidine (C or U).

	Codon									
	1	**2**	**3**	**4**	**5**	**6**	**7**	**8**	**9**	**10**
Wild type =	Met	Tyr	Ile	Thr	Trp	Asp	Glu	Pro	Val	Lys	Stop
Codon =	AUG	UAY	AUY	ACN	UGG	GAY	GAR	CCN	GUN	AAR	UAR
			AUA								UGA

We can now examine each mutant and methodically try to determine the identity of the N, Y, and R bases.

Mutant 1: The only difference between this mutant and the wild-type sequence is the presence of an isoleucine instead of a lysine at position number 10. This is the result of a missense mutation that transformed an AA(A/G) triplet into AU(A/C/U). The only possibility for the single base difference is an AAA → AUA point mutation. This can be caused by an AT-to-TA transversion in the DNA. Therefore, the wild-type lysine codon (10) is AAA.

Mutant 2: The amino acid sequence shows three missense mutations (codons 4–6), which are immediately followed by a premature termination signal. This suggests the presence of a frameshift mutation in the DNA near codons 3 and 4. A comparison of the mutant and wild-type mRNA sequences should reveal which single mutation (addition or deletion) can explain these amino acid differences.

	Codon									
	1	2	3	4	5	6	7	8	9	10
Wild type =	Met	Tyr	Ile	Thr	Trp	Asp	Glu	Pro	Val	Lys	Stop
Codon =	AUG	UAY	AUY	ACN	UGG	GAY	GAR	CCN	GUN	AAA	UAR
			AUA								UGA
Mutant 2 =	Met	Tyr	Ile	His	Val	Gly	Stop				
Codon =	AUG	UAY	AUY	CAY	GUN	GGN	UAR				
			AUA				UGA				

Let's closely examine the wild-type sequence of codons 4–6 = ACN UGG GAY. If the A in codon 4 is deleted, the resulting codons 4 and 5 would be CNU GGG. CNU can be rewritten as CAY, if N = A, and so it will specify histidine (amino acid 4 in mutant 2). However, the codon GGG specifies glycine and not valine (amino acid 5 in mutant 2). Therefore, the deletion of the A cannot cause mutant 2. A similar reasoning demonstrates that the deletions of the second or third bases of the wild-type codon 4 (ACN) cannot account for the amino acid differences between mutant 2 and the wild-type protein. Therefore, mutant 2 must have been the result of an insertion. The mRNA sequence of mutant 2 shows the first base in its codon 4 is a C. This suggests that the insertion of a C between the wild-type codons 3 and 4 could account for mutant 2. Inserting a C in that position yields the following mutant mRNA sequence:

	Codon									
	1	2	3	4	5	6	7	8	9	10
Codon =	AUG	UAY	AUY	CAC	NUG	GGA	YGA	RCC	NGU	NAA	AUA
			A								G

This sequence can indeed explain the origin of mutant 2. The new codon 4, CAC, specifies histidine. Codon 5, NUG, would encode valine, if N = G. GGA (codon 6) specifies glycine. Codon 7, YGA, can be a termination signal, if Y = U. Therefore, mutant 2 resulted from the insertion of a C between codons 3 and 4 in the wild-type DNA.

Mutant 3: The amino acid sequence shows three missense mutations (codons 6–8), which are immediately followed by a premature termination signal. This suggests the presence of a frameshift mutation in the DNA near codons 5 and 6. A comparison of the mutant and wild-type mRNA sequences should reveal which single mutation (addition or deletion) can explain these amino acid differences.

	Codon									
	1	**2**	**3**	**4**	**5**	**6**	**7**	**8**	**9**	**10**
Wild type =	Met	Tyr	Ile	Thr	Trp	Asp	Glu	Pro	Val	Lys	Stop
Codon =	AUG	UAY	AUY	ACG	UGG	GAU	GAR	CCN	GUN	AAA	UAR
			AUA								UGA

	Codon									
	1	**2**	**3**	**4**	**5**	**6**	**7**	**8**	**9**	**10**
Mutant 3 =	Met	Tyr	Ile	Thr	Trp	Met	Asn	Leu	Stop
Codon =	AUG	UAY	AUY	ACN	UGG	AUG	AAY	CUN	UAR
			AUA					UUR	UGA

(*Note:* The identity of the underlined bases in the wild-type sequence was obtained from the sequences of mutants 1 and 2.)

Let's closely examine the wild-type sequence of codons 6–8 = GAU GAR CCN. Any single base insertion in codon 6, or between codons 5 and 6, would generate an immediate stop codon (UGA) at position 7. Since mutant 3 is longer than six amino acids, it could not have resulted from a base insertion. Therefore, mutant 3 must have been produced by a deletion in the wild-type DNA sequence. Indeed, the deletion of any of the two G's in codon 5 or the G in codon 6 would account for the mRNA sequence of mutant 3. If such a deletion occurs, the mRNA and amino acid sequences of mutant 3 would be

Codon									
1	**2**	**3**	**4**	**5**	**6**	**7**	**8**	**9**	**10**
AUG	UAY	AUY	ACG	UGG	AUG	ARC	CNG	UNA	AAU
		A							
Met	Tyr	Ile	Thr	Trp	Met	Asn	Leu	Stop	

The new codon 6, AUG, specifies methionine. Codon 7, ARC, would encode asparagine, if R = A. CNG (codon 8) would specify leucine, if N = U. Codon 9, UNA, can be a termination signal, if N = A or G. Therefore, mutant 3 resulted from the deletion of one of the three Gs in codons 5 and 6 in the wild-type DNA.

b. Based on the sequences of mutants 1–3, the wild-type mRNA sequence is

	Codon									
	1	**2**	**3**	**4**	**5**	**6**	**7**	**8**	**9**	**10**
Codon =	AUG	UAC	AUY	ACG	UGG	GAU	GAA	CCU	GUA	AAA	UAR
		U	A							G	GA

CHAPTER INTEGRATION PROBLEM

a. The 30-bp DNA fragment shows the sequence of codons 136–145 of the *fb*⁺ gene.
The schematic of the *fb*⁺ gene is shown here (with distances in base pairs):

The entire coding sequence of the *fb*⁺ gene is 375 (exon 1) + 1971 (exon 2) + 672
(exon 3) = 3018 bp. Because each 3 bp constitute a codon, this coding sequence
corresponds to 3018/3 = 1006 codons (including the start and stop codons). Codons
136–145 must be located in exon 2, and so the 30-bp sequence should not contain a
stop codon in its coding or nontemplate strand. Let's start by examining the top
strand. This strand contains termination codons in both directions: a TGA (codon 5,
going from left to right), and a TAA (codon 3, going from right to left). Therefore,
the top strand cannot be the coding strand, and so it must be the template strand.
Examination of the bottom strand reveals a TAG stop codon (number 2) from right
to left but none from left to right. Therefore, the bottom strand is the coding strand
and its polarity is 5' → 3' from left to right. The polarity of the double-stranded DNA
sequence is shown here:

```
3 –GAC CTT CGA GTT TGA AAA AAA AAT CTA ATG–5
5 –CTG GAA GCT CAA ACT TTT TTT TTA GAT TAC–3
```

The top strand is the template strand and is transcribed from left to right.

b. Remember that the mRNA sequence is identical to the DNA coding strand with U's
replacing T's. Therefore, the mRNA and polypeptide sequences of codons 136–145 of the
fb⁺ gene are

```
5 –CUG GAA GCU CAA ACU UUU UUU UUA GAU UAC–3
   Leu Glu Ala Gln Thr Phe Phe Leu Asp Tyr
```

c. There are 30 bases in the coding sequence. Each of these can undergo three different base-
substitution mutations. For example, the first C (DNA coding strand) can be substituted for a
T (transition) or either an A or a G (transversions). Therefore, 30 × 3 = 90 different single
base substitutions are possible.

d. The DNA sequence contains nine consecutive AT base pairs. This repeated sequence is
prone to addition and deletion mutations because of potential misalignment of the template
and daughter strands during DNA replication (refer to figure 18.16). In addition, the run of
T's on the bottom strand makes the DNA prone to the formation of thymine dimers.

e. i. Codon 145 is 5'-TAC-3'. The deletion of the C makes the new codon 5'-TAN-3', where N is the base immediately downstream of the deleted C. N can be an A, C, G, or T. For 5'-TAN-3' to be a termination codon, N would have to be either an A or a G (5'-TAA-3' and 5'-TAG-3' are two of the three stop codons). Because the four bases are found in equal frequency, the probability that N is A or G is 50%. Therefore, the probability that the deletion of C will generate a translation termination signal in codon 145 is 1/2.

ii. The insertion of a G–C base pair immediately following codon 145 creates a new codon 146 that has the sequence 5'-GNN-3'. None of the three stop codons has a G as its 5' base. Therefore, the probability that the new codon 146 is a termination triplet is 0. Let's now consider the new codon 147. We don't know anything about its sequence. However, we do know that the insertion of the G–C base pair produced a random codon sequence downstream. We also know that there are 64 possible codons, three of which signal the end of protein synthesis. Therefore, the probability that codon 147 is a termination signal is 3/64, or approximately 4.7%.

f. i. The DNA sequence in question is that of codon 136:

```
3 -GAC-5
5 -CTG-3
```

Tautomerization to the imino form of the A will make it pair like a G, causing an A-to-G transition at that spot. Consequently, there will be a T-to-C transition in the nontemplate or coding strand. The mutant DNA sequence would be

```
3 -GGC-5
5 -CCG-3
```

The transcribed mRNA codon would be 5'-CCG-3', which encodes the amino acid proline. Therefore, the mutation is a missense mutation at the level of the protein.

ii. The DNA sequence in question is that of codon 138:

```
3 -CGA-5
5 -GCT-3
```

Tautomerization to the enol form of the T will make it pair like a C, causing a T-to-C transition at that spot. Consequently, there will be an A-to-G transition in the template or transcribed strand. The mutant DNA sequence would be

```
3 -CGG-5
5 -GCC-3
```

The transcribed mRNA codon would be 5'-GCC-3', which also encodes the amino acid alanine. Therefore, the mutation is a silent mutation at the level of the protein.

iii. The DNA sequence in question is that of codon 139:

```
3 -GTT-5
5 -CAA-3
```

Deamination of the C will convert it to a U, thereby causing a C-to-T transition at that spot. Consequently, there will be an G-to-A transition in the template or transcribed strand. The mutant DNA sequence would be

```
3 -ATT-5
5 -TAA-3
```

The transcribed mRNA codon would be 5'-UAA-3', which is a termination signal. Therefore, the mutation is a nonsense mutation that would cause a premature end to protein synthesis.

iv. Nucleotide number −173 is located in the promoter region of the fb^+ gene, upstream of the transcribed sequence. It is likely not within any of the known conserved elements of promoters. Therefore, the A-to-G transition will have no effect on the mRNA or its encoded protein.

v. The triplet TGA encodes a stop codon. However, it will cause premature termination *only* if it is inserted in the correct reading frame (that is, in between two wild-type codons). Therefore, to find the effect of this insertion, we have to determine the exact location of nucleotide +208.

First, it is important to remember that the first transcribed base is given the designation +1. With this in mind, let us determine the exact locations of all the regions of the fb^+ gene transcript. The 5'-UTR (5' untranslated region) is 153 bases, and so it will stretch from nucleotides +1 to +153. Therefore, exon 1 will begin at nucleotide +154 and end 375 bases later at +528. Similar reasoning yields the following table:

Region	Nucleotides (Coding Strand)
5' Untranslated region (5'-UTR)	+1 to +153
Exon 1	+154 to +528
Intron 1	+529 to +639
Exon 2	+640 to +2610
Intron 2	+2611 to +2900
Exon 3	+2901 to +3572
3' untranslated region (3'-UTR)	+3573 to +3751

Note that the translation initiation codon spans nucleotides +154 to +156, while the termination codon is located at nucleotides +3570 to +3572.

Now back to nucleotide +208. It is located 208 − 156 = 52 bases downstream of the translation initiation codon. This is equivalent to 52/3 = 17.33 codons. This simply means that there are 17 full codons *between* that nucleotide and the translational start site, and so nucleotide +208 is the 5' end of the very next codon (number 19, including the start codon). Therefore, insertion of the TGA triplet after this nucleotide will not cause premature termination of protein synthesis. However, this insertion will alter the meaning of the next two codons, after which the correct reading frame is maintained. It does not matter what the identities of bases +208 to +210 are. The insertion of TGA between bases 208 and 209 will only influence this codon and the next one. Moreover, this insertion cannot generate a premature termination signal. It is hard to determine the effect of this mutation on the wild-type protein. The mutation adds one amino acid, while potentially modifying another. If these changes are in the active site of the protein, its function may be significantly affected.

vi. Nucleotides +529 to +538 constitute the first 10 bases of intron 1. Their deletion will not affect the primary mRNA transcript. However, because the deleted region contains conserved sequences required for intron removal, intron 1 cannot be excised from the primary transcript. This will most likely produce a nonfunctional Fb^+ protein.

vii. Nucleotides +1840 to +1842 fall in exon 2. The TTG to CTG conversion is a transition at the level of the DNA. If the three nucleotides constitute a single codon, the mutation would be silent at the level of the protein since both TTG and CTG encode leucine. However, the three nucleotides may span two different codons, and so the change may affect the protein. Let's determine the exact location of these nucleotides.
     Base +1840 falls $1840 - 639 = 1201$ bases downstream of the intron 1–exon 2 junction. This is equivalent to $1201/3 = 400.33$ codons. This simply means that there are 400 full codons *between* that nucleotide and the start of exon 2, and so nucleotide +1840 is the 5' end of the very next codon. Therefore, nucleotides +1840 to +1842 do make up a single codon, and thus the TTG to CTG mutation has no effect on the Fb^+ protein.

viii. Nucleotide +3570 is the 5' base of the stop codon, so it is not a coincidence that it is a T. (Remember the three stop codons all start with a T: 5'-TAA-3', 5'-TAG-3', 5'-TGA-3'.) Therefore, a T-to-C transition in the DNA coding strand will convert the stop codon into a sense codon, which can only encode one of two amino acids: glutamine (5'-CAA-3' and 5'-CAG-3') or arginine (5'-CGA-3'). Regardless, this mutation will cause run-on protein synthesis. The result is a longer and most likely nonfunctional Fb^+ protein.

g. Mutant 1. The deletion of nucleotides –35 to –17 removed the TATA (Hogness) box which is usually centered at –25. This box is important for promoter recognition by RNA polymerase and its associated transcription factors. Therefore, the *fb*$^+$ gene in this mutant could not be transcribed resulting in the absence of a sense of humor.

Mutant 2. The deletion contains nine base pairs. Nucleotide +760 falls $760 - 639 = 121$ bases downstream of the intron 1–exon 2 junction. This is equivalent to $121/3 = 40.33$ codons. This simply means that there are 40 full codons *between* that nucleotide and the start of exon 2, and so nucleotide +760 is the 5' end of the very next codon. Therefore, nucleotides +760 to +768 account for three consecutive codons, and so the 9-bp deletion does not cause a frameshift mutation. The fact that the sense of humor of this mutant is OK suggests that the missing amino acids are not critical for the function of the Fb^+ protein. They may be in a region of the protein that is distant from the active site.

Mutant 3. Nucleotides +1005 to +1007 fall in exon 2. The CTC-to-CTA conversion is a transversion at the level of the DNA. If the three nucleotides constitute a single codon, the mutation at nucleotide +1007 would be silent at the level of the protein since both CTC and CTA encode leucine. However, there is no sense of humor in this mutant, so the mutation must have had an effect at the level of the protein. Nucleotide +1005 is located 1005 − 639 = 366 bases into exon 2. This corresponds to the 3' base of the 122nd codon in exon 2. Therefore, nucleotide +1007 occupies the middle position of the 123rd codon. The wild-type codon must be 5'-TCN-3', which encodes serine. The mutant codon has the sequence 5'-TAN-3'. If the N is an A or G, this would create a premature termination codon. This in turn would produce a much shorter and nonfunctional Fb protein and therefore no sense of humor in the mutant.

Mutant 4. There are 2440 − 639 = 1801 bases or 1801/3 = 600.33 codons between nucleotide +2440 and the beginning of exon 2. Therefore, nucleotides +2440 to +2442 constitute codon number 601 of exon 2. The AAG-to-GAG conversion is a transition at the level of the DNA, and a missense (lysine to glutamic acid) at the level of the protein. This particular amino acid must be critical to the function of the wild-type Fb protein. Changing the amino acid from a lysine (positively charged) to a glutamic acid (negatively charged) must have disrupted the function of the protein. This produced a mutant with no sense of humor.

Mutant 5. Nucleotide +3558 is located at a distance of 15 nucleotides from the end of exon 3. Therefore, nucleotides +3558 to +360 make up a single codon. The CGA-to-TGA conversion is a nonsense mutation, changing an arginine codon into a stop signal. While this would cause a premature termination of translation, the Fb protein would only be shortened by four amino acids. The wild-type protein has a primary structure of 1005 amino acids, and so the mutant protein would be 1001 amino acids long. This mutant protein is still capable of folding into its active 3-D structure and still able to perform its function. The missing amino acids must not have a noticeable effect on the protein's activity, and thus the mutant individual still possesses a sense of humor.

h. The wild-type nucleotide +692 is located 692 − 639 = 53 bases into exon 2. But what is it? Well, the exact nature of this base can be deduced by analysis of the codon structure of the fb^+ gene.
Exon 1 contains 375 bp/3 bp = 125 codons, and so exon 2 begins with codon 126. But exon 2 starts at nucleotide +640; therefore, codon 126 has nucleotide +640 at its 5' end. Nucleotide +692 is 53 bases downstream of the start of exon 2, or 53/3 = 17.66 codons. This simply means that nucleotide +692 is the second base of the 18th codon in exon 2. This particular codon is actually the 143rd codon in the fb^+ gene (125 from exon 1 plus 18 from exon 2). The 30-bp DNA sequence provided in this problem spans codons 136–145. Therefore, nucleotide +692 is the middle T of the codon 5'-TTA-3' (coding strand of DNA).
Transversion of this T produces an A or a G, resulting in either a 5'-TAA-3' or a 5'-TGA-3' codon. Both of these are nonsense codons which cause the synthesis of the Fb^+ protein to terminate prematurely. The mutant protein would only contain 142 out of the wild-type 1005 amino acids and would surely be nonfunctional. Therefore, unfortunately, Zorkan Tarazan lacks a sense of humor!

i.   Zorkan and Zelda are homozygous for the transversion mutant and wild-type alleles, respectively. Their child Zultan will inherit each of these two alleles and will therefore be heterozygous. He in turn will have a 50:50 chance of transmitting the mutant allele to his son. Therefore, the probability that Zultan's son will inherit the transversion mutant allele from his grandfather Zorkan is 1/2.

For Zoltan's son to be homozygous for the mutant transversion allele, Zoltan's wife would have to be a "carrier" for this allele (P = 1/12,500), and she would also have to transmit it to her son (P = 1/2). The probability that Zoltan will transmit the mutant allele to his son is 1/2. Therefore, the probability that Zoltan's son will be homozygous for the transversion mutant allele is 1/2 × 1/12,500 × 1/2 = 1/50,000.

CHAPTER 19 EXTRANUCLEAR INHERITANCE

Chapter Goals
1. Describe the various functions of chloroplasts and mitochondria in the eukaryotic cell.
2. What is the endosymbiosis theory? Compare the similarities and differences in the origin of chloroplasts and mitochondria according to this theory.
3. Compare and contrast the genomes associated with chloroplasts and mitochondria. How do these organellar genomes compare with the nuclear genomes in the same cell? Why do these organelles require their own genome?
4. What are the different patterns of organeller inheritance? How can you distinguish between these different patterns of inheritance?
5. What are the general features of inherited human mitochondrial diseases?

Key Chapter Concepts
19.1 Physiological Roles of Mitochondria and Chloroplasts
- **Mitochondria and Chloroplasts:** Mitochondria and chloroplasts are **organelles** found in eukaryotic cells that exhibit **extranuclear inheritance**. Both organelles contain DNA and ribosomes allowing transcription and translation to be carried out within the organelle.
- **Structure and Function of Mitochondria:** Mitochondria are found in all eukaryotes. The mitochondria of a cell consist of an outer and inner membrane that is folded into structures called cristae. The DNA, RNA, and ribosomes of the mitochondria are contained within the mitochondrial matrix. The Krebs cycle and oxidative phosphorylation occur in the mitochondria of the cell. The Krebs cycle is involved in the breakdown of glucose to CO_2 in the mitochondrial matrix and also produces electrons that are transferred to the electron transport chain creating a proton gradient across the inner mitochondrial membrane. This gradient allows for the aerobic production of ATP by the mitochondria.
- **Structure and Function of Chloroplasts:** Chloroplasts are organelles that are found in all plants and some protists. Chloroplasts are **plastids** that develop from **proplastids** and are the sites of photosynthesis in the leaves and stems of green plants. They contain both an inner and an outer membrane and a third membrane arranged in flattened disks called **thylakoids** that are arranged in stacks called **grana**. The fluid environment surrounding the thylakoids is the **stroma**. During photosynthesis, ions from the stroma are transported into the thylakoid space within the thylakoid disks. This proton gradient is used to produce ATP and reduced cofactors that allow the fixation of CO_2 into sugars and fatty acids.

19.2 Origin of Mitochondria and Chloroplasts
- **Endosymbiotic Theory:** The endosymbiotic theory suggested that eukaryotic organelles were originally derived by a cell that engulfed either DNA or a prokaryotic cell. Mitochondria and plastids were thought to be the result of endosymbiosis. Mitochondria are believed to have originated in archaebacterium from the endocytosis of an aerobic α-proteobacterium. Chloroplasts are believed to have originated in this cell by the subsequent endocytosis of a cyanobacterium.
- **Origin of Mitochondria:** Evidence of the bacterial ancestry of mitochondria has been found. Mitochondrial genes, like bacterial genes, lack introns (except in yeast). In addition, mitochondrial gene sequences have been found to be similar to related genes from α-proteobacteria; α-proteobacteria also have been found to form symbiotic relationships with eukaryotes. Like bacterial translation initiation, translation initiation of mitochondrial mRNAs uses $tRNA_f^{Met}$ and the mitochondrial ribosomes are more similar to prokaryotic ribosomes, including their rRNAs and sensitivity to antibiotics. Mitochondrial genomes are similar to bacterial genomes in that they are circular and lack chromosomal proteins. Finally, the mitochondrial genome is replicated using bacterial-like mechanisms and mitochondria divide by a process similar to bacterial fission.
- **Origin of Chloroplasts:** Chloroplasts appear to share a common ancestor with cyanobacteria. Both possess internal membranes and carry out photosynthesis. In addition, the ribosomal RNA of chloroplasts shares significant DNA sequence similarity with the DNA from cyanobacteria. Plastids, like bacteria and mitochondria, divide by a process similar to bacterial fission. The leucine tRNA gene

from the chloroplast genome contains a single intron in the same location as related genes from five diverse cyanobacteria, and there is significant sequence similarity between the introns.

19.3 Structure and Organization of the Organellar Genomes

- **Mitochondrial Genome:** Mitochondrial DNA is located in the nucleoid region of the mitochondria. Since there are several mitochondria per cell, there are numerous copies of the mitochondrial genome in every cell. There is great variation in the content and gene organization of different mitochondrial DNAs, ranging in size from 4 to 360 kb and containing 3 to 97 genes. In humans, mitochondria contain 2 to 10 copies of the genome per organelle with very few noncoding regions and no introns. They contain the genes for 13 of the 69 proteins required for oxidative phosphorylation; the others are transcribed in the nucleus and imported into the mitochondria.

- **Mitochondrial Gene Transfer to the Nucleus:** Many of the genes required by the mitochondria were lost from the original symbiotic bacteria and transferred into the host genome. This transfer cannot continue to occur in fungi and animals because these organisms use a different genetic code for mitochondrial and nuclear-encoded genes. In plants, however, the genetic code is universal.

- **Chloroplast Genome:** The chloroplast genome is similar between different algae and plants. The malarial parasite *Plasmodium falciparum* was found to contain an organelle with a degenerate plastid genome indicating that it likely evolved from parasitic unicellular algae but lost the need to perform photosynthesis as eukaryotes evolved.

19.4 Mechanisms of Organellar Inheritance

- **Inheritance of Organelles from a Single Parent:** Unlike the nuclear chromosomes, parents typically contribute unequal volumes of cytoplasm and cytoplasmic components during sexual reproduction. Chloroplasts and mitochondria are two cytoplasmic components that not synthesized de novo but are passed to a new individual intact from a sperm or an egg. In general, maternal inheritance is observed when mitochondria and/or chloroplasts are partitioned randomly into the egg cell during cytokinesis. An organism or cell may have genetically identical mitochondria or chloroplasts (**homoplasmy**), or an organism or cell may have two or more genetically different mitochondria or chloroplasts (**heteroplasmy**).

- **Complex Patterns of Inheritance:** In *Chlamydomonas*, the chloroplast genes are preferentially inherited from the mt^+ parent, while the mitochondrial genes are preferentially inherited from the mt^- parent. Yeast **petite mutants** result from mutations in nuclear genes (**segregational petites**) or in mitochondrial genes (**neutral petites** and **suppressive petites**) that prevent oxidative phosphorylation from occurring in the mitochondria. Petite mutants appear as small, anaerobic-like colonies when growing aerobically.

- **Somatic Segregation of Organelles:** Organelles tend to increase in number as the cell volume increases, and they are divided randomly into the daughter cells. When a heteroplasmic cell divides, the daughter cells may be either heteroplasmic or homoplasmic. If a heteroplasmic cell has organelles with different genotypes, the resulting daughter may have a variety of phenotypes.

19.5 Inherited Human Mitochondrial Diseases

- **General Features of Human Mitochondrial Diseases:** More than 200 human genetic disorders have been associated with mitochondrial defects. In humans, these diseases include **Luft disease, Leber hereditary optic neuropathy** (LHON), and **myoclonic epilepsy and ragged red fiber disease** (**MERRF**). The general features of these diseases are that they affect oxidative phosphorylation in the mitochondria, they are inherited in a maternal manner, individuals typically contain some percentage of wild-type mitochondrial genomes due to heteroplasmy which affects the severity of the disease, and random segregation of mitochondria during cell division affects the severity of the disease in different tissues of an organism.

Key Terms

- maternal inheritance
- mitochondrial Eve hypothesis
- organelles
- extranuclear inheritance
- plastids
- proplastids

- endosymbiotic theory
- mitochondrial DNA (mtDNA)
- homoplasmy
- heteroplasmy

- petite mutations
- segregational petite
- neutral petite
- suppressive petite

- Leber hereditary optic neuropathy (LHON)
- myoclonic epilepsy and ragged red fiber disease (MERRF)

Understanding the Key Concepts
Use the key terms or parts of their definitions to complete the following sentences.

Mitochondria and (1)_____ are two membrane-bound (2)_____ that each contain their own (3)_____ and exhibit extranuclear inheritance. According to the (4)_____ theory, (5)_____ are believed to have originated in eukaryotes from the (6)_____ of an aerobic (7)_____ by an archaebacterium. This cell evolved into a present-day (8)_____ cell. (9)_____ originated later than (10)_____ when a cell already containing a (11)_____ engulfed a (12)_____. This cell evolved into a present-day (13)_____ cell. The DNA in the (14)_____ and (15)_____ is similar to bacterial DNA in that it is (16)_____ and essentially free of (17)_____. Many of the genes from the original symbiotic bacteria were lost or (18)_____ into the (19)_____. This type of (20)_____ is not likely to continue in many organisms because of differences in the (21)_____ found in the (22)_____ genes and in the (23)_____ genes.

Unlike the inheritance of the (24)_____ genome in which two parents contribute an equal number of (25)_____ to an offspring, (26)_____ are (27)_____ partitioned into daughter cells at (28)_____. In humans, the (29)_____ are inherited almost exclusively from the (30)_____ parent due to the unequal ratios of (31)_____ contributed by the egg and the sperm. Both Leber hereditary optic neuropathy (LHON) and myoclonic epilepsy and ragged red fiber disease (MERRF) are (32)_____ inherited diseases. Both of these diseases are caused by (33)_____ in the (34)_____ DNA. These (35)_____ affect the ability of the mitochondria to carry out (36)_____. The (37)_____ of these diseases is dependent upon the percentage of (38)_____ mitochondrial (39)_____ that are present in a cell. (40)_____ of the mitochondria during cell division causes some tissues to be affected by (41)_____ more than other tissues. Furthermore, tissues with (42)_____ requirements, such as neurons and muscles, experience more detrimental defects from mitochondrial mutations.

Figure Analysis
Use the indicated figure or table from the textbook to answer the questions that follow.

1. **Figure 19.3**
   (a) Describe the membrane structure of mitochondria.

   (b) Where is the DNA in the mitochondria located?

   (c) What important cellular processes take place in the mitochondria?

   (d) Where is the proton gradient in the mitochondria established as a result of electron transport?

   (e) What is this proton gradient used for?

2. **Figure 19.4**
   (a) Describe the membrane structure of a chloroplast.

   (b) Where is the DNA in the chloroplast located?

   (c) What important cellular process takes place in the chloroplast?

(d) How is this process similar to the process of oxidative phosphorylation that takes place in the mitochondria?

(e) Where is the proton gradient in the chloroplast established as a result of electron transport?

3. **Figure 19.5**
   (a) Describe how mitochondria are thought to have originated according to the endosymbiotic theory.

   (b) When did mitochondria first originate? How was this deduced?

   (c) What are the origins of the inner and outer mitochondrial membranes according to the endosymbiotic theory?

   (d) Describe how chloroplasts are thought to have originated according to the endosymbiotic theory.

   (e) When did chloroplasts first originate? How was this deduced?

   (f) What are the origins of the inner and outer membranes of the chloroplast according to the endosymbiotic theory?

4. **Table 19.2**
   (a) Two codons that specify the amino acid arginine in nuclear-encoded genes specify different amino acids in the mitochondrial genomes of which organisms?

   (b) What amino acid is specified by the codons AGA or AGG in the mitochondrial genome of vertebrates?

   (c) If a gene containing either of these codons was transferred from the vertebrate mitochondrial genome into the nuclear genome, what effect would this have on the protein encoded by the gene?
   (d) What amino acid is specified by the codon UGA in the nuclear genome of vertebrates? What amino acid is specified by this codon in the mitochondrial genome of vertebrates?

   (e) What would be the effect of transferring a gene containing this codon from the vertebrate mitochondrial genome into the nuclear genome?

5. **Figure 19.24**
   (a) What is the difference between the normal wild-type colonies and the petite yeast colonies morphologically?

   (b) What defect in the petite mutants accounts for this morphological difference?

   (c) What is the specific defect in a segregational petite mutant?

   (d) What is the phenotypic ratio of the progeny that result from a cross between a segregational petite mutant and a wild-type yeast strain?

   (e) Why do the progeny of this cross exhibit a Mendelian pattern of inheritance while the progeny of the crosses generated from a cross between the other petite mutants and a wild-type strain do not?

   (f) What is the specific defect in a neutral petite mutant?

   (g) What is the phenotypic ratio of the progeny that result from a cross between a neutral petite mutant and a wild-type yeast strain?

   (h) Why aren't any of the progeny that result from this cross petite?

(i) What is the specific defect in a suppressive petite mutant?

(j) What is the phenotypic ratio of the progeny that result from a cross between a suppressive petite mutant and a wild-type yeast strain?

(k) What are the possible explanations for how the suppressive petite mutants could exert their influence over the normal wild-type mitochondria?

6. **Figure 19.28**
   (a) How many of the children of the affected mother (I-1) have myoclonic epilepsy and ragged red fiber disease (MERRF)?

   (b) How severe are the MERRF symptoms of the I-1 female? Why?

   (c) What is the difference between the individuals that display severe MERRF symptoms and those that display only mild symptoms?

   (d) What is the most likely explanation for why none of the sons of the affected II-6 male had MERRF disease symptoms?

General Questions

Q19-1. Jesse James, the notorious bank robber, was allegedly killed in 1882. Since that time, rumors that the dead man was not Jesse James circulated widely and were supported by several men who claimed to be the dead "legend." Recently, a group of scientists exhumed the body of Jesse James in an effort to confirm his death. The research team obtained DNA from James' tooth pulp. They also obtained DNA samples from the descendants of James' sister. Why were these samples chosen to confirm the identity of Jesse James?

Q19-2. Describe the mitochondrial features that suggest the mitochondria have a bacterial ancestry.

Q19-3. What evidence is there to suggest that chloroplasts and cyanobacteria share a common ancestor?

Q19-4. Explain why mitochondrial genomes exhibit more variability than chloroplast genomes.

Q19-5. Describe the two different ways in which heteroplasmy can arise.

Q19-6. What are the two main principles that govern organellar inheritance?

Multiple Choice
For each of the following, circle the letter of the choice that most appropriately answers the question.

1.) In which of the following organisms is gene transfer between a mitochondrion and the nucleus still most likely to occur?
   a. ascidians
   b. vertebrates
   c. invertebrates
   d. plants
   e. yeast

2.) Which of the following organisms has the largest number of genes in its mitochondrial genome?
   a. *Schizosaccharomyces*
   b. *Homo sapiens*
   c. *Reclinomonas americana*
   d. *Marchantia polymorpha*
   e. *Plasmodium falciparum*

3.) Which of the following modifications has *not* been shown to occur in organellar RNAs?
   a. splicing together RNAs transcribed from opposite DNA strands
   b. polyadenylation
   c. removal of introns from some RNAs
   d. RNA editing
   e. None of the above; all of these modifications have been shown to occur in organellar RNAs.

4.) Which of the following organisms was found to have undergone degenerative evolution and lost the genes required for oxidative phosphorylation?
   a. *Giardia intestinalis*
   b. *Marchantia polymorpha*
   c. *Arabidopsis thaliana*
   d. *Plasmodium falciparum*
   e. *Spizellomyces punctatus*

5.) Which of the following yeast petite mutants when crossed to a wild-type yeast strain will produce progeny that are all wild type?
   a. segregational petite mutants
   b. suppressive petite mutants
   c. neutral petite mutants
   d. All of the above.
   e. None of the above; all of the yeast petite mutants will produce some mutant progeny.

6.) Suppressive petites are mutant yeast colonies with which of the following defects?
   a. They contain small deletions in their mitochondrial genomes.
   b. They lack mitochondria altogether.
   c. They have mitochondria with no DNA.
   d. They have defects in their mitochondrial ribosomes preventing the translation of proteins that are required for oxidative phosphorylation.
   e. They have a mutation in a nuclear-encoded gene required for mitochondrial function.

7.) The organelles in animals and plants are inherited almost exclusively from the female parent due to which of the following?
   a. The organelles found in the sperm disintegrate after gamete fusion.
   b. Only the egg contributed by the female parent has organelles; the sperm never has organelles.
   c. The egg contributed by the female parent contains a larger amount of cytoplasm than the sperm contributed by the male parent.
   d. The egg contains the maternal proteins necessary for the *de novo* synthesis of the organelles.
   e. Organelles from the female parent are preferentially maintained because they are genetically favorable.

8.) The severity of mitochondrial diseases is determined by which of the following?
   a. progression of the disease at the time of detection
   b. degree of heteroplasmy
   c. age of the affected individual
   d. nuclear-encoded genes that are being expressed
   e. which parent the disease is inherited from

9.) Which of the following patterns of inheritance is consistent with strict maternal inheritance?
   a. The mother passes a trait on to all of her children, while the father passes it on only to his daughters.
   b. The father passes a trait inherited from his mother on to all of his children.
   c. The mother passes a trait on to all of her daughters, while the father never passes it on to any of his children.

d. The mother passes a trait on to all of her children, while the father never passes it on to his children.

e. The mother passes a trait on to all of her sons, while the father only passes it on to his daughters.

10.) Which of the following statements comparing the human and yeast mitochondrial genomes is *not* true?

a. The human mitochondrial genome is smaller than the yeast mitochondrial genome.

b. Both the human and the yeast mitochondrial genomes encode a large number of tRNAs.

c. The human mitochondrial genome contains large intergenic regions not present in the yeast mitochondrial genome.

d. The human mitochondrial genome contains more genes than the yeast mitochondrial genome.

e. The human mitochondrial genome lacks introns which are found in the yeast mitochondrial genome.

Practice Problems

P19-1. In *Drosophila*, sensitivity or resistance to carbon dioxide displays maternal inheritance. Predict the outcomes of reciprocal crosses involving matings between sensitive and resistant flies.

(*Hint 1*: Extranuclear inheritance is suggested if the results of reciprocal crosses produces progeny with the phenotype of only one parent.
Hint 2: If a trait is maternally *inherited*, genetic transmission is only through the maternal cytoplasm. The male makes no contribution to the progeny phenotype. Regardless of the number of generations examined, offspring will have the same phenotype as their mother if the trait is maternally inherited.)

P19-2. Three individual crosses were made between killer and sensitive *Paramecium*. Examine the results of each cross and use this information to assign genotypes to the parental strains. Cross 1: all killer; cross 2: 75% killer and 25% sensitive; and cross 3: 50% killer and 50% sensitive.

P19-3. A male from a true-breeding strain of nervous zeebogs is crossed to a calm female, also from a true-breeding strain. The F_1 and F_2 offspring are all calm. Propose an explanation for these findings. If the F2 generation information were unavailable, what conclusions would you have drawn?

(*Hint 3*: The phenotypes of F_1 and F_2 progeny will be the same as those of the female parent if the trait is maternally influenced *or* maternally inherited. Several generations must be examined to conclusively conclude that a trait is maternally inherited.)

P19-4. Complete the following pedigree such that maternal inheritance is suggested and dominant sex-linked inheritance can be excluded.

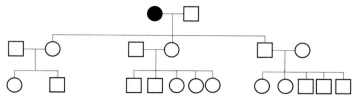

P19-5. Examine the pedigree below, and determine how the trait is likely inherited.

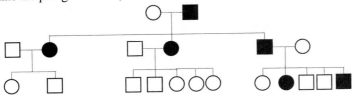

P19-6. In four-o'clock plants, leaves can be green, white, or variegated. Flowers from branches of each type were used to perform all possible crosses (as outlined in the following). Predict the progeny phenotypes expected if (a) inheritance is biparental (codominance) or (b) inheritance is maternal.

	Female Parent	Male Parent	(a) Codominance	(b) Maternal Inheritance
1	White	White		
2	White	Green		
3	White	Variegated		
4	Green	White		
5	Green	Green		
6	Green	Variegated		
7	Variegated	White		
8	Variegated	Green		
9	Variegated	Variegated		

Assessing Your Knowledge

Understanding the Key Concepts—Answers

1.) chloroplasts; 2.) organelles; 3.) DNA; 4.) endosymbiotic; 5.) mitochondria; 6.) endocytosis or phagocytosis; 7.) α-proteobacterium; 8.) animal; 9.) Chloroplasts; 10.) mitochondria; 11.) α-proteobacterium; 12.) cyanobacterium; 13.) plant; 14–15.) mitochondria; chloroplasts; 16.) circular; 17.) chromosomal proteins; 18.) transferred; 19.) host genome; 20.) transfer; 21.) genetic code; 22–23.) mitochondrial; nuclear-encoded; 24.) nuclear; 25.) chromosomes; 26.) organelles; 27.) randomly; 28.) cytokinesis; 29.) organelles; 30.) female; 31.) cytoplasm; 32.) maternally; 33.) mutations; 34.) mitochondrial; 35.) mutations; 36.) oxidative phosphorylation; 37.) severity; 38.) wild-type; 39.) genomes; 40.) Random segregation; 41.) mitochondrial defects or mutations; 42.) high energy.

Figure Analysis—Answers

1a.) Mitochondria have an outer and an inner membrane. The outer membrane is smooth and surrounds the entire organelle, while the inner membrane is highly folded into cristae that surround the mitochondrial matrix; b.) The mitochondrial genomic DNA is located in the central mitochondrial matrix; c.) The Krebs cycle and oxidative phosphorylation are two important cellular processes that take place in the mitochondria; d.) A proton gradient is established across the mitochondrial inner membrane as a result of electron transport; e.) This proton gradient drives the production of ATP.

2a.) A chloroplast has two peripheral membranes that are both smooth. These are the inner and outer membranes. In addition, a chloroplast has a third membrane arranged in flattened disks called thylakoids; b.) The DNA in the chloroplast is located in the stroma, or the fluid environment surrounding the thylakoids; c.) Photosynthesis takes place in the chloroplast; d.) Similar to the electron transport that takes place in the mitochondria, photosynthesis involves electron transfer in the chloroplast that transports ions from the stroma into the thylakoid space creating a proton gradient that drives the production of ATP; e.) A proton gradient is established across the thylakoid membrane as a result of electron transfer in the chloroplast.

3a.) According to the endosymbiotic theory, mitochondria are thought to have originated by the endocytosis of an aerobic α-proteobacterium by an early eukaryotic cell. This α-proteobacterium evolved into a mitochondrion; b.) Mitochondria are thought to have first appeared shortly after eukaryotic cells evolved from prokaryotes. This was deduced from the fact that all eukaryotes have mitochondria, but all prokaryotes lack mitochondria; c.) According to the endosymbiotic theory, the inner membrane of the mitochondrion is derived from the α-proteobacterium that was endocytosed, while the outer membrane is derived from the endocytosing eukaryotic cell; d.) According to the endosymbiotic theory, chloroplasts are thought to have originated by the endocytosis of a cyanobacterium by an early eukaryotic cell that previously endocytosed an α-proteobacterium to form a mitochondrion; e.) Chloroplasts originated later than the appearance of the mitochondria. This was deduced because only some eukaryotes have chloroplasts, while all eukaryotes have mitochondria; f.) According to the endosymbiotic theory, the inner membrane of the chloroplast is derived from the cyanobacterium that was endocytosed, and the outer membrane is derived from the endocytosing eukaryotic cell.

4a.) The AGA and AGG codons that specify the amino acid arginine in nuclear-encoded genes specify different amino acids in the mitochondrial genomes of vertebrates, invertebrates, and ascidians; b.) In the mitochondrial genome of vertebrates, the codons AGA or AGG specify a stop codon; c.) If a gene containing either of these codons was transferred from the mitochondrial genome into the nuclear genome, an arginine residue would be incorporated into the protein encoded by the gene rather than a stop codon. This would result in the failure to properly terminate translation at this site and would lead to the production of a larger protein from the nuclear gene; d.) In the nuclear genome of vertebrates, the codon UGA specifies a stop codon. In the mitochondrial genome of vertebrates, the codon UGA specifies the incorporation of a tryptophan residue; e.) If a gene containing this codon was transferred from the mitochondrial genome into the nuclear genome, a premature stop codon would be incorporated into the protein encoded by the gene. This premature translation termination codon would lead to the production of a truncated protein.

5a.) The petite yeast mutants are morphologically smaller than the wild-type yeast colonies; b.) Petite yeast colonies have defects in the function of their mitochondria; c.) The segregational petite mutant has a mutation in a nuclear gene that encodes a protein required for mitochondrial function; d.) A cross between a segregational petite mutant and a wild-type yeast strain results in progeny with a phenotypic ratio of 2 wild types:2 petites; e.) The progeny of this cross exhibit a Mendelian pattern of inheritance because the segregational petite mutants have a nuclear mutation such that it segregates 2:2 when crossed to a wild-type strain. This is different than the other yeast petite mutants (neutral and suppressive) that have mutations in the mitochondrial genome and therefore do not segregate according to the Mendelian patterns of inheritance; f.) A neutral petite mutant has mitochondria that entirely lack DNA; g.) A cross between a neutral petite mutant and a wild-type yeast strain results in progeny with a phenotypic ratio of 4 wild types:0 petites; h.) None of the progeny from the cross between the neutral petite mutant and a wild-type strain are petite because mitochondria from both parents are inherited in the progeny. In diploid cells, the mitochondria can fuse, and mitochondria lacking mtDNA can fuse with wild-type mitochondria. When the fused mitochondria divide, they will all contain at least one copy of the mtDNA, so only wild-type mitochondria are present. Meiosis produces haploid spores that all have normal mitochondria; i.) The suppressive petite mutants contain small deletions in the mitochondrial genome; j.) A cross between a suppressive petite mutant and a wild-type yeast strain results in progeny with a phenotypic ratio of 0 wild type:4 petites; k.) Suppressive petite mutants could exert their influence over normal wild-type mitochondria by replicating more rapidly than the wild-type mitochondria such that the number of suppressive petite mitochondria increases in the cells. Secondly, they could exert their influence over normal wild-type mitochondria in fused mitochondria having both types of genomes. If the suppressive petite mutant mtDNA replicates more rapidly than the wild-type mtDNA, a greater percentage of the mtDNA in the fused mitochondria will be petite. When the fused mitochondrion divides, it will produce organelles that contain both petite and wild-type mtDNA. The percentage of wild-type mtDNA will determine if the cell will have a petite or a wild-type phenotype. Finally, in the fused mitochondria, recombination could occur between the wild-type mtDNA and the suppressive petite mtDNA. This could alter the relative abundance of the mutant petite mtDNA, and when the mitochondrion divides, the percentage of wild-type mtDNA will again determine if the cell will have a petite or a wild-type phenotype.

6a.) All four of the children (thee daughters and one son) of the affected mother (I-1) have myoclonic epilepsy and ragged red fiber disease (MERRF); b.) The I-1 female has mild MERRF symptoms because she has only 73% mutant mtDNA, while greater than 90% of the mitochondria must be mutant to develop the severe MERRF symptoms; c.) The individuals with severe MERRF symptoms all have greater than 90% mutant mtDNA, while the individuals with mild MERRF symptoms have 73%, 85%, or 90% mutant mtDNA; d.) The most likely explanation for why none of the sons of the II-6 male have MERRF disease symptoms is that this disease is inherited in a strictly maternal manner such that an affected woman passes the disease on to all of her children, but an affected man does not pass the disease on to his children.

General Questions—Answers

Q19-1. Mitochondrial DNA was used to identify the remains of Jesse James. While James and his sister shared only 50% (on average) of their nuclear genes, they both received their mitochondrial DNA from their mother. James' sister in turn passed her mitochondrial DNA to her children and so on. With the exception of mutation, the mitochondrial DNA sequence of the descendants of James' sister should be identical to that of James. The identity of the remains of Jesse James was finally confirmed by mitochondrial DNA analysis.

Q19-2. Several mitochondrial features suggest that it has a bacterial ancestry. Mitochondrial genes, like bacterial genes lack introns. Mitochondrial gene sequences are similar to related genes from α-proteobacteria. Translation initiation of mitochondrial mRNAs utilizes $tRNA_f^{Met}$ as does bacterial translation initiation, and mitochondrial ribosomes share similarities with bacterial ribosomes in both their rRNAs and in their sensitivity to antibiotics. Finally, like the bacterial genome, mitochondrial DNA is circular and free of chromosomal proteins, both genomes are replicated using similar mechanisms, and mitochondria divide by a process similar to bacterial fission.

Q19-3. Chloroplasts and cyanobacteria seem to share a common ancestor because they both possess similar structural features and carry out photosynthesis. The ribosomal RNA of chloroplasts shares significant DNA sequence similarity with the DNA from cyanobacteria. In addition, the leucine tRNA gene from the chloroplast genome contains a single intron in the same location as related genes from five diverse cyanobacteria and there is significant sequence similarity between the introns.

Q19-4. Mitochondrial genomes exhibit more variability than chloroplast genomes because the endosymbiotic event that created mitochondria occurred earlier than that which created chloroplasts, so mitochondrial genomes have had more time than chloroplast genomes for their sequences to diverge from the original bacterial ancestor.

Q19-5. Heteroplasmy is the presence of a mixture of organellar genomes within a single cell. This happens when a fertilized egg contains an organelle type that is composed of two different genotypes. This can occur either by a mutation arising within the organellar DNA or by both the male and the female parents contributing cytoplasmic organelles to the fertilized egg.

Q19-6. The two main principles that govern organellar inheritance are that the organelles are inherited preferentially from only one of the two parents and organelles exhibit somatic segregation.

Multiple Choice—Answers

1.) d; 2.) c; 3.) e; 4.) a; 5.) c; 6.) a; 7.) c; 8.) b; 9.) d; 10.) c.

Practice Problems—Answers

P19-1. When a trait is inherited maternally, genetic transmission is only through the maternal cytoplasm. If we crossed sensitive females with resistant males, we would expect all of the offspring to be sensitive. If we cross resistant females with sensitive males, we would expect all of the offspring to be resistant.

P19-2. In cross 1, all of the offspring display the killer phenotype, suggesting that both parents also had the killer phenotype. Genotypes would be $KK \times KK$ or $KK \times Kk$. The 3:1 ratio of killer to sensitive phenotypes in cross 2 suggests that both individuals were heterozygous killer (*Kk* × *Kk*). In cross 3, crossing a heterozygous killer and a sensitive strain (*kk*) would result in the observed 1:1 ratio.

P19-3. In this cross, we see that the F_1 and F_2 phenotypes are the same as the P_1 female, which suggests maternal inheritance. If we had looked at the F_1 phenotypes only we probably would have concluded that the trait was likely inherited as an autosomal with calm dominant to nervous. This problem illustrates that it is important to examine multiple generations resulting from a cross before drawing conclusions.

P19-4. In maternally inherited traits, all individuals have the phenotype of the female parent. In generation III, we can distinguish maternal inheritance from sex-linked dominant inheritance by examining the offspring of affected males and females. For example, approximately one-half of the daughters of the affected male would be affected if the trait were sex-linked dominant. No affected daughters will result from the affected father if the trait is maternal.

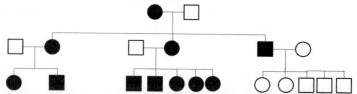

P19-5. The pedigree is probably showing paternal influence or paternal inheritance. Because affected males produce both affected daughters and affected sons, we can rule out Y-linked inheritance as a likely explanation. Sex-linked inheritance is eliminated because none of the affected females have affected children.

P19-6. Cells in variegated regions of the plant contain both green and white chloroplasts. Depending on how the chloroplasts are partitioned in the egg, a variegated plant can give rise to green, white, or variegated offspring phenotypes.

Cross (Female × Male)	(a) Codominance	(b) Maternal Inheritance
1. White × White	White	White
2. White × Green	Variegated	White
3. White × Variegated	1 Variegated : 1 White	White
4. Green × White	Variegated	Green
5. Green × Green	Green	Green
6. Green × Variegated	1 Variegated : 1 Green	Green
7. Variegated × White	1 Variegated : 1 White	Green, White, Variegated
8. Variegated × Green	1 Variegated : 1 Green	Green, White, Variegated
9. Variegated × Variegated	1 Green : 2 Variegated : 1 White	Green, White, Variegated

19

EXTRANUCLEAR INHERITANCE

CHAPTER SUMMARY QUESTIONS

1. 1. b, 2. d, 3. f, 4. h, 5. j, 6. a, 7. c, 8. e, 9. g, 10. i.

2. The Mitochondrial Eve Hypothesis proposes that all human mitochondrial genomes evolved from a "single" original genome approximately 200,000 years ago. The human mitochondrial genome is maternally inherited. Therefore, the original genome must have been present in the first *Homo sapiens* female—hence, the name "Eve."

3. Human mitochondrial DNA (mtDNA) is circular, consisting of 16,569 base pairs. It has very few noncoding regions and no introns. The genome carries genes for 22 tRNAs, 2 rRNAs (16S and 12S), and 13 mRNAs that encode polypeptides. The human mitochondrial genome is present in 2 to 10 copies per organelle. Because every cell contains many mitochondria, there are numerous copies of the mitochondrial genome per cell.

4. The mitochondrion is the site of electron transport, a process that produces high levels of ATP via oxidative phosphorylation. A mitochondrion possesses two main features that make it ideal for ATP production: (1) It contains membrane-enriched cristae which provide a large surface area for ATP production. (2) Its inner membrane is impermeable to ions and small molecules.

   The process of energy production begins when sugars are broken down in the cytoplasm to produce pyruvate and a small amount of ATP. The pyruvate enters the mitochondria, where it is degraded to CO_2 via the Krebs cycle. The Krebs cycle also produces reduced NADH and $FADH_2$ molecules which are the source of the electrons that pass through the respiratory electron transport chain, with oxygen as the terminal electron transport. This electron transfer creates a proton gradient across the inner mitochondrial membrane that drives ATP biosynthesis.

5. Chloroplasts are the sites of photosynthesis in plants and green algae. Photosynthesis generates ATP and reducing power in the form of reduced ferredoxin and NADPH. These facilitate several other basic functions: (1) CO_2 fixation, (2) starch synthesis,

(3) fatty acid synthesis, (4) nitrogen fixation, (5) sulfur fixation, and (6) amino acid synthesis.

6. Both mitochondria and chloroplasts contain DNA and ribosomes with prokaryotic affinities. The two organelles are not generated de novo when a eukaryotic cell divides. Rather, they arise from pre-existing chloroplasts and mitochondria via binary fission. Defects in either organelle can produce similar phenotypic effects of reduced growth.

7. Transcription in animal mitochondria differs from both bacterial and nuclear transcription in that each of the two strands is transcribed into a single RNA product that is then cut into smaller transcripts for individual genes. Unlike nuclear mRNAs in humans, mitochondrial mRNAs lack introns, and unlike bacterial mRNAs and the vast majority of nuclear mRNAs, mitochondrial mRNAs may undergo RNA editing (where bases are replaced, modified, removed, or inserted). Furthermore, mitochondria use a slightly different genetic code from the universal code used by bacteria and nuclei. For example, the codon 5'-UGA-3' is a termination signal in both bacterial and nuclear genes; however, it encodes tryptophan in mitochondria.

8. The endosymbiotic theory proposes an explanation for the origin of mitochondria and chloroplasts. The theory holds that ancestral eukaryotic cells were archaeal cells that grew anaerobically and lacked mitochondria and chloroplasts. These early cells engulfed by endocytosis an aerobic α-proteobacterium, which evolved into a mitochondrion. Chloroplasts were formed when this cell (or its descendants) subsequently endocytosed a photosynthetic cyanobacterium.

    Data that supports this theory include: (1) Each organelle is surrounded by two membranes—the inner membrane is likely derived from the proteobacterium or cyanobacterium, and the outer membrane from the endocytosing host cell. (2) Like bacteria, both organelles divide by binary fission. (3) Many mitochondrial genomes and all chloroplast genomes are circular, just like the vast majority of bacterial genomes. (4) Mitochondrial gene sequences are similar to related genes from α-proteobacteria. (5) Many α-proteobacteria form symbiotic relationships with eukaryotic cells. (6) Mitochondrial ribosomes are similar in structure to bacterial ribosomes (even though the mitochondrial ribosome is constructed of imported cellular proteins, it is sensitive to antibiotics that inhibit translation from bacterial ribosomes but not from eukaryotic ribosomes). (7) Like bacterial cells, mitochondria use $tRNA_f^{Met}$ for translation initiation (it is charged with N-formyl methionine instead of regular methionine).

9. Petite mutants of *Saccharomyces cerevisiae* have abnormalities in mitochondrial function, specifically in oxidative phosphorylation. They all appear as small, anaerobic-like colonies when grown under aerobic conditions. Segregational petites are caused by a mutation in a nuclear gene that encodes a protein required for mitochondrial function. In contrast, both neutral and suppressive petites are due to mitochondrial genome mutations. Neutral petites lack the entire mitochondrial

genome, while suppressive petites contain only small deletions in their mitochondrial genome.

10. The two terms refer to the presence of multiple mitochondria in a single cell and more than one genome per organelle. Homoplasmy is the presence of a single common genotype in the organellar genome, while heteroplasmy is the presence of a mixture of organellar genomes in a cell.

11. **a.** 2 normal:2 petite. A nuclear gene should exhibit a 2:2 Mendelian segregation pattern.

　　b. 0 normal:4 petite. Suppressive petites contain small deletions in their mitochondrial genome. This permits a shorter length of time for replication, giving them an advantage in heteroplasmy with wild-type mitochondria.

　　c. 4 normal:0 petite. Neutral mitochondria are lost in heteroplasmy to wild-type mitochondria.

12. The shrinking of mitochondrial genomes has occurred through two basic mechanisms: (1) transfer of genes to the nuclear genome, and (2) loss of genes that were no longer required (degenerative evolution).

13. Human mitochondrial diseases are characterized by four general features:
　　1. They exhibit strict maternal inheritance.
　　2. They can vary in severity between individuals because of heteroplasmy.
　　3. They can vary in severity in different tissues of the same individual because of random segregation of mitochondria during cell division.
　　4. They primarily affect the nervous and muscular systems.

EXERCISES AND PROBLEMS

14. By looking at different species, it is clear that few genes for oxidative phosphorylation are found in all mitochondrial genomes.

15. Human mitochondrial DNA does not have introns. Finding an intron would suggest that the mitochondria had acquired a nuclear gene.

16. A mating between an affected male and normal female would produce normal children. The children inherit the mitochondria of their normal mother and not those of their affected father. In contrast, the children of a normal father and affected mother could develop blindness. If the mother is homoplasmic, all children will be blind. If she is heteroplasmic, the children may or may not develop blindness. Their exact phenotype would depend on how many defective mitochondria they inherit from their mother.

17. The simplest way to determine the nature of the lesion resulting in the petite phenotype is to make a cross of the petite strain with a wild-type strain. After meiosis, isolate the four products (spores) and allow them to grow separately under normal, aerobic conditions. If the ratio of petite to wild type is 1:1, the mutation is of a nuclear gene. If progeny are wild type, the mutation is in the mitochondrial genome and is of the neutral type. If progeny are mostly petites, the mutation is also in the mitochondrial genome, but it is of the suppressive type.

18. When a trait is inherited maternally, genetic transmission is only through the maternal cytoplasm. If we crossed sensitive females with resistant males, we would expect all of the offspring to be sensitive. In contrast, a cross of resistant females with sensitive males would produce offspring that are resistant.

19. None of the couple's four children will develop the disease because they all inherit the mitochondria of their unaffected mother.

20. In this cross, we see that the F_1 and F_2 phenotypes are the same as the parental female, which suggests maternal inheritance. If we had looked at the F_1 phenotypes only, we probably would have concluded that the trait was likely inherited as autosomal with calm dominant to nervous. This problem illustrates that it is important to examine multiple generations resulting from a cross before drawing conclusions.

21. Mitochondrial DNA was used to identify the remains of Jesse James. While James and his sister shared only 50% (on average) of their nuclear genes, they both received their mitochondrial DNA from their mother. James's sister in turn passed her mitochondrial DNA to her children, and so on. With the exception of mutations, the mitochondrial DNA sequence of the descendants of James's sister should be identical to that of James. Using the descendants of James's brother would not have been useful. These descendants would not have the same mitochondrial DNA as their ancestors, James, or his brother.

22. In X-linked inheritance both parents can transmit the trait to their children, while in cytoplasmic inheritance only one does.

23.

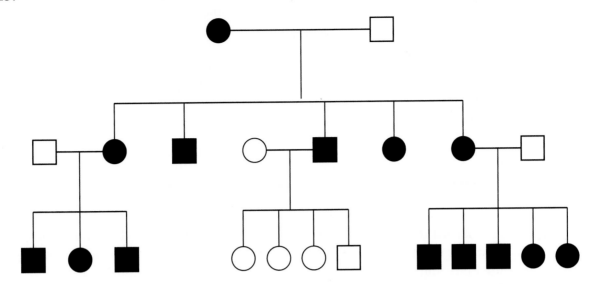

24. Nuclear genes would show a Mendelian segregation pattern, whereas chloroplast genes would be inherited strictly from the female.
   a. $C^T N^T N^t$—normal leaves.
   b. $C^t N^T N^t$—normal leaves.
   c. The cross is: female $C^T N^T N^t$ × male $C^t N^T N^t$. The offspring would be 1/4 $C^T N^T N^T$, 2/4 $C^T N^T N^t$, 1/4 $C^T N^t N^t$. All would have normal leaves.
   d. The cross is: female $C^t N^T N^t$ × male $C^T N^T N^t$. The offspring would be 1/4 $C^t N^T N^T$ (normal leaves), 2/4 $C^t N^T N^t$ (normal leaves), 1/4 $C^t N^t N^t$ (twisted leaves).

25. In humans, mitochondrial DNA exhibits strict maternal inheritance. Therefore, each of these individuals could have inherited only the top molecule, only the bottom molecule, or both. If an individual is homoplasmic for the top mitochondrial DNA, the probe will hybridize to two *Pst*I restriction fragments: 7 kb and 4 kb. If an individual is homoplasmic for the bottom mitochondrial DNA, the probe will hybridize to three *Pst*I restriction fragments: 7, 3, and 1 kb. Finally, heteroplasmic individuals will show four fragments in their Southern blot: 7, 4, 3, and 1 kb.

26. Y-linked inheritance shows a strict male-to-male transmission, while paternal inheritance occurs from male to all offspring (male and female).

27. a. A circular molecule that contains *n* sites for restriction enzymes will produce *n* fragments when digested. In contrast, a linear molecule will generate *n* + 1 fragments. Digestion of the mitochondrial genome with *Pst*I, *Eco*RI, and *Bgl*II individually produced two fragments in each lane. If the genome is circular, there must be two restriction sites for each enzyme, whereas a linear genome would only contain one site for each restriction enzyme. Therefore, a triple digest of this mitochondrial genome would yield four fragments if the genome is

linear (3 restriction sites + 1), and six fragments if the genome is circular (6 restriction sites). We see that the actual triple digest produced four fragments, and so the mitochondrial genome must be linear.

b.   Each enzyme has only one restriction site in the mitochondrial DNA, so it should not be hard to locate them. Let's start with *Pst*I, which produces 3- and 13-kb fragments. We can place the *Pst*I site arbitrarily on the left to produce the following map:

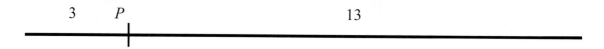

The *Eco*RI restriction fragments are 4- and 12-kb in length. The *Eco*RI site could be on the opposite end of the mtDNA or the same end as the *Pst*I site. The first possibility predicts that a 4-kb fragment should be seen in the triple digestion. This fragment would result because the 4-kb *Eco*RI fragment would be too small to contain a restriction site for the third enzyme, *Bam*HI, and so it should appear if all three enzymes are used simultaneously. Since we don't see a 4-kb fragment in the triple digest, the *Eco*RI site must be adjacent to the *Pst*I site:

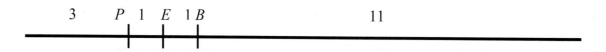

Finally, digestion by *Bam*HI alone produces 5- and 11-kb fragments. If the *Bam*HI site is adjacent to the *Eco*RI, the map would be

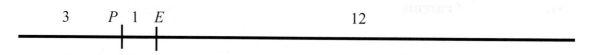

The triple digest clearly refutes this restriction map. For instance, there is no 11-kb fragment. Therefore, the *Bam*HI site must be on the opposite end of the mtDNA relative to the *Eco*RI site. The restriction map and location of the rRNA gene are as follows:

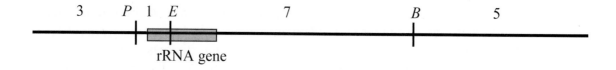

28. Segregation can occur in one of two ways: (1) wild-type (A) mitochondria to the same pole and mutant (a) mitochondria to the opposite pole, or (2) a wild-type (A) and a mutant (a) mitochondrion migrate to each of the two poles. Therefore, the probabilities are 1/4 AA:2/4 Aa:1/4 aa.

29. In *Chlamydomonas reinhardtii*, chloroplasts are inherited from the mt^+ parent, and mitochondria are inherited from the mt^- parent. The mt^+ parent in this cross has the genotype c^+d^- for the two chloroplast genes; therefore, all of the meiotic products will inherit c^+d^-. The two nuclear genes will segregate according to normal Mendelian ratios.
   a. Zygote = $a^+a^- \; b^+b^- \; c^+c^- \; d^+d^-$
   b. Meiotic products = $a^+b^+c^+d^-$; $a^+b^-c^+d^-$; $a^-b^+c^+d^-$; $a^-b^-c^+d^-$.

30. In strain 1, both crosses produced offspring that have the same prescottine phenotype as the mt^+ parent. Therefore, the gene for prescottine resistance is located in the chloroplast genome. On the other hand, the reciprocal crosses involving strain 2 yielded a 1:1 phenotypic ratio in the offspring. Because this is a Mendelian segregation pattern, eduardomycin resistance must be encoded by a nuclear gene. Both strain 3 crosses produced offspring with the same brownicillin phenotype as the mt^- parent. Therefore, the gene for brownicillin resistance is found in the mitochondrial genome.

31.

	Parents		**Inheritance**	
	Female	**Male**	**(a) Biparental**	**(b) Maternal**
1.	green	× green	green	green
2.	green	× white	variegated	green
3.	green	× variegated	1 green:1 variegated	green
4.	white	× green	variegated	white
5.	white	× white	white	white
6.	white	× variegated	1 white:1 variegated	white
7.	variegated	× green	1 variegated:1 green	green, white, variegated
8.	variegated	× white	1 variegated:1 white	green, white, variegated
9.	variegated	× variegated	1 green:2 variegated:1 white	green, white, variegated

32. a. The labeled *rbcL* probe hybridized to chloroplast but not nuclear DNA. This indicates that the large subunit of RuBisCO is encoded in the chloroplast genome. The *rbcS* probe on the other hand hybridized to nuclear but not chloroplast DNA. Indeed, two bands appeared in the Southern blot, suggesting the presence of two *rbcS* genes. Therefore, the small subunit of RuBisCO is encoded by two nuclear genes.

b.   It could be that the large and small subunit polypeptides have to undergo posttranslational modification before they can form the RuBisCO holoenzyme. *E. coli* cells do not have the proper proteins for this modification.

CHAPTER INTEGRATION PROBLEM

a.   Autosomal dominance is possible, but not an attractive explanation. Let A = disease allele and a = normal allele. All unaffected individuals are homozygous recessive (*aa*), while all affected individuals, with the possible exception of I-1, are heterozygous (*Aa*). The mating between individuals II-1 and II-2 is *Aa* × *aa* and is expected to produce offspring in a 1:1 phenotypic ratio. However, the mating actually produced children in an 8 affected:1 unaffected ratio. Therefore, individual II-1 would have to pass the *A* allele to eight of her nine children. Girls III-2, III-7, and III-13 would then go on to transmit the *A* allele to all of their children, while boys III-4 and III-10 would transmit the normal *a* allele to their children. Therefore, although possible, autosomal dominant inheritance is highly unlikely.

The same goes for autosomal recessive inheritance. Let A = normal allele and a = disease allele. All affected individuals are homozygous recessive (*aa*). Individuals I-2, II-2, III-1, III-6, and III-12 would all have to be carriers. This is not likely since the disease is rare. Moreover, II-2 (genotype *Aa*) would have to transmit the *a* allele to eight of his nine children. The probability of this happening is $(1/2)^8$ = 1/256. Therefore, this disease is unlikely to be inherited in an autosomal recessive manner.

Y-linkage is impossible because the disease is found in females.

X-linkage is also impossible. The disease cannot be inherited in a dominant fashion because affected males III-4 and III-10 should have transmitted the disease to all of their daughters. Furthermore, the disease cannot be inherited recessively because males I-2 and III-1, who are normal, have daughters with the disease.

Sex-limited inheritance is impossible because both males and females are affected.

Finally, sex-influenced inheritance can also be excluded. Let A = disease allele and a = normal allele. If the trait was dominant in males, female I-1 and male I-2 would have to be homozygous dominant (*AA*) and homozygous recessive (*aa*), respectively. Their daughter (II-1) would have to be *Aa*, which means she should not have the disease. Therefore, the disease cannot be sex-influenced dominant in males. On the other hand, if the trait was dominant in females, males III-4 and III-10 would have to be homozygous dominant (*AA*), because males would not express the trait in the heterozygous state. These two fathers would transmit the *A* allele to all of their daughters, who will be affected with the disease. The presence of five normal daughters (IV-5 and IV-8,9,10,11) rules out this mode of inheritance.

b.   The findings eliminate the possibility of autosomal recessive inheritance because two unaffected individuals cannot have affected offspring (in the absence of mutations). Autosomal dominant inheritance is still possible. The disease could be incompletely penetrant, which would explain why individual III-14 was unaffected, and it could show variable expressivity, which would explain the varying levels of disease expression in different individuals.

c.   Affected female II-1 produced only affected offspring (with the exception of daughter III-14). In addition, she had three daughters (III-2, III-7, and III-13) who got married and all of them had affected children. However, her two married sons (III-4 and III-10) had only normal children—who inherited the normal alleles of their mothers III-3 and III-9, respectively. So it is clear that the gene responsible for this neurodegenerative disease is transmitted only by female parents. Because human mitochondrial DNA is transmitted by a mother to all her children, mitochondrial inheritance is very likely for this disease.

The variable expressivity of the disease can be explained by the fact that affected individuals are heteroplasmic, with both mutant and wild-type mitochondria. These organelles are randomly distributed during cell division, resulting in cells with varying proportions of mutant and wild-type mitochondria. The severity of the symptoms is dependent on the proportion of mutant mtDNAs in the zygote and cells derived from it—the higher the proportion, the more severe the symptoms.

d.   The mutant phenotype is expressed only when the proportion of mutant mitochondrial DNA reaches a threshold level. In individual III-14, this threshold was clearly not reached. However, because she has two affected daughters, individual III-14 must have inherited mutant mitochondria from her mother. Therefore, just by chance female II-1 must have transmitted predominantly normal mitochondria to her daughter III-14. On the other hand, if random mitochondrial segregation in III-14 was heavily slanted in the "opposite" direction of that of her mother, III-14 would transmit predominantly mutant mitochondria to both of her children.

e.   Let's first write the complementary strand to produce the double-stranded DNA region:

```
AATGATCTGC TGCAGTGCTC TGAGCCCTAG GATTCATCTT TCTTTTCACC GTAGGTGGCC
TTACTAGACG ACGTCACGAG ACTCGGGATC CTAAGTAGAA AGAAAAGTGG CATCCACCGG
```

The polarity is 5' → 3' for the top strand and 3' → 5' for the bottom strand.
Type II restriction endonucleases recognize and cut at palindromic sequences. Palindromes in DNA are sequences that are identical when read in the 5' → 3' direction on each strand. Careful scanning of this double-stranded DNA reveals the following four restriction sites:

```
AATGATCTGC TGCAGTGCTC TGAGCCCTAG GATTCATCTT TCTTTTCACC GTAGGTGGCC
TTACTAGACG ACGTCACGAG ACTCGGGATC CTAAGTAGAA AGAAAAGTGG CATCCACCGG
```

f.   The restriction digest does validate the conclusions of parts (c) and (d). First the restriction patterns clearly rule out nuclear inheritance because there is no evidence of equal segregation.

All normal individuals have two fragments of size 14.5 and 2.1 kb, while all affected individuals possess a 16.6-kb DNA molecule. Individuals IV-1 and IV-2 exhibit the restriction pattern of their affected mother (and not that of their normal father), while the restriction pattern of individuals IV-3, IV-4, and IV-5 is identical to that of their normal mother (and not that of their affected father). Furthermore, individual III-14 appears to have very little mutant mitochondrial DNA (based on the relative intensity of the bands), which would explain why she was not affected with the disease. However, her children have a large proportion of mutant mitochondrial DNA, and hence are affected.

g.   Wild-type human mitochondrial DNA is a double-stranded circle of 16,569 bp. When the DNA of normal individuals was subjected to the 6-bp cutter used in this experiment, two fragments of sizes 14.5 and 2.1 kb were produced. This indicates that there are two recognition sites for this restriction enzyme. (Remember that a circular molecule that contains *n* sites for restriction enzymes will produce *n* fragments when digested.) The location of one of these two restriction sites can be obtained from part **(e)**. The 6-bp palindromic sequence is situated 10 nucleotides from the beginning of the given DNA region. Therefore, one of the two restriction sites for the 6-bp cutter is at nucleotide 6910 of the wild-type mitochondrial DNA sequence. Because one of the two digestion fragments is 2.1 kb in size, the second restriction site must be about 2.1 kb away from the first. This can be visualized by the following map:

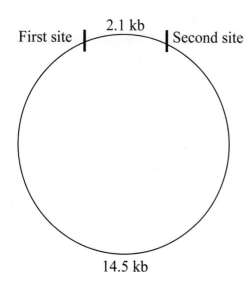

The disease appears to correlate with a mutation that eliminated one of the two restriction sites. The remaining wild-type restriction site was digested by the 6-bp cutter leading to a single molecule of 16.6 kb. It is not clear from this experiment which of the two sites was mutated.

Furthermore, we cannot deduce the exact nature of this mutation. The gel is not sensitive enough to detect the actual size of this mutant 16.6-kb fragment. The intact wild-type mitochondrial DNA is 16,569 bp in length. A base-pair substitution could have eliminated one of the restriction sites and still produced no change in the overall length of the mitochondrial genome. Similarly, a small (one or a few base pairs) deletion or even insertion could eliminate the restriction site without a detectable change in the length of the mitochondrial genome. However, we can deduce that the region encompassing the mutated restriction site must be crucial to mitochondrial function.

h.   The restriction pattern of individual III-8 shows only the 16.6-kb fragment, indicating that she is homoplasmic for mutant mitochondria. Therefore, all of her children would be expected to be affected by the disease.

CHAPTER 20 MUTATIONAL ANALYSIS

Chapter Goals

1. Describe the various classes of mutations and how they differ.
2. Understand the genetic crosses used to generate different types of mutants.
3. Describe the differences between forward and reverse genetics and the advantages of both approaches.
4. Discuss the various approaches used to isolate mutants in forward and reverse genetic approaches.
5. Understand the different ways that enhancers and suppressors are isolated and the ways they may operate.
6. Discuss the advantages of isolating recessive mutations in a mosaic screen and how it is performed.
7. Describe two different ways to phenocopy a gene and the advantages of each.

Key Chapter Concepts

- Mutations naturally occur in populations in a random manner, leading to the vast genetic diversity that we see today. On the other hand, researchers create random and specific mutations within the genes of many organisms to help elucidate the nature of cellular mechanisms and the functions of particular proteins. This chapter focuses on the methods used by researchers to create these mutations and information that can be gained from the phenotypes they generate.

20.1 Types of Mutations

- **Types of Mutations: Loss-of-function** mutations produce proteins that either have less activity (**hypomorph**) than the normal protein or no activity at all (**null** or **amorph**). **Gain-of-function** mutations produce proteins that either have more activity (**hypermorph**) than the normal protein or a new activity distinct from the normal protein's activity (**neomorph**). Any of these mutations can be **conditional mutants** that express the wild-type phenotype under **permissive conditions** but display a mutant phenotype under **restrictive conditions**.
- **Types of Mutagens:** Chemical mutagens, energy mutagens such as X-rays, and transposable element insertions are the three common methods that researchers employ to create random mutations in a gene or genome. The molecular effects of each of these are described in chapter 18 of your textbook. Chemical-induced mutagenesis usually results in single nucleotide changes, thus resulting in missense or nonsense gene mutations. X-ray–induced mutagenesis usually produces DNA breaks that result in some of the chromosomal rearrangements that are discussed in chapter 8 of your textbook. Finally, random insertion of foreign DNA into a genome will induce mutant phenotypes by disrupting open reading frames. Thus, chemical mutagens often create amorphs and hypomorphs via single amino acid changes, X-rays produce amorphs or hypermorphs by breaking open reading frames or changing their expression, and DNA insertion mutagenesis creates amorphs by disrupting open reading frames.

20.2 Forward Genetics

- **Types of Forward Genetic Screens:** In an attempt to determine what genes in an organism perform what functions, researchers randomly mutate a genome, select for mutant phenotypes of the process they wish to learn about, and then determine which gene was mutated to create the phenotype.
  - **Dominant Mutation Screen**: This is a straightforward screen because heterozygous individuals will display the mutant phenotype. Thus, an **F_1 screen** is used in which individuals are randomly mutagenized, crossed to a wild-type organism to dilute out the mutations, and the F_1 offspring screened for the desired mutations.
  - **Recessive Viable Mutation Screen:** Screens for recessive mutations are much more complex than those for dominant mutations in diploid organisms since a homozygous mutant individual must be created in order to observe the mutant phenotype. Thus, an **F_3 screen** is required in which the F_1 progeny are mated to a wild-type individual to produce families of F_2 progeny. Siblings within F_2 families are then mated to create F_3 progeny, some of which will be homozygous for a mutation. These F_3 progeny are then screened for the desired mutant phenotype. To enrich the amount of homozygous mutant F_3 progeny, an individual carrying an easily identifiable recessive mutation is mutagenized and the wild-type individual that the mutants are crossed to contain a

balancer chromosome (to prevent recombination of the mutation away from the recessively marked chromosome) marked with an easily identifiable dominant mutation.

- o **Recessive Lethal Mutation Screen:** An F_3 screen is performed as in the screen for recessive viable mutations; however, the F_3 generation is screened differently. Instead of looking for a mutant phenotype in the F_3 generation, researchers look for F_2 families that fail to produce *any* viable F_3 progeny that do not display the dominant marker on the balancer chromosome since they are the homozygous mutant individuals and they are apparently inviable. The mutation is then collected from the heterozygous F_2 individuals.
- o **Transposable Element Insertion Screen:** A transposable element is a piece of DNA that can insert (with the help of enzymes) into random places in a genome. If it inserts in the middle of a gene, it usually disrupts the expression of that gene and a mutant phenotype can result. In these screens, transposable elements are allowed to insert randomly, homozygous mutants are then generated as in recessive screens, and the mutants are screened for the desired phenotype. The disrupted gene is then identified by sequencing since primers can be annealed to the transposable element that is inserted near or within the gene.

- **Methods to Screen for Mutants:** The easiest system available for isolating the desired mutants in a genetic screen from the wild-type individuals or other unrelated mutants is through **selection** which uses clever methods to eliminate all of the individuals except the mutants that are of interest. This strategy, however, is not always feasible, and other means must be employed for isolating the desired mutants. In some cases, **detection** methods are used to simplify the identification of the desired mutants by causing them to produce a visible marker such as a blue pigment.

20.3 Reverse Genetics

- **Types of Reverse Genetic Screens:** The identity of a significant number of genes in many organisms has now been identified through forward genetics and genome sequencing. Therefore, the most popular genetic screening currently involves **reverse genetics**, a method for the identification of the function of a specific gene. In reverse genetics, a specific gene is mutated and the resulting mutant phenotype demonstrates something about the function of that gene. This is the opposite of forward genetics in which a mutant phenotype is formed and then the gene that produces it is identified.
    - o **Targeted Screening**: In this type of screen, a cloned gene is randomly or site-specifically mutated in vitro. The mutant gene is then reintroduced into the organism to either replace the genomic versions of the gene or to express an extra copy of the gene.
    - o **Deletion Screening:** This screen uses pseudodominance to circumvent the necessity of reintroducing a mutant version of a gene and creating a homozygous mutant in order to identify recessive mutant phenotypes. This is accomplished by randomly mutagenizing the genome of an organism that is heterozygous for a deletion spanning the gene you wish to screen. The researcher can then screen for recessive mutations that cause a phenotype by pseudodominance. Recessive mutations in any other gene will not display a phenotype. The downfall of this type of screening is that dominant mutations in other genes *will* be found, so a method for eliminating them must be considered.
    - o **TILLING:** This is a second method for screening for mutant phenotypes resulting from mutations in a specific gene. In this type of screen, DNA is isolated from dispensable tissue of mutated individuals and the gene of interest is sequenced. Only individuals whose DNA show that a mutation is present in the gene of interest are crossed to produce homozygous mutants.
    - o **Conditional Screening:** In this screen, a promoter that is able to be regulated by a drug is cloned in front of a gene of interest and that recombinant gene is then reintroduced into the organism. The expression of that gene can then be "turned on" or "turned off" at will by including the regulating drug in the food eaten by the organism. This is especially beneficial for the determination of the function of genes essential for development of an embryo. In this situation, the gene can be expressed until the embryo matures and is born. It can then be "turned off" and the function of the gene's product determined by the phenotype observed.

20.4 Advanced Genetic Analysis

- **Suppressors and Enhancers:** In order to determine the function of a protein, it is usually helpful to determine which other proteins in the cell either bind to it or participate in a similar function. **Suppressor** and **enhancer** screens help to identify such proteins. In these screens, mutations are sought in other genes that either suppress or enhance a hypomorphic phenotype produced by a mutant gene of interest. The mutations usually occur in the genes of proteins that either directly bind to the protein of interest or participate in a common biochemical pathway. Recessive suppressors and enhancers are identified in an F_3 screen, while dominant suppressors or enhancers are easily identified in an F_1 screen. In a version of these screens called a sensitized screen, only a *slight* suppression or enhancement of the phenotype is necessary. Thus, isolated mutations are usually dominant and are very easy to select.

- **Mosaic Organisms:** Recessive mutations of some genes are inviable in the homozygous state making it difficult to ascertain the function of the protein when a null mutant cannot be born. Therefore, genetic methods have been designed that create regions (called patches) of tissue with homozygosity in a living heterozygous individual. The organism can survive since it is mostly heterozygous, and the phenotype of the patches can be examined to determine the function that the protein of interest normally provides. The mosaic individuals are created by promoting *mitotic* recombination randomly via radiation or specifically by expressing a recombinase enzyme in specific tissues.

20.5 Phenocopying

- **Phenocopying:** Two other methods commonly used for the study of cells or tissue that lack a particular protein without creating a homozygous mutant individual are RNA interference and chemical genetics. These methods help to copy the phenotypes of mutations without altering the genome. In **RNA interference**, small double-stranded RNA that is complementary to the mRNA of the gene to be studied is synthesized and injected into cells or tissue. These RNAs, like endogenous miRNAs, use cellular proteins to reduce the expression of the complementary gene. In the second method, small chemical molecules are used to either activate or inhibit the activity of the protein of interest. Chemicals that affect the activity of a particular protein can be found by testing libraries of compounds for one that binds to the protein. An alternative approach to chemical genetics is to test the libraries of compounds on cells and screen for one that elicits a desired phenotype. The protein being affected by the chemical is then determined.

Key Terms

- missense mutation
- nonsense mutation
- frameshift mutation
- conditional mutants
- permissive condition
- restrictive condition
- nutritional mutants
- temperature-sensitive mutants
- null allele
- amorph
- hypomorph
- loss-of-function
- hypermorphic allele

- gain-of-function alleles
- neomorphic allele
- forward genetics
- reverse genetics
- saturation
- F_1 screen
- F_3 screen
- balancer chromosomes
- marker
- inverse circular PCR
- genetic screen
- detection
- enhancer trap screens
- dominant-negative mutations

- TILLING
- Tet-Off system
- Tet-On system
- suppressor
- enhancer
- sensitized screen
- mosaic
- site-specific recombination
- RNAi (RNA interference)
- short interfering RNA (siRNA)
- morpholinos

Understanding the Key Concepts
Use the key terms or parts of their definitions to complete the following sentences.

In (1)_____ genetics, researchers often generate random mutations in the genome of an organism and then select individuals with specific mutant phenotypes. The identity of the mutated gene is then determined, informing the researcher of what genes are involved in a particular cellular process. The specific role of the gene product from the identified gene is then determined through (2)_____ genetics which involves creating specific mutations in that gene and then searching for proteins that interact with it or phenotypes that result from the mutations. A particularly helpful type of mutation is a (3)_____ mutation that allows organisms to grow under (4)_____ conditions but not under (5)_____ conditions. One type of this mutation is the (6)_____ mutant which allows researchers to switch the gene product from an active state to an inactive state simply by switching the growth temperature. These mutants allow researchers to analyze the phenotype immediately before and after the temperature shift, thus exposing acute effects of the mutation and limiting all other variables.

Several methods are used to induce mutations in genetics. (7)_____ such as EMS and ENU generally create (8)_____ changes that result in missense or nonsense mutations, while (9)_____ such as X-rays generally induce (10)_____ that lead to (11)_____. Another type of mutagenesis strategy involves allowing (12)_____ to insert into random locations in the genome, thus on random occasions, inactivating nearby genes.

Once random mutagenesis has been performed, the desired mutant progeny must be identified. This can be performed through either a (13)_____ screen or an improved detection screen, the former being more helpful since less progeny will need to be inspected. To isolate dominant mutations, an (14)_____ screen can be performed, while an (15)_____ screen is required to identify recessive mutations that produce the desired mutant phenotype. The latter screen for recessive mutations requires the mating of (16)_____ in order to create individuals that are homozygous for the recessive mutation. If the desired phenotype is embryonic lethality, then the mutation must be recovered from the heterozygous F_2 individuals that were mated to create the lethal progeny. In order to simplify the creation of homozygous recessive progeny needed for this screen, a recessive mutation that acts as a (17)_____ can be included on the mutagenized chromosome. To ensure that this mutation remains on the mutagenized chromosome, a (18)_____ chromosome is included in the nonmutagenized individuals.

Alternative genetic methods utilize mutation of the gene of interest and then study the phenotypes produced. Often this requires the generation of recombinant organisms that contain the mutated gene. The recombinant organism can be engineered so that the expression of the gene can be regulated using the (19)_____ or the (20)_____ system, or it can be engineered so that the mutant is always expressed. If the mutation is dominant, the phenotype of the mutant can then be assessed. However, if the mutation is recessive, the endogenous copies of the gene will need to be replaced by the mutant in order to create a phenotype. Once a mutant organism is generated, (21)_____ or (22)_____ screens that reduce or exacerbate the phenotype of the mutation can be performed in order to find other genes whose gene products are related to the gene of interest. An alternative approach to studying the function of a gene is to use (23)_____ to reduce the activity of an endogenous protein or (24)_____ to reduce the expression of an endogenous gene.

Figure Analysis
Use the indicated figure or table from the textbook to answer the questions that follow.

1. **Figure 20.10**

   (a) What type of mutations are being identified in this F_3 screen?

   (b) Mutations created on the male chromosomes are designated with an "m." Why don't all of the F_1 progeny contain such a mutation?

   (c) Which individuals will display the desired mutant phenotype?

   (d) Why are siblings mated in this screen?

2. **Figure 20.11**
   (a) In this figure, a modified version of a recessive genetic screen is depicted. This screen utilizes a balancer chromosome, a dominant curly-wing mutation, and a recessive cinnabar-eye mutation? What is the purpose of each of these components?

   Balancer chromosome:

   Curly-wing mutation:

   Cinnabar-eye mutation:

   (b) Why are the straight-winged F_1 progeny discarded?

   (c) Why are the wild-type eye F_2 progeny discarded?

   (d) Why are the curly-winged F_3 progeny discarded?

   (e) Which individuals are screened for the desired mutations?

3. **Figure 20.14**
   (a) What is the goal of performing the PCR amplification outlined in figure 20.14?

   (b) Why does only one circular piece of DNA produce a PCR product?

   (c) What is the PCR product used for?

4. **Figure 20.30**
   (a) What type of cell division is outlined in figure 20.30 (mitosis or meiosis), and at what stage of development is this occurring?

   (b) Describe how the recombination event depicted in steps 2 and 3 occurs.

   (c) In step 4, in addition to the two homozygous daughter cells pictured, what other daughter cells are possible? What determines which of the two sets (those pictured or the other possibility) results?

General Questions

Q20-1. Discuss the differences between amorphic, hypomorphic, hypermorphic, and neomorphic mutations.

Q20-2. In general terms, describe how the method employed in forward genetics is different than the method employed in reverse genetics.

Q20-3. Why are recessive mutations more difficult to identify in diploid organisms than dominant mutations?

Q20-4. In performing genetic screens, what is the important advantage of creating a *selection* screen instead of an improved *detection* screen?

Q20-5. What is the purpose of including a red-eye-color allele in an FLP–FRT system for the generation of mosaic fruit flies with patches of their eyes that are homozygous for a lethal mutation?

Multiple Choice

For each of the following, circle the letter of the choice that most appropriately answers the question.

1.) Which of the following methods of mutagenesis is most likely to produce chromosomal rearrangements?
    a. ethyl methanesulfonate (EMS) treatment
    b. *N*-ethyl-*N*-nitrosourea treatment
    c. X-ray treatment
    d. transposable element insertion
    e. All of the above are equally likely to
        produce chromosomal rearrangements.

2.) Which of the following is most likely to produce a gain-of-function mutation?
    a. chromosomal translocation that moves a gene in close proximity to a very active promoter
    b. chromosomal translocation whose breakpoint is in the middle of an open reading frame
    c. nonsense mutation
    d. missense mutation
    e. temperature-sensitive mutation

3.) Why is a balancer chromosome often used in recessive mutation screens?
    a. It marks the chromosome that contains the mutation.
    b. It marks the chromosome that does not contain the mutation.
    c. It promotes mutagenesis by creating chromosomal breaks.
    d. It prevents the mutation from recombining off of the marked chromosome.
    e. It allows for selection of mutant phenotypes under restrictive conditions.

4.) In addition to the desired mutations, which of the following types of mutations would also be discovered in a deletion screen?
    a. recessive mutations in other genes
    b. dominant mutations in other genes
    c. deletion mutations in other genes
    d. DNA insertions in other genes
    e. null mutations of other genes

5.) In a TILLING experiment, how can you be certain that all of the individuals that are screened for a phenotype actually do contain a mutation in the gene of interest?
    a. Individuals not containing a mutation will be dead due to selection in restrictive conditions.
    b. Individuals containing a mutation will produce a blue color that improves detection.
    c. The gene is first sequenced from many individuals, and only individuals with a mutation in that gene are pursued.
    d. The mutations identified in a TILLING experiment are always dominant.
    e. A dominant mutation on the mutagenized chromosome indicates the presence of an induced mutation.

6.) Genetic screens for (i)_____ mutations can be accomplished with F_1 screens, while genetic screens for (ii)_____ mutations require F_3 screens.
    a. i.) recessive   ii.) dominant
    b. i.) dominant   ii.) recessive
    c. i.) suppressor ii.) enhancer
    d. i.) lethal   ii.) viable
    e. i.) viable   ii.) lethal

7.) Which method is used to create mosaic individuals that have patches of tissue that are homozygous for a lethal mutation?
    a. Mitotic recombination is induced after birth in individuals *heterozygous* for the lethal mutation through the use of X-rays or other high-energy radiation.
    b. An enzyme that catalyzes a specific mitotic recombination is induced in developing heterozygous individuals only in the tissue to be studied.
    c. Mitotic recombination is induced after birth in *homozygous* wild-type individuals through the use of X-rays or other high-energy radiation.
    d. Heterozygous F_2 siblings are mated with each other to create homozygous individuals.
    e. Both a and b

8.) Mutation of which of the following proteins is likely to suppress the phenotype of a mutation in the gene that produces protein A?
- a. a protein that directly binds to protein A
- b. a protein that functions in the *same* biochemical pathway as protein A
- c. a protein that functions in a *different* biochemical pathway as protein A
- d. a protein that is expressed in different tissues than those in which protein A is expressed
- e. Both a and b

9.) Which of the following is a difference between siRNAs and miRNAs?
- a. siRNAs can silence *transcription* of a complementary gene, while miRNAs cannot.
- b. siRNAs utilize the RISC proteins, while miRNAs utilize the Dicer protein instead.

- c. miRNAs can silence *transcription* of a complementary gene, while siRNAs cannot.
- d. siRNAs utilize the Argonaute protein, while miRNAs utilize the Dicer protein instead.
- e. Both a and b

10.) Phenocopying is advantageous over genetic screens that create homozygous null mutants for which of the following reasons?
- a. Phenocopying can usually create a completely null phenotype, while mutagenesis cannot.
- b. Phenocopying can create a greater spectrum of phenotypes than mutagenesis can.
- c. Some organisms are not very amenable to mutagenic screens or gene replacements.
- d. Phenocopying creates permanent phenotypes that are continually inherited.
- e. Phenocopying can be performed on living organisms, while mutagenesis cannot.

Practice Problems

P20-1. A Southern blot was performed to determine what type of chromosomal aberration occurred to produce a particular mutant phenotype that was produced by X-ray mutagenesis. The following Southern blot was obtained using the probes that anneal to the genomic region depicted. Describe the chromosomal aberration that is present in mutant 1 and in mutant 2.

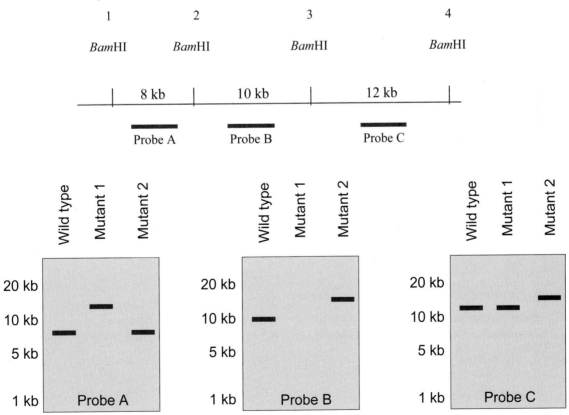

P20-2. A forward genetic screen was performed to isolate recessive lethal mutations in *Drosophila*. Male flies with cinnabar eyes were randomly mutagenized with the mutagen EMS. These males were then mated to female flies heterozygous for the SM1 balancer chromosome. The F₃ progeny shown in the following table were produced from eight F₂ sibling matings. Which flies should be kept in order to isolate the mutations that cause the desired mutations?

Mutant	Curly Wings Cinnabar Eyes	Straight Wings Cinnabar Eyes
1	31	9
2	29	12
3	57	17
4	41	0
5	81	26
6	58	0
7	19	7
8	21	0

P20-3. Design a *selection* method to select for the following desired mutants.
(*Hint 1*: Remember that a selection screen is different from a detection screen. In a selection screen, the desired mutant phenotypes are collected without having to observe every individual.)

      a. A mutation in a bacterial gene that creates a new gene whose protein product can degrade a compound called Toxo that is normally toxic to bacteria.

      b. A mutation in flightless fruit flies that restores their ability to fly.

P20-4. Homozygosity for a null allele of a gene known as *abc1* was recently discovered to be lethal due to a failure in embryonic development. Describe two experiments that could be performed to elucidate the molecular functions that the Abc1 protein performs in cells. The first experiment should be designed to identify other proteins that function with Abc1 in cells, and the second experiment should be designed to determine what happens inside cells lacking the Abc1 protein.

Assessing Your Knowledge

Understanding the Key Concepts—Answers

1.) forward; 2.) reverse; 3.) conditional; 4.) permissive; 5.) restrictive; 6.) temperature-sensitive; 7.) Chemicals; 8.) base or nucleotide; 9.) energy; 10.) DNA breaks; 11.) chromosome rearrangements; 12.) transposable elements; 13.) selection; 14) F₁; 15.) F₃; 16.) F₂ siblings; 17.) marker; 18.) balancer; 19.) Tet-On; 20.) Tet-Off; 21.) suppressor; 22.) enhancer; 23.) chemical genetics; 24.) RNA interference.

Figure Analysis—Answers

1a.) Recessive viable mutations are being identified; b.) The mutagen administered to the flies causes mutations *randomly*. Therefore, some individuals will have more than one mutation while other individuals will not have any.

c.) The F_3 individuals in the lower right corner: $\dfrac{+\ m\ +}{+\ m\ +}$

d.) Mating of siblings is the only way to generate individuals that are *homozygous* for the mutation. Homozygosity is necessary because the desired mutations are recessive.

2a.) **Balancer chromosome:** The balancer chromosome prevents the mutation from recombining off of the "marked" chromosome and onto the unmarked chromosome. Thus, it ensures that the "marked" chromosome contains the mutation. **Curly-wing mutation:** This mutation is dominant and thus indicates which of the progeny contain the balancer chromosome. It serves as a marker for the balancer chromosome. **Cinnabar-eye mutation:** This mutation is recessive and "marks" the mutagenized chromosome. Cinnabar-eyed flies must either contain both the balancer chromosome and a mutagenized chromosome (if they also have the curly-wing mutation) or two copies of the mutagenized chromosome. Thus, this marker helps to identify which individuals should be screened; b.) The straight-winged F_1 progeny are discarded because they did not receive the balancer chromosome. Therefore, even though some of them have a desirable mutation, it will be difficult to create homozygous individuals since recombination can move the mutation off of the marked chromosome; c.) The wild-type eye F_2 progeny are discarded because they either do not contain a mutagenized chromosome or they do not contain a balancer chromosome. The only F_2 progeny that do have a mutagenized chromosome paired with a balancer chromosome will have cinnabar eyes since one recessive allele for this trait is on the mutagenized chromosome and one is on the balancer chromosome. Again, the balancer chromosome is necessary in order to prevent recombination of the mutant off of the marked chromosome; d.) The curly-winged F_3 progeny are discarded because they are *not* homozygous for the mutation. They must contain the balancer chromosome since that is where the curly-wing mutation resides; e.) Straight-winged F_3 progeny are examined for the desired mutation since they are the only flies that can be homozygous for a mutation. It is important to note, however, that not all of them will be homozygous for a mutation since some mutagenized chromosomes did not receive any mutations.

3a.) The goal is to identify the location in the genome where a transposable element inserted, causing a particular phenotype. Amplifying the DNA by PCR creates enough DNA to allow it to be sequenced. Determining this sequence and what gene(s) surround the insertion will suggest which gene normally prevents the phenotype observed; b.) Only circular pieces of DNA that contain the DNA of the transposable element will produce a PCR product. This is because the primers for the PCR reaction are complementary only to the transposable element DNA; c.) The PCR product is sequenced using primers that anneal to the transposable element in order to determine what genes are disrupted by the insertion.

4a.) Mitosis. This mitosis is occurring during development of the organism after the essential function of the gene of interest has been performed; b.) Energy such as X-rays or irradiation, cause breaks in the DNA of nonsister chromatids. Enzymes rejoin the free ends of the chromatids, and by random chance they are rejoined to the wrong chromatids; c.) Two heterozygous daughter cells identical to the parental cells can also be generated. Whether these are produced or the recombinant daughter cells are produced depends on how the homologous chromosomes align during metaphase.

General Questions—Answers

Q20-1. Amorphic mutations express gene products that have *no* activity, while hypomorphic mutations express gene products that have *reduced* activity but still retain some activity. Hypermorphic mutations express gene products that have *more* activity than the wild-type allele, and neomorphic mutations express gene products with a *new* function.

Q20-2. In forward genetics, the genes that carry out a particular process are identified. Mutagenesis is performed in order to randomly mutate genes in the genome of the selected organism, and mutants that have a desired phenotype are selected and the mutant genes identified. In contrast, reverse genetics seeks to identify the cellular process that is carried out by a particular gene. Mutagenesis is performed on the specific gene, and the resulting phenotype is determined.

Q20-3. Recessive mutations must be in a homozygous state in order to produce a phenotype. Thus, an F_3 screen is used; many matings are required to produce these individuals that are homozygous for each mutation. Dominant mutations will display their phenotype in the F_1 generation circumventing the need for all of the matings required for F_3 screens.

Q20-4. A selection screen will only allow the desired mutant phenotypes to be seen, while detection screens require that all individuals (even those who do not contain any mutations) be observed. If a selection screen is devised, the work involved in finding and recovering the desired mutants is dramatically reduced.

Q20-5. The red-eye-color allele will mark the patches of the fly eye that are not homozygous for the lethal mutation being investigated. The white patches in the eye will, however, be homozygous for the lethal mutation being investigated. Thus, the eye color allele makes it easy to know where to look in the eye for a mutant phenotype resulting from the mutation being investigated.

Multiple Choice—Answers

1.) c; 2.) a; 3.) d; 4.) b; 5.) c; 6.) b; 7.) e; 8.) e; 9.) a; 10.) c.

Practice Problems—Answers

P20-1. Mutant 1: Chromosomal deletion. The left break point must lie somewhere between the left end of probe A and the second *Bam*HI site. The right break point must lie somewhere between the left end of probe B and the third *Bam*HI site.
Mutant 2: Chromosomal deletion; The left break point must lie somewhere between the left end of probe B and the third *Bam*HI site. The right break point must lie somewhere between the third BamHI site and the right end of probe C.

P20-2. The heterozygous F_2 flies from mutants 4, 6, and 8 should be kept in order to isolate a recessive lethal mutation. Since no straight-winged, cinnabar-eyed flies were born, the mutation present in these flies must be lethal in the homozygous state.

P20-3. a.) The mutagenized bacteria should be grown on a medium containing Toxo. The wild-type bacteria will die, and only the bacteria containing the desired mutation will live.
b.) The mutagenized flies should be placed in a container with an opening near the top. The opening should lead into a second container through a one-way funnel. Only the flies that can fly will be able to get up to the opening and across to the second container. Thus, the desired mutants will accumulate in the second container.

P20-4. Experiment 1: A forward genetic screen can be used to search for a suppressor mutation that suppresses the lethal phenotype caused by *abc1* mutations. The proteins whose mutant genes suppress *abc1* mutations are likely to interact or function with the Abc1 protein.
Experiment 2: A mosaic screen can be performed to generate live organisms that contain patches that are homozygous for the lethal *abc1* mutations. The phenotype of these patches can then be determined in order to discover the normal function of the Abc1 protein. Alternatively, RNAi can be used in cultured cells or a living organism to reduce the expression of the *abc1* gene. The resulting phenotype may then indicate the normal function of the Abc1 protein.

20

Mutational Analysis

CHAPTER SUMMARY QUESTIONS

1. 1. g, 2. d, 3. e, 4. a, 5. h, 6. j, 7. i, 8. c, 9. b, 10. f.

2. Generation of specific mutants allows scientists to study the detailed structure of the gene and how it functions, such as how its expression is regulated. Mutations can also elucidate information about the encoded protein, such as its function in the cell and how it may interact with other proteins in a biological process.

3. A molecular null allele does not produce any functional gene (protein) product. This molecular null allele would not transcribe any mRNA or either the transcribed mRNA would not translate a protein or the translated protein would be completely nonfunctional. A genetic null allele produces a phenotype that corresponds to a complete absence of any functional gene product. However, low levels of functional protein may be present if they are insufficient to alter the null phenotype. In this case, the genetic null allele would not be a molecular null allele.

4. Forward genetics is mutant first because a mutant is randomly isolated and then used to identify the gene that has been altered. Reverse genetics is gene first because a specific gene is identified to be altered and then the resulting phenotype of the mutant is determined.

5. Short generation time, large numbers of offspring, relatively easy screens for mutants, pre-existing and genetic knowledge about the organism.

6. Multiple inversions must prevent crossing over between chromosomes during meiosis. They contain a dominant mutation that also has a recessive lethal phenotype, which allows the identification of individuals that either contain the balancer chromosome (dominant phenotype) or lack it (wild-type phenotype). Individuals that are homozygous for the balancer chromosome would die due to the recessive lethal phenotype. The balancer chromosome also has recessive visible mutations that are also present on the chromosome to be mutagenized. This allows individuals that contain both the balancer chromosome and the mutagenized chromosome to be easily identified.

7.   Yes, because FM7 includes the recessive *w* mutant allele. Animals that contain an insertion of the transposon carrying the w^+ allele would be easily identified by red patches in their eyes.

8.   A selection screen would be preferred because it will eliminate a large number of nonmutagenized individuals. This will reduce the total number of viable individuals that will have to be analyzed for the desired mutant.

9.   The major advantage of a *P*-element mutagenesis is that the transposon insertion that causes the mutant phenotype can also be used as a tag to identify and clone the affected gene. It is also more flexible—there is no need to use multiple screens with separate balancers for each chromosome.

10. The *P* element must contain a weak promoter that controls the expression of the reporter gene within the *P* element. This weak promoter is usually insufficient to transcribe the reporter gene at a level that is high enough to observe its expression phenotypically. However, when the *P* element inserts near an enhancer element, the enhancer can increase the expression from the weak promoter to phenotypically observe the expression of the reporter gene. This enhancer presumably also controls the expression of an endogenous promoter in the genome. This enhancer should also activate the expression of both the reporter and endogenous genes simultaneously in the same tissues.

11. Expression of the dominant negative allele is either on or off in *all* tissues of the organism in response to tetracycline. Addition of tetracycline will then have the same effect on the expression of the dominant-negative transgene in all tissues. While addition of tetracycline will lead to the expression of the dominant-negative transgene in the desired tissue to generate the potential phenotype, it will also be expressed in other tissues that could lead to undesirable side effects.

12. A neomorph is defined as a mutation with a function that was not previously present. Dominant-negative mutations are one type of neomorph—the new function is that the mutant protein interacts with wild-type protein to prevent the normal wild-type activity. Other neomorphs may not interact with the wild-type protein at all.

13. In TILLING, mutants are identified based on a DNA sequence. In forward genetics, mutants are identified based on a phenotype. TILLING may allow identification of unexpected phenotypes, but only in known genes. Traditional forward genetics may identify unknown genes associated with a phenotype but may not identify unexpected functions of those genes.

14. Genetic mosaics will have tissues with more than one different genotype. This could occur due to mutations in the somatic tissues that occur during development of the individual (such as errors in DNA synthesis during development) or as a

result of mutations in the somatic tissues in the adult (such as induced mutations of skin cells that lead to cancerous tumors).

15. Both techniques allow down-regulation of gene expression after transcription. In antisense RNA technology, a single-stranded RNA complementary to the gene in question must be present in the tissue of interest. In siRNA techniques, the RNA produced can be much smaller than an antisense RNA, and indirectly triggers endogenous cellular mechanisms that either block the translation of the targeted mRNA, degrade the targeted mRNA, or repress the transcription of the targeted mRNA.

EXERCISES AND PROBLEMS

16. EMS and 5-bromouracil would both be useful to generate hypomorphic and null alleles because they will generate single-base changes, which can result in both null (nonsense and missense mutations) and hypomorphic (missense) mutations. X-rays are unlikely to be useful because they tend to generate DNA rearrangements such as deletions, which usually produce null alleles, but not hypomorphs.

17. EMS generates single base-substitutions. Since the protein product has a greater molecular weight than expected, we can anticipate that the mutation interferes with the stop codon. A substitution changing the stop codon to one encoding an amino acid will allow translation through the codon that corresponded to the stop codon, leading to a larger protein than usual. The sequencing gel confirms that a T-to-C transition has occurred in the *chk* mutant and that depending on the reading frame, this mutation could change a TAG stop codon into a CAG codon, resulting in read-through.

18. Northern blots would allow analysis of the mRNAs that are transcribed from the gene and could potentially identify transcripts of incorrect length due to deletions or splicing errors, mutations that affect the amount of gene transcription, or mutations that affect when and where the gene is transcribed. Western blots allow the analysis of proteins directly. Western blotting used with denaturing gels would allow analysis of protein size, the amount of protein, and where and when the protein is expressed. Western blotting can be combined with other types of gels, such as isoelectric focusing gels, to analyze protein characteristics such as folding and ionic character. DNA sequencing can identify base changes directly, and those sequence changes can allow scientists to predict an effect on the protein.

Allele 1 produces a normal band on the genomic Southern blot, indicating that it is not due to a large deletion. However, the mRNA and protein are absent on the Northern and Western blots, respectively. Examination of the DNA sequence reveals that allele 1 contains an A-to-G substitution in the promoter region. This substitution prevents transcription of the mRNA of this gene.

Allele 2 has a larger than normal mRNA on the Northern blot and a smaller than normal protein on the Western blot. The DNA sequence of the promoter/mRNA start and exon 2 are both wild type. The most likely defect in this allele is a splice site mutation. The incorrect splicing in the mutant results in a larger than normal mRNA but presumably introduces a stop codon that produces a truncated protein.

Allele 3 appears normal on the Southern, Northern, and Western blots. However, there is a point mutation near the beginning of exon 2 (G-to-C mutation). This must be a missense mutation because it does not affect the size of the mRNA or protein but does result in a mutant phenotype.

Allele 4 produces a normal transcript on the Northern blot but a smaller than normal protein on the Western blot. The DNA sequence of the promoter/mRNA start and exon 2 are both normal. Allele 4 most likely contains a nonsense mutation downstream of the sequenced region.

Allele 5 produces a normal transcript but no protein on the Western blot. Inspection of the DNA sequence reveals a G-to-T substitution in exon 2. The lack of protein suggests that this is a nonsense mutation. Indeed, if the reading frame normally results in translation of GAA (glutamate), then the G-to-T substitution would change this codon to TAA (stop).

19. **a.** Both phenotypes are affected by the same mutation. In the F_2 generation, 517 wild-type offspring and 498 offspring with both reduced bristle number and eye abnormalities are observed. If different mutations caused the two phenotypes, then they would have segregated in the F_2 generation.

   b. The mutation here is dominant. The F_1 offspring has both mutant phenotypes; if the mutation were recessive, the F_1 offspring would have been wild type in appearance, and the mutant phenotype would not have appeared until two F_1 heterozygotes were mated to each other. Furthermore, mating of a heterozygous F_1 fly with a wild-type female results in a 1:1 ratio of wild-type:mutant, which corresponds to the predicted ratio for an autosomal dominant mutation.

20.   The expected genotypes are as follows:

Among the surviving offspring, one-third will have curly wings and cinnabar eyes, one-third will be wild type for both wings and eyes, and one-third will have curly wings and wild-type eyes.

21. The minimum number of inversions required to generate this balance chromosome is 3. Ideally, a balancer chromosome should contain a large paracentric inversion as well as a pericentric inversion on each of the two chromosome arms.

22. Because the snout, ears, and extremities of the rabbit are normally pigmented, and you expect these areas to be cooler than the body core, you can hypothesize that the permissive temperature is slightly cooler than the rabbit's normal body temperature. If this is the case, you would expect the patch with the cold pack to grow back pigmented fur like the cooler extremities, while the untreated patch would grow back white like the rest of the body.

23.

Genotype	Expected Phenotype
AA	Wild type
Aa	Wild type
aa	Albino
Aa^h	Wild type
aa^h	Himalayan
$a^h a^h$	Himalayan

24. **a.** A deletion heterozygote is heterozygous for a mutant allele and for a known deletion. It is useful for characterization of the mutant allele. For example, if a homozygous mutant has the same phenotype as the deletion heterozygote, then the mutation is a null allele. If the phenotype of the homozygous mutant is more severe than that of the deletion heterozygote, then the mutation is a hypomorphic allele. If the phenotype of the homozygous mutant is less severe than that of the deletion heterozygote, then the mutation is a hypermorphic or neomorphic allele (see figure 20.3).

b. For mutant 1, the phenotype of the homozygous mutant is more severe than that of the deletion heterozygote and is identical to the phenotype of the *mutant 1/curled* heterozygote. Therefore, mutant 1 is a hypomorphic allele of the *curled* gene. For mutant 2, the phenotype of the *mutant 2/mutant 2* homozygote, the *mutant 2/curled* heterozygote, and the *mutant 2/deletion* heterozygote are identical. Mutant 2 is a null allele of *curled*. Mutant 3 complements the *curled* mutation and the deletion of *curled*; therefore, mutant 3 affects a different gene. In mutant 4, the phenotype of the *mutant 4/mutant 4* homozygote and the *mutant 4/curled* heterozygote are less severe than the phenotype of the deletion heterozyote, which has very curly wings. Mutant 4 is either a hypomorphic or neomorphic allele of *curled*. A *mutant 4/mutant 4/+* fly would have to be generated to distinguish between these two possibilities.

25. **a.** A conservative mutation may still allow normal folding but will not be recognized by the kinase (phosphorylation enzyme), so this is likely to be a genetic null. Although the protein will be present, it will likely not be able to be activated.

b. A nonconservative mutation will probably disrupt folding of at least the hinge region. If only the hinge region is affected, the mutant could be a hypermorph—constitutively active. If the folding error also affects the active site, the mutation could be a hypomorph or a genetic null.

c. A missense mutation in the active site will probably reduce or eliminate protein function, creating a hypomorphic or null mutant.

d. Deletion of the regulatory region will lead to a constitutively active protein—a hypermorphic allele.

26. Mutations expected to enhance the Ras hypomorph would be additional mutation in Ras itself, further limiting the signaling. Hypomorphic mutations in the Raf protein could further limit Ras signaling, by disrupting the normal interaction between Ras and Raf. Suppression of the Ras mutation could be accomplished by hypermorphic mutations in Raf, such that Raf is more easily activated by Ras, or even may be active in the absence of Ras activity.

27. **a.** In the F_2 generation, the flies will be heterozygous for the mutations that were generated. Therefore, dominant mutations will be apparent in F_2 flies and can be directly studied at that generation. Identification of recessive alleles requires at least crosses between F_2 siblings to produce homozygous mutant progeny. Use of balancer chromosomes simplifies identification of recessive and lethal recessive alleles.

b. If a deletion exists for the gene that you are interested in studying, the potential mutants can be crossed to the deletion. For example, if either the mutagenized male or the F_1 progeny are crossed to individuals with the deletion, the resulting progeny will be deletion heterozygotes. The difference is that when the mutagenized male is crossed to females with the deletion, a small percentage of the F_1 progeny will be deletion heterozygotes because only a small percentage of the mutagenized male's sperm will carry the desired mutation. In contrast, if the F_1 progeny are crossed to an individual with the deletion, then 50% of the progeny will be deletion heterozygotes if the F_1 individual has the desired mutation.

28. The *P* element contains a wild-type copy of the *w* gene (w^+). This allele encodes a *trans*-acting factor that, if present, will allow normal pigment production of eye cells regardless of its chromosomal location. If the starting *Drosophila* strain contains the w^- allele on the X chromosome, it will exhibit the recessive white-eyed phenotype. The presence of the $P[w^+]$ element, which inserts anywhere in the *Drosophila* genome, will produce a pigmented eye due to the presence of the w^+ allele on the *P* element.

29. **a.** A chimeric mouse will exhibit the phenotype that is associated with the genotypes of both the blastocyst and the ES cells. In this example, the chimeric mouse would be albino with patches of agouti coloration. The albino blastocyst should incorporate the ES cells randomly; offspring from the ES cells will create agouti coloration if located in the skin.

b. To generate mice that are homozygous for the YFG knockout, you would cross chimeric mice to an albino mouse and screen for agouti offspring. These agouti mice will not be chimeric. However, they will be heterozygous for the agouti allele (Aa). Half of the agouti mice will also be heterozygous for the recessive knockout mutation. Remember that the ES cells were homozygous for the wild-type agouti allele but heterozygous for the knockout allele. Unless *YFG* is linked to the coat color gene, they should assort independently. Thus, you would need to identify knockout heterozygotes from the agouti offspring and then cross two of the heterozygous agouti siblings; one-fourth of the progeny will be homozygous for the knockout YFG allele.

c. The probe will hybridize to two bands of the wild-type allele and to one larger band in the knockout allele. Mice 1 and 4 are homozygous for the wild-type allele. Mice 3 and 6 are heterozygous for the wild-type allele and the knockout allele. Mice 2 and 5 are homozygous for the knockout allele.

30. When studying complex processes, or processes that are not well understood, a traditional screen can help identify all the genes involved in the process. For situations where the genes involved are already known, TILLING can help reveal the intricate function of each specific gene by generating multiple alleles (null, hypomorphic, hypermorphic, neomorphic). For example, if you are interested in *Drosophila* wing development, a traditional screen for wingless flies will identify the assortment of genes that are required for the generation of a wing. Assume that you identify the recessive no wings (*nw*) mutant that completely lacks wings. If you then use TILLING to identify additional mutations in the *nw* gene, you may isolate five new alleles based on the DNA sequence. When you look at the phenotypes of these five new alleles, you discover that one is recessive lethal, one is dominant for a small eye (but it has normal wings), one is a recessive no wings phenotype, and two are phenotypically wild type.

31. The most straightforward method would be to make a transgenic animal containing an antisense RNA transgene to your gene that is expressed after the critical embryonic period. This expression could be driven by a fairly broad regulatory element, because the antisense RNA is unlikely to harm the organism after embryonic development. The regulatory elements to be used should be tested with a reporter gene to verify when and in what tissues the antisense RNA can be expected to be produced.

32. Both systems require two different strains of transgenic mice. Tet-On and Tet-Off systems are focused on temporal regulation, while CRE-loxP can generate tissue-specific knockouts.

33. We expect these genes to be recessive mutations, so we will need a detection system for recessive mutations. Our phenotype (detection screen) is sevenless (lack of R7). Mutagenized males are crossed to wild-type females for R7 specification.

Individual F$_1$ flies could contain different mutations, so individual F$_1$ flies are backcrossed to wild-type flies. F$_2$ siblings are interbred, and the F$_3$ offspring are examined for a sevenless phenotype.

34. Flies homozygous for each new mutation are crossed to flies homozygous for a null mutation in sevenless. If the compound heterozygote displays the sevenless phenotype, we expect that mutation to be in the sevenless gene. If the compound heterozygote is wild type with respect to R7 formation, the mutations are in different genes.

35. Mutation 1 affects the start codon and could theoretically result in a null allele. However, a second ATG occurs two codons downstream; use of this start codon could produce a protein missing the first two amino acids, which could have normal function.

Mutation 2 is a deletion of three bases. This mutation would delete one amino acid in the protein. CCGAAATTT (Pro Lys Phe) becomes CCATTT (Pro Phe). Depending on the normal role of the deleted lysine, this could be a hypomorphic, hypermorphic, or neomorphic allele. It is unlikely that it would be a null allele.

Mutation 3 deletes a large segment of the protein. It also changes the reading frame so that a stop codon is introduced immediately after the deletion. This mutation would result in a null allele.

Mutation 4 changes an AAG codon to a TAG stop codon. This would result in a truncated protein, which would most likely be a null mutation.

Mutation 5 affects the normal stop codon, causing read-through and synthesis of a longer than normal protein. Depending on whether a stop codon occurs shortly after this mutation and the effect of the extra amino acids, this mutation could produce a hypomorphic, hypermorphic, or neomorphic protein.

36. Genotypes that would be useful for characterization of this mutant are *fz/fz*, *fz/deletion,* and *fz/fz/+*. These three genotypes would allow characterization of the mutation as a hypomorphic, hypermorphic, neomorphic, or null allele. Appropriate molecular techniques for further characterization are Southern, Northern, and Western blots; PCR; and DNA sequencing.

37. Assuming that the deletion (*del*) uncovers part of the *fz* gene, we will not be able to generate recombinants between the *fz* gene and the deletion. Thus, we must cross the *fz/del* heterozygotes with wild-type flies, which will produce equal numbers of *fz/+* and *+/del* heterozygotes for this experiment. These offspring could then be irradiated at various stages of development to activate mitotic recombination. The irradiated *fz/+* individuals would contain patches of *fz/fz* tissue, which could then be analyzed for the potential function of the Fz protein. The irradiated *+/del* individuals would contain patches of *del/del* tissue, which would either be phenotypically identical to the *fz/fz* patches or more severe.

38. a. This female must be mated to a male that is homozygous for the following second chromosome:

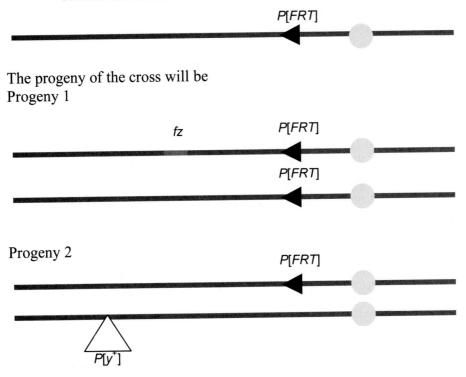

The progeny of the cross will be
Progeny 1

Progeny 2

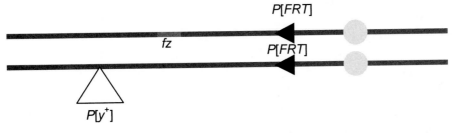

We will then mate progeny 1 males with progeny 2 females. The females will produce recombinant chromosomes that will place the $P[y^+]$ transposon onto the $P[FRT]$ chromosome. The resulting progeny will be

b. The males must supply a source of flippase enzyme that will then mediate the recombination event between the *FRT* elements after DNA replication in the somatic cells. The males (and females) must also be y^- (yellow) on the X chromosome. This will allow the selection of yellow patches on the cuticle that will correspond to mitotic recombinant clones that are *fz/fz*.

39. a. The kinase domain is part of the intracellular domain. This mutant would not have any kinase activity, so it would be inactive. In a wild-type background, this would probably be a hypomorphic allele. If a wild-type receptor dimerized with a mutant receptor, neither monomer would be activated. However, wild-type proteins could still dimerize with each other.

b. The extracellular domain is required for ligand binding. Again, in a wild-type background, this mutation would likely be hypomorphic.

c. This mutant would be constitutively active and would be a hypermorphic allele.

d. This mutant would fail to be activated and would be hypomorphic.

40. Mutant (c) should demonstrate rescue. This mutant will be constitutively active. The other mutations are hypomorphic and will not rescue a null mutation.

CHAPTER INTEGRATION PROBLEM

a. Cross a true-breeding female mellow cat with a true-breeding male spritely cat. Cross the F_1 offspring with one another to generate the F_2 generation. If the gene is autosomal, a 3:1 ratio should be observed, and the dominant allele can be identified.

Perform the reciprocal cross of a true-breeding male mellow cat with a true-breeding female spritely cat, and again generate the F_2 generation. If the same results are obtained as in the first cross, the trait is autosomal. If different results are obtained and there are differences in the male:female ratio in either of these crosses, then the trait is X-linked.

b. Mutagenize cats and identify spritely mutants. Cross the new mutants with the original spritely mutant. Complementation indicates mutations in different genes. Through complementation analysis, the number of genes in the pathway can be identified.

c. Use the following symbols to represent the genes in this cross:
sp, sprite; sp^+, wild type (mellow)
W, Wiskerless; W^+, wild type
T, Tailless; T^+, wild type

The first cross can be represented as
$sp/sp\ W^+/W^+\ T^+/T^+ \times sp^+/sp^+\ W/W\ T/T$

The offspring are heterozygous at all three loci.

In the second cross, a triply heterozygous female is crossed with a homozygous *sprite* male, as follows:

$sp^+/sp\ W^+/W\ T^+/T$ female $\times\ sp/sp\ W^+/W^+\ T^+/T^+$ male

We know that the *Wiskerless* gene is on the third chromosome and the *Tailless* gene is on the fifth chromosome, so they will assort independently. To determine whether the *sprite* gene is on one of these chromosomes, we must examine the offspring from this cross. The male parent will always contribute the three recessive alleles sp, W^+, and T^+. By examining the offspring, we can determine the alleles contributed by the female parent's gamete. The parental allele combinations for the female parent are *sp* W^+ T^+ and sp^+ *W T*. We need to determine which gametes contain recombinant genotypes. The data are summarized in the following table:

Phenotype	Female Gamete Genotype	Number	Recombinant for *sp-W*?	Recombinant for *sp-T*?
Wild type	$sp^+W^+T^+$	4	Yes	Yes
sprite, Wiskerless, Tailless	*sp W T*	5	Yes	Yes
Wiskerless	$sp^+W\,T^+$	6	No	Yes
sprite, Tailless	$sp\,W^+T$	3	No	Yes
Tailless	sp^+W^+T	19	Yes	No
sprite, Wiskerless	$sp\,W\,T^+$	20	Yes	No
sprite	$sp\,W^+\,T^+$	22	No	No
Wiskerless, Tailless	$sp^+\,W\,T$	21	No	No

From the data in this table, we see that there are 4 + 5 + 19 + 20 = 48 offspring that are recombinant for *sprite* and *Wiskerless* and 6 + 3 + 22 + 21 = 52 offspring that are not recombinant for *sprite* and *Wiskerless*. This is very close to the 1:1 ratio that is expected if these genes assort independently. Therefore, we can conclude that *sprite* is not linked to *Wiskerless*.

There are 4 + 5 + 6 + 3 = 18 recombinants for *sprite* and *Tailless* and 19 + 20 + 22 + 21 = 82 offspring that are not recombinant for *sprite* and *Tailless*. The data do *not* fit the 1:1 ratio predicted under independent assortment. We can conclude that *sprite* is linked to *Tailless* and therefore must map to the fifth chromosome. The distance between these genes is (4 + 5 + 6 + 3) / (4 + 5 + 6 + 3 + 19 + 20 + 22 + 21) = 18/100 = 0.18, or 18.0 map units.

CHAPTER 21 DEVELOPMENTAL GENETICS

Chapter Goals

1. Distinguish between determination and differentiation.
2. Understand maternal-effect genes, and describe the roles of cytoplasmic determining factors in pattern formation.
3. Explain the levels at which zygotic gene activity can be regulated.
4. Gain a basic knowledge of the regulatory molecules and their roles in cell–cell signaling during development.
5. Discuss the role of the homeotic genes in development and their conservation between invertebrates and vertebrates.
6. Explain the differences between master regulator genes and the genes required for development of the R7 photoreceptor.
7. Describe the differences between necrosis and apoptosis and the roles of apoptosis in development.

Key Chapter Concepts

21.1 Basic Concepts in Animal Development

- **Gene Expression in a Developing Embryo:** Sexual reproduction involves the fusion of a male-produced gamete (sperm) and a female-produced gamete (egg) to produce a diploid zygote. The egg contributes not only a haploid complement of chromosomes to the zygote as does the sperm, but also the majority of the cytoplasmic material including RNA and proteins expressed from **maternal-effect genes**. The products of these genes are the only RNA and proteins present in the early embryo and are responsible for carrying out all of the initial steps of development. When the rate of cell division slows, transcription of the zygotic genome is induced (**zygotic induction**).

- **Differential Gene Activity:** Development is controlled by the differential expression of genes. In eukaryotic organisms, any nucleus can theoretically give rise to any cell type (**totipotent**). As the embryo develops, most cells become committed to a particular cell fate by the process of **determination**. Each determined cell must then take on the characteristics of its adult fate by the process of **differentiation**. Cell differentiation results in the production of different cell types without permanently changing the DNA.

- **Internal and External Determination Cues:** Cell determination can be affected by **intrinsic** factors that originate from within the cell or by **extrinsic** factors in which signals from neighboring cells lead to determination. The maternal-effect gene products are intrinsic factors that determine the future identity of a cell. As transcription factors or regulators of transcription factors, the maternal-effect gene products control zygotic gene activity. Diffusable proteins act as extrinsic signals that allow cells to communicate with each other. These signals indicate the location of the cell within the embryo and can determine the fate of the cell.

21.2 Pattern Formation During *Drosophila* Embryonic Development

- **Maternal-Effect Factors and Generation of the Initial Axis Polarities:** *Drosophila* development is controlled by genes from maternal cells and genes of the zygote. In many cases, gene expression is also controlled by external environmental factors via signal transduction. A body plan must be initiated early in development to construct a complete multicellular organism with a variety of tissues and systems. In *Drosophila* about 30 maternally derived gene products (RNA or protein) are introduced into the egg and are the first involved in determining the general axes (anterior–posterior, dorsal–ventral) of the body plan. Gradients of maternal RNA or protein across the egg play a large role. A protein whose gradient determines two or more cell fates is called a **morphogen**.
  - **Anterior:** Maternally derived *bicoid* RNA is translated into the bicoid protein that is localized toward the end of the egg that will become the anterior end of the embryo. This protein, which is a morphogen, is also a specific transcription factor that acts to suppress translation of the *caudal* mRNA in the anterior pole.
  - **Posterior:** Localizing the product of the maternal-effect gene *nanos* determines the posterior region. This protein, which is also a morphogen, prohibits translation of *hunchback* mRNA in the posterior pole.

      o   **Dorsal–ventral:** Dorsal–ventral development is also initiated by the deposition of a maternal factor in the unfertilized egg. Unlike the anterior–posterior axis, the maternal factor that determines the dorsal–ventral axis is a protein rather than mRNA. The Spätzle protein that is deposited in the egg is cleaved by maternally deposited proteases following fertilization. This cleavage releases Spätzle from the vitelline membrane forming a gradient with the highest concentration at the ventral midline decreasing toward the dorsal midline. The Spätzle protein is bound by the Toll receptor protein, activating the Pelle kinase which phosphorylates the Dorsal protein. Phosphorylation of the Dorsal protein allows it to enter the nucleus and activate the transcription of zygotic genes.

- **Generation of the Segmented Pattern in the *Drosophila* Embryo:** In a developing *Drosophila* embryo, the **segment genes** determine the number and fate of segments that will eventually give rise to the various parts of the body. These genes, which are activated sequentially, are of three major types: **gap**, **pair-rule**, and **segment-polarity** genes. For example, correct gene expression of the gap genes depends on the gradients of maternal-effect genes, and the pair-rule genes depend on the expression of the gap genes, and so on. Each additional level acts to narrow and focus the body plan until 14 segments exist.

      o   **Gap Genes:** Transcription of *hunchback*, a gap gene, only at the anterior end is determined by bicoid and nanos concentrations (see earlier). Specific concentrations of hunchback, in turn, determine the transcription of three additional gap genes: *Krüppel*, *knirps,* and *giant*. These proteins, in combination with other maternal-effect proteins or proteins derived from their signals, construct a gene expression pattern with which the pair-rule genes can interact.

      o   **Pair-Rule Genes:** The pair-rule gene products, whose expression is determined by the gap genes, are transcription factors that affect odd- or even-numbered segments.

      o   **Segment Polarity Genes:** The segmentation genes are activated by the combinatorial effects of all the preceding proteins to further narrow and define the 14 segments. These genes are required for individual cells to recognize their relative position within a parasegment.

21.3 Generation of Cell Identity During Development

- **Homeotic Genes and Segment Identity:** The products of the **homeotic genes** control the developmental fate of a cell type within the 14 segments. Cells with a mutation in a homeotic gene mistakenly follow the developmental pathway of another cell type, without changing the number of segments. Genes within the *bithorax* complex control the fate of posterior thorax and abdomen structures, while genes within the *Antennapedia* complex control the fate of anterior structures. The **homeobox** is a consensus DNA sequence found in promoters of *Drosophila* homeotic genes and other developmentally important genes including those of human beings. The homeobox is a 180-bp sequence encoding the **homeodomain**, a 60-amino-acid helix-loop-helix DNA-binding motif found in the homeotic proteins.

- **Homeobox Genes and Vertebrate Development:** Mammalian homeobox genes (***Hox* genes**), like the *Drosophila* homeotic genes, are arranged in complexes in which the expression of the gene correlates with its position in the complex. These genes control the body regions along the anterior–posterior axis and the proximal–distal axis in vertebrate limb development. According to the **Hox code**, the segments of an animal are specified based on their unique pattern of *Hox* gene expression.

21.4 Sex Determination

- **Sex Determination in *Drosophila*:** Sex is determined in *Drosophila* by the X/A ratio. After this ratio is determined, the *Sex lethal* gene is transcribed leading to the establishment of a specific sex. In females, transcription of this gene from the P_F promoter generates a functional protein in the early embryo. This does not occur in males because they lack the proteins required for transcription. Later in development, the *Sex lethal* gene is transcribed in both males and females from the PL promoter. Because of the presence of the Sex lethal protein in females, the PL transcript undergoes alternative splicing in males and females. In females, the Sex lethal protein prevents the inclusion of the L3 exon in the mature mRNA generating a functional protein. In males, lack of the Sex lethal protein allows the inclusion of the L3 exon in the mature transcript. L3 contains a translation termination codon that leads to the production of a truncated, nonfunctional Sex lethal protein. The presence of the Sex lethal protein in females also affects the alternative splicing of the *transformer* and the *msl-2* primary transcripts. In females, a functional Tra protein is produced, while in males, alternative splicing produces a truncated, nonfunctional Tra protein. The presence or absence of the Tra protein affects the splicing of the *doublesex* and *fruitless* primary transcripts which generates female- or male-specific functional Doublesex and Fruitless proteins. In males, the absence of the Sex lethal protein leads to the production of a functional Msl-2 protein, while females produce a truncated, nonfunctional Msl-2 protein by alternative splicing. The Msl-2 protein functions in males to double the transcription of the X-linked genes.

- **Sex Determination in Mammals:** Mammalian sex is determined by the presence or absence of the *SRY* gene located on the Y chromosome. This gene encodes a DNA-binding protein with an HMG domain. The SRY protein activates expression of the autosomal *SOX9* gene which is involved in the activation of genes that encode testis-determining factors and anti-Müllerian hormone. In addition, the SRY protein blocks the expression of the *WNT4* gene which activates the expression of genes that encode the ovary-determining factors and the *DAX1* gene which blocks the expression of proteins that activate the expression of the testis-determining factors. Dosage compensation in mammals is accomplished by the inactivation of one of the X chromosomes in females. This occurs by a blocking factor binding to the counting element (CE) located adjacent to the *XIST* gene on one X chromosome which becomes the active X chromosome in the cell. The other X chromosome is bound by *XIST* RNA which spreads toward the ends of the chromosome, followed by histone modifications which inactivate it. The *XIST* promoter on the active X chromosome becomes methylated to terminate *XIST* transcription. Recent studies show that up to 25% of the genes on the inactive X chromosome may still be expressed.

21.5 *Drosophila* Eye Development

- **Master Regulator Genes:** Master regulator genes are genes that commit a cell to a particular cell fate. In the development of the eye in *Drosophila*, the two homeobox-containing genes, *eyeless* and *twin of eyeless* are master regulators. These two genes can induce eye formation in different parts of a developing fly when they are expressed in the wrong cells. Many master regulator genes are conserved in different organisms. *Pax6* is the mammalian homolog of the *Drosophila eyeless* and *twin of eyeless* genes.

- **Generation of the R7 Photoreceptor Cell:** The Sevenless signaling cascade is required for the determination of the R7 photoreceptor cell in *Drosophila*. This was determined by the isolation and characterization of a number of mutations in different genes in this pathway. The *Drosophila sevenless* mutant was isolated as a fly that could not respond to UV light. This mutant lacked the R7 photoreceptor and was found to encode a receptor tyrosine kinase that activated a signaling cascade presumed to activate the transcription of other genes. To identify other factors in this signaling cascade, investigators identified other mutants with a similar phenotype. By this method, the *bride of sevenless* gene which encodes an integral membrane protein expressed on the R8 photoreceptor was identified. This protein serves as a ligand for the adjacent cell expressing a Sevenless receptor to become an R7 photoreceptor. A sensitized screen using a temperature-sensitive mutant of the *sevenless* gene was also carried out to identify mutations in other components of the Sevenless signaling pathway. This screen identified a *ras* gene encoding a GTP-binding protein, the *Son of sevenless* gene encoding a guanine exchange factor, and the *GRB2* gene encoding an adaptor protein. A second screen to isolate mutants that suppress or enhance the phenotype of flies expressing a *ras^D* transgene, which results in extra R7 cells being produced in each ommatidium, identified additional components in the Sevenless signaling cascade.

21.6 Programmed Cell Death in Development

- **Discovery of Apoptosis:** Apoptosis is programmed cell death that is genetically controlled. It was first discovered in *C. elegans* and is known to be a normal part of organismal development. Apoptosis, unlike necrosis, involves the activation of a signaling cascade that results in the engulfment and removal of a specific cell without significant damage to surrounding cells. This process is characterized by condensation of the cytoplasm and nucleus, blebbing of the plasma membrane, and fragmentation of the nuclear DNA. Commitment to apoptosis leads to the activation of caspases which are cysteine proteases that cleave cytoskeletal proteins and activate a specific DNase.

Key Terms

- developmental genetics
- maternal-effect genes
- zygotic induction
- undifferentiated
- stem cells
- totipotent
- determination
- pluripotent
- differentiation

- intrinsic
- extrinsic
- morphogens
- gap genes
- parasegment
- pair-rule gene
- segment polarity genes
- homeotic genes
- homeobox

- homeodomain
- *Hox* genes
- Hox code
- *XIST* RNA
- *TISX* RNA
- master regulator gene
- suppressor mutation
- enhancer mutation
- apoptosis

Understanding the Key Concepts

Use the key terms or parts of their definitions to complete the following sentences.

The oocytes of most mammals are arrested at (1)_____ of meiosis until they are (2)_____. Following (3)_____, (4)_____ genes carry a zygote through many rounds of cell division. During this time, cells spend only a short period in the (5)_____ and (6)_____ phases of the cell cycle, such that the RNAs and proteins produced by the (7)_____ genes are the only ones that are (8)_____ and subsequently (9)_____.

Cells become determined as they are (10)_____ to develop into a particular type of cell by (11)_____ or (12)_____ signals. A cell takes on the (13)_____ of its adult fate by the process of differentiation. Only (14)_____ retain their (15)_____ nature in that they do not take on the fate of any particular adult cell type, but instead remain (16)_____.

Anterior–posterior axis formation in *Drosophila* relies on the products of the (17)_____ genes. These genes generate (18)_____ protein concentration gradients in the developing embryo by the sequestration of the (19)_____ and (20)_____ mRNAs at opposite poles and their ability to block the (21)_____ of the (22)_____ and (23)_____ mRNAs, respectively.

(24)_____ axis formation in *Drosophila* also requires (25)_____, such as (26)_____, (27)_____, and (28)_____ to be deposited and evenly distributed in the unfertilized egg. Once Spätzle is (29)_____ in the (30)_____ space, it is activated and can then bind to the (31)_____ receptor located in the (32)_____ which activates the (33)_____ kinase to phosphorylate the (34)_____ protein allowing it to move into the (35)_____. Therefore, a gradient of activated (36)_____ creates a gradient of activated (37)_____ in the nucleus that can activate the transcription of (38)_____ genes required for the development of the (39)_____ side of the *Drosophila* embryo. Four classes of genes transcribed in the *Drosophila* embryo direct anterior–posterior pattern formation. The (40)_____, such as *hunchback*, (41)_____, *giant*, and *knirps* are transcribed during the (42)_____ stage and define regions along the anterior–posterior axis that will ultimately consist of several different (43)_____. The locations of these (44)_____ are determined by the (45)_____ and (46)_____ genes. Most of the (47)_____ genes are transcription factors that define the number of (48)_____ and the location of the (49)_____ boundary of each. (50)_____ genes determine the (51)_____ and polarity of each segment. The (52)_____ genes in *Drosophila* are the final class of anterior–posterior patterning genes required for assigning unique identities to each cell within a parasegment. The (53)_____ gene cluster directs the development of the (54)_____ and the first (55)_____ thoracic segments, while the (56)_____ gene cluster directs the proper development of the posterior region of the (57)_____ thoracic segment, the (58)_____ thoracic segment, and all of the (59)_____ segments. These genes all contain a (60)_____-base-pair sequence that encodes a 60-amino-acid DNA-binding motif called a (61)_____. This motif binds conserved sequences in the (62)_____ of specific genes to regulate gene expression. Significantly, this DNA-binding motif is conserved in vertebrate (63)_____ genes. Like the *Drosophila* genes, the

vertebrate (64)_____ genes display (65)_____ in which the order of the genes along the
chromosome correlates with (66)_____ and (67)_____ the gene is expressed along the
anterior–posterior axis.

      (68)_____ in *Drosophila* depends on the ratio of X chromosomes to (69)_____, while
in mammals it depends on the presence of the (70)_____ gene on the (71)_____ chromosome.
(72)_____ is achieved in *Drosophila* by (73)_____ the transcription of (74)_____ genes
in (75)_____ flies due to the presence of a functional (76)_____ protein. In contrast,
(77)_____ in mammals involves the (78)_____ of one of the (79)_____ in
(80)_____. This is accomplished by (81)_____ binding to X_i and a (82)_____ binding
to X_a.

      Apoptosis is a normal part of development in many organisms. Unlike (83)_____ which
removes (84)_____ cells, apoptotic cell death is (85)_____ controlled by the (86)_____
and (87)_____ gene families. Apoptotic cells are removed by (88)_____ following
condensation of the cytoplasm and nucleus and (89)_____ of the DNA.

Figure Analysis

Use the indicated figure or table from the textbook to answer the questions that follow.

1. **Figure 21.10**

    (a) Where is the *bicoid* mRNA deposited? What is the function of the Bicoid protein?

    (b) What is the phenotype of a *bcd⁻/bcd⁻* female fly? What is the phenotype of the progeny of this fly?
    Explain why.

    (c) Where is the *nanos* mRNA deposited? What is the function of the Nanos protein?

    (d) How are the *bicoid* and the *nanos* mRNAs localized and maintained at the anterior and posterior poles
    of the egg, respectively?

    (e) How do mRNAs like *hunchback* and *caudal* which are evenly distributed throughout the unfertilized
    egg play a role in anterior–posterior axis formation?

    (f) What protein affects the expression of the *hunchback* mRNA? How does it affect its expression?

    (g) What protein affects the expression of the *caudal* mRNA? How does it affect its expression?

2. **Figure 21.19**

    (a) What is the function of the Hedgehog protein at the parasegment boundary?

    (b) What cell secretes the Hedgehog protein? How is Hedgehog expression directly regulated?

    (c) What is the effect of the Hedgehog protein binding to the Patched receptor on the posterior-most cell?

    (d) What is the effect of the Wingless protein binding to the Frizzled receptor on the anterior-most cell?

3. **Figure 21.29**

    (a) Why is the Sex lethal protein different in male and female fruit flies?

    (b) How does the presence of a functional Sex lethal protein in female fruit flies affect the Tra protein?
    How does its absence in male flies affect this protein?

    (c) How does the presence of a functional Tra protein affect the Doublesex and Fruitless proteins?

    (d) How does the Sex lethal protein affect the expression of the Msl-2 protein?

(e) What is the function of the Msl-2 protein in male fruit flies?

4. **Figures 21.33 and 21.34**

   (a) What are two factors that affect the binding of the *XIST* RNA to the X chromosomes?

   (b) What effect do these two factors have on the *XIST* RNA?

   (c) Which X chromosome is typically selected as the X_i chromosome?

   (d) Once the X_i chromosome has been selected, how is inactivation of the chromosome carried out?

5. **Figure 21.39**

   (a) What developmental process is the Sevenless signaling cascade involved in?

   (b) What is the phenotype of a *sevenless* mutant? What does this mutant lack?

   (c) What is the function of the Boss protein in this signaling pathway?

   (d) How was the Boss protein identified?

   (e) What are the functions of the Son of sevenless, Ras, and GRB2 proteins in the Sevenless signaling pathway?

   (f) How were these proteins identified to be part of the Sevenless signaling cascade?

   (g) How was it confirmed that the activated Ras protein was required to generate an R7 cell?

General Questions

Q21-1. What are the two primary methods of determination? What is the difference between cell determination and cell differentiation?

Q21-2. Describe the function of the homeotic genes in *Drosophila*. What vertebrate genes are related to the homeotic genes? What similarities do these genes share with the *Drosophila* homeotic genes?

Q21-3. Describe how sex is determined in *Drosophila* and in mammals? How is dosage compensation achieved in these two groups of organisms?

Q21-4. What is a master regulator gene? What is the effect of a loss-of-function mutation in a master regulator gene? What is the effect of expressing a master regulator gene in cells where it is not typically expressed?

Q21-5. Explain how BMP signaling differs in the chicken and duck hind limbs. What is the phenotypic effect of this difference?

Multiple Choice

For each of the following, circle the letter of the choice that most appropriately answers the question.

1.) Which of the following proteins is *not* evenly distributed in an unfertilized *Drosophila* egg?
   a. Toll
   b. Easter protease
   c. Spätzle
   d. Dorsal
   e. None of the above; all of these proteins are evenly distributed.

2.) Which of the following groups of genes are the first to be transcribed following zygotic induction?
   a. segment polarity genes
   b. maternal-effect genes
   c. pair-rule genes
   d. gap genes
   e. sex determination genes

3.) Walter Gehring demonstrated that the *eyeless* and *twin of eyeless* genes in *Drosophila* are which of the following?
   a. maternal-effect genes
   b. gap genes
   c. pair-rule genes
   d. master regulator genes
   e. segment polarity genes

4.) When is the first time during development when paternally contributed genes can affect the phenotype of an embryo?
   a. after the first cell division
   b. during the first meiotic division
   c. after zygotic induction
   d. during cellular differentiation
   e. immediately following fertilization

5.) Removal of one of the cells of a sea urchin embryo at the two-cell stage produces a larval sea urchin with which of the following characteristics?
   a. normal size and possessing all cell types
   b. normal size but possessing only half as many cell types
   c. smaller than normal and possessing only half as many cell types
   d. smaller than normal but possessing all cell types
   e. unable to develop beyond the gastrula stage

6.) Activated Spätzle protein is involved in the transcription of genes that direct the development of which part of the *Drosophila* embryo?

   a. ventral
   b. dorsal
   c. anterior
   d. posterior
   e. compound eye

7.) For the following classes of *Drosophila* genes, which are *not* acted on by the pair-rule genes?
   a. other pair-rule genes
   b. homeotic genes
   c. segment polarity genes
   d. gap genes
   e. None of the above; all of these genes are acted on by the pair-rule genes.

8.) Which of the following defects is produced as a result of mutations in the homeotic genes in *Drosophila*?
   a. loss of alternating segments in the embryo
   b. one body part developing as another body part
   c. loss of anterior or posterior structures
   d. tandem duplication of the anterior or posterior region of a parasegment
   e. changes in rhombomere identity

9.) The *Polycomb* and *Trithorax* genes regulate homeotic gene expression by which of the following mechanisms?
   a. initiating a signal transduction cascade
   b. promoting alternative splicing
   c. altering the chromatin conformation
   d. suppressing their translation
   e. sequestering maternal effect mRNAs

10.) An enhancer mutation in the $P[ras^D]$ transgenic fly produces a fly with which of the following?
   a. orderly and smoothly packed ommatidia each containing an R7 photoreceptor cell
   b. phenotypically normal eyes lacking R7 photoreceptor cells
   c. unpatterned compound eyes due to the random duplication of different photoreceptor cells
   d. extremely rough eye surfaces due to extra R7 photoreceptor cells in each ommatidia
   e. additional compound eyes in aberrant locations

Practice Problems

P21-1. If the expression of the *rotator* gene is dependent on the presence of the murky protein and the murky protein is expressed as shown in a concentration gradient, where would you expect the highest expression of *rotator* to be found? Where would the highest expression of *rotator* be found if the murky protein inhibits *rotator* transcription?

a. posterior
b. anterior
c. dorsal
d. ventral

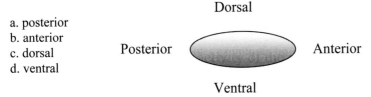

Dorsal

Posterior   Anterior

Ventral

P21-2. Actinomycin D is an antibiotic that inhibits RNA synthesis. If this antibiotic is added to fertilized frog eggs, protein synthesis continues normally until the gastrula stage. From this point on, the embryo stops synthesizing protein. Why?

P21-3. Identify the source (mutated maternal-effect gene, segmentation gene, or homeotic gene) of each *Drosophila* mutant described.
a. embryos are missing heads and thoracic structures
b. affected individuals are missing two abdominal segments
c. an individual has legs in place of antennae on the head
d. an individual has two thoracic segments
e. an individual lacks all abdominal segments

P21-4. A hypothetical gene in mouse development, *devoid,* is expressed at embryonic day 12. A second gene, *bilateral*, is expressed at embryonic day 14. You hypothesize that *devoid* is a transcription factor that turns on the transcription of *bilateral*, so you construct individual organisms that are homozygous mutant for these genes and look for the presence or absence of each RNA at day 15. Fill in the following tables if your hypothesis is true or not true.

Results that show my hypothesis is *true*.

Embryo genotype	*devoid* RNA	*bilateral* RNA
Wild-type embryo	Present	Present
devoid mutant		
bilateral mutant		

Results that show my hypothesis is not true.

Embryo genotype	*devoid* RNA	*bilateral* RNA
Wild-type embryo	Present	Present
devoid mutant		
bilateral mutant		

P21-5. Grumpy and Sleepy are two hypothetical proteins in a signal transduction cascade to direct anterior development in *Drosophila*. Flies with mutations in either of these genes develop a head that lacks the typical anterior structures. Hypothetical *noseless*, *pouty*, and *weepy* mutations have been identified to cause a phenotype similar to *grumpy* and *sleepy* suggesting that they may also be involved in the same signaling pathway. Use the following information to determine the order in which these proteins act assuming that they are all part of the same hypothetical signaling pathway.

Treatment	Mutant *pouty*	*noseless*	*weepy*
Activate Grumpy	No rescue	Rescue	No rescue
Excess Sleepy	No rescue	Rescue	Rescue

(*Hint 1:* If the addition of activated Grumpy protein or the injection of Sleepy protein rescues a mutant phenotype to restore the wild-type phenotype, the mutation must disrupt a step prior to the action of Grumpy or Sleepy in the signaling pathway.)

P21-6. The order in which a series of proteins act in a signal transduction cascade is as follows:
$$\text{Sevenless} \rightarrow \text{GRB2} \rightarrow \text{Sos} \rightarrow \text{Ras} \rightarrow \text{Raf} \rightarrow \text{MAP kinase} \rightarrow \text{AP-1}$$

Determine whether the mutant phenotypes associated with the following four mutations listed would be "Rescued" or "Not rescued" with the treatments listed on the left:

Treatment	Mutant *sevenless*	*grb2*	*raf*	*ap-1*
Activated Ras				
Activated Map kinase				
Activated Raf				
Active Sevenless				

Assessing Your Knowledge

Understanding the Key Concepts—Answers

1.) metaphase II; 2.) fertilized; 3.) fertilization; 4.) maternal-effect; 5–6.) G_1; G_2; 7.) maternal-effect; 8.) transcribed; 9.) translated; 10.) committed; 11–12.) intrinsic; extrinsic; 13.) characteristics; 14.) stem cells; 15.) undifferentiated; 16.) totipotent; 17.) maternal-effect; 18.) four; 19.) *bicoid*; 20.) *nanos*; 21.) translation; 22.) *caudal*; 23.) *hunchback*; 24.) Dorsal–ventral; 25.) maternal factors; 26–28.) Spätzle; Toll; Dorsal; 29.) cleaved; 30.) previtelline; 31.) Toll; 32.) plasma membrane; 33.) Pelle; 34.) Dorsal; 35.) nucleus; 36.) Spätzle; 37.) Dorsal; 38.) zygotic; 39.) ventral; 40.) gap genes; 41.) *Krüppel*;

42.) syncitial blastoderm; 43.) segments; 44.) segments; 45–46.) pair-rule; segment polarity; 47.) pair-rule; 48.) parasegments; 49.) anterior; 50.) Segment-polarity; 51.) identity; 52.) homeotic; 53.) *Antennapedia*; 54.) head; 55.) two; 56.) *Bithorax*; 57.) second; 58.) third; 59.) abdominal; 60.) 180; 61.) homeodomain; 62.) promoters; 63.) homeobox or *HOX*; 64.) homeobox or *HOX*; 65.) colinearity; 66–67.) when; where; 68.) Sex determination; 69.) autosomes; 70.) *SRY*; 71.) Y; 72.) Dosage compensation; 73.) doubling; 74.) X-linked; 75.) male; 76.) Msl-2; 77.) dosage compensation; 78.) inactivation; 79.) X chromosomes; 80.) females; 81.) *XIST* RNAs; 82.) blocking factor; 83.) necrosis; 84.) injured or damaged; 85.) genetically; 86–87.) *BCL-2*; *P53*; 88.) macrophages; 89.) fragmentation.

Figure Analysis—Answers

1a.) The *bicoid* mRNA is deposited in the anterior end of the egg. The Bicoid protein directs the proper development of the anterior region of the embryo; b.) A *bcd⁻/bcd⁻* female fly is phenotypically normal because its mother is *bcd⁻/bcd⁺*, and therefore this fly has the necessary Bicoid protein to develop normally. The progeny of this fly will be lethal because the mother will be unable to deposit the *bicoid* mRNA into the unfertilized egg. Upon fertilization, the egg will develop into an embryo lacking anterior structures; c.) The *nanos* mRNA is deposited in the posterior end of the egg. The Nanos protein directs the proper development of the posterior region of the embryo; d.) The *bicoid* and *nanos* mRNAs are properly localized to the anterior and posterior poles of the egg, respectively, because they are deposited in the egg at this location by the mother. In addition, the 3′ UTR of the *bicoid* and *nanos* mRNAs bind proteins associated with the microtubules. The 3′ UTR of the *bicoid* mRNA binds to a protein associated with the minus ends of the microtubules which are located in the anterior end of the egg, and the *nanos* mRNA binds to a protein associated with the plus ends of the microtubules which are located in the posterior end of the egg. This binding controls the localization of the mRNAs in the egg; e.) The *hunchback* and *caudal* mRNAs which are evenly distributed throughout the unfertilized egg play a role in anterior–posterior axis formation because these mRNAs are only translated in either the anterior or the posterior pole again generating a protein concentration gradient of both the Hunchback and Caudal proteins in the embryo; f.) The Nanos protein suppresses the translation of the *hunchback* mRNA in the posterior pole of the embryo; g.) The Bicoid protein suppresses the translation of the *caudal* mRNA in the anterior pole of the embryo.

2a.) The Hedgehog protein binds to the Patched receptor on the posterior-most cell and activates it to initiate a signal transduction cascade that activates the transcription of the *wingless* gene; b.) The anterior-most cell at the parasegment boundary secretes the Hedgehog protein. The expression of Hedgehog is directly regulated by the Engrailed protein which acts as a transcriptional activator to promote transcription of the *hedgehog* gene; c.) Binding of the Hedgehog protein to the Patched receptor on the posterior-most cell activates the Patched receptor which stabilizes the Cubitus interruptus transcription factor. Once stabilized, this factor activates the transcription of the *wingless* gene which then produces Wingless protein that is secreted by the cell; d.) Wingless binding to the Frizzled receptor on the anterior-most cell activates Frizzled which stabilizes the Armadillo transcription factor and promotes the transcriptional activation of the *engrailed* gene. The Engrailed protein can then bind to the promoter of the *hedgehog* gene to activate its transcription resulting in the production of the Hedgehog protein by the anterior-most cell.

3a.) The Sex lethal protein is different in male and female flies due to alternative splicing. The female fly contains the Sex lethal protein in the early embryo which is transcribed from the *PE* promoter, but the male fly lacks this protein because it lacks the proteins necessary for transcription of the *Sex lethal* gene from the *PE* promoter. The early Sex lethal protein prevents the inclusion of exon L3 in the late *Sex lethal* transcript in the female flies, but the lack of the early Sex lethal protein in males permits the inclusion of this exon in the late *Sex lethal* transcript. This exon contains a premature translation termination codon, so the *late Sex lethal* transcript in male flies produces a truncated, nonfunctional protein. Since this exon is not included in the late *Sex lethal* transcript in female flies, a functional Sex lethal protein is produced; b.) The Sex lethal protein in female flies affects the splicing of the *transformer* primary transcript and leads to the production of a functional Tra protein. Lack of the Sex lethal protein in male flies leads to an alternatively spliced *tra* transcript which includes an exon containing a premature translation termination codon leading to the production of a truncated, nonfunctional protein; c.) In females, the Tra protein interacts with the Tra-2 protein which regulates the splicing of the *Doublesex* and the *Fruitless* primary transcripts and produces female-specific Doublesex and Fruitless proteins; d.) Presence of the Sex lethal protein in female flies leads to an alternatively spliced *msl-2* transcript that is translated to produce a truncated nonfunctional

Msl-2 protein. Absence of the Sex lethal protein in males leads to the production of an mRNA that is translated into a functional Msl-2 protein; e.) The Msl-2 protein is required in male fruit flies to properly balance the expression of the X-linked genes to achieve dosage compensation among the male and female flies. The Msl-2 protein is required to double the transcription of the X-linked genes in male flies.

4a.) Binding of the blocking factor and the presence of the *TSIX* RNA affect the binding of the *XIST* RNA to the X chromosomes; b.) Both binding the blocking factor and the presence of *TSIX* RNA affect which X chromosome is chosen as the active X chromosome. Binding of the blocking factor to the counting element in a region adjacent to the *XIST* gene prevents the spreading of the *XIST* RNA and activates the transcription of the *TSIX* RNA. The *TSIX* RNA base-pairs with the *XIST* RNA and decreases its stability resulting in less *XIST* RNA being associated with a particular X chromosome. This frees the counting element on one of the X chromosomes to be bound by the blocking factor; c.) The X chromosome that produces the greater amount of *XIST* RNA is usually selected as X_i; d.) Once the X_i chromosome has been selected, the inactivation of this chromosome is carried out by the spreading of the *XIST* RNA to the ends of the X_i chromosome. This is followed by increased methylation and decreased acetylation of the histones as well as DNA methylation to keep this X chromosome inactive.

5a.) The Sevenless signaling cascade is involved in the generation of the R7 photoreceptor cell; b.) A *sevenless* mutant is unable to respond to ultraviolet light. This mutant lacks the R7 photoreceptor in every ommatidium of its eye; c.) The Boss protein encodes an integral membrane protein on the R8 photoreceptor that serves as a ligand for an adjacent cell containing the Sevenless receptor to initiate the signal transduction cascade necessary to direct the cell to become an R7 photoreceptor; d.) The Boss protein was identified in a screen for additional mutants with the *sevenless* phenotype; e.) The Son of sevenless protein removes the GDP from the inactive Ras and replaces it with GTP to reactivate the protein. The Ras protein activates a series of kinases in a signal transduction cascade that leads to the transcriptional activation of AP-1 and PNT and suppresses the activation of Yan. The GRB2 protein serves as an adaptor protein that binds both the phosphorylated Sevenless receptor and the Son of Sevenless protein; f.) These proteins were identified in a screen using a temperature-sensitive mutant of the *sevenless* gene at a temperature intermediary to its permissive and restrictive temperature to identify second site mutations that would reduce the activity of the Sevenless signaling pathway and prevent the R7 cell from developing at this semi-permissive temperature; g.) The requirement for the activated Ras protein in the Sevenless signaling pathway was confirmed by using in vitro mutagenesis of the *ras* gene to generate a dominant form of Ras that was constitutively active. This mutant was transposed into the genome of a *sevenless* mutant to determine if it was sufficient to rescue the development of the R7 photoreceptor cell.

General Questions—Answers

Q21-1. A cell can become determined by either intrinsic or extrinsic signals. For intrinsic determination, one or more factors present within the cell determine the fate of the cell, such as transcription factors that regulate the expression of specific genes that determine a cell's fate. For extrinsic determination, signaling molecules and receptor proteins allow cells to communicate with other cells. By controlling the expression of the genes encoding the receptors for signaling molecules, the fate of some cells can be determined. Binding of signaling factors by receptors on adjacent cells identifies the location of the cell in the embryo and allows its fate to be properly determined. Cell determination is the commitment of a cell to a particular fate, while cell differentiation is the process by which a cell undergoes changes in structure and function to reach its determined cell fate.

Q21-2. The homeotic genes in *Drosophila* include the *Antennapedia* and the *Bithorax* gene clusters. The genes in the *Antennapedia* complex are required for the proper development of the head and the first two thoracic segments. The genes in the *Bithorax* complex are required for the development of the posterior region of the second thoracic segment, the third thoracic segment, and all of the abdominal segments. The mammalian homeobox genes are related to the *Drosophila* homeotic genes. The mammalian homeobox genes share sequence similarity to the *Drosophila* homeotic genes in that both the homeotic and the homeobox genes contain a homeobox that encodes a DNA-binding domain that is used to regulate the expression of a specific set of genes in *Drosophila* and vertebrates. The homeobox genes in vertebrates, like the *Drosophila* homeotic genes, control the specification of body segments along the anterior–posterior axis. Like the *Drosophila* homeotic genes, the homeobox genes in vertebrates also exist in regulated complexes in which the order of the genes along the chromosome correlates with their expression along the anterior–posterior axis and the time at which each gene is expressed.

Q21-3. In *Drosophila*, sex is determined by the X/A ratio in which a ratio of 0.5 produces a male and a ratio of 1.0 produces a female fly. In mammals, the presence or absence of the *SRY* gene on the Y chromosome determines the sex of an organism. An organism with a Y chromosome is male, and an organism lacking a Y chromosome is female. Dosage compensation in *Drosophila* is achieved in male fruit flies by the Msl-2 protein functioning to double the transcription of the X-linked genes. In mammals, dosage compensation is achieved by the inactivation of one of the X chromosomes in females by binding of *XIST* RNAs to the X chromosome to be inactivated leading to histone modifications, and binding of a blocking factor to the active X chromosome.

Q21-4. A master regulator gene is a gene that commits a cell to a particular cell fate. The action of a master regulator gene typically involves the initiation of a cascade of gene activations. A loss-of-function mutation in a master regulator gene prevents initiation of the signaling cascade and, ultimately, prevents the formation of the corresponding organ or tissue whose development the master gene directs. Expression of a master regulator gene in cells where it is not typically expressed creates a tissue or organ in a location where it is not typically found on an organism.

Q21-5. BMP signaling in the chick and duck hind limbs differs in that BMPs act as ligands to stimulate a signaling cascade that promotes apoptosis in the interdigital regions of the chick. This signaling cascade produces separate digits in the chicken. In the duck hind limb, an inhibitor of BMPs prevents this apoptotic cascade from occurring. Inhibition of the BMPs leads to loss of the increased rates of apoptosis in the interdigital regions and results in interdigital webbing in the duck hind limb.

Multiple Choice—Answers

1.) b; 2.) d; 3.) d; 4.) c; 5.) d; 6.) a; 7.) d; 8.) b; 9.) c; 10.) d.

Practice Problems—Answers

P21-1. The highest expression of murky in the diagram is at the ventral side, so you would expect to find the highest expression of *rotator* in this same location: the ventral side (d). If murky inhibited the expression of *rotator*, then you would expect to find rotator protein only where murky was at the lowest concentration, the dorsal side (c).

P21-2. Actinomycin D inhibits the embryo's ability to synthesize RNA. Prior to gastrulation, the mRNAs existing in the cell are already translated. Because they already exist in the cell at the time of treatment, early protein synthesis is unaffected.

P21-3. Phenotypes (a) and (e) are likely caused by maternal-effect genes; phenotype (b) results from a mutated segmentation gene; phenotypes (c) and (d) result from homeotic gene mutants.

P21-4.
Results that show my hypothesis is *true*.

Embryo genotype	*devoid* RNA	*bilateral* RNA
Wild-type embryo	Present	Present
devoid mutant	Absent	Absent
bilateral mutant	Present	Absent

Results that show my hypothesis is *not true*.

Embryo genotype	*devoid* RNA	*bilateral* RNA
Wild-type embryo	Present	Present
devoid mutant	Absent	Present
bilateral mutant	Present	Absent

If devoid activates transcription of *bilateral* (devoid → bilateral) and we mutate *devoid*, bilateral will also be absent because devoid is not available to turn on *bilateral* transcription. However, if devoid is not necessary to turn on *bilateral* expression, then we would expect the results shown at the right. Some other protein activates *bilateral* transcription (X → bilateral), not devoid.

P21-5. Since the injection of activated Grumpy rescued the mutant phenotype of *noseless*, but failed to rescue *pouty* or *weepy*, Grumpy must act prior to Weepy and Pouty, but downstream of Noseless in the signaling pathway. Likewise, since the addition of excess Sleepy rescued the mutant phenotypes of *noseless* and *weepy*, but not of *pouty*, Sleepy must act prior to Pouty, but downstream of Noseless and Weepy. Therefore, the proteins must act in the following order: Noseless → Grumpy → Weepy → Sleepy → Pouty.

P21-6. To restore the wild-type phenotype and rescue the mutant, the proteins being added as treatment must be required downstream of the mutation in question because the mutations listed disrupt the normal activation of the receptor or the kinases in the signaling cascade. If this is true, addition of the activated receptor or activated kinase will rescue the normal signaling pathway. If this is not true and the mutation is in a component in the pathway that acts downstream of the receptor, addition of the activated receptor or kinase will fail to rescue the mutant phenotype. Similarly, if the mutation is in a component in the pathway that acts downstream of one of the kinases being added as treatment, addition of the activated kinase will fail to rescue the mutant phenotype.

	Mutant			
Treatment	*sevenless*	*grb2*	*raf*	*ap-1*
Activated Ras	Rescued	Rescued	Not rescued	Not rescued
Activated Map kinase	Rescued	Rescued	Rescued	Not rescued
Activated Raf	Rescued	Rescued	Rescued	Not rescued
Active Sevenless	Rescued	Not rescued	Not rescued	Not rescued

21

Developmental Genetics

CHAPTER SUMMARY QUESTIONS

1. 1. i, 2. g, 3. f, 4. h, 5. j, 6. c, 7. b, 8. a, 9. d, 10. e.

2. Sperm cells are haploid, containing one complete copy of the genetic material. The number of chromosomes depends on the species.

3. A maternal effect gene controls production of one or more components of the oocyte, or otherwise influences the maternal ability to produce viable oocytes.

4. The polarity axes are the first pattern element formed during *Drosophila* embryonic development. The anterior–posterior axis runs from head to tail; the dorsal–ventral axis is from back to belly.

5. A syncytial blastoderm is seen in early *Drosophila* development, where repeated nuclear divisions give rise to many embryonic nuclei, but cytokinesis has not occurred in concert. The result is that there are many nuclei located within a common cytoplasm—a syncytium.

   A cellular blastoderm is formed from the syncytial blastoderm once the nuclei migrate to the periphery. Invagination of the plasma membrane and production of new membrane material combine to generate individual cells at the surface of the embryo all at once.

6. Because all embryonic nuclei are located in a common cytoplasm, diffusion of material through that cytoplasm will easily establish a gradient that can have differential effects on the nuclei. Once cellularization has occurred, existing gradients have become intrinsic cues to individual cells; new signals cannot easily diffuse, and other mechanisms of patterning become more important.

7. A morphogen is a diffusable substance in a developing embryo that controls gene expression depending on the concentration of the morphogen at that point in its gradient. The bicoid gene, expressed in maternal tissue, results in a region of the messenger RNA concentrated at the anterior tip of the embryo at fertilization. The messenger is then translated into a protein that diffuses into the embryo, creating a gradient that controls expression of other genes such as *hunchback*.

8. Gap (specify large chunks of pattern); pair-rule (specify every other segment); Segment polarity (specify orientation of segment along AP axis); homeotic (specify location of segment along AP axis).

9. *Hunchback.*

10. The four genes are *Bicoid*, *Giant*, *Hunchback*, and *Krüppel*. *Bicoid* and *Hunchback* activate transcription of *even-skipped* in parasegment 3. *Giant* represses transcription of *even-skipped* in the anterior portion of the embryo, and *Krüppel* represses transcription of *even-skipped* in the posterior portion of the embryo.

11. No. One unique feature of parasegment 3 (the location of eve stripe 2) is that Krüppel expression is very low. Loss of function of Krüppel will probably have many other effects on gene expression in the *Drosophila* embryo, but little or no effect in this specific interaction.

12. Regulation of splicing of the sex lethal transcript from the late promoter (LP; splicing produces a productive transcript if functional sex lethal protein is present) and regulation of splicing of transformer (again, functional transformer protein is not produced unless sex lethal protein is already present).

13. Hyperactivation of gene expression of X-linked genes is important because male *Drosophila* have only one X chromosome. In order to keep up with transcription of X-linked genes in females (considering they can transcribe from both X chromosomes at the same time and thus have a higher transcriptional rate), males must have factors to increase X-linked gene transcription.

14. A portion of TSIX is complementary to the XIST RNA sequence, so the two molecules can base-pair. This destabilizes XIST RNA interaction with the X chromosome and allows blocking factor to reach the CE more efficiently, and thus increase the chances of that particular X remaining active.

15. Activation of a master regulator gene sets off a cascade of gene actions resulting in specification of a particular tissue. Loss of function mutants prevent development of that specific tissue, while gain of function mutations lead to production of the specific tissue in locations where it would not otherwise exist.

16. R7 precursor cells receive the Boss signal; their response to this signal is to develop into R7 rather than a cone cell.

17. Apoptosis is a specific type of cell death that is under genetic control.

EXERCISES AND PROBLEMS

18. For a cross between a heterozygous female ($X^A X^a$) and a male carrying the dominant allele $X^A Y$, all offspring would be unaffected: 25% would be $X^A Y$ males, 25% $X^a Y$ males, 25% $X^A X^A$ females, 25% $X^A X^a$ females. None of these would be affected themselves. In a testcross, the tester parent by definition has only recessive alleles for the trait in question. A homozygous recessive female is unable to produce viable oocytes, so no viable offspring will be produced regardless of the male's genotype. Of the two possible female genotypes, both will produce all unaffected offspring when crossed to a recessive male. However, half of the female offspring from the heterozygous female will be homozygous recessive and unable to produce any offspring.

For a cross between a heterozygous female and a male carrying the recessive allele, the same two possibilities for male offspring exist, since the male parent's X chromosome does not influence his male offspring. Female offspring will either be heterozygous (and thus half of the female offspring in a testcross will be able to produce offspring) or homozygous recessive and unable to produce offspring themselves.

19. If one cell is killed but the material that formed the cell remains in place, the other cell is still receiving extrinsic cues and "believes" that the other half of the embryo is still there developing apace. Separation of the two cells prevents extrinsic cues from the "other half" of the embryo, so the separated cells regulate and establish a complete body plan.

20. The embryo, like the future adult, is bilaterally symmetrical. Intrinsic cues are probably distributed in the zygote and early embryo in a bilaterally symmetrical pattern. Division of the embryo along the plane of symmetry maintains the basic pattern of intrinsic cues, while division in any other plane results in imbalanced patterns of intrinsic cues being distributed to the two parts.

21. **a.** and **c.** are loss-of-function mutations, probably nulls. Bicoid mRNA, and thus protein, will be evenly distributed through the oocyte. *Hunchback* translation will be evenly activated at a low level. Lack of high levels of bicoid leads to failure of activation of zygotic *hunchback* gene expression, and ultimately prevents formation of normal anterior structures.

b. may or may not have a significant phenotype; this mutation may cause a mild hypomorph or a complete null depending on the specific amino acid affected.

d. will probably cause an allele that has normal function in regard to Bicoid localization but will be a complete null relative to localization of the other transcripts.

22. A complete loss of function mutation in cactus would prevent sequestration of Dorsal protein in the cytoplasm. Thus, Dorsal can enter the nucleus and possibly activate gene transcription regardless of activation of the Toll pathway. (*Note:* This explanation ignores evidence that phosphorylation of Dorsal is required for efficient translocation to the nucleus.) Because Toll acts upstream of Dorsal, a gain of function mutation in Toll will not rescue the mutant phenotype of the cactus mutant.

23. The expression of genes at the boundaries of parasegments is controlled in part by a feedback loop that stabilizes the boundary. Two rows of cells (either expressing wingless or engrailed) must communicate with each other—the "sending" cell displays a cell surface ligand (wingless or hedgehog) recognized by a receptor on the "receiving" cell (frizzled or patched); the signal must then be transduced across the membrane and converted into transcriptional regulation.

Null mutations of patched will prevent cells from responding to the hedgehog signal, and will allow expression of wingless, leading to loss of the parasegment boundary. Hypermorphic alleles of armadillo will likely lead to increased expression of engrailed and excessive activation of hedgehog. Depending on the interactions with other players in the pathway, this could have no phenotypic effect or could disrupt maintenance of parasegment boundaries as engrailed expression becomes deregulated.

24. a. Pair-rule genes are expressed in seven stripes—in alternating parasegments. If one of the pair-rule genes was responsible for engrailed activation, loss of that pair-rule gene would lead to loss of the seven stripes regulated by that pair-rule gene.

   b. The gene in question is probably under transcriptional control. The regulatory gene is probably a transcription factor.

   c. The seven stripes give us a clue—we can start by sequencing the pair-rule genes present in our mutant and see if any obvious mutations are present in those candidate genes. We could try to identify specific changes in the proteins expressed in cells that should express engrailed but do not, compared to similar cells in wild-type embryos. If the mutation were generated by transposon insertion, we could clone genomic DNA adjacent to the transposon and look to see what known (or unknown) genes may be present. Other options are also possible.

25. A reasonable working hypothesis would be that there is no difference in the expression of homeotic genes between segments of a centipede. In addition to direct detection of gene expression in centipedes, you could consider tinkering with expression of homeotic genes (probably by targeted mutagenesis) and see what changes are able to change (even if it is a disruption) the body pattern.

26. Wings and legs are characteristic of thoracic segments rather than abdominal segments, so we need to consider mutations in genes expressed in the thorax. Specifically, *Ultrabithorax* and *Antennapedia* gene expression would need to be "adjusted." In *Drosophila*, *Antennapedia* differentiates the second thoracic segment from the first. We might expect *Antennapedia* expression not to be present in the second thoracic segment of the dragonfly, allowing production of wings (rather than the halteres of *Drosophila*), shifting the anterior boundary of *Antennapedia* to the

T1/T2 boundary rather than the T2/T3 boundary might be a reasonable idea. *Ultrabithorax* expression should also be considered.

27. Deletion of a portion of one of the two homeotic complexes, and thus deletion of multiple genes; mutations in Polycomb family members; and mutations in Trithorax family members. Sequencing of the homeotic complexes, *Polycomb*, and/or *Trithorax* genes would be one start, but requires a lot of work for not much information. Observation of gene expression over *time* would help; if the problem lies in misactivation of homeotic genes, then mutations in the complexes (deletion of genes or changes in regulatory elements that control several genes) would be expected. If the problem lies in a failure to maintain correctly initiated homeotic gene expression, *Polycomb* and/or *Trithorax* become much more attractive candidates.

28. Dramatic homeotic transformations are unlikely in mammals in part due to a lack of obviously segmented structures in the adult. Subtle transformations are possible in structures that display segmentation either during embryonic development (for example, rhombomeres) or in adult structures (for example, vertebrae, ribs). Other subtle transformations may be interpreted as homeotic if the abnormal structures resemble structures normally present at a different anterior–posterior level of the animal; one example would be a pattern of spinal neurons similar to that expected at the level of a limb occurring at cervical or thoracic regions. The presence of multiple copies of each gene raises the possibility that functions may be redundant (that is, more than one gene is responsible for each function), so loss of *one* gene does not lead to complete loss of that specific function. Also, the presence of multiple copies of each gene has freed those copies to gain unique functions not seen in species with less complex sets of homeotic genes, increasing the likelihood of nonhomeotic effects of mutations in these genes.

29. In both cases, males have only one X chromosome and females have two. In mammals the "extra" X in the female is shut down, so only one X chromosome functions in both sexes. This leads to mosaic expression of X-linked genes in females. If a female is heterozygous for an X-linked gene (ignoring for the moment the fact that many X-linked genes are not fully inactivated), she will be mosaic in terms of expression of X-linked genes. On average, half of her cells will be expressing each specific allele. In *Drosophila*, the female continues to express genes from both her X chromosomes. Male *Drosophila* upregulate the levels of gene expression from their single X to compensate.

30. 1. Blocking factor. This gene is not yet characterized but probably encodes a *trans*-acting protein factor that must bind the CE. A null mutation will lead to failure to specify an X to remain active, and thus inactivation of both X chromosomes. A hypermorph may produce enough blocking factor to bind the CE of both X chromosomes, and thus the cell will attempt to maintain them both as active.

2. XIST. The transcription product of XIST is thought to associate with the inactive X chromosome and apparently initiates and maintains a "closed" chromatin configuration. A loss of XIST function will prevent X-inactivation from occurring. A hypermorph might overwhelm normal mechanisms for selecting Xa and inactivate both chromosomes.

3. TSIX. The transcription product of TSIX can bind with XIST and prevent it from blocking factor binding to CE. A loss of function may prevent the cells from efficiently recognizing a CE and selecting Xa. A hypermorph may be *too* efficient in destabilizing XIST RNA and prevent X inactivation entirely.

31. The most efficient system would be if Xist and blocking factor would not be synthesized in males, since they aren't needed. However, we know that males with an extra X chromosome will undergo X inactivation, so the basic elements of the machinery must be present. If blocking factor is produced, it will efficiently bind to the single CE present. The single CE is the only possible target, so amounts of blocking factor are not competing for multiple sites. It's pretty hard to make a guess what XIST will do in males. Excessive activation of XIST may overcome the blocking of CE and lead to the fatal inactivation of the cell's only X chromosome.

32. Embryonic stem cells will be subjected to homologous recombination using a disrupted CE element. Selected cells demonstrated to lack CE will be transplanted into blastocysts, which will be transferred into receptive females for gestation. You expect chimeric offspring to be generated. Within each chimera, individual cells which lack a functional CE entirely may not be able to select an X to remain active; failure to maintain gene expression from the X chromosome will lead to death of the cell. Some cells will contain one X chromosome with a functional CE and one X chromosome with a nonfunctional CE, leading to selective inactivation of the latter. Germ-line chimeras may be able to pass the CE-inactivated X to their offspring.

33. Yes, although generation of homozygous strains will be difficult or impossible. A single mutant generated by knockout technology will be balanced by the wild-type CE element on the other X chromosome; that X chromosome with a functional CE will remain active while the other(s) are inactivated. Males may possibly be viable even if their only X chromosome contains the nonfunctional CE; if so, the line could be maintained by crossing such males with heterozygous or nonmutant females. If not, the line could be maintained by crossing heterozygous females to nonmutant males. In either case, selection of CE mutant progeny would be required every generation. Homozygous females without functional CEs may inactivate both chromosomes instead of only one, likely with a lethal result.

34. An important line of evidence would be the specific pattern of expression of the gene. If the gene is *not* expressed in fin precursor tissues, or if the gene is expressed in many tissues in addition to fin precursors, we may become suspicious of our initial idea. If the gene is expressed in the right place and time, we have one supporting piece of evidence. Knocking the gene out should prevent the formation of fins;

expressing the gene where fins are not normally present should lead to production of fins in those locations.

35. Double mutants that are homozygous for both mutations should be considered. If the two genes operate in the same pathway, there should be no or very little effect on the phenotype in the double mutant—disruption and inactivation of the pathway by one mutant results in the same phenotype even if the pathway is also disrupted somewhere else and thus is inactive. If they operate in different pathways, we might expect a more severe or broader phenotype in the double mutant, since their effects essentially add together. Molecular analysis of the structure and probable function of the two genes may also be useful.

36. Eyeless should be able to partially rescue *Pax6* mutants. However, the success of potential rescue will be determined by how closely the "rescue" expression of eyeless replaces the normal pattern of *Pax6* function, and thus how effectively eyeless is able to substitute for the missing *Pax6*.

37. There must be other factors at work in R1, R3, R4, and R6, to prevent the signal transduction cascade that activates R7 specific development, or to change the result of that cascade. Either another gene must be present in the four cells that "ignore" the Boss signal (in fact, this has been shown; the gene is called seven-up and is a transcription factor expressed only in R1, R3, R4, and R6), or a gene present in the future R7 must be lacking from the other cells (proposals might include downstream components of the pathway).

38. Since the Bcl-2 and Bcl-x_i are known to suppress apoptosis, delivery of these gene products to the injured region may be a potential therapy. Conversely, blocking the action of Bax, Bad, and/or Bac could prevent their normal function to promote apoptosis. In the neural development example, we see that neurotrophic factors are required for survival of correctly connected neurons. In the limb development example, we see that inhibition of BMP signaling prevents apoptosis in the duck hindlimb. Delivery of neurotrophic factors or activation of BMP inhibitors in the injured regions may also be potential therapies.

In all of these examples, we must consider what effect the treatment will have on healthy neurons and glial cells in the area. If the treatment is to be delivered via the bloodstream, we must also be concerned about potential effects on all other cell types in the body, especially for situations where apoptosis is required to maintain healthy tissue.

Secondly, delivery mechanism is a major concern. If a local injury such as a blow to the spinal cord is to be treated, ideally the therapy could be delivered only to that region. However, we must also ensure that the treatment penetrates the tissue to reach all cells where it would be helpful. In some cases of well-localized brain tumors, direct injection into the tumor is possible, but physical methods of delivery in general run a significant risk of generating further injury.

This is an area of active study, both for treatment of spinal cord injury and brain injury, where neural loss may be irreversible, and the reverse goal to enhance apoptosis in tumors.

CHAPTER INTEGRATION PROBLEM

a.

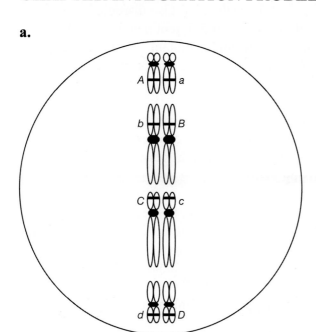

This is one of several possible arrangements.

b. There are $2^4 = 16$ possible allelic combinations in gametes.

c.

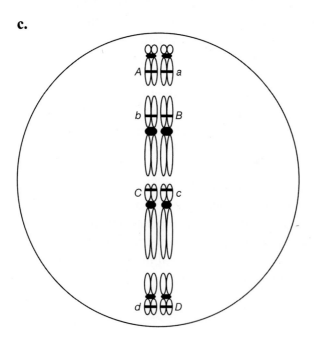

Starting from this arrangement at meiosis I, and assuming that no crossing over occurs between *C* and *E* loci (estimated 100% frequency), the combinations will be *ABCd*, and *abcDEF* at approximately equal frequencies.

Crossing over will generate additional genotypes. For example, if a crossover occurs between gene *A* and its centromere, four types of gametes will be produced: *ABCd*, *aBCd*, *AbcD*, and *abcD*.

d. The diagram of female meiosis should include polar bodies and one viable oocyte. Color coding should indicate the alleles from the male parent [one gamete from part (c)] and the alleles from the female parent (the preceding oocyte) in different colors.

e. Any zygote with the *AA* genotype will be embryonic lethal and therefore unable to survive to adulthood. *Aa* and *aa* zygotes are expected to be viable.

f. *Aa* females are expected to display the maternal effect, which is presumably incompatible with viable offspring. *aa* females will be fertile and produce viable offspring.

g. Because the phenotype of this gene is a maternal effect phenotype, a traditional mutant screen will require at least four generations of breeding to generate homozygous mutant females and screen for embryonic lethals. A genetic mosaic screen, for instance, using the FLP–FRT technique, will be useful to reduce generations of breeding. If the gene sequence is understood, a TILLING approach may possibly be a richer source of mutations in this specific gene than a random mutagenesis approach.

CHAPTER 22 CANCER GENETICS

Chapter Goals
1. Understand the various stages of cancer progression and the cellular basis of each stage.
2. Describe different types of carcinogens and how they function to modify DNA.
3. Explain how the cell cycle is regulated by checkpoints and understand how various types of mutations disrupt the regulation of the cell cycle.
4. Diagram the various pathways involved in cell death (apoptosis) and describe how apoptosis can be altered during cancer progression.
5. Describe the similarities and differences between oncogenes and tumor suppressor genes, the types of mutations associated with each class, and their roles in cancer development.
6. Compare and contrast the development and types of mutations associated with sporadic cancers and familial cancers.
7. Explain the cellular events involved in tumor progression, the reason behind the two-hit hypothesis, and how this relates to colorectal cancer progression.

Key Chapter Concepts
22.1 What Is Cancer?
- **Cancer:** Cancer is now recognized as a genetic disease that results from the inappropriate activity of certain genes. Cancers arise from a single aberrant cell that proliferates and does not respond to normal mitotic controls. This inappropriate growth and division results when genes are mutated or when genes are imported, activated, or inactivated by viruses. Because DNA mutations can initiate tumor formation, a predisposition for cancer formation can be inherited. When a cell begins to proliferate out of control, it forms a tumor that is usually benign. However, some cells may break away and generate tumors in other parts of the body (**metastasis**). To become metastatic, the cells must acquire new characteristics by turning on other normal genes inappropriately. Metastasic tumor formation is believed to be a multistep process that involves both DNA instability and natural selection. Generally, there are two types of genes that balance cell division and cell death: tumor-suppressor genes and protooncogenes.
- **Tumor-Suppressor Genes:** Tumor-suppressor genes normally act to suppress malignant growth. However, if mutated, the homozygous recessive genotype results in a loss of gene function and the formation of cancer. Examples of tumor-suppressor genes include the genes for **retinoblastoma** (Rb1) and **p53**. A normally functioning *p53* gene will trigger programmed cell death (**apoptosis**) if damage to the DNA is severe. Approximately 50% of cancers involve loss of *p53* activity.
- **Protooncogenes and Oncogenes:** A protooncogene normally works in the cell to stimulate growth and division under appropriate conditions. These genes are found in normal, nontransformed cells. A protooncogene can be converted to an oncogene by mutation, by movement of the gene within the genome to a region with a strong promoter or enhancer, or by amplification of the gene. One copy of an oncogene is capable of transforming a cell, so it is considered a dominant mutation. Oncogenes are found in retroviruses and in transformed cells.

22.2 Carcinogens
- **Chemical Carcinogens:** More than 90% of all human tumors arise spontaneously and are induced by human-created or natural carcinogens. Chemical carcinogens include alkylating agents, aralkylating agents, and arylhydroxylamines. These compounds form stable altered nucleotides or **DNA adducts** typically leading to a point mutation that is inherited by the daughter cell. Most of these carcinogens are absorbed through the lungs and intestines and are oxidized in the liver by enzymes in the cytochrome P450 family. These enzymes generate **proximate carcinogens** that can be more readily excreted by the kidneys, but that can also be transported from the liver to target cells where they can be further metabolized to form active **ultimate carcinogens**.
- **Environmental Carcinogens:** Tobacco smoke, containing benzo[a]pyrene, dimethylbenzanthracene, arylhydroxylamines, and other polyaromatic hydrocarbons (PAH) is the most important environmental carcinogen accounting for 30% of all human cancer deaths. Aflatoxin B1 is a naturally occurring compound found in the mold *Aspergillus flavus* that also causes the formation of DNA adducts. Consumption of red meat and animal fat has also been

implicated in certain forms of cancer. Asbestos is inhaled and disrupts normal segregation leading to mesothelioma.

- **Radiation:** Ultraviolet radiation (particularly UV-B) induces the formation of pyrimidine cyclobutane dimers in the DNA which induce G–C to A–T transition mutations or frameshift mutations. UV radiation can induce skin cancers such as squamous cell carcinoma, basal cell carcinoma, or the most aggressive form of skin cancer, malignant melanoma. Ionizing radiation (X-rays, γ-radiation) produces transversion mutations, single- and double-strand breaks in the DNA. Cells with large chromosomal aberrations are often tumorigenic.

- **Viral Carcinogens:** Viral infections account for 5–10% of all cancers worldwide. Some forms of the **human papilloma virus** (**HPV**) can cause cervical cancer. The viral proteins E6 and E7 bind to and inactivate the cellular p53 and Rb1 proteins, respectively, resulting in **hyperplasia**, dysplasia, and cancer. **Adenovirus** also encodes two proteins, E1A and E1B, that bind RB1 and p53, respectively, and disrupt their activities. In the **polyomavirus**, **SV40**, the large T antigen binds to both p53 and Rb1 and blocks their regulation of the cell cycle. **Epstein–Barr virus** (**EBV**) infects primary B cells and induces hyperplasia. This virus encodes the EBNA-2 protein that increases the expression of cellular Myc. Hepatitis B and hepatitis C infection are associated with liver cancer.

22.3 The Cell Cycle

- **Regulation of the Cell Cycle:** Different **cyclin/CDK complexes** regulate the progression through the G_1/S transition, S phase, and the G_2/M transition of the cell cycle. Cyclins act as regulatory subunits that bind and activate the cyclin-dependent kinases. The kinase activity of each complex is also regulated by cyclin kinase inhibitory proteins that are induced in response to DNA damage. Growth factors and cytokines promote cell cycle progression by binding to specific **transmembrane receptors** and initiating signaling cascades within the cell that result in the transcriptional activation of specific target genes.

22.4 Apoptosis

- **Induction of Apoptosis:** Cells initiate apoptosis because of the loss of paracrine survival signal or because of the presence of extrinsic or intrinsic cell death signals such as intracellular stress or DNA damage. The mitochondria serve as sensors of the general health of the cell.

- **Regulation of Apoptosis:** Cleavage of the Bax protein to t-Bax and its insertion into the outer mitochondrial membrane promotes apoptosis. In normal cells, the cleavage of Bax is blocked by the Bcl-2 and BclX1 proteins which form heterodimers with Bax to prevent the formation of t-Bax. When the levels of the Bad protein increase in the cytoplasm, Bcl-2 preferentially heterodimerizes with Bad, releasing Bax and allowing it to be cleaved. Signaling through receptor tyrosine kinases can also regulate apoptosis. Activation of phospholipase C as a result of ligand binding to specific receptor tyrosine kinases ultimately results in the activation of Akt which phosphorylates Bad and sequesters it by binding to the protein 14-3-3. This pathway is regulated by PTEN which is a phosphatase that can block Akt activation.

22.5 Spontaneous Cancers

- **Activation of Oncogenes:** Ninety percent of tumors occur as a result of spontaneous random mutations. Protooncogenes encode proteins that promote cell cycle progression. Gain-of-function mutations in these genes can result in the creation of an oncogene. Overexpression of cyclins, growth factors, or of the Myc protein can lead to a loss of cell cycle regulation and continuous stimulation of cell division. Mutations that result in constitutively activated Ras can not only block the induction of cell death but also can promote cell cycle progression and stimulate cell division.

- **Mutations in Tumor Suppressor Genes:** Tumor suppressor genes regulate cell cycle progression, or cell survival and cell death. Inactivating mutations in both alleles of a tumor suppressor gene leads to an increase in cell number and cancer. *RB1*, *p53*, *BrCA-1*, *BrCA-2*, and *PTEN* are all tumor suppressor genes. Loss of function of the Rb1 protein shortens the cell cycle and eliminates the opportunity to repair DNA damage. Loss of p53 leads to an inability to inhibit cell cycle progression when DNA damage is present and loss of activation of apoptosis. BRCA-1 and BRCA-2 mutations result in the loss of DNA repair and accumulation of DNA double-strand breaks, while mutations in the PTEN protein result in the proliferation of cells with DNA damage.

22.6 Cancer Predisposition and Familial Cancers

- **Genetic Predisposition to Cancer:** Most gain-of-function oncogenic mutations that affect the regulation of the cell cycle are embryonic lethal. However, individuals that inherit a single mutant copy of a tumor suppressor gene are usually not affected during development, but instead have a genetic predisposition to cancer. In these individuals, a second spontaneous mutation in the second copy of the tumor suppressor gene will lead to cancer. The likelihood of this occurring is greater than two spontaneous mutations occurring in both alleles of an individual that inherits two wild-type copies of a tumor suppressor gene. Loss of heterozygosity in an individual with a mutation in one allele of *RB1*, *p53*, *PTEN*, or another tumor suppressor gene will lead to cancer because of loss of cell cycle regulation or the inhibition of apoptosis.

22.7 Tumor Progression

- **Multiple-Hit Hypothesis:** Statistical analysis of sporadic and hereditary cancers suggests that two genetic "hits" or mutations are required for the initiation of tumor progression. These two mutations lead to the formation of a benign hyperplasia. Gain-of-function mutations in oncogenes or loss-of-function mutations in both alleles of a tumor suppressor gene increase cell cycle progression and reduce the opportunity for DNA repair leading to the accumulation of additional mutations that affect cell division, DNA repair proteins, and apoptotic pathways. Additional mutations lead to the generation of a malignant metastatic disease. The generation of mutations in different individuals and in different forms of cancer varies and does not seem to follow a specific pattern. Colorectal cancer is an exception to this statement because mutations in this cancer do seem to occur in a predictable sequence.

Key Terms

- predisposition
- cancer
- protooncogenes
- tumor suppressor genes
- hyperplasia
- carcinogens
- metastasis
- malignant
- DNA adducts
- human papilloma virus (HPV)
- adenovirus
- polyomavirus
- Epstein-Barr virus (EBV)
- hepatitis B virus
- hepatitis C virus
- M phase
- interphase
- G_1 phase
- S phase
- G_2 phase

- cyclin-dependent kinase (CDK)
- paracrine signaling
- transmembrane receptor
- GTPases
- mitogen-activated protein kinases (MAPKs)
- apoptosis
- precondensation stage
- condensation phase
- nuclear condensation phase
- fragmentation phase
- phagocytosis stage
- mitochondrial membrane permeability transition
- oncogene
- cyclins
- growth factor receptor genes
- Ras proteins
- Myc protein
- retinoblastoma 1 (*Rb1*)

- genomic instability
- *p53*
- *BRCA-1*
- *PTEN*
- retinoblastoma
- loss of heterozygosity (LOH)
- Li-Fraumeni syndrome
- multiple-hit hypothesis

Understanding the Key Concepts

Use the key terms or parts of their definitions to complete the following sentences.

Most human tumors arise (1)_____, while only a small percentage of tumors are (2)_____. (3)_____ mutations in protooncogenes or (4)_____ mutations in tumor suppressor genes can lead to (5)_____ and primary tumor formation. Individuals that inherit a mutant copy of a (6)_____ or a (7)_____ have a (8)_____ to cancer because the generation of a single (9)_____ mutation in the wild-type allele of these genes will lead to a loss of (10)_____ and (11)_____ formation. Most primary tumors are (12)_____ because their cells do not possess the ability to (13)_____ and (14)_____ healthy tissues. As these cells accumulate additional (15)_____, they become more aggressive and are more likely to become (16)_____.

Most spontaneous mutations arise as a result of exposure to (17)_____ found in (18)_____, (19)_____, or (20)_____. These chemicals are either intrinsically reactive with (21)_____ or are metabolized by the (22)_____ in the (23)_____ to generate (24)_____ that can then be further metabolized by tissue-specific (25)_____ to form (26)_____. These (27)_____ can react with (28)_____ to form a (29)_____. If this (30)_____ is not repaired, a (31)_____ mutation will be generated that will be inherited by the daughter cell. (32)_____ from the sun can also act as a (33)_____ to introduce (34)_____ or (35)_____ mutations in the DNA that can lead to (36)_____ cancer. Some (37)_____, such as HPV 16, (38)_____, and polyomavirus can also cause cancer because they encode proteins that inactivate the tumor suppressor genes, (39)_____ and (40)_____. (41)_____ virus encodes a protein that induces the expression of the (42)_____ gene which regulates the expression of the (43)_____ genes that control cell cycle progression.

Regulation of (44)_____ and (45)_____ are critical to the maintenance of homeostasis within the tissues of an organism. Protooncogenes encode proteins that stimulate (46)_____ progression, while tumor suppressor genes encode proteins that regulate (47)_____ or activate (48)_____. Mutations that result in the loss of these regulatory mechanisms can lead to an increase in (49)_____ and primary tumor formation.

Apoptosis is induced by the loss of extracellular (50)_____ signaling or by intracellular (51)_____ or (52)_____. The induction of apoptosis results in the increase in cytoplasmic levels of the (53)_____ protein which disrupts (54)_____ heterodimers by binding to (55)_____, releasing (56)_____ which is then cleaved by (57)_____ to form (58)_____. (59)_____ inserts into the outer (60)_____ membrane leading to the release of (61)_____ which binds to Procaspase-9 and APAF to form the (62)_____. Signaling through receptor tyrosine kinases also regulates apoptosis because it activates (63)_____ which phosphorylates and sequesters the (64)_____ protein, promoting cell (65)_____. The (66)_____ protein can block (67)_____ activation to induce cell (68)_____. Mutations in the (69)_____ gene that inhibit apoptosis, like overexpression of (70)_____, can result in tumor formation.

Figure Analysis

Use the indicated figure or table from the textbook to answer the questions that follow.

1. **Figure 22.1**

    (a) How could a single mutation in a cell allow the mutant cell to grow more rapidly than the wild-type cells?

    (b) Why are most primary tumors typically benign?

    (c) What are the mechanisms by which additional mutations can be generated in the cells of a primary tumor?

(d) What is the likely effect of additional mutations being generated in the cells of a primary tumor?

(e) When is a tumor considered to be malignant?

2. **Figure 22.2**
(a) How can exposure to ultraviolet light cause cancer? What types of cancers can exposure to UV light cause? Which form is the most severe?

(b) Why doesn't exposure to UV light lead to other forms of cancer?

(c) Describe two different types of DNA mutations that can be caused by ionizing radiation. How can these mutations lead to cancer?

(d) How can excessive alcohol use lead to cancer?

(e) Describe one way in which sexual behavior has been linked to cancer.

(f) Explain how asbestos, a commonly used industrial product, can cause cancer.

3. **Figure 22.12**
(a) How is Ras activated? How is it inactivated?

(b) Generally describe two different Ras mutations that could keep the protein constitutively activated?

(c) What two pathways are stimulated by Ras activation?

(d) What would be the effect on these two pathways if Ras was constitutively activated?

4. **Figure 22.14**
(a) What are the two mechanisms by which the intrinsic apoptotic pathway can be triggered?

(b) In normal cells, how is the intrinsic apoptotic pathway blocked?

(c) What two proteins heterodimerize with Bcl-2? Which heterodimer promotes life? Which heterodimer promotes cell death?

(d) When can Bax be cleaved to form t-Bax? What protein cleaves Bax?

(e) How are Bad protein levels in the cytoplasm regulated?

(f) What is the effect of t-Bax inserting into the outer mitochondrial membrane?

(g) What proteins form the apoptosome?

(h) What are the functions of caspase-9 and caspase-3?

(i) How can chemotherapeutic therapies affect the intrinsic apoptotic pathway?

(j) What is the function of Akt? How is Akt activated?

(k) What is the effect of Akt activation on apoptosis?

(l) How does PTEN regulate this pathway?

(m) Explain how loss-of-function mutations in *PTEN* could lead to cancer.

(n) What would be the effect of Bcl-2 overexpression on the intrinsic apoptotic pathway?

General Questions

Q22-1. Most spontaneous mutations do not affect the phenotype of a cell. Explain this statement. What types of mutations are likely to cause cellular defects? What types of mutations can lead to cancer?

Q22-2. How are chemical carcinogens metabolized? What is the difference between a proximate and an ultimate carcinogen?

Q22-3. Explain how viral infections can cause cancer.

Q22-4. What is the function of the cyclin-dependent kinases? How is their activity regulated? Why do cyclins D and E act as oncogenes when they are overexpressed?

Q22-5. What functions of *Rb1* and *p53* make these tumor suppressor genes?

Q22-6. Explain why overexpression or constitutive activation of growth factors, Ras, and Myc has oncogenic potential.

Multiple Choice

For each of the following, circle the letter of the choice that most appropriately answers the question.

1.) What percent of human tumors are familial or hereditary?
   a. 5%
   b. 10%
   c. 20%
   d. 40%
   e. 60%

2.) Burkitt lymphoma is caused by which of the following?
   a. translocation of the *myc* gene
   b. excess mitogenic signaling
   c. inactivation of p53 and Rb1
   d. Ras mutations
   e. inhibition of apoptosis

3.) Which of the following carcinogens is *not* found in cigarette smoke?
   a. dimethylbenzanthracene
   b. benzo[a]pyrene
   c. aflatoxin B1

   d. arylhydroxylamines
   e. polyaromatic hydrocarbons

4.) What form of cancer are *PTEN* mutations commonly associated with?
   a. melanoma
   b. ovarian
   c. prostate
   d. mesothelioma
   e. pancreatic

5.) Which of the following genes is mutated in over 50% of all cancers?
   a. *Brca1*
   b. *Rb1*
   c. *fos*
   d. *PTEN*
   e. *p53*

6.) Tobacco smoke accounts for what percentage of all human cancer deaths?
 a. 5%
 b. 10%
 c. 20%
 d. 30%
 e. 50%

7.) Multistep progression of which of the following types of cancer appears to follow a predictable sequence of mutations?
 a. breast
 b. colorectal
 c. lung
 d. prostate
 e. ovarian

8.) Which of the following viruses cause changes in the growth factor environment within the cell?
 a. adenovirus
 b. HBV

c. polyomavirus
 d. EBV
 e. HPV 16

9.) Which of the following cyclin/CDK complexes is required for the initiation of chromosome segregation?
 a. B/CDK1
 b. A/CDK2
 c. E/CDK2
 d. D/CDK4
 e. D/CDK6

10.) Individuals with inactivating mutations in both alleles of *PTEN* develop tumors as a result of which of the following?
 a. constitutively activated Ras
 b. rapid cell cycle progression
 c. failure of cells to undergo apoptosis
 d. an inability to repair DNA damage
 e. an inability to inactivate E2F

Practice Problems

P22-1. You are given the task of investigating a new gene that may be involved in cancer. You have: cells that are normal, cells that are mutant for your gene of interest, the normal gene contained on a plasmid, and the mutant gene on a plasmid. You transfect the normal and mutant gene into both sets of cells (mutant and normal) and find the following results concerning whether the cells continue to undergo cell division indefinitely (+) or not (−). Does the ***normal*** gene encode a tumor suppressor gene or a protooncogene? Explain.

Cell type	+ No DNA	+ Normal gene	+ Mutant gene
Normal	−	−	−
Mutant	+	−	+

[*Hint 1*: When determining whether a cancer gene of interest is a mutant tumor-suppressor gene or an oncogene, examine whether adding a normal version corrects the aberrant cell division phenotype of the mutant cell. Adding the normal version of an oncogene (proto-oncogene) to a cell that is already mutant for that gene will not correct the dominant mutation that already exists in the cell. However, adding the normal version of a tumor suppressor gene to a cell mutant for that gene will correct the cell.]

P22-2. Categorize each of the following as a protooncogene or as a tumor suppressor gene.
 a. *myc*   _____
 b. *BRCA-1*   _____
 c. *Rb1*   _____
 d. *ras*   _____
 e. *bcl-2*   _____
 f. *p53*   _____
 g. *Wt1*   _____
 h. *fos*   _____
 i. *raf*   _____
 j. *Apc*   _____

P22-3. For each mutation listed, indicate how it would affect the cell. Then indicate the cancer associated with each of the mutations.

a. Translocation of the cyclin E gene adjacent to the IgH promoter region on chromosome 14

b. Hypermorphic mutation in *myc*

c. Null mutation of *BRCA-1*

d. Mutation in a growth factor receptor that prevents ligand dissociation and keeps Ras constitutively activated

P22-4. For each cell cycle phase, indicate the critical events that take place.

a. G_1 phase:

b. S phase:

c. G_2 phase:

d. M phase:

P22-5. For each cyclin, indicate the CDK protein it binds to and activates. Indicate which phase of the cell cycle each cyclin/CDK complex is required for.

Cyclin	CDK	Cell Cycle Requirement
a. Cyclin D		
b. Cyclin E		
c. Cyclin A		
d. Cyclin B		

P22-6. Match the five stages of apoptosis with the hallmark events that occur during that stage on the right.

_____ a. precondensation   i. Cell is subdivided into several apoptotic bodies.

_____ b. condensation   ii. Interaction between the dying cell and neighbors is lost.

_____ c. nuclear condensation   iii. Apoptotic cell is engulfed by neighboring cells.

_____ d. fragmentation   iv. DNA is cleaved and redistributed to nuclear margins.

_____ e. phagocytosis   v. Cell receives the signal to induce cell death.

P22-7. *K-ras* mutations that change the Gly residue at codon 12 (5′-GGT-3′) to Asp (5′-GAT-3′), Val (5′-GTT-3′), or Ser (5′-AGT-3′) are associated with pancreatic cancer. These three mutations result in constitutive activation of the K-ras enzyme. In addition, all three mutations disrupt the *Mva*I restriction site at codon 12 creating a restriction fragment length polymorphism.

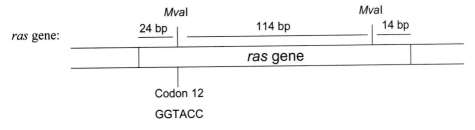

a. Given two tissue samples from different biopsies, explain how you could determine if the pancreatic tissue has a *K-ras* mutation indicating that it is likely cancerous.   What two controls should be used? Explain why.

b. Draw what the agarose gel would look like if sample 1 does not contain a *K-ras* mutation, while sample 2 does contain a mutation at this site. Indicate the size of the fragments. Indicate what the two control lanes would look like as well.

Control 1   Sample 1   Sample 2   Control 2

Assessing Your Knowledge

Understanding the Key Concepts—Answers

1.) spontaneously; 2.) familial or inherited; 3.) Gain-of-function; 4.) loss-of-function; 5.) hyperplasia; 6–7.) protooncogene; tumor suppressor gene; 8.) predisposition; 9.) spontaneous; 10.) heterozygosity; 11.) tumor; 12.) benign; 13.) metastasize or migrate; 14.) invade; 15.) mutations; 16.) malignant; 17.) carcinogens; 18–20.) air; food; tobacco smoke; 21.) DNA; 22.) cytochrome P450 oxidases; 23.) liver; 24.) proximate carcinogens; 25.) cytochrome P450s; 26.) ultimate carcinogens; 27.) ultimate carcinogens; 28.) DNA; 29.) DNA adduct; 30.) DNA adduct; 31.) point; 32.) UV radiation; 33.) carcinogen; 34–35.) transition; frameshift; 36.) skin; 37.) viruses; 38.) adenovirus; 39–40.) *p53*; *Rb1*; 41.) Epstein–Barr; 42.) *c-myc*; 43.) cyclin; 44–45.) cell division; apoptosis; 46.) cell cycle; 47.) cell cycle checkpoints; 48.) apoptosis; 49.) cell number; 50.) growth factor; 51.) stress; 52.) DNA damage; 53.) Bad; 54.) Bcl-2/Bax; 55.) Bcl-2; 56.) Bax; 57.) caspase-8; 58.) t-Bax; 59.) t-Bax; 60.) mitochondrial; 61.) cytochrome c; 62.) apoptosome; 63.) Akt; 64.) Bad; 65.) survival; 66.) PTEN; 67.) Akt; 68.) death; 69.) *PTEN*; 70.) *bcl-2*.

Figure Analysis—Answers

1a.) A gain-of-function mutation in a protooncogene or a loss-of-function mutation in a tumor suppressor gene could allow a mutant cell to grow more rapidly than a wild-type cell. A mutation in a protooncogene could lead to constitutive activation which could either disrupt the regulation of the cell cycle or the apoptotic pathway. In addition, if an individual is heterozygous for a mutation in a tumor suppressor gene, a single mutation would result in a loss of heterozygosity that would lead to an increase in cell cycle progression or disruption of the normal regulation of the apoptotic pathway. Either of these types of mutations would allow the cell containing the loss-of-function mutation in the tumor suppressor gene to grow more rapidly than the wild-type cells; b.) Most primary tumors are typically benign because a single mutation in a protooncogene or in a tumor suppressor gene in a heterozygous individual can result in the formation of a primary tumor. However, additional mutations are required for a cell to migrate from its original location to colonize another site and for the cancer to progress. Typically only a few cells in a primary tumor can acquire this ability. Additional mutations are also required for the migrating cells to be able to invade healthy tissues and become malignant; c.) Additional mutations in the cells of a primary tumor can be generated by the loss of DNA repair mechanisms in the cell, the loss of cell cycle regulatory mechanisms, inappropriate chromosomal segregation, or environmental carcinogens; d.) The generation of additional mutations in the cells of a primary tumor will cause most cells to die, but some cells will acquire the ability to metastasize to a new location from these mutations. Additional mutations may allow cells to invade a new site and proliferate there. The more mutations there are, the more likely the cancer cells will be to invade and become malignant; e.) A tumor is malignant if the cancerous cells have adopted the ability to invade healthy tissues within the body.

2a.) Exposure to UV radiation causes the formation of pyrimidine cyclobutane dimers between adjacent pyrimidines in the DNA. If these distortions in the DNA are not repaired, they can cause G–C to A–T transitions or frameshift mutations. If these mutations affect protooncogenes or tumor suppressor genes, or if they disrupt cell cycle regulation or the apoptotic pathway, they could lead to cancer. Exposure to UV light can cause different forms of skin cancer, including squamous cell carcinoma, basal cell carcinoma, or malignant melanoma. Malignant melanoma is the most severe of these because it is the most aggressive form of skin cancer. b.) Exposure to UV light can only cause skin cancer, but not other forms of cancer because UV light can only penetrate the first few layers of skin. Other cells are protected from exposure to UV light by the skin; c.) Ionizing radiation can produce oxidized nucleotides from the reaction of oxygen free radicals with the DNA. If not repaired, these oxidized nucleotides can produce transversion mutations. If these mutations affect protooncogenes or tumor suppressor genes, or if they disrupt cell cycle regulation or the apoptotic pathway, they could lead to cancer. Ionizing radiation can also cause double-strand breaks that can result in chromosomal rearrangements such as deletions, translocations, or insertions. Cells with these types of mutations often lead to tumor formation because of disruption of the regulation of the cell cycle; d.) Excessive alcohol use can lead to scarring of the liver that leads to chronic inflammation. These inflammatory cells continually secrete cytokines leading to a loss of cell cycle regulation in the hepatocytes over a long period of time. This loss of cell cycle regulation can initiate tumor formation and lead to additional mutations that cause hepatocellular carcinoma; e.) Sexual behavior has been linked to cancer because two strains of human papilloma virus that are known to cause cervical cancer are sexually transmitted. Therefore, one of the risk factors for cervical cancer is dictated by sexual behavior and the number of sexual partners an individual has; f.) Asbestos is inhaled and taken up by the lung epithelial cells. It physically disrupts normal chromosome segregation producing chromosomal aberrations that lead to lung cancer.

3a.) Ras is activated by binding to GTP, and it is inactivated by the hydrolysis of GTP to GDP; b.) One mutation that could keep Ras constitutively activated is a mutation in the active site of the protein that prevents the hydrolysis of GTP to GDP. A second constitutively activating Ras mutation would be a mutation in the GTP-binding pocket that prevents the release of the phosphate from GTP following hydrolysis; c.) Ras activation stimulates the Akt pathway and stimulates the activation of MAP kinases that activate transcription factors which regulate the expression of genes required for cell cycle progression, such as cyclin D and Myc; d.) If Ras was constitutively activated, it would continually stimulate the Akt pathway causing Bad to be maintained in a phosphorylated state and blocking the induction of cell death by the intrinsic apoptotic pathway. Constitutively activated Ras would also increase cyclin D levels which would lead to loss of cell cycle regulation, promoting the cells to rapidly progress through G_1 and into S phase and stimulating them to divide.

4a.) The intrinsic apoptotic pathway can be triggered by the loss of extracellular growth factor signaling or by an increase in intracellular stress or DNA damage; b.) The intrinsic apoptotic pathway is blocked in normal cells by Bcl-2 and BclX1 which heterodimerize with the Bax protein to block its cleavage to t-Bax; c.) Bad and Bax both heterodimerize with Bcl-2. The Bcl-2/Bax heterodimer promotes life, while the Bcl-2/Bad heterodimer promotes cell death; d.) Bax can be cleaved to form t-bax when Bcl-2 heterodimerizes with the Bad protein and releases the Bax protein. t-Bax is generated by cleavage of the Bax protein by caspase-8; e.) Bad protein levels in the cytoplasm are regulated by ligand binding to specific receptor tyrosine kinases; f.) t-Bax inserting into the outer mitochondrial membrane forms pores in the membrane that allows the release of cytochrome c from the mitochondria; g.) Cytochrome c binding to procaspase-9 and the APAF protein form the apoptosome; h.) Caspase-9 cleaves procaspase-3 to form caspase-3. Caspase-3 activates DNA fragmentation; i.) Chemotherapeutic therapies can affect the intrinsic apoptotic pathway by causing intracellular stress that leads to an increase in the levels of Bcl-2 in the cells that fail to undergo apoptosis. This makes the cells that are capable of surviving more resistant to subsequent treatments; j.) Akt phosphorylates Bad. Akt is activated by binding of PIP_3 to subsequent phosphorylation of Akt in the Akt/PKB protein complex; k.) Akt activation prevents apoptosis because it keeps Bad phosphorylated and bound by the protein 14-3-3 so that it cannot disrupt the Bcl-2/Bax heterodimers; l.) PTEN regulates this pathway by metabolizing PIP_3 and blocking Akt activation; m.) Loss-of-function mutations in *PTEN* could cause cancer because these mutations would keep PIP_3 from being metabolized which would prevent Akt from being deactivated. This would keep the Bad protein phosphorylated and sequestered preventing the removal of cells with DNA damage or cellular stress by the normal apoptotic pathway. Proliferation of these cells could lead to tumor progression; n.) Bcl-2 overexpression also blocks the normal apoptotic pathway because in this situation, there is plenty of Bcl-2 protein to bind all of the Bax protein preventing its cleavage to t-Bax. Even if cytoplasmic levels of the Bad protein were to increase, there would be enough free Bcl-2 to bind to both Bax and Bad such that the Bcl-2/Bax heterodimers would not be disrupted. Therefore, Bax could not be cleaved to induce apoptosis.

General Questions—Answers

Q22-1. Most spontaneous mutations do not affect the phenotype of a cell because they are typically point mutations located in regions of the DNA outside of the genes, in the introns of genes, or in the wobble position of codons, or they result in conservative amino acid substitutions in the affected protein. Spontaneous mutations in the exons of genes that make nonconservative amino acid changes in the proteins that these genes encode are likely to affect the activity of the proteins and potentially lead to cellular defects. Loss-of-function mutations in tumor suppressor genes, gain-of-function mutations in oncogenes, or mutations that disrupt the regulation of the cell cycle or apoptosis can cause an increase in cell number and lead to hyperplasia.

Q22-2. Chemical carcinogens are metabolized by the P450 mixed-function oxidases in the liver. This alters the chemical structure of the carcinogen to generate a proximate carcinogen. The proximate carcinogen can be more readily excreted by the kidneys, but is also more likely to enter target cells. It is less reactive than an ultimate carcinogen because it is not intrinsically reactive with the DNA. However, in some cells the proximate carcinogen will be further metabolized by tissue-specific cytochrome P450s. This will lead to the production of an ultimate carcinogen that is extremely reactive because it can readily form covalent bonds with nucleotides in the DNA.

Q22-3. Viruses can cause cancer by producing proteins that interfere with cell cycle control. Human papillomaviruses, adenoviruses, and polyomaviruses all encode proteins that interfere with p53 and Rb1 activity, disrupting cell cycle regulation and initiating tumor formation. Epstein–Barr virus encodes a protein that interacts with transcription factors to increase the expression of c-Myc, a protein that regulates the expression of cell cycle proteins. Hepatitis B and C infections cause inflammation of the liver that causes inflammatory cells to secrete growth factors which drives hepatocytes through the cell cycle disrupting its normal regulation and promoting tumor formation.

Q22-4. Cyclin-dependent kinases (CDKs) are cellular proteins that promote progression through the different phases of the cell cycle when they are complexed with cyclin proteins. Cyclin-dependent kinases are regulated by cyclin proteins which are transiently expressed at different times during the cell cycle. Cyclins heterodimerize with CDKs to activate them, promoting progression through specific cell cycle transitions. Cyclin/CDK complexes are also modulated by cyclin kinase inhibitory proteins which bind to the complexes in response to DNA damage to arrest the cell cycle prior to S phase allowing DNA repair to take place prior to DNA replication. Overexpression of cyclins D and E keeps the cyclin/CDK complexes that they participate in, in an active state. Cyclins D and E regulate progression through the G_1 restriction point and the G_1/S transition, respectively. In normal cells, degradation of these cyclins returns the CDK to an inactive state. When cyclin D or E is overexpressed, however, cells are continually permitted to transit the G_1 restriction point or the G_1/S transition, disrupting normal cell cycle control.

Q22-5. *Rb1* is considered to be a tumor suppressor gene because of its involvement in the cell cycle. The Rb1 protein binds to the E2F transcription factor in G_1 of the cell cycle and inactivates it. Release of E2F from Rb1 is a critical regulatory mechanism in the progression of the cell cycle through G_1 and the G_1/S transition because E2F acts as a transcription factor to activate the expression of genes required for cell cycle progression at this stage. Loss of the Rb1 protein shortens the cell cycle and reduces or eliminates the opportunity to repair DNA damage which can lead to genomic instability and ultimately cancer. *p53* is considered to be a tumor suppressor gene because the p53 protein regulates the transcription of cyclin kinase inhibitory proteins which bind to cyclin/CDK complexes to inhibit the cell cycle before S phase when DNA damage is present. This permits DNA repair to take place prior to DNA replication. In addition, *p53* is a tumor suppressor gene because it promotes apoptosis of damaged cells by activating the transcription of the *bax* gene and repressing the transcription of the *bcl-2* gene.

Q22-6. Growth factor overexpression or constitutive activation has oncogenic potential because this situation can lead to the continuous stimulation of cell division. Constitutively activated Ras has oncogenic potential because it continually stimulates the Akt pathway which blocks the induction of cell death through the intrinsic apoptotic pathway. In addition, it increases the level of cyclin D so cells are stimulated to rapidly progress through G_1 and enter S phase, and are continuously stimulated to divide. Overexpression of Myc also leads to rapid cell cycle progression due to the consistent elevation of the transcription of Myc-responsive genes that promote cell division. Myc can also act with constitutively activated Ras to up-regulate cyclin expression and down-regulate CK1 expression.

Multiple Choice—Answers

1.) b; 2.) a; 3.) c; 4.) c; 5.) e; 6.) d; 7.) b; 8.) b; 9.) a; 10.) c.

Practice Problems—Answers

P22-1. Since adding the normal version of the gene of interest does correct the aberrant cell division of the mutant cell, the gene of interest is a tumor suppressor gene.

P22-2. Protooncogenes: *myc, ras, bcl-2, fos, raf.* Tumor suppressor genes: *BRCA-1, Rb1*, p53, *Wt1, Apc*.

P22-3. a.) Translocation of the cyclin E gene adjacent to the IgH promoter on chromosome 14 would lead to the overexpression of cyclin E. When overexpressed, cyclin E acts as an oncogene because it results in the deregulation of the progression from G_1 to S eliminating the time that the cell normally has to repair damaged DNA prior to S phase and promoting cell cycle progression. In addition, cyclin E/CDK2 remains active promoting the phosphorylation of the Rb1 protein and keeping the E2F transcription factor active. Cyclin E overexpression is associated with breast and colon cancers and some leukemias; b.) A hypermorphic mutation of *myc* will lead to Myc overexpression resulting in the overexpression of Myc-responsive genes including the cyclin genes. This will lead to a loss of cell cycle control and rapid cell division. Mutations in *myc* are associated with leukemia, breast, colon, stomach, and lung cancers, as well as neuroblastoma and glioblastoma; c.) A null mutation of *BRCA-1* will result in the loss of DNA repair and the accumulation of DNA double-strand breaks in the surviving daughter cells. *BRCA-1* is a tumor suppressor gene, and its loss results in significant genomic instability. Inactivating mutations in *BRCA-1* are associated with breast and ovarian cancers; d.) A mutation in a growth factor receptor that prevents ligand dissociation and keeps Ras constitutively activated will lead to the continuous stimulation of the Akt pathway, keeping Bad in a phosphorylated state and preventing the induction of apoptosis. In addition it will lead to increased levels of cyclin D such that there is a loss of cell cycle regulation and cells are continually stimulated to divide. Constitutively activating mutations in Ras are associated with bladder, breast, lung, head, neck, ovarian, and pancreatic cancers.

P22-4. a.) G_1 phase: The cell synthesizes components required for cell division; b.) S phase: DNA is replicated; c.) G_2 phase: The cell completes the synthesis of the nucleosomes and packaging of the newly synthesized DNA into chromatin; d.) M phase: Cell division takes place.

P22-5.

Cyclin	CDK	Cell Cycle Requirement
a. Cyclin D	CDK4 and/or CDK6	Transit through the G_1 restriction point
b. Cyclin E	CDK2	Progression through the G_1/S transition
c. Cyclin A	CDK2	Progression through S phase
d. Cyclin B	CDK1	Progression through the G_2/M transition

P22-6. a.) v; b.) ii; c.) iv; d.) i; e.) iii.

P22-7. a.) To determine if either of the tissue samples are cancerous, isolate genomic DNA and use PCR to amplify a region of the *ras* gene using primers that flank the two *Mva*I restriction sites. Digest the PCR products with *Mva*I and run the digests on an agarose gel. Determine if the *Mva*I restriction site at codon 12 is present or not. If the *Mva*I site is present, the pancreatic tissue does not contain a *K-ras* mutation. If the *Mva*I site is not present, the pancreatic tissue does contain a *K-ras* mutation. DNA isolated from normal pancreatic tissue and DNA isolated from cancerous tissue should be used as controls to be able to easily determine if the other tissue samples that are being analyzed are cancerous or not.
b.) Sample 1: Fragments of 14 bp, 24 bp, 114 bp.
   Sample 2: Fragments of 14 bp, 138 bp.

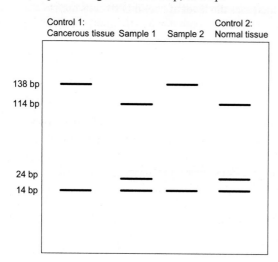

22

Cancer Genetics

CHAPTER SUMMARY QUESTIONS

1. 1. e, 2. h, 3. b, 4. j, 5. d, 6. a, 7. i, 8. f, 9. c, 10. g.

2. Cancer is a group of diseases characterized by uncontrolled cell growth and the spread of abnormal cells.

3. Protooncogenes are normal genes that stimulate cell division; mutants in these genes may be oncogenic (cancer-causing). Tumor suppressor genes encode proteins that either slow the cell cycle or induce apoptosis.

4. Hyperplastic cells accumulate additional mutations and become dysmorphic (abnormal in addition to uncontrolled division); dysmorphic cells may acquire the ability to separate from the mass of cells, migrate between cells, and enter the bloodstream. A metastatic tumor is one where the cells have gained the ability to travel throughout the body and establish secondary tumor locations.

5. Carcinogens generally interact with DNA to either modify the structure of a nucleotide or physically substitute for a nucleotide; in either case modified base-pairing properties cause mispairing and subsequent base substitutions during DNA synthesis.

6. They may occur in intragenic regions or introns, or they may affect wobble positions; in these examples they will have no effect on protein structure. Mutations that cause a change in amino acid sequence may not affect protein function if they are conservative missense and/or affect nonessential amino acids. Mutations that destroy protein function will not affect a cell's phenotype if that function is still performed—for example, by proteins encoded by the other allele.

7. Benzo[*a*]pyrene, arylhydroxylamine, dimethyl benzanthracene, aflatoxin B1, asbestos, radiation (X, UV, gamma), cigarette smoke, and so forth.

8. The role of these proteins is to regulate the cell cycle by causing the cell to arrest at a cell cycle checkpoint. If Rb and p53 are inactivated, a cell may divide inappropriately, resulting in hyperplasia.

9. In both cases, c-myc is misregulated.

10. Cyclin levels vary throughout the cell cycle, while cdk levels are fairly constant. Expression of cdk's at the level of *functional* protein requires an additional activation step (association with a cyclin protein), which is reversible.

11. G1 restriction point. The point at which a cell either goes into G0 or moves on toward S.
    G1/S transition. Transition from the first growth phase, beginning of DNA synthesis. Any induced errors in DNA structure (chromosomal breaks, chemical modification, and so forth) need to have been repaired and sufficient resources available to complete division.

G2/M transition. Following completion of DNA replication, any errors need to be repaired before cell division. High levels of cellular stress or DNA damage trigger apoptosis.

12. Precondensation. The period after the cell has received a signal to induce cell death but before there are any visible signs.
Condensation. Loss of interactions between the dying cell and its neighbors; degradation of the extracellular matrix (if any) and a decrease in the cytoplasmic volume.
Nuclear condensation. DNA cleavage and redistribution to nuclear margins.
Fragmentation. Division of the apoptotic cell into apoptotic bodies.
Phagocytosis. Neighboring cells engulf the apoptotic bodies.

13. Tumor cells become more and more aggressive as they accumulate additional mutations.

14. APC, KRAS, SMAD4/DCC, p53. This is unusual because in most cases the order of mutations is random rather than following a typical sequence.

15. If a person is heterozygous for a loss-of-function mutation in a tumor suppressor gene, a second mutation will mean that cell will now display the recessive (tumorigenic) phenotype. Once a cell has lost the tumor suppressor function, additional mutations may begin to accumulate.

EXERCISES AND PROBLEMS

16. a. Hypomorphic or null alleles of tumor suppressor genes are recessive mutations but will cause errors to accumulate in the cellular genome. Hypomorphic or null mutations of protooncogenes are unlikely to cause cancer.
    b. Hypermorphic alleles of tumor suppressor genes are unlikely to cause cancer. Hypermorphic alleles of protooncogenes are likely to be oncogenic, because they increase the amount of cell signaling that stimulated cell division.

17. a. The virus could increase the amount of messenger RNA for the growth factor.
    b. Use Northern blots to quantify the amount of oncogenic messenger RNA present in infected and uninfected cells. If a growth factor is overproduced, the excess molecules could cause cells to divide more rapidly and become cancerous. Many retroviruses contain regions that behave as enhancers of transcription of cellular oncogenes. If the virus inserts near an oncogene, the enhancer could stimulate transcription of that gene. An increase in transcription of a specific gene can be followed by quantifying the amount of specific RNA present in infected and uninfected cells. The intensity of the hybridized band should be higher in cells that are making more RNA.

18. We see one band that is common to both cell lines; this band must represent the normal protooncogene. The fact that this band is present in both lines indicates that the insertion of a virus has occurred in only one of the two copies of the gene present in the clone 1 cell. If it had inserted within both genes, we should not have seen the normal band. We see a larger fragment in clone 1, indicating that the DNA of the virus does not contain a site for the restriction enzyme used and that the virus has inserted within the restriction sites that define the band, lengthening the region probed. Alternatively, the virus could contain a restriction site and has still inserted in such a way as to lengthen the band probed by inserting between the sequence probed and the original restriction site.

19. Adenovirus attacks normal cells by binding to p53; it also attacks cancerous cells that lack p53. If we remove the gene for the protein that binds p53, the *E1B* gene, then the modified adenovirus will not be able to attack normal cells but will be able to attack cancerous ones lacking p53. This modification of adenovirus as a potential tool in treating cancer was published in 1996.

20. If the viral proteins completely prevent RB1 protein function, this would be very similar to a homozygous loss of function mutation. If the viral proteins only reduce the level of RB1 function, the situation would be more like a hypomorph. In either case, the RB1 protein is still produced normally by the cell. A loss of function mutation is different because it affects the production or function of the protein directly.

21. The major determining factors would be exposure to mutagens, and the normal rate of cell division in various tissues. Tissues that are more likely to be exposed to mutagens, and/or where there is a high rate of cell division are more likely to accumulate a mutation in their functional *RB1* allele. Especially for a tissue like the developing retina where cell division is relatively rapid, tumor development is likely.

22. Contact inhibition requires signals from nearby cells to repress cell division, or influence the G1 restriction point. Signals that allow cells to ignore repressive signals from neighbors will allow transition to S rather than G0. For anchorage dependence, it seems the cells must normally require some sort of signal from a cell:cell or cell:matrix connection in order to prevent apoptosis. Mutations in the pathway through which this signal is mediated could prevent stimulation of apoptosis even if not connected; such a cell could continue dividing and pass that trait on to daughter cells.

23. Viral "fitness" is all about making as many copies of the virus as possible, with a secondary goal of ensuring that those copies have access to cells to infect. If a virus inserts its genetic material into the genome of an infected cell, replication of the cell will ensure replication of the virus, and uncontrolled replication ensures even more viral replication. Even if the virus has not inserted its genetic material, replication of the host cell will assist in viral replication. If the cell also gains the ability to travel, the virus will have wider access to additional host cells.

24. The specific checkpoints affected will determine the nature of the cell cycle of each type of cancerous cell. The key comparison is that the cell cycle functions without normal controls—not necessarily that cell division is rapid. For instance, in many tumors, a G1-S signal is excessive or constant and G1 phase will be very short, leading to rapid division. In tissues such as the prostate, which normally have very low rates of cell division, a mutation that interferes with G2-M arrest will allow cancerous cells to divide but will not affect the overall rate of cell division and thus will result in a slow-growing tumor.

   If the spindle assembly checkpoint in M has been disrupted, metaphase may be shorter as the cell does not have to wait to ensure all chromosomes are attached to spindle fibers. This type of mutation may not directly affect the cell cycle per se, but will cause genomic instability and rapid development of abnormalities in the tumor cells.

25. Loss of function mutations would probably be less common in tumor cells than gain of function mutations. Loss of function mutations in cyclins D, E, or A will "merely" prevent the cell from future division—this is not necessarily lethal but will prevent the cell from forming a tumor. Gain of function mutations in those cyclins increase the rate of progression through the cell cycle. Cyclin B loss of function mutations may allow cells to move *too* quickly through M, leading to inappropriate chromosomal segregation; cyclin B loss of function mutations would be more likely in tumor cells than cyclin B gain of function mutations.

26. The extracellular signal binds to a receptor protein located in the cell membrane. The intracytoplasmic portion of the receptor then interacts with other proteins such as adapter proteins and G proteins to initiate a signal cascade composed of protein kinases. Ultimately, phosphorylation of transcription factors leads to a specific pattern of transcription factor binding at the upstream regulatory sites of cyclin D, activating transcription.

27. Allele 1. This one base-pair insertion will cause a frameshift, disrupting the rest of the coding sequence and eliminating protein activity. This will be a null allele.
Allele 2. This is a silent mutation and will have no effect on protein function or cellular phenotype.
Allele 3. This missense mutation will cause an arginine to threonine substitution. This is a nonconservative substitution and will probably cause a loss of function mutation.
Allele 4. This mutation changes a TCG codon (serine) to a TGG codon (tryptophan). The cell would be unable to progress through the cell cycle and would enter G0. If the mutant protein binds E2F but cannot be released by phosphorylation, this is likely to be a dominant negative mutation.
Allele 5. This is a silent mutation and will have no effect on protein function or cellular phenotype.

28. a. The initial Myc mutation is a hypomorph; along with a wild-type copy, this allele produces an adequate amount of functional Myc for wild-type protein activity. However, two hypomorphs produce so little functional Myc that Myc's normal role to maintain proliferation is compromised and cell division fails, resulting in embryonic lethality. The suppressor mutation is likely to be another gene that normally competes with Myc. If levels of Myc are normal, loss of function at this gene results in relative overactivity of Myc and increased proliferation. The reduced levels of Myc activity in the homozygous hypomorph balance the levels of activity of the mutant allele at the new gene, returning the cell cycle control to its usual balance.

b. Since these two proteins do not directly interact, there must be some other protein that mediates their interaction. A straightforward method would be to identify known targets of Myc regulation and study potential protein:protein interactions between Myc, the target genes, and the unknown gene identified earlier. If no known genes are identified, the yeast two-hybrid system could help identify other proteins involved in this pathway.

29. Cytochrome C is a key component of the electron transfer cascade used by the mitochondrion to capture energy into a hydrogen ion gradient for production of ATP. Removal of cytochrome C will prevent the cell from making ATP and will certainly be lethal.

30. a. If free Bad is present, it will bind with Bcl-2, preventing Bcl-2 from binding Bax. Free Bax will be cleaved and lead to apoptosis. Overexpression of Bcl-2 will allow binding to both Bax and Bad, preventing apoptosis unless Bax levels are very high. This is potentially an oncogenic mutation, as cells that should undergo apoptosis will be allowed to survive and continue through the cell cycle.

b. Reduced levels of Bad will reduce the amount of Bcl-2 required to sequester Bad protein and, conversely, increase the relative amount of Bcl-2 available to bind Bax, preventing apoptosis unless Bax levels are very high. Again, this is likely to have an oncogenic effect because cells that should undergo apoptosis are allowed to survive and divide.

31. The active p53 protein increases transcription of CKIs, which inactivate cyclin/CDK complexes. Once the cellular stress subsides and DNA damage is repaired, p53 is inactivated. CKIs are degraded, and the cell is allowed to continue through the cell cycle.

32. 20,000 lesions per day $\times$ 3 days = 60,000 lesions. If 99% of these lesions are repaired, $60,000 \times 0.01 = 600$ remain.

33. The natural function of the immunoglobulin promoter is to be spliced together with DNA segments during development of lymphocytes in order to control expression of the immunoglobulin gene. While other factors may also influence the frequency of rearrangements, errors in splicing during lymphocyte development would be expected to be more common than in other tissues; those errors will sometimes lead to overexpression of an oncogene under the control of the immunoglobulin promoter. This rearrangement will lead to inappropriate expression of the oncogene only in B lymphocytes, where the immunoglobulin heavy chain promoter is normally active.

34. Tissue-specific expression (for example, DCC—colon) of the wild-type gene will often result in tissue-specific expression of the mutant allele, or tissue-specific responses to loss of gene function. Some tissues are also more likely to be exposed to mutagens, or more likely to accumulate a mutation due to rapid cell division or inefficient DNA repair in that tissue. Normal levels of oxidative damage caused by the presence of ROS will also vary between tissues, with tissues that have high energy demands having higher levels of ROS.

35. Breast tissue has a relatively high level of CYP enzymes, resulting in high levels of proximate carcinogens produced. These carcinogens are drained at least in part via the lymphatic system. It is possible that cells in or near the lymphatic system are more likely to accumulate mutations as a result of this exposure. However, this is probably not the entire story, as most breast cancers arise from cells lining the milk ducts of the breast.

CHAPTER INTEGRATION PROBLEM

a. A first step would be sectioning and staining of the tissue. Various staining techniques would allow visualization of cellular and extracellular structures. An increased frequency of mitotic cells than would be expected for that tissue indicates hyperplasia. In the epidermal reference sample provided, the only mitotic cells are the basal layer. Abnormally shaped cells, presence of abnormal cytoskeletal structures, and evidence of unusual cell behavior would be indicators of potential malignancy. A further indication is proliferation of a single cell type, rather than the mix of cell types typically present in the epithelial tissue. If a portion of the edge of the growth is included, observation of the cells at the border would provide an indication if the growth is fully contained or has potentially become migratory.

b. Preparation of a karyotype with FISH will provide detailed information regarding chromosomal rearrangements or changes in the number of chromosomes present. Alternative staining procedures such as G-banding may allow identification of deletions, but are unlikely to add significant information to a FISH karyotype. PCR amplification and analysis of genes that are known or suspected to play a role in development of carcinomas from epithelial tissues is a more time intensive procedure, and it may or may not lead to identification of specific mutations or rearrangements. However, if specific mutations are suspected, PCR amplification and individual gene analysis may provide additional useful information.

c. The best method for studying gene expression for a large number of genes is microarray analysis. mRNA from healthy tissue can be compared to mRNA from the biopsy to identify any changes. Subtractive cDNA hybridization and library screening for differentially active genes is a second method that could be used. This method is useful if cloning the genes of interest and further study is needed. For diagnostic purposes, microarray analysis is faster and more accurate. A third technique to consider is Northern blot analysis of mRNA. This procedure requires a large amount of mRNA and is not technically feasible for a small amount of tissue such as a typical biopsy. However, after removal of a large tumor, Northern blot analysis of mRNA from the tumor could provide detailed information about the expression of specific genes.

d. Genes 1, 4, and 8 are only active during tumor growth or late in the process of carcinogenesis. As these genes seem to promote growth, they are potentially involved in growth factor response. Genes that are inactivated in the process of tumorigenesis, such as 2, 5, 6, and 7, are more likely to be tumor suppressors. If they are down-regulated, tumor formation is allowed.

e. This will depend on the specific function of these genes. Sunburn 2 may be involved in precancerous changes, but does not appear to play a role in late development of tumors, as it is inactive both in healthy skin and in late or metastatic tumors. Sunburn 5 may play a role in *preventing* precancerous changes, as it is activated only in acute exposure. One possible mechanism would be participation in DNA repair. Preventing activation of these genes during the normal response to acute UV exposure could potentially increase risk of future cancer development.

f. Inhibition of genes that play a role in tumor development or that increase rates of cell division could potentially reduce cancer development. Inhibition of genes that protect against tumor development could have the opposite effect.

g. In an adult patient without a history of skin cancer, new epidermal growths are more likely to be caused by environmental exposure to mutagens such as UV radiation. Such exposures are not heritable. However, a complete answer would require information about the patient's past history as well as a complete family history. Heritable factors such as an underlying predisposition or genetic disease related to DNA repair could be involved and would likely be identified in a thorough discussion of personal and family history. No, the stage of development of the tumor would not affect this reasoning. A relatively common type of skin tumor would support it; if the growth were very rare and/or almost always seen only in cases of an underlying predisposition, the possibility of a family history becomes more relevant.

h. No. The most common risk factors for breast cancer are sex (female) and age (most breast cancers are diagnosed after the age of 70). Only about 15% of breast cancer patients have a family history of breast cancer; of those only about 85% have an identifiable genetic mutation.

CHAPTER 23 POPULATION GENETICS

Chapter Goals

1. Learn the forces that determine the amount and kind of genetic variation found in a population.
2. Describe the way the five major evolutionary forces change the amount and kind of genetic variation found in a population.
3. Understand the principle behind the Hardy–Weinberg equilibrium and apply it in a problem.
4. Use the different types of departures from Hardy–Weinberg equilibrium to predict the causes of disequilibrium.
5. Discuss the differences between the effects of random genetic drift and inbreeding.

Key Chapter Concepts

23.1 What Is Population Genetics?

- **Population genetics** is the study of genetic variation at the population level that results from **mutation**, **selection**, **random genetic drift**, **migration**, and **nonrandom mating**. Populations are the units of evolutionary change consisting of an interbreeding group of individuals that exist together. The complement of alleles available among all of the reproductive members of a population comprises the **gene pool**.

23.2 Frequencies of Genotypes and Alleles

- **Calculating Genotypic Frequencies:** The proportion of individuals in a population with a specific genotype (that is, *AA*, *Aa*, or *aa*) is the genotypic frequency. If the homozygous individuals can be distinguished from the heterozygous individuals, the genotypic frequency can be calculated directly. The sum of the genotypic frequencies is equal to 1.0.
- **Calculating Allelic Frequencies:** The proportion of each allele in a gene pool is the allelic frequency. In cases where heterozygotes can be identified, allele frequencies are calculated by dividing the number of alleles of that type in the population by the total number of alleles in the population. For example, in a sample of 200 individuals there were 50 with genotype *AA*, 100 with genotype *Aa*, and 50 with genotype *aa*. The frequency of the *A* allele can be calculated as: $(100 + 100) \div 400 = 0.5$, which corresponds to twice the number with the *AA* genotype + the number with heterozygous genotype ÷ twice the total number of individuals.

23.3 Hardy–Weinberg Equilibrium

- Allele frequencies determine the genotypic frequencies if the population has an infinite number of members, random mating, no mutation, no migration, and no selection. When these conditions are met, the population is in **Hardy–Weinberg equilibrium**. If the frequencies of two alleles in the parental groups are *p* and *q*, then the genotypic frequencies of the two homozygous genotypes are p^2 and q^2 and the heterozygote is $2pq$. The Hardy–Weinberg equation for one gene with two alleles is $p^2 + 2pq + q^2 = 1$. The Hardy–Weinberg theory or rule states that allele frequencies of an autosomal gene will not change from one generation to the next and that the genotypic frequencies in the population can be predicted from allele frequencies. The Hardy–Weinberg law further states that if a population is perturbed, equilibrium will be reestablished after one generation of random mating.
- **Application of Hardy–Weinberg Equilibrium:** The Hardy–Weinberg equilibrium uses allele frequencies (*p* and *q*) to predict genotypic frequencies. The sum of the allelic frequencies (*p* + *q*) is equal to 1.0. The sum of the genotypic frequencies (*P* + *H* + *Q*) is also equal to 1.0, where *P* = the frequency of genotype A_1A_1, *H* = the frequency of genotype A_1A_2, and *Q* = the frequency of genotype A_2A_2. The frequency of the first allele, *p*, includes all of the A_1A_1 homozygotes plus half the A_1A_2 heterozygotes [$p = P + (0.5)H$] and the frequency of the second allele, *q*, includes all of the A_2A_2 homozygotes plus half the A_1A_2 heterozygotes [$q = Q + (0.5)H$]. Allelic frequencies can also be used to calculate the phenotypic frequencies. If the frequency of one allele, *M*, is 0.6 and the frequency of the second allele, *N*, is 0.4, the phenotypic frequencies are $MM = P = p^2 = 0.36$, $MN = H = 2pq = 0.48$, and $NN = Q = q^2 = 0.16$. If the phenotypic frequencies are known, the frequency of the recessive allele (*q*) can be obtained by taking the square root of the recessive phenotype frequency. The frequency of the dominant allele (*p*) is obtained from the equation $p + q = 1$, by algebraic rearrangement ($p = 1 - q$).

- **Testing Whether the Genotypic Frequencies in a Population Are in Hardy–Weinberg Equilibrium:** Chi-square analysis can be used to determine if the observed genotypic frequencies at a locus are statistically similar to the expected genotypic frequencies indicating that a population is in Hardy–Weinberg equilibrium at this locus.
- **Hardy–Weinberg Equilibrium with More Than Two Alleles:** For the three-allele system with alleles A_1, a_2, and a_3 where the frequency of the A_1 allele is p, the frequency of the a_2 allele is q, and that of the a_3 allele is r, the allelic frequencies $p + q + r$ must equal 1.0, and the genotypic frequencies $(p + q + r)^2$ must also equal 1.0. The trinomial expansion can be used to calculate allele frequencies and expected genotypic frequencies for single genes with multiple alleles. The Hardy–Weinberg equation for a gene with three alleles (p, q, and r) is: $p^2 + 2pq + 2pr + q^2 + 2qr + r^2 = 1$.

23.4 Selection: Shifting Allele Frequencies

- **The Concept of Relative Fitness:** Under Hardy–Weinberg equilibrium we assume that all members of a population are equally likely to reproduce. In reality, some phenotypes may leave more offspring than other phenotypes. This difference in **reproductive success** results from selection. **Fitness** is a measure of reproductive success: a genotype that leaves more offspring that survive and reproduce has a higher fitness than the alternative genotypes. **Absolute fitness** refers to the average reproductive rate of individuals with the same genotype, while **relative fitness** refers to a genotype's ability to survive and reproduce relative to other genotypes in the population.
- **Simple Models of Selection:** Selection is the process by which one allele or allele combination is favored over another and results in allele frequencies changing away from Hardy–Weinberg equilibrium. Selection acts upon the phenotype and in doing so changes the genotypic makeup of a population. In general, selection may shift the population mean toward one end of the phenotypic distribution (**directional selection**), it may favor individuals with phenotypes near the mean of the population (**stabilizing selection**), or it may favor individuals with extreme phenotypes (**disruptive selection**). **Additive effects** occur when the phenotypic value for the heterozygote is the mean of the phenotypic value for each of the homozygotes.
- **Response to Selection:** The speed by which allelic frequencies change depends on the strength of selection, the level of dominance, and the allelic frequencies themselves. In general, selection at a single locus tends to be weak, and therefore changes in allelic frequencies from selection alone typically occur very slowly.
- **Maintaining Genetic Variability Through Selection:** Selection is a necessary component of evolutionary change. **Frequency-dependent selection**, **density-dependent selection**, and **genotype-environment interactions** are forms of selection that maintain genetic variation. Selection may change allelic and genotypic frequencies by removing alleles associated with a particular phenotype (for example, a recessive phenotype). Once removed, they can be reintroduced into the population through mutation. Selection can also act to maintain two alleles in a population (**heterozygote advantage**).
- **Mutation-Selection Balance:** Because these two forces are essential for a population to adapt to new environments and evolve, a balance must be maintained between the two. A mutation is more likely to generate a deleterious allele, whereas selection usually decreases the presence of deleterious alleles.

23.5 Random Genetic Drift

- **Effect of Random Genetic Drift: Random genetic drift** refers to changes in allelic frequencies that occur by chance alone. Random genetic drift results in the random fixation of an allele at a genetic locus, whether the allele has advantageous or deleterious effects. The effect of drift increases as the number of effective breeders decreases. The introduction of additional breeders into a population provides the opportunity to decrease the frequency of the deleterious alleles and reduce the effect of random genetic drift.
- **Interactions Between Selection, Random Genetic Drift, and Mutation:** Genetic variation is dependent on the interactions between selection, random genetic drift, and mutation rates. Without selection, mutation, or migration, every allele will eventually become fixed or lost by continuous genetic drift. The amount of genetic drift is inversely proportional to the size of a population. Selection exerts a more powerful effect than random genetic drift as the population size increases.

23.6 Migration and Population Structure

- **Gene flow** occurs when individuals migrate from one population into another, interbreed, and reproduce. Genetically isolated populations become genetically differentiated from each other by mutations, random genetic drift, and differential selection.
- **The Continent-Island Model:** The rate of change of allelic frequency in a small population subject to immigration depends on the rate of individuals moving into the small population and the difference in allelic frequencies. The greater the immigration rate and the more the two populations initially differ from each other, the faster the allelic frequencies will change in the small population.
- **Migration, Selection, and Drift:** Mutation and drift are often relatively minor forces acting on populations relative to gene flow and selection. Immigrant alleles can face both strong positive or strong negative selection. Directional changes in allelic frequencies (**clines**) can often be correlated with environmental factors which provide evidence of adaptation.

23.7 Nonrandom Mating

- **Nonrandom mating** refers to mating between two individuals based on the level of physical resemblance or the level of genetic relatedness.
- **The Wahlund Effect:** Migration involves the adding or removing of alleles from populations as individuals move from one population to another. When migrants enter a native population, the resulting subdivided group will not be in equilibrium because of a decrease in heterozygotes (the **Wahlund effect**). Hardy–Weinberg equilibrium is established in the new conglomerate population after one generation of random mating.
- **Inbreeding:** The assumption of random mating underlying Hardy–Weinberg equilibrium is violated when genetic relatedness (**inbreeding** or **common ancestry**) influences mating patterns. Inbreeding has several effects. First, it leads to an increased expression of hidden recessive alleles. Inbreeding increases the likelihood that a zygote will receive two copies of a deleterious allele. Second, inbreeding increases homozygosity. Inbred individuals are more likely to receive alleles that are copies of the same ancestral allele or identical by descent, including deleterious recessive alleles. This can lead to a reduction in fitness known as **inbreeding depression**. A statistic referred to as the **inbreeding coefficient** (F) provides a measure of the probability that two alleles in an individual are identical because of their inheritance for a single allele in an ancestor. An F value of zero is obtained when there is no inbreeding, while an F value of one is obtained if everyone in a population is homozygous for an allele that is derived from a common ancestral allele. Pedigree analysis can be used to calculate an inbreeding coefficient for an individual. The inbreeding coefficient for a population can also be calculated by measuring the reduction in heterozygosity resulting from inbreeding.
- **Relationship Between Inbreeding and Drift:** Inbreeding, unlike genetic drift, does not change allelic frequencies and can increase the strength of selection against deleterious recessive alleles.
- **Assortative–Disassortative Mating:** These are two other forms of nonrandom mating. **Assortative mating** occurs when mates are chosen because they share a common trait, while **disassortative mating** occurs when mates are chosen because they are different in some respect.

Key Terms

- population genetics
- population
- metapopulation
- gene pool
- mutation
- fitness
- selection
- random genetic drift
- migration
- nonrandom mating
- Hardy–Weinberg equilibrium

- absolute fitness
- relative fitness
- directional selection
- stabilizing selection
- disruptive selection
- selection coefficient
- additive effects
- heterozygote advantage
- frequency-dependent selection
- density-dependent selection

- effectively neutral
- clines
- Wahlund effect
- inbreeding
- outbreeding
- inbreeding coefficient (F)
- inbreeding depression
- purging
- assortative mating
- disassortative mating

Understanding the Key Concepts

Use the key terms or parts of their definitions to complete the following sentences.

A population consists of individuals that exist together and (1)_____. The genetic (2)_____ of a (3)_____ is affected by five major forces: mutation, (4)_____, random genetic drift, (5)_____, and (6)_____ mating. Mutations introduce new (7)_____ into a population and either decrease or have no effect on (8)_____. (9)_____ increases the frequency of a favored allele leading to changes in the (10)_____ and (11)_____ frequencies. (12)_____ also increases the frequency of a particular allele in a (13)_____, but the allele that increases in different populations is (14)_____. Migration leads to (15)_____ between two different populations that makes the two populations more genetically similar. Nonrandom mating, such as inbreeding, results in a decrease in the number of (16)_____ and an increase in the number of (17)_____. This can lead to an increase in the expression of deleterious (18)_____ alleles.

An infinitely large (19)_____ that is not acted on by any of these forces is in (20)_____. The (21)_____ and (22)_____ frequencies in this (23)_____ are not changing. Under (24)_____, the genotypic frequency is represented by the equation (25)_____, where p is the (26)_____ of (27)_____ and q is the (28)_____ of (29)_____. The genotypic frequency of a homozygote is the (30)_____ of the (31)_____ frequency, where p^2 is the genotypic frequency of the (32)_____ homozygote and q^2 is the genotypic frequency of the (33)_____ homozygote. The genotypic frequency of a heterozygote is (34)_____ the (35)_____ of the two (36)_____ frequencies.

(37)_____ and (38)_____ are two opposing forces that are crucial to evolution. (39)_____ can introduce deleterious alleles, while (40)_____ can remove these alleles because it is the process by which one allele is favored over another. (41)_____ changes allele frequencies away from (42)_____ and maintains genetic (43)_____. Directional selection continuously removes individuals from one end of the (44)_____ distribution, while (45)_____ selection removes individuals from both ends of the (46)_____ distribution. It occurs when the (47)_____ is the fittest genotype. In contrast, in (48)_____ selection the (49)_____ trait is selected against increasing the frequency of both (50)_____ traits. Selection at a single locus tends to be (51)_____, but some rapid changes in (52)_____ frequencies have occurred in instances of (53)_____ selection. Frequency-dependent selection occurs when the (54)_____ of one allele depends on the frequency of (55)_____ in the population. Density-dependent selection occurs when the (56)_____ of a genotype depends on the (57)_____ of the population. (58)_____ and (59)_____ variation in an (60)_____ also maintains genetic variation because some genotypes are favored at one time in a specific (61)_____, while another may be favored at a different time in a different place. In some cases, individuals settle in (62)_____ where they have the highest (63)_____.

(64)_____ and (65)_____ are different because the allele that becomes fixed at a locus by (66)_____ is random, and the effect of this process (67)_____ as the number of effective breeders (68)_____. The introduction of additional breeders into a population reduces the effect of (69)_____. As the population size (70)_____, (71)_____ exerts a more powerful effect than random genetic drift.

For Hardy–Weinberg equilibrium to exist in a population, mating must be (72)_____. Nonrandom mating occurs when mates are chosen based on a common (73)_____ or genetic (74)_____. When two populations of equal size in Hardy–Weinberg equilibrium migrate and intermingle, there is initially a low number of (75)_____ called the (76)_____ effect. As random mating begins, (77)_____ becomes established within the new combined population. (78)_____ within a population will lead to an increase in homozygosity that can lead to an increase in deleterious (79)_____ alleles, leading to a decrease in (80)_____ known as inbreeding (81)_____.

Figure Analysis

Use the indicated figure or table from the textbook to answer the questions that follow.

1. **Figure 23.1**
   (a) What is predicted in figure 23.1?

   (b) Between what allele frequencies is there predicted to be more heterozygous individuals than either of the homozygous individuals in a population?

   (c) Between what frequencies of allele *A* are *AA* individuals predicted to be the most abundant genotype in a population?

   (d) At what allele frequency are the *AA* and *aa* individuals predicted to be found in equal proportions in a population?

   (e) What must be true in order for these genotypic frequency predictions to be true?

2. **Figure 23.3**
   (a) In the middle row of graphs, what do the shaded areas represent?

   (b) Which selection mechanism increases the risk of an allele becoming fixed in a population?

   (c) Which of the three selection mechanisms will preserve genetic diversity?

   (d) Which selection method will increase the percentage of heterozygous individuals in a population?

   (e) Which selection method will decrease the percentage of heterozygous individuals in a population?

3. **Figure 23.11**
   (a) How many populations are shown in each graph? How many total populations are shown in figure 23.11?

   (b) How many individuals are there in each of the populations shown in the top graph?

   (c) What is the starting frequency of allele b^+ in each of the populations?

   (d) In which set of populations ($N = 10$, $N = 20$, $N = 50$, or $N = 100$) does genetic drift have the largest effect?

   (e) Why does the frequency of allele b^+ drift upward (toward 1.0)?

   (f) Why does the frequency of allele b^+ become fixed at zero in one of the populations in the top graph but not in any of the populations shown in the other graphs?

   (g) Which set of populations ($N = 10$, $N = 20$, $N = 50$, or $N = 100$) are almost effectively neutral?

General Questions

Q23-1. Since Hardy–Weinberg equilibrium assumptions are rarely met in nature, of what value are these assumptions?

Q23-2. Tongue rolling is believed to be a dominant trait in human beings. A student of genetics was asked the following question: "Will the frequency of tongue rolling increase until 75% of the population expresses the dominant phenotype?" How would you answer this question?

Q23-3. For each force discussed in this chapter (genetic drift, migration, mutation, directional selection, stabilizing selection, and disruptive selection), speculate whether each would act to increase or decrease variation within populations and between populations.

Q23-4. Would selection against a dominant allele or against a recessive allele be most effective in removing the allele from the population?

Q23-5. Describe the methods by which genetically isolated populations become genetically differentiated from other populations. What is the effect of gene flow between populations?

Multiple Choice

For each of the following, circle the letter of the choice that most appropriately answers the question.

1.) Which of the following is *not* a required condition for Hardy–Weinberg equilibrium to exist in a population?
   a. Individuals mate randomly.
   b. There is no migration into or out of the population.
   c. The introduction of mutations is balanced by selection against these mutations.
   d. The genotypes in the population exhibit equal fitness.
   e. The population is maintained at a very large size.

2.) If a population consists of 65 A_1A_1 individuals, 95 A_1A_2 individuals, and 20 A_2A_2 individuals, the frequency of the A_1 allele is which of the following?
   a. 0.313
   b. 0.375
   c. 0.444
   d. 0.625
   e. 0.889

3.) In a particular population of *Drosophila*, vestigial wings (allele *vg*) are recessive to the wild-type condition of normal wings (allele vg^+). You isolate a population of 20 vestigial-winged flies, 80 flies that are heterozygous for normal wings, and 140 flies that are homozygous for normal wings. What is the frequency of alleles in this population?
   a. 30% *vg*; 70% vg^+
   b. 70% *vg*; 30% vg^+
   c. 40% *vg*; 60% vg^+

   d. 25% *vg*; 75% vg^+
   e. 75% *vg*; 25% vg^+

4.) Using the information given in question 3, what is the predicted number of each genotype if the population of 240 flies were in Hardy–Weinberg equilibrium?
   a. 135 $vg^+ vg^+$;   90 $vg\ vg^+$;   15 $vg\ vg$
   b. 150 $vg^+ vg^+$;   73 $vg\ vg^+$;   17 $vg\ vg$
   c. 180 $vg^+ vg^+$;   48 $vg\ vg^+$;   12 $vg\ vg$
   d. 175 $vg^+ vg^+$;   51 $vg\ vg^+$;   14 $vg\ vg$
   e. 120 $vg^+ vg^+$;   101 $vg\ vg^+$;   19 $vg\ vg$

5.) In a population that has 100 *AA*, 200 *Aa*, and 100 *aa* individuals, calculate the frequency of allele *A* after one generation of natural selection if the following relative fitness values apply: *AA* = 1.0, *Aa* = 1.0, *aa* = 0.5.
   a. 0.43
   b. 0.50
   c. 0.57
   d. 0.60
   e. 0.63

6.) As populations become endangered, which of the following increases the probability that deleterious alleles will be fixed in the population?
   a. differential selection
   b. random genetic drift
   c. spontaneous mutations
   d. reduction of relative fitness
   e. None of the above

7.) The scale-eating fish *Perrisodus microlepis* has an asymmetrical joint in its jaw that allows it to open its mouth to only one side. These fish exhibit what type of selection at this locus?
   a. stabilizing selection
   b. spatial variation
   c. frequency-dependent selection
   d. disruptive selection
   e. directional selection

8.) Which of the following types of selection favors only one extreme of a phenotypic distribution?
   a. disruptive
   b. directional
   c. assortative
   d. stabilizing
   e. disassortative

9.) The speed with which allelic frequencies change is dependent on all but which of the following?
   a. immigration rate in small populations
   b. strength of selection
   c. allelic frequencies themselves
   d. level of dominance
   e. None of the above; the speed at which allelic frequencies change is dependent on all of these.

10.) Which of the following is the major force acting to decrease the frequency of deleterious alleles in a population?
   a. reverse mutations
   b. natural selection
   c. genetic drift in large populations
   d. gene flow
   e. nonrandom mating

Practice Problems

P23-1. Consider a population of cattle composed of 50 individuals with red hair (*RR*), 48 roan individuals (*Rr*), and 12 white individuals (*rr*). Calculate allele frequencies, and determine if the population is in equilibrium.

(*Hint 1*: If heterozygotes can be identified, the direct count method should be used to calculate allele frequencies. *Never* assume equilibrium for a population in which heterozygotes can be identified. *Hint 2*: Chi-square analysis can be used to determine if a population is in equilibrium. When using chi-square analysis, the degrees of freedom are obtained by subtracting *2* from the number of classes.)

P23-2. In a human population, a recessive autosomal disorder occurs in approximately 1 of 2500 births. Calculate the frequency of the recessive allele and the percentage of heterozygous carriers.

P23-3. In most states, newborn children are required by law to be tested for PKU, an autosomal recessive genetic disorder. If untreated, individuals homozygous for PKU suffer severe mental retardation. In a recent year, four phenylketonuric babies were detected out of 126,000 tested. Assuming Hardy–Weinburg conditions, what is the frequency of the PKU gene in this population?

P23-4. Consider the following data for allozyme phenotypes in the plant *Phlox drummondii*: 7 *AA*, 15 *AB*, 38 *BB*, 30 *BC*, 16 *CC*, and 4 *AC*. Calculate the frequency of each allele and determine if the population is in equilibrium.

P23-5. Assume that 30% of the human beings on campus have type A blood, 20% have type B blood, 10% have type AB blood, and 40% have type O blood. Calculate allele frequencies for this population.

[*Hint 3*: Hardy–Weinberg equilibrium must be assumed to calculate allele frequencies when heterozygotes cannot be identified. Obtain the frequency of the recessive allele (q) by taking the square root of the frequency of the recessive phenotype.]

P23-6. Red-green color blindness is a sex-linked recessive trait in humans. Assume that the frequency of color-blind males on campus is 0.008. Calculate the frequency of all phenotypes.

[*Hint 4*: For a sex-linked recessive trait, the frequency of affected males is equal to the frequency of the recessive allele (q) because males possess only one copy of the × chromosome. The frequency of affected females is q^2.]

P23-7. Answer problem P23-6 assuming that the frequency given (0.008) pertains to color-blind females rather than males. What is the frequency of each phenotype in the population?

P23-8. Assume that the frequency of heterozygotes in a population is 0.35. Assume further that $p = 0.65$ and $q = 0.35$. Estimate the inbreeding coefficient (F). Interpret the calculated value.

[*Hint 5*: An excess of homozygotes in a population *suggests* inbreeding. Inbreeding alters genotypic but not allelic frequencies.]

P23-9. Consider a gene that shows overdominance: the heterozygote has a higher fitness in the environment than the homozygous normal individual (AA), and the homozygous affected individual (aa) has reduced fitness. Given the fitness values provided, calculate allele frequencies.

Genotype	Fitness
AA	0.75
Aa	1.0
aa	0.1

P23-10. Assume that a recessive phenotype is selected against and that the selection coefficient is 0.1. The mutation rate (A to a) is 4×10^{-6}. What is the selection–mutation equilibrium frequency of the recessive allele?

P23-11. For each major type of selection (directional, stabilizing, and disruptive), provide the relative fitness values associated with each genotype (AA, Aa, and aa).

Assessing Your Knowledge

Understanding the Key Concepts—Answers

1.) interbreed; 2.) variation; 3.) population; 4–5.) selection; migration; 6.) nonrandom; 7.) alleles; 8.) fitness; 9.) Selection; 10–11.) allelic; genotypic; 12.) Random genetic drift; 13.) population; 14.) different; 15.) gene flow; 16.) heterozygotes; 17.) homozygotes; 18.) recessive; 19.) population; 20.) Hardy–Weinberg equilibrium; 21–22.) allelic; genotypic; 23.) population; 24.) Hardy–Weinberg; 25.) $p^2 + 2pq + q^2 = 1$; 26.) frequency; 27.) allele A_1; 28.) frequency; 29.) allele A_2; 30.) square; 31.) allelic; 32.) dominant; 33.) recessive; 34.) double; 35.) product; 36.) allelic; 37–38.) Mutation; selection; 39.) Mutations; 40.) selection; 41.) Selection; 42.) Hardy–Weinberg equilibrium; 43.) variation or diversity; 44.) phenotypic; 45.) stabilizing; 46.) phenotypic; 47.) heterozygote; 48.) disruptive; 49.) intermediate; 50.) homozygous; 51.) weak; 52.) allelic; 53.) strong; 54.) fitness; 55.) other alleles; 56.) fitness; 57.) density; 58–59.) Spatial; temporal; 60.) environment; 61.) habitat; 62.) habitats; 63.) fitness; 64–65.) Selection; random genetic drift; 66.) genetic drift; 67.) increases; 68.) decreases; 69.) random genetic drift; 70.) increases; 71.) selection; 72.) random; 73.) trait; 74.) relatedness; 75.) heterozygotes; 76.) Wahlund; 77.) Hardy–Weinberg; 78.) Inbreeding; 79.) recessive; 80.) fitness; 81.) depression.

Figure Analysis—Answers

1a.) The frequency of each of the genotypes that will result from a particular allele frequency is predicted from the Hardy–Weinberg equation; b.) Between allele frequency 0.333 and 0.667, the heterozygous individuals are predicted to be more prevalent than either of the homozygous individuals; c.) Between *A* allele frequency 0 and 0.333, the *AA* individuals are predicted to be the most prevalent genotype in a population; d.) At an allele frequency of 0.5, the two homozygote genotypes are expected to be in equal proportions; e.) The population must be in Hardy–Weinberg equilibrium.

2a.) The shaded areas represent phenotypes of individuals who are selected *against*. These individuals may be the primary target of predators, for example; b.) Directional selection will increase the risk of allele fixation. This is because selection will cause a continuous loss of one of the alleles; c.) Stabilizing selection and disruptive selection will both maintain genetic variation in a population; d.) Stabilizing selection will generally select for heterozygous individuals since they can display an intermediate phenotype. Thus, heterozygous individuals will have an advantage in the population and will be more prevalent; e.) Disruptive selection will generally decrease the percentage of heterozygotes in a population since the intermediate phenotype will be selected against.

3a.) There are 13 populations shown in each graph for a total of 52 populations in the figure; b.) There are 10 individuals in each of the 13 populations shown in the top graph; c.) 0.5; d.) Genetic drift has the greatest effect on the smallest population, $N = 10$; e.) There is selection in favor of the b^+ allele; f.) The frequency of allele b^+ became fixed at zero in the top graph because the effect of genetic drift was great enough in that small population to overcome the selection pressure. The populations shown in the lower graphs were large enough that genetic drift could not overcome the selection pressure; g.) The smallest population ($N = 10$) is almost effectively neutral because genetic drift can overcome the selection pressure.

General Questions—Answers

Q23-1. The Hardy–Weinberg equilibrium assumptions are valuable for several reasons. First, the concept provides us with a null hypothesis that can be tested. We cannot easily examine the forces of evolutionary change without knowing if the population is experiencing change or not. Second, assumptions of Hardy–Weinberg are important in human genetics because they allow the geneticist to calculate the frequency of a recessive allele in the general population. This method is the only one available for disorders that are caused by as yet-unidentified genes.

Q23-2. Many people mistakenly believe that 75% of the population will express a dominant phenotype given Mendelian ratios. Mendelian ratios are achieved in a population only when the frequency of each allele is 0.5. Substitute these numbers into the Hardy–Weinberg equation and you obtain Mendel's

1:2:1 genotypic and 3:1 phenotypic ratios. It is unreasonable to assume that alleles for all genes occur equally frequently.

Q23-3. Within a population, you would expect mutation, migration, and stabilizing selection to increase variation. Mutation and migration introduce new or rare alleles into a population, while stabilizing selection maintains both alleles in a population. Genetic drift, directional selection, and disruptive selection act to remove alleles (directly or accidentally) and would therefore decrease variation within a population. Between populations, genetic drift, disruptive selection, and directional selection would act to increase differences. Mutation, migration, and stabilizing selection would tend to increase the similarity between different populations.

Q23-4. Selection against a dominant phenotype would quickly remove the dominant allele from the population. In contrast, selection against a recessive allele occurs *very* slowly. This is because heterozygotes carry a "hidden" recessive allele. Remember that selection acts on phenotypes; heterozygotes express the dominant phenotype and would not be selected against. Heterozygotes would live and be able to pass the recessive allele to their offspring. As the recessive allele becomes rare, fewer individuals will be homozygous recessive and selected against. With each generation it becomes harder to remove the recessive allele from the population.

Q23-5. Genetically isolated populations become genetically differentiated from other populations by mutations, random genetic drift, and differential selection. Mutations are rare events that are not likely to be duplicated in isolated populations; therefore the generation of mutations increases the genetic differences between the population incurring the mutation and other populations. Random genetic drift can fix one allele in one population, but another allele in a second population increasing the genetic variation between the two populations. Differential selection is a product of populations living in different environments. These populations will be acted on by different selective pressures, favoring some alleles and selecting against other. This will serve to increase the genetic variation between the population being acted on by certain selective pressures and other populations being acted on by different selective pressures. The effect of gene flow between populations is to make the populations more genetically similar.

Multiple Choice—Answers

1.) c; 2.) d; 3.) d; 4.) a; 5.) c; 6.) b; 7.) c; 8.) b; 9.) e; 10.) b.

Practice Problems—Answers

P23-1. In this problem, heterozygotes can be identified and counted, so we should use the direct method to calculate allele frequencies. The allele frequencies can be used to predict the expected genotypic frequency distribution at equilibrium.

Total number of alleles	=	$2(50 + 48 + 12) = 220$
Total number of r alleles	=	$(12 \times 2) + 48 = 72$ $f(r) = q = 72/220 = 0.327$
Total number of R alleles	=	$(50 \times 2) + 48 = 148$ $f(R) = p = 148/220 = 0.673$

H_0 = population is in Hardy–Weinberg equilibrium

Genotype	Observed	Expected	$(O - E)^2$	$(O - E)^2 / E$
RR	50	$p^2 \times 110$ = 49.82	0.0324	0.0007
Rr	48	$2pq \times 110$ = 48.42	0.1764	0.004
rr	12	$q^2 \times 110$ = 11.76	0.0576	0.005
				0.0097

Critical chi-square value (1 df, 0.05 probability) = 3.841   Fail to reject H_0.

P23-2. In order to calculate allele frequencies, we must assume that the population is in Hardy–Weinberg equilibrium. The frequency of the recessive allele (q) is estimated from the frequency of the recessive phenotype. The percentage of heterozygous carriers is obtained by substituting the allele frequencies in the term $2pq$.

$f(rr) = 1/2500 = 0.0004$ $q^2 = $ square root of $0.0004 = 0.02$ $p = 1 - q = 0.98$
$f(Rr) = 2pq = 2(0.02)(0.98) = 0.0392$ or 3.92%

P23-3. Approach this problem as you approached problem P23-2.
$q^2 = 4/126{,}000 = 0.0000317$ $q = 0.006$

P23-4. There are three alleles for this gene, and we are able to identify all genotypic classes. Therefore, the direct count method is used to determine allele frequencies.
$f(A) = p = [(7 \times 2) + 15 + 4] / 220 = 0.15$
$f(B) = q = [(38 \times 2) + 15 + 30] / 220 = 0.55$
$f(C) = r = [(16 \times 2) + 30 + 4] / 220 = 0.3$

$H_0 = $ population is in Hardy–Weinberg equilibrium

Genotype	Observed	Expected			$(O - E)^2$	$(O - E)^2 / E$
AA	7	$p^2 \times 110$	=	2.475	20.4756	8.273
AB	15	$2pq \times 110$	=	18.15	9.9225	0.547
BB	38	$q^2 \times 110$	=	33.275	22.3256	0.671
BC	30	$2qr \times 110$	=	36.3	39.69	1.093
CC	16	$r^2 \times 110$	=	9.9	37.21	3.759
AC	4	$2pr \times 110$	=	9.9	34.81	3.516
						17.859

Critical chi-square value (4 df, 0.05 probability) = 9.488   Reject H_0.

P23-5. In this problem, we must assume equilibrium to estimate allele frequencies. Obtain an estimate of r from the frequency of the recessive phenotype (type O blood, ii). Next, examine the A and O, or B and O, alleles simultaneously to estimate a second allele frequency. The third frequency is obtained by subtracting from one.

$\quad\quad\quad\quad\quad = \quad \sqrt{0.4} = 0.633$
$(p + r)^2 \quad = \quad p^2 + 2pr + r^2$
$(p + r)^2 \quad = \quad f(I^A I^A) + f(I^A I^O) + f(ii)$
$(p + r)^2 \quad = \quad 0.30 + 0.4$
$(p + r)^2 \quad = \quad 0.7$
$(p + r) \quad = \quad \sqrt{0.7} = 0.837$
$p \quad\quad = \quad 0.837 - r$
$p \quad\quad = \quad 0.837 - 0.633 = 0.204$
$q \quad\quad = \quad 1 - 0.633 - 0.204 = 0.163$

P23-6. For a sex-linked recessive trait, the frequency of affected males is equal to q. In this problem, $q = 0.008$. The frequency of unaffected males is $1 - 0.008 = 0.992$. The frequency of color-blind females is q^2, or 0.0064%. The frequency of females with normal vision is $p^2 + 2pq$, or 99.99%.

P23-7. For a sex-linked recessive trait, the frequency of affected females is equal to q^2. In this problem, $q^2 = 0.008$ so $q = 0.089$. The frequency of color-blind males is q, or 8.9%, and the frequency of males with normal vision is 91.1%. The frequency of females with normal vision is $p^2 + 2pq$, or 99.2%.

P23-8. The inbreeding coefficient (F) is calculated by dividing the observed frequency of heterozygotes by the expected frequency of heterozygotes and subtracting this number from one.
$F = 1 - [0.35/(2pq)]$
$\quad = 1 - [0.35/0.455]$
$\quad = 1 - 0.769$
$\quad = 0.231$
This value indicates that there is an excess of homozygotes that may be attributable to inbreeding.

P23-9. First, calculate the selection coefficients for the homozygous normal and the homozygous affected genotypes. Remember that the selection coefficient is equal to 1 minus fitness. In this problem, the selection coefficient for the homozygous normal genotype (s_1) is $1 - 0.75 = 0.25$. The selection coefficient for the homozygous affected genotype (s_2) is $1 - 0.1 = 0.9$. Next, calculate the frequency of q expected at equilibrium $[= s_1 / (s_1 + s_2)]$. Substituting, we find that the frequency of q expected at equilibrium = 0.25/ (0.9 + 0.25) or 0.217. The frequency of p would therefore be $1 - 0.217$ or 0.783.

P23-10. The frequency of q at equilibrium equals: $\sqrt{(\mu/s)} = \sqrt{4 \times 10^{-6} / 0.1} = 0.006$. The frequency of the selected against allele (a) is very low, but it is maintained in the population by mutation.

P23-11. Directional selection can be modeled as selection against a recessive homozygote. Stabilizing selection can be modeled as heterozygote advantage, while disruptive selection is modeled as heterozygote disadvantage. The following table presents the fitness values for each type of selection.

Selection Type	AA	Aa	aa
Directional	1	1	$1 - s$
Stabilizing	$1 - s_1$	1	$1 - s_2$
Disruptive	1	$1 - s$	1

23

Population Genetics

CHAPTER SUMMARY QUESTIONS

1. 1. f, 2. d, 3. g, 4. a, 5. h, 6. c, 7. b, 8. e.

2. The assumptions of Hardy–Weinberg equilibrium (HWE) are: (1) diploid organism engaging in sexual reproduction, (2) nonoverlapping generations, (3) mating is at random, (4) infinite population size, and (5) no mutation, migration, or selection occurs. The first two assumptions are made for mathematical simplicity and to make HWE relevant. These assumptions can be relaxed. The third assumption allows us to be able to detect inbreeding (or outbreeding) independently of changes in allele frequencies themselves. The fourth assumption eliminates the stochastic evolutionary force, random genetic drift. The fifth eliminates other evolutionary forces that might change allele frequencies. The purpose of HWE is to construct a null hypothesis of random mating and no changes in allele frequencies.

3. Population geneticists use the term *fitness* to describe the ability of a given genotype (individual, population, and so forth) to pass on its alleles to the next generation. The more copies of the genetic information passed to future generations, the higher the fitness.

4. The five major evolutionary forces are mutation, selection, random genetic drift, migration, and nonrandom mating. (i) Mutation changes allele frequencies very slowly and is the only evolutionary process capable of creating new alleles. (ii) Selection is generally believed to be the strongest evolutionary force and increases the frequency of the allele with the highest fitness. (iii) Random genetic drift consists of stochastic changes in allele frequencies. It causes the loss of genetic variation and increases the variation in allele frequencies in now genetically isolated lineages. (iv) Migration can potentially change allele frequencies rapidly depending on the rate of gene flow and the differences in allele frequencies between the populations among which gene flow is occurring. Effective gene flow can be much greater than the migration rate if the recipient population is highly inbred and migrants have high levels of fitness compared with natives, or much lower than the migration rates if strong local adaptation means that migrants have low mean fitness. (v) Nonrandom mating most often takes the form of inbreeding due to structuring of populations. The effect of inbreeding is to decrease the heterozygosity levels of individuals, but expected heterozygosity (under HWE) for the overall population remains the same. Assortative mating has the same effect, assuming that the trait being selected on has a genetic basis. Disassortative mating has the opposite effect, increasing heterozygosity levels in individuals while the overall population's expected heterozygosity remains the same.

5. **a.** Differential reproduction is a necessary but not sufficient condition for evolution by natural selection. If all individuals reproduce equal numbers of offspring regardless of phenotype or if individuals produce different numbers of offspring based solely on differences in environment, no adaptation occurs. However, stochastic changes in allele frequencies (evolution) can still occur.

b. Environmental change is not necessary for natural selection to occur. Even in a stable environment, mutations that increase fitness can and will occur. However, the rate of evolution is expected to be much slower in a stable environment that the organism is already well adapted to.

c. Heritable variation in fitness is a necessary but not sufficient condition for natural selection to occur.

d. Sexual reproduction is not necessary for natural selection. Asexual organisms have genotypes that differ in fitness and the most fit genotypes increase in frequency over time.

6. The seven assumptions of the selection models we used are: (i) differences in fitness are due only to differences in viability, (ii) fitness differences are due to genotypic differences at a single locus, (iii) mating is at random, (iv) selection is identical in both sexes, (v) the selection coefficient is constant through time, (vi) population size is infinite, (vii) and no mutation occurs at the locus. All of these assumptions are made in order to make the mathematics simple. None of the assumptions is likely to be true, and many models have been built to explore what result relaxing these assumptions has on the outcome of selection.

7. The rate at which allele frequencies change under selection is determined primarily by the selection coefficient, the degree of dominance or recessiveness of the allele, and the frequency of the allele at the onset of selection.

8. Directional selection favors high (or low) values for a trait. Selection for extreme values for the trait also means selection for homozygosity at the loci contributing to the trait (except where there is heterozygote advantage). Disruptive selection favors both extremes of a trait and is believed to be rather rare. If the genetically based differences in phenotype are also linked to genetically based preferences for mating, disruptive selection may lead to speciation. Stabilizing selection occurs when intermediate phenotypic values have the highest fitness; this type of selection is thought to be very common (for example, the survival rate for human infants is highest at intermediate birth weights).

9. Selection can maintain genetic variation whenever the fitness of alleles can vary spatially or temporally, for example, with density- or frequency-dependent selection or with genotype-environment interactions. When the heterozygote has higher fitness than either homozygote, selection will act to maintain a polymorphism at this locus. Stabilizing selection can also maintain genetic variation. Most genetic variation is probably maintained through a balance between mutation, selection, and drift.

10. The two major effects of random genetic drift are to reduce genetic variation and to potentially fix alleles other than the one with the highest fitness despite their deleterious nature. Drift is stronger in smaller populations because it is a sampling process, and variation among samples is largest when samples are smallest.

11. Figure 23.11 demonstrates the interaction between drift and selection. At any given population size, the fate of an allele with a given selection coefficient might be purely stochastic (determined by drift), deterministic (determined by selection), or determined in part by both (more likely to fix the beneficial allele).

12. The major impact of gene flow among populations is to make those populations more genetically similar to each other, by countering the effects of random genetic drift, unique mutations, and differential selection. Gene flow can have different effects on population fitness depending on the details of the populations among which gene flow is occurring. Highly inbred populations usually benefit from gene flow from other populations as it relieves inbreeding depression. High-profile examples of this have been seen in populations of adders in Sweden, prairie chickens in Illinois, the Florida panther, and a population of wolves in Norway. Conversely, if the populations differ phenotypically because of differential selection to the local environment or have undergone chromosomal rearrangements that affect the fertility of hybrids, gene flow can be detrimental to the population. This is especially true where gene flow rates

exceed selection's ability to eliminate particular alleles that are deleterious in the new environment. An example of this is the breeding of escaped domestic (captive bred) salmon with wild salmon, which has been shown to have deleterious effects on population numbers of wild salmon due to the low fitness of hybrids.

13. Inbreeding almost universally decreases mean fitness, and this phenomenon is termed inbreeding depression. Inbreeding per se, in the absence of drift, increases the homozygosity levels of individuals without decreasing the total genetic diversity available to the population. The two facts are thought to be cause and effect. The increase in homozygosity for recessive deleterious alleles across numerous loci throughout the genome is thought to be the primary cause of inbreeding depression.

14. Inbreeding coefficients for an individual can be determined by using a pedigree to determine the likelihood of an allele being homozygous by descent (that is, inherited from a common ancestor). Inbreeding coefficients for a population are usually determined by comparing heterozygosity levels for a single population at different times or by comparing heterozygosity levels of different populations against each other.

15. Random genetic drift decreases heterozygosity levels. It does so by the stochastic loss of alleles at a locus until, given enough time, one allele is fixed at random. Random genetic drift decreases fitness by sometimes fixing deleterious alleles, despite their being detrimental to fitness. Inbreeding also decreases heterozygosity, but does so at the individual level and does not change allele frequencies as drift does. Inbreeding also decreases fitness, but does so through increasing homozygosity levels for deleterious recessive alleles within individuals in a population.

16. By *kind* of genetic variation, population geneticists are referring to whether the genetic variation is found in the form of beneficial, deleterious, or neutral alleles, whether the alleles are dominant or recessive to the wild type, whether they are of large or small effect, whether selection coefficients are constant across environments, and so forth.

EXERCISES AND PROBLEMS

17. You are given that $M = 0.59$ and $N = 0.41$. Therefore, under HWE, $MM = p^2 = P = 0.3481$; $MN = 2pq = H = 0.4838$; and $NN = q^2 = Q = 0.1681$. Since there are 3772 people we expect 1313 to be genotype MM (0.3481×3772), 1825 to be MN (0.4838×3772), and 634 to be NN (0.1681×3772).

18. Allele frequencies can be calculated as follows: 500 individuals with the MM genotype contribute 1000 M alleles, and 300 individuals with the MN genotype contribute 300 M alleles, for a total of 1300 M alleles. Then 700 individuals with the NN genotype contribute 1400 N alleles, and 300 individuals with the MN genotype

contribute 300 N alleles, for a total of 1700 N alleles. The relative frequency of the M allele is 0.4333 (1300/3000), and the frequency of the N allele is 0.5667 (1700/3000).

Under HWE, the following genotypic frequencies are expected: $MM = p^2 = 0.1878$, $MN = 2pq = 0.4911$, and $NN = p^2 = 0.3211$. The population consists of 1500 people, so expected numbers of individuals with each genotype can be calculated by multiplying each of the genotypic frequencies by the number of individuals.

	Observed	Expected	χ^2
MM	500	281.70	169.2
MN	300	736.65	258.8
NN	700	481.65	98.9
Total	1500	1500	526.9

The calculated chi-square value exceeds the chi-square critical value for two degrees of freedom immensely. Thus, you can reject the null hypothesis that this population is in HWE with great confidence. The excess of both types of homozygotes leads you to suspect strong inbreeding in this population.

19. To answer this question you must recognize that the 1% of girls that are naughty are homozygous for the n allele. Thus, the frequency of the recessive allele can be calculated as the square root of the frequency of the homozygote ($q^2 = 0.01$). Thus, the naughty allele (n) exists at a frequency of 0.10. Since males receive only one copy of the X chromosome, the frequency of the recessive genotype will be equal to the frequency of the allele in the population. Alice has a 10% chance of meeting a naughty boy and being disappointed.

20. From the data given we know that 24% of the students had dimples. Thus, they must be either heterozygous or homozygous for the dominant allele. Thus, 76% must be homozygous for the recessive allele. If $q^2 = 0.76$, then q must equal the square root of this, so the recessive allele, d, exists at a frequency of 0.8718. This makes p equal to $1 - 0.8718$, so the dominant allele, D, exists at a frequency of 0.1282.

21. The frequency of the recessive genotype is 0.7887. If $q^2 = 0.7887$, then the frequency of d is equal to 0.8881. The frequency of D must then be 0.1119. Given this, the frequency of individuals that are homozygous for the D allele (and will give dimples to all their children) is $p^2 = (0.1119)^2 = 0.0125$ or about 1.25% of the population.

22. The proportion of the population that is homozygous for the recessive allele is 0.3023. If we assume HWE, the frequency of the recessive allele must be 0.5498, making the dominant allele occur with a frequency of 0.4502.

23. The estimated allele frequencies are: $A = 0.45$ (27/60), $B = 0.1667$ (10/60), $C = 0.0833$ (5/60), $D = 0.15$ (9/60), and $E = 0.15$ (9/60). To test for HWE, it might seem obvious to arrange the chi-square analysis in the following way:

	Observed	Expected	χ^2
AA	9	6.07	1.41
AB	2	4.50	1.39
AC	0	2.25	2.25
AD	5	4.05	0.22
AE	2	4.05	1.04
BB	3	0.83	5.67
BC	0	0.83	0.83
BD	1	1.50	0.17
BE	1	1.50	0.17
CC	2	0.21	15.25
CD	0	0.75	0.75
CE	1	0.75	0.08
DD	1	0.68	0.15
DE	1	1.35	0.09
EE	2	0.68	2.56
Total	30	30	32.03

This provides a strong rejection of the null hypothesis, and we reach the conclusion that the population is not in HWE. However, the sample sizes are very small and any expected values less than 5 should be lumped. This leaves us only two categories, which will be rather uninformative. Examining the data, we see that the departure from HWE is due mostly to an excess of *BB* and especially *CC* individuals. In fact, all the homozygous genotypes are in excess. Thus, a better test might be to see if homozygous genotypes at all loci are more frequent than expected under HWE.

	Observed	Expected	χ^2
Homozygotes	17	8.47	8.39
Heterozygotes	13	21.53	3.38
Total	30	30	11.77

With one degree of freedom, the chi-square critical value is 3.84. Even at a *p* value of 0.005, the critical value is only 7.88. Thus, we have very strong evidence against the null hypothesis and reject that this population is in HWE. Further, we show that the departures from HWE are due to an excess of homozygous individuals, likely due to inbreeding within the population.

24. The fitness of the three genotypes are

W_1W_1 $144.0 \times 0.016 = 2.30$
W_1W_2 $133.2 \times 0.025 = 3.33$
W_2W_2 $102.9 \times 0.034 = 3.50$

The relative fitnesses are then

W_1W_1 $2.30/3.50 = 0.6571$
W_1W_2 $3.33/3.50 = 0.9514$
W_2W_2 $3.50/3.50 = 1.0000$

25. From the sample, we calculate that the frequency of the A_1 allele is 0.675 and the A_2 allele is 0.325. HW expectations and observed values are

Genotype	Expected	Observed	χ^2
A_1A_1	182.25	180	0.0278
A_1A_2	175.5	180	0.1154
A_2A_2	42.25	40	0.1198
Total	400	400	0.263

The calculated chi-square value is very small compared to the chi-square critical value for two degrees of freedom ($0.263 < 5.991$). Thus, we accept the null hypothesis that the population is still in HWE. This suggests that the three genotypes survived the 15 years equally well.

26. The information you were given and calculations from that information can be summarized in the following form:

Genotype	Frequency	Relative Fitness	Contribution
A_1A_1	0.25	1.00	0.25
A_1A_2	0.50	1.00	0.50
A_2A_2	0.25	0.00	0.00

The fitness of this population is equal to the sum of the contributions of the three genotypes, which in this case is 0.75. Thus, the population has its fitness reduced by 25% due to the segregation of a recessive lethal at a frequency of 0.50. The relative frequencies of the genotypes in the next generation are

A_1A_1 $0.25/0.75 = 0.333$
A_1A_2 $0.50/0.75 = 0.667$
A_2A_2 $0.00/0.75 = 0.000$

The frequency of the A_1 allele $= [0.333 + (0.5)(0.667)] = 0.667$. The frequency of the A_2 allele is then 0.333. Thus, the frequency of the allele with the higher fitness (A_1) increased in frequency from 0.5 to 0.667 in one generation.

27. To answer this question, we must first calculate the current allele frequencies. There are 1484 R alleles and 516 Y alleles. This corresponds to frequencies of $R = 0.742$ and $Y = 0.258$. Expectations under HWE are 0.551, 0.383, and 0.067 for RR, RY, and YY, respectively. Multiplying expected frequencies by 1000 flowers gives us our expected values. Let's first test to see if the population is still in HWE.

Genotype	Expected	Observed	χ^2
RR	531	551	0.753
RY	383	422	3.971
YY	67	47	5.970
Total	400	400	10.694

The calculated chi-square value exceeds the chi-square critical value for two degrees of freedom (5.991) by enough that the *p* value is slightly smaller than 0.005. Thus, we reject the null hypothesis that the population is currently in HWE. What has caused the population to go out of HWE? The chi-square values suggest that the departures from HWE are due primarily to a deficiency in *YY* individuals and also to an excess of heterozygotes. However, the real test is to see what has happened to the frequency of genotypes over time. The proportion of *RR* homozygotes has increased, and the proportion of *YY* homozygotes has decreased. This suggests that the yellow flowers are being selected against, and the frequency of the *R* allele is increasing (from 0.70 to 0.742).

28. To answer this question, we must first calculate the current allele frequencies. There are 1400 *R* alleles and 600 *Y* alleles. This corresponds to frequencies of $R = 0.70$ and $Y = 0.30$. Expectations under HWE are 0.49, 0.42, and 0.09 for *RR*, *RY*, and *YY*, respectively. Multiplying expected frequencies by 1000 flowers gives us our expected values. Let's first test to see if the population is still in HWE.

Genotype	Expected	Observed	χ^2
RR	490	495	0.051
RY	420	410	0.238
YY	90	95	0.278
Total	400	400	0.567

The calculated chi-square value is very small compared to the chi-square critical value for two degrees of freedom (5.991). Thus, we accept the null hypothesis that the population is currently in HWE. The genotypic frequencies have changed so slightly that it is not unexpected by chance alone, and the allele frequencies are unchanged.

29. From the information given, we can form the following table:

	Frequency	Relative Fitness	Contribution
A_1A_1	0.50	0.80	0.40
A_1A_2	0.40	1.00	0.40
A_2A_2	0.10	0.40	0.04

The relative contributions are then
A_1A_1 0.40/0.84 = 0.476
A_1A_2 0.40/0.84 = 0.476
A_2A_2 0.04/0.84 = 0.048

In the next generation, the frequency of A_1 will be $0.476 + (0.5)(0.476) = 0.714$. The frequency of A_2 will be $0.048 + (0.5)(0.476) = 0.286$.

Thus, the A_1 allele increased from a frequency of 0.70 to 0.714. The equilibrium frequency of the A_1 allele can be calculated by noting that s (the selection coefficient against individuals homozygous for the A_1 allele) is 0.2 and t (the selection coefficient against individuals homozygous for the A_2 allele) is 0.6. The equilibrium frequency of the A_1 allele in the absence of any forces except selection is $t/(s + t) = 0.75$. Thus, the equilibrium frequency for the A_2 allele is 0.25. In this example we expect the A_1 allele to continue to increase in frequency until it reaches 0.75. At that point population fitness is maximized and selection will act to maintain this equilibrium.

30. From the information provided:

	Frequency	Relative Fitness	Contribution
A_1A_1	0.4	1	0.4
A_1A_2	0.5	0.8	0.4
A_2A_2	0.1	0.6	0.06

The relative contribution to the next generation is

A_1A_1	$0.40/0.86 = 0.465$
A_1A_2	$0.40/0.86 = 0.465$
A_2A_2	$0.06/0.86 = 0.070$

The frequency of the A_1 allele is $0.465 + (0.5)(0.465) = 0.698$. Thus, it increased from 0.65 to 0.698 in one generation.

31. Since selection against the two homozygous genotypes is equal ($s = t$), fitness is maximized when heterozygosity is maximized (that is, the allele frequencies are even). Formally, $s = 0.15$ and $t = 0.15$. The equilibrium frequency of the M_1 allele is $t/(s + t)$, which is equal to 0.50.

32. To solve the problem one must first recognize that the homozygous recessive individuals have died. The population is not in HWE, and the proportion of individuals that are AA is 0.1111 (0.04/0.36) and the proportion of individuals with the Aa genotype is 0.8889(0.32/0.36). With random mating among the survivors, the expected genotypic frequencies are again p^2, $2pq$, and q^2 for the AA, Aa, and aa genotypes, respectively. The A allele exists at a frequency of $0.1111 + (0.5)(0.8889) = 0.5555$. Thus, the a allele exists at a frequency of 0.4445. The proportion of individuals that will die because they are homozygous for the now deleterious recessive allele is 0.1976 or about 19.76% of the population.

33. The larger population is fixed for the single most advantageous allele. Thus, its relative fitness is 1.0. In the second population the D_1 allele has a frequency of 0.8367.

The population fitness is:

Frequency	Relative Fitness	Contribution
0.7000	1.00	0.7000
0.2733	0.99	0.2706
0.0267	0.95	0.0254

The relative fitness of the smaller population is approximately 99.6% of that of the larger population. The difference in the allele frequencies is most likely due to random genetic drift being greater in the smaller population.

34. The formula for calculating the frequency of a completely recessive deleterious allele, q, in mutation selection balance is $q = (\mu/s)^{1/2}$, where s is the selection coefficient and μ is the mutation rate. In this case, $q_e = (0.000005/0.01)^{1/2} = 0.0224$.

35. Because each source population was fixed for its respective alleles and the new population was initiated from equal numbers from each of the source populations, the initial allele frequencies are equal. It would normally take only one generation to reach HWE, but because different sexes migrated from different source populations, mating in the "parental" generation is not really at random. In the first generation, all individuals will be heterozygous for MN. However, after the initial generation, if the conditions for HWE hold, the genotypic frequencies will be in HWE. Thus, since $M = 0.5$ and $N = 0.5$, the corresponding genotypic frequencies are 0.25, 0.50, and 0.25 for the MM, MN, and NN genotypes, respectively.

36. The trick to solving this problem rapidly is to recognize that the speed at which migration changes allele frequencies depends only on the rate of migration and the difference between the native and migrant populations. Thus,

   A. Forty migrants in a total population of $100 = 0.40$. The difference between the allele frequencies is $1.00 - 0.85 = 0.15$. Total effect $= 0.40 \times 0.15 = \mathbf{0.06}$.
   B. Eight migrants in a total population of $80 = 0.10$. The difference between the allele frequencies is $0.70 - 0.50 = 0.20$. Total effect $= 0.10 \times 0.20 = \mathbf{0.02}$.
   C. Seventy migrants in a total population of $700 = 0.10$. The difference between the allele frequencies is $0.50 - 0.40 = 0.10$. Total effect $= 0.10 \times 0.10 = \mathbf{0.01}$.
   D. Sixty migrants in a total population of $120 = 0.50$. The difference between the allele frequencies is $1.00 - 0.05 = 0.95$. Total effect $= 0.50 \times 0.95 = \mathbf{0.475}$.
   E. Seventy-five migrants in a total population of $750 = 0.10$. The difference between the allele frequencies is $0.70 - 0.25 = 0.45$. Total effect $= 0.10 \times 0.45 = \mathbf{0.045}$.

   Situation D provides the largest change in allele frequencies with migration. This is obvious in retrospect, because it has both the highest migration rate and the largest difference in initial allele frequencies between the migrant and native populations.

37. The new allele frequency in the combined population is the weighted frequencies of the allele in the source populations. Thus, natives comprise 70% of the population

and migrants 30%. The allele frequencies are 0.08 and 0.005, for the native and migrant populations, respectively. The frequency in the combined population is $[(0.70)(0.08) + (0.30)(0.005)] = 0.0575$. Assuming random mating, the frequency of the affected individuals would have been $0.08^2 = 0.0064$ or 64 individuals in 10,000 births. The new frequency, assuming random mating, is $0.0575^2 = 0.0033$ or 33 individuals in 10,000 births. Thus, the proportion of individuals affected by the disease is approximately half of what it was prior to migration.

38. There are two paths, each with seven ancestors (see the following figure). Therefore, the inbreeding coefficient F is equal to $\sum[(1/2)^n(1 + F_J)] = 2(1/2)^7 = 0.016$.

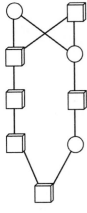

39. The inbreeding coefficient F is a measure of inbreeding within a population. It can be estimated by the reduction in heterozygosity through time or by comparing heterozygosity levels between populations. In this case, $F = (H_L - H_S)/H_L$, where H_L is the amount of heterozygosity in the larger population and H_S is the amount of heterozygosity in the smaller population. Thus, $F = (0.76 - 0.608)/0.76 = 0.20$ in this case. The smaller population is 20% more inbred (homozygous by descent) than the larger.

40. The four cases are outlined here:

a. **Additive with No Inbreeding**

Genotype	Frequency	Relative Fitness	Contribution
A_1A_1	0.81	1.0	0.810
A_1A_2	0.18	0.8	0.144
A_2A_2	0.01	0.6	0.006
Population fitness			0.96

b. **Additive with Inbreeding**

Genotype	Frequency	Relative Fitness	Contribution
A_1A_1	0.828	1.0	0.828
A_1A_2	0.144	0.8	0.115
A_2A_2	0.028	0.6	0.017
Population fitness			0.96

c.   **Dominance with No Inbreeding**

Genotype	Frequency	Relative Fitness	Contribution
A_1A_1	0.81	1.0	0.810
A_1A_2	0.18	1.0	0.180
A_2A_2	0.01	0.6	0.006
Population fitness			0.996

d.   **Dominance with Inbreeding**

Genotype	Frequency	Relative Fitness	Contribution
A_1A_1	0.828	1.0	0.828
A_1A_2	0.144	1.0	0.144
A_2A_2	0.028	0.6	0.017
Population fitness			0.989

The relative fitness (relative to an ideal population without the deleterious allele) for each population is listed in each table. In particular, it is worth noting that inbreeding does not decrease fitness when the alleles act additively. Inbreeding only decreases fitness when there is dominance. Thus, inbreeding depression is indirectly the result of selection, as selection efficiently removes deleterious alleles that are not recessive.

41. This question is asking about mutation-selection balance. In an earlier problem we stated that the $q_e = (\mu/s)^{1/2}$. In this case we are given the equilibrium frequency and the mutation rate μ and asked to estimate the selection coefficient s. The equation can be arranged so that $s = 1/(q_e^2/\mu)$. It was stated that 1 in 50,000 people have this form of dwarfism. This is not the equilibrium allele frequency q_e, but rather the number of individuals homozygous for the allele. It is, however, under random mating, the expected value for q^2, which is what we are actually looking for. The mutation rate μ is given as 9.13×10^{-6}. Thus, we can plug into the equation that $s = 1/(0.00002/0.00000913) = 0.4565$.

42. The two forces are random genetic drift and selection. If the population were made even smaller, drift should become more powerful relative to selection. If the populations are made small enough, drift should become so powerful that the alleles act as if they were completely neutral. Thus, you expect half the populations to eventually fix the beneficial allele and half the deleterious allele.

CHAPTER INTEGRATION PROBLEM

a. Methionine is one of those rare amino acids represented by only one codon, AUG. An AUG codon could mutate into AAG (lysine), ACG (threonine), AGG (arginine), any of the three isoleucine codons (AUA, AUC, or AUU), CUU (leucine), GUG (valine), or UUG (leucine).

b. The equilibrium allele frequency q_e for a totally recessive deleterious allele is $(\mu/s)^{1/2}$, where μ is the forward mutation rate and s is the selection coefficient. In this case, $\mu = 2.5 \times 10^{-5}$. The selection coefficient against the B_2 allele in the absence of disease is 0.03 and 0.05 against the B_3 allele, as provided in the question. For the B_2 allele, $q_e = (0.000025/0.03)^{0.5} = 0.029$. For the B_3 allele, $q_e = (0.000025/0.05)^{0.5} = 0.022$.

c. The allele frequencies are $B_1 = 0.756$, $B_2 = 0.173$, and $B_3 = 0.071$. Expected genotypic frequencies under HWE are $B_1B_1 = 0.572$, $B_1B_2 = 0.261$, $B_1B_3 = 0.107$, $B_2B_2 = 0.030$, $B_2B_3 = 0.025$, and $B_3B_3 = 0.005$. Relative fitness for each genotype is 1.0, 1.0, 1.0, 0.97, 0.96, and 0.95, respectively.

d. The following tables take you through the steps of calculating the new allele frequencies. As expected, the frequency of the B_1 allele increased while the other two decreased, since the other two alleles were deleterious in this environment.

Genotype	Frequency	Relative Fitness	Contribution
B_1B_1	0.572	1.00	0.5720
B_1B_2	0.261	1.00	0.2610
B_1B_3	0.107	1.00	0.1070
B_2B_2	0.030	0.97	0.0291
B_2B_3	0.025	0.96	0.0240
B_3B_3	0.005	0.95	0.0048
Total	1.000		0.9979

Relative genotypic contribution:
$B_1B_1 = 0.5720/0.9979 = 0.5732$
$B_1B_2 = 0.2610/0.9979 = 0.2615$
$B_1B_3 = 0.1070/0.9979 = 0.1072$
$B_2B_2 = 0.0291/0.9979 = 0.0292$
$B_2B_3 = 0.0240/0.9979 = 0.0241$
$B_3B_3 = 0.0048/0.9979 = 0.0048$

New allele frequencies:
$B_1 = [0.5732 + (0.5)(0.2615) + (0.5)(0.1072)] = 0.7576$
$B_2 = [0.0292 + (0.5)(0.2615) + (0.5)(0.0241)] = 0.1720$
$B_3 = [0.0048 + (0.5)(0.1072) + (0.5)(0.0241)] = 0.0704$

e. The allele frequencies are $B_1 = 0.756$, $B_2 = 0.173$, and $B_3 = 0.071$. Expected genotypic frequencies under HWE are $B_1B_1 = 0.572$, $B_1B_2 = 0.261$, $B_1B_3 = 0.107$, $B_2B_2 = 0.030$, $B_2B_3 = 0.025$, and $B_3B_3 = 0.005$. Relative fitness for each genotype is 0.0, 0.0, 0.0, 0.20, 1.00, and 0.50, respectively.

f. The following tables take you through the steps of calculating the new allele frequencies. Not surprisingly, this level of intense selection radically changes the allele frequencies in one generation.

Genotype	Frequency	Relative Fitness	Contribution
B_1B_1	0.572	0.00	0.0000
B_1B_2	0.261	0.00	0.0000
B_1B_3	0.107	0.00	0.0000
B_2B_2	0.030	0.20	0.0060
B_2B_3	0.025	1.00	0.0250
B_3B_3	0.005	0.50	0.0025
Total	1.000		0.0335

Relative genotypic contribution:
$B_1B_1 = 0.0/0.0335 = 0.0000$
$B_1B_2 = 0.0/0.0335 = 0.0000$
$B_1B_3 = 0.0/0.0335 = 0.0000$
$B_2B_2 = 0.006/0.0335 = 0.1791$
$B_2B_3 = 0.0250/0.0335 = 0.7463$
$B_3B_3 = 0.0025/0.0335 = 0.0746$

New allele frequencies:
$B_1 = 0.0$
$B_2 = [0.1791 + (0.5)(0.7463)] = 0.5523$
$B_3 = [0.0746 + (0.5)(0.7463)] = 0.4477$

g. The equilibrium frequency of two alleles when there is heterozygote advantage depends on the relative selection coefficients when the alleles are homozygous. In the presence of disease X, 80% of individuals homozygous for the B_2 allele die, so the selection coefficient s is 0.8. In the presence of disease X, 50% of individuals homozygous for the B_3 allele die, so the selection coefficient t is 0.5. The equilibrium frequency of the B_2 allele is $t/(s + t) = 0.5/(0.8 + 0.5) = 0.385$. The B_3 allele then exists at a frequency of 0.615.

h. The following tables take you through the steps of calculating the new allele frequencies. The trick is realizing that the relative frequencies are now a weighted average of the fitness in each environment. Thus, the B_1B_1 homozygote has a fitness of one 60% of the time and a fitness of zero 40% of the time. The B_3 allele increases fastest, the B_2 allele also increases, and the B_1 allele decreases in frequency.

Genotype	Frequency	Relative Fitness	Contribution
B_1B_1	0.035	0.600	0.0210
B_1B_2	0.161	0.600	0.0966
B_1B_3	0.143	0.600	0.0858
B_2B_2	0.185	0.662	0.1225
B_2B_3	0.329	0.976	0.3211
B_3B_3	0.147	0.770	0.1132
Total	1.000		0.7602

Relative genotypic contribution:

$B_1B_1 = 0.0210/0.7602 = 0.0276$

$B_1B_2 = 0.0966/0.7602 = 0.1271$

$B_1B_3 = 0.0858/0.7602 = 0.1129$

$B_2B_2 = 0.1225/0.7602 = 0.1611$

$B_2B_3 = 0.3211/0.7602 = 0.4224$

$B_3B_3 = 0.1132/0.7602 = 0.1489$

New allele frequencies:

$B_1 = [0.0276 + (0.5)(0.1271) + (0.5)(0.1129)] = 0.1476$

$B_2 = [0.1611 + (0.5)(0.1271) + (0.5)(0.4224)] = 0.4359$

$B_3 = [0.1489 + (0.5)(0.1129) + (0.5)(0.4224)] = 0.4165$

i. The slowest moving band must be the B_1 allele, because all individuals with that band died. Surviving individuals all had the same genotype, so it is impossible to differentiate the other two alleles.

j. Each child has an allele not found in both parents. Child 1 had to inherit the B_1 allele maternally, because the father does not have a copy of that allele. The other then must come from the father. Child 2 has two alleles distinct to each parent; the B_1 allele came from the mother, and the fastest moving band had to come from the father. The two youngest children have the same genotype and must have inherited the fastest moving allele paternally; therefore, the other allele came from the mother.

k. Half of the children are expected to inherit the B_1 allele from the mother and be 100% susceptible to the disease. Regardless of which band represents which of the two mutant alleles, you expect one-fourth of the offspring to be heterozygous and completely immune and one-fourth to be homozygous for the mutant allele and partially immune.

CHAPTER 24 QUANTITATIVE GENETICS

Chapter Goals

1. Determine the proportion of variation in a trait that results from the environment and the genotype.
2. Understand how the mean phenotype of a given genotype is altered with changes in the environment.
3. Describe how and to what extent loci and alleles interact with each other to influence a phenotype.
4. Explain the differences observed due to natural and artificial selection.
5. Understand how the theory of quantitative genetics may affect the management of endangered species of plants and animals.
6. Estimate the number of loci involved in a trait and any limitations to this approach.

Key Chapter Concepts

24.1 What Is Quantitative Genetics?

- **Overview:** Traits such as height, intelligence, or crop seed size are all traits that do not have discrete phenotypes. Instead, they vary continuously so that they display a continuum of possible phenotypes. The inheritance of these traits is more complicated than the simple patterns described by Mendel. They often result from the influence of many genes *and* the influence of the environment. Researchers studying quantitative genetics study how many genes influence a trait, what percentage of the trait is determined by each of these genes, and what the consequences of environmental changes are on the trait.

24.2 Polygenic Traits

- **Polygenic Inheritance: Polygenic traits** are traits that are controlled by more than one gene. These traits are usually influenced by several pairs of alleles and by the environment. There are three types of polygenic traits.
  - **Continuous traits**: Rather than discrete phenotype possibilities such as black or white, these traits can exhibit an infinite number of values. An example is human height.
  - **Threshold traits:** These traits are those that are either present or absent in an individual and where a threshold value must be met in order to produce the trait. There does, however, exist an infinite number of *probabilities* that an individual will reach the threshold value and produce the trait. An example of a threshold trait is heart disease.
  - **Meristic traits:** These traits have a large number of possible values, but each possible value is always a whole number. An example is the number of scales on a fish.
- **Approximating the Number of Genes Influencing a Polygenic Trait:** The number of genes controlling a polygenic trait can be approximated by obtaining an F_2 generation with individuals that display an extreme phenotype. The number of F_2 progeny with a phenotype as extreme as one of the parents equals $(1/4^n)$, where n equals the number of pairs of alleles. For example, an F_2 generation with approximately 8 out of 500 individuals (0.016) resembling one of the parents closely approximates the value of $1/4^3$ ($1/64 = 0.0156$), indicating that three genes (each with two alleles) influence this trait.
- **Environmental Effects on Continuous Variation:** A polygenic trait controlled by a limited number of genes may still have discrete phenotypes that result from specific genotypes. If the environment, however, can cause the phenotype of each genotype to vary slightly, then the phenotypic possibilities of different genotypes may overlap and a continuous trait will result. Thus, as the number of genes controlling a trait and/or the influence of the environment increases, so does the number of possible phenotypes.

24.3 Statistical Tools in Quantitative Genetics

- **Parameters That Describe a Distribution of Phenotypes:** More complex methods of analysis are necessary to describe and understand polygenic inheritance. Important statistics include the **mean**, **median**, **mode**, **variance**, and **standard deviation**. Each of these statistics can be used to describe the distribution of individuals in a population and to compare the distributions of different populations. In a truly normal distribution, the mean, median, and mode will all be equal. The variance describes the extent of the extreme phenotypes in a population, while the standard deviation describes the mean distance of any value from the mean of the population. A population can be described using only the mean and standard deviation.

- **Measurements That Describe Covariation: Covariance**, **correlation**, and the **correlation coefficient** are used to determine if there is a relationship between two characteristics in a series of individuals or a relationship between a characteristic in a series of two types of individuals (parents versus offspring, for example). Correlation coefficients vary between –1 and 1. A coefficient of –1 indicates that as one trait increases, the other *always* decreases. A coefficient of 0 indicates that no correlation exists at all, and a coefficient of 1 indicates that as one trait increases in value, the other trait *always* increases also. A correlation does not necessarily identify a cause.

24.4 The Nature of Continuous Variation

- **Variation:** Continuous variation results from small variances that result from one or more of several contributing factors. This phenotypic variance can be broken into its component parts that include both gene and environmental influences. Within the genetic component of variance, **additive genetic variance** results from genetic traits inherited from the parents, while **dominance variance** and **epistatic variance** are genotype-specific and are therefore not inherited but relate to the combination of alleles that are inherited. Thus, only additive variance experiences natural selection. A final component of the phenotypic variance is the **norm of reaction** which relates the interaction between a gene and the environment. In other words, this describes the variation between different genotypes with respect to how they change in changing environments. This is important for the development of agricultural crops with genotypes that display a suitable variation in phenotype across the entire range of expected conditions.

24.5 Heritability

- **Overview of Heritability:** Individuals within a population often vary in the characteristics that they possess. For years, geneticists have taken advantage of this natural variation to create varieties of organisms, such as domestic animals and corn plants that possess characteristics of interest (such as increased milk production in domestic animals or increased seed production in corn). In order for the characteristics of a population to change, the trait must have a genetic basis. **Heritability** is a statistic that allows a geneticist to determine if the observed differences in the population (variation) are due to genetic differences among the individual members of the population or due to the influence of the environment. A heritability of one (1.0) implies that all of the variation observed in a population is due to genetic factors, while a heritability of zero implies that all of the variation observed is due to environmental influences. Remember that heritability describes the percentage of the *variation* observed in a population and not the percentage of the phenotype. **Broad-sense heritability** (H^2) describes the proportion of the observed variation that is due to all genetic factors (additive, dominance, and epistatic variance), while **narrow-sense heritability** (h^2) estimates the proportion of the observed variation that is due only to additive genetic effects. Narrow-sense heritability is generally a more useful statistic since it results from factors that can experience selection in selective breeding experiments in order to produce desired traits in plants and animals.

- **Measuring Heritability:** Controlled experiments can be conducted to estimate broad-sense and narrow-sense heritability values for quantitative traits. To estimate heritability, either the genotype or the environment is fixed and the phenotypic variation is measured. This value is then compared when both components vary freely. **Cross-fostering** experiments are useful in determining heritability in animals. In these experiments, siblings, ideally identical twins, are separated and raised in differing environments. Since nonidentical siblings share 50% of their genes and identical twins share 100% of their genes, the variation in the phenotypes of identical twins will be equal to the environmental variation, and the variation in nonidentical siblings will be equal to the environmental variation plus half of the genetic variation. Thus, the genetic variation can be computed by subtracting the environmental variation from the variation in the nonidentical siblings and then multiplying by 2.

- **Realized Heritability:** Narrow-sense heritability can be estimated by applying a selection (through selective breeding) and determining how much change is gained from a given selection pressure. This is known as **realized heritability**. To perform this, two individuals are chosen for mating. The **selection differential** is the difference between the mean difference of the two selected individuals and the mean phenotype of the population.

24.6 Quantitative Genetics in Ecology

- **Selection Pressures:** Evolution of a species in a changing environment requires natural selection of quantitative traits. Selection works on genetically inherited traits; in order for evolution to occur, the population must display genetic variation for quantitative traits. As you learned in chapter 23, selection can sometimes lead to a loss of genetic diversity, or variation. Population size, the strength and nature of selection, and the rate of mutation (new allele formation) all affect the maintenance of genetic variation. Selection may shift the population mean toward one end of the phenotypic distribution (**directional selection**), it may favor individuals with phenotypes near the mean of the population (**stabilizing selection**), or it may favor individuals with extreme phenotypes (**disruptive selection**). Understanding these principles is necessary for our society in order to ensure that genetic diversity is maintained in wild populations so that extinctions or catastrophic loss of biodiversity does not occur.

- **Lessons from Selection:** Most natural populations have a sufficient amount of segregating genetic diversity for most phenotypic traits to ensure that genetic variation is maintained even in selective environments (the alleles for a trait do not become fixed). Thus, when the environment changes, most populations are able to adapt. Furthermore, many traits do not ever seem to reach a limit when experiencing strong selective pressure. A controlled experiment involving 100 generations of corn plant selection demonstrated both of these concepts; oil content continued to rise during 100 generations of selection, and even after 48 generations, reverse selection was able to restore the original oil content.

- **Conservation:** The exception to the idea that most natural populations can adapt indefinitely lies in small populations that are more susceptible to random genetic drift. Thus, maintaining sufficiently large natural populations is an important concern for our society if we wish to prevent catastrophic loss of biodiversity. Successes have been made in saving small, wild populations by breeding individuals in captivity where the offspring have an increased chance of survival or by introducing genetically similar, but more genetically diverse, individuals into a population to increase the diversity of the small population. The former was successful in increasing the population size of the California condor and the latter was successful in reestablishing the Florida panther.

24.7 Quantitative Trait Loci Mapping

- **Overview: Quantitative trait loci (QTL) mapping** is used to determine the number of loci that affect a quantitative trait, the relative contribution of each loci to the phenotype, and the location of each loci within the genome. Interestingly, experiments have discovered that only 4 to 12 QTLs account for the majority of the phenotypic variation in many traits.

- **Method:** QTL mapping takes advantage of the linkage of QTLs and DNA markers such as SNPs, RFLPs, and microsatellites. These markers were described in chapter 13 of your textbook. In QTL mapping, genetic crosses are examined for the **cosegregation** of a phenotype and a DNA marker. A high incidence of cosegregation suggests that a QTL that contributes to the phenotype must be physically near the DNA marker such that recombination between the two does not occur very frequently. Computer programs have been designed to analyze the data from such a cross and determine DNA markers that frequently cosegregate with a phenotype.

Key Terms

- quantitative trait loci (QTL)
- polygenic trait
- continuous traits
- threshold traits
- meristic traits
- sample
- central tendency
- arithmetic mean
- median

- mode
- variance (s^2)
- standard deviation (s)
- covariation
- covariance (cov)
- correlation
- correlation coefficient (r)
- additive genetic variance (V_A)

- dominance variance (V_D)
- epistatic genetic variance (V_I)
- norm of reaction
- heritability
- broad-sense heritability (H^2)
- narrow-sense heritability (h^2)

- cross-fostering
- concordance
- discordance
- response to selection

- selection differential
- directional selection
- stabilizing selection
- disruptive selection

- quantitative trait loci mapping
- cosegregation

Understanding the Key Concepts

Use the key terms or parts of their definitions to complete the following sentences.

Quantitative genetics involves examining how genes interact with (1)_____ in order to produce a continuous distribution of (2)_____ . Most of the traits that are quantitative in nature are (3)_____, meaning that they are influenced by several genes. The *number* of genes affecting the trait can be approximated by comparing the number of individuals in the F_2 generation that display one of the two extreme phenotypes of the (4)_____ generation. Still, a finite number of genes can only produce a finite number of phenotypic values unless each phenotype can vary slightly; for example, the (5)_____ might have the ability to influence each phenotype. This can occur by a slight shift toward either extreme. Quantitative traits can be classified as (6)_____, representing traits with an infinite number of values; (7)_____, representing traits that are either present or not present; or (8)_____, representing traits that vary by a whole number. The distribution of phenotypes of quantitative traits usually generate a bell-shaped curve with the average value occurring near the middle. If the distribution is normal, then the (9)_____ , the (10)_____, and the (11)_____ will all be at or very near the center of the curve. The variance of the population indicates the extent of the (12)_____ which relates to how wide the curve is. The (13)_____ of the variance is known as the standard deviation. Approximately (14)_____ % of a population should lie between ± one standard deviation of the mean, while approximately (15)_____ % of the population should lie between ± two standard deviations of the mean.

Variation in the phenotypes of quantitative traits can arise from differences in genetics, differences in the (16)_____, or differences in the way a genotype interacts with the (17)_____ . The genetic component of the variance results from three different subcomponents. (18)_____ results from the additive effects of accumulating active alleles of the genes that produce the trait. (19)_____ is the proportion of the variance that results from dominant relationships between alleles of a single locus, while (20)_____ results from gene interactions between alleles at different loci. (21)_____ is used to describe the amount of phenotypic variation that is determined by genetics. This can be broken into two categories; (22)_____ describes the variance that is due to all genetic contributions, while (23)_____ describes only the variance that is due to (24)_____ . To estimate the latter value of a population, a selection differential can be applied to the population and the populations (25)_____ can be divided by the degree of selection (selection differential) to estimate (26)_____.

Figure Analysis

Use the indicated figure or table from the textbook to answer the questions that follow.

1. **Figure 24.3**

    (a) The horizontal axis of these graphs represents a continuum of color between red and white. What areas of the graphs represent the shades of color that can arise if the environment does *not* affect the phenotype at all? What areas represent the shades of color that can arise if the environment *does* cause some variation in the phenotype of a particular genotype to occur?

    (b) In the top graph, what is meant by the labels 2-1 alleles, 1-1 alleles, and 0-1 alleles?

    (c) In order to produce a continuous variation of color in wheat, would a larger or smaller environmental effect be required if the number of genes controlling seed color were increased from 2 to 3?

    (d) Give some examples of changes in environmental conditions that could change the wheat color of a given genotype in order to produce the distributions shown in figure 24.3.

2. **Figure 24.9**

(a) In order to get the earliest flowering plants, which genotype should be planted and at what density should it be planted?

(b) In order to get the most seeds (ovules per flower) as possible, which genotype should be planted and at what density should it be planted?

(c) Which two genotypes show the *least* variation in petal area in response to changes in planting density? Which genotype shows the *most* variation?

(d) Which trait yields the least average variation (regardless of genotype) in response to changes in planting density?

(e) Which trait yields the least variation between genotypes in response to changes in planting density?

3. **Figure 24.11**

(a) What were some of the goals of performing this selection experiment?

(b) How was high oil content selected for? How was low oil content selected for?

(c) At generation 48, how was the reverse selection performed?

(d) What two major findings were discovered from this experiment?

General Questions

Q24-1. Compose a list of traits in human beings that are likely to be inherited quantitatively.

Q24-2. Virtually all human beings are born with two eyes: there is little to no phenotypic variation in this population. The heritability of eye number is zero. Explain.

Q24-3. Explain two reasons why a calculated correlation coefficient may be misleading.

Q24-4. Explain the meaning of the norm of reaction for a plant and its usefulness in agriculture.

Q24-5. Explain why a heritability value is only an accurate description of the population from which it was derived and it cannot be applied to a different population of individuals in a different environment.

Multiple Choice

For each of the following, circle the letter of the choice that most appropriately answers the question.

1.) Which of the following human traits is not a quantitative trait?
  a. height
  b. development of breast cancer
  c. eye color
  d. intelligence
  e. heart disease

2.) A quantitative trait is being studied, and two true-breeding plants are cross-fertilized and the resulting F_1 progeny are self-fertilized. If 3 plants out of 200 displayed the phenotype of one of the parents, how many genes would you predict control this trait?

    a. one
    b. two
    c. three
    d. four
    e. five

3.) What is the most likely explanation for why four genes, each having two alleles, that work together to produce a trait can produce more than nine different phenotypes?

    a. The genotype of an organism can change through mutation.
    b. The environment can influence the phenotypes of individuals.
    c. Mendel's law of independent assortment demonstrates 16 phenotypic classes.
    d. Some genes may contribute more to the phenotype than other genes.
    e. None of the above are correct; only nine phenotypes can be produced.

4.) A correlation coefficient of 0.9 was calculated for two traits (body mass and hair length) in wolf spiders. Which of the following is a correct interpretation of this finding?

    a. A small body on a wolf spider causes its hair to grow longer.
    b. A large body on a wolf spider causes its hair to grow longer.
    c. Larger wolf spiders usually have shorter hairs than smaller spiders.
    d. Larger wolf spiders usually have longer hairs than smaller spiders.
    e. Larger wolf spiders always have longer hairs than smaller spiders.

5.) Which of the following is typically *not* a major source of phenotypic variance (V_P)?

    a. combination of alleles of an individual (genotype)
    b. mutation of genes
    c. environment
    d. genetic inheritance
    e. interaction between genotype and environment (norm of reaction)

6.) Which of the following correctly describes heritability?

    a. proportion of variation in a population of phenotypes that is attributable to genetics
    b. proportion of a phenotype that is attributable to genetics

    c. proportion of the variation in a population of phenotypes that is attributable to the environment
    d. proportion of a phenotype that is attributable to the environment
    e. proportion of alleles in a population that contribute to a phenotype

7.) What heritability value would you expect to see for a population of corn plants that are all grown in an identical, controlled environment?

    a. 0
    b. 0.25
    c. 0.5
    d. 1.5
    e. 2.0

8.) Which of the following is an example of concordance?

    a. Two identical twins have identical Iqs.
    b. Neither one of two identical twins have brown hair.
    c. Of two identical twins, one has brown hair and one has black hair.
    d. All of the above
    e. Both a and b

9.) Which of the following is described by heritability?

    a. percentage of a phenotype that can be attributed to genetics
    b. percentage of a phenotype that can be attributed to the environment
    c. percentage of phenotypic variation that can be attributed to genetics
    d. percentage of a genotype that is inherited from a parent
    e. percentage of individuals in a population that express a particular phenotype

10.) Which of the following suggests that a quantitative trait locus is located near a particular molecular marker?

    a. Recombination occurs frequently between the molecular marker and the quantitative trait.
    b. The quantitative trait masks the phenotype associated with the molecular marker.
    c. The molecular marker masks the phenotype associated with the quantitative trait.
    d. There is frequent cosegregation of the phenotype associated with the quantitative trait and the molecular marker.
    e. The phenotype associated with the quantitative trait and the molecular marker never cosegregate.

Practice Problems

P24-1. A pure-breeding strain of hypothetical zeebogs with long tails (14 cm) is crossed with a strain of short-tailed (2 cm) zeebogs. All of the F_1 zeebogs have tails of intermediate length (8 cm). The F_1 generation is interbred to produce an F_2 generation ($N = 1000$) composed of individuals with tails ranging from 2 to 14 cm in length. Many of the F_2 individuals have 8-cm-long tails, while approximately three individuals have tails as long as the parental zeebogs.

[*Hint 1*: The number of polygenes influencing a trait can be estimated by determining the fraction of an F_2 generation that express one of the parental (P_1) phenotypes.
Hint 2: Each contributing allele (capital letter) adds a set amount to the phenotype.]
a.) Predict the number of genes that influence the polygenic trait of tail length in zeebogs.

b.) Provide the genotypes of individuals in the parental and F_1 generations.

P24-2. Consider the following set of data. Then calculate the correlation coefficient between trait 1 and trait 2.

Trait 1	Trait 2		Trait 1	Trait 2		Trait 1	Trait 2
8	5		3	5		6	4
10	6		8	7		5	7
5	4		4	7		6	7
8	10		9	4		8	9

[*Hint 3*: Calculation of a correlation coefficient (r) first requires the determination of covariances (cov) and standard deviations (s).]

P24-3. An Easter bunny farmer wants to increase the chocolate content of his Easter bunnies. In his population, the heritability of chocolate content is 0.69. The farmer selectively breeds bunnies with high chocolate content: the selected bunnies have, on average, 12 mg more chocolate than the unselected bunnies. What gain in chocolate content is expected in the next chocolate bunny generation?

P24-4. Assume that height in humans is controlled by six genes with no environmental influence. In this model, each additive allele contributes 4 inches to a base height of 4 feet. What is the range of heights possible in this population? If a woman with the minimal height specified marries a man of intermediate height, what is the maximum height the child that they produce could achieve?

[*Hint 4*: Each contributing allele (capital letter) adds a set amount to the phenotype.]

P24-5. What are the expected correlations between the following pairs of relatives: a woman and her (a) mother, (b) brother, (c) husband, (d) grandmother, (e) uncle, and (f) first cousin? If this woman's mother has a monozygotic twin, what is the expected correlation between the woman and the twin aunt's child (g)?

Assessing Your Knowledge

Understanding the Key Concepts—Answers

1.) the environment; 2.) phenotypes; 3.) polygenic; 4.) parental; 5.) environment; 6.) continuous; 7.) threshold; 8.) meristic; 9.) mean; 10.) median; 11.) mode; 12.) extreme phenotypes; 13.) square root; 14) 68; 15.) 95; 16.) environment; 17.) environment; 18.) Additive genetic variance; 19.) Dominance variance; 20.) epistatic genetic variance; 21.) heritability; 22.) broad-sense heritability; 23.) narrow-sense heritability; 24.) additive genetic variance; 25.) response to selection; 26.) realized heritability.

Figure Analysis—Answers

1a.) Very little environmental influence: The areas *under* the blue bars represent the possible colors that can arise from each genotype. The width of the bar represents the amount of variation that can arise from the corresponding genotype. Large environmental influence: The area under each bell-shaped line represents the possible colors that can arise from each genotype. The wider the bell-shaped line, the more environmental influence and the more possible phenotypes that can result. If the bell-shaped lines overlap (as in the lower three graphs), then continuous variation will be observed. If they do not overlap, then some phenotypes cannot be created; b.) 2-1: This means that the individuals depicted by the bar below have two copies of allele a^1. 1-1: This means that the individuals depicted by the bar below have one copy of allele a^1. 0-1: This means that the individuals depicted by the bar below do not have any copies of allele a^1; c.) With more genes, a smaller environmental effect would be necessary in order to produce a continuum of variation since there would be more possible genotypes in between the two extremes; d.) A change in the levels of a particular nutrient in the soil, a change in the amount of sunlight (more sunny days, for example), or a change in the average temperature may shift the color produced by each genotype to either more red or less red.

2a.) Genotypes 3 or 4 should be planted at low density in order to get flowers at the earliest possible date; b.) Genotype 2 should be planted at low density in order to get the most ovules per flower; c.) Genotypes 3 and 14 show the least variation in petal area, while genotype 4 shows the most variation; d.) Flowering date; e.) Germination date

3a.) The overall goal of this experiment was to determine the long-term response of an organism to selection. Is there a maximum value that can be achieved? After long-term selection does enough genetic diversity remain to select for other traits? Does selection prevent the phenotype from ever returning to the original state? b.) High oil content was selected by choosing the plants that had the highest oil content in each generation and crossing them to produce the next generation. Low oil content was selected by crossing the plants with the lowest oil content in each generation; c.) After 48 generations of selecting and crossing the plants with the *highest* oil content, the plants producing the *lowest* oil content were selected and crossed for each subsequent generation. The opposite was also done; d.) It was demonstrated that long-term selection for a trait (high oil content in this case) will not necessarily reach a maximum value, even under very high selection pressure. Secondly, it was demonstrated that even after 48 generations of selection, enough genetic diversity remained in the population that low oil content could still be produced by selective breeding.

General Questions—Answers

Q24-1. Some potentially polygenic characteristics in human beings include height, weight, longevity, hair color, eye color, skin color, limb length, IQ, personality, aggression, memory, Alzheimer's disease, diabetes, body fat content, heart disease, blood pressure, lung capacity, and heart rate.

Q24-2. If there is no phenotypic variation in a population, then there is no measurable genetic variation ($V_P = V_G + V_E$; $0 = 0 + 0$). Remember that heritability *measures the proportion of phenotypic variation* that is due to genetic factors. It does not determine to what extent a trait is genetic.

Q24-3. First, the correlation coefficient is subject to sampling error; a different sample may yield a different correlation coefficient. Secondly, the correlation coefficient does not demonstrate a cause-and-effect relationship even in the instance of a very large correlation coefficient.

Q24-4. The norm of reaction describes how a particular genotype will vary under a variety of conditions. Although different genotypes may produce the same phenotype under a specific condition, a different environmental condition may affect each genotype differently. Therefore, different genotypes will respond differently to environmental changes. In agriculture, it is imperative to plant crops that will thrive under all of the conditions that may be experienced. For example, it is impossible to predict whether a year will be warmer than average or cooler than average. Therefore, it is important to plant a crop with a particular genotype that will perform well in either warm or cool temperatures.

Q24-5. A different environment will have a different contribution to the variation in phenotypes. Since the environmental contribution will be different, the genetic contribution (heritability) will also be different.

Multiple Choice—Answers

1.) c; 2.) c; 3.) b; 4.) d; 5.) b; 6.) a; 7.) c; 8.) e; 9.) c; 10.) d.

Practice Problems—Answers

P24-1. a.) To estimate the number of polygenes influencing a trait, determine the average number of F_2 individuals that have the parental phenotype. In this example, an average of 3/1000 individuals have 14-cm-long tails, like the P_1 zeebogs. This value approximates the value expected if four pairs of alleles are involved ($1/4^n = 1/256$). Thus four genes likely control this trait.

b.)   Parentals: *AABBCCDD* (14 cm)   *aabbccdd* (2 cm)
        F_1:   *AaBbCcDd* (8 cm)

P24-2. Calculation of a correlation coefficient will provide an indication of the relationship between values for traits 1 and 2. If there is no relationship, the correlation will be 0. If there is a direct correspondence between traits 1 and 2, the correlation coefficient will be +1 or –1.

Trait 1	Trait 2	(Trait 1 – mean 1)	(Trait 1 – mean 2)^2	(Trait 2 – mean 2)	(Trait 2 – mean 2)^2	(Trait 1 – mean 1) (Trait 2 – mean 2)
8	5	1.33	1.769	–1.25	1.5625	–1.663
10	6	3.33	11.089	0.25	0.625	0.833
5	4	–1.67	2.789	–2.25	5.0625	3.758
8	10	1.33	1.769	3.75	14.063	4.988
3	5	–3.67	13.469	–1.25	1.563	4.586
8	7	1.33	1.769	0.75	0.563	0.998
4	7	–2.67	7.129	0.75	0.563	–2.003
9	4	2.33	5.429	–2.25	5.063	–5.243
6	4	–0.67	0.449	–2.25	5.063	1.508
5	7	–1.67	2.789	0.75	0.563	–1.253
6	7	–0.67	0.449	0.75	0.563	–1.253
8	9	1.33	1.769	2.75	7.563	3.658
Σ	Σ		Σ = 50.667		Σ = 42.813	Σ = 8.915

Mean1 = 6.67   Mean2 = 6.25
$s_1 = \sqrt{50.667 / 11} = 2.146$ $s_2 = \sqrt{42.813 / 11} = 1.973$
Cov $(x,y) = 8.915/11 = 0.811$
$r = 0.811/(2.146)(1.973) = \mathbf{0.191}$

P24-3. Substitute the information provided into the formula $h^2 = R/S$ where h^2 = the realized heritability (0.69), R = the response to selection (what you are looking for), and S = the selection differential (12 mg) in order to obtain the increase in chocolate content expected:
0.69 = offspring gain / 12 mg   →   (0.69)(12 mg) = offspring gain = **8.28 mg**

P24-4. If human height is controlled by the three pairs of additive alleles that add 4 inches to a base height of 4 feet, then there would be seven different phenotypes possible. An individual with no additive alleles would have the minimum height of 4 feet. An individual with any one contributing allele will be 4 feet, 4 inches. An individual with any two contributing alleles will be 4 feet, 8 inches. An individual with any three contributing alleles will be 5 feet. An individual with any four contributing alleles will be 5 feet, 4 inches. An individual with any five contributing alleles will be 5 feet, 8 inches. An individual with all six contributing alleles will be 6 feet. A woman with the minimum height would be genotype *aabbcc*, while a male of intermediate height would have genotype *AaBbCc* (or some other combination of three contributing and three noncontributing alleles). The woman will pass only noncontributing alleles to her offspring. Therefore, the maximum height attainable for a child of this couple will be determined by the male parent. In this example, the maximum number of contributing alleles that can be passed to a child is three (*ABC*). Thus, the offspring of this couple cannot be taller than 5 feet.

P24-5. The expected correlations between relatives reflect the number of genes they share, on average. The correlations expected are: (a) 0.5; (b) 0.5; (c) 0; (d) 0.25; (e) 0.25; (f) 0.125; (g) 0.25

24
Quantitative Genetics

CHAPTER SUMMARY QUESTIONS

1. 1. j, 2. i, 3. a, 4. h, 5. f, 6. b, 7. k, 8. l, 9. e, 10. c, 11. g, 12. D.

2. Quantitative genetics differs from previous topics in the book in that it deals with the complex inheritance patterns of traits determined by multiple loci and the environment.

3. Broad-sense heritability can be defined as the proportion of phenotypic variation among individuals in a population (in a given environment) that is due to genotypic differences among individuals. Narrow-sense heritability is the proportion of phenotypic variation among individuals in a population (in a given environment) that is due to additive genetic differences among individuals. These two measures differ in that the additive portion of the genotypic differences is the only part that contributes to the resemblance of offspring to their parents and to the evolutionary response to selection.

4. R. A. Fisher first showed how traits determined by multiple loci could be inherited in Mendelian fashion (particulate inheritance) but still show a distribution of phenotypic values typical of a quantitative trait. Adding an environmental component makes such distributions even easier to imagine. Prior to Fisher's synthesis of the evidence for polygenic traits, most "Mendelians" thought that Darwin must have been incorrect in his assumption of evolution by gradual change because they could only imagine discrete changes in a phenotype resulting from discrete changes in the single gene determining the trait. The "discovery" of polygenic traits opened the door for smaller changes accumulating over many genes each contributing only a small amount to the trait.

5. Much of the study of ecology and evolution is focused on how plants and animals adapt to their environments. Since most traits important to adaptation are quantitative traits, quantitative genetics is highly relevant to these fields.

6. The three leading causes of mortality in industrialized nations, heart disease, cancer, and diabetes, are all quantitative traits. Further, in less-industrialized nations, infectious diseases are major causes of mortality, and the response to infectious diseases is also a quantitative trait.

7. Quantitative genetics has always been central to improving food production. Since most traits of importance in agriculture and livestock production are quantitative traits and such conditions lend themselves to controlled breeding programs, quantitative genetics has had more of an impact in this field than any other.

8. Clinical depression and diabetes (as two examples) are often treated as a dichotomous trait (presence or absence). However, each is in fact a quantitative trait for which a threshold exists for which we define the symptoms as severe enough to be described as an illness. Both clinical depression and diabetes are determined by multiple genes and their interactions with the environment, and both are continuous traits with differing severity and lengths of depression or differing levels of control over glucose levels in the blood, for depression and diabetes, respectively.

9. The amount of genetic variation for quantitative traits is determined by most of the same forces that determine genetic variation at individual loci, a balance between mutation, selection, and drift. There are typically large amounts of genetic variation for quantitative traits because of the large number of loci that contribute to the trait and the often weak selection on any particular locus.

10. Changes in phenotype due to selection often reach an asymptote due to the loss of all additive genetic variation associated with that trait or due to negative genetic or phenotypic correlations with fitness.

11. Conservation genetics is the application of genetic principles and theory to issues concerning the conservation of endangered species and the conservation of genetic varieties for food security and future medicinal value. It is important to conserve genetic resources because genetic variation is the precursor to biological diversity, which forms the foundation of human existence on the earth.

12. A minimum viable population size is the population size that gives a certain probability of persistence (not going extinct) over a given amount of time. The three most pressing genetic concerns facing endangered species of plants and animals are inbreeding depression, random genetic drift, and the input of beneficial mutations.

13. Conserving species of plants with agricultural potential but not currently used widely in food production allows their genes to be saved for future agricultural purposes. These plant species could be either used as agricultural products themselves or by

incorporating their genes into currently used plants via hybridization or genetic engineering.

14. The mapping of loci involved in the determination of quantitative traits relies on the linkage of those loci to molecular markers such as microsatellite loci or RFLPs.

15. A norm of reaction describes how the phenotype of a given genotype changes across an environmental gradient. Genotype–environment interaction occurs when two or more genotypes respond in a different way to an environmental gradient. Genotype–environment interactions are important to evolutionary ecology, medicine, and food production. In wild populations these interactions may help maintain genetic variation because one genotype is not always the most fit through space and/or time, preventing fixation of the single most fit allele. Genotype–environment interactions for drugs are gaining more attention from medical researchers. In agriculture and livestock breeding, these interactions are crucial for maximizing production and profit in changing, specific, or variable rearing conditions.

16. Stabilizing selection on a quantitative trait can potentially help maintain genetic variation for quantitative traits because values closer to the mean for the trait are favored. Therefore, phenotypic variation is lost, but because there are multiple loci and multiple ways to arrive at the mean trait value there need not be directional selection on a specific allele.

17. The number of loci affecting a quantitative trait is typically moderate but highly variable. In *Drosophila melanogaster* there have been 28 loci identified as affecting sensory bristle number, 19 loci affecting longevity, and 21 affecting wing shape. These account for a large portion of the variance in these traits, but some variance remains unexplained. Thus, there are more loci affecting these traits than have been found, possibly large numbers of loci with very small effects. In pigs, five loci have been found influencing teat number in females and six for determining age at puberty. However, the molecular markers for pigs are less dense than in *D. melanogaster*, and the studies explain less of the total phenotypic variance. Thus, the difference in the number of loci is likely a statistical artifact rather than a real difference.

    Other information from QTL studies indicates that dominance, epistasis, and pleiotropy are at least as widespread as we believed prior to the development of QTL mapping methods. Genotype–environment interactions and sex-specific effects are also common, and their importance may have been underestimated in the past.

18. QTL mapping can be performed on artificially selected populations or on natural populations. The first is a more statistically powerful technique with great application in agriculture and medicine, because it identifies QTLs of large effect that are selected for strongly in the two lines being selected for different phenotypic extremes (or comparing symptomatic versus asymptomatic individuals). Using natural populations is far less statistically powerful but finds QTLs responsible for naturally occurring variation within a population.

19. a. Heritability decreases and the phenotypic mean remains the same.
  b. Heritability remains the same and the phenotypic mean changes (probably decreases depending on the trait).
  c. Heritability decreases and phenotypic mean increases.

20. Because pairs of mono- and dizygotic twins are expected to share the same level of similarity in their environments, but monozygotic twins share twice as much of their DNA, traits with similar concordance values for pairs of twins are expected to be due mostly to the environment. When mono- and dizygotic twins differ significantly in their concordance rates, we expect that there is a significant genetic component to variance for the trait (that is, the trait is heritable).

EXERCISES AND PROBLEMS

21. The mean heterozygosity is 0.5964, and the median is 0.6085. Given that none of the values is repeated, they each represent the mode in this sample. A hint that the distribution is at least approximately normal can be found in the similarity of the mean and median values. Plotting a frequency distribution shows the values to be very roughly bell-shaped.

22. The variance is calculated as

$$s^2 = \frac{\sum(x_i - \bar{x})^2}{(n-1)}$$

The standard deviation, s, is the square root of the variance. Begin by setting up a data table.

Mean Heterozygosity (x_i)	Mean ($\bar{x}$)	($x_i - \bar{x}$)	($x_i - \bar{x}$)^2
0.526	0.5964	−0.0704	0.00496
0.638	0.5964	0.0416	0.00173
0.637	0.5964	0.0406	0.00165
0.476	0.5964	−0.1204	0.01450
0.568	0.5964	−0.0284	0.00081
0.603	0.5964	0.0066	0.00004
0.629	0.5964	0.0326	0.00106
0.431	0.5964	−0.1654	0.02736
0.614	0.5964	0.0176	0.00031
0.569	0.5964	−0.0274	0.00075
0.576	0.5964	−0.0204	0.00042
0.524	0.5964	−0.0724	0.00524
0.730	0.5964	0.1336	0.01785
0.620	0.5964	0.0236	0.00056

0.647	0.5964	0.0506	0.00256
0.804	0.5964	0.2076	0.04310
0.643	0.5964	0.0466	0.00217
0.456	0.5964	−0.1404	0.01971
0.750	0.5964	0.1536	0.02359
0.487	0.5964	−0.1094	0.01197

$$\sum(x_i - \bar{x})^2 = 0.18033$$

The variance $s^2 = 0.18033/(20 - 1) = 0.00949$. The standard deviation of the heterozygosities is the square root of the variance, or 0.09742.

23. The mean body mass is 327.6 kg, the median is 333.0 kg, and the modal values are 306, 333, and 344 kg, which each appear twice in the distribution. The phenotypic variance is calculated as

$$V_p = \frac{\sum(P_i - P_M)^2}{n - 1}$$

To calculate the estimated phenotypic variance, begin by setting up a data table.

Weight in kg (P_i)	Mean (P_M)	$P_i - P_M$	$(P_i - P_M)^2$
344	327.6	16.4	268.96
492	327.6	164.4	27027.36
342	327.6	14.4	207.36
209	327.6	−118.6	14065.96
267	327.6	−60.6	3672.36
465	327.6	137.4	18878.76
304	327.6	−23.6	556.96
303	327.6	−24.6	605.16
333	327.6	5.4	29.16
255	327.6	−72.6	5270.76
354	327.6	26.4	696.96
334	327.6	6.4	40.96
219	327.6	−108.6	11793.96
297	327.6	−30.6	936.36
336	327.6	8.4	70.56
407	327.6	79.4	6304.36
343	327.6	15.4	237.16
277	327.6	−50.6	2560.36
323	327.6	−4.6	21.16
344	327.6	16.4	268.96
333	327.6	5.4	29.16
337	327.6	9.4	88.36
341	327.6	13.4	179.56
275	327.6	−52.6	2766.76

Weight in kg (P_i)	Mean (P_M)	$P_i - P_M$	$(P_i - P_M)^2$
345	327.6	17.4	302.76
306	327.6	−21.6	466.56
315	327.6	−12.6	158.76
306	327.6	−21.6	466.56
252	327.6	−75.6	5715.36
316	327.6	−11.6	134.56
273	327.6	−54.6	2981.16
395	327.6	67.4	4542.76
434	327.6	106.4	11320.96
302	327.6	−25.6	655.36
367	327.6	39.4	1552.36
350	327.6	22.4	501.76
			$\sum(P_i - P_M)^2 = 125{,}376.36$

The estimated phenotypic variance for body mass is $125{,}376.36/(36 − 1) =$ 3582.18 kg.

24. Begin by constructing a data table.

Eggs (x_i)	Mean ($\bar{X}$)	$x_i - \bar{X}$	Follicles (y_i)	Mean ($\bar{Y}$)	$y_i - \bar{Y}$	$[(x_i - \bar{X})(y_i - \bar{Y})]$
39	38.4	0.6	37	37.5	−0.5	−0.3
29	38.4	−9.4	34	37.5	−3.5	32.9
46	38.4	7.6	52	37.5	14.5	110.2
28	38.4	−10.4	26	37.5	−11.5	119.6
31	38.4	−7.4	32	37.5	−5.5	40.7
25	38.4	−13.4	25	37.5	−12.5	167.5
49	38.4	10.6	55	37.5	17.5	185.5
57	38.4	18.6	65	37.5	27.5	511.5
51	38.4	12.6	44	37.5	6.5	81.9
21	38.4	−17.4	25	37.5	−12.5	217.5
42	38.4	3.6	45	37.5	7.5	27.0
38	38.4	−0.4	26	37.5	−11.5	4.6
34	38.4	−4.4	29	37.5	−8.5	37.4
47	38.4	8.6	30	37.5	−7.5	−64.5
	s_x	10.80		s_y	12.77	$\sum[(x_i - \bar{X})(y_i - \bar{Y})]$ $= 1471.50$

First calculate the mean number of eggs, the mean number of follicles, and the standard deviation for both eggs and follicles. Use these values to complete the rest of the data table. Next calculate $[(x_i - \bar{X})(y_i - \bar{Y})] = 1471.50$. The covariance $\text{cov}_{xy} =$ $1471.51/(14 − 1) = 113.19$. The correlation coefficient r is $\text{cov}_{xy}/(s_x s_y)$, or $1471.51/(10.80 \times 12.77) = 0.82$.

Graph the data as shown here.

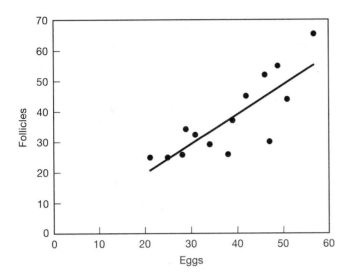

25. There will be eight phenotypic classes ranging from being homozygous for the high-intensity allele at all loci to being homozygous at the null allele at all four loci.

26. The regression coefficients using father–child, mother–child, and midparent–child are 0.26, 0.46, and 0.50, respectively. The narrow-sense heritabilities estimated from these are 0.53, 0.91, and 0.50, respectively. You expect the midparent value to provide the highest correlation coefficient because it includes more information (genes of both parents). This is true, though the difference between mothers alone and midparent values is rather small. This suggests very large maternal effects.

27. The narrow-sense heritability estimate is calculated as R/S, where R is the difference between the mean phenotypic value of the offspring and the mean phenotypic value of the original population, and S is the difference between the mean phenotypic value of the selected parents and the mean phenotypic value of the original population. In this case, $R = 0.62 - 0.44 = 0.18$ and $S = 0.68 - 0.44 = 0.24$. The narrow-sense heritability estimate is therefore $0.18/0.24 = 0.75$.

28. **i.** To answer this question, we must first calculate the narrow-sense heritability for each trait. Narrow-sense heritability is V_A/V_P. These values are calculated as follows:

Average daily growth $h^2 = 0.013/0.040 = 0.325$
Backfat thickness $h^2 = 0.036/0.050 = 0.720$
Loin eye area $h^2 = 0.268/0.413 = 0.649$

Backfat thickness has the highest narrow-sense heritability; therefore, it should respond best to selection.

ii. We know that $h^2 = R/S$. We have already calculated the narrow-sense heritability in part (i). The selection differential S is the difference between the phenotypic value of the parents and the mean phenotypic value of the population. The response to selection R is the difference between the mean phenotypic value of the offspring and the mean phenotypic value of the original population. The only value that is unknown is the mean of the offspring. Putting these numbers into the realized heritability equation, we determine that 0.325 = (mean offspring – 0.474)/(0.65 – 0.474). By rearranging this equation and solving for the mean of the offspring, we determine that the mean after one generation of selection will be 0.531 kg/day.

iii. Such selection will not be met with much success because of the negative genetic correlation between them.

29. The key to this problem is to remember that narrow-sense heritability is $h^2 = R/S$. We need to solve for R. We are given h^2, and we can determine S from the mean and the standard deviation. We can calculate the standard deviation as the square root of the phenotypic variance. The individuals selected for breeding differ from the mean by two standard deviations. For the *Phil* strain, $V_P = 100$, so the standard deviation is 10. Therefore, the individuals selected for breeding have a mean of 42.0 + (2 × 10) = 62.0. The selection differential S is then 62.0 – 42.0 = 20.0, and $R = h^2 × S$, or 20.0 × 0.75 = 15.0.

Therefore, the new mean is 42.0 + 15.0 = 57.0. The other values can be calculated in a similar manner. The results are summarized here.

	Mean	V_P	h^2	Response	New Mean
Phil (normal)	42.0	100	0.75	+15.0	57.0
Thai (normal)	50.0	144	0.50	+12.0	62.0
Phil (drought)	37.0	182	0.67	+18.0	55.0
Thai (drought)	30.0	506	0.44	+19.8	49.8

In the normal environment the *Phil* genotype has a larger response to one generation of selection than does the *Thai* genotype. The opposite is true in the drought environment, with the Thai genotype having a larger response (despite having a smaller heritability). The two genotypes do not change their ranks in the two environments; in other words, the *Phil* strain will have a higher yield in either environment after one generation of selection.

30. Begin by constructing a data table and calculating the correlation coefficient.

Number of Offspring (x_i)	Mean ($\bar{X}$)	$x_i - \bar{X}$	Mean Weight (y_i)	Mean ($\bar{Y}$)	$y_i - \bar{Y}$	$[(x_i - \bar{X})(y_i - \bar{Y})]$
177	123.8	53.2	1.41	1.36	0.05	2.66
132	123.8	8.2	1.29	1.36	−0.07	−0.574
194	123.8	70.2	1.44	1.36	0.08	5.616
199	123.8	75.2	1.16	1.36	−0.2	−15.04
138	123.8	14.2	1.38	1.36	0.02	0.284
132	123.8	8.2	1.14	1.36	−0.22	−1.804
92	123.8	−31.8	1.63	1.36	0.27	−8.586
120	123.8	−3.8	1.33	1.36	−0.03	0.114
126	123.8	2.2	1.51	1.36	0.15	0.33
197	123.8	73.2	1.27	1.36	−0.09	−6.588
100	123.8	−23.8	1.6	1.36	0.24	−5.712
121	123.8	−2.8	1.65	1.36	0.29	−0.812
207	123.8	83.2	1.16	1.36	−0.2	−16.64
102	123.8	−21.8	1.27	1.36	−0.09	1.962
13	123.8	−110.8	1.52	1.36	0.16	−17.728
155	123.8	31.2	1.29	1.36	−0.07	−2.184
134	123.8	10.2	1.27	1.36	−0.09	−0.918
8	123.8	−115.8	1.25	1.36	−0.11	12.738
159	123.8	35.2	1.26	1.36	−0.1	−3.52
31	123.8	−92.8	1.48	1.36	0.12	−11.136
158	123.8	34.2	1.33	1.36	−0.03	−1.026
89	123.8	−34.8	1.35	1.36	−0.01	0.348
74	123.8	−49.8	1.45	1.36	0.09	−4.482
107	123.8	−16.8	1.21	1.36	0.15	2.52
152	123.8	28.2	1.18	1.36	−0.18	−5.076
123	123.8	−0.8	1.38	1.36	0.02	−0.016
99	123.8	−24.8	1.52	1.36	0.16	−3.968
83	123.8	−40.8	1.33	1.36	−0.03	1.224
85	123.8	−38.8	1.41	1.36	0.05	−1.94
207	123.8	83.2	1.26	1.36	−0.1	−8.32
	s_x	52.97		s_y	0.141	$\sum[(x_i - \bar{X})(y_i - \bar{Y})]$ = −88.27

The covariance is −88.27/29 = −3.04. The correlation coefficient is −3.04/(52.97 × 0.141) = −0.41. It is impossible to tell from the data provided whether you can select for both more and larger offspring, because these are phenotypic correlations not genetic correlations.

31.

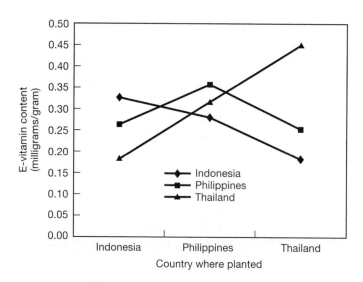

There appears to be strong genotype–environment interactions for the trait of interest.

32. The selection differential is 3 inches, and $h^2 = 0.5$. Therefore, the response to selection should be $h^2 \times S$, or $0.5 \times 3 = 1.5$. In the next generation, the height of humans should **(a)** increase by 1.5 inches.

33. **i.** $h^2 = V_A/V_P$. Rearranging, $V_P = V_A/h^2$. Since $h^2 = 0.25$ and $V_A = 140$, then $V_P = 140/0.25 = 560$.

   ii. $V_P = V_A + V_D + V_I + V_E$, so $V_E = 560 - (140 + 113 + 176) = 131$

   iii. $H^2 = V_G/V_P = (140 + 113 + 176)/560 = 0.766$

   iv. The phenotypic variance is 560, so the standard deviation is 23.6. The plants chosen for breeding differ from the mean by 1.5 standard deviations, or $23.6 \times 1.5 = 35.5$ kg/acre. Therefore, the selection differential $S = 35.5$, and $R = 0.25 \times 35.5 = 8.9$ kg/acre. After one generation, the mean grain yield will be $136.5 + 8.9 = 145.4$ kg/acre.

34. $V_G = V_A + V_D + V_I = 4.0 + 1.8 + 0.5 = 6.3$. $V_P = V_G + V_E = 6.3 + 2.5 = 8.8$. The broad-sense heritability equals $V_G/V_P = 6.3/8.8 = 0.716$ and the narrow-sense heritability equals $V_A/V_P = 4.0/8.8 = 0.455$.

35. Genetic variance is the sum of additive genetic variance + dominance genetic variance + epistatic variance. We know that epistatic variance = 0. To determine the dominance variance, we need to calculate the additive genetic variance. $V_A = V_P \times h^2 = 44 \times 0.15 = 6.6$. The dominance variance is equal to $22 - 6.6 = 15.4$ kg^2. The phenotypic variance is the sum of the genetic variance and the environmental variance, so the environmental variance is $44 - 22 = 22$ kg^2.

36. The selection differential is $84 - 76 = 8$ kg. The responses to selection will be $0.4 \times 8 = 3.2$ kg. Therefore, the mean weight of 4-month-old hogs should increase by 3.2 kg.

37. Heritability can be estimated as twice the difference between the concordance rate for monozygotic twins versus dizygotic twins. Based on the data given, the following values can be calculated.

Trait	MZ	DZ	MZ–DZ	$(MZ - DZ)^2$
Adult schizophrenia	0.44	0.12	0.32	0.64
Breast cancer	0.19	0.10	0.09	0.18
Hypertension	0.41	0.13	0.28	0.56
Insulin-dependent diabetes	0.53	0.11	0.42	0.84
Nontraumatic epilepsy	0.58	0.09	0.49	0.98

Thus, nontraumatic epilepsy has the highest heritability ($h^2 = 0.98$) and breast cancer the lowest ($h^2 = 0.18$).

38. This suggests that in the field 56.25% (0.09/0.16) of the total variance was due to differences in the environment experienced by individuals. (This assumes environmental variation in the greenhouse to be zero.) Thus, the total genetic contribution to phenotypic variance could be estimated as 0.07 and the broad-sense heritability of lysine content would be 0.4375.

39. In the F_1 generation, all individuals would be heterozygous at all loci and genetically identical to one another. Therefore, genetic variance would be 0, and the phenotypic variance would be due only to environmental variance. Therefore, $V_E = 20$. In the F_2 generation, environmental variance is the same, but genetic variance has increased. $V_G = 60 - 20 = 40$. The broad-sense heritability (H^2) for plant height in this population is 40/60 = 0.67.

40. i. The estimated heritabilities are calculated as R/S. For the upward selection line, in the first generation, $R_1 = 14.6 - 12.1 = 2.5$ and $S_1 = 17.0 - 12.1 = 4.9$. Therefore, $h_1^2 = 2.5/4.9 = 0.51$. For the second generation, the mean of the population is now 14.6, and the mean of the parents is 19.2. Now $R_2 = 16.6 - 14.6 = 2.0$ and $S_2 = 19.2 - 14.6 = 4.6$, so $h_2^2 = 2.0/4.6 = 0.43$. The other values are calculated in a similar fashion. The estimated heritabilities for the upward selection line are $h_1^2 = 0.51$, $h_2^2 = 0.43$, $h_3^2 = 0.35$, $h_4^2 = 0.31$. The estimated heritabilities for the downward selection line are: $h_1^2 = 0.50$, $h_2^2 = 0.31$, $h_3^2 = 0.16$, $h_4^2 = 0.03$. The heritabilities for each selection line decrease with time. This is expected as the additive genetic variation should be depleted with strong selection. The heritabilities start out being nearly identical for upward and downward selection, but genetic variation for the downward selection line appears to disappear faster. This could be due to several things but is most likely caused by a negative correlation between lower values of the trait and fitness.

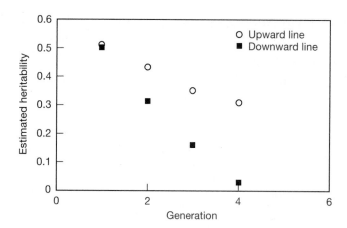

ii. The estimated selection differentials for the upward selection line are $S_1 = 4.9$, $S_2 = 4.6$, $S_3 = 4.8$, $S_4 = 4.9$. The estimated selection differentials for the upward selection line are $S_1 = 4.6$, $S_2 = 4.2$, $S_3 = 3.7$, $S_4 = 3.2$. For the upward selection lines the selection differentials remained fairly constant. For the downward selection lines there is a steady decrease in the selection differentials. Since the proportion of the population selected for breeding was constant, this shows that either (1) phenotypic variation was being eroded in the downward selection lines or (2) fitness may have dropped so low that they may not have been producing enough female spiders to maintain the population at 100 females (or a combination of both).

Because of lowered heritability and reduced selection differential the total response in four generations of downward selection was only 4.3 days as compared to a 11.1-day change in the positive direction with the same strength of selection.

CHAPTER INTEGRATION PROBLEM

a. The broad-sense heritability $H^2 = 0.90$, and $V_P = 28,561$. Therefore $V_G = H^2 \times V_P = 0.90 \times 28,561 = 25,704.9$. We know that $V_G = V_A + V_D + V_I$. We can calculate V_A once we have determined the narrow-sense heritability from the selection response and selection differential. Based on the data provided, $R = 1450 - 1300 = 150$ and $S = 1600 - 1300 = 300$, so $h^2 = 150/300 = 0.5$. $V_A = h^2 \times V_P = 0.5 \times 28,561 = 14,280.5$. The nonadditive genetic variance ($V_D + V_I$) for bite force in this population of alligators is $25,704.9 - 14,280.5 = 11,424.4$ psi^2.

b. The environmental variance V_E for bite force in the described population is $28,561 - 25,704.9 = 2856.1$ psi^2.

c. The narrow-sense heritability h^2 of bite force in this population is 0.50, as calculated in part (a).

d. High levels of narrow-sense (much less broad-sense) heritability do not rule out substantial changes in the mean phenotype with a change in environment.

Heritabilities deal with the proportion of the phenotypic *variation* attributable to different causes in a given population in a given environment. They do not explain mean values, and a change in the environment can change the heritability in any event. Thus, the mean bite force of the alligators could potentially be changed greatly by the new food regardless of heritability.

e. More uniform environmental conditions should increase heritability, by making the proportion of the phenotypic variance due to genetic causes (and additive genetic causes specifically).

f. Despite the increase in the heritability of the trait, you may not get a stronger response to selection because the total phenotypic variance should decrease. However, selection will become more efficient because you can more easily identify the genetically superior individuals.

g. For traits controlled by many genes of very small effect, traditional breeding programs often perform better than genetic engineering. The ability to locate many loci of small effect using QTL methods may be more trouble than it is worth if several dozen loci contribute to the trait and none of them explain more than 5% of the variation in the trait.

CHAPTER 25 EVOLUTIONARY GENETICS

Chapter Goals

1. Describe the differences between allopatric and sympatric speciation and the role of reproductive isolation in both mechanisms.
2. Compare the following terms as they relate to speciation: *anagenesis, cladogenesis, phyletic gradualism,* and *punctuated equilibrium.*
3. Explain the neutral theory and how it relates to the molecular clock and phylogenetic analysis.
4. Describe the main features identified when genomes or proteomes are compared across species.
5. Explain how phylogenetic analysis has been used to examine the origins of *Homo sapiens.*
6. Summarize the different ways that phylogenetic studies have been used in studying HIV.

Key Chapter Concepts

25.1 Charles Darwin and the Principles of Natural Selection

- **Darwinian Evolution:** Evolution describes the changes in allele frequencies in populations over generations. While Charles Darwin did not discover evolution, he did propose a mechanism of evolution, **natural selection**. The requirements for evolution by natural selection include variation among individuals, inheritance of these differences, overproduction of offspring by organisms that compete for available resources, and the survival and reproduction of individuals possessing favorable characteristics.

25.2 What Is Evolutionary Genetics?

- **Evolutionary genetics:** Evolutionary genetics looks at how genetic processes, such as mutation, migration, natural selection, and random genetic drift have affected the variation of species throughout time. Biogeography, comparative anatomy, embryology, and paleontology support the idea that evolution has occurred. Inferences of common ancestry based on comparisons of the genes of living species are supported by direct fossil evidence of evolutionary transitions.
- **The Role of Natural Selection in Evolution:** Evolution has two major components: changes within a genetic lineage and branching of the lineages into different species. Natural selection is the ultimate cause of **adaptation** and is primarily responsible for the diversity of life on Earth. For natural selection to occur two conditions must be met, reproductive excess must be accompanied by competition to survive and reproduce, and there must be a correlation between the differences in reproductive success and the heritable differences in phenotype.

25.3 Speciation

- **Speciation:** Speciation refers to a process whereby an ancestral species evolves into a new species (**anagenesis**) or whereby one population diverges to form two, simultaneously existing species (**cladogenesis**). In order for speciation to occur, populations must be isolated and members must be unable to reproduce with each other (**reproductive isolation**).
- **Classification of Species:** The definition of a species is not simple. Several different definitions are used by evolutionary geneticists. The **morphological species concept** classifies two organisms as belonging to the same species if they are morphologically similar and belonging to two different species if their differences are as great as two organisms already classified to belong to two different species. The **biological species concept** classifies two organisms as belonging to the same species if they can interbreed to produce fertile offspring. The **species recognition concept** is based on the idea that organisms belonging to the same species have the ability to locate each other and successfully mate. The **species phylogeny concept** is based on the idea that organisms of the same species share one or more DNA sequences or molecular features such that they belong to the same **clade**.
- **Mechanisms of Reproductive Isolation:** Environmental, behavioral, mechanical, or physiological barriers prevent isolated populations of individuals from interbreeding and exchanging genes (gene flow). Over evolutionary time, the separated members of one species become two distinct species incapable of reproducing with each other. A number of **prezygotic** (before zygote formation) and **postzygotic** (after zygote formation) **mechanisms** have been identified that allow reproductive

isolation to be achieved. Speciation can occur while populations are separated by geographical barriers (**allopatric speciation**) or by divergences, either genetic or behavioral within the population (**sympatric speciation**), that allow a group of organisms to enter a new niche in the same area.

- **Gradualism Versus Punctuated Equilibrium:** Evolution, as envisioned by Darwin, is a gradual process (**phyletic gradualism**). Examination of the fossil record suggests that evolution occurs in rapid bursts separated by long periods of stasis (**punctuated equilibrium**).

25.4 Evolution at the Molecular Level
- **Phylogenetic Analysis:** The examination of the nucleotide sequence of orthologous genes allows the number of nucleotide differences to be determined and the degree of relatedness among the members of a taxonomic group to be established. **Phylogenetic trees** diagram the relationship between different **orthologs** or different species based on several genes. They are based on the idea that groups that are more similar in nucleotide sequence share a more recent common ancestor than those that are less similar.
- **Gene Trees Versus Species Trees:** Gene trees and species trees are two different kinds of phylogenetic trees. A **gene tree** shows the evolutionary relationships for a single homologous gene, both **paralogs** and **homologs**, while a **species tree** shows the evolutionary relationships among species by examining multiple independent genes.
- **Neutral Theory:** The neutral theory suggests that most of the genetic diversity at the molecular level is due to random genetic drift, rather than selection. Most mutations have a small effect on the fitness of an organism and are therefore essentially neutral.
- **The Molecular Clock:** A molecular evolutionary clock measures evolutionary time in the number of nucleotide substitutions per million years. Statistical methods can be used to estimate the number of nucleotide substitutions that have taken place over evolutionary time if the current amino acid differences in a protein between species is known.

25.5 Comparison of Genomes and Proteomes
- **Comparison of Genomic Organization Between Species:** A change in the ploidy of an organism can result in instantaneous speciation. Reciprocal translocations can also lead to sympatric speciation. Therefore, related species should exhibit similar types of chromosomal rearrangements throughout their genomes. Chromosomal banding patterns (low resolution) and fluorescent in situ hybridization (high resolution) can be used to look at the chromosome organization of different species. Large blocks of genes that remain together in different species such as what is observed in humans and mice are a phenomenon known as **synteny**.
- **Comparison of the Proteome Between Species:** Comparison of the proteomes from different species has highlighted an increase in complexity from yeast to invertebrates to humans. This is not simply due to an increase in the number of proteins, but also an increase in the number of paralogs and an increase in the number of proteins with multiple functions. Only about 1% of the proteins in the human proteome are unique to humans, and less than 1% of mouse proteins are present only in rodents.

25.6 Applications of Evolutionary Genetics
- **Origin of Modern Humans:** Mitochondrial DNA was used to trace the maternal origins of *Homo sapiens*, and Y chromosomal DNA sequences were used to trace the paternal origins. Mitochondrial DNA was useful for this analysis because it accumulates mutations 10-fold faster than nuclear DNA allowing scientists to examine evolution over a short period of time. *Homo sapiens* originated from a population in Africa. One group migrated out of Africa, along the southern coast of Asia to Australia, and later a second group migrated up the Nile River to Eurasia and fanned out across Europe and Asia.
- **Analysis of Viral Origin:** Genetics can be used to describe the origin of a virus, how it is propagated, the amount of genetic variation it contains, and how these factors may influence its infectivity. Phylogenetic trees indicate that there are two types of HIV: HIV-1 and HIV-2, where HIV-1 is more common and virulent and HIV-2 is primarily found in Africa. Subtypes of HIV are determined based upon genetic variation. This allows subtypes to be organized into phylogenetic trees based on their DNA sequence variations. The route of transmission between individuals in a population can also be determined using a phylogenetic tree. The DNA sequence of specific genes in the viral genome can be determined from different viral strains obtained from different sources. A phylogenetic tree can then

be constructed by comparing the sequences of the genes from different sources. Viral strains containing fewer DNA mutations are more closely related to each other than viral strains that have a greater number of differences in their DNA sequences. Using this type of analysis, the relationship between different strains of the virus can be determined.

Key Terms

- vestigial features
- natural selection
- evolutionary genetics
- adaptation
- anagenesis
- cladogenesis
- stasis
- morphological species concept
- biological species concept
- species recognition concept
- species phylogeny concept
- clade
- reproductive isolating mechanisms
- prezygotic mechanism
- residential isolation
- seasonal or temporal isolation
- ethological isolation
- mechanical isolation
- postzygotic mechanism
- hybrid zygotes
- F_1 hybrid breakdown
- developmental hybrid sterility
- segregational hybrid sterility
- F_2 breakdown
- gene flow
- allopatric speciation
- depauperate fauna
- sympatric speciation
- phyletic gradualism
- punctuated equilibrium
- homologs
- orthologs
- paralogs
- phylogenetic trees
- gene tree
- species tree
- molecular clock
- synteny

Understanding the Key Concepts

Use the key terms or parts of their definitions to complete the following sentences.

Changes in a genetic lineage and the branching of lineages lead to (1)_____. Darwin proposed that evolution occurs by the process of (2)_____. This theory is based on the idea that (3)_____ differences are genetically based and heritable. In addition, it is based on the idea that organisms reproduce in excess although (4)_____ are limited. Therefore, organisms must (5)_____ for resources. Those that can (6)_____ and (7)_____ these resources the best will survive. The (8)_____ characteristics of the surviving organisms are passed onto the next (9)_____. (10)_____ is the study of how mutation, migration, random genetic drift, natural selection, and nonrandom mating affect the genetic (11)_____ of populations over long periods of time. (12)_____ is the ultimate cause of adaptation. As an organism's (13)_____ changes or as beneficial (14)_____ arise, the characteristics of a population can be altered which can lead to the generation of a new species.

New species can arise by (15)_____ or by (16)_____. (17)_____ is the (18)_____ of one species into a new species, while (19)_____ occurs when one species (20)_____ into two new species. Species maintain their separate identities and gene pools by (21)_____. This can be achieved by (22)_____ barriers which prevent the (23)_____ of an egg and inhibit the production of a (24)_____ either by (25)_____ isolation, (26)_____ or (27)_____ isolation, or by incompatible (28)_____ behaviors or differences in (29)_____ structures. (30)_____ can also be achieved by (31)_____ mechanisms which affect the (32)_____ of the F_1 (33)_____, or their fertility, or affect the F_2 generation in these ways. Both of these mechanisms prevent (34)_____. A (35)_____ that separates a population into two or more distinct groups leads to (36)_____ speciation, while (37)_____ speciation occurs when a group of organisms diverges from a parental population due to a (38)_____ or a (39)_____ change that allows the group to occupy a new (40)_____.

Which organisms belong to the same species is not easy to determine. Some evolutionary biologists define a species according to (41)_____ similarities, while others define a species based on the organisms' ability to (42)_____ and produce (43)_____. Still others rely on molecular data such as similarities in the sequence of the organisms' (44)_____ or other molecular features. (45)_____ sequence analysis has led to the identification of (46)_____ genes, which are genes found in the same, or different species that share a single ancestral origin. By examining the nucleotide sequence of (47)_____ genes from different species and determining the nucleotide changes, a (48)_____ can be constructed to diagram the evolutionary relationships for a single gene, or the evolutionary relationships among (49)_____ if multiple independent genes are analyzed. Groups that have more similar nucleotide sequences share a more recent (50)_____ than those that are less similar. This type of analysis has been used to determine the relatedness of different HIV (51)_____. In addition, it has been used to examine the (52)_____ of the viral infection in different individuals.

Chromosomal banding patterns and fluorescent in situ hybridization (FISH) can be used to analyze chromosomal (53)_____ that could have led to "instantaneous" speciation. Comparison of the human and the mouse genomes has highlighted the fact that clusters of (54)_____ genes are conserved in both species, a phenomenon called (55)_____. By comparing the human (56)_____ to that of all other species, it was revealed that only (57)_____ of all human proteins are unique to humans.

Figure Analysis
Use the indicated figure or table from the textbook to answer the questions that follow.

1. **Figure 25.2**
   (a) What happens to species 1 as it undergoes stasis?

   (b) How does species 2 evolve from species 1?

   (c) How are species 3 and 4 generated?

2. **Figure 25.3**
   (a) What are the two mechanisms by which reproductive isolation is achieved?

   (b) What is the difference between these two mechanisms?

   (c) What type of divergence in a population can lead to sympatric speciation?

   (d) What type of barriers can result in allopatric speciation?

3. **Figure 25.4**
   (a) How did allopatric speciation take place in the finches of the Galápagos Islands?

   (b) What three groups of finches evolved?

   (c) How were the insect-eating finches that feed on insects under the bark of trees able to exist in the Galápagos Islands when they could not exist in South America?

   (d) What are the major differences in the beaks of the different species of seed-eating finches?

4. **Figure 25.8**
   (a) What is the difference between phyletic gradualism and punctuated equilibrium?

   (b) How do two species evolve from one by punctuated equilibrium?

   (c) How do two species evolve from one by phyletic gradualism?

   (d) Why are these two models difficult to distinguish in practice?

5. **Figure 25.13 and Table 25.1**
   (a) What is the neutral theory?

   (b) Which protein has the highest rate of amino acid substitutions?

   (c) Which protein has the lowest rate of amino acid substitutions?

   (d) What types of proteins typically have the lowest rates of neutral substitution?

   (e) What types of proteins typically have higher rates of neutral substitutions?

   (f) What is the likely explanation for why histone H4 has undergone such a low level of amino acid substitutions?

6. **Figures 25.15 and 25.16**
   (a) What are the differences among chromosome 1 from the four different organisms?

   (b) What are the differences among chromosome 2 from the four different organisms?

   (c) What are the differences among chromosome 3 from the four different organisms?

   (d) When did the inversion of chromosome 3 likely occur in the orangutan?

   (e) When did the fusion that resulted in human chromosome 2 likely occur?

7. **Figure 25.23**
   (a) How was it determined if the infected patients acquired HIV from their dentist?

   (b) What are the critical controls in the experiment? Why are these important?

   (c) How can this type of analysis identify patients that likely acquired HIV from the dentist?

   (d) Which HIV strains diverged prior to the dentist's strain?

   (e) Which HIV strains diverged later than the dentist's strain?

   (f) Which patients did the dentist likely infect with HIV?

   (g) Which patients likely acquired HIV from a source other than the dentist?

General Questions

Q25-1. Describe the different requirements used by evolutionary biologists to define distinct species.

Q25-2. Explain the problems that arise in using the biological species concept to define distinct species.

Q25-3. Describe the two major mechanisms by which reproductive isolation can be achieved to prevent two subpopulations from interbreeding. Explain the different ways in which these mechanisms are accomplished.

Q25-4. The three requirements for evolution by natural selection are variation, selection, and inheritance. Elaborate on the necessity of each factor.

Q25-5. What two different DNA sequences were used to determine the origin of *Homo sapiens*? Why were these sequences particularly useful?

Multiple Choice

For each of the following, circle the letter of the choice that most appropriately answers the question.

1.) Which species most recently diverged from *Homo sapiens*?
   a. *Pan troglodytes*
   b. *Hylobates syndactylus*
   c. *Pongo pygmaeus*
   d. *Gorilla gorilla*
   e. *Mus musculus*

2.) Which of the following is an example of orthologs?
   a. human α- and β-globin genes
   b. human α-globin and mouse β-globin genes
   c. human β-globin and mouse β-globin genes
   d. human α_1- and α_2-globin genes
   e. the ancestral α-globin gene and any duplications of this gene in any organism

3.) Which region of nucleotide sequence among orthologous genes is most highly conserved?
   a. 5' UTR
   b. coding sequences
   c. introns
   d. 3' UTR
   e. exon–intron splice junctions

4.) According to the neutral theory, most diversity at the molecular level is due to which of the following?
   a. spontaneous nucleotide substitutions
   b. gene duplications
   c. random genetic drift
   d. chromosomal rearrangements
   e. selection

5.) The number of species of cichlid fish in Lake Victoria is an example of which of the following?
   a. anagenesis
   b. sympatric speciation
   c. cladogenesis
   d. allopatric speciation
   e. stasis

6.) Which of the following is *not* one of the requirements for Darwin's theory of natural selection?
   a. genotypically based phenotypic variation
   b. heritable phenotypic differences
   c. competition to survive and reproduce
   d. reproductive excess
   e. acquired characteristics resulting in an increase in fitness

7.) The lowest percentage of sequence identity among orthologous genes is found in which of the following?
   a. 200 bp upstream of the transcriptional start site
   b. coding sequences
   c. 3' UTR
   d. introns
   e. exon–intron splice junctions

8.) Which of the following is *not* a prediction or an implication of the neutral theory?
   a. Divergence among species is more rapid for genes with a smaller effect on fitness.
   b. Proteins with variable activities among different species have a larger number of conserved domains.
   c. Most genetic diversity is not dependent upon selection.
   d. Most mutations do not affect the fitness of an organism.
   e. Greatest diversity in the region of a gene occurs at sites that are functionally less important.

9.) By comparing the sequences of hemoglobin, cytochrome c, and fibrinopeptide A, which of the following organisms are most distantly related?
   a. chicken and frog
   b. cow and horse
   c. monkey and chicken
   d. dog and chicken
   e. sheep and cow

10.) Which of the following human
chromosomes shares the largest syntenic region
with a single mouse chromosome?

    a. chromosome 9
    b. chromosome 11
    c. chromosome 15
    d. chromosome 17
    e. chromosome 22

Practice Problems

P25-1. Identify each of the reproductive isolating mechanism described.
    a. Two species of insect produce different chemicals used to locate mates.
    b. Two species of insect produce the same mating signal, but one mates at dusk and the other mates at dawn.
    c. Two species produce phenotypically normal offspring that are sterile.
    d. The pedipalp of a male spider (a sperm transfer structure) is shaped differently than the female reproductive tract.
    e. Two species produce compatible gametes, but one species lives in the forest canopy while the other species occurs on the forest floor.

P25-2. The sequence of the same region of six orthologous genes from six different organisms is depicted in the following. An asterisk below a nucleotide indicates that an identical nucleotide is found in that position in all six organisms. Highlighted nucleotides indicate deviations from the most common nucleotide found at that position. (a) Which two genes are most closely related? (b) Which two genes are most distantly related?

```
1. GGCACTGCGGGTGCTCAGAGGCAAGTTCAGGCCGTGTTGTTCTCTTTTCA
2. GGCCAGGCGGGTTCTCAGATGCAAGTTCAGGCCGTGAAGTTTTCTTTTGG
3. GGCCCTGAGGGTTCTCAGAGGCAAGTTTTTGCCGTGTTGTTTTCTGTACA
4. GGCCCTGCGGGTTCTCGCAGGCAAGTTCAGGCCGTGTTGTTTTCTTTTCA
5. GGCCCTGCGGGTTCTCAGAGGCAAGTTCAGGCCGTGTTGTTTTCTTTTCA
6. GGCCCTGCGGGTTCTCAGAGGCTAGTTCAGGCCGTGTTGTTTTCTTTTCA
   ***   * **** ***  *  ** ****   ****** *** *** *
```

P25-3. On an expedition to Madagascar, you find three isolated, phenotypically identical populations of zeebogs. Upon cytological examination, you find that each population differs with regard to the sequence of genes found on chromosome number 10. Specifically, the genes *Z1, Z2, Z3, Z4, Z5,* and *Z6* are all present, although their sequence differs between populations. Which population is likely ancestral to the others? Note the centromere is represented by the letter "c."

(*Hint 1:* When inferring evolutionary relationships, the fewest number of chromosomal changes should be considered.)

Population 1: *Z1*	*Z2*	c	*Z4*	*Z3*	*Z5*	*Z6*
Population 2: *Z1*	*Z3*	*Z4*	c	*Z2*	*Z5*	*Z6*
Population 3: *Z1*	*Z2*	c	*Z4*	*Z6*	*Z5*	*Z3*

P25-4. Consider the two orthologous proteins whose amino acid sequence is given in the following. The top protein is found in humans, and the bottom protein is present in *Drosophila*. Sequence identity is denoted by an asterisk (*), and sequence similarity is denoted by a period or colon (. or :). What region (between which amino acids) of this protein is most likely to contribute an important function to this protein? In what region should researchers create mutations in order to discover the function of this protein?

```
human   1 DLLQAAQTCRYWRILAEDNLLWREKCKEEGIDEPLHIKRRKVIKPGAGDF  50
fly     1 DLLRAAQTCRSWRFLCDDNLLWKEKCRKAQILAEPRSDRPKRGRDGNMPP  50
          ***:****** **:*.:*****:***::   *      :.* *   : *

human  51 NAETGECIHTLYGHTSTVRCMHLHEKRVVSGSRDASLRVWDIETGQCLHV 100
fly    51 DMDSGACVHTLQGHTSTVECMHLHGSKVVSPSRKATLRVWDIEQGSCLHV 100
          : ::* *:*** ****** ***** .:*** ** *:******* *.****

human 101 LMGHVAAVRPVQYDGRRVVSGAYDFMVKVWDPETETCLHTLQGHTNRVYS 150
fly   101 LVGHLAAVRCVQYDGKLIVSGDYDYMVKIWHPERQECLHTLQGHTNRVYS 150
          *:**:**** *****: :*** **:***:*.** : *************

human 151 LQFDGIHVVSGSLDTSIRVWDVETGNCKHTLTGHQSLTSGMELKDNILVS 200
fly   151 LQFDGLHVVSGSLDTSIRVWDVETGNCKHTLMGHQSLTSGMELRQNILVS 200
          *****:*********************** ***********::*****

human 201 VPCSATPTTFGDLRAANGQGQQRRITSVQPPTGLQEWLKMFQSWSGPEK  250
fly   201 SVTPSSHLTSSTPGSALGRRTPRSVPSRDNPPPELQHWLAQFQRWSHVER 250
          .::  *  .   :* *:     *    :  :**. **.**  ** **  *:
```

P25-5. Use the information given for each of the following proteins to determine its evolutionary rate in mutations per 100 amino acids per 10 million years. Then specify which of the proteins likely carries out a function that is extremely essential to life.

Protein	Mutations per 100 Amino Acids Since Divergence 200 Million Years Ago	Evolutionary Rate
Protein A	50	
Protein B	80	
Protein C	150	
Protein D	2	
Protein E	20	

Assessing Your Knowledge

Understanding the Key Concepts—Answers

1.) evolution; 2.) natural selection; 3.) phenotypic; 4.) resources; 5.) compete; 6–7.) obtain; utilize; 8.) favorable; 9.) generation; 10.) Evolutionary genetics; 11.) diversity; 12.) Natural selection; 13.) environment; 14.) mutations; 15–16.) anagenesis; cladogenesis; 17.) Anagenesis; 18.) transformation; 19.) cladogenesis; 20.) splits; 21.) reproductive isolation; 22.) prezygotic; 23.) fertilization; 24.) zygote; 25.) residential; 26–27.) spatial; temporal; 28.) courtship; 29.) reproductive; 30.) Reproductive isolation; 31.) postzygotic; 32.) viability; 33.) hybrid zygotes; 34.) gene flow; 35.) physical barrier; 36.) allopatric;

37.) sympatric; 38–39.) genetic; behavioral; 40.) niche; 41.) morphological; 42.) interbreed or mate; 43.) fertile offspring; 44.) DNA; 45.) DNA; 46.) homologous; 47.) orthologous; 48.) phylogenetic tree; 49.) species; 50.) common ancestor; 51.) subtypes or strains; 52.) origin; 53.) rearrangements; 54.) orthologous; 55.) synteny; 56.) proteome; 57.) 1%.

Figure Analysis—Answers

1a.) Essentially nothing happens to species 1 as it undergoes stasis. This is a period of time when the species is undergoing little change and is not undergoing either anagenesis or cladogenesis; b.) Species 2 evolves from species 1 by anagenesis which is a process by which species 1 undergoes a number of changes transforming it into a new species that is significantly different than its progenitor. Species 2 and species 1 are not present at the same time; c.) Species 3 and 4 are generated by cladogenesis of species 2 in which it splits into 2 new species that differ from each other and from species 2 itself. Species 3 and 4 are present at the same time as each other but not at the same time as species 2.

2a.) Reproduction isolation is achieved either by physical separation due to a geographic barrier or by the divergence of organisms within a population that allows them to enter a new niche in the same area; b.) The difference between these two mechanisms is that reproductive isolation by allopatric speciation involves the physical separation of groups of organisms, but reproductive isolation by sympatric speciation does not; c.) Genetic changes such as a specific mutation, a change in chromosome number or a translocation, or behavioral changes such as the use of a different food source are divergences in a population that can lead to sympatric speciation; d.) Physical barriers caused by bodies of water, mountain ranges, or the construction of a highway or buildings can lead to reproductive isolation and allopatric speciation.

3a.) The finches of the Galápagos Islands underwent allopatric speciation because they had limited abilities to move from island to island; b.) The three groups of finches that evolved were bud eaters, seed eaters, and insect eaters; c.) Finches that feed on the insects under the bark of trees were not able to exist in South America because many woodpecker species that were feeding on these insects were already living there. In the Galápagos Islands, there were no woodpeckers, so insects living under the bark of trees were an unused food resource. Finches that could feed on these insects were favored by natural selection; d.) The beaks of the different species of seed-eating finches differ in their length and depth depending upon what type of seeds that the finches eat.

4a.) By phyletic gradualism, evolution of a new species occurs through small genetic changes that accumulate over time, while evolution of a new species by punctuated equilibrium occurs by the rapid accumulation of changes that leads to rapid morphological changes in short periods of time separated by long periods of time when no or little changes occur; b.) The generation of two species from one by punctuated equilibrium occurs by the rapid divergence of two distinct groups from a common progenitor; c.) The generation of two species from one by phyletic gradualism occurs by the gradual divergence of two different populations from a common progenitor; d.) These two models of evolutionary change are difficult to distinguish in practice because to determine if phyletic gradualism was taking place, a scientist would need to recover fossils that exhibit numerous different gradual changes. Since this recovery of fossils is very difficult, it is often difficult to determine which evolutionary process has taken place.

5a.) The neutral theory states that most of the genetic diversity that we observe at the molecular level is not due to mutations but is due to random genetic drift such that most mutations have a small enough effect on the fitness of an organism that they are essentially neutral. It is the basis for the molecular clock which states that neutral substitution is constant over time for a given homolog across species; b.) The fibrinopeptide protein has the highest rate of amino acid substitutions; c.) Histone H4 has the lowest rate of amino acid substitutions; d.) Proteins encoded by genes that have highly conserved essential functions are expected to have the lowest rates of neutral substitutions; e.) Proteins encoded by genes that have highly variable activities in different species are expected to have higher rates of neutral substitutions; f.) Histone H4 has likely undergone such a low rate of amino acid substitutions because its function in different species is highly conserved and is essential for the survival of the cell.

6a.) Chromosome 1 is similar among the chimp, gorilla, and orangutan. It is also similar to the human chromosome 1 except that the banding pattern flanking the centromere on the human chromosome is different; b.) The banding pattern for human chromosome 2 is present on 2 different chromosomes in the chimp, gorilla and orangutan. Otherwise, the banding patterns are nearly identical in the four different organisms; c.) Chromosome 3 in humans, chimps, and gorillas have essentially an identical banding pattern with only a slight difference in the gorilla chromosome at the very end. Chromosome 3 from the orangutan has a pericentric inversion; d.) The inversion of chromosome 3 in the orangutan likely occurred at or after the split of the orangutan lineage; e.) The fusion of the two smaller chromosomes to form human chromosome 2 likely occurred at or after the split between the human and chimpanzee lineages.

7a.) To determine if the infected patients acquired HIV from their dentist, viral samples were collected from all of the patients, the dentist, and HIV-positive individuals in the area. The DNA was extracted and regions of the viral genome were PCR-amplified and then sequenced. The DNA sequences were then analyzed for similarities and differences that would indicate which strains were most closely related to each other; b.) The critical controls in the experiment are DNA samples from other individuals infected with HIV in the area. These samples are important because they allow a phylogenetic tree to be established and offer a comparison for the strains obtained from the patients and that obtained from the dentist. If the patients' strains are more closely related to the area controls than they are to the dentist's strain, it suggests that the source of their infection is likely not the dentist; c.) This type of analysis can identify the patients that likely acquired the virus from the dentist because a phylogenetic tree can be generated that indicates which strains are most closely related to one another and when the strains likely diverged relative to each other. Strains that are most closely related to the dentist's strain containing mutations that demonstrate that they diverged later than the dentist's strain indicate that the dentist was the likely source of the infection; d.) Strains D and F diverged prior to the dentist's strain; e.) Strains A, B, C, E, and G diverged later than the dentist's strain; f.) The dentist likely infected the patients with strains A, B, C, E, and G with HIV; g.) The patients carrying strains D and F likely acquired the HIV virus from a source other than the dentist.

General Questions—Answers

Q25-1. Some evolutionary biologists define a species by the morphological species concept in which two organisms are classified as belonging to the same species if they are morphologically similar, and classified as belonging to two different species if their differences are as great as two organisms already classified to belong to two different species. Other evolutionary biologists define a species by the biological species concept in which two organisms are classified to belong to the same species if they can interbreed and produce fertile offspring. Still other evolutionary biologists define a species by the species recognition concept in which two organisms belong to the same species if they are capable of locating each other and successfully mating with each other. Finally, some evolutionary biologists define a species by the species phylogeny concept in which two organisms belong to the same species if they share one or more DNA sequences or molecular features.

Q25-2. There are several problems in using the biological species concept to define distinct species. Since this concept is based on the ability of organisms to successfully mate and produce fertile offspring, it can not be used to classify extinct organisms and is difficult to use to classify haploid or asexual species. In addition, it is difficult to classify species that do not normally interbreed but do in fact do so in a laboratory setting. Likewise, it is difficult to classify organisms that are geographically isolated from each other because these groups of organisms can frequently interbreed when they are brought together.

Q25-3. Reproductive isolation can be achieved by prezygotic or postzygotic mechanisms. Prezygotic reproductive isolation prevents the fertilization of the egg and the formation of a zygote. Prezygotic reproductive isolation can be accomplished by residential isolation in which two populations exist in different habitats, by seasonal or temporal isolation in which two populations are sexually mature at different times, by ethological isolation in which two populations have incompatible courtship behavior, or by mechanical isolation in which the reproductive structures of two populations are incompatible. Postzygotic reproductive isolation affects the zygote after fertilization has taken place. It is accomplished by the F_1 zygote being inviable or having reduced viability or by the F_1 zygote being sterile if it is viable due to abnormal gonad development or failures during gamete formation, or due to an abnormal distribution of chromosomes or chromosomal segments. If the F_1 zygote is viable and fertile, postzygotic reproductive isolation is accomplished by the F_2 generation containing many compromised or sterile individuals.

Q25-4. Natural selection can act only on phenotypic variation present in a population. If variation is not present, there is no selective advantage to any one phenotype. Natural selection also requires that the source of the variation selected for or against have a genetic component. Allele frequencies in a population cannot change over generations if there is no mechanism for transmitting the observed differences from parents to offspring. Natural selection is the mechanism of change. Without selection, all phenotypes would be equally likely to reproduce and leave offspring.

Q25-5. Mitochondrial DNA (mtDNA) and Y chromosomal DNA sequences were used to determine the origin of *Homo sapiens*. Since mitochondrial DNA is maternal in origin, while Y chromosomal DNA is paternal in origin, these sequences were useful in tracing both the maternal and paternal human lineages. In addition, mtDNA was particularly useful for this analysis because it accumulates mutations 10-fold faster than nuclear DNA allowing scientists to examine evolutionary relationships based on the presence or the absence of particular mutations in the mtDNA. This allowed the relationships among organisms to be determined using a much shorter period of time.

Multiple Choice—Answers

1.) a; 2.) c; 3.) e; 4.) c; 5.) b; 6.) e; 7.) d; 8.) b; 9.) c; 10.) d.

Practice Problems—Answers

P25-1. To identify which reproductive isolating mechanism is being described you must determine how the two populations are being separated. For example, the two species described in (a) use different types of mating signals which suggests behavioral isolation. (b) Different times of mating suggest temporal isolation. (c) Production of sterile offspring results from hybrid sterility. (d) Incompatibility of reproductive structures is a type of mechanical isolating mechanism. (e) Physical separation of two compatible species is an example of geographical isolation.

P25-2.

Number of Nucleotide Differences Between Two Different Organisms

	1	2	3	4	5	6
1		10	9	5	3	4
2			13	9	7	8
3				8	6	7
4					2	3
5						1
6						

(a) Genes 5 and 6 are the most closely related genes.
(b) Genes 2 and 3 are the most distantly related genes.

P25-3. To infer the evolutionary relationships between the three populations, you need to determine how many changes would be necessary if each of the species were ancestral. The order requiring the smallest number of changes is the most likely. Population 2 is probably ancestral. One inversion (of the *Z3 Z4 Z2* sequence) would produce population 1. An inversion in population 2 (of the *Z3 Z5 Z6* sequence) would produce population 3.

P25-4. An important functional region of the protein most likely resides between amino acid 137 and amino acid 200 since this region is extremely conserved between these two distantly related organisms. This region would be the region that researchers should mutate in order to determine what the normal function of this protein is.

P25-5. Protein D is likely to be the most essential of these five proteins since it has experienced the smallest evolutionary rate.

Protein	Mutations per 100 Amino Acids Since Divergence 200 Million Years Ago	Evolutionary Rate
Protein A	50	2.5
Protein B	80	4.0
Protein C	150	7.5
Protein D	2	0.1
Protein E	20	1.0

25

Evolutionary Genetics

CHAPTER SUMMARY QUESTIONS

1. 1. j, 2. l, 3. n, 4. c, 5. e, 6. d, 7. b, 8. g, 9. m, 10. f, 11. k, 12. i, 13. a, 14. h.

2. There is no universally agreed upon definition of a species. Species and speciation may be better thought of as a continuum rather than discrete units. Probably the two most widely used definitions of species are the morphological and the biological species concepts. The morphological species concept supports the widely held view that members of a species are individuals that look similar to one another. This concept became criticized by biologists because it was arbitrary. Many examples were found in which individuals of two populations were very hard to tell apart but would not mate with one another, suggesting that they were in fact different species. Mimicry complexes supplied further evidence against the concept, as organisms of the same species can look very different, depending upon where they are reared or their life-cycle stage (some insects produce a spring brood that looks like one host plant and a summer brood that looks like another).

The biological species concept states that a species is a group of actually or potentially interbreeding individuals who are reproductively isolated from other such groups. It suggests that two individuals belong to the same species if their gametes can unite with each other under natural conditions to produce fertile offspring, emphasizing that a species is an evolutionary unit. Members share genes with other members of their species, and not with members of other species. Problems with this definition are that it cannot be used for extinct species and does not apply to asexually reproducing organisms.

3. Allopatric speciation occurs when populations physically isolated by an external barrier evolve reproductive (genetic) isolating mechanisms. Most evolutionary biologists agree that allopatric speciation is common. The evolution of reproductive isolation is generally thought to be passive, the product of differential selection (adaptation to different environmental conditions) as well as random genetic drift and unique mutations.

Sympatric speciation occurs among populations not geographically isolated from one another. How frequently sympatric speciation occurs is somewhat contentious; however, well-documented empirical evidence does exist. This speciation is most often thought to occur under disruptive selection where there is a significant genotype–environment interaction for two genotypes in two very discrete environments (for example, specialization on different host plants). Linked alleles for superiority in one environment and a preference for that environment can quickly lead to genetic isolation of the genotypes and a type of reinforcing selection for that isolation.

4. The data suggest that the fitness of hybrids may be very low compared to the parental species. This could be due to genetic divergence between the populations and could lead to speciation.

5. *QR* hybrids are likely sterile because of the difficulties during meiosis presented by the hybridization of a diploid and polyploid parental strain.

6. Evolution can be defined as a change in the genetic composition of a population over time or as phenotypic changes within an evolutionary lineage over time. Evolution and adaptation are not synonymous because there are nonadaptive evolutionary changes that occur.

7. The major mechanisms of evolutionary change are mutation, selection, and random genetic drift.

8. Natural selection is the mechanism by which an allele that provides a fitness advantage increases in frequency as the result of the advantage it provides.

9. Many things limit the scope of phenotypic change, such as finite population size limiting the input of new mutations and causing the fate of some of those mutations to be stochastic rather than deterministic, or negative genetic correlations among two or more traits that need to be increased simultaneously. Perhaps the biggest limitation is the developmental and phylogenetic constraints imposed on evolution. In other words, evolution must build incrementally on what already exists. It cannot deconstruct an organism and start its design from scratch.

10. **a.** Differential reproduction is a necessary but not a sufficient condition for evolution by natural selection. Without differential reproduction, changes in allele frequencies would be stochastic and due to random genetic drift and the vagaries of mutation.

   b. Environmental change is not necessary for evolution by natural selection. Even in a constant environment beneficial mutations will sometime occur and they can be selected for. However, the rate of evolution might be slow.

   c. Heritable variation for fitness is a necessary but not a sufficient condition for evolution by natural selection. If differential reproduction is linked to heritable variation for those differences, we have sufficient conditions for evolution by natural selection.

   d. Sexual reproduction is not a necessary condition. Asexual genotypes can have both differential reproductive rates and heritable genetic variation of those differences.

11. The cause of this phenomenon is natural selection. At first the antibiotic killed a large proportion of the bacteria and reduced their density to the point that the infected individual became symptom-free. However, a small number of bacteria contained an allele (or alleles) that provided resistance to the antibiotic. Over time the resistance genotype multiplied in number until the symptoms returned. Because of the very strong selection imposed by the antibiotic, all of the bacteria now causing the infection are resistant to the antibiotic.

12. The riddle is solved by recognizing that although most mutations are deleterious, natural selection acts like a filter removing the deleterious mutations from the population and increasing the frequency of the beneficial ones.

13. The fate of a new mutation is determined by whether it is beneficial or deleterious; the size of the selection coefficient for or against the allele relative to population size; whether it is dominant, recessive, and so forth; and its linkage associations with other alleles.

14. It is not proof. This demonstrates that individuals in human populations have differential reproduction but does not link these differences to heritable genetic differences.

15. Because some parts of the protein are more essential to carrying out the normal function of that protein, these parts will undergo stronger selection against nonsynonymous substitutions and are constrained from changing.

16. This suggests that the 1% difference in nucleotide sequences between humans and chimpanzees is due mostly to random genetic drift and is of little consequence to the differences in behavior, morphology, and so forth. The phenotypic differences may be due to small differences in regulatory genes caused by natural selection.

17. The third nucleotide position in a codon should contain the most genetic variation because it is the least likely to actually change the amino acid coded for by the mRNA.

EXERCISES AND PROBLEMS

18. This is due to convergent evolution. Dolphins and many sharks have a similar biology (marine fish predators) and therefore have converged on somewhat similar morphologies due to similar selection pressures. However, because they evolved from very different evolutionary lineages there are very large differences between them as well (air breathing versus water breathing).

19. Such designs result from the fact that mutations can only build on what already exists. A simpler form of the eye that preceded our current eye evolved in a way that maximized its performance under the current constraints. New innovations introduced by mutation and selection had to work within the framework that existed when they occurred.

20. Evolutionary genetics is a predictive science with relevance to many areas of modern biology. As one example, evolutionary biologists as early as the 1950s warned that the indiscriminate use of antibiotics would quickly lead to resistant strains of pathogens. This scenario has played out and multiple-drug resistant strains of tuberculosis and many other pathogens are now one of the greatest health crises facing the world today.

21. One way to test them is to artificially shorten the life span and see if this indirectly results in increased reproduction early in life or merely allows mutations to accumulate during the later stages of life and shorten life span further without reproductive compensation early in life.

22. All of these taken together argue strongly for evolution by descent from a common ancestor. These similarities are unlikely to have arisen by chance.

23. The necessary conditions for evolution by natural selection are differential reproduction among individuals due to heritable differences in fitness. This leads to adaptation by increasing the frequency of the advantageous allele(s) and moving the phenotype closer to the optimum for that environment.

24. This situation could potentially result in sympatric speciation. Because the hybrids have much higher fitness in the midshore zones, they could occupy this niche and exclude the parental types. However, for speciation to occur some type of isolating mechanism must evolve that stops gene flow from the two parental species into the hybrid zone.

25. The species recognition concept of a species relies on the identification of specific mate recognition systems. Species are defined as organisms that share a common fertilization system. The recognition concept suffers from all of the flaws of the biological species concept and differs from it primarily by focusing on a subset of isolating mechanisms.

26. Natural selection will "reward" individuals that discriminate against mating with individuals of a genotype that produces nonviable or low fitness offspring. Thus, natural selection might be expected to make prezygotic reproductive isolating mechanisms more prevalent among two diverging genotypes where postzygotic isolating mechanisms already exist.

27. Allopatric speciation is often thought of as passive because the two populations have no chance to hybridize and isolating mechanisms evolve as the consequence of differential selection, mutation (especially chromosomal rearrangements), and random genetic drift, whereas sympatric speciation probably involves active selection for prezygotic isolating mechanisms.

28. The large number of species on islands and isolated lakes is probably due to the large number of empty niches available for the rare colonists to reach those areas.

29. Under neutral theory those changes are mostly neutral and may not reflect the behavioral, ecological, and morphological changes usually associated with most definitions of species.

30. The α-globin gene in a modern human should be much more similar to the α-globin gene in a Neanderthal than to a β-globin gene from a human, because the α- and β-globin genes have been evolving independently for a much longer time.

31. They have diverged for an estimated 100,000,000 years. 0.2 substitutions per site/2 $\times$ 10^{-9} substitutions per site per year = 100,000,000 years.

32. Evolution appears to be much better at modifying existing structures than it is at coming up with de novo designs. Duplication of genes may allow the old gene to continue to its normal function, while the new copy accumulates mutations and can "experiment" with new functions at little or no cost to the organism.

33. Chimpanzees appear to be most closely related to humans, and yeast the least closely related. We make this assumption based on the fact that the chimpanzee gene has the fewest number of differences and the yeast gene the most. The evolutionary relationships suggested by the amino acid differences in this gene agree very well with the evolutionary relationships suggested by morphology, physiology, and behavior.

34. An evolutionarily conserved gene is one that is very similar even in organisms that are only distantly evolutionarily related. It is common for some parts of a gene (protein) to be more highly conserved than others. This is because not all parts of a protein are equally important for its function.

35. The mutation rate is the rate at which mutations actually occur. The substitution rate is the rate at which one nucleotide (codon, allele) goes to fixation, replacing another. The substitution rate reflects the impact of mutation, selection, and drift interacting. The mutation rate is rarely known and is not used for phylogenetic comparisons; the substitution rate is used for phylogenetic comparisons. That is why it is important to use neutral changes and to focus on homologous genes among organisms.

36. For one thing, the specific genetic changes causing virulence might be identified.

37. If a plant is discovered with weak cancer fighting properties, the most likely place to find similar chemical compounds with superior cancer fighting properties is in that plant's closest evolutionary relatives. This often cannot be known from morphology or geographical proximity. Phylogenetic analysis would be more useful for identifying these evolutionary relatives.

CHAPTER INTEGRATION PROBLEM

a. The rate of synonymous substitutions is more than 40 times the nonsynonymous substitution rate. This agrees with expectations. The synonymous substitutions are thought to be neutral with respect to fitness and will go to fixation at a rate equal to neutral expectations. Nonsynonymous substitutions are typically not neutral and are usually selected against. Those that are fixed are fixed through random genetic drift overpowering weak selection or because the mutation was beneficial in one species and not the other.

b. The genes do seem to differ in their substitution rates. This is expected as some genes are more important or the structure of their proteins more susceptible to disruption of function.

c. Nonsynonymous substitution rates are far more variable among genes, the largest rate being more than 400 times the smallest. For synonymous substitutions the largest is less than 1.4 times as large as the smallest. This is also in agreement with theory. Synonymous substitutions are expected to be the same or similar for all genes. However, the nonsynonymous substitution rate can vary greatly depending on the function of the gene and its importance to fitness.

d. Under the phylogenetic species concept most would consider these different species. However, one of the weaknesses of this approach is that the cutoff for different species is rather arbitrary and differs within different taxonomic groups.

e. You would want to know about the viability of offspring from a cross between these two species (assuming there are not prezygotic isolation mechanisms that prevent them from mating at all), their ecological role in the environment, and their morphology.

Notes

Notes